PROINFLAMMATORY AND ANTIINFLAMMATORY PEPTIDES

LUNG BIOLOGY IN HEALTH AND DISEASE

Executive Editor

Claude Lenfant
Director, National Heart, Lung and Blood Institute
National Institutes of Health
Bethesda, Maryland

ADDITIONAL VOLUMES IN PREPARATION

Five-Lypoxygenase Products in Asthma, *edited by Jeffrey M. Drazen, Sven-Erik Dahlén, and Tak H. Lee*

Complexity in Structure and Function of the Lung, *edited by Michael P. Hlastala and H. Thomas Robertson*

Human Immunodeficiency Virus and the Lung, *edited by Mark J. Rosen and James M. Beck*

The opinions expressed in these volumes do not necessarily represent the views of the National Institutes of Health.

PROINFLAMMATORY AND ANTIINFLAMMATORY PEPTIDES

Edited by

Sami I. Said

State University of New York Health Sciences Center
Stony Brook
and Northport Veterans Affairs Medical Center
Northport, New York

CRC Press
Taylor & Francis Group
Boca Raton London New York

CRC Press is an imprint of the
Taylor & Francis Group, an **informa** business

CRC Press
Taylor & Francis Group
6000 Broken Sound Parkway NW, Suite 300
Boca Raton, FL 33487-2742

First issued in paperback 2019

ISBN-13: 978-0-367-40067-5

**Visit the Taylor & Francis Web site at
http://www.taylorandfrancis.com**

**and the CRC Press Web site at
http://www.crcpress.com**

To Viktor Mutt and Miklos Bodanszky,
who introduced me to the study of peptides,
and to my family:
Mufeed and Nadia
and the memory of Wadie and my father and mother,
for their love and support

SERIES INTRODUCTION

The rapid progress that true science now makes occasions my regretting sometimes that I was born too soon. It is impossible to imagine the height to which may be carried, in a thousand years, the power of man over matter.

Benjamin Franklin
February 1770

Among the "matter" that has received considerable scientific attention since Franklin's statement is that of the peptides. This history of these molecules is a relatively recent one, and, indeed, much has been learned in far less than a thousand years. It seems that it began with the crystallization of L-asparagine from asparagus shoots by two French chemists, L. N. Vauquelin and P. Robinet, in 1806. Emil Fisher, however, was the one who really focused interest on peptide chemistry and synthesis following years of work on amino acids, as he realized that before understanding the peptides, it was critical to know their components. Although not for his work on peptides per se, Fisher's discoveries led to the receipt, in 1902, of the second Nobel prize ever awarded.

Since then, the world of peptides has seemingly become limitless. Scores of brilliant scientists successfully spent years on the isolation, characterization, and synthesis of peptides. At the same time, the potency of their biological function was recognized by biologists. It is, therefore, not surprising that eventually peptides became part of our therapeutic arsenal. The list of these biologically active compounds is impressive; all organ systems are affected by one or many peptides that can either be the cause of a deleterious effect or the agent that will correct a dysfunction.

Since its inception, the Lung Biology in Health and Disease series has addressed many aspects of the biological functions of peptides. Many of its volumes have chapters discussing them. Recognizing the therapeutic power of peptides, the series even included a volume (Volume 107), *Inhalation Delivery of Therapeutic Peptides and Proteins*, edited by Drs. Adjei and Gupta, but never before did the series present one entirely and exclusively devoted to peptides and their biological function.

Dr. Sami Said, who accepted the invitation to serve as editor of this volume, assembled an international cast of established leaders in their field. Those interested in understanding disease processes or devoted to treating disease will find this book to be state of the art.

As the Executive Editor of this series, I am grateful to Dr. Said and to the contributors for the opportunity to present this volume.

Claude Lenfant, M.D.
Bethesda, Maryland

FOREWORD

A 1953 paper in the *Journal of the American Chemical Society* on the structure and synthesis of oxytocin, by Vincent du Vigneaud and his associates, inaugurated a new age of peptide chemistry. On reading this seminal publication, I thought that oxytocin was just the first biologically active peptide that became amenable to chemical studies, and other biologically active peptides would follow it in the foreseeable future. Therefore, I decided to address my research to peptide synthesis. And indeed, new publications soon appeared from the same laboratory on the vasopressins, and not much later, studies on angiotensins and on bradykinin were published by other groups of investigators. Today, a glance through the catalogs of research supply houses specializing in peptides reveals that hundreds of biologically active peptides are commercially available. Their still rapidly increasing number far exceeds my most sanguine expectations.

Two obvious questions can be asked: Why did nature choose peptides to play such an important role in the regulation of life processes? Will the discovery of more and more biologically active peptides ever come to an end?

There are also other groups of compounds with conspicuous biological activities, for instance, steroids and prostaglandins. These groups, however, merely represent merely variations on single themes. For example, the steroid nucleus carries substituents such as the hydroxyl group in one or several positions, or double bonds appear in the structure at various areas of the four rings. The variations thus created are quite numerous, but their possible number is not limitless. In contrast, peptide chains—built of the amino acids that are the constituents of proteins—have various lengths and, more importantly, various amino acid sequences. Therefore, a vast number of different peptides can be constructed from the same 20 building stones. Their number is further increased by posttranslational modifications, such as the conversion of the phenolic hydroxyl in the tyrosine side chain to a sulfate ester in the molecule of cholecystokinin. Hence, there is a virtually unlimited array of possible structures, a large enough number of potential peptides for the regulation of a multitude of biological processes.

Yet the seemingly simple building components of peptides, the amino acids themselves, are also worthy of some thought. Although the generation of large molecules from small entities can be accomplished by various means, the presence of an amino group and a carboxyl group within the same molecule is eminently suitable for chain formation through amide (peptide) bonds. It is remarkable, however, that both groups are attached to the same carbon atom. This circumstance is not merely a question of economy in design, it also lends to these small molecules a quite unusual stability, including chiral stability, even under fairly drastic conditions. Chirality follows from the fact that the central (alfa) carbon atom, in addition to the amino and the carboxyl group, carries a so-called side chain as well. This side chain is different for each amino acid, thereby lending individual character to each. Also, the side chains are functional, and are the source of attraction between distant parts of the peptide chain via ion pairs, hydrogen bonds, and nonpolar or hydrophobic inter-actions. The sum of these attracting forces results in a preferred conformation (or secondary–tertiary structure). The architecture of a peptide is a feature fundamental to biological activity: it allows a firm attachment between an agonist and its receptor with complementary geometry.

In spite of the conformational freedom in many open-chain peptides, certain repeating architectural motives can still be discerned in certain areas. Thus, in dilute aqueous solutions, vasoactive intestinal peptide (VIP) and related peptides, such as secretin, glucagon, or helodermin, reveal a three-dimensional, folded structure with low helix content. Addition of organic solvents or micelles, and also aggregation during crystallization, induce a major conformational change: the molecules now appear in an extended form with much higher helix content. It is tempting to speculate that a similar conformational change takes place on attachment of these peptides to their individual receptors and that the resulting extended chains penetrate a membrane to induce the biological activity, for instance, by forming channels for certain inorganic ions. This speculation is supported by our experiment (carried out in collaboration with Professor Said) (1), in which the 15th residue of secretin (Asp) was replaced by the residue (Lys) in the corresponding position of VIP. In various biological actions the resulting analog is less secretin-like and more VIP-like than the parent hormone. Position 15 is in the area of the turn and hence exposed. This allows contact with a specific site of a membrane protein, a contact that should trigger the dramatic conformational change and the concomitant biological event.

Strategically positioned interactive side chains participate in the attraction between agonists and their specific receptors as well. Binding to a receptor, often reaching binding constants with nigh astronomical numbers, is again the consequence of interactions between amino acid side chains, although in this case not between residues belonging to two distant parts of the same chain, but between residues in two separate molecules. Thus, we are witnessing an unparalleled sophistication in the structures of nature's most important functional molecules, peptides, and proteins. Understanding the development of this machinery transcends present-day chemistry, physics, and molecular biology. In fact, the many questions that remain unanswered in our quest for explanations suggest that Schrödinger was right in postulating (in *What Is Life?*; Cambridge: Cambridge University Press, 1967) a yet nonexistent science, comparable to chemistry or physics.

The second question raised in this Foreword is whether or not the so-far unabated stream of newly discovered biologically active peptides will continue to flow or will diminish to a trickle, as happened in the field of steroids. It may be safe to say that discoveries will continue at a rapid pace for a long time. The earliest biologically active peptides were isolated by methods based on their physiological or pharmacological properties. This is true for the hormones of the neurohypophysis, oxytocin and the vasopressins, and for the gastrointestinal hormones glucagon, secretin and VIP (2) as well. Medical research seeking remedies for high blood pressure yielded the angiotensins. The proinflammatory peptides kallidin and bradykinin were discovered through their pharmacological activities. It is interesting to note that the recognition (3) of a carboxydipeptidase cleaving bradykinin, a nonapeptide between the eighth and ninth residues, was instrumental in the study of the analogous conversion of angiotensin I to angiotensin II by cleavage catalyzed by a specific carboxydipeptidase, the angiotensin converting enzyme (ACE). A systematic investigation of ACE inhibitors (4) resulted in extremely potent inhibitory pseudopeptide substrates. Since then, several more enzyme inhibitors have been developed, some of them with major significance in medicine, including the treatment of AIDS.

Isolation of biologically active peptides is, however, not limited to known biological or pharmacological activity. The mere fact that many of the active peptides have the amide of an amino acid as their C-terminal residue rather than an amino acid with a free carboxyl group (probably as protection against degradation by carboxypeptidases) offered a most interesting alternative route (5) to the isolation of new peptides; their biological activities were determined subsequently.

The active peptides are generally synthesized in the ribosomes as parts of protein molecules from which they are cleaved by specific proteolytic enzymes. Therefore, taking into account the vast but not unlimited number of proteins with sequences encoded in the nucleic acids of the cell, we should conclude that the number of newly discovered peptides can continue to grow, but it is not infinite. Yet posttranslational transformation can further increase their number. Beyond nature's work, one should also consider human ingenuity in molecular manipulations. These can be quite practical, as exemplified by the long-acting vasopressin preparation used for the control of diabetes insipidus. Synthesis can provide peptides with striking new properties.

Articles, and even books, have appeared in recent years suggesting that research already has covered almost all that can be discovered in nature, that no further major scientific breakthroughs should be expected. This volume can be regarded as evidence against this view; its chapters on new results in the field of inflammatory peptides are convincing testimony on continued rapid expansion. We should, therefore, conclude that scientific research is alive and well, and that in spite of impressive advances, new discoveries should be expected from generations of investigators to come.

Miklos Bodanszky
Mabery Professor Emeritus of Chemistry
Case Western Reserve University
Cleveland, Ohio

References

1. Bodanszky M, Natarajan S, Gardner JD, Makhlouf G, Said SI. Synthesis and some pharmacological properties of the 23-peptide 15-lysine-secretin (5-27): special role of the residue in position 15 in biological activity of the vasoactive intestinal polypeptide. J Medicinal Chem 1978; 21:1171–1172.
2. Said SI, Mutt V. Polypeptide with broad biological activity: isolation from small intestine. Science 1970; 169:1217–1218.
3. Erdös EG, Sloane EM. An enzyme in human blood plasma that inactivates bradykinin and kallidins. Biochem Pharmacol 1962; 11:585–592.
4. Ondetti MA, Rubin B, Kushman DW. The design of specific inhibitors of angiotensin-converting enzyme; a new class of orally active anti-hypertensive agents. Science 1977; 196:441–444.
5. Tatemoto K, Mutt V. Isolation of two novel candidate hormones using a chemical method for finding naturally occurring polypeptides. Nature 1980; 285:417–418.

PREFACE

My introduction to the study of peptides began with a sabbatical year I spent at the Karolinska Institute, in Stockholm, Sweden, in 1968 and 1969. My associates and I, at the Medical College of Virginia, in Richmond, had found that extracts of mammalian lungs caused systemic vasodilation and hypotension upon injection into other animals. This vasoactivity could not be explained solely by the compounds then known to be present in lung tissue, including histamine and prostaglandins, and we concluded that we were dealing with one or more vasoactive peptides. Lacking experience in this area, I decided to seek expert help with the isolation of these peptides.

Through Professor Sune Bergström, then rector of the Karolinska Institute, I met Professor Viktor Mutt, who accepted me as a guest researcher in his laboratory. Together, we extracted and partially purified a vasodilator peptide from porcine lung, but complete purification was delayed by the inadequate supply of fresh lung tissue. Reasoning that the same peptide might occur in the upper intestine, which has the same embryological derivation as the lung, we turned our attention to duodenal extracts, which were readily available in the laboratory and had served as the source material for Professors Mutt and Erik Jorpes, in their preparation of secretin and cholecystokinin. Using the same bioassay that had guided our work on the lung, in which vasodilation was assessed as increased peripheral blood flow together with decreased arterial blood pressure in anesthetized dogs, we successfully isolated a vasodilator peptide and named it vasoactive intestinal polypeptide (VIP) (1). The peptide was quickly synthesized by Miklos Bodanszky (2).

A few years later, we, and others, determined that VIP was richly present in the brain and peripheral nerves (3,4), and redefined it as a neuropeptide. Its ready isolation from intestine was due to its high concentration in the dense networks of enteric nerves.

Today, VIP is widely viewed as a physiological regulator of major body functions (5–11), including brain metabolism and blood flow, gastrointestinal motility and secretion, cardiovascular and respiratory function, neuroendocrine secretion, cell survival and proliferation, immune and inflammatory responses, and sexual activity and reproduction. Hypersecretion of VIP by certain tumors results in a distinct clinical entity, and VIP deficiency has been linked to several other diseases, including bronchial asthma, cystic fibrosis, and the acquired immunodeficiency syndrome (AIDS).

Over the years, my work with VIP has introduced me to several new and exciting vistas in science and biology. And I have gained valued new acquaintances and friendships with colleagues around the world. This volume, based on a FASEB symposium that focused on the role of peptides in inflammation (12), represents a cooperative effort by an international group of scientists at the forefront of research in this field. Several of the authors were largely responsible for the discovery, characterization, and development of the peptides they discuss. The authoritative and up-to-date information provided here, not available in any other single volume, includes discussions of the chemistry, biochemistry, pharmacology, enzymatic degradation, molecular biology, and therapeutic potential of key peptides that can influence inflammation and cell injury, as well as their analogs, agonists, and antagonists.

Sami I. Said

References

1. Said SI, Mutt V. Polypeptide with broad biological activity: isolation from small intestine. Science 1970; 169:1217–1218.
2. Bodanszky M, Klausner YS, Said SI. Biological activities of synthetic peptides corresponding to fragments of and to the entire sequence of the vasoactive intestinal peptide. Proc Natl Acad Sci (USA) 1973; 70:382–384.
3. Said SI, Rosenberg RN. Vasoactive intestinal polypeptide: abundant immunoreactivity in neural cell lines and normal nervous tissues. Science 1976; 192:907–908.
4. Larsson L-I, Fahrenkrug J, Schaffalitzky de Muckadell O, Sundler F, Håkanson R, Rehfeld JF. Localization of vasoactive intestinal polypeptide (VIP) to central and peripheral neurons. Proc Natl Acad Sci (USA) 1976; 73:3197–3200.
5. Said SI, ed. Vasoactive Intestinal Peptide. New York: Raven Press, 1982.
6. Bataille D, Said SI, eds. VIP and Related Substances: 2nd International Symposium. Peptides 1986; 7(Suppl 1):1–297.

7. Said SI, Mutt V, eds. Vasoactive Intestinal Peptide and Related Peptides. New York: Ann NY Acad Sci, 1988; 527:1–527.

8. Yanaihara, N. Vasoactive Intestinal Peptide and Related Peptides. 5th International Symposium, Shizuoka, Japan, November 12–15, 1991. Biomedical Research 1992; 13 (Suppl 2):1–393.

9. Fahrenkrug J. Transmitter role of vasoactive intestinal peptide. Pharmacol Toxicol 1993; 72:354–363.

10. Rosselin G, ed. VIP, PACAP, and Related Regulatory Peptides. River Edge, NJ: World Scientific, 1994.

11. Arimura A, Said SI. VIP, PACAP, and related peptides. Ann NY Acad Sci 1996; 805: 1–792.

12. Said SI, Wei ET. Symposium on Pro-Inflammatory and Anti-Inflammatory Peptides. Exp Biol Meetings, Atlanta, Georgia, April 9–13, 1995.

CONTRIBUTORS

Peter Baluk, Ph.D.   Assistant Research Anatomist, Cardiovascular Research Institute and Department of Anatomy, University of California, San Francisco, San Francisco, California

G. Boccoli   Third Division of Internal Medicine, IRCCS Ospedale Maggiore di Milano, Milan, Italy

Douglas E. Brenneman, Ph.D.   Chief, Section on Developmental and Molecular Pharmacology, Laboratory of Developmental Neurobiology, National Institute of Child Health and Human Development, National Institutes of Health, Bethesda, Maryland

Patricia K. Byrd, B.A.   Staff Research Associate, Department of Medicine, University of California, San Francisco, and Veterans Affairs Medical Center, San Francisco, California

Girolamo Caló, M.D., Ph.D.   Assistant Professor, Section of Pharmacology, Department of Experimental and Clinical Medicine, University of Ferrara, Ferrara, Italy

Anna Catania   Third Division of Internal Medicine, IRCCS Ospedale Maggiore di Milano, Milan, Italy

Giuliana Ceriani   Third Division of Internal Medicine, IRCCS Ospedale Maggiore di Milano, Milan, Italy

Dipak K. Das, Ph.D.   Professor and Director, Cardiovascular Division, Department of Surgery, University of Connecticut School of Medicine, Farmington, Connecticut

Emanuel DiCicco-Bloom, M.D.   Associate Professor, Department of Neuroscience and Cell Biology and Department of Pediatrics, University of Medicine and Dentistry of New Jersey/Robert Wood Johnson Medical School, Piscataway, New Jersey

Costanza Emanueli, Ph.D.   Investigator, Section of Pharmacology, Department of Experimental and Clinical Medicine, University of Ferrara, Ferrara, Italy

Richard M. Engelman, M.D.   Professor and Chief of Cardiac Surgery, Department of Surgery, Baystate Medical Center, Springfield, Massachusetts

Ervin G. Erdös, M.D.   Professor, Departments of Pharmacology and Anesthesiology, University of Illinois College of Medicine at Chicago, Chicago, Illinois

Alexander Faussner, Ph.D.   Researcher Associate, Johns Hopkins Asthma and Allergy Center, Baltimore, Maryland

Michela Figini, Ph.D.   Investigator, Section of Pharmacology, Department of Experimental and Clinical Medicine, University of Ferrara, Ferrara, Italy

J. Stephen Fink   Department of Pharmacology, Uniformed Services University of the Health Sciences, Bethesda, Maryland

Thomas Flüge, M.D.   Lower Saxony Institute for Peptide Research, Hannover, Germany

Wolf-Georg Forssmann, M.D.   Professor, Lower Saxony Institute for Peptide Research, Hannover, Germany

Pierangelo Geppetti, M.D.   Professor, Section of Pharmacology, Department of Experimental and Clinical Medicine, University of Ferrara, Ferrara, Italy

Edward J. Goetzl, M.D.   Professor of Medicine and Microbiology, Department of Medicine, University of California, San Francisco, San Francisco, California

Roy G. Goldie, B.Sc., Ph.D.   Professor and Senior Principal Research Fellow, Department of Pharmacology, University of Western Australia, Perth, Western Australia, Australia

Illana Gozes, Ph.D.   Professor, Department of Clinical Biochemistry, The Lily and Avraham Gildor Chair for the Investigation of Growth Factors, Sackler School of Medicine, Tel Aviv University, Tel Aviv, Israel

Pierre Gressens, M.D., Ph.D.   Chargé de Recherche Inserm, Service de Neurologie Pédiatrique and INSERM CRI-9603, Hôpital Robert-Debré, Paris, France

Christopher Haslett, B.Sc.(Hons.), M.B.Ch.B.(Hons.), F.R.C.P(E), F.R.C.P.
Professor, Department of Medicine, University of Edinburgh, Edinburgh, Scotland,
United Kingdom

Douglas W. P. Hay, B.Sc., Ph.D. Director, Department of Pulmonary Pharmacology, SmithKline Beecham Pharmaceuticals, King of Prussia, Pennsylvania

Maria G. M. O. Henriques, M.Sc., Ph.D. Senior Researcher, Department of
Applied Pharmacology, Institute of Drug Technology, Fundação Oswaldo Cruz, Rio
de Janeiro, Brazil

Peter J. Henry, B.Sc., Ph.D. Senior Research Officer, Department of Pharmacology, University of Western Australia, Perth, Western Australia, Australia

Joanna M. Hill, Ph.D. Staff Scientist, Section on Developmental and Molecular
Pharmacology, Laboratory of Developmental Neurobiology, National Institute of
Child Health and Human Development, National Institutes of Health, Bethesda,
Maryland

Ernst Jaeger Max-Planck-Institut für Biochemie, Munich, Germany

H. Benfer Kaltreider, M.D. Professor in Residence, Medicine, and CVRI, Department of Medicine, University of California, San Francisco, and Veterans Affairs
Medical Center, San Francisco, California

Craig M. Lilly, M.D. Assistant Professor, Department of Medicine, Harvard Medical School, Boston, Massachusetts

J. M. Lipton University of Texas Southwestern Medical Center at Dallas, Dallas,
Texas

Nicholas W. Lukacs, Ph.D. Assistant Research Scientist, Department of Pathology, University of Michigan Medical School, Ann Arbor, Michigan

Carlo Alberto Maggi, M.D. Director of Discovery, Menarini Richerche, Florence,
Italy

Nilanjana Maulik, Ph.D. Assistant Professor, and Director, Molecular Cardiology
Laboratory, Department of Surgery, University of Connecticut School of Medicine,
Farmington, Connecticut

Donald M. McDonald, M.D., Ph.D. Professor of Anatomy, Cardiovascular Research Institute and Department of Anatomy, University of California, San Francisco, San Francisco, California

Markus Meyer, M.D. Lower Saxony Institute for Peptide Research, Hannover,
Germany

J. Morley, Ph.D.   Honorary Senior Lecturer, The Sackler Institute of Pulmonary Pharmacology and King's College School of Medicine and Dentistry, London, England

C. P. Page, Ph.D.   Professor of Pharmacology, The Sackler Institute of Pulmonary Pharmacology and King's College School of Medicine and Dentistry, London, England

Sudhir Paul, Ph.D.   Professor, Department of Anesthesiology, University of Nebraska Medical Center, Omaha, Nebraska

Giles A. Rae, M.Sc., Ph.D.   Full Professor, Department of Pharmacology, Biological Sciences Centre, Universidade Federal de Santa Catarina, Florianópolis, Brazil

Domenico Regoli, M.D.   Professor, Section of Pharmacology, Department of Experimental and Clinical Medicine, University of Ferrara, Ferrara, Italy

Nilum Rajora   University of Texas Southwestern Medical Center at Dallas, Dallas, Texas

Adelbert A. Roscher, M.D.   Professor of Biochemistry, Department of Clinical Chemistry, Biochemistry, and Metabolism, Children's Hospital, University of Munich, Munich, Germany

Adriano Giorgio Rossi, B.Sc.(Hons.), Ph.D.   Senior MRC Research Scientist, Department of Medicine, University of Edinburgh, Edinburgh, Scotland, United Kingdom

Sami I. Said, M.D., B.Ch.   Professor of Medicine and Physiology, State University of New York Health Sciences Center, Stony Brook, and Chief of Pulmonary and Critical Care Medicine, Northport Veterans Affairs Medical Center, Northport, New York

Stephanie A. Shore, Ph.D.   Associate Professor of Physiology, Department of Environmental Health, Harvard School of Public Health, Boston, Massachusetts

Randal A. Skidgel, Ph.D.   Associate Professor, Departments of Pharmacology and Anesthesiology, University of Illinois College of Medicine at Chicago, Chicago, Illinois

D. Spina, Ph.D.   Lecturer, The Sackler Institute of Pulmonary Pharmacology and King's College School of Medicine and Dentistry, London, England

Sunil P. Sreedharan, Ph.D.   Assistant Professor, Division of Immunology and Allergy, Department of Medicine, University of California, San Francisco, San Francisco, California

R. A. Star   Department of Internal Medicine, University of Texas Southwestern Medical Center at Dallas, Dallas, Texas

Aviva J. Symes, Ph.D.   Assistant Professor, Department of Pharmacology, Uniformed Services University of the Health Sciences, Bethesda, Maryland

S. Taherzadeh   University of Texas Southwestern Medical Center at Dallas, Dallas, Texas

Peter A. Ward, M.D.   Professor and Chairman, Department of Pathology, University of Michigan Medical School, Ann Arbor, Michigan

Menghang Xia, Ph.D.   Amgen Inc., Thousand Oaks, California

CONTENTS

1

Perspective

Peptides in Relation to Tissue Injury and Inflammation

SAMI I. SAID

State University of New York Health Sciences Center
Stony Brook
and Northport Veterans Affairs Medical Center
Northport, New York

I. The Multiple Roles of Peptides

Numerous biologically active peptides have been identified in mammalian and nonmammalian systems. Most of these peptides are synthesized and released by neuronal cells and hence are known as *neuropeptides*. As *neurotransmitters* in the central and peripheral nervous systems, neuropeptides are in a position to exert wide-ranging and important regulatory influences on almost all body functions. Peptides may also act as *endocrine* or *neuroendocrine* messengers, or by *paracrine, neurocrine*, or *autocrine* mechanisms (1). Functions that are influenced by peptides include digestion, absorption, and gastrointestinal motility; metabolism; electrolyte balance; blood pressure, cardiac performance and blood flow to vital organs; vascular and nonvascular smooth muscle responses; cellular and humoral immune mechanisms; hormonal and neurohormonal secretion; and cell proliferation and differentiation. Peptides may also have antibacterial (2) and other activities.

1

II. Peptides and Inflammation

In addition to their physiological regulatory roles, peptides may be major players in a variety of pathophysiological and pathological processes (1). Until recently, the role of peptides in the response to inflammation and cell injury had received sporadic attention. Several groups of peptides have been recognized for their potential or proven contributions to pathophysiological events: the tachykinins, released from sensory nerves, for their mediation of "neurogenic" inflammation (Chapters 4, 5, 6, 22); bradykinin, for its role in local and systemic inflammation (Chapter 10); and the renin-angiotensin system, for its importance in some forms of hypertension (3). However, the full participation of peptides in the production, potentiation, or amelioration of inflammation, and the enhancement or protection against cell-tissue injury, has remained insufficiently appreciated.

Designed to defend against noxious stimuli, limit injury, and promote tissue repair and regeneration, the inflammatory response often ends in resolution and recovery. If unchecked, however, it may lead to excessive local damage, and potentially life-threatening systemic complications (4). The complex events of the inflammatory response—its initiation, perpetuation, accentuation, and modulation—are brought about by a host of chemicals, including biogenic amines, lipids, peptides, enzymes, reactive oxygen metabolites, and nitric oxide. These compounds are released by, and interact with, a variety of inflammatory cells, including neutrophils, T lymphocytes, macrophages, eosinophils, basophils, and mast cells (5).

The focus of this book, based in large part on a special symposium devoted to the subject (6), is the role of peptides in inflammation and cell death—both in promoting the inflammatory-injury reaction, and in effectively moderating inflammation and defending cells against injury and death. The larger polypeptide cytokines and chemokines, which exert considerable influence on all aspects of inflammation and cell death, have been reviewed elsewhere (e.g., Ref. 7).

III. Experimental Models for the Study of Inflammation and Tissue Injury

Different experimental models have been employed for demonstrating the phenomena of inflammation and their modification by peptides and other agents. Among these models are the hamster cheek pouch preparation, in which intravital microscopy is used to quantify changes in the microcirculation, particularly vascular leakage due to increased microvascular permeability (8). Useful information on inflammation and its modification by drugs has also been derived from studies of inflammation produced in joints and other tissues by adjuvant, carrageenin, and other agents (9), in the rat paw by exposure to heat, in skeletal muscle by a knife cut, in the cerebral cortex by a freeze lesion (10), and in the lung by a variety of procedures in which toxic agents are introduced either by inhalation or via the bloodstream (11).

Because of inherent differences in structure and function between various systemic circulations and the pulmonary circulation, investigations of the pharmacological and physiological defenses against lung injury and inflammation are best learned from studies on the lung itself, rather than by extrapolation of results from nonpulmonary tissues and organs. Because of these differences, certain compounds may have divergent effects on the production of systemic and pulmonary edema. Thus, vasoactive intestinal peptide (VIP), which has been reported to enhance cutaneous edema (12), actually suppresses the development of high-permeability pulmonary edema (Chapter 15). Further, the same agent, e.g., prostacyclin, may, as a vasodilator, promote cutaneous edema when injected locally, but reduce the edema when given systemically (13).

IV. Two Modes of Cell Death: Necrosis Versus Apoptosis

Apoptosis, sometimes also known as programmed cell death, is a fundamental homeostatic mechanism that serves to maintain a balance between cell death and proliferation during development and in other physiological processes (14). Accomplished through genetically encoded and carefully regulated pathways, apoptosis is also emerging as a potentially important mechanism in the pathogenesis of varied pathological disorders, including neurodegenerative diseases such as Alzheimer's and Parkinson's diseases (15), heart failure (16), and acute lung injury (17). Elucidation of the role of apoptosis in pathological cell death may have therapeutic implications, through appropriate manipulation of proapoptotic and antiapoptotic pathways.

V. Peptides That Promote Inflammation and Injury (Table 1)

When peptides and polypeptides are discussed in the context of inflammation, it is most often with reference to their proinflammatory effects. Among the earliest and best characterized are *activated complement peptides*. The complement system, and its component proteins and peptides, play a critical role in the immunological defense against microbial agents. During complement activation, several peptides are generated, especially C3 and C5, which can trigger such features of inflammation as in-

Table 1 Proinflammatory Peptides and Polypeptides

Bradykinin	Certain cytokines and chemokines
	(e.g., tumor necrosis factor, interleukin-1, interleukin-8)
Endothelins	Complement peptides
Tachykinins	Peptido-leukotrienes

creased vascular permeability; leukocyte activation, chemotoxis, and adhesion; phagocytosis; and activation of the proinflammatory lipoxygenase pathway of arachidonic acid metabolism. The activation of the complement system is closely linked to the pathways leading to activation of blood coagulation, fibrinolysis, and kinin formation (18).

Bradykinin a powerful vasodilator released from kininogen by the action of kallikrein, produces pain, inflammation, systemic vasodilation and hypotension, nonvascular smooth muscle contraction, and increased vascular permeability. The recent development of selective antagonists should help define the contribution of bradykinin to the abnormal inflammatory response (Chapter 10).

Cytokines, originally called lymphokines or monokines, as products of lymphocytes or monocytes/macrophages, respectively, may be secreted by a variety of cells and can profoundly influence inflammation, immunity, and growth and differentiation (7). Of special relevance to the inflammatory response are *tumor necrosis factor-α* and *interleukin-(IL)-1* (Chapter 3), both of which are major mediators of septic shock (19).

Chemokines, a novel supergene family of chemotactic cytokines, serve a key role in the recruitment of leukocytes to an areas of inflammation (Chapter 3). Chemokines exhibit a relative specificity for the recruitment and activation of subpopulations of blood leukocytes. Examples are *IL-8*, or neutrophil activating protein 1, a potent chemoattractant for neutrophils, and *eotaxin*, particularly active toward eosinophils.

The *peptido-leukotrienes* (LTs) are biologically active products of arachidonic acid metabolism via the lipoxygenase system. LTC_4 and LTD_4 increase systemic and pulmonary vascular permeability, and cause bronchoconstriction and pulmonary vasoconstriction (20).

So named because of the fast onset of their action, as opposed to the slower-acting bradykinin, *tachykinins*, also known as *neurokinins*, are a family of neuropeptides with structural and biological features in common. Mammalian tachykinins comprise substance P and neurokinin A, derived from a single gene, and neurokinin B, derived from another. Tachykinins continue to be the focus of active investigation as to their distribution, localization in sensory nerves, co-localization with other neuropeptides such as *calcitonin gene-related peptide (CGRP)*, biosynthesis, mode of action, specific receptors and receptor antagonists, degradation, and possible role in bronchial asthma and other inflammatory states (Chapters 4, 5, 6, 22).

Bursting on the scene less than 10 years ago, the *endothelins* have quickly gained a prominent place among biologically active peptides. Located primarily in vascular endothelium, respiratory epithelium, and certain other cells, the endothelin family of peptides exerts numerous important actions on multiple systems. These actions include smooth muscle contraction, microvascular leakage and edema, mitogenesis, and mucous gland hypersecretion. Investigation of the possible contributions of the endothelins to a variety of human diseases in progressing at a rapid pace (Chapter 7, 8).

VI. Peptides That Protect Against Injury and Cell Death (Table 2)

The concept that certain peptides may have antiinflammatory or antiinjury properties is relatively new—and one that is not yet widely known. The concept is founded primarily on work with *VIP and related peptides, corticotropin-releasing factor (CRF) and related peptides,* α*-melanocyte-stimulating hormone,* and *lipocortins.*

Over the past decade, a series of reports has demonstrated the VIP has potent antiinflammatory, and antiinjury activity in the lung (Chapter 15). This conclusion is based on evidence that VIP can prevent or attenuate acute edematous lung injury in several experimental models of the adult respiratory distress syndrome (ARDS). VIP dose-dependently protected perfused, ventilated rat or guinea pig lungs in which the injury was induced by intratracheal instillation of HCl; infusion of platelet-activating factor (PAF), of the oxidant herbicide paraquat, or of xanthine and xanthine oxidase, which produce injury through the generation of toxic oxygen species; and prolonged perfusion of the lungs ex vivo, a form of injury that also results from free oxygen radicals. Of the VIP-related peptides, only helodermin gave equal protection, while secretin and glucagon were ineffective (21), and results with PACAP were inconclusive.

The tissue-protective effect of VIP is not limited to the lung: VIP promotes the survival of neuronal cells and their precursors in culture, and protects neurons from the toxic effect of envelope protein of the human immunodeficiency virus (reviewed in Chapter 17). VIP also protects the heart against reperfusion injury (Chapter 14), a common cause of oxidant injury of cardiac muscle.

Several publications have provided evidence that CRF, and other compounds with CRF activity can prevent or attenuate increased vascular permeability in a variety of experimental situations, including the anesthetized rat's paw skin after exposure to heat or to extreme cold, tracheal and esophageal mucous membrane after subcutaneous substance P injection or exposure to formaldehyde vapor, skeletal muscle after surgical incision or local injection of substance P, brain cortex after a

Table 2 Antiinflammatory Peptides and Polypeptides

Corticotropin-releasing factor (CRF) and related peptides; neurotensin, xenopsin, and related peptides; certain opioid peptides

Vasoactive intestinal peptide (VIP); helodermin; pituitary adenylate cyclase-activating peptide (PACAP)

Melanocyte-stimulating hormone

Certain cytokines and chemokines (e.g., interleukin-10, transforming growth factor-β)

Lipocortins

freeze lesion, and pulmonary edema in rats caused by intravenous injection of epinephrine or intratracheal instillation of formalin (10).

Similar antiinflammatory activity has been reported for several other peptides, or groups of peptides, with structural similarities to CRF. Such peptides include the frog skin peptide *sauvagine*, the sucker fish peptide *urotensin I*, *neurotensin*, and the synthetic, CRF-modified "mystixins" (10). It is noteworthy that sequence similarities exist between CRF-related peptides and VIP (10).

Other peptides with antiinflammatory potential include α-melanocyte-stimulating hormone (Chapter 11), *CGRP* (10), *natriuretic peptides* (Chapter 23), and the lipocortins and the related *antiflammins* (10,22), initially reported to act by inhibiting phospholipase A_2 activity. Of the polypeptide cytokines/chemokines, IL-4, IL-10, and transforming growth factor-β (TGF-β) have generally antiinflammatory properties (Chapter 3).

VII. Interactions Among Peptides, and with Other Mediators/Modulators of Cell Injury

In the course of the inflammatory response, peptides often interact not only among other peptides and polypeptides, but also with other pro- and antiinflammatory agents. Examples of such interactions are the complement system with the kallikrein-kinin, clotting, and fibrinolytic systems; the preceding systems with the arachidonate metabolic pathways; cytokines and chemokines with other inflammatory mediators and the cells that release them; bradykinin with the tachykinins (Chapter 9); various peptides with biogenic armines and acetylcholine; VIP with other mast cell mediators, with substance P, and with nitric oxide (23). Future research will no doubt uncover numerous other instances where peptides can modify the inflammatory/injury responses in cells and tissues, either directly or indirectly by interacting with other mediators and modulators. The coexistence of peptides and other neurotransmitters within the same neurons (24) multiplies the opportunities for such interactions.

References

1. Maggi CA, Giachetti A, Dey RD, Said SI. Neuropeptides as regulators of airway function: vasoactive intestinal peptide and the tachykinins. Physiol Rev 1995; 75: 277–322.
2. Mursh J, Goode JA. Antimicrobial Peptides. Ciba Foundation Symposium #186. New York: J. Wiley, 1994.
3. Frohlich ED. Angiotensin converting enzyme inhibitors: present and future. Hypertension 1989; 13 (suppl 1):1125–1130.
4. Cotran RS, Kumar V, Robbins SL. Inflammation and repair. In: Cotran RS, Kumar V, Robbins SL, eds. Robbins: Pathologic Basis of Disease. 5th ed. Philadelphia: Saunders, 1994:51–92.

5. Gallin JA, Goldstein IM. Snyderman R, eds. In: Inflammation: Basic Principles and Clinical Correlates. 2d ed. New York: Raven Press, 1992.
6. Said SI, Wei ET. Symposium on Pro-Inflammatory and Anti-Inflammatory Peptides. Exp Biol Meetings, Atlanta, Georgia, April 9–13, 1995.
7. Remick DG, Friedland JS, eds. In: Cytokines in Health and Disease. 2d ed. New York: Marcel Dekker, 1997.
8. Svensjo E. The hamster cheek pouch as a model in microcirculation research. Eur Respir J 1990; 3:595s–601s.
9. Aspinall RL, Cammarata PS. Effect of prostaglandin E_2 on adjuvant arthritis. Nature (Lond) 1969; 224:1320–1321.
10. Wei ET, Thomas HA. Anti-inflammatory peptide agonists. Annu Rev Pharmacol Toxicol 1993; 33:91–108.
11. Said SI (ed.) The Pulmonary Circulation and Acute Lung Injury. 2d ed. Mount Kisco, NY: Futura, 1991.
12. Brain SD, Williams TJ. Neuropharmacology of peptides in skin. Sem Dermatol 1988; 7:278–283.
13. Rampart M, Williams TJ. Polymorphonuclear leukocyte-dependent plasma leakage in the rabbit skin is enhanced or inhibited by prostacyclin, depending on the route of administration. Am J Pathol 1986; 124:66–73.
14. Rudin CM, Thompson CB. Apoptosis and disease: regulation and clinical relevance of programmed cell death. Annu Rev Med 197; 48:267–281.
15. Margolis RL, Chuang D-M, Post RM. Programmed cell death: implications for neuropsychiatric disorders. Biol Psychiatr 1994; 35:946–956.
16. Olivetti MD, Abbi R, Quaini F, et al., Apoptosis in the failing human heart. N Engl J Med 1997; 336:1131–1141.
17. Said SI, Pakbaz H, Berisha HI, Raza S, Goel S, DeStefanis P. Apoptosis is a major component of excitotoxic lung injury. Am J Respir Crit Care Med 1997; 154:A95.
18. Ghebrehiwet B. The complement system: mechanisms of activation, regulation, and biologic functions. In: Kaplan A, ed. Allergy. 2d ed. Philadelphia: Saunders, 1997: 219–234.
19. Tracey KJ, Beutler B, Lowry SF, et al. Shock and tissue injury induced by recombinant human cachectin. Science 1986; 234:470–474.
20. Samuelsson B. Leukotrienes: mediators of immediate hypersensitivity and inflammation. Science 1983; 220:568–575.
21. Pakbaz H, Foda HD, Berisha HI, Trotz M, Said SI. Paraquat-induced lung injury: prevention by vasoactive intestinal peptide (VIP) and the related peptide helodermin. Am J Physiol 265 (Lung Cell Mol Physiol 9):L369–L373, 1993.
22. Perretti M, Flower RJ. Anti-inflammatory lipocortin-derived peptides. In: Pruzanski W, Vadas P. Eds. Novel Molecular Approaches to Anti-Inflammatory Theory. Basel: Birkhäuser, 1995:131–138.
23. Said SI. Vasoactive intestinal peptide and nitric oxide: Divergent roles in relation to tissue injury. Ann NY Acad Sci 1996; 805:379–388.
24. Lundberg J, Hökfelt T. Coexistence of peptides and classical neuro-transmitters. Trends Neurosci 1983; 6:325–333.

Part One

CELL INJURY AND INFLAMMATION

2

Inflammation, Cell Injury, and Apoptosis

ADRIANO GIORGIO ROSSI and CHRISTOPHER HASLETT

University of Edinburgh
Edinburgh, Scotland, United Kingdom

I. Introduction

Inflammation is normally a highly complex, interdependent biochemical and physiological protective response to tissue injury (e.g., due to bacterial infection or trauma).
This process serves to destroy, dilute, partition off, or remove the injurious agent and
the injured tissues, thereby promoting tissue repair. When this crucial and normally
beneficial response occurs in an uncontrolled or exaggerated manner, the result is
inappropriate or excessive tissue damage that often results in chronic inflammation.
The products liberated by leukocytes recruited to sites of inflammation are primarily
responsible for the tissue damage associated with these chronic inflammatory diseases. For example, leukocyte-derived reactive oxygen species (e.g., O_2-, H_2O_2,
OH^-, NO) and proteases (e.g., elastase and collagenase) can be deleterious to cells
and tissues. Thus, the inflammatory response is also a fundamental process involved
in the pathogenesis of many diseases, including asthma, chronic bronchitis, emphysema, rheumatoid arthritis, glomerulonephritis, and myocardial infarction and reperfusion injury, most of which are highly prominent, and on the increase, in developed
societies. Understanding the mechanisms underlying the induction and resolution of
the inflammatory response will have far-reaching implications for the prevention and
treatment of inflammatory diseases.

II. The Inflammatory Response

The gross clinical hallmarks of the acute inflammatory response were described at the beginning of the first millennium and include redness and swelling, with heat, pain, and often loss of function. The processes responsible for these changes have only really became apparent during the last century. The redness (erythema) and increased warmth occurring at sites of inflammation result mainly from vasodilatation leading to increased blood flow to the tissues. Vasodilatation of capillaries and arterioles produced by agents such as prostaglandins (e.g., PGE_2, PGI_2, PGD_2), neuropeptides (e.g., calcitonin gene-related peptide, vasoactive intestinal peptide, substance P), and nitric oxide cause an augmented intraluminal hydrostatic pressure downstream in postcapillary venules where plasma leakage occurs. These mediators act in concert with agents that cause an increase in postcapillary venule microvascular permeability, resulting in tissue fluid exudation (swelling or edema). Two main types of agents cause changes in microvascular permeability, those which act directly on endothelial cells (e.g., histamine, vasopressin, bradykinin, PAF) and those which cause increased microvascular permeability that depends on the recruitment of inflammatory cells. These latter mediators include complement-derived peptides (e.g., C5a and C3a), bacterial products (e.g., formylated peptides), lipids (e.g., LTB_4), and cytokines (e.g., IL-8) (1–3). The concerted action of these processes leads to the classical responses observed during inflammation.

III. Mechanisms Underlying Inflammatory Cell Recruitment

Over the last two decades a great deal of effort has been devoted to elucidating the mechanisms underlying recruitment of inflammatory cells to sites of inflammation (2,3). Inflammatory cells, in order to reach sites of inflammation, have to (a) adhere loosely (tethering and rolling) to the postcapillary endothelial cells, a process mediated primarily by the selectins (E-, L- and P-selectin) and their respective ligands; (b) flatten and undergo firm adhesion to the endothelial cells, where the adhesive events are mediated mainly by the integrins (e.g., β_2 and β_1 integrins) and the immunoglobulin superfamily (e.g., ICAM-1, ICAM-2, VCAM-1); (c) traverse the endothelial cell layer of the blood vessel wall; (d) penetrate the basement membrane; and (e) migrate through the interstitial medium to reach the affected area (see Fig. 1). The site or the cause of the inflammatory response will often determine which type of inflammatory cell is recruited, possibly as a result of mediators generated (e.g., IL-8 for neutrophils and eotaxin for eosinophils) and adhesion molecules expressed (e.g., VLA-4 [$\alpha_4\beta_1$] is highly expressed on eosinophils and lymphocytes but not on neutrophils). In addition, the structure and physiology of the vasculature may influence leukocyte recruitment to an inflammatory site. For example, in rheumatoid arthritis and in respiratory distress syndromes of the adult and neonate, the predominant leukocyte is the neutrophil; whereas in parasitic infection and in asthma, there is

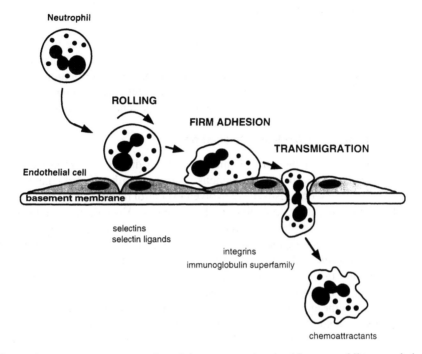

Figure 1   Schematic representation of the processes involved in neutrophil accumulation.

an abundance of recruited eosinophils. Evidence indicates that capillaries of the lung are the major site of leukocyte emigration, whereas postcapillary venules are the principal site in the systemic circulation (including the bronchial circulation). Intravascular pressure in the pulmonary circulation is approximately 10-fold lower than in the systemic circulation, the blood flow is pulsatile, and the diameter of the pulmonary capillaries is somewhat smaller than the mean diameter of granulocytes. These cells therefore have to deform to pass through the pulmonary capillaries. Agents such as FMLP and C5a reduce neutrophil deformability, making them more rigid and favoring retention in the capillaries of the lung (see Refs. 3 and 4 for reviews).

    Once inflammatory cells have been recruited to an inflammatory focus, if the concentration of secretagogue stimuli is sufficient, they become fully activated and, for example, phagocytose invading bacteria and liberate into the phagosome digestive enzymes and bactericidal oxygen metabolites. This scenario is perhaps akin to the phenomenon of "priming" observed in vitro, where cells such as the neutrophil and eosinophil in the presence of appropriate agents are rendered hyperresponsive to further activation by the same or other stimuli (5–7). The accumulation of leukocytes therefore depends on the complex interaction between inflammatory mediators, the expression of adhesion molecules and extracellular matrix proteins, and the physiology of the vasculature.

IV. Resolution of Inflammation

Although much attention has been directed toward understanding the mechanisms underlying the induction of inflammation, it is now apparent that the inflammatory process is regulated by a balance between its induction and its resolution (8). The remainder of this review will focus on this latter, rapidly developing area, with particular emphasis on the process of apoptosis and the removal of apoptotic cells.

A. Inflammatory Mediator Dissipation

At inflammatory foci, a plethora of inflammatory mediators are produced endo-genously and/or introduced from exogenous sources. In order for inflammation to resolve, a number of mechanisms have evolved to control removal or inactivation of these pleiotropic mediators, including the following. (a) Vasodilatation and edema formation not only bring plasma proteins and leukocytes to inflammatory foci but also dilute, thereby reducing the effective concentrations, of inflammatory stimuli at the site. (b) The primary function of the accumulated leukocytes is to remove or destroy invading organisms; for example, eosinophils will neutralize parasites and neutrophils will phagocytose infecting bacteria. Removal of the source of exogenous mediators reduces the effective concentration of the inflammatory products. (c) Some inflammatory mediators will spontaneously become deactivated; for example, throm-boxane A_2 and nitric oxide can rapidly decompose to their relatively inactive me-tabolites, whereas others such as C5a, PAF, and LTB_4 are enzymatically rendered impotent. (d) A reduction in the synthesis and release, or an increase in the metabo-lism of a particular mediator, will also limit the concentration of the mediator. (e) Mediator-induced cellular or tissue responsiveness may also be reduced due to less well defined phenomena such as cellular or tissue desensitization. (f) Many mediators, generated at inflammatory sites, have opposing effects on effector cells and/or tissues; for example, C5a and IL-8 are potent neutrophil chemotactic agents, whereas prostaglandins can exert inhibitory effects on these cells by acting on specific cell surface receptors (9). Some inflammatory mediators (e.g., GM-CSF) prolong the survival of neutrophils by inhibiting apoptosis, whereas others (e.g., TNF-α) can induce apoptosis (see below).

B. Termination of Inflammatory Cell Recruitment

Factors that control cessation of inflammatory cell accumulation remain ill defined. A number of in vivo studies have investigated the kinetics of accumulation of radiolabeled inflammatory cells in order to obtain information on the temporal emigration profiles of these cells from the blood to experimentally induced inflamed tissue. For example, in the acutely inflamed skin, lung, or joints, neutrophils accumu-late within the first few hours (usually between 30 min and 4 hr), followed by a later, more protracted accumulation of monocytes. It was observed that the neutrophil influx in these acute inflammatory situations ceased remarkably quickly (10–13). In chronic inflammatory models or diseases where there is tissue damage and scarring, neutrophils and macrophages continue to accumulate for several days before resolution occurs (13). The mechanisms responsible for early cessation are obscure, but our study (14) suggests that dissipation of mediators may be a key element.

C. Control of Inflammatory Cell Responses

Undoubtedly, a number of mechanisms exist to limit the destructive capacity of inflammatory cells and allow resolution to occur, including: (a) a reduction in the effective concentration of proinflammatory stimuli; (b) an increase in antiinflammatory mediators; (c) an exhaustion of the responsiveness of inflammatory cells—i.e., cells may no longer have the capability to liberate destructive enzymes or toxic oxygen metabolites due to depletion of cellular resources or internal energy stores; (d) receptor or cellular desensitization; and (3) ultimately, the death and removal of the cell itself.

D. Apoptosis of Inflammatory Cells

Before the seminal work of Wyllie and colleagues (15,16), who described a physiological mode of cell death termed apoptosis (programmed cell death), it was widely believed that cells simply died by necrosis in an uncontrolled manner (see Ref. 17). If this were the case, most cells, and especially inflammatory cells, would liberate their intracellular contents, many of which have major histotoxic and proinflammatory properties and would thus be expected to provoke an exaggerated inflammatory response. Virtually all cells under normal conditions undergo controlled physiological death by apoptosis, a process known to play a fundamental role in almost all aspects of life (16), including the resolution of the inflammatory response (8,18). Cells that have undergone apoptosis show remarkably similar structural, morphological, and biochemical changes. These similarities are perhaps indicative of a common underlying series of molecular mechanisms. Apoptotic cells are often smaller (due to cytoplasmic shrinkage), vacuolated, and exhibit major cell surface changes. Importantly, apoptotic cells remain intact, retaining their cytoplasmic granules and maintaining plasma membrane integrity such that they exclude vital dyes (e.g., trypan blue). It is the ultrastructural changes observed in the nucleus that are strikingly characteristic of an apoptotic cell; there is condensation of nuclear chromatin into dense, crescent-shaped aggregates, with the nucleolus becoming more conspicuous. In cells such as the neutrophil and eosinophil, where the nucleus is multilobed or bilobed, there is nuclear coalescence. Biochemically, apoptosis is typically characterized by endogenous endonuclease activation resulting in internucleosomal cleavage of chromatin. When DNA is extracted from apoptotic cells and separated electrophoretically on an agarose gel, there is a characteristic "ladder" pattern of DNA fragments, representing multimers of the 180–200 base pairs of DNA associated with nucleosomes. Most apoptotic cells express phosphatidylserine molecules on the external surface of their plasma membranes, a process which appears to be highly regulated (19–22), important in phagocyte recognition (see below), and useful for assessing apoptosis per se (23). Other cell surface changes may be unique to a particular cell system—for example, shedding of FcγRIII (CD16) from neutrophils (24,25). It has also been shown that apoptotic neutrophils are less responsive to external receptor-directed stimuli, in that neutrophil functions such as chemotaxis, phagocytosis, and degranulation are downregulated due to loss of cell surface receptors (e.g., FMLP receptors) and molecules important in phagocytosis (CD16) and adhesion (L-selectin) (24–27). These processes are likely to be important mechanisms in reducing the potential of cells to cause tissue damage, isolating apoptotic neutrophils from proinflammatory stimuli that would normally trigger effector function.

E. Regulation of Inflammatory Cell Apoptosis

It is now clear that apoptosis in granulocytes is regulated differently than in lymphocytes and thymocytes. Indeed, there are differences in the regulatory mechanisms for the closely related neutrophilic and eosinophilic granulocytes (28,29). It is well established that certain inflammatory mediators prolong survival of inflammatory cells by inhibiting the rate of apoptosis (30–32). Furthermore, inflammatory stimuli such as GM-CSF, LPS, and C5a, by profoundly inhibiting neutrophil apoptosis, greatly prolong the functional longevity of these cells as assessed by chemotaxis and degranulation assays (26). Eosinophil apoptosis is also suppressed by GM-CSF and Il-5 (31), and by eotaxin (Rossi, Hannah, and Haslett, unpublished observations); IL-5 and eotaxin having no effect on neutrophil longevity. Suppression of apoptosis is not a universal phenomenon associated with all proinflammatory mediators; treatment of isolated human neutrophils in culture with TNF-α for 6 hr will accelerate the rate of apoptosis; whereas at later time points ($>$18 hr), apoptosis is inhibited by this cytokine (33,34). The mechanisms underlying TNF-α induction of apoptosis is unknown, but likely involves TNF receptor-associated proteins such as TRADD, FADD, and FLICE and may involve Fas (CD95/APO-1) and/or the sphingomyelin-ceramide pathways (35,36). It has recently been demonstrated that anti-Fas-activating antibodies or Fas-L promote neutrophil and eosinophil apoptosis, an effect blocked by anti-Fas-blocking antibodies (37,38). Whether Fas ligation plays an important role in inflammatory disease remains to be fully determined, although this pathway has been reported to induce apoptosis in experimental models in vivo (39). We have recently reported that nitric oxide donors induce granulocyte apoptosis via a cGMP-independent pathway, an observation that may explain part of the beneficial influence of nitric oxide in inflammation (40). Hypoxic conditions, which are found in chronically inflamed sites, inhibit neutrophil apoptosis, whereas apoptosis is induced in some cell lines (41). Other agents, such as the glucocorticoids (e.g., dexamethasone), inhibit neutrophil apoptosis concentration and steroid receptor dependently, but promote eosinophil apoptosis (28). Although the mechanisms underlying this divergent effect of glucocorticoids on granulocyte apoptosis are currently unknown, the effect may explain why steroid therapy is so efficacious in eosinophilic inflammation such as asthma. Interestingly, asthmatic patients who show an improvement in lung function after corticosteroid treatment also have an increased proportion of apoptotic eosinophils in their airway secretions (42).

Using a pharmacological approach, the intracellular signaling pathways that control granulocyte apoptosis are slowly being unraveled. Neutrophil apoptosis is inhibited by elevation of intracellular free Ca^{2+} levels by calcium ionophores such as A23187 and ionomycin or by mobilization of intracellular stores using thapsigargin (29,43), whereas eosinophil (29) and thymocyte (44) apoptosis is promoted. A key role for protein kinases is indicated by experiments in which apoptosis is promoted by inhibition of protein kinase C by staurosporine and Ro-31,8220 (29) or delayed by activation of protein kinase A following treatment by agents that elevate cAMP (25,45). Interestingly, elevation of cAMP promotes the rate of apoptosis in thymocytes (46) and in leukemic cell lines (47,48). In addition, phosphatase inhibitors, okadaic acid and calyculin, inhibit or promote granulocytes apoptosis depending on the concentration used (49). Although intracellular kinase activity plays a key regulatory role in granulocyte apoptosis, the precise kinase(s) involved remain to be established. Why the intracellular signaling pathways that regulate neutrophil apoptosis are different from other cells, including the eosinophil, remains elusive but may be due to the differential expression of levels or types of protooncogenes shown to be involved in other cells systems (50,51). Although we and others have demonstrated that neutrophils do not normally express Bcl-2 protein, it is possible that other family members may be involved since it has been reported that overexpression of Bcl-2 in neutrophils prolongs neutrophil survival (52). It remains to be established whether the recently characterized novel Bcl-2-like proteins are also expressed or play a role in neutrophil apoptosis (53). Protein synthesis inhibitors such as cycloheximide and actinomycin D, while inhibiting thymocyte apoptosis, promote neutrophil apoptosis, indicating endogenous synthesis of neutrophil survival proteins (54). Recent data also suggest that CD11b/CD18-dependent neutrophil phagocytosis can induce neutrophil apoptosis, a response that appears to be regulated by the formation of reactive oxygen intermediates (55). Despite the progress made during the last few years in identifying signaling mechanisms that control granulocyte apoptosis, it is obvious that much remains to be elucidated.

F. Macrophage Phagocytosis of Senescent Apoptotic Cells

Apoptotic cells at the inflammatory site would inevitably undergo disintegration (secondary necrosis) if macrophages were absent in situ, resulting in the release of their histotoxic contents and thereby causing further inflammation and the destruction of tissue cells. This highly undesirable scenario may occur in uncontrolled chronic inflammation, where the normally highly effective macrophage recognition and clearance mechanisms (see below) could be overwhelmed or defective. Indeed, it was once generally assumed that the majority of extravasated leukocytes met their ultimate fate at the inflamed site by necrosis, with removal of cell debris by tissue macrophages (17). Such a removal pathway would result in release of large quantities of leukocyte granular contents into the surrounding milieu, resulting in an exaggerated, potentially injurious inflammatory response. The majority of senescent cells at sites of inflammation are recognized by nearby phagocytes. Over 100 years ago, using recently developed intravital light microscopic techniques on a number of transparent invertebrates, the Russian biologist Elie Metchnikoff (56) first described phagocytosis of bacteria and senescent leukocytes by macrophages. Of particular relevance to the resolution of inflammation is the description of Reiter's cells in the inflamed joint fluid from people with Reiter's disease and other forms of acute arthritis. These Reiter's cells were shown to be macrophages containing apoptotic neutrophils (57,58). Furthermore, in experimental inflammatory models where there is a large influx of neutrophils, the predominant mode of extravasated intact neutrophil removal is engulfment by the macrophages (59,60), which subsequently leave the inflammatory site via the draining lymph nodes (61).

Although there have been numerous accounts of macrophage phagocytosis of neutrophils over the last century, it is only relatively recently that major developments regarding recognition of apoptotic cells have been described. Newman et al. (62) first showed that isolated neutrophils which had been aged in culture were recognized and ingested by inflammatory macrophages. The macrophages engorged themselves with senescent neutrophils but did not ingest freshly isolated neutrophils, indicating that aged neutrophils had undergone a fundamental process leading to recognition by macrophages. We subsequently showed that cells (e.g., neutrophils and eosinophils) had to undergo apoptosis before they were recognized and ingested by the macrophage—an extremely rapid and efficient process (31,63,64). The macrophage can ingest several apoptotic cells at a time and, importantly, there is no detectable release of potentially injurious neutrophil contents (e.g., granule enzyme markers) into the surrounding medium (65,66), effectively neutralizing any proinflammatory potential of apoptotic cells. Furthermore, when macrophages ingest particles such as zymosan (yeast cell walls) in vitro, the response is to liberate proinflammatory mediators (e.g., eicosanoids, granular enzymes, and cytokines). However, ingestion of apoptotic neutrophils does not result in such a proinflammatory mediator release from macrophages (67). We subsequently showed that this noninflammatory clearance mechanism was due to the molecular mechanisms by which macrophages recognize and ingest apoptotic cells (see 8,18).

G. Mechanisms for Recognition and Clearance of Apoptotic Cells

Phagocytes (usually macrophages) have developed highly efficient recognition mechanisms for the removal of apoptotic cells. Although it is the macrophage that is believed to remove the majority of apoptotic cells, other cells, such as fibroblasts (68), hepatocytes (69), endothelial cells (70), and glomelular mesangial cells (71), have a limited capacity for apoptotic cell recognition and removal. Perhaps the removal of apoptotic cells by these "semiprofessional" phagocytes plays a crucial role when local macrophages are overwhelmed by major waves of apoptotic cells. Surface changes on the apoptotic cell that may lead to phagocytic recognition remain poorly understood. The known molecular mechanisms of recognition of apoptotic cells are listed in Table 1, together with appropriate references [for a more detailed

Table 1   Molecular Mechanisms of Recognition of Apoptotic Cells by Phagocytes

	References
Lectin-like receptors	
Specific carbohydrates on the surface of apoptotic cells are recognized by phagocytic cells containing lectin-like receptors such as the asialoglyco-protein, mannose or mannose/fucose receptor.	68–70,73
$\alpha_v\beta_3$/CD36/thrombospondin (TSP)	
It is hypothesized that a TSP-binding, moiety expressed on the surface of apoptotic cells binds to nearby TSP, which in turn acts as a bridging molecule between the apoptotic cell and the ingesting phagocyte. The phagocyte expresses two receptors on its surface, the $\alpha_v\beta_3$ "vitronectin receptor" integrin and an 88-kDa monomer termed CD36, which cooperates to bind TSP.	63,74–77
Phosphatidylserine (PS) and PS receptors	
Exposure of phosphatidylserine on the surface of apoptotic cells is believed to be recognized by putative phosphatidylserine receptors located on the surface of the ingesting phagocyte. Recently, it has been suggested that members of the scavenger receptor family may act as PS receptors.	19–22,78–84
61D3 antigen	
The mAb 61D3 can specifically attenuate the recognition of apoptotic cells by human monocyte-derived macrophages. The 61D3 antigen has not been fully characterized.	85

description of these mechanisms, we recommend the reviews by Savill et al. (18) and Hart et al. (72)]. Given that multiple genes are involved in the disposal of apoptotic bodies in the nematode *Caenorhabditis elegans*, it is likely that other unknown mechanisms exist in the recognition and phagocytosis of mammalian inflammatory cells.

V.   Relevance to Inflammatory Diseases: The Utilization of Apoptotic Mechanisms and Macrophage Clearance Mechanisms as Potential New Antiinflammatory Strategies

Given that in chronic inflammatory diseases there is persistent accumulation of inflammatory cells which likely liberate their toxic intracellular products and thereby lead to tissue injury, scarring and architectural disruption with catastrophic loss of organ function, it seems likely that effective removal of these tissue-damaging inflammatory cells would be desirable. It is possible that in these diseases the inflammatory cell clearance mechanisms are either defective or overwhelmed, since in other examples of massive inflammatory reactions, such as lobar streptococcal pneumonia and acute gout, the lesions clear completely. The mechanisms by which inflammation normally resolves are not only essential for understanding of the persistent inflammatory states that characterize chronic inflammatory diseases, but should also provide novel approaches to antiinflammatory therapy. Significant differences in the intracellular regulation of inflammatory cells provide a special opportunity to induce apoptosis in inflammatory cell types selectively. Thus, by utilizing mechanisms of receptor-mediated induction of apoptosis of specific inflammatory cells (e.g., Fas ligation and glucocorticoid receptor activation) and by inhibition of the powerful survival signals found at inflammatory foci (e.g., GMCSF, LPS, prostaglandins, IL-5), it may be possible to accelerate apoptosis for therapeutic gain. In

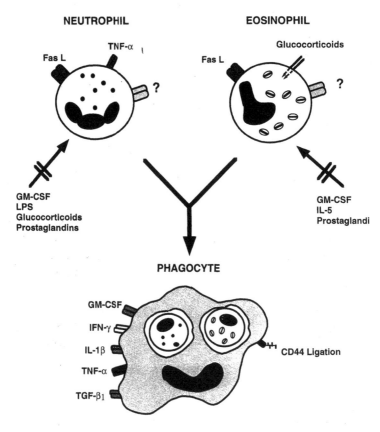

Figure 2   Strategies for promotion of granulocyte apoptosis and removal of apoptotic cells by phagocytes (GM-CSF, granulocyte/macrophage-colony stimulating factor; LPS, lipopolysaccharide; TNF-α, tumor necrosis factor-α; IL-5, interleukin-5; INF-γ, interferon-γ, IL-1β, interleukin-1β; TGF-β_1; transforming growth factor-β_1.

combination with an effective therapeutic strategy for inducing apoptosis, development of mechanisms to specifically enhance apoptotic cell clearance by phagocytes such as the macrophage, or by "semiprofessional" phagocytes, may be equally desirable or indeed a prerequisite to preventing proinflammatory secondary necrosis (Fig. 2). It has recently been reported that certain cytokines (77), activation of macrophage protein kinase C (McCutcheon, Haslett, and Dransfield, unpublished observations), and ligation of macrophage CD44 (86) can enhance macrophage clearance of apoptotic cells.

VI. Concluding Remarks and Future Directions

Although much progress has been made in understanding the mechanisms controlling the accumulation of inflammatory cells to sites of inflammation, we believe that elucidation of the mechanisms regulating the resolution of the inflammatory process can lead to novel approaches for therapeutic gain. Specifically, development of strategies to induce selective inflammatory cell apoptosis, together with augmentation of macrophage clearance mechanisms, will have far-reaching implications for the control of the many inflammatory diseases that are responsible for the heavy burden of morbidity and untimely deaths observed throughout the world.

Acknowledgments

We thank Dr. Ian Dransfield for helpful discussion and are grateful to the Medical Research Council (UK) for financial support.

References

1. Wedmore CV, Williams TJ. Control of vascular permeability by polymorphonuclear leukocytes in inflammation. Nature 1981; 289:646–650.
2. Showell HJ, Williams TJ. The neutrophil in inflammation. In: Roger TJ, Gilman SC, eds. Immunopharmacology. New York: Telford, 1989:23–63.
3. Rossi AG, Hellewell PG. Mechanisms of neutrophil accumulation in tissues. In: Hellewell PG, Williams TJ, eds. Immunopharmacology of Neutrophils. London: Academic Press, 1994:223–243.
4. MacNee W, Selby C. Neutrophil kinetics in the lung. Clin Sci 1990; 79:97–107.
5. O'Flaherty JT, Rossi AG. 5-Hydroxyicosatetraenoate stimulate neutrophils by a stereospecific, G-protein linked mechanism. J Biol Chem 1993; 268:14708–14714.
6. Rossi AG, MacIntyre DE, Jones CJP, McMillan RM. Stimulation of human neutrophil polmorphonuclear leukocytes by leukotriene B$_4$ and platelet-activating factor: an ultrastructural and pharmacological study. J Leuk Biol 1993; 53:117–125.
7. Kitchen E, Rossi AG, Condliffe AM, Haslett C, Chilvers ER. Demonstration of reversible priming of human neutrophils using platelet-activating factor. Blood 1996; 88:4330–4337.

8. Haslett C, Savill JS, Whyte MKB, Stern M, Dransfield I, Meagher LC. Granulocyte apoptosis and the control of inflammation. Phil Trans R Soc 1994; 345:327–333.
9. Rossi AG, O'Flaherty JT. Prostaglandin binding sites in human polymorphonuclear neutrophils. Prostaglandins 1989; 37:641–653.
10. Colditz IG, Movat HZ. Densensitisation of acute inflammatory lesions to chemotaxins and endotoxin. J Immunol 1984; 133:2163–2168.
11. Saverymuttu SH, Phillips G, Peters AM, Lavender JP. Indium-111 autologous leucocyte scanning in lobar pneumonia and lung abscesses. Thorax 1985; 40:925–930.
12. Clark RJ, Jones HA, Rhodes CG, Haslett C. Non-invasive assessment in self-limited pulmonary inflammation by external scintigraphy of ¹¹¹indium-labelled neutrophil influx and by measurement of the local metabolic response with positron emission tomography. Am Rev Respir Dis 1989; 139:A58.
13. Haslett C, Shen AS, Feldsien DC, Allen D, Henson PM, Cherniack RM. ¹¹¹Indium-labelled neutrophil flux into the lungs of bleomycin-treated rabbits assessed non-invasively by external scintigraphy. Am Rev Respir Dis 1989; 140:756–763.
14. Haslett C, Jose PJ, Giclas PC, Williams TJ, Henson PM. Cessation of neutrophil influx in C5a-induced acute experimental arthritis is associated with loss of chemoattractant activity from joint spaces. J Immunol 1989; 142:3510–3517.
15. Kerr JFR, Wyllie AH, Currie AR. Apoptosis: a basic biological phenomenon with wide ranging implications in tissue kinetics. Br J. Cancer 1972; 26:239–257.
16. Wyllie AH, Kerr JFR, Currie AR. Cell death: the significance of apoptosis. Int Rev Cytol 1980; 68:251–306.
17. Hurley JV. Termination of acute inflammation. I. Resolution. In: Hurley JV, ed. Acute Inflammation. 2d ed. London: Churchill Livingstone, 1983:109–117.
18. Savill JS, Haslett C. Fate of neutrophils. In: Hellewell PG, Williams TJ, eds. Immunopharmacology of Neutrophils. London: Academic Press, 1994:295–314.
19. Fadok VA, Savill JS, Haslett C, Bratton DL, Doherty D, Campbell PA, Henson PM. Different populations of macrophages use either the vitronectin receptor or the phosphatidylserine receptor to recognize and remove apoptotic cells. J Immunol 1992; 149:4029–4035.
20. Fadok VA, Voelker DR, Campbell PA, Cohen JJ, Bratton DL, Henson PM. Exposure of phosphatidylserine on the surface of apoptotic lymphocytes triggers specific recognition and removal by macrophages. J Immunol 1992; 148:2207–2216.
21. Fadok VA, Laszlo DJ, Noble PW, Weinstein L, Riches DW, Henson PM. Particle digestibility is required for induction of the murine phosphatidylserine recognition mechanism used by murine macrophages to phagocytose apoptotic cells. J Immunol 1993; 151:4274–4285.
22. Fadok VA, Laszlo DJ, Noble PW, Weinstein L, Riches DWH, Henson PM. Particle digestibility is required for induction of the phosphatidylserine recognition mechanism used by murine macrophages to phagocytose apoptotic cells. J Immunol 1993; 151:4274–4285.
23. Koopman G, Reutelingsperger CPM, Kuijten GAM, Keehnen RMJ, Pals ST, Van Oers MHJ. AnnexinV for flow cytometric detection of phosphatidylserine expression on B cells undergoing apoptosis. Blood 1994; 84:1415–1420.
24. Dransfield I, Buckle A-M, Savill JS, McDowall A, Haslett C, Hogg N. Neutrophil apoptosis is associated with a reduction in CD16 (FcγRIII) expression. J Immunol 1994; 153:1254–1263.

25. Rossi AG, Cousin JM, Dransfield I, Lawson MF, Chilvers ER, Haslett C. Agents that elevate cAMP inhibit human neutrophil apoptosis. Biochem Biophys Res Commun 1995; 217:892–899.

26. Whyte MKB, Meagher LC, MacDermot J, Haslett C. Impairment of function in aging neutrophils is associated with apoptosis. J Immunol 1993; 150:5124–5134.

27. Dransfield I, Stocks SC, Haslett C. Regulation of cell adhesion molecule expression and function associated with neutrophil apoptosis. Blood 1995; 85:3264–3273.

28. Meagher L, Cousin JM, Seckl JR, Haslett C. Opposing effects of glucocorticoids on the rate of apoptosis in neutrophilic and eosinophilic granulocytes. J Immunol 1996; 156: 4422–4428.

29. Cousin JM, Haslett C, Rossi AG. Regulation of granulocyte apoptosis by PKC inhibition and elevation of [Ca^{2+}]i. Biochem Soc Trans 1997; 25:243S.

30. Lee A, Whyte MBK, Haslett C. Prolonged *in vitro* lifespan and functional longevity of neutrophils induced by inflammatory mediators acting through inhibition of apoptosis. J Leuk Biol 1993; 54:283–288.

31. Stern M, Meagher L, Savill JS, Haslett C. Apoptosis in human eosinophils. Programmed cell death in the eosinophil leads to phagocytosis by macrophages and is modulated by IL-5. J Immunol 1992; 148:3543–3549.

32. Ward C, Hannah S, Chilvers ER, Farrow S, Haslett C, Rossi AG. Transforming growth factor-β increases the inhibitory effect of GM-CSF and dexamethasone on neutrophil apoptosis. Biochem Soc Trans 1997; 25:244S.

33. Takeda Y, Watanabe H, Yonehara S, Yamashita T, Saito S, Sendo F. Rapid acceleration of neutrophil apoptosis by tumor necrosis factor. Int Immunol 1993; 5:691–694.

34. Murray J, Barbara JAJ, Dunkley SA, Lopez AF, Ostade XF, Condliffe AM, Dransfield I, Haslett C, Chilvers ER. Regulation of neutrophil apoptosis by tumour necrosis factor α: requirement for TNFR55 and TNFR-75 for induction of apoptosis *in vitro*. Blood 1997. In press.

35. Whyte MKB. A link between cell-surface receptors and ICE proteases. Trends Cell Biol 1996; 6:418.

36. McConkey DJ, Orrenius S. Signal transduction pathways to apoptosis. Trends Cell Biol 1996; 4:370–374.

37. Liles WC, Kiener PA, Ledbetter JA, Aruffo A, Klebenoff SJ. Differential expression of Fas (CD95) and Fas ligand on normal human phagocytes: implications for the regulation of apoptosis in neutrophils. J Exp Med 1996; 184:429–440.

38. Matsumoto K, Schleimet RP, Saito H, Iikura Y, Bochner BS. Induction of apoptosis in human eosinophils by anti-Fas antibody treatment *in vitro*. Blood 1995; 86:1437–1443.

39. Anderson GP. Resolution of chronic inflammation by therapeutic induction of apoptosis. TIPS 1996; 17:438–442.

40. Chilvers ER, Wong TH, Rahman I, Haslett C, Rossi AG. Nitric oxide stimulates neutrophil apoptosis via a cyclic-GMP-independent mechanisms. Keystone Symposia: Apoptosis and Programmed Cell Death, 1997; A109.

41. Hannah S, Mecklenburgh K, Rahman I, Bellingan GJ, Greening A, Haslett C, Chilvers ER. Hypoxia prolongs neutrophil survival *in vitro*. FEBS Lett 1995; 372:233–237.

42. Woolley KL, Gipson PG, Carty K, Wilson AJ, Twaddell SH, Woolley MJ. Eosinophil apoptosis and the resolution of airway inflammation in asthma. Am J Respir Crit Care 1996; 154:237–243.

43. Whyte MKB, Meagher LC, Hardwick SJ, Savill JS, Haslett C. Transient elevations of

cytosolic free calcium retard subsequent apoptosis in neutrophils *in vitro.* J Clin Invest 1993; 92:446–455.

44. McConkey DJ, Nicotera P, Hartzell P, Bellomo G, Wyllie AH, Orrenius S. Glucocorticoids activate a suicide process in thymocytes through an elevation of cytosolic Ca^{2+} concentration. Arch Biochem Biophys 1989; 269:365–370.

45. Hallsworth MP, Geimbycz MA, Barnes PJ, Lee TH. Cyclic AMP-elevating agents prolong or inhibit eosinophil survival depending on prior exposure to GM-CSF. Br J Pharmacol 1996; 117:79–86.

46. McConkey DJ, Orrenius S, Okret S, Jondal M. Cyclic AMP potentiates glucocorticoid-induced endogenous endonuclease activation in thymocytes. FASEB J 1993; 7:580–585.

47. Lanotte M, Riviere JB, Hermouet S, Houge G, Vintermyr OK, Gjertsen BT, Doskeland SO. Programmed cell death (apoptosis) is induced rapidly and with positive cooperativity by activation of cyclic adenosine monophosphate-kinase I in a myeloid leukemic cell line. J Cell Phys 1991; 146:73–80.

48. Gjertsen BT, Cressey LI, Ruchard S, Houge G, Lanotte M, Doskeland SO. Multiple apoptotic death types triggered through activation of separate pathways by cAMP and inhibitors of protein phosphatases in one (IPC leukemia) cell line. J Cell Sci 1994; 107: 3363–3377.

49. Cousin JM, Haslett C, Rossi AG. Effect of protein phosphatase inhibitors on dexamethasone-mediated inhibition of neutrophil apoptosis. Biochem Soc Trans 1997; 25:246S.

50. Yang E, Korsmeyer SJ. Molecular thanatopsis: a discourse on the BCL2 family and cell death. Blood 1996; 88:386–401.

51. Martins LM, Earnshaw WC. Apoptosis: alive and kicking in 1997. Trends Cell Biol 1997; 7:111–114.

52. Lagasse E, Weissman IL. Bcl-2 inhibits apoptosis of neutrophils but not their engulfment by macrophages. J Exp Med 1994; 179:1047–1051.

53. Gibson L, Holmgreen SP, Huang DCS, Bernard O, Copeland NG, Jenkins NA, Sutherland GR, Baker E, Adams JM, Cory S. bcl-w, a novel member of the bcl-2 family, promotes cell survival. Oncogene 1996; 13:665–675.

54. Whyte MKB, Meagher LC, Lee A, Haslett C. Coupling of neutrophil apoptosis to recognition by macrophages: co-ordinated acceleration by protein synthesis inhibitors. J Leuk Biol 1997. In press.

55. Coxon A, Rieu P, Barkalow FJ, Askari S, Sharpe AH, von Andrian UH, Arnaout MA, Mayadas TN. A novel role for the β2 integrin CD11b/CD18 in neutrophil apoptosis: a homeostatic mechanism in inflammation. Immunity 1996; 5:653–666.

56. Metchnikoff E. Lectures on the comparative pathology of inflammation. Lecture VII. Delivered at the Pasteur Institute in 1891 (transl Starling FA, Starling EH). London: Kegan, Paul, Trench, Trubner, 1968.

57. Pekin T, Malinin TI, Zwaifler R. Unusual synovial fluid findings in Reiter's syndrome. Ann Intern Med 1967; 66:677–684.

58. Spriggs RS, Boddington MM, Mowat AG. Joint fluid cytology in Reiter's syndrome. Ann Rheum Dis 1978; 37:557–560.

59. Sanui H, Yoshida S-I, Nomoto K, Ohhara R, Adachi Y. Peritoneal macrophages which phagocytose autologous polymorphonuclear leucocytes in guinea-pigs. Br J Exp Pathol 1982; 63:278–285.

60. Chapes SK, Haskill S. Evidence for granulocyte-mediated macrophage activation after *C. parvum* immunization. Cell Immunol 1983; 75:367–377.

61. Bellingan GJ, Caldwell H, Howie SEM, Dransfield I, Haslett C. *In vivo* fate of the inflammatory macrophage during the resolution of inflammation: inflammatory macrophages do not die locally but emigrate to the draining lymph nodes. J Immunol 1996; 157: 2577–2585.
62. Newman SL, Henson JE, Henson PM. Phagocytosis of senescent neutrophils by human monocyte-derived macrophages and rabbit inflammatory macrophages. J Exp Med 1982; 156:430–442.
63. Savill JS, Henson PM, Haslett C. Phagocytosis of aged human neutrophils by macrophages is mediated by a novel "charge-sensitive" recognition mechanism. J Clin Invest 1989; 84:1518–1527.
64. Stern M, Savill JS, Haslett C. Human monocyte-derived macrophage phagocytosis of senescent eosinophils undergoing apoptosis. Mediation by $\alpha_v\beta_3$/CD36/ thrombospondin recognition mechanism and lack of phlogistic response. Am J Pathol 1996; 149:911–921.
65. Savill JS, Wyllie AH, Henson JE, Walport MJ, Henson PM, Haslett C. Macrophage phagocytosis of aging neutrophils in inflammation: programmed cell death in the neutrophil leads to its recognition by macrophages. J Clin Invest 1989; 83:865–875.
66. Kar S, Ren Y, Savill JS, Haslett C. Inhibition of macrophage phagocytosis *in vitro* of aged neutrophils increases release of neutrophil contents (abstr). Clin Sci 1995; 85:27p.
67. Meagher LC, Savill JS, Baker A, Fuller RW, Haslett C. Phagocytosis of apoptotic neutrophils does not induce macrophage release of thromboxane B_2. J Leuk Biol 1992; 52:269–273.
68. Hall SE, Savill JS, Henson PM, Haslett C. Apoptotic neutrophils are phagocytosed by fibroblasts with participation of the fibroblast vitronectin receptor and involvement of a mannose/fucose-specific lectin. J Immunol 1994; 153:3218–3227.
69. Dini L, Autuori F, Lentini A, Oliverio S, Piacentini M. The clearance of apoptotic cells in the liver is mediated by the asialoglycoprotein receptor. FEBS Lett 1992; 296:174–178.
70. Dini L, Lentini A, Diez Diez G, Rocha M, Falasca L, Serafino L, Vidal-Vanaclocha F. Phagocytosis of apoptotic bodies by liver endothelial cells. J Cell Sci 1995; 108:967–973.
71. Savill JS, Smith J, Sarraf C, Ren Y, Abbott F, Rees A. Glomerular mesangial cells and inflammatory macrophages ingest neutrophils undergoing apoptosis. Kidney Int 1992; 42: 924–936.
72. Hart SP, Haslett C, Dransfield I. Recognition of apoptotic cells by phagocytes. Experientia 1996; 52:950–956.
73. Duvall E, Wyllie AH, Morris RG. Macrophage recognition of cells undergoing programmed cell death (apoptosis). Immunology 1985; 56:351–358.
74. Savill JS, Dransfield I, Hogg N, Haslett C. Vitronectin receptor-mediated phagocytosis of cells undergoing apoptosis. Nature 1990; 342:170–173.
75. Savill JS, Hogg N, Ren Y, Haslett C. Thrombospondin cooperates with CD36 and the vitronectin receptor in macrophage recognition of neutrophils undergoing apoptosis. J Clin Invest 1992; 90:1513–1522.
76. Ren Y, Silverstein RL, Allen J, Savill JS. CD36 gene transfer confers capacity for phagocytosis of cells undergoing apoptosis. J Exp Med 1995; 181:1857–1862.
77. Ren Y, Savill JS. Proinflammatory cytokines potentiate thrombospondin-mediated phagocytosis of neutrophils undergoing apoptosis. J Immunol 1995; 154:2366–2374.
78. Martin SJ, Reutelingsperger CPM, McGahon AJ, Rader JA, van Schie RCAA, LaFace DM, Green DR. Early redistribution of plasma membrane phosphatidylserine is a general feature of apoptosis regardless of the initiating stimulus: inhibition by overexpression of bcl-2 and Abl. J Exp Med 1995; 182:1545–1556.

79. Verhoven B, Schlegel RA, Williamson P. Mechanisms of phosphatidylserine exposure, a phagocyte recognition signal, on apoptotic T lymphocytes. J Exp Med 1995; 182:1597–1601.

80. Platt N, Suzuki H, Kurihara Y, Kodama T, Gordon S. Role of the class A macrophage scavenger receptor in the phagocytosis of apoptotic thymocytes *in vitro*. Proc Natl Acad Sci (USA) 1997; 93:12456–12460.

81. Ramprasad MP, Fischer W, Wiztum JL, Sambrano GR, Quehenberger O, Steinberg D. The 94-to 97-kDa mouse macrophage membrane protein that recognises oxidised low density lipoprotein and phosphatidylserine-rich liposomes is identical to macrosialin, the mouse homologue of human CD68. Proc Natl Acad Sci (USA) 1995; 92:9580–9584.

82. Rigotti A, Acton SL, Krieger M. The class B scavenger receptors SR-BI and CD36 are receptors for anionic phospholipids. J Biol Chem 1995; 270:16221–16224.

83. Sambrano GR, Parsatharathy S, Steinberg D. Recognition of oxidatively damaged erythrocytes by a macrophage receptor with specificity for oxidized low density lipoprotein. Proc Natl Acad Sci (USA) 1994; 91:3265–3269.

84. Sambrano GR, Steinberg D. Recognition of oxidatively damaged and apoptotic cells by an oxidized low density lipoprotein receptor on mouse peritoneal macrophages: role of membrane phosphatidylserine. Proc Natl Acad Sci (USA) 1995; 92:1396–1400.

85. Flora PK, Gregory GD. Recognition of apoptotic cells by human macrophages: inhibition by a monocyte/macrophage-specific monoclonal antibody. Eur J Immunol 1994; 24:2625–2632.

86. Hart SP, Dougherty GJ, Haslett C, Dransfield I. CD44 regulates phagocytosis of apoptotic neutrophil granulocytes, but not apoptotic lymphocytes, by human macrophages. J Immunol 1997; 159:919–925.

3

Proinflammatory and Antiinflammatory Cytokines in the Inflammatory Response

NICHOLAS W. LUKACS and PETER A. WARD

University of Michigan Medical School
Ann Arbor, Michigan

I. Introduction

The induction of immune/inflammatory responses is initiated by a complex series of events which can deal efficiently with a multitude of infectious agents, such as bacteria and fungi. However, an overzealous inflammatory reaction may lead to harmful responses within the tissues and organ systems, often leading to irreversible damage and dysfunction. Uncontrolled inflammation can be initiated by either nonspecific stimuli, such as endotoxin (lipopolysaccharide, LPS) or by specific stimuli, such as bacterial, viral, parasitic, or allergic antigen, any one of which may have a propensity for inducing chronic inflammatory responses. A common thread which runs among these latter outcomes is an intense leukocyte recruitment into inflamed areas (1,2). The recruitment of leukocyte populations into inflamed tissues is thought to be initiated by cytokine-induced expression of adhesion molecules on vascular endothelium. Endothelial adhesion molecules, which can be upregulated by TNF-α and IL-1, include ICAM-1, VCAM-1, and E-selectin (3–5). These adhesion molecules are considered to play an essential role in facilitating leukocyte adhesion to the endothelium, a necessary step to leukocyte transmigration. Subsequently, leukocyte spreading along the activated vascular endothelium is followed by leukocytic migration into the inflamed tissue. presumably directed by chemotactic molecules at the

25

site of inflammation. Rapid advances in the understanding of the sequence of events during leukocyte accumulation is only now beginning to shed light on the complex mechanisms which control the steps occurring during inflammatory events. In addition, regulation and resolution of inflammation relies on the production of antiinflammatory peptides which can modulate the intensity and duration of an inflammatory response. We will examine the involvement of proinflammatory and chemotactic cytokines/chemokines, which consist of peptides that maintain the leukocyte influx into sites of inflammation, as well as antiinflammatory peptides which regulate the inflammatory reaction.

II. Proinflammatory Cytokines

Cytokines produced during an inflammatory response allow the cell-to-cell communication which orchestrates the recruitment and activation of leukocyte populations. TNF-α and IL-1 production have been associated with both acute and chronic inflammatory disorders, including septic shock, pulmonary and hepatic injury, rheumatoid arthritis, and parasitic and viral infections. TNF-α was originally described by its ability to inhibit and/or directly kill tumor cells (6). However, the direct toxic effects of TNF-α appear to represent only a minor role when compared to its homeostatic and inflammatory functions. Both TNF-α and IL-1 appear to have overlapping effects on an inflammatory response. These effects depend on a number of parameters, including cytokine concentration, duration of cell exposure to cytokine, and the presence of other mediators that may act in a synergistic manner (1,2,6). The production of these two "early-response" cytokines, TNF-α and IL-1, during initial phases of a response appears to be crucial for subsequent leukocyte recruitment. IL-1 and TNF-α appear to have a primary role in induction of adhesion molecule expression and inflammatory cytokine production (Table 1). The ability of the system to produce IL-1 and TNF-α quickly during infectious stimuli allows a rapid upregulation of vascular adhesion molecules, a necessary step for leukocyte recruitment. Subsequently, the production of chemotactic peptides, such as chemo-

Table 1 Proinflammatory Cytokines and Their Functions

Cytokine	Function during inflammation
IL-1, TNF	Vascular adhesion molecule expression
	Initiation of cytokine cascades
	Activation of chemokine production
C-X-C chemokines	Recruitment of neutrophils
	Angiogenic/angiostatic activity
C-C chemokines	Recruitment of monocytes, T cells, eosinophils, basophils
	Costimulation of T lymphocytes
	Antibody isotype switching

kines, appears to induce the transendothelial migration of leukocytes into the developing inflammatory response. A cytokine network can be induced by the presence of TNF-α and IL-1, affecting surrounding immune and nonimmune cell populations and leading to the production of chemokines which then mediate recruitment and localization of leukocytes to the site of inflammation (7). The continuous activation and production of the leukocyte-derived early response mediators, as well as associated cytokines, can maintain the activation of the surrounding stromal cell populations sustaining leukocyte recruitment. This process may be maintained for extended periods, often after the initiating stimuli have been removed.

In recent years much focus has been directed toward understanding the cytokines involved in leukocyte accumulation at a site of inflammation. The elucidation of these pathways may offer an avenue for therapeutic intervention which could attenuate the intensity of the leukocyte infiltration in many inflammatory diseases and therefore alleviating the tissue injury and fibrosis. Chemokine molecules, unlike complement activation products (C3a, C5a) and lipid mediator metabolites (leukotriene B4, platelet activating factor), appear to induce specific chemotactic responses of leukocytes (8,9). Based upon the first two cysteine residues in their sequence, the chemokines have been divided into two distinct families, the C-X-C (alpha) and the C-C (beta) families (10–12) (Table 2). The functional activities of these family members also have distinct properties. A third family of chemokines have been described which contain a single cysteine. The C-X-C family of chemokines is predominantly chemotactic for neutrophils, whereas the C-C family members are chemotactic for mononuclear leukocytes, monocytes and lymphocytes, as well as for eosinophils and basophils, depending on the particular C-C chemokine. To an extent, these attributes appear to divide chemokine-driven functions between acute (C-X-C) and chronic (C-C) inflammatory responses. Not unexpectedly, the specificity of

Table 2 Chemokines

C-X-C (alpha)	C-C (beta)
ELR-containing	
IL-8	MCP-1, 2, 3, 4
ENA-78	MIP-1α
GRO-α,β,γ	MIP-1β
GCP-2	RANTES
b-thromboglobulin	Eotaxin
CTAP III	C10[a]
NAP-2	I-309[a]
Non-ELR-containing	
PF-4	
IP-10	
MIG	

[a]Found only in mouse systems.

biological activity for chemokines is not completely clear-cut, since multiple studies now have identified cross-over functions for C-X-C and C-C family members related to both neutrophil and mononuclear chemotactic responses. The C-X-C family includes IL-8, ENA-78, GRO-α,β,γ, CTAP III, NAP-2, MIP-2, PF-4, IP-10, and MIG. This family can be divided according to the presence of the amino acid motif glutamate-leucine-arginine (E-L-R), which appears to confer the ability to recruit neutrophils as well as provide an angiogenic signal (i.e., endothelial cell growth and vessel formation) (13,14). The C-C family contains several related proteins: MCP-1,-2,-3, MIP-1α, MIP-1β, eotaxin, and RANTES. As discussed above, these proteins are considered to have a role in promoting the recruitment of specific leukocyte populations during inflammatory events. The ability of chemokines to bind to heparin present on endothelial cells has led to the suggestion that chemokines operate in a "solid-phase" gradient, together with adhesion molecule expression, to cause leukocytes to accumulate specifically at a site of inflammation. The solid-phase bound chemokines would allow a chemokine gradient to localize the response to a very distinct area without being easily perturbed. This concept is attractive, since the chemotactic gradient can be readily maintained on a stable entity. The chemokine family members appear to have other functions as well (Table 1). For example, the C-X-C family chemokines (E-L-R containing) not only can recruit and activate neutrophils but also have angiogenic/angiostatic activity (13), whereas the C-C chemokines may have potent costimulatory functions on T-lymphocyte activation, either directly (15–17) or possibly through activation of antigen-presenting cells and adhesion (18,19). In addition, it appears that particular C-C chemokines can augment/mediate antibody isotype switching in B cells (20). More recently, new insights into the role of chemokine receptors as cofactors for the binding of HIV to CD4 molecules on monocytes and lymphocytes have suggested that they may be important for inhibition of viral entry into susceptible cells through blocking of permiscuous receptors (CCKR-3 and CCKR-5) using synthetic products which resemble chemokines (21–23). Especially relevant chemokines are MIP-1α, MIP-1β, and RANTES, which have already been shown to block viral entry and inhibit infection in vitro. The chemokines may be reasonable targets during inflammatory responses in human disease for several reasons: (a) They are produced later and in a more prolonged time frame than TNF-α and IL-1; (b) targeting the chemokines may inhibit not only recruitment but also activation of leukocytes; and (c) because the various family (C-X-C or C-C) members share permiscuous receptors, a single competitive receptor antagonist may block multiple chemokines with overlapping functions. This information has made the chemokines an attractive option for controlling the inflammatory response in a variety of conditions.

A. Role of Cytokines in the Acute Inflammatory Response

The role of cytokines in inflammation is suggested by their appearance in several disease states (Table 3). High levels of cytokines have been detected particularly in patients with septicemia, and the cytokines appear not only to activate the inflamma-

Table 3 Models of Acute Inflammation

Models	Relevant cytokines	
	Activating	Chemotactic
Septicemia		
Endotoxin	TNFα, IL-1	KC, MIP-2, MIP-1α, RANTES
Cecal ligation and puncture	TNF-α	MIP-2
Immune complex		
IgG	TNF-α, IL-1	MIP-2
IgA	TNF-α	MCP-1

tory pathways but also to alter physiological parameters throughout the body. The increase in systemic levels of TNF-α and IL-1 may be associated with development of multiorgan failure, including liver, lung, and kidney. Continued elevation of these early-response mediators systemically may lead to leukocyte infiltration into tissues, activation and release of damaging proteases and oxidants, extensive tissue damage and dysfunction, and finally multiorgan failure. Once this pathway has been initiated, current treatment protocols appear to be ineffective in altering morbidity and mortality outcomes. Several reagents have been established to target the early-response mediators including, TNF-α and IL-1, but none of the interventions involving the use of blocking antibodies or receptor antagonists has proven to be of therapeutic benefit.

The evidence for the role of TNF-α and IL-1 in septic responses comes primarily from animal models of septicemia. Different types of models have been employed: (a) endotoxin infusion, which utilizes a bolus injection of endotoxin to activate cytokine pathways; (b) bacterial infusion (*Escherichia coli*); or (c) cecal ligation and puncture models which allow a slow, continuous leak of bacteria into the peritoneum. A great deal of our knowledge on TNF-α and IL-1 activation pathways in vivo has come from studies utilizing the endotoxin models. Intraperitoneal injection of endotoxin in mouse models results in the sequential release of TNF-α, IL-1β, and IL-1α peaking at 1, 2.5, and 6 hr, respectively (26). In addition, it appears that the primary source of these early-response cytokines after intraperitoneal injection is Kupffer cells within the liver, with subsequent transport of cytokines to the lung where inflammation follows. This was one of many studies using animal models where a relationship of activation from the gut to the liver to the lung has been established. One of the primary events post-early-response cytokine (TNF-α, IL-1) production is adhesion molecule expression and leukocyte localization. In cynomologous monkeys, upregulation of E-selectin expression after LPS infusion has been found to correlate with the severity of the response (27). E-selectin in the vascular endothelium was induced by 2 hr and peaked at 4 hr after LPS infusion. This rapid increase in E-selectin may mediate the initial neutrophil adherence or "rolling" response following endotoxin infusion. The second step in leukocyte adherence is the firm adhesion to vascular endothelium via binding to endothelial ICAM-1 via neutrophil β-integrins (CD11/CD18). In the same model, administration of a blocking antibody to CD18 significantly attenuated neutrophil accumulation within the lung as well as alveolar capillary membrane injury, with reduced capillary leak (28). Blockade of TNF-α and IL-1 during septic responses in animal models has suggested that these adhesion molecules may be relevant targets for blockade. In particular, the targeting of these two early-response cytokines using soluble TNF receptor constructs or IL-1 receptor antagonist proteins has shown success in animal models of endotoxemia (29) but not in clinical trials with patients with septic syndromes. Likewise, using a cecal ligation and puncture model of septicemia, in-vivo neutralization of TNF-α did not significantly attenuate mortality (30). More recent data using a similar cecal ligation and puncture model have suggested that the level of TNF may not be as important as the persistent prolonged production of the suppressive cytokine, IL-10 (31–33). These latter data may explain why in human clinical trials neutralization of TNF-α in patients with septic shock did not alter the clinical outcome (34).

The primary cell populations which appear to be active in in-vivo responses to endotoxin appear to be phagocytic cells (neutrophils and macrophage/monocytes). The production of specific chemokines during septic responses have been documented. In human sepsis the production of IL-8 during septic responses correlates to the severity of the damage (35). Studies in rodents have been performed by examining a C-X-C chemokine, macrophage inflammatory protein-2 (MIP-2), which is chemotactic for neutrophils and a functional murine homolog for IL-8. The production of MIP-2 during a septic response in mice appears be a pivotal stimulus for the recruitment of neutrophils, and inhibition of MIP-2 significantly attenuated mortality in endotoxemia (36). Likewise, other C-X-C family members also appear to be involved in the septic responses, such as KC, the murine homolog of human GRO-α (37). During septic responses the mononuclear phagocyte accumulation may also play a role in tissue damage. Studies examining the production as well as the effects of neutralization of C-C chemokine family members suggest that both MIP-1α and RANTES appear to be important mediators of tissue damage and mortality during the endotoxin responses in animal models (38,39). However, using a cecal ligation and puncture model of peritonitis, blockade of MIP-2, but not MIP-1α or MCP-1, reduced mortality, suggesting that the neutrophil was the primary cell inducing damage (Walley, unpublished observation). The identification of the specific chemokines responsible for the recruitment of various populations of leukocytes during septic responses may be useful in determining targets for therapeutic applications (Table 3).

Another model of acute inflammation which has been useful for the elucidation of specific cytokine/chemokine mechanisms is immune complex-induced lung injury. This model, which has been studied extensively in the lung, may represent responses observed in both pulmonary and nonpulmonary diseases, such as in rheumatoid arthritis, Wegner's granulomatous, and systemic lupus erythmatosis (SLE). Intrapulmonary deposition of IgG or IgA immune complexes in rats leads to acute injury to lung vascular and alveolar epithelium (40,41), related to the participation of either neutrophils or tissue macrophages, depending on the antibody isotype used. The activation of residential macrophages appears to be a distinguishing feature between the two models. In the IgG immune complex model a substantial upregulation of TNF-α and IL-1 is associated with the expression of adhesion molecules, E-selectin and ICAM-1, resulting in recruitment of neutrophils into the alveolar compartment. The activation and degranulation of migrated neutrophils and activation of residential macrophages releases multiple oxidants (O2·, HO·, ·NO, ONOO·, etc.) and proteases (serine proteases and metalloproteases), which cause structural damage of cells and connective tissue matrix. The IgG immune complex-induced injury can be significantly attenuated by the administration of blocking antibodies specific for TNF-α (42) and subsequent production of neutrophil chemoattractants, such as MIP-2 (unpublished data). In the IgA immune complex model the activation of resident pulmonary macrophage populations leads to the production of MCP-1, a member of the C-C chemokine family, which appears to cause the activation of residential macrophages. This is followed by the generation of oxidants and release of metalloproteases which in turn damage cells and connective tissue matrix in a similar manner to that in the IgG immune complex model. Similar to the endotoxin models (Table 3), it appears that coordinated mechanisms consisting of activation of residential cells and production of early-response cytokines and chemokines lead to leukocyte accumulation, activation, and tissue damage.

B. Role of Cytokines in the Chronic Inflammatory Response

The prolonged accumulation of leukocytes at a site of inflammation can have lasting deleterious effects. These effects can be observed in several human inflammatory diseases, including sarcoidosis, asthma, idiopathic pulmonary fibrosis (IPF), rheumatoid arthritis, and many other diseases for which the inciting agent is not known. In all of these diseases a prolonged, unregulated, and persistent accumulation of leukocytes appears to be the driving force behind the progressive tissue/organ damage. As with acute inflammatory events, the initiating stimuli appear to drive the production of early-response mediators such as IL-1 and TNF-α. The subsequent activation and upregulation of adhesion molecules (ICAM-1, E-selectin, and VCAM-1) are necessary to recruit leukocytes. Although chronic inflammation may primarily resemble acute inflammation and cells, such as neutrophils, may be present at the site of chronic inflammatory reactions, there appears to be an antigen-specific response which is the driving force behind the persistent inflammatory response.

T-lymphocyte specific responses have been classified into two distinct categories, depending on the type of cytokines produced (43,44). Th1-type responses are characterized by lymphocytes that produce primarily IFN-γ, whereas Th2-type responses are associated with production of IL-4, IL-5, IL-10, and IL-13. It is likely that during in-vivo responses these classifications are rarely exclusive. Most immune responses utilize a complex mix of cytokines inducing both phenotype subsets. However, the predominance of one cytokine, IFN versus IL-4, during a reaction allows classification of responses as being predominantly Th1 or Th2 type. The

T-lymphocyte specific response likely determines the type of inflammation which accompanies the reaction. Intracellular infectious agents, such as mycobacteria, tend to induce a Th1-type (IFN-γ mediated), mononuclear cell inflammation, whereas extracellular organisms, such as parasites or antigens related to allergic responses, tend to have an eosinophilic component and cause a predominant Th2-type (IL-4 driven) response. Evolutionarily, the Th2-type responses may have been developed to combat extracellular parasitic organisms, infectious agents, and bacteria.

Granulomatous Inflammation

Granulomatous inflammation is characterized by an accumulation of leukocytes consisting of macrophages, lymphocytes, mast cells, epithelioid cells, multinucleated giant cells, eosinophils, and/or fibroblasts around an infectious or noninfectious agent (45,46). Each of these cell populations likely contributes to the overall pathology and tissue damage associated with granulomatous reactions. The residual fibrosis which accompanies the resolution phase of granulomas may result in irreversible tissue damage and organ dysfunction. Noninfectious foreign body-type granulomas can be induced by industrial agents such as talc, silica, or beryllium, while infectious granulomas are induced by bacteria (mycobacterial sp.) fungi, viruses, or parasites (leishmania, schistosoma). In some instances the initiation of the response is unclear, as in sarcoidosis and Wegner's disease. In cases of progressive granulomatous inflammation, options for treatment are limited. A better understanding of the mechanisms of induction, maintenance, and resolution of the disease would likely result in more effective treatment strategies. Several models of granulomatous inflammation exist and cover the entire spectrum of disease observed in humans, including non-specific, Th1-type and Th2-type inflammatory processes (Table 4). The evaluation of the role of inflammatory peptides in these models is incomplete, but several key cytokines which mediate disease progression have been identified.

Early-response cytokines have been found to be an important initiating factor in chronic granulomatous inflammation, similar to the role they play in acute inflammation. TNF-α was first identified as a major participant in granulomatous inflammation using a severe combined immunodeficient (SCID) mouse system and the *Schistosoma mansoni* egg granuloma model (47). In these experiments an embolized egg to the lungs of SCID mice induced little granuloma formation or leukocyte accumulation. However, after the animals were treated with exogenous TNF-α, significant mononuclear cell accumulation and granuloma formation could be observed. This classic study set the stage for the examination of the role of TNF in chronic inflammation. The use of nonreactive synthetic beads coated with various cytokines demonstrated that TNF-α, IL-1, but not IL-6, could induce granulomatous responses in the lungs of mice (48). TNF-α has been shown to induce the expression of vascular ICAM-1 in the lung, which is required to mediate the granulomatous response and accumulation of leukocytes around the egg nidus (49). In-vivo neutralization of TNF-α inhibits development of mycobacterial granulomas (50), suggesting its role in eradication of intracellular organisms. In a glucan-induced foreign body pulmonary

Table 4 Models of Chronic Inflammation in Lung

Models	Relevant cytokines	
	Activating	Chemotactic
Granulomatous inflammation		
Foreign body (glucan)	TNF-α, IL-1	MCP-1
TH1 type	IFN-γ, TNF-α	MIP-1α
TH2 type	IL-4, IL-5	MCP-1
Allergic inflammation	IL-4, TNF-α	Eosinophils: MIP-1α, RANTES, eotaxin; lymphocytes: MCP-1

granuloma model in rats, a significant mononuclear cell infiltrate (present at an angiocentric location) has been associated with the expression and production of IL-1β and TNF-α (51). Neutralization of either of these cytokines significantly alters the development of the glucan-induced granulomatous lesion.

The use of pulmonary granuloma models to determine the mechanisms of leukocyte recruitment has yielded a great deal of information about specific chemotactic molecules. In foreign-body granuloma formation, it appears that MCP-1 may play a primary role in the recruitment of mononuclear phagocytic cells (52). The role of specific chemokines during different phases of granulomatous inflammation has also been investigated. During initial stages of schistosome egg granuloma formation, a Th1-type (IFN-γ) response predominates. During this stage of granuloma formation depletion of C-C family chemokines demonstrate that MIP-1α, but not MCP-1, is crucial in mononuclear cell accumulation (53,54). In contrast, in the secondary, anemnestic egg granulomas, a Th2-type (IL-4) response predominates, with changes in leukocyte population (50% eosinophils) and significantly larger lesion sizes. In these secondary responses the depletion of MCP-1 (but not MIP-1α) reduces the size of the lesion and appears to control the inflammatory response (Table 4). Although other chemokines and factors likely contribute to the overall accumulation of leukocytes within the granulomas, these data demonstrate that particular chemokines may function under specific phases or within certain types (Th1 versus Th2) of inflammation.

The regulation of granuloma formation by inflammatory/immune cytokines has been investigated extensively. In human disease, the prevailing concept is that granulomatous responses, as in sarcoidosis, are mediated by Th1-type, (IFN-γ-mediated) classical delayed-type hypersensitivity (DTH) responses. However, research in animal models has provided conflicting information. The initial identification of lymphokine differences in various types of granulomas (BCG-mediated → IFN-γ versus *Schistosoma* egg-mediated → IL-4) has indicated that a strict cytokine phenotype is not required for granuloma formation (55–57). It appears that Th2-type (IL-4-mediated) granulomatous inflammation is more intense and ultimately may facilitate end-stage fibrosis. Certain cytokine phenotypes may alter the outcome of an inflammatory response. IFN-γ has demonstrated a function as an antiproliferative reagent which inhibits the maturation of Th2-type lymphocytes, fibroblast proliferation, and collagen production (58–60). Several studies, in both granulomatous and nongranulomatous diseases, have also suggested that IFN-γ is a potent antifibrotic factor (61,62). In contrast, recent data have suggested that Th2-type responses play a direct role in fibroblast proliferation and connective-tissue-matrix gene activation (63,64). Depletion of IL-4 in the schistosome egg granuloma inflammatory model significantly decreases granuloma size (65), and deposition of matrix proteins and fibrosis. IL-4 may also contribute to the overall fibrotic response by increasing the number of appropriate leukocytes (i.e., eosinophils) through upregulation of VCAM-1 (66) or through the induction of additional fibrotic mediators, such as TGF-β (67). Altogether these studies suggest that the regulation of end-stage disease (fibrosis) may depend ultimately on the types and profile of cytokines produced during the inflammatory response.

Allergic Airway Inflammation

Inflammation induced during allergic airway responses is mediated by the coordination of several immune-specific events, including production of relevant inflammatory cytokines and chemokines resulting in infiltration of leukocytes in a sequential manner (68) (Table 4). The long-term pathophysiological events can have devastating outcome, especially as related to acute life-threatening asthmatic attacks. The production of cytokines and altered pathophysiological lung responses have shown significant correlation, both in humans and in animal models of allergic airway inflammation. The correlation of increased TNF production and pathological events in asthma has been supported both in vivo and in vitro using cellular isolates from asthmatic patients. The most convincing evidence that TNF-α plays a role in the pathophysiology of asthma was recently observed in normal subjects receiving inhaled recombinant TNF-α (69). These studies demonstrated a significant increase in airway hyperresponsiveness and decreases in FEV_1 within those subjects receiving TNF-α compared to the placebo group. Production of TNF in asthmatic patients after IgE-dependent stimulation from both mast cells (70) and alveolar macrophages (71) demonstrate the importance of these two cellular populations in contributing to the early activation of the response. In addition, elevated release of TNF-α was observed in airway leukocytes isolated from asthmatics compared to leukocytes from non-asthmatic patients (72).

TNF-α has also been demonstrated to play a significant role in airway hyperresponsiveness in other animal models of airway inflammation. Rats exposed to LPS demonstrated elevated airway hyperreactivity which was alleviated by antibodies to TNF-α (73). In addition, direct exposure by aerosolized TNF also increased airway hyperreactivity. In both instances a correlation with neutrophil influx was observed, suggesting that this cell population can contribute to airway hyperresponsiveness. Using a guinea pig model of ovalbumin-induced airway hyperresponsiveness, the administration of IL-1 receptor antagonist protein inhibited airway hyperreactivity, eosinophil accumulation, and TNF production (74). More recently, in a mouse model of allergic airway inflammation, passive immunization with a soluble TNF receptor construct significantly reduced both early neutrophil and later eosinophil influx in and around the airways of the challenged mice (75). In addition, the neutralization of TNF-α significantly diminished the level of eosinophil-specific chemokines, MIP-1α and RANTES, produced after an allergic challenge (76). Altogether, it appears that TNF-α may play a significant role during allergic airway inflammation as an overall inflammatory mediator, perhaps by its ability to upregulate pathways specific for leukocyte recruitment.

The accumulation of leukocytes into and around the airways in allergic asthmatics appears to correlate with expression of a number of chemotactic mediators. The production of specific chemokines, which can preferentially recruit particular leukocyte populations, has been a major focus of many laboratories. Initial investigations in human populations have centered around eosinophil-specific chemokines, since the eosinophil appears to be an important cell in the pathology of asthma. One of the most potent chemokines which promotes eosinophil recruitment in vitro is RANTES (77,78). In vivo, the intradermal injection of RANTES into dogs caused the accumulation of eosinophils and mononuclear leukocytes, whereas other chemokines did not (79), while the addition of RANTES to cultures of eosinophils caused activation and degranulation of these cells (80). As with all C-C chemokines, RANTES is also a potent chemoattractant for lymphocytes and monocytes and may contribute to the influx of these cells during allergic responses. As stated above, MIP-1α has chemotactic activity for eosinophils and has been observed to be increased in BAL fluids of asthmatics as compared to BAL fluids from nonasthmatic patients (81). Increased MCP-1 expression can also be observed in bronchial epithelial cells (82) and may play a role as a basophil-degranulating agent (83) and lymphocyte chemoattractant (84). Other C-C chemokines may participate in the eosinophil-specific infiltration observed in allergic asthmatic responses, such as MCP-3 and eotaxin, which have potent eosinophil chemotactic properties (85,86). Eotaxin has demonstrated eosinophil-specific chemoattraction, with little activity for neutrophils or mononuclear cells (87,88). Eotaxin was first identified in allergic guinea pig models but has more recently been isolated from human and murine sources and may serve as the most specific chemoattractant for eosinophils during allergic responses. Finally, IL-8, the "flagship" C-X-C chemokine, appears to have some activity in mediating eosinophil recruitment (89). Altogether, it appears that multiple chemokines with overlapping functions are expressed during allergic asthmatic responses, and targeting a single chemokine may not prove effective in altering the clinical outcome of this disorder.

The use of animal models to investigate allergic airway inflammation may support a role of chemokines involved in allergic airway responses. During allergic responses, several C-C chemokines are produced which have overlapping functions in vitro, such as monocyte and lymphocyte chemotactic properties (90). The in-vivo production of chemokines may occur in a regulated manner, causing migration of particular leukocytes into the airway. When allergic mice were passively immunized with antibodies specific for MIP-1α, MCP-1, or RANTES, distinct patterns of leukocyte infiltration were interrupted. In-vivo neutralization of MIP-1α or RANTES significantly affected the eosinophil populations (40–60% decrease) but did not significantly alter the influx of mononuclear cells. These latter data have been confirmed in vitro using eosinophil chemotactic assays with samples from allergic mice (91). In contrast, when MCP-1 was neutralized in vivo, the recruitment of mononuclear cells, monocytes and lymphocytes, was significantly reduced. In further studies, when changes in airway physiology were examined in the allergic model, significant decreases in SEA-induced hyperreactivity were observed in anti-MCP-1-, but not anti-MIP-1α- or anti-RANTES-treated animals. These studies imply that in this model of allergic airway inflammation, monocytes and T lymphocytes play a more important role in changes in airway pathophysiology than do eosinophils. These studies suggest that T lymphocytes and their products may provide appropriate stimuli to induce airway responsiveness, possibly with the aid of effector cells, such as eosinophils.

III. Antiinflammatory Peptides

An overzealous inflammatory response is often the cause of organ dysfunction. Modulation of a response can often be more critical than upregulation of effector mechanisms which can cause irreversible side effects. The identification of regulatory peptides during inflammatory responses may serve a critical role in controlling the cellular damage and tissue injury induced during a reaction (Table 5). The use of antiinflammatory peptides as therapeutics may serve well to keep various acute and chronic inflammatory reactions "in check." This may especially be critical given the limited success of blockade strategies specific for inflammatory peptides, such as TNF-α and IL-1, as discussed above. Interruption of any one of a number of mediators may be sufficient to alleviate inflammation-induced damage within tissues.

Table 5 Antiinflammatory cytokines

Cytokine	Function during disease
IL-10	Downregulation of inflammatory cytokines (TNF-α, IL-1, IFN, IL-6, etc.)
	Homeostatic control of inflammation
	Downregulation of adhesion molecules
TGF-β	Antiproliferative for lymphocytes
	Downregulation of cytokine production
	Upregulation of collagen deposition
	Fibroblast activation factor
IL-4	Inhibition of TNF, IL-1 production
	Upregulation of VCAM-1 expression
	Mediation of IgE ab isotype switching
	Fibroblast activation and collagen gene activation

The most successful therapy to date for controlling inflammation has been the use of steroidal compounds which nonspecifically inhibit the production of nearly all inflammatory cytokines. Reliance on this strategy, although often effective, demonstrates our lack of understanding of the mechanisms which govern the inflammatory response. In many cases the use of an appropriate antiinflammatory peptide may provide better-suited regulation in the absence of steroidal side effects. Several antiinflammatory peptides have been investigated, including IL-10, TGF-β, and IL-1 receptor antagonist (IL-1ra), as well as IL-4 and IL-13.

Perhaps the most attractive antiinflammatory cytokine with broad-spectrum activity is IL-10. Originally described as a cytokine synthesis inhibitory factor, it has been assessed in many types of inflammation. IL-10 has a high degree of homology (~80%) with an Epstein-Barr virus-encoded protein, and its expression during viral infection likely modulates the host response (92). IL-10 is predominantly produced by macrophage populations, Th2-type lymphocytes, and B cells, but has recently been shown to be produced constitutively by airway epithelial cells (93) and by many types of cancerous cell populations (94,95). The function of IL-10 appears to be important during normal physiological events, as IL-10 gene knockout mice have lethal inflammatory bowel dysfunction (96). The potency of this cytokine in disease has been observed in studies which have administered IL-10 to mice and protected them from a lethal endotoxin challenge (97,98). These latter observations can likely be attributed to the ability of IL-10 to downregulate multiple inflammatory cytokines, including TNF, IL-1, IL-6, and IFN-γ, expression of adhesion molecules, and production of nitric oxide (99). IL-10 therapy was also effective in cecal ligation and puncture models of septicemia in which anti-TNF therapy exacerbated lethality (33). The downregulation of the inflammatory cytokine mediators by IL-10 therapy suggest an extremely potent antiinflammatory agent to be used for intervention of inflammation-induced injury. Several other disease states have demonstrated the ability of IL-10 to act as a potent antiinflammatory cytokine. In animal models of inflammatory arthritis, the administration of IL-10 modulates disease progression (100), while in-vivo neutralization of IL-10 exacerbates disease severity (101). In other studies, IL-10 has been used to modulate allograft rejection episodes and has significantly prolonged allograft survival (102). It appears that IL-10 is effective as an antiinflammatory compound, is well tolerated, and to date demonstrates no undesirable effects.

Other cytokines also function to suppress the inflammatory response. Transforming growth factor-beta (TGF-β) has antiproliferative effects in lymphocytes, downregulates cytokine production, and inhibits endotoxin-induced inflammation (103). In TGF-β gene knockout mice a progressive inflammatory response was observed early on (day 14 after birth) throughout the body, including heart, lung, salivary glands, and pancreas, and in virtually every organ at later time points (104). Thus TGF-β, like IL-10, plays a critical role in homeostatic regulation of inflammatory response within the body. However, the use of this cytokine as a therapeutic must be viewed cautiously, as TGF-β, as stated above, may promote fibrogenic outcomes (105). Therefore, the exogenous administration may alter the inflammatory response and yet shift it into a fibrotic event. Another cytokine which has demonstrated antiinflammatory functions is IL-4. IL-4 is a member of the Th2-type cytokine family and has the ability to downregulate the production of inflammatory cytokines from macrophages (106). However, like TGF-β, IL-4 has a number of alternative functions which should be viewed carefully, including upregulation of VCAM-1, IgE antibody isotype switching, and fibroblast activation (63,64,107). For example, IL-4 has demonstrated an intimate role in progression of pulmonary granulomatous responses (55), in allergic airway eosinophilia (108,109), and in proliferation and collagen-gene expression in pulmonary fibroblast populations (64). The use of either TGF-β or IL-4 as a therapeutic antiinflammatory agent will likely have to be carefully evaluated.

IV. Summary

The regulation of inflammatory responses is dependent on the production of various pro- and antiinflammatory cytokines/chemokines which together regulate and control the intensity of the reaction. The intimate coordination and the balance of these peptides will dictate the initiation, maintenance, and resolution phases of antiinflammatory responses. The excessive production of these proinflammatory peptides can cause an intense leukocyte accumulation and activation, leading to tissue damage and organ dysfunction. Improper or premature regulation of the response may lead to incomplete clearance of an infectious agent and eventual overgrowth of the invading organism, leading to infectious pathology. The continued investigation into the mechanisms and functions of pro- and antiinflammatory cytokines will expand our understanding of the reactions and lead to development of better pharmacological interventions.

References

1. Kunkel SL, Remick DG, Strieter RM, Larrick JW. Mechanisms that regulate the production and effects of tumor necrosis factor-alpha. Crit Rev Immunol 1989; 9(2): 93–117.
2. Strieter RM, Kunkel SL, Bone RC. Role of tumor necrosis factor-alpha in disease states and inflammation. Crit Care Med 1993; 21(10 suppl):S447–S463.
3. Springer TA, Dustin ML, Kishimoto TK, Martin SD. The lymphocyte function-associated LFA-1, CD2, and LFA-3 molecules: cell adhesion receptors of the immune system. Ann Rev Immunol 1987; 5:223.
4. Zimmerman GA, Prescott SM, McIntyre TM. Endothelial cell interactions with granulocytes: tethering and signaling molecules. Immunol Today 1992; 13:93.
5. Imhof BA, Dunon D. Leukocyte migration and adhesion. Adv Immunol 1995; 58:345–416.
6. Beutler B. The tumor necrosis factors: cachectin and lymphotoxin. Hosp Pract 1990; 2:45–56.
7. Standiford TJ, Kunkel SL, Basha MA, Chensue SW, Lynch JP, Toews GB, Strieter RM. Interleukin-8 gene expression by a pulmonary epithelial cell line: a model for cytokine networks in the lung. J Clin Invest 1990; 86:1945–1953.

8. Ward PA, Warren JS, Johnson K. Oxygen radicals, inflammation, and tissue injury. Free Radical Biol Med 1988; 5(5–6):403–408.

9. Shimizu T, Mutoh H, Waga I, Nakamura M, Honda Z, Izumi T. Structure and function of platelet-activating factor receptor. In: Navarro J, Molecular Basis of Inflammation. 1994.

10. Oppenheim JJ, Zachariae COC, Mukaida N, Matsushima K. Properties of the novel proinflammatory "intercrine" cytokine family. Annu Rev Immunol 1991; 9:617.

11. Schall TJ. Biology of the RANTES/SIS cytokine family. Cytokine 1991; 3:165–183.

12. Baggiolini M, Dewald B, Moser B. Interleukin-8 and related chemotactic cytokines—CXC and CC chemokines. Adv Immunol 1994; 55:97–179.

13. Strieter RM, Polverini PJ, Arenberg DA, et al. Role of C-X-C chemokines as regulators of angiogenesis in lung cancer. J Leukocyte Biol 1995; 57:752–762.

14. Chuntharapai A, Kim KJ. Regulation of the IL-8 receptor A/B by IL-8: possible functions of each receptor. J Immunol 1995; 155:2587–2594.

15. Bacon KB, Premack BA, Gardner P, Schall TJ. Activation of dual T cell signaling pathways by the chemokine RANTES. Science 1995; 269:1727–1730.

16. Turner L, Ward SG, Westwick J. RANTES-activated human T lymphocytes. A role for phosphoinositide 3-kinase. J Immunol 1995; 155:2437–2444.

17. Taub DD, Ortaldo JR, Turcovski-Corrales SM, Key ML, Longo DL, Murphy WJ. Beta chemokine costimulate lymphocyte cytolysis, proliferation, and lymphokine production. J Leukocyte Biol 1996; 59:81–89.

18. Jiang Y, Beller DI, Frendl G, Graves DT. Monocyte chemoattractant protein-1 regulates adhesion molecule expression and cytokine production in human monocytes. J Immunol 1992; 148:2423–2428.

19. Lloyd AR, Oppenheim JJ, Kelvin DJ, Taub DD. Chemokines regulate T cell adherence to recombinant adhesion molecules and extracellular matrix proteins. J Immunol 1996; 156:932–938.

20. Kimata H, Yoshida A, Ishioka C, Fujimoto M, Lindley I, Furusho K. RANTES and macrophage inflammatory protein 1 selectively enhance immunoglobulin E (IgE) and IgG4 production by human B cells. J Exp Med 1996; 183:2397–2402.

21. Dragic T, Litmin V, Allaway GP, Martin SR, Huang Y, Nagashima KA, Cayanan C, Maddon PJ, Koup RA, Moore JP, Paxton WA. HIV-1 entry into CD4+ cells is mediated by the chemokine receptor CC-CKR-5. Nature 1996; 381:667–673.

22. Deng H, Liu R, Ellmeier W, Choe S, Unutmaz D, Burkhart M, Di Marzio P, Marmon S, Sutton RE, Hill CM, Davis CB, Peiper SC, Schall TJ, Littman DR, Landau NR. Identification of a major co-receptor for primary isolates of HIV-1. Nature 1996; 381: 661–666.

23. Alkhatib G, Combadiere C, Broder CC, Feng Y, Kennedy PE, Murphy PM, Berger EA. CC CKR5: a RANTES, MIP-1alpha, MIP-1beta receptor as a fusion cofactor for macrophage-tropic HIV-1. Science 1996; 272:1955–1958.

24. Parrillo JE. The cardiovascular pathophysiology of sepsis. Annu Rev Med 1989; 40: 469–485.

25. Strieter RM, Lynch JP 3d, Basha MA, Standiford TJ, Kasahara K, Kunkel SL. Host responses in mediating sepsis and adult respiratory distress syndrome. Semin Respir Infect 1990; 5:233–247.

26. Chensue SW, Terebuh PD, Remick DG, Scales WE, Kunkel SL. In vivo biologic and immunohistochemical analysis of interleukin 1 alpha, beta, and tumor necrosis factor during experimental endotoxemia: kinetics, Kupffer cell expression, and glucocorticoid effects. Am J Pathol 1991; 138:395–402.

27. Engelberts I, Samyo SK, Leeuwenberg JF, van der Linden CJ, Buurman WA. A role for ELAM-1 in the pathogenesis of MOF during septic shock. J Surg Res 1992; 53:136–144.
28. Walsh CJ, Carey PD, Cook DJ, Bechard DE, Fowler AA, Sugarman HJ. Anti-CD18 antibody attenuates neutropenia and alveolar capillary-membrane injury during gram-negative sepsis. Surgery 1991; 110:205–211.
29. Ulich TR, Yi ES, Smith C, Remick D. Intratracheal administration of endotoxin and cytokines. VII. The soluble interleukin-1 receptor antagonist and the soluble tumor necrosis factor receptor II (p80) inhibit acute inflammation. Clin Immunol Immunopathol 1994; 72:137–140.
30. Eskandari MK, Bolgos G, Miller C, Nguyen DT, DeForge LE, Remick DG. Anti-tumor necrosis factor antibody therapy fails to prevent lethality after cecal ligation and puncture or endotoxemia. J Immunol 1992; 148:2724–2730.
31. Kato T, Murata A, Ishida H, Toda H, Tanaka N, Hayashida H, Monderm M, Matsuura N. Interleukin 10 reduces mortality from severe peritonitis in mice. Antimicrob Agents Chemother 1995; 39:1336–1340.
32. van der Poll T, Marchant A, Buurman WA, Berman L, Keogh CV, Lazarus DD, Nguyen L, Goldman M, Moldawer LL, Lowry SF. Endogenous IL-10 protects mice from death during septic peritonitis. J Immunol 1995; 155:5397–5401.
33. Walley KR, Lukacs NW, Standiford TJ, Strieter RM, Kunkel SL. Balance of inflammatory cytokines related to severity and mortality of murine sepsis. Infect Immun 1996; 64:4733–4738.
34. Fisher CJ, Agosti JM, Opal SM, Lowry SF, Balk RA, Sadoff JC, Abraham E, Schein RM, Benjamin E. Treatment of septic shock with the tumor necrosis factor:Fc fusion protein. The Soluble TNF Receptor Sepsis Study Group. N Engl J Med 1996; 334:1667–1702.
35. Hack CE, Hart M, van Schijndel RJ, Eerenberg AJ, Nuijens JH, Thijs LG, Aarden LA. Interleukin-8 in sepsis: relation to shock and inflammatory mediators. Infect Immun 1992; 60:2835–2842.
36. Standiford TJ, Strieter RM, Lukacs NW, Kunkel SL. Neutralization of IL-10 Increases lethality in endotoxemia: cooperative effects of macrophage inflammatory protein-2 and tumor necrosis factor. J Immunol 1995; 155:2222–2229.
37. Huang S, Paulauskis JD, Godleski JJ, Kobzik L. Expression of macrophage inflammatory protein-2 and KC mRNA in pulmonary inflammation. Am J Pathol 1992; 141:981–988.
38. VanOtteren GM, Strieter RM, Kunkel SL, Paine R, Greenberger MJ, Danforth JM, Burdick MD, Standiford TJ. Compartmentalized expression of RANTES in a murine model of endotoxemia. J Immunol 1995; 154:1900–1908.
39. Standiford TJ, Kunkel SL, Lukacs NW, Greenberger MJ, Danforth JM, Kunkel RG, Strieter RM. Macrophage inflammatory protein-1α mediates lung leukocyte recruitment, lung capillary leak, and early mortality in murine endotoxemia. J Immunol 1995; 155:1515–1524.
40. Johnson KJ, Chapman WE, Ward PA. Immunopathology of the lung: a review. Am J Pathol 1979; 95:795–844.
41. Johnson KJ, Ward PA, Kunkel RG, Wilson BS. Mediation of IgA induced lung injury in the rat. Role of macrophages and reactive oxygen products. Lab Invest 1986; 54:499–506.
42. Johnson KJ, Ward PA. Role of oxygen metabolites in immune complex injury of the lung. J Immunol 1981; 126:2365–2369.
43. Mosmann TR, Moore KW. The role of IL-10 in crossregulation of Th1 and Th2 response. Immunol Today 1989; 12:A49–A58.

44. Mosmann TR, Cerwinski H, Bond MW, Giedlin MA, Coffman RL. Two types of murine helper T cell clone. I. Definition according to profiles of lymphokine activities and secreted proteins. J Immunol 1986. 136:2348–2359.

45. Boros DL. Immunopathology of *Schistosoma mansoni* infection. Clin Microbiol Rev 1989; 2:250–269.

46. Kunkel SL, Strieter RM, Lukacs NW, Chensue SW. Molecular aspects of granulomatous inflammation. EOS J Immunol Immunopharm 1994; 14:71–77.

47. Amiri P, Locksley RM, Parslow TG, Sadick M, Rector E, Ritter D, McKerrow JH. Tumor necrosis factor-a restores granulomas and induces parasite egg laying in schistosome-infected SCID mice. Nature 1992; 356:604–610.

48. Kasahara K, Kobayashi K, Shikama Y, Yoneya I, Kaga S, Hashimoto M, Odagiri T, Soejima K, Ide H, Takahashi T. The role of monokines in granuloma formation in mice: the ability of interleukin 1 and tumor necrosis factor-alpha to induce lung granulomas. Clin Immunol Immunopathol 1989; 51:419–425.

49. Lukacs NW, Chensue SW, Strieter RM, Warmington K, Kunkel SL. Inflammatory granuloma formation is mediated by TNF-a-inducible intracellular adhesion molecule-1. J Immunol 1994; 152:5883–5889.

50. Tumang MC, Keogh C, Moldawer LL, Helfgott DC, Teitelbaum R, Hariprashad J, Murray HW. Role and effect of TNF-alpha in experimental visceral leishmaniasis. J Immunol 1994; 153:768–775.

51. Flory CM, Jones ML, Miller BF, Warren JS. Regulatory roles of tumor necrosis factor-alpha and interleukin-1 beta in monocyte chemoattractant protein-1-mediated pulmonary granuloma formation in the rat. Am J Pathol 1995; 146:450–462.

52. Jones ML, Warren JS. Monocyte chemoattractant protein 1 in a rat model of pulmonary granulomatosis. Lab Invest 1992; 66:498–503.

53. Lukacs NW, Kunkel SL, Strieter RM, Warmington K, Chensue SW. The role of macrophage inflammatory protein-1 alpha in *Schistosoma mansoni* egg-induced granulomatous inflammation. J Exp Med 1993; 177:1551–1559.

54. Chensue SW, Warmington KS, Lukacs NW, Lincoln PM, Burdick MD, Strieter RM, Kunkel SL. Monocyte chemotactic protein expression during schistosome egg granuloma formation: sequence of production, localization, contribution, and regulation. Am J Pathol 1995; 146:130–138.

55. Chensue SW, Terebuh PD, Warmington KS, Hersey SD, Evanoff HL, Kunkel SL, Higashi GI. Role of IL-4 and IFN-gamma in *Schistosoma mansoni* egg-induced hypersensitivity granuloma formation. Orchestration, relative contribution, and relationship to macrophage function. J Immunol 1992; 148:900–911.

56. Grzych JM, Pearce E, Cheever A, Caulada ZA, Caspar P, Henry S, Lewis F, Sher A. Egg deposition is the major stimulus for the production of Th2 cytokines in murine schistosomiasis mansoni. J Immunol 1991; 146:1322–1329.

57. Henderson GS, Lu X, McCurley TL, Colley DG. In vivo molecular analysis of lymphokines involved in the murine immune response during *Schistosoma mansoni* infection. II. Quantification of IL-4 mRNA, and IL-2 mRNA levels in the granulomatous livers, mesenteric lymph nodes, and spleens during the course of modulation. J Immunol 1992; 148:2261–2269.

58. Gajewski TF, Pinnas M, Wong T, Fitch FW. Murine Th1 and Th2 clones proliferate optimally in response to distinct antigen-presenting cell populations. J Immunol 1991; 146:1750–1758.

59. Amento EP, McCullagh AK, Krane SM. Influence of gamma interferon on synovial

fibroblast-like cells:Ia induction and inhibition of collagen synthesis. J Clin Invest 1985; 76:836–848.

60. Rosenbloom J, Feldman G, Freundlich B, Jiminez SA. Transcriptional control of human diploid fibroblast collagen synthesis by gamma interferon. Biochem Biophys Res Commun 1984; 123:365–372.

61. Giri SN, Hyde DM, Marafino BJJ. Amelioration effect of murine interferon gamma on bleomycin-induced lung collagen fibrosis in mice. Biochem Med Metabol Biol 1989; 36:194–197.

62. Czaja MJ, Weiner FR, Takahashi S, Gimbrone MA, van der Meide PH, Schellekens H, Biempica L, Zern MA. Gamma-interferon treatment inhibits collagen deposition in murine schistosomiasis. Hepatology 1989; 10:795–800.

63. Monroe JG, Haldar S, Prystowsky MB, Lammie P. Lymphokine regulation of inflammatory processes: interleukin-4 stimulates fibroblast proliferation. Clin Immunol Immunopathol 1988; 49:292–298.

64. Sempowski GD, MP, Derak S, Phipps RP. Subsets of murine lung fibroblasts express membrane-bound and soluble IL-4 receptors. Role of IL-4 enhancing fibroblast proliferation and collagen synthesis. J Immunol 1994; 152:3606–3614.

65. Joseph AL, Boros DL. TNF plays a role in *Schistosoma mansoni* egg-induced granulomatous inflammation. J Immunol 1993; 151:5461–5468.

66. Schleimer RP, Sterbinsky SA, Kaiser J, Bickel CA, Klunk DA, Tomioka K, Newman W, Luscinskas FW, Gmbrone MA Jr, McIntyre BW, Bochner BS. IL-4 induces adherence of human eosinophils and basophils but not neutrophils to endothelium. J Immunol 1993; 148:1086.

67. Gauldie J, Jordana M, Cox G. Cytokines and pulmonary fibrosis. Thorax 1993; 48:931–935.

68. Corrigan CJ, Kay AB. T cells and eosinophils in the pathogenesis of asthma. Immunol Today 1992; 13:501–507.

69. Walsh CJ, Sugerman HJ, Mullen PG, Carey PD, Leeper-Woodford SK, Jesmok GJ, Ellis EF, Fowler AA. Monoclonal antibody to TNF attenuates cardiopulmonary dysfunction in porcine gram-negative sepsis. Arch Surg 1992; 127:138–144.

70. Bradding P, Roberts JA, Britten KM, Montefort S, Djukanovic R, Mueller R, Heusser CH, Howarth PH, Holgate ST. IL-4, -5, -6, and TNF in normal and asthmatic airways; evidence for the human mast cell as a source. Am J Respir Cell Mol Biol 1994; 10: 471–480.

71. Gosset P, Tsicopoulos A, Wallaert B, Joseph M, Capron A, Tonnel AB. Am Rev Respir Dis 1992; 146:768–774.

72. Cimbryzynska-Nowak M, Szklarz E, Inglot AD, Teodorczyk-Injeyan JA. Am Rev Respir Dis 1993; 147:291–295.

73. Kips JC, Tavernier J, Paumels RA. Am Rev Respir Dis 1992; 145:332–336.

74. Watson ML, Smith D, Bourne AD, Thompson RC, Westwick J. Cytokines contribute to airway dysfunction in antigen-challenged guinea pigs: inhibition of airway hyperreactivity, pulmonary eosinophil accumulation, and tumor necrosis factor generation by pretreatment with an interleukin-1 receptor antagonist. Am J Respir Cell Mol Biol 1993; 8:365–369.

75. Lukacs NW, Strieter RM, Chensue SW, Widmer M, Kunkel SL. TNFα mediates recruitment of neutrophils and eosinophils during allergic airway inflammation. J Immunol 1995; 154:5411–5417.

76. Lukacs NW, Strieter RM, Chensue SW, Kunkel SL. Activation and regulation of chemokines in allergic airway inflammation. J Leukocyte Biol 1996; 59:13–18.

77. Rot A, Krieger M, Brunner T, Bischoff SC, Schall TJ, Dahinden CA. RANTES and macrophage inflammatory protein 1α induce the migration and activation of normal human eosinophil granulocytes. J Exp Med 1992; 176:1489–1495.
78. Ebisawa M, Yamada T, Bickel C, Klunk D, Schleimer RP. Eosinophil transendothelial migration induced by cytokines. II. Potentiation of eosinophil transendothelial migration by eosinophil-active cytokines. J Immunol 153:2153–2160.
79. Meurer R, Van Riper G, Feeney W, Cunningham P, Hora D, Springer MS, MacIntyre DE, Rosen H. Formation of eosinophilic and monocytic intradermal inflammatory sites in the dog by injection of human RANTES but not monocyte chemoattractant protein 1, human macrophage inflammatory protein 1 alpha, or human interleukin 8. J Exp Med 1993; 178:1913–1921.
80. Chihara J, Hayashi N, Kakazu T, Yamamoto T, Kurachi D, Nakajima S. RANTES augments radical oxygen products from eosinophils. Int Arch Allergy Immunol 1994; 104:52–53.
81. Cruikshank WW, Long A, Tarpy RE, Kornfeld H, Carroll MP, Teran L, Holgate ST, Center DM. Early identification of IL-16 and MCP-1 alpha in BAL of asthmatics. Am J Respir Cell Mol Biol 1995; 13:738–747.
82. Sousa AR, Lane SJ, Nakhosteen JA, Yoshimura T, Lee TH, Poston RN. Increased expression of MCP-1 in bronchial tissue from asthmatic subjects. Am J Respir Cell Mol Biol 1994; 10:142–147.
83. Bischoff SC, Krieger M, Brunner T, Rot A, von Tscharner V, Baggiolini M, Dahinden CA. Eur J Immunol 1993; 23:761–767.
84. Carr MW, Roth SJ, Luther E, Ross SS, Springer TA. MCP-1 is a T cell chemoattractant. Proc Natl Acad Sci (USA) 1994; 91:3652–3663.
85. Dahinden CA, Geiser T, Brunner T, von Tscharner V, Caput D, Ferrara P, Minty A, Baggiolini M. Monocyte chemotactic protein 3 is a most effective basophil- and eosinophil-activating chemokine. J Exp Med 1994; 179:751–760.
86. Jose PJ, Griffiths-Johnson DA, Collins PD, Walsh DT, Moqbel R, Totty NF, Truong O, Hsuan JJ, Williams TJ. Eotaxin: a potent eosinophil chemoattractant cytokine detected in a guinea pig model of allergic airways inflammation. J Exp Med 1994; 179:881–887.
87. Rothenberg ME, Luster AD, Lilly CM, Drazen JM, Leder P. Constitutive and allergen-induced expression of eotaxin mRNA in the guinea pig lung. J Exp Med 1995; 181:1211–1216.
88. Collins PD, Marleau S, Griffiths-Johnson DA, Jose PJ, Williams TJ. Cooperation between interleukin-5 and the chemokine eotaxin to induce eosinophil accumulation in vivo. J Exp Med 1995; 182:1169–1174.
89. Schweizer RC, Welmers BAC, Raaijmakers JAM, Zanen P, Lammers JWJ, Koenderman L. RANTES and IL-8-induced responses in eosinophils: effect of IL-5 priming. Blood 1994; 83:3697–3704.
90. Lukacs NW, Strieter RM, Warmington K, Lincoln P, Chensue SW, Kunkel SL. Differential recruitment of leukocyte populations and alteration of airway hyperreactivity by C-C family chemokines in murine allergic airway inflammation. Submitted.
91. Lukacs NW, Standiford TJ, Strieter RM, Chensue SW, Kunkel RG, Kunkel SL. C-C chemokine-induced eosinophil chemotaxis during allergic airway inflammation. J Leukocyte Biol. In press.
92. Fiorentino DF, Zlotnik A, Mosmann TR, Howard M, O'Garra A. IL-10 inhibits cytokine production by activated macrophages. J Immunol 1991; 147:3815–3822.
93. Bonfield TL, Konstan MW, Burfiend P, Panuska JR, Hilliard JB, Berger M. Normal

bronchial epithelial cells contitutively produce the anti-inflammatory cytokine inter-leukin-10, which is downregulated in cystic fibrosis. Am J Respir Cell Mol Biol 1995; 13:257–261.

94. Smith DR, Kunkel SL, Burdick MD, Wilke CA, Orringer MB, Whyte RI, Strieter RM. Production of IL-10 by human bronchogenic carcinoma. Am J Pathol 1994; 145:18–25.

95. Maeurer MJ, Martin DM, Castelli C, Elder E, Leder G, Storkus WJ, Lotze MT. Host immune response in renal cell cancer: interleukin 4 (IL-4) and interleukin 10 (IL-10) mRNA are frequently detected in freshly collected tumor-infiltrating lymphocytes. Cancer Immunol Immunother 1995; 41:111–112.

96. Rennick D, Davidson N, Berg D. Interleukin-10 gene knock-out mice: a model of chronic inflammation. Clin Immunol Immunopathol 76:S174–178.

97. Howard M, Muchamuel T, Andrade S, Menon S. Interleukin 10 protects mice from lethal endotoxemia. J Exp Med 1993; 177:1205–1208.

98. Rogy MA, Auffenberg T, Espat NJ, Philip R, Remick D, Wollenberg GK, Copeland EM, Moldawer LL. Human tumor necrosis factor receptor (p55) and interleukin 10 gene transfer in the mouse reduces mortality to lethal endotoxemia and also attenuates local inflammatory responses. J Exp Med 1995; 181:2289–2293.

99. Romani L, Puccetti P, Mencacci A, Cenci E, Spaccepalo R, Tonnetti L, Grohmann U, Bistoni F. Neutralization of IL-10 up-regulates nitric oxide production and protects sus-ceptible mice from challenge with *Candida albicans*. J Immunol 1994; 152:3514–3521.

100. van Roon JA, van Roy JL, Gmelig-Meyling FH, Lafeber FP, and Bijlsma JW. Preven-tion and reversal of cartilage degradation in rheumatoid arthritis by interleukin-10 and interleukin-4. Arthritis Rheum 1996; 39:829–835.

101. Kasama T, Strieter RM, Lukacs NW, Lincoln PM, Burdick MD, Kunkel SL. Inter-leukin-10 expression and chemokine regulation during the evolution of murine type II collagen-induced arthritis. J Clin Invest 1995; 95:2868–2876.

102. Bromberg JS. IL-10 immunosuppression in transplantation. Curr Opin Immunol 7:639–643.

103. Ulich TR, Yin S, Guo K, Yi ES, Remick D, del Castillo J. Intratracheal injection of endotoxin and cytokines. II. Interleukin-6 and transforming growth factor beta inhibit acute inflammation. Am J Pathol 138:1097–1101.

104. Boivin GP, O'Toole BA, Orsmby IE, Diebold RJ, Eis MJ, Doetschman T, Kier AB. Onset and progression of pathological lesions in transforming growth factor-beta 1-defi-cient mice. J Immunol 1995; 146:276–288.

105. Border WA, Noble NA. Transforming growth factor in tissue fibrosis. J Immunol 1994; 331:1286–1292.

106. Sone S, Yanagawa H, Nishioka Y, Orino E, Bhaskaran G, Nil A, Mizuno K, Heike Y, Ogushi F, Ogura T. Interleukin-4 as a potent down-receptor for human alveolar macro-phages capable of producing tumor necrosis factor-alpha and interleukin-1. Eur Respir J 1992; 5:174–181.

107. Coffman RL, Lebman DA, Rothman P. Mechanism and regulation of immunoglobulin isotype switching. Adv Immunol 1993; 54:229–270.

108. Brusselle G, Kips J, Joos G, Bluethmann H, Pauwels R. Allergen-induced airway inflammation and bronchial responsiveness in wild-type and interleukin-4-deficient mice. Am J Respir Cell Mol Biol 1995; 12:254–259.

109. Lukacs NW, Strieter RM, Chensue SW, Kunkel SL. Interleukin-4-dependent pulmonary eosinophil infiltration in a murine model of asthma. Am J Respir Cell Mol Biol 1994; 10:526–532.

Part Two

PROINFLAMMATORY PEPTIDES

4

Proinflammatory Peptides in Sensory Nerves of the Airways

PETER BALUK and DONALD M. McDONALD

University of California, San Francisco
San Francisco, California

I. Introduction

The aim of this chapter is to review the inflammatory effects of sensory nerve peptides in airways. Most of our knowledge of the mechanism of action of inflammatory peptides comes from studies of animals, particularly rodents. The extent to which the data from experimental animals applies to human airways in health and disease is much less clear, and is only now being systematically investigated.

Traditionally, the four classical signs of inflammation are redness, heat, swelling, and pain. The first two signs result from the dilatation of arterioles and increased blood flow, whereas swelling, or tissue edema, results from plasma leakage, mainly from postcapillary venules. Pain results from the stimulation of nociceptive sensory nerve fibers, which are mostly unmyelinated C-fibers and small-diameter myelinated A-δ fibers. Many inflammatory stimuli and conditions can cause these physical signs to be manifested to different extents in various organs. A role for the sensory nervous system itself in inflammation has been recognized for more than a century, but for many years the only inflammatory sign believed to result from stimulation of sensory nerves was vasodilatation, as evidenced by redness observed in the skin after electrical or chemical stimulation of sensory nerves. For interesting historical accounts of the early studies of the involvement of sensory nerves in inflammation, see Refs. 1 and 2.

Major progress in our understanding of the inflammatory functions of sensory nerves came in the 1960s through the work of Miklós Jancsó. He introduced both the term and the concept of neurogenic inflammation (3–5). In experiments performed mainly in the skin, Jancsó clearly recognized that "the pain sensory nerve endings are actually the receptors of [the] inflammatory reaction," and anticipating the identification of proinflammatory sensory nerve peptides by more than a decade, he proposed that a neurotransmitter or "neurohumor" was released from sensory nerve fibers, thus laying the basis for modern concepts of the dual afferent and efferent nature of some peripheral sensory nerves (3,6). Jancsó distinguished inflammatory mediators that acted via a neurogenic mechanism—for example, mustard oil or capsaicin, the extract of hot peppers—from those like histamine, which could cause inflammation without the involvement of sensory nerves. Although he noted that vasodilatation and hypersensitivity to noxious stimuli occurred in inflamed skin, for Jancsó the cardinal sign of neurogenic inflammation was the abnormally increased vascular permeability that resulted in plasma leakage. Jancsó's pioneering studies have been extended to other organs, and many other cellular effects of sensory nerve peptides have been described. In addition to the skin, neurogenic inflammation has been documented in the nasal mucosa and airways, conjunctiva, meninges and sheaths of peripheral nerves, middle ear, tooth pulp and gums, salivary gland ducts, esophagus, stomach, gall bladder, bile and pancreatic ducts, the ureter and urinary bladder, joints and periosteum of bones, and in the female reproductive tract (1,7). A common feature of many of these tissues is a dense innervation by nerves sensitive to noxious stimuli. The first study of neurogenic inflammation in the airways reported the leakage of Evans blue dye in the trachea and main stem bronchi of rats upon electrical stimulation of the cut peripheral ends of the vagus nerves (8).

This review focuses on the lower airways, i.e., the trachea and bronchi. For a detailed consideration of the neural control of the nasal passages, see Ref. 9.

II. Mechanism of Proinflammatory Effects of Sensory Nerve Peptides

A. Anatomical Basis for Proinflammatory Effects

The airways are innervated by three sets of anatomically distinct pathways of sympathetic, parasympathetic, and sensory nerves. All three pathways contain neuropeptides, sometimes coexisting with nonpeptide transmitters, such as acetylcholine, norepinephrine, and nitric oxide. The postganglionic cell bodies of the sympathetic nerves are located in the superior cervical and stellate ganglia, whereas the postganglionic cell bodies of the parasympathetic nerves are located in small, local intrinsic ganglia embedded in the walls of the airways (10). Nerve degeneration studies and retrograde tracing studies show that the cell bodies of the tracheal sensory nerves are located in the jugular and nodose ganglion, while the upper thoracic dorsal root ganglia also contribute sensory fibers to the lungs (11–14). It has also been suggested recently that local interneurons or sensory neurons may exist within the intrinsic airway ganglia (15), similar to the intrinsic sensory neurons in enteric ganglia (16). Most sensory nerve fibers in the airways are long, branching, unmyelinated fibers with many varicosities or beads approximately 1–2 μm in diameter, connected by thinner intervaricose portions approximately 0.1–0.5 μm in diameter, thus making the entire nerve fiber within the resolution of the light microscope. Synaptic vesicles within the varicosities contain various transmitters. Some sensory nerve fibers, including those for stretch receptors, are myelinated (17).

The anatomical complexity of the airway innervation and the frequent coexistence of neurotransmitters makes it difficult to assign a functional role to particular nerve fibers without additional supporting evidence (18). Fortunately, in addition to tracing experiments and nerve degeneration studies mentioned above, nerve pathways can be identified by the use of pharmacological tools for the selective degeneration of subsets of nerves. When given in high enough doses, capsaicin (19), and its ultrapotent analog, resiniferatoxin (20), eliminate a subset of sensory nerves containing tachykinins and calcitonin gene-related peptide (CGRP), and 6-hydroxydopamine eliminates postganglionic sympathetic nerve fibers (21,22). Obviously, because it is impossible to use these approaches in humans, the precise pattern of innervation of human airways is less clear.

Substance P and neurokinin A (NKA) are the two members of the tachykinin family of peptides that are most important in mammalian airways. The other main tachykinin, neurokinin B (NKB), is not present in detectable concentrations in the peripheral nervous system, although it is co-localized with the other tachykinins and CGRP in the central nervous system (23,24). Two distinct genes for tachykinins have been cloned, termed preprotachykinins I and II. By alternate splicing, the preprotachykinin I gene gives rise to three mRNA's and protein products, termed α-PPT, β-PPT, and γ-PPT, which by posttranslational processing give rise to substance P, NKA, and peptides expressed in smaller amounts, a short form of NKA termed NKA(3-10), and extended forms termed neuropeptide K and neuropeptide γ. The preprotachykinin II gene codes only for NKB. The mammalian tachykinins share a common carboxyl terminal (Phe-X-Gly-Leu-Met-NH$_2$, where X is Phe or Val), which confers their activity at neurokinin receptors. Entirely different tachykinins are present in fish and amphibians. The tachykinins frequently coexist in the same sensory nerve fibers with CGRP (see below). The molecular biology of the tachykinins is reviewed in more detail in earlier publications (18,25,26), and Chapter 6. In mammals, substance P was identified as the major mediator of neurogenic inflammation, first in the skin (27,28), and then in the airways (29). The tachykinins remain the best characterized of the proinflammatory sensory nerve peptides, perhaps because of their availability as very effective tools—capsaicin to selectively deplete tachykinins from nerve fibers, and many nonpeptide receptor antagonists to eliminate tachykinin-mediated effects.

Sensory nerve fibers containing substance P or CGRP have been localized immunohistochemically in the lower airways of animals and humans, as listed in Table 1.

Table 1 Immunohistochemical Localization of Tachykinins and CGRP in Airway Sensory Nerves

Species	Tachykinins	CGRP
Rat	(30–37)	(11,12,32–34,38–44)
Guinea pig	(14,23,30–33,35,36,45)	(14,32,33,35,40,46)
Cat	(31–33,35,36,47,48)	(32,33,40,48)
Mouse	(30,35)	—
Hamster	—	(49)
Dog	(50,51)	—
Rabbit	(52)	—
Ferret	(35,53,54)	(53)
Pig	(30,33,36,55)	(33,55)
Horse	(56,57)	(56,57)
Sheep	(36)	—
Monkey	(58,59)	(58,59)
Human	(31–33,36,60–65)	(32,33,40,60,61,63–66)

Overall, a consistent picture has emerged for the distribution of substance P-and CGRP-containing nerve fibers in the airways: (a) most densely in the mucosa, often close to or in the epithelium (Figs. 1A and 1B), (b) around submucosal glands, (c) on arterioles, but rarely on postcapillary venules (Fig. 2A), (d) in airway smooth muscle (Figs. 2B and 2C), and (e) around cell bodies of postganglionic neurons in intrinsic airway ganglia (Fig. 2D). Sensory fibers also innervate bronchial associated lymphoid tissue (67,68). Few quantitative studies have been done, and the methods used have often differed in tissue sampling (cryostat sections, whole mounts, biopsies), and have used different fixation procedures and antibodies. Although it is difficult to compare the results of different studies directly, it is clear that there are significant differences among species. All studies of human airways indicate that the density of innervation by substance P and CGRP-containing nerves is much lower than in rodents. A quantitative immunohistochemical study of the relative density of substance P nerve fibers compared to the total number of nerves labeled by a panneuronal marker, Protein Gene Product 9.5 (PGP 9.5), revealed that substance P fibers constituted the following percentages of nerves in the airway mucosa: 60% in guinea pigs, 26% in cats, 15% in pigs, 12% in sheep, and only 1% in humans (36). Indeed, some studies of human airways have failed to find any substance P or CGRP-containing fibers, perhaps because of sampling limitations (52,69). The question arises if human sensory nerves do not contain substance P or CGRP, then what transmitters do they contain, and what is the action of the transmitters?

Some exceptions to the rule that substance P is present only in extrinsic nerve fibers have been reported in the airways of cats, dogs, ferrets, and monkeys. In these species, substance P is present in the cell bodies of some intrinsic airway neurons, in addition to sensory nerve fibers of extrinsic origin (47,48,51,59). In the superficial

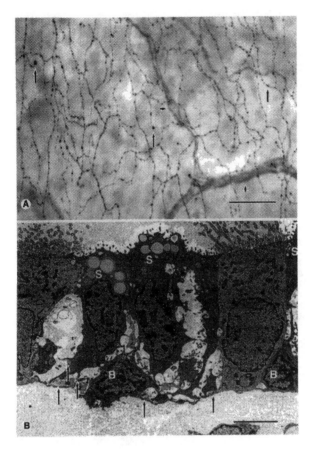

Figure 1 (A) A low-magnification view of substance P-immunoreactive axons in the mucosa of a whole mount of rat trachea. Many single varicose axons (arrows) are located in a plexus at the base of the epithelium. Nerve bundles (arrowheads) in the lamina propria contain preterminal nonvaricose substance P-immunoreactive nerve fibers. Bar = 50 µm. (From Ref. 37.) (B) Transmission electron micrograph showing intraepithelial axons (arrows) located at the base of the epithelium in the rat trachea. Also visible are changes resulting from 2 min of vagal nerve stimulation: widening of extracellular spaces next to secretory cells (S) and basal cells (B), and clustering of secretory granules at the apex of secretory cells. Bar = 2 µm. (From Ref. 13.)

nerve plexus of the ferret trachea, substance P is co-localized with VIP in about 40% of the intrinsic neurons, some of which also contain nitric oxide synthase (54). In bronchial ganglia of the monkey, substance P is co-localized with CGRP, VIP, and choline acetyltransferase, the enzyme that synthesizes acetylcholine (59). These observations suggest that such tachykinin-containing intrinsic neurons are likely to

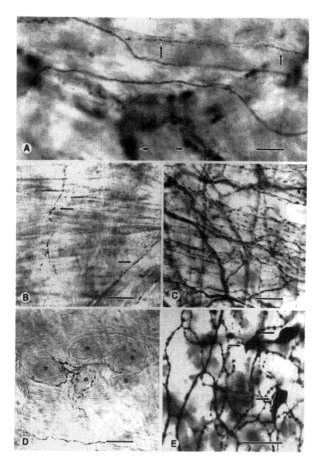

Figure 2   Substance P and CGRP immunoreactivity in whole mounts of rat trachea. (A) Substance P-immunoreactive axons (arrows) run along the adventitial surface of an arteriole in the mucosa, but are more distant from a leaky venule (arrowheads) labeled with Monastral blue. (B) A whole mount of a region of the posterior membrane stained for substance P immunoreactivity. Only a few substance P-immunoreactive axons are present (arrows), and these do not run parallel to the tracheal smooth muscle cells (horizontal orientation). (C) A region of posterior membrane similar to (B), but stained for PGP 9.5 immunoreactivity, to demonstrate all of the axons. A dense plexus of axons is present, many of which run parallel to the smooth muscle (horizontal direction). (D) A small parasympathetic ganglion in the adventitial nerve plexus of the posterior membrane stained for substance P immunoreactivity. None of the neuronal cell bodies (asterisks) is stained, but intraganglionic substance P-immunoreactive axons are present near them. (E) CGRP immunoreactivity is present in varicose axons and three neuroendocrine cells (arrows) located at the base of the epithelium. Bar = 25 μm for all micrographs. [A–D from Ref. 37; E from Ref. 44.]

be motor and not sensory in nature. There are also wide differences in the density of innervation of tracheal muscle by substance P nerve fibers, from sparse in pig (55) and rat (37) (Figs. 2B and 2C), to dense in guinea pig (14) and ferret (53).

B. Physiological Proinflammatory Effects of Tachykinins and CGRP

Substance P and NKA in the airways have multiple effects that are mediated chiefly through three distinct neurokinin receptors, termed NK-1, NK-2, or NK-3 receptors (70). A notable exception is mast cell degranulation, which appears not to be mediated by NK-1, NK-2, or NK-3 receptors, but instead is a direct activation of mast cell G proteins by the basically charged amino terminal of substance P (71,72). The inflammatory effects of tachykinins have been studied in great detail in rodents, and have been frequently reviewed (73–78).

Plasma leakage in response to stimulation of sensory nerves is the defining criterion of neurogenic inflammation, according to the original concept of Jancsó (4). Plasma leakage in response to substance P or capsaicin has been extensively documented in rats and guinea pigs; see Refs. 77 and 79–81. It also occurs in mice, particularly when the enzymes that degrade substance P are inhibited (82). Neurogenic plasma leakage has been reported not to occur in the airways of cats and dogs in response to antidromic stimulation of the sensory nerves (83), or in hamsters (unpublished observations). The same techniques cannot be used in humans, as in experimental animals, and there is no clear evidence that neurogenic plasma leakage occurs. However, there is much circumstantial evidence from bronchial lavage and histopathogical studies that tissue edema contributes to airway pathology in asthma, and the concentration of albumin is greater in asthmatics than in normal subjects; see Ref. 81. If plasma leakage does occur in humans, it could contribute to airway pathophysiology by (a) directly narrowing airways by thickening the mucosa, or exaggerating the effects of smooth muscle contraction; (b) facilitating the movement of inflammatory mediators from plasma into tissues; (c) acting as a "mucosal defense" to dilute or flush out irritants into the airway lumen (84); (d) cause hypersecretion of mucus from submucous glands, adding to airway blockage.

Based on numerous studies using selective tachykinin agonists and antagonists, there is general agreement that neurogenic plasma leakage is mediated by NK-1 receptors; see Refs. 18,78 and 85, and Chapters 6 and 22 in this volume. In guinea pig distal bronchi, NK-2 receptors have been reported as contributing to the leakage (86). Plasma leakage occurs as a result of the increased permeability of postcapillary venules and collecting venules by the formation of transient and reversible gaps between endothelial cells. The gaps have been visualized by transmission electron microscopy, silver nitrate staining of endothelial cell borders, lectin staining, and scanning electron microscopy (87–91) (Figs. 3A and 3B). In rats, few substance P-containing nerves innervate directly, or even come close to the endothelial cells of venules that leak (37) (Fig. 2A). However, venular endothelial cells express NK-1

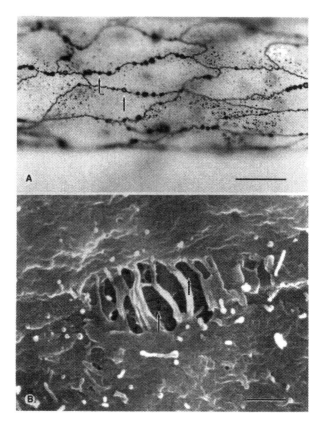

Figure 3 (A) Endothelial cells of a postcapillary venule stained with silver nitrate in a whole mount of rat tracheal mucosa 3 min after inflammation has been induced by an intravenous injection of substance P. Endothelial gaps are represented by black dots (arrows) on the cell borders. Bar = 25 μm. (From Ref. 338.) (B) Scanning electron micrograph of lumenal aspect of postcapillary venule in rat tracheal mucosa fixed 1 min after inflammation has been induced by an intravenous injection of substance P. The endothelial cells at top and bottom of the picture are separated by multiple cytoplasmic fingers and oval-shaped gaps. The underlying basement membrane is visible through some of these gaps (arrows). Bar = 1 μm. (From Ref. 91.)

receptors, suggesting that the permeability increase by substance P is a direct effect (92) (Fig. 4). One possibility is that during neurogenic inflammation, enough transmitter diffuses from the epithelial nerves near the epithelium to produce an effective concentration at the target cells.

Another effect mediated by the release of sensory nerve peptides is leukocyte

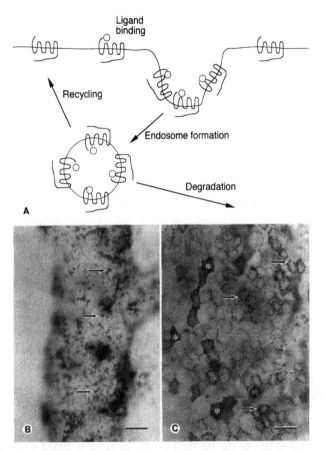

Figure 4 (A) Schematic representation of NK-1 receptor internalization and endosome formation after binding its ligand, substance P. The seven-transmembrane domain receptors are first internalized together with the bound substance P via clathrin-coated pits and then concentrated in endosomes that are recognizable by light microscopic immunohistochemistry. Ultimately, the receptors are recycled or degraded. (B) NK-1 receptor immunoreactivity in endothelial cells of postcapillary venule in rat trachea 3 min after injection of substance P. Immunoreactive endosomes (arrows) are abundant. (From Ref. 92.) (C) NK-1 receptor immunoreactivity in a whole mount of epithelium of rat trachea 5 min after vagal nerve stimulation. NK-1 immunoreactive endosomes clearly visible in otherwise faintly stained epithelial cells, but some darkly stained ciliated cells (asterisks) have few endosomes. (From Ref. 122.) Bar = 10 μm for (B) and (C).

adhesion. Neutrophils, eosinophils, and monocytes adhere to the endothelium of venules prior to their migration into inflamed tissue. The leukocyte adhesion and plasma leakage induced by stimulation of sensory nerves have several points in common: (a) Both phenomena are triggered by the same stimuli, including vagal stimulation, substance P, capsaicin, hypertonic saline, cigarette smoke, and selective NK-1 agonists (87,93–96), and are inhibited by NK-1 receptor antagonists (95,96), indicating that they are mediated by NK-1 receptors; (b) both occur in postcapillary venules and collecting venules, although these phenomena are not entirely coincident, and leakage occurs preferentially from smaller venules, whereas leukocyte adhesion tends to occur in larger venules (88,95); (c) both phenomena are inhibited by antiinflammatory drugs (97,98), and are increased by inhibition of enzymes that degrade tachykinins (97,99).

There is strong evidence that substance P and capsaicin induce vasodilatation and increase tracheobronchial blood flow (100–102). In contrast to plasma leakage, which has only been convincingly shown in the airways of rodents, vasodilatation in response to stimulation of sensory nerves or application of tachykinins has been reported in cats (103,104), dogs (105), pigs (106), sheep (107), and rats (108). Also, in contrast to plasma leakage, which is exclusively mediated by tachykinins released from sensory nerves, vasodilatation has a reflex parasympathetic component mediated by acetylcholine and VIP, in addition to a sensory nerve component mediated by tachykinins and CGRP (102). Again, there are species and organ differences in the balance of the different components. In the pig, the blood flow of the upper trachea is under both parasympathetic and sensory control, whereas the blood flow of peripheral airways is exclusively under sensory control (109). In the dog trachea, the relative order of potency of peptides in increasing tracheal blood flow in vivo is NKA > VIP > CGRP > SP > Peptide histidine isoleucine (PHI) (105), whereas for vasodilatation of the dog bronchial artery in vitro the order of potency is SP > NKA > CGRP > NKB (110). In the rat trachea, the increase in blood flow induced by capsaicin was mimicked by substance P, but not by CGRP (108). The neurogenic increased blood flow induced in rat airways is mediated by NK-1 receptors (111). In agreement with a similar observation in rat skin (27), in rat trachea the threshold of stimulation of sensory nerves required to increase blood flow is much lower than that required to produce plasma leakage (108). Perhaps mild irritants can be eliminated by increased blood flow, whereas stronger noxious stimuli may require additional defense mechanisms such as plasma leakage and leukocyte activation. In airways of other species—for example, pig—and in other vascular beds—for example, in the rat skin—CGRP appears to be the main mediator of vasodilatation (see below). The physiologically observed increase in blood flow mediated by sensory nerve activation is in excellent agreement with the intimate anatomical relationship of sensory nerves with arterioles (Fig. 2A).

Tachykinins produce bronchoconstriction in several species, including humans (76,78,112,113). NKA is considerably more potent than substance P in inducing airway smooth muscle contraction, and evidence from experiments using selective neurokinin receptor agonists and antagonists indicates that bronchoconstriction is mediated by NK-2 receptors (114,115). Furthermore, the contractile effect of tachykinins is greater in the more distal airways. Substance P can contract human bronchi in vitro (116), but does not do so in vivo, when given either by inhalation (117,118) or by intravenous infusion (117,119). NKA induces bronchoconstriction in normal humans after intravenous infusion (119), and in asthmatics after inhalation (118). In some species, e.g., guinea pig, both NK-1 and NK-2 receptors are present in the airway muscle (120), and in others, e.g., rat, substance P can contract the tracheal muscle indirectly, by production of arachidonic acid metabolites in the epithelium (121). This last is effect is consistent with the observed distribution of nerves and neurokinin receptors in the rat trachea (37,122) (Figs. 1 and 4).

Secretion of products into the airway lumen occurs from mucous and serous cells both on the epithelial surface and in the submucosal glands, and constitutes a defense against lumenal irritants (123). All of these secretory cells are stimulated by tachykinins in humans and in animals (124–126), and also by other neuropeptides, notably VIP and related peptides; see Ref. 127 and Chapter 15. The secretomotor effect is mediated by NK-1 receptors (128). Tachykinins and capsaicin also increase ciliary beat frequency by NK-1 receptors (129,130). The dense distribution of tachykinin-containing sensory nerves in the epithelium is well placed to mediate these effects (Fig. 1).

Tachykinin-containing nerve endings are found within the intrinsic parasympathetic ganglia of airways (Fig. 2D). Tachykinins increase the release of acetylcholine from parasympathetic nerves, thereby facilitating ganglionic transmission. This effect appears to be mediated by NK-1 receptors (131–133), although recently evidence has been presented from desensitization experiments that in bronchial ganglia there is also an NK-3 component (134).

In addition to the above rapid effects on resident cells of the airways, substance P and tachykinins have been reported to have various activating, priming, or mitogenic effects on an intriguing collection of inflammatory and immune cells that can migrate into inflamed airways (135). They may also have slower trophic actions, such as angiogenesis and wound healing (136). These fields promise to be expanding areas of research. In many cases, the effects have only been studied in vitro, and their relevance to normal physiology or disease remains to be determined. Substance P and other tachykinins can also cause eosinophil activation (137,138), neutrophil phagocytosis (139,140), macrophage activation and phagocytosis (141,142), B-lymphocyte activation (143,144), T-lymphocyte activation (145,146), monocyte cytokine production (147,148), fibroblast chemotaxis (149,150), and endothelial cell proliferation, migration, and angiogenesis (136,151).

As mentioned above, NKA is co-localized with substance P in peripheral sensory nerves (23,24,152). NKA can activate NK-1, NK-2, and NK-3 receptors, but is the preferred ligand for NK-2 receptors. Accordingly, the main effect of NKA in the airways appears to be bronchoconstriction (73,114,153).

Neuropeptide K (NPK) and neuropeptide γ (NPγ) are N-terminally extended forms of NKA, arising by alternate posttranslational processing of β and γ pre protachykinin I. They are present in smaller quantities in sensory neurons and are colocalized with substance P, neurokinin A, and CGRP (23,153). Although there are as yet fewer reports on the role of these extended tachykinins than of their shorter relatives, it is known that they are potent constrictors of guinea pig and human airways (154–156).

CGRP is a 37-amino acid peptide that results from alternative splicing of the mRNA of the precursor coded by the calcitonin gene. Human CGRP has at least two isoforms, termed CGRPα and CGRPβ, that differ by three amino acids, with other slight differences in other species. Both isoforms are present in peripheral sensory neurons (157). The molecular biology and pharmacology of CGRP has been reviewed elsewhere (18,158,159). CGRP is co-localized with tachykinins in sensory nerves (40,46), and is present in the airways of several species, including humans. In fact, in some cases CGRP appears to be present in a greater number of sensory nerves than substance P, and is detected in greater amounts biochemically (160). This property may be due partly to the greater resistance of CGRP than substance P to enzymatic degradation (161), and makes CGRP a useful assay of sensory nerve peptide release (160,162). In some species, especially rodents, CGRP is also present in airway neuroendocrine cells (34,42,44) (Fig. 2E). In rats, CGRP is also located in some serous-type epithelial cells (44).

The main effect of CGRP in skin and airways of various species including humans appears to be vasodilatation (46,163,164). The vasodilatation induced by CGRP in the microvasculature is independent of the endothelium, and is long-lasting (165). In some vascular beds, CGRP has a better case than tachykinins for being the primary sensory nerve peptide responsible for vasodilatation. The vasodilatation induced by electrical nerve stimulation or capsaicin or resiniferatoxin is mimicked by CGRP, but not by substance P or NK-1 agonists, and is blocked by the CGRP receptor antagonist CGRP (8-37), but not by NK-1 antagonists (165–168).

CGRP does not induce plasma leakage by itself, but can act synergistically with substance P to potentiate leakage in skin and airways (169–171).

For reasons that are unclear, CGRP can constrict human bronchi in vitro (66), but in vivo it appears to have no direct effects on airway smooth muscle (40). CGRP appears to induce only a small increase in submucosal gland secretion, consistent with the apparent lack of CGRP receptors on glands (127,172).

In addition to the above effects, CGRP may also have activating or trophic effects on various cells. In human skin, CGRP can cause accumulation of leukocytes (173) and is chemotactic for T lymphocytes (174). Degradative enzymes such as neutral endopeptidase can cleave CGRP to produce chemotactic factors for eosinophils (175,176). In addition, CGRP is mitogenic for human endothelial cells in vitro (177).

On the other hand, CGRP has been reported to have antiinflammatory actions. Somewhat paradoxically, pretreatment with CGRP reduces plasma leakage induced by histamine, leukotriene B_4 and serotonin in hamster cheek pouch (178). CGRP also inhibits proliferation of mouse T lymphocytes (179), and inhibits macrophage function (180).

C. Co-localization of Neuropeptides and Co-transmission

In addition to the tachykinins and CGRP, many other neuropeptides have been reported in the airways, but their possible roles have yet to be determined (78,181). To date, three other peptides have been localized to a subset of sensory nerves in the airways. One is VIP, in the tracheal epithelium of young rats (182). These nerves degenerate after capsaicin treatment or vagal nerve transsection, and decrease in number with age. The second peptide is dynorphin, which has been co-localized to a subset of substance P/CGRP-containing nerves in the guinea pig airway lamina propria and smooth muscle, but not in the epithelium or on blood vessels (14). The third peptide is pituitary adenylate cyclase activating peptide (PACAP) co-localized with substance P/CGRP in the tracheal epithelium of rats (183). All three of these peptides are also present in other populations of airway nerves that are not sensory. Their roles in the sensory nerves have yet to be determined.

The other neuropeptides reported in the airways cannot be considered as sensory nerve peptides because they are localized either in sympathetic or parasympathetic motor nerves, or their distribution or sensitivity to capsaicin is not known. These peptides also cannot be considered as inflammatory, and indeed evidence has been presented that some may be antiinflammatory. The peptides localized in airway nerves are listed in Table 2.

Table 2 Neuropeptides Other Than Tachykinins or CGRP Localized Immunohistochemically in Airway Nerves

Peptide	Localization	Function	Reference
VIP	Sensory	Unknown	(182)
Dynorphin	Sensory	Unknown	(14)
PACAP	Sensory	Unknown	(183)
VIP, PHI/PHM, PACAP	Parasympathetic	Vasodilator Bronchodilator Secretomotor	(18,59,64,78,184)
VIP	Sympathetic	Unknown	(185)
Neuropeptide Y	Sympathetic	Vasoconstrictor Antiinflammatory	(18) (186,187)
Dynorphin	Sympathetic	Unknown	(14)
Enkephalins	Parasympathetic	Unknown	(188,189)
Endogenous opioids	Not determined	Antiinflammatory	(190)
Galanin	Parasympathetic	Unknown	(58,64,191,192)

III. Natural and Experimental Stimuli for Release of Sensory Nerve Peptides

One of the special features of peripheral sensory nerve fibers is that they are polymodal; i.e., they will respond to a wide range of noxious irritant chemicals, to osmotic stimuli, and to mechanical irritation, by transmitting action potentials to the CNS, and by releasing transmitters at the peripheral nerve endings (193,194). These stimulatory actions on nerves can be either direct or indirect, with the production of intermediates, e.g., arachidonic acid metabolites. In many cases it is not known exactly which cells or mechanisms are involved in the excitation process, although epithelial cells are likely to be involved, because of their proximity to sensory nerves and the external environment. Some mediators that excite sensory nerves, e.g., bradykinin, histamine, and serotonin, can produce inflammatory effects such as plasma leakage by direct actions on endothelial cells.

Capsaicin merits a special place among the chemicals that excite C-fiber sensory nerves, not only for historical reasons (195), but because it has been used as a pharmacological tool to *define* the population of nerves in question, which are termed *capsaicin-sensitive nerves*. Low concentrations of capsaicin depolarize these nerve fibers, probably by interaction with a specific capsaicin receptor coupled to a non-selective cation channel through which calcium and sodium ions enter (196–198). The binding of capsaicin is competitively antagonized by capsazepine (199), and the nonspecific cation channel is blocked by the inorganic dye ruthenium red (200,201). Higher doses of capsaicin lead to long-lasting structural damage of the nerves. In adult animals, the sensory nerve fibers are depleted of transmitter (19,202), and can physically retract, but the neuronal cell bodies are spared (203,204). In neonatal animals, entire neurons can degenerate and are permanently eliminated (204,205). However, caution should be used in interpreting the results of experiments using capsaicin, for several reasons. The populations of capsaicin-sensitive and "small dark" neuronal cell bodies with unmyelinated nerve fibers are largely overlapping *but are not identical*, and little is known about the identity of the exceptions (204–206). Even after very high doses of capsaicin, not all of the cell bodies and nerve fibers degenerate, and the surviving fibers can sprout and regrow (204,207). Many, but certainly not all, capsaicin-sensitive nerves contain substance P (206,208). In addition to NKA and CGRP, which are largely co-expressed with substance P, other capsaicin-sensitive nerve fibers in the airways contain somatostatin or VIP (182). Little is known about the function of the capsaicin-sensitive nerves that *do not* contain substance P.

In experimental animals, administration of capsaicin produces many signs of inflammation in the airways, including plasma extravasation, vasodilatation, bronchoconstriction, leukocyte adhesion, and mucus secretion. These effects are particularly evident in rodents, but both central and peripheral nervous reflex components of vasodilatation have also been documented in sheep, dogs, and pigs (209–211). The inflammatory effects of cigarette smoke or citric acid are reduced or eliminated in capsaicin-pretreated rodents (29,212). Furthermore, in several animal models, responses of airways to allergen are eliminated by pretreatment with high doses of capsaicin, indicating a role for capsaicin-sensitive nerves (213–216). Ever since the introduction by Jancsó of capsaicin as a sensory nerve stimulant and neurotoxin, many investigators have been struck by the large differences among species in the sensitivity, physiological responses, and toxicity to it (4,77,197,217,218). For example, although capsaicin will cause vasodilatation in the airways of dogs and pigs, there is no evidence for neurogenic plasma leakage in these species; see Ref. 77. Furthermore, the effects of capsaicin seem to be poorly correlated with the content of substance P in sensory nerves (215,219). In humans, the main effects of capsaicin in the airways are cough (220) and reflex bronchoconstriction (221). Clearly, further studies are needed to explain the species differences in the sometimes paradoxical effects of capsaicin, and to elucidate the role of capsaicin-sensitive nerves in human airways.

A further development that has advanced our understanding of capsaicin-sensitive nerves is the introduction of resiniferatoxin, an ultrapotent analog of capsaicin (222). Although capsaicin and resiniferatoxin are derived from a different genus of plants (*Capsicum* and *Euphorbia*, respectively), they share a common homovanillic acid motif, and hence their binding site has been termed the vanilloid receptor (223). Use of radiolabeled resiniferatoxin has permitted the direct visualization of binding sites on sensory nerve fibers in the dorsal horn of the spinal cord (224). Resiniferatoxin has a similar spectrum of actions to capsaicin, for example, it causes the release of peptides from sensory nerves in the airways (162,225), and desensitization and destruction of sensory nerves [20], but it is up to several orders of magnitude more potent. Interestingly, the relative potency of resiniferatoxin compared to capsaicin varies for different actions. Resiniferatoxin is 1000 times more potent at inducing neurogenic inflammation, but does not stimulate pulmonary chemoreceptors or cause apnea, as does capsaicin (20,223). This suggests the possible existence of subtypes of vanilloid receptors. The strong binding of radiolabeled resiniferatoxin to the vanilloid receptor has permitted the direct visualization of binding sites on sensory nerve fibers in the dorsal horn of the spinal cord (224), and offers hope for the isolation and molecular characterization of the receptor (222). It is not yet clear why vanilloid receptors should exist on sensory nerves at all, or whether there is an endogenous ligand that stimulates these receptors.

Antidromic stimulation of sensory nerves induces release of peptides and effects such as plasma leakage and vasodilatation in the airways and in other tissues (4,8,13,29). The parameters of electrical stimulation at frequencies of 1–10 Hz indicate that A-δ (thinly myelinated) and C-fiber (unmyelinated) axons are involved. There is evidence from skin that the A-δ fibers may be preferentially responsible for vasodilatation, whereas the C fibers are responsible for plasma leakage (226). Antidromic electrical stimulation of sensory nerves can be prevented by application of local anesthetics such as lidocaine that block axonal conduction. Interestingly, these same anesthetics do not block the release of transmitter induced by the direct stimulation of sensory nerve endings by capsaicin or other inflammatory stimuli (4,6). The consequence of this aspect of the efferent nature of sensory nerves is that local anesthetics block vasodilatation, which requires an *axon reflex* and nerve conduction, but not plasma leakage, which requires only an *axon response*, i.e., by direct activation of nerve endings and release of sensory nerve peptides independent of nerve conduction (6).

Bradykinin can be generated endogenously in airways injured by mechanical trauma, cold air, anoxia, low pH, allergen, virus infection, or various other irritants by the action of the proteolytic enzyme kallikrein on the plasma precursor kininogen. Among the many inflammatory actions of bradykinin (reviewed in Ref. 227 and Chapters 10 and 20) is a stimulatory effect on sensory nerves, causing the release of peptides (228). This action appears be indirect, dependent on the production of prostaglandins as intermediates, and is reduced by indomethacin (225,229). However, the excitatory action of bradykinin is mediated by B2 receptors, and can be eliminated by treatment with B2 receptor antagonists, e.g., HOE 140 (230,231).

Exposure of the airway mucosa to low-pH solutions, as may occur after aspiration of gastric juice as a possible complication of anesthesia, induces the release of sensory nerve peptides (225,232). The mechanism is not quite clear, but may involve a direct activation of the capsaicin (vanilloid) receptor, or generation of an endogenous ligand, because the effect is attenuated by capsazepine (193,194,233).

Breathing dry or cold air can be an important trigger of asthma (74). These stimuli may act by concentrating the thin film of liquid on the apical surface of the airway epithelium, and can be mimicked by breathing hypertonic saline. These stimuli release sensory nerve peptides and cause plasma leakage and bronchoconstriction in rats, guinea pigs, and dogs (93,234–237). Electrical recording from single nerve fibers in the guinea pig trachea shows that the nerves sensitive to hypertonic saline are A-δ and C fibers (238).

Jancsó noted in his early studies on the conjunctiva that nicotine produced plasma leakage that was reduced by prior surgical denervation of the trigeminal nerve (4,195). These results were extended by others to the airways, who showed that acetylcholine, nicotine, and nicotinic agonists can release peptides from capsaicin-sensitive nerves (162,228,239,240). The release of CGRP from sensory nerve terminals in the rat trachea is reduced by indomethacin pretreatment, or by chemical destruction of sympathetic nerves by 6-hydroxydopamine treatment, suggesting that production of prostaglandins by sympathetic nerves may be an intermediate step in this excitatory pathway (225).

Histamine and serotonin (5-HT), inflammatory mediators released from mast cells, can stimulate release of peptides from sensory nerves that cause plasma leakage in rodent airways (29,228,241). These mediators also cause plasma leakage by a direct, nonneurogenic action on endothelial cells (242,243).

Exposure to allergens can increase the amount of several of the inflammatory mediators mentioned above. In experimental animals, it has been convincingly demonstrated that inflammatory responses to allergens are reduced by the elimination of capsaicin-sensitive nerves or blockade of NK-1 receptors: plasma leakage and bronchoconstriction in guinea pigs (241,244–246); bronchial hyperreactivity in mice (247); and vasodilatation in pigs (213).

In addition to the above stimuli, many gaseous noxious chemicals produce inflammatory effects in the airways of experimental animals that are at least partially dependent on capsaicin-sensitive nerves. These include cigarette smoke (29,248,249); formaldehyde (248,250); acrolein (a component of tear gas) (41); mustard oil vapor, xylene, and ether (4,29); toluene diisocyanate (251); and sulfur dioxide (252,253).

IV. Receptors for Sensory Nerve Peptides

Receptors for sensory nerve peptides, especially the tachykinins, have recently been a very active research area; see Refs. 25 and 254, and the chapter by Regoli. The genes for tachykinin NK-1, NK-2, and NK-3 receptors have been cloned, and DNA and RNA probes are becoming available (255,256). Many stable, nontoxic, and potent nonpeptide antagonists to all three neurokinin receptors have become available, and their use may help to elucidate the role of tachykinin-containing nerves in humans (see Chapter 22). Here, the focus is on neurokinin receptor localization.

Three approaches have been used to localize neurokinin receptors histologically in tissues: (a) autoradiography with radioactively labeled ligands, (b) immunohistochemical staining with antiidiotypic antibodies to tachykinin receptors (antibodies whose binding specificities resemble the natural binding site of the selected ligand), and more recently, (c) immunohistochemical staining using antibodies to specific peptide sequences of the receptors themselves.

The principles and practice of autoradiographic localization of neurokinin receptors using labeled substance P, and selective agonists and antagonists, have been reviewed (257,258). In general, these methods are effective at localizing binding sites to tissues, and give an idea of the binding intensity, but generally they lack the resolution to localize receptors to the cellular or subcellular level. Nevertheless, substance P-binding sites have been localized to airway smooth muscle and mucosa, especially epithelium and glands of rat, guinea pig, rabbit, and human airways (172,259–262). Some autoradiographic studies have found high-affinity substance P-binding sites in small blood vessels in rat, guinea pig, and human airways (263, 264), although it has generally been difficult to identify the precise vessel type, or exactly which cells bind the ligand. More recently, NK-1 receptors have been more selectively localized using radiolabeled antagonists FK-888 and CP-96,345 (265,266).

Immunohistochemistry using antiidiotypic antibodies detected substance P receptors on the lumenal aspect of all epithelial cells, in submucosal glands, and in the tracheal muscle, but did not detect labeling of blood vessels, and did not distinguish between subtypes of receptors (267–269).

Immunohistochemistry using antibodies to receptors has allowed a more precise cellular localization of NK-1 receptors (Fig. 4). In rat trachea, abundant NK-1 receptor immunoreactivity is present on endothelial cells of postcapillary venules that are the sites of tachykinin-induced plasma leakage (92) (Fig. 4B) and on epithelial cells and submucosal glands (122) (Fig. 4C). However, there is no detectable staining of arterioles, even though these vessels are directly innervated by substance P-containing nerves. This approach also permits the subcellular localization of receptors and visualization of the intracellular trafficking of receptors. When NK-1 receptors are stimulated by substance P, they are internalized into endosomes together with their ligand (Figs. 4A–4C). The internalization is blocked by the NK-1 receptor antagonist SR 140333 (92). The number of endosomes gives a quantitative index of receptor number and activation, and this number parallels the degree of leakage observed in different vessels (Fig. 5). As a consequence the pool of cell surface receptors is reduced, and the cell is desensitized to further stimulation (92,270). Thus, the process of receptor internalization, together with associated phosphorylation of stimulated receptors (271), contributes to tachyphylaxis. Eventually the internalized receptors are recycled to the cell surface, and the cells are resensitized to substance P (92,272).

More recently, similar antibodies have been generated to all three subtypes of neurokinin receptors (273). In guinea pig airways, NK-2 receptor immunoreactivity is found in the smooth muscle of airways and pulmonary arteries, but no staining for NK-3 receptor has yet been detected (274).

Receptors for CGRP are members of the superfamily of seven transmembrane-domain G-protein-coupled receptors. There is pharmacological evidence for at least two subtypes of CGRP receptors, termed $CGRP_1$ and $CGRP_2$ receptors, which are selectively antagonized by CGRP (8-37) fragment and by the CGRP analog [Cys(ACM2,7)CGRP], respectively (275,276). The human $CGRP_1$ receptor cDNA has been cloned and sequenced, and it has been deduced that it encodes for a 461-amino acid peptide that is coupled to adenylate cyclase (277). The receptor is expressed in high amounts in the lung, and has been localized in the lung by autoradiography using radiolabeled CGRP (172,278), and by in-situ gene amplification for $CGRP_1$ mRNA (277). These receptors appear to be more densely distributed on bronchial blood vessels than on muscle or epithelium, but the precise cellular localization of $CGRP_1$ receptors is not yet known.

V. Modulation of Proinflammatory Effects of Sensory Nerve Peptides

The inflammatory effects of sensory nerve peptides can be potentiated or inhibited by (a) modulating the synthesis or release of sensory nerve peptides, (b) modulating the number of sensory nerves or their proximity to target cells, (c) modulating the number of peptide receptors, or the volume of target tissue, or (d) modulating the amount or activity of enzymes that degrade the peptides. All of these phenomena can contribute to the overall severity of inflammation, as exemplified in rats with chronic *Mycoplasma pulmonis* airway infections. Each factor will be considered in turn for the potentiation, and then for the inhibition of inflammation.

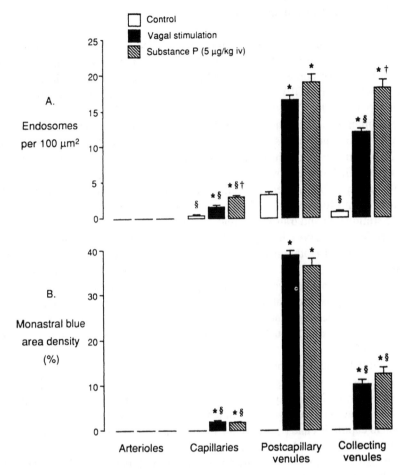

Figure 5 Bar graph comparing effects of vagal stimulation, injection of substance P (5 μg/kg, iv), and no stimulation (control) on number of NK-1 receptor immunoreactive endosomes (A) and amount of Monastral blue extravasation (B) in blood vessels of rat tracheal mucosa. Number of endosomes is expressed per 100 μm² of endothelial surface. Monastral blue area density (amount of extravasated Monastral blue in vessel wall expressed as percent of vessel surface area) was determined by stereological point counting. Values are means ± SEM, *n* = 4 rats per group. *Significantly different from corresponding control value. §Significantly different from corresponding value for postcapillary venules. †Significantly different from corresponding value for vagal stimulation. (From Ref. 122.)

A. Increased Inflammation

The possible upregulation of tachykinins has attracted attention as a mechanism of regulating the response to stimuli in conditions of chronic inflammation and pain (279,280). Sensory nerves, like sympathetic nerves, depend for their survival on nerve growth factor (NGF), and other neurotrophins produced by the target cells that they innervate (281). Cytokines also regulate the expression of substance P (282). Chronic inflammation or pain, such as the injection of formalin into the hindpaws of rats, can increase the synthesis of neurokinins in the spinal cord or dorsal root ganglia (283,284). Guinea pigs chronically exposed to cigarette smoke had increased amounts of CGRP in their lower airways (285), and rats chronically exposed to sulfur dioxide had threefold higher concentrations of substance P in their tracheas (286). The mechanisms for these increases are not yet clear. However, one possibility is that chronic stimulation and transmitter release may necessitate and facilitate transmitter synthesis. Sensory nerves, like sympathetic nerves, depend for their survival on nerve growth factor (NGF), other neurotrophins, and cytokines produced by the target cells they innervate (281,282,287). The chronic presence of other inflammatory mediators, such as bradykinin and cyclooxygenase products (288), or infections can cause sensory hyperalgesia, so that the threshold for release of peptides is decreased. For example, plasma leakage can be induced in the airways of rats with *Mycoplasma pulmonis* infections with much lower doses of capsaicin than required in pathogen-free rats (289).

To our knowledge, there is as yet no direct evidence for an increase of the number of nerves containing sensory peptides in animal models of chronic inflammation. However, in the airways of rats with *Mycoplasma pulmonis* infection there is extensive tissue remodeling, and a much greater proportion of sensory nerves are adjacent to capillaries and postcapillary venules than in pathogen-free controls (207).

In the formalin-induced inflammation model of the rat hindpaw, not only the tachykinins themselves, but the mRNA for their receptors is upregulated (280,284). In the *Mycoplasma pulmonis* infection model in rats, many vessels, particularly angiogenic capillaries, become more responsive to substance P than in controls. There is an increase in NK-1 receptor immunoreactivity in these vessels that parallels the increased leakage of vascular tracers (290). In response to the infection, there is a large thickening of the airway mucosa, infiltration with inflammatory cells, and angiogenesis, or proliferation of existing blood vessels. It is precisely the newly formed vessels that show the greatest increase in leakage and expression of NK-1 receptors.

Modulation of inflammation by the enzymes neutral endopeptidase (NEP) and angiotensin-converting enzyme (ACE), which degrade sensory nerve peptides and bradykinin, have been frequently reviewed (see Refs. 291 and 292, and several chapters in this volume). Here it will suffice to say that many inhaled irritant and natural infections such as respiratory viruses can decrease the activity of neutral endopeptidase (293–295). Inhibition of both NEP and ACE in rats results in an elevated baseline plasma leakage, suggesting that there is a tonic baseline release of inflammatory mediators (296). In the case of *Mycoplasma pulmonis*-infected rats, NEP activity in the epithelium is only half of that in pathogen-free controls (297); i.e., in this model, the decrease in NEP activity contributes only partially to the much larger increase in plasma leakage.

B. Decreased Inflammation

Inhibiting the synthesis or release of sensory nerve peptides is an effective way of reducing the inflammatory effects of sensory nerve stimulation (298). Reduced synthesis of sensory peptides is achieved in capsaicin-depletion procedures in adult animals, where the sensory nerve fibers persist structurally but are depleted of transmitter (204), and in tachyphylaxis, where nerves are overstimulated and become refractory (299). Blockade of capsaicin-receptor-mediated activation of sensory nerves can be achieved by pretreatment with capsazepine (300) or ruthenium red (301,302).

Presynaptic inhibition of sensory nerve transmitter release can be induced by a wide range of agents that can hyperpolarize the sensory nerves (303). Several different mechanisms and ion channels may be involved, because not all inflammatory stimuli are uniformly inhibited (18,162,304). Examples of agents that can block effects of sensory nerve stimulation in airways in vivo through a presumptive action on the nerves themselves include opioids acting via μ receptors (190,305), catecholamines acting via α_2-adrenoceptors (306), neuropeptide Y acting via Y_2 receptors (186,187), GABA acting via $GABA_2$ receptors (307), and histamine acting via H_3 receptors (308,309), corticotrophin-releasing factor and related peptides (310), and potassium channel openers (311,312). In view of this wide variety of inhibitory agents, an interesting question is whether there is a naturally occurring antiinflammatory agent that reduces sensory peptide release, and if so, whether advantage could be taken of it therapeutically. Two potential advantages of doing so would be the ability to inhibit inflammation caused by a wide variety of stimuli, and blocking the effects of release of multiple sensory nerve peptides more effectively than could be achieved by a single receptor antagonist.

Degeneration of sensory nerves by chronic capsaicin or resiniferatoxin treatment in newborn animals eliminates sensory nerve-mediated inflammatory effects in later life (204), but this approach is unlikely to be of therapeutic value in humans.

In recent years, many potent selective antagonists to neurokinin receptors have become available, and their future use promises to define the role of sensory nerves in humans. These drugs are reviewed in Refs. 313 and 314, and in Chapter 22. On the other hand, the study of the regulation of the expression of receptors to the sensory peptides is still in its infancy, and as yet there are no effective ways of manipulating the number of receptors that are expressed in vivo. However, it has been reported that dexamethasone reduced the concentration of NK-1 receptors in human asthmatic lung tissue (315). In the *Mycoplasma pulmonis*-infection model of rat airways, dexamethasone not only reversed the potentiation of neurogenic inflammation, but also caused a regression of angiogenic changes, i.e., eliminated newly formed vessels bearing NK-1 receptors (316).

In principle, it should be possible to inhibit inflammation by increasing the amount or activity of enzymes that degrade sensory peptides in vivo. In practice, this approach is limited by the difficulty of administering large molecules into the airways. However, aerosolized recombinant human neutral endopeptidase inhibits neuropeptide-induced cough in guinea pigs (317). It has also been reported that dexamethasone treatment increases neutral endopeptidase and angiotensin-converting enzyme activity in the airways in vivo (292).

VI. Role of Sensory Nerve Peptides in Disease

The hypothesis that asthma may be a neural disease has stimulated an enormous amount of research on autonomic and sensory nerves in humans and animals (318). In animals, especially in rodents, the evidence is clear that sensory nerve peptides are involved in many aspects of airway inflammation. As our knowledge of the animal models and inflammatory mechanisms increases, so does our ability to manipulate these models, and ultimately to control inflammation in humans.

Many relevant issues involving human airway sensory nerves and their transmitters in health and disease remain unresolved. Key questions for establishing the role of a particular chemical substance as a neurotransmitter are: (a) Is it present in nerves? (b) Is it released by nerve stimulation? (c) Does the exogenously applied substance mimic the effects of nerve stimulation? (d) Can the effects of nerve stimulation be blocked by antagonists of the substance? (e) Can the effects of nerve stimulation be potentiated by drugs that prevent removal of the substance—for example, by preventing its enzymatic breakdown? Other questions are also relevant: Are the numbers of sensory nerves and their receptors different in healthy and diseased airways? What transmitters do human sensory nerves contain? Is the sensitivity of sensory nerves changed in disease?

To date, it has proved extraordinarily technically difficult to assess these issues in human tissues with the same precision as in animals, and most of the above questions have not been answered satisfactorily. On the brighter side, appropriate techniques and tools are now becoming available to address these issues systematically.

There is no doubt that sensory nerves are present in human airways, and that some nerves express substance P or CGRP, but in contrast to rodents, such nerves are sparse. This should not come as too great a surprise in view of the diversity of chemical coding of airway nerves documented in animals. Efforts should be made to identify other transmitters in human sensory nerves and to determine the action of these transmitters. Comparisons of the number or density of nerves containing substance P, CGRP, and other peptides in airways from "healthy controls" and asthmatics have produced conflicting results, with some investigators reporting large differences in nerve number (62,319) and others reporting no differences in nerve number (65,69) or transmitter content (320). Perhaps these discrepancies can be partly explained by the differences in the severity of disease assessed, patient age and statistical sample size, tissues sampled, and techniques used. Similar considerations may apply to discrepant reports where increased amounts of neurokinin NK-1 receptors were observed in asthma (315), or no differences were observed (321).

Evidence has been presented that substance P and CGRP are released into nasal or bronchial lavage fluids after allergen challenge with allergen or with hypertonic saline, and the release of substance P is greater in allergic individuals (322–324). In humans, inhaled capsaicin can produce cough, sneezing, mucus secretion, nasal congestion, and a relatively small and transient bronchoconstriction (221). However, the sensitivity of sensory nerves to capsaicin is known to vary enormously among different animals species (217,218). Capsaicin is undoubtedly a very useful tool, but may not necessarily be the appropriate "gold standard" against which to measure human sensory nerves. Human beings are not alone among mammals in failing to respond to capsaicin by activation of all categories of their sensory nerves.

The inflammatory effects of exogenously applied tachykinins in human airways have been reviewed (76,85,325). Substance P causes an increased amount of albumin in nasal secretions, suggesting plasma leakage in the nasal mucosa (326). However, it is not yet known if plasma leakage, originally defined as the key characteristic of neurogenic inflammation, really occurs in the lower airways (77,80). Neurokinin A is more effective at producing bronchoconstriction than substance P, particularly in allergic individuals (325). Three other proinflammatory effects of sensory peptides and capsaicin (pain, vasodilatation, and mucus secretion) occur in human nasal mucosa (327), but have not yet been convincingly demonstrated in the lower airways. The effects of sensory nerve peptides on inflammatory cell influx in human airways have yet to be carefully assessed in vivo.

The effects of the new generation of tachykinin receptor antagonists on inflammation in human airways are reviewed in Ref. 314 and in Chapter 22. At present few studies have been published, but more information is expected as clinical studies are completed. In one study, the neurokinin antagonist FK-224, which is equally potent against NK-1 and NK-2 receptors, blocked the bradykinin-induced bronchoconstriction in asthmatics, suggesting the involvement of tachykinins (328). The same group also reported that pretreatment with the selective NK-1 receptor antagonist FK-888 protected against airway narrowing in exercise-induced asthma (329). However, in another study, the NK-1 receptor antagonist CP-99,994 had no effect on bronchoconstriction induced in asthmatics by hypertonic saline challenge (330). In monkeys, dual antagonisms of NK-1 and NK-2 receptors protected against allergen-induced airway hyperresponsiveness (331).

From animal studies, it would be expected that inhibition of neutral endopeptidase would reveal or potentiate effects of sensory nerve peptide release in human airways. So far, such experiments in humans have produced mixed results. In some studies, inhibition of NEP by thiorphan potentiates the bronchoconstriction induced in normal humans by neurokinin A (332), and inhaled sodium metabisulfite (333). Similarly, inhibition of NEP by thiorphan or phosphoramidon potentiates the airway narrowing induced in asthmatics by leukotriene D_4 and by bradykinin, two mediators believed to be produced in allergen challenge (334,335). However, in other studies, inhibition of NEP does not potentiate the bronchoconstriction induced in asthmatics by sodium metabisulfite inhalation (336), or by allergen challenge (337).

VII. Conclusions

Tachykinins and CGRP are the main proinflammatory peptides contained within sensory nerves in the airways. They have multiple actions, including vasodilatation, plasma leakage, leukocyte adhesion, bronchoconstriction, gland secretion, epithelial cell ion transport, and ciliary beat activity, and various chemotactic and trophic effects on immune cells. These phenomena have been studied extensively in animal models, and their cellular and molecular mechanisms have been well characterized, particularly in rodents. Sensory nerves undoubtedly also occur in human airways, although fewer of these nerves appear to contain substance P than in rodents. The role of airway sensory nerves in humans is only now starting to be addressed systematically. These experiments should be facilitated by the development of techniques and specific, effective, and nontoxic reagents that can be used in the clinical setting. In view of the multiple proinflammatory effects of sensory nerve peptides in animal models, it is possible that sensory nerves also play an important role in human health and disease.

Acknowledgments

This work was supported in part by National Heart, Lung, and Blood Institute Program Project Grant HL-24136.

References

1. Chahl LA. Antidromic vasodilatation and neurogenic inflammation. Pharmacol Ther 1988; 37:275–300.
2. Szolcsányi J. Antidromic vasodilatation and neurogenic inflammation. Agents Actions 1988; 23:4–11.
3. Jancsó N. Role of the nerve terminals in the mechanism of inflammatory reactions. Bull Millard Fillmore Hosp (Buffalo, NY) 1960; 7:53–77.
4. Jancsó N, Jancsó-Gábor A, Szolcsányi J. Direct evidence for neurogenic inflammation and its prevention by denervation and by pretreatment with capsaicin. Br J Pharmacol Chemother 1967; 31:138–151.
5. Jancsó N, Jancsó-Gábor A, Szolcsányi J. The role of sensory nerve endings in neurogenic inflammation induced in human skin and in the eye and paw of the rat. Br J Pharmacol Chemother 1968; 33:32–41.
6. Szolcsányi J. Neurogenic inflammation: reevaluation of axon reflex theory. In: Geppetti P, Holzer P, eds. Neurogenic Inflammation. Boca Raton, FL: CRC Press, 1996:35–45.
7. Geppetti P, Holzer P, eds. Neurogenic Inflammation. Boca Raton, FL: CRC Press, 1996.

8. Szolcsányi, J, Jancsó-Gábor A, Salomon I. Vascular permeability-increasing effect of electrical stimulation of peripheral nerves, sensory ganglia and spinal roots. Acta Physiol Hung 1976; 47:255.

9. Baraniuk JN. Neural control of the upper respiratory tract. In: Kaliner MA, et al., eds. Neuropeptides in Respiratory Medicine. New York: Marcel Dekker, 1994:79–123.

10. Canning BJ, Undem BJ. Parasympathetic innervation of airways smooth muscle. Airways Smooth Muscle: Structure, Innervation and Neurotransmission 1994:43–78.

11. Cadieux A, Springall DR, Mulderry PK, Rodrigo J, Ghatei MA, Terenghi G, Bloom SR, Polak JM. Occurrence, distribution and ontogeny of CGRP immunoreactivity in the rat lower respiratory tract: effect of capsaicin treatment and surgical denervations. Neuroscience 1986; 19:605–627.

12. Springall DR, Cadieux A, Oliveira H, Su H, Royston D, Polak JM. Retrograde tracing shows that CGRP-immunoreactive nerves of rat trachea and lung originate from vagal and dorsal root ganglia. J Auton Nerv Syst 1987; 20:155–166.

13. McDonald DM, Mitchell RA, Gabella G, Haskell A. Neurogenic inflammation in the rat trachea. II. Identity and distribution of nerves mediating the increase in vascular permeability. J Neurocytol 1988; 17:605–628.

14. Kummer W, Fischer A, Kurkowski R, Heym C. The sensory and sympathetic innervation of guinea-pig lung and trachea as studied by retrograde neuronal tracing and double-labelling immunohistochemistry. Neuroscience 1992; 49:715–737.

15. Canning BJ, Myers AC, Undem BJ. Evidence for interganglionic communication between guinea pig tracheal (GPT) ganglion neurons. Am J Respir Crit Care Med 1996; 153:A844.

16. Furness JB, Costa M. The Enteric Nervous System. Edinburgh: Churchill Livingstone, 1986.

17. Coleridge JCG, Coleridge HM. Afferent vagal C fibre innervation of the lungs and its functional significance. Rev Physiol Biochem Pharmacol 1984; 99:1–110.

18. Lundberg JM. Pharmacology of cotransmission in the autonomic nervous system: integrative aspects on amines, neuropeptides, adenosine triphosphate, amino acids and nitric oxide. Pharmacol Rev 1996; 48:113–178.

19. Jancsó G, Király E, Jancsó-Gábor A. Pharmacologically induced selective degeneration of chemosensitive primary sensory neurones. Nature 1977; 270:741– 743.

20. Szolcsányi J, Szallasi A, Szallasi Z, Joó F, Blumberg PM. Resiniferatoxin: an ultrapotent selective modulator of capsaicin-sensitive primary afferent neurons. J Pharmacol Exp Ther 1990; 255:923–928.

21. Tranzer J, Thoenen H. An electron microscope study of selective acute degeneration of sympathetic nerve terminals after administration of 6-hydroxydopamine. Experientia 1968; 24:155–156.

22. Sulakvelidze I, Baluk P, McDonald DM. Plasma extravasation induced in the rat trachea by 6-hydroxydopamine is mediated by sensory nerves, not by sympathetic nerves. J Appl Physiol 1994; 76:701–707.

23. Hua XY, Theodorsson-Norheim E, Brodin E, Lundberg JM, Hökfelt T. Multiple tachykinins (neurokinin A, neuropeptide K and substance P) in capsaicin-sensitive sensory neurons in the guinea-pig. Regul Pept 1985; 13:1–19.

24. Moussaoui SM, Le PN, Bonici B, Faucher DC, Cuiné F, Laduron PM, Garret C. Distribution of neurokinin B in rat spinal cord and peripheral tissues: comparison with neurokinin A and substance P and effects of neonatal capsaicin treatment. Neuroscience 1992; 48:969–978.

25. Buck SH, ed. The tachykinin receptors. In: Bylund DB, ed. The Receptors, Totawa, NJ: Humana Press, 1994:630.

26. Fong TM. Molecular biology of tachykinins. In: Geppetti P, Holzer P, eds. Neurogenic Inflammation. Boca Raton, FL: CRC Press, 1996:3–14.

27. Lembeck F, Holzer P. Substance P as neurogenic mediator of antidromic vasodilation and neurogenic plasma extravasation. Naunyn-Schmiedeberg's Arch Pharmacol 1979; 310:175–183.

28. Lembeck F. The 1988 Ulf von Euler Lecture. Substance P: from extract to excitement. Acta Physiol Scand 1988; 133:435–454.

29. Lundberg JM, Saria A. Capsaicin-induced desensitization of airway mucosa to cigarette smoke, mechanical and chemical irritants. Nature 1983; 302:251–253.

30. Wharton J, Polak JM, Bloom SR, Will JA, Brown MR, Pearse AGE. Substance P-like immunoreactive nerves in mammalian lung. Invest Cell Pathol 1979; 2:3–10.

31. Lundberg JM, Hökfelt T, Martling CR, Saria A, Cuello C. Substance P-immunoreactive sensory nerves in the lower respiratory tract of various mammals including man. Cell Tissue Res 1984; 235:251–261.

32. Martling CR. Sensory nerves containing tachykinins and CGRP in the lower airways. Functional implications for bronchoconstriction, vasodilatation and protein extravasation. Acta Physiol Scand Suppl 1987; 563:1–57.

33. Lundberg JM, Martling C-R, Hökfelt T. Airways, oral cavity and salivary glands: classical transmitters and peptides in sensory and autonomic motor neurons. In: Björklund A, Hökfelt T, Owman C, eds. Handbook of Chemical Neuroanatomy. Vol. 6. The Peripheral Nervous System. Amsterdam: Elsevier, 1988:391–444.

34. Domeij S, Dahlqvist A, Forsgren S. Regional differences in the distribution of nerve fibers showing substance P-related and calcitonin gene-related peptide-like immunoreactivity in the rat larynx. Anat Embryol 1991; 183:49–56.

35. Uddman R, Sundler F. Neuropeptides in the airways: a review. Am Rev Respir Dis 1987; 94:53–58.

36. Bowden J, Gibbins IL. Relative density of substance P-immunoreactive (SP-IR) nerve fibres in the tracheal epithelium of a range of species. FASEB J, 1992; 6:A1276.

37. Baluk P, Nadel JA, McDonald DM. Substance P-immunoreactive sensory axons in the rat respiratory tract: a quantitative study of their distribution and role in neurogenic inflammation. J Comp Neurol 1992; 319:586–598.

38. Springall DR, Polak JA, Ghatei MA, Lackie P, Bloom SR. Calcitonin gene-related peptide (CGRP), a new regulatory peptide widely distributed in lung. Lung 1984; 143: 306–307.

39. Uddman R, Luts A, Sundler F. Occurrence and distribution of calcitonin-gene-related peptide in the mammalian respiratory tract and middle ear. Cell Tissue Res 1985; 241: 551–555.

40. Martling CR, Saria A, Fischer JA, Hökfelt T, Lundberg JM. Calcitonin gene-related peptide and the lung: neuronal coexistence with substance P, release by capsaicin and vasodilatory effect. Regul Pept 1988; 20:125–139.

41. Springall DR, Edginton JAG, Price PN, Swanston DW, Noel C, Bloom SR, Polak JM. Acrolein depletes the neuropeptides CGRP and substance P in sensory nerves in rat respiratory tract. Environ Health Perspect 1990; 85:151–157.

42. Shimosegawa T, Said SI. Pulmonary calcitonin gene-related peptide immunoreactivity: nerve-endocrine interrelationships. Am J Respir Cell Mol Biol, 1991; 4:126–134.

43. Terada M, Iwanaga T, Takahashiiwanaga H, Adachi I, Arakawa M, Fujita T. Calcitonin gene-related peptide (CGRP)-immunoreactive nerves in the tracheal epithelium of rats: an immunohistochemical study be means of whole mount preparations. Arch Histol Cytol 1992; 55:219–233.

44. Baluk P, Nadel J, McDonald D. Calcitonin gene-related peptide in secretory granules of serous cells in the rat tracheal epithelium. Am J Respir Cell Mol Biol 1993; 8:446–453.

45. Nilsson G, Dahlberg K, Brodin E, Sundler F, Strandberg K. Distribution and constrictor effect of substance P in guinea pig tracheobronchial tissue. In: von Euler US, Pernow B, eds. Substance P. New York: Raven Press, 1977:75–81.

46. Lundberg JM, Franco CA, Hua X, Hökfelt T, Fischer JA. Co-existence of substance P and calcitonin gene-related peptide-like immunoreactivities in sensory nerves in relation to cardiovascular and bronchoconstrictor effects of capsaicin. Eur J Pharmacol 1985; 108:315–319.

47. Dey RD, Hoffpauir J, Said SI. Co-localization of vasoactive intestinal peptide- and substance P-containing nerves in cat bronchi. Neuroscience 1988; 24:275–281.

48. Dey RD, Altemus JB, Zervos I, Hoffpauir J. Origin and colocalization of CGRP- and SP-reactive nerves in cat airway epithelium. J Appl Physiol 1990; 68:770–778.

49. Keith IM, Ekman R. Calcitonin gene-related peptide in hamster lung and its coexistence with serotonin: a chemical and immunocytochemical study. Regul Pept 1988; 22:315–323.

50. Shin T, Wada S, Maeyama T, Wtanabe S. Substance P immunoreactive sensory nerve fibers of the canine laryngeal mucosa. In: Fujimura O, ed. Vocal Physiology: Voice Production, Mechanisms and Functions. New York: Raven Press, 1988:115–127.

51. Nohr D, Weihe E. Koexistenz von Substanz P (SP) und vasoaktivem intestinalem Polypeptid (VIP) in intrinsichen Neuronen der Lunge von Katze und Hund. In: Pfannenshil H-D, ed. 82 Jahresversammlung der Deutchen Zoologischen Gesellschaft. Düsseldorf: Gustav Fischer Verlag, 1989.

52. Laitinen LA, Laitinen A, Panula PA, Partanen M, Tervo K, Tervo T. Immunohistochemical demonstration of substance P in the lower respiratory tract of the rabbit and not of man. Thorax 1983; 38:531–536.

53. Luts A, Sundler F. Peptide-containing nerve fibers in the respiratory tract of the ferret. Cell Tissue Res 1989; 258:259–267.

54. Dey RD, Altemus JB, Rodd A, Mayer B, Said SI, Coburn RF. Neurochemical characterization of intrinsic neurons in ferret tracheal plexus. Am J Respir Cell Mol Biol 1996; 14:207–216.

55. Martling CR, Matran R, Alving K, Hökfelt T, Lundberg JM. Innervation of lower airways and neuropeptide effects on bronchial and vascular tone in the pig. Cell Tissue Res 1990; 260:223–233.

56. Sonea IM, Bowker RM, Robinson NE, Holland RE. Distribution of SP- and CGRP-like immunoreactive nerve fibers in the lower respiratory tract of neonatal foals: evidence for loss during development. Anat Embryol 1994; 190:469–477.

57. Sonea IM, Bowker RM, Robinson NE, Broadstone RV. Substance P and calcitonin gene-related peptide-like immunoreactive nerve fibers in lungs from adult equids. Am J Vet Res 1994; 55:1066–1074.

58. Ghatei MA, Springall DR, Richards IM, Oostveen JA, Griffin RL, Cadieux A, Polak JM, Bloom SR. Regulatory peptides in the respiratory tract of *Macaca fascicularis*. Thorax 1987; 42:431–439.

59. Nohr D, Eiden LE, Weihe E. Coexpression of vasoactive intestinal peptide, calcitonin gene-related peptide and substance p immunoreactivity in parasympathetic neurons of the rhesus monkey lung. Neurosci Lett 1995; 199:25–28.

60. Springall DR, Polak JM, Howard L, Power RF, Krausz T, Manickam S, Banner NR, Khagani A, Rose M, Yacoub MH. Persistence of intrinsic neurones and possible phenotypic changes after extrinsic denervation of human respiratory tract by heart-lung transplantation. Am Rev Respir Dis 1990; 141:1538–1546.

61. Hislop AA, Wharton J, Allen KM, Polak JM, Haworth SG. Immunohistochemical localization of peptide-containing nerves in human airways—age-related changes. Am J Respir Cell Mol Biol 1990; 3:191–198.

62. Ollerenshaw SL, Jarvis D, Sullivan CE, Woolcock AJ. Substance P immunoreactive nerves in airways from asthmatics and nonasthmatics. Eur Respir J 1991; 4:673–682.

63. Komatsu T, Yamamoto M, Shimokata K, Nagura H. Distribution of substance P-immunoreactive and calcitonin gene-related peptide-immunoreactive nerves in normal human lungs. Int Arch Allergy Appl Immunol 1991; 95:23–28.

64. Luts A, Uddman R, Alm P, Basterra J, Sundler F. Peptide-containing nerve fibers in human airways—distribution and coexistence pattern. Int Arch Allergy Immunol 1993; 101:52–60.

65. Howarth PH, Springall DR, Redington AE, Djukanovic R, Holgate ST, Polak JM. Neuropeptide-containing nerves in endobronchial biopsies from asthmatic and non-asthmatic subjects. Am J Respir Cell Molec Biol 1995; 13:288–296.

66. Palmer J, Cuss F, Mulderry P, Ghatei M, Springall D, Cadieux A, Bloom S, Polak JM, Barnes PJ. Calcitonin gene-related peptide is localized to human airway nerves and potently constricts human airway smooth muscle. Br J Pharmacol 1987; 91:95–101.

67. Nohr D, Weihe E. The neuroimmune link in the bronchus-associated lymphoid tissue (BALT) of cat and rat: peptides and neural markers. Brain Behav Immun 1991; 5:84–101.

68. Nilsson G, Alving K, Ahlstedt S, Hökfelt T, Lundberg JM. Peptidergic innervation of rat lymphoid tissue and lung: relation to mast cells and sensitivity to capsaicin and immunization. Cell Tissue Res 1990; 262:125–133.

69. Howarth PH, Djukanovic R, Wilson JW, Holgate ST, Springall DR, Polak JM. Mucosal nerves in endobronchial biopsies in asthma and non-asthma. Int Arch Allergy Appl Immunol 1991; 94:330–330.

70. Regoli D, Drapeau G, Dion S, D'Orléans JP. Receptors for substance P and related neurokinins. Pharmacology 1989; 38:1–15.

71. Foreman JC. Substance P and calcitonin gene-related peptide: effects on mast cells and in human skin. Int Arch Allergy Appl Immunol 1987; 82:366–371.

72. Mousli M, Bronner C, Landry Y, Bockaert J, Rouot B. Direct activation of GTP-binding regulatory proteins (G-proteins) by substance P and compound 48/80. FEBS Lett 1990; 259:260–262.

73. Barnes PJ, Baraniuk JN, Belvisi MG. Neuropeptides in the respiratory tract. Part I. Am Rev Respir Dis 1991; 144:1187–1198.

74. Solway J, Leff AR. Sensory neuropeptides and airway function. J Appl Physiol 1991; 71:2077–2087.

75. Frossard N, Advenier C. Tachykinin receptors and the airways. Life Sci 1991; 49:1941–1953.

76. Joos GF, Germonpré PR, Kips JC, Peleman RA, Pauwels RA. Sensory neuropeptides and the human lower airways—present state and future directions. Eur Respir J 1994; 7:1161–1171.

77. McDonald DM. Neurogenic inflammation in the airways. In: Barnes P, ed. Autonomic Control of the Respiratory System. Harwood Academic Publishers, 1997: 249–290.

78. Maggi CA, Giachetti A, Dey RD, Said SI. Neuropeptides as regulators of airway function: vasoactive intestinal peptide and the tachykinins. Physiol Rev 1995; 75:277–322.

79. Chung KF, Rogers DF, Barnes PJ, Evans TW. The role of increased airway microvascular permeability and plasma exudation in asthma. Eur Respir J 1990; 3:329–337.

80. McDonald DM. The concept of neurogenic inflammation in the respiratory tract. In: Kaliner M, et al., eds. Neuropeptides in the Respiratory Medicine. New York: Marcel Dekker, 1994:321–349.

81. Bowden JJ, McDonald DM. The microvasculature as a participant in inflammation. In: Holgate ST, ed. Immunopharmacology of the Respiratory System. New York: Academic Press, 1995:147–168.

82. Figini M, Emanueli C, Grady EF, Kirkwood K, Payan DG, Ansel J, Gerrard C, Geppetti P, Bunnett N. Substance P and bradykinin stimulate plasma extravastion in the mouse gastrointestinal tract and pancreas. Am J Physiol 1997; 272:9785–9793.

83. Lundberg JM, Lundblad L, Änggård A, Martling C-R, Theodorsson-Norheim E, Stjärne P, Hökfelt TG, Saria A. Bioactive peptides in capsaicin-sensitive C-fiber afferents of the airways: functional and pathophysiological implications. In: Kaliner MA, Barnes PJ, eds. The Airways, Neural Control in Health and Disease. New York, Basel: Marcel Dekker, 1988:417–445.

84. Persson CGA. Plasma exudation from tracheobronchial microvessels in health and disease. In: Butler J, ed. The Bronchial Circulation. New York: Marcel Dekker, 1991: 443–473.

85. Barnes PJ. Sensory neuropeptides and airway disease. In: Geppetti P, Holzer P, eds. Neurogenic Inflammation. Boca Raton, FL: CRC Press, 1996:169–185.

86. Tousignant C, Chan C-C, Guevremont D, Brideau C, Hale JJ, MacCoss M, Rodger IW. NK_2 receptors mediate plasma extravasation in guinea-pig lower airways. Br J Pharmacol 1993; 108:383–386.

87. McDonald DM. Neurogenic inflammation in the rat trachea. I. Changes in venules, leukocytes, and epithelial cells. J Neurocytol 1988; 17:583–603.

88. McDonald DM. Endothelial gaps and permeability of venules in rat tracheas exposed to inflammatory stimuli. Am J Physiol 1994; 266:L61–L83.

89. Hirata A, Baluk P, Fujiwara T, McDonald DM. Location of focal silver staining at endothelial gaps in inflamed venules examined by scanning electron microscopy. Am J Physiol 1995; 269:403–418.

90. Thurston G, Baluk P, Hirata A, McDonald DM. Permeability-related changes revealed at endothelial cell borders in inflamed venules by lectin binding. Am J Physiol 1996; 271:H2547–H2562.

91. Baluk P, Hirata A, Thurston G, Fujiwara T, Neal CR, Michel CC, McDonald DM. Endothelial gaps: time course of formation and closure in inflamed venules of rats. Am J Physiol 1997; 272:L155–L170.

92. Bowden JJ, Garland A, Baluk P, Lefevre P, Grady E, Vigna SR, Bunnett NW, McDonald DM. Direct observation of substance P-induced internalization of NK_1 receptors at sites of inflammation. Proc Natl Acad Sci (USA) 1994; 91:8964–8968.

93. Umeno E, McDonald DM, Nadel JA. Hypertonic saline increases vascular permeability in the rat by producing neurogenic inflammation. J Clin Invest 1990; 85:1905–1908.

94. Umeno E, Nadel JA, McDonald DM. Neurogenic inflammation of the rat trachea: fate of neutrophils that adhere to venules. J Appl Physiol 1990; 69:2131–2136.

95. Baluk P, Bertrand C, Geppetti P, McDonald DM, Nadel JA. NK-1 receptors mediate leukocyte adhesion in neurogenic inflammation in the rat trachea. Am J Physiol 1995; 268:L263–L269.

96. Baluk P, Bertrand C, Geppetti P, McDonald DM, Nadel JA. NK$_1$ receptor antagonist CP-99,994 inhibits cigarette smoke-induced neutrophil and eosinophil adhesion in rat tracheal venules. Exp Lung Res 1996; 22:409–418.

97. Katayama M, Nadel JA, Piedimonte G, McDonald DM. Peptidase inhibitors reverse steroid-induced suppression of neutrophil adhesion in rat tracheal blood vessels. Am J Physiol 1993; 264:L316–L322.

98. Bowden JJ, Sulakvelidze I, McDonald DM. Inhibition of neutrophil and eosinophil adhesion to venules of the rat trachea by the b$_2$-adrenergic agonist formoterol. J Appl Physiol 1994; 77:397–405.

99. Umeno E, Nadel JA, Huang HT, McDonald DM. Inhibition of neutral endopeptidase potentiates neurogenic inflammation in the rat trachea. J Appl Physiol 1989; 66:2647–2652.

100. Laitinen LA, Laitinen A, Widdicombe J. Effects of inflammatory and other mediators on airway vascular beds. Am Rev Respir Dis 1987; 135:1959.

101. Salonen RO, Webber SE, Widdicombe JG. Effects of neurotransmitters on tracheobronchial blood flow. Eur Respir J Suppl 1990; 12:636s–637s.

102. Widdicombe JG, Webber SE. Neuroregulation and pharmacology of the tracheobronchial circulation. In: Butler J, ed. The Bronchial Circulation. New York: Marcel Dekker, 1992:249–289.

103. Martling C-R, Änggård A, Lundberg JM. Non-cholinergic vasodilation in the tracheobronchial tree of the cat induced by vagal nerve stimulation. Acta Physiol Scand 1985; 125:343–346.

104. Martling CR, Gazelius B, Lundberg JM. Nervous control of tracheal blood flow in the cat measured by the laser Doppler technique. Acta Physiol Scand 1987; 130:409–417.

105. Salonen RO, Webber SE, Widdicombe JG. Effects of neuropeptides and capsaicin on the canine tracheal vasculature in vivo. Br J Pharmacol 1988; 95:1262–1270.

106. Martling CR, Matran R, Alving K, Lacroix JS, Lundberg JM. Vagal vasodilatory mechanisms in the pig bronchial circulation preferentially involves sensory nerves. Neurosci Lett 1989; 30:306–311.

107. Parsons GH, Nichol GM, Barnes PJ, Chung KF. Peptide mediator effects on bronchial blood velocity and lung resistance in conscious sheep. J Appl Physiol 1992; 72:1118–1122.

108. Piedimonte G, Hoffman JIE, Husseini WK, Hiser WL, Nadel JA. Effect of neuropeptides released from sensory nerves on blood flow in the rat airway microcirculation. J Appl Physiol 1992; 72:1563–1570.

109. Matran R, Alving K, Martling C-R, Lacroix JS, Lundberg JM. Vagally mediated vasodilatation by motor and sensory nerves in the tracheal and bronchial circulation of the pig. Acta Physiol Scand 1989; 135:29–37.

110. McCormack DG, Salonen RO, Barnes PJ. Effect of sensory neuropeptides on canine bronchial and pulmonary vessels in vitro. Life Sci 1989; 45:2405–2412.

111. Piedimonte G, Hoffman JI, Husseini WK, Snider RM, Desai MC, Nadel JA. NK1 receptors mediate neurogenic inflammatory increase in blood flow in rat airways. J Appl Physiol 1993; 74:2462–2468.

112. Lundberg JM, Saria A. Bronchial smooth muscle contraction induced by stimulation of capsaicin-sensitive sensory neurons. Acta Physiol Scand 1982; 116:473–476.

113. Frossard N, Barnes PJ. Effect of tachykinins in small human airways. Neuropeptides 1991; 19:157–161.

114. Advenier C, Naline E, Drapeau G, Regoli D. Relative potencies of neurokinins in guinea pig and human bronchus. Am Rev Respir Dis 1987; 136:50–54.

115. Advenier C, Naline E, Toty L, Bakdach H, Emonds-Alt X, Vilain P, Breliere JC, Le Fur G. Effects on the isolated human bronchus of SR 48968, a potent and selective non-peptide antagonist of the neurokinin A (NK2) receptors. Am Rev Respir Dis 1992; 146:1177–1181.

116. Lundberg JM, Martling CR, Saria A. Substance P and capsaicin-induced contraction of human bronchi. Acta Physiol Scand 1983; 119:49–53.

117. Fuller RW, Maxwell DL, Dixon CM, McGregor GP, Barnes VF, Bloom SR, Barnes PJ: Effect of substance P on cardiovascular and respiratory function in subjects. J Appl Physiol 1987; 62:1473–1479.

118. Joos G, Pauwels R, Van der Straeten M. Effect of inhaled substance P and neurokinin A on the airways of normal and asthmatic subjects. Thorax 1987; 42:779–783.

119. Evans TW, Dixon CM, Clarke B, Conradson TB, Barnes PJ. Comparison of neurokinin A and substance P on cardiovascular and airway function in man. Br J Clin Pharmacol 1988; 25:273–275.

120. Maggi CA, Patacchini R, Rovero P, Santicioli P. Tachykinin receptors and noncholinergic bronchoconstriction in the guinea pig isolated bronchi. Am Rev Respir Dis 1991; 144:363–367.

121. Joos GF, Kips JC, Pauwels RA. In vivo characterization of the tachykinin receptors involved in the direct and indirect bronchoconstrictor effect of tachykinins in two inbred rat strains. Am J Respir Crit Care Med 1994; 149:1160–1166.

122. Bowden JJ, Baluk P, Lefevre PM, Vigna SR, McDonald DM. Substance P (NK_1) receptor immunoreactivity on endothelial cells of the rat tracheal mucosa. Am J Physiol 1996; 270:L404–L414.

123. Rogers DF. Airway goblet cells: responsive and adaptable front-line defenders. Eur Respir J 1994; 7:1690–1706.

124. Coles SJ, Neill KH, Reid LM. Potent stimulation of glycoprotein secretion in canine trachea by substance P. J Appl Physiol 1984; 57:1323–1327.

125. Rogers DF, Aursudkij B, Barnes PJ. Effects of tachykinins on mucus secretion in human bronchi in vitro. Eur J Pharmacol 1989; 174:283–286.

126. Gentry SE. Tachykinin receptors mediating airway macromolecular secretion. Life Sci 1991; 48:1609–1618.

127. Ramnarine SI, Rogers DF. Non-adrenergic, non-cholinergic neural control of mucus secretion in the airways. Pulmon Pharmacol 1994; 7:19–33.

128. Rogers DF. Neurokinin receptors subserving airways secretion. Can J Physiol Pharmacol 1995; 73:932–939.

129. Wong LB, Miller IF, Yeates DB. Pathways of substance P stimulation of canine tracheal ciliary beat frequency. J Appl Physiol 1991; 70:267–273.

130. Lindberg S, Dolata J. NK1 receptors mediate the increase in mucociliary activity produced by tachykinins. Eur J Pharmacol 1993; 231:375–380.

131. Martling CR, Saria A, Andersson P, Lundberg JM. Capsaicin pretreatment inhibits vagal cholinergic and non-cholinergic control of pulmonary mechanics in the guinea pig. Naunyn Schmiedebergs Arch Pharmacol 1984; 325:343–348.

132. Undem BJ, Myers AC, Barthlow H, Weinreich D. Vagal innervation of guinea pig bronchial smooth muscle. J Appl Physiol 1990; 69:1336–1346.

133. Watson N, Maclagan J, Barnes PJ. Endogenous tachykinins facilitate transmission through parasympathetic ganglia in guinea-pig trachea. Br J Pharmacol 1993; 109: 751–759.

134. Myers AC, Undem BJ. Electrophysiological effects of tachykinins and capsaicin on guinea-pig bronchial parasympathetic ganglion neurones. J Physiol 1993; 470: 665–470.

135. Payan DG. The role of neuropeptides in inflammation. In: Gallin JI, Goldstein IM, Snyderman R, eds. Inflammation: Basic Principles and Clinical Correlates. 2d ed. New York: Raven Press, 1992:177–192.

136. Ziche M. Sensory neuropeptides: mitogenic and trophic functions. In: Geppetti P, Holzer P, eds. Neurogenic Inflammation. Boca Raton, FL: CRC Press, 1996:253–263.

137. Kroegel C, Giembycz MA, Barnes PJ. Characterization of eosinophil cell activation by peptides. Differential effects of substance P, melittin, and FMET-Leu-Phe. J Immunol 1990; 145:2581–2587.

138. Fajac I, Braunstein G, Ickovic MR, Lacronique J, Frossard N. Selective recruitment of eosinophils by substance P after repeated allergen exposure in allergic rhinitis. Allergy 1995; 50:970–975.

139. Helme RD, Eglezos A, Hosking CS. Substance P induces chemotaxis of neutrophils in normal and capsaicin-treated rats. Immunol Cell Biol 1987; 65:267–269.

140. Serra MC, Bazzoni F, Della BV, Greskowiak M, Rossi F. Activation of human neutrophils by substance P. Effect on oxidative metabolism, exocytosis, cytosolic Ca^{++} concentration and inositol phosphate formation. J Immunol 1988; 141:2118–2124.

141. Brunelleschi S, Vanni L, Ledda F, Giotti A, Maggi CA, Fantozzi R. Tachykinins activate guinea-pig alveolar macrophages: involvement of NK_2 and NK_1 receptors. Br J Pharmacol 1990; 100:417–420.

142. Boichot E, Lagente V, Paubert-Braquet M, Frossard N. Inhaled substance-P induces activation of alveolar macrophages and increases airway responses in the guinea-pig. Neuropeptides 1993; 25:307–313.

143. Bost KL, Pascual DW. Substance-P: a late-acting B lymphocyte differentiation cofactor. Am J Physiol 1992; 262:C537–C545.

144. Takeda Y, Chou KB, Takeda J, Sachais BS, Krause JE. Molecular cloning, structural characterization and functional expression of the human substance-P receptor. Biochem Biophys Res Commun 1991; 179:1232–1240.

145. Payan DG, Brewster DR, Goetzl EJ. Specific stimulation of human T lympocytes by substance P. J Immunol 1983; 131:1613–1604.

146. Nio DA, Moylan RN, Roche JK. Modulation of T lymphocyte function by neuropeptides. Evidence for their role as local immunoregulatory elements. J Immunol 1993; 150:5281–5288.

147. Lotz M, Vaughan JH, Carson DA. Effect of neuropeptides on production of inflammatory cytokines by human monocytes. Science 1988; 241:1218–1221.

148. Jeurissen F, Kavelaars A, Korstjens M, Broeke D, Franklin RA, Gelfand EW, Heijnen CJ. Monocytes express a non-neurokinin substance P receptor that is functionally coupled to MAP kinase. J Immunol 1994; 152:2987–2994.

149. Nilsson J, Von Euler AM, Dalsgaard CJ. Stimulation of connective tissue cell growth by substance P and substance K. Nature 1985; 315:61–63.

150. Harrison NK, Dawes KE, Kwon OJ, Barnes PJ, Laurent GJ, Chung KF. Effects of neuropeptides on human lung fibroblast proliferation and chemotaxis. Am J Physiol 1995; 12:L278–L283.

151. Fan TPD, Hu DE, Guard S, Gresham GA, Watling KJ. Stimulation of angiogenesis by substance P and interleukin-1 in the rat and its inhibition by NK1 or interleukin-1 receptor antagonists. Br J Pharmacol 1993; 110:43–49.

152. Hua XY, Saria A, Gamse R, Theodorsson-Norheim E, Brodin E, Lundberg JM. Capsaicin induced release of multiple tachykinins (substance P, neurokinin A and eledoisin-like material) from guinea-pig spinal cord and ureter. Neuroscience 1986; 19:313–319.

153. Martling CR, Theodorsson-Norheim E, Lundberg JM. Occurrence and effects of multiple tachykinins; substance P, neurokinin A and neuropeptide K in human lower airways. Life Sci 1987; 20:1633–1643.

154. Burcher E, Alouan LA, Johnson PR, Black JL. Neuropeptide γ, the most potent contractile tachykinin in human isolated bronchus, acts via a "non-classical" NK2 receptor. Neuropeptides 1991; 20:79–82.

155. Shore SA, Sharpless C, Drazen JM. Bronchoconstrictor activities of NP-γ and NPK in anaesthetized guinea-pigs—effect of NEP inhibition. Pulmon Pharmacol 1993; 6: 143–147.

156. Qian Y, Advenier C, Naline E, Bellamy JF, Emonds-Alt X. Effects of SR 48968 on the neuropeptide γ-induced contraction of the human isolated bronchus. Fundam Clin Pharmacol 1994; 8:71–75.

157. Noguchi K, Senba E, Morita Y, Sato M, Tohyama M. Coexistence of a-CGRP and β-CGRP mRNAs in rat dorsal root ganglion cells. Neurosci Lett 1990; 108:1–5.

158. Emeson RB. Posttranscriptional regulation of calcitonin gene-related peptide (CGRP) mRNA production. In: Geppetti P, Holzer P, eds. Neurogenic Inflammation. Boca Raton, FL: CRC Press, 1996:15–32.

159. Hall JM, Brain SD. Pharmacology of calcitonin gene-related peptide. In: Geppetti P, Holzer P, eds. Neurogenic Inflammation. Boca Raton, FL: CRC Press, 1996:101–114.

160. Hua XY, Yaksh TL. Release of calcitonin gene-related peptide and tachykinins from the rat trachea. Peptides 1992; 13:113–120.

161. Katayama M, Nadel JA, Bunnett NW, Di Maria GU, Haxhiu M, Borson DB. Catabolism of calcitonin gene-related peptide and substance P by neutral endopeptidase. Peptides 1991; 12:563–567.

162. Lou YP. Regulation of neuropeptide release from pulmonary capsaicin-sensitive afferents in relation to bronchoconstriction. Acta Physiol Scand 1993; 149:1–88.

163. Brain SD, Williams TJ, Tippins JR, Morris HR, MacIntyre I. Calcitonin gene-related peptide is a potent vasodilator. Nature 1985; 313:54–56.

164. Franco-Cereceda A, Henke H, Lundberg JM, Petermann JB, Hökfelt T, Fischer JA. Calcitonin gene-related peptide (CGRP) in capsaicin-sensitive substance P-immunoreactive sensory neurons in animals and man: distribution and release by capsaicin. Peptides 1987; 8:399–410.

165. Franco-Cereceda A, Rudehill A, Lundberg JM. Calcitonin gene-related peptide but not substance P mimics capsaicin-induced coronary vasodilation in the pig. Eur J Pharmacol 1987; 13:235–243.

166. Franco-Cereceda A. Resiniferatoxin-evoked, capsaicin-evoked and CGRP-evoked porcine coronary vasodilatation is independent of EDRF mechanisms but antagonized by CGRP (8-37). Acta Physiol Scand 1991; 143:331–337.

167. Hughes SR, Brain SD. A calcitonin gene-related peptide (CGRP) antagonist (CGRP8-37) inhibits microvascular responses induced by CGRP and capsaicin in skin. Br J Pharmacol 1991; 104:738–742.

168. Rinder J, Lundberg JM. Effects of hCGRP 8-37 and the NK1-receptor antagonist SR

140.333 on capsaicin-evoked vasodilation in the pig nasal mucosa in vivo. Acta Physiol Scand 1996; 156:115–122.

169. Gamse R, Saria A. Potentiation of tachykinin-induced plasma protein extravasation by calcitonin gene-related peptide. Eur J Pharmacol 1985; 114:61–66.

170. Brain SD, Williams TJ. Inflammatory oedema induced by synergism between calcitonin gene-related peptide (CGRP) and mediators of increased vascular permeability. Br J Pharmacol 1985; 86:855–860.

171. Brokaw JJ, White GW. Calcitonin gene-related peptide potentiates substance P-induced plasma extravasation in the rat trachea. Lung 1992; 170:85–93.

172. Mak J, Barnes PJ. Autoradiographic localization of calcitonin gene-related peptide (CGRP) binding sites in human and guinea-pig lung. Peptides 1988; 9:957–963.

173. Piotrowski W, Foreman JC. Some effects of calcitonin gene-related peptide in human skin and on histamine release. Br J Dermatol 1985; 114:37–46.

174. Foster CA, Mandak B, Kromer E, Rot A. Calcitonin gene-related peptide is chemotactic for human lymphocytes. Ann NY Acad Sci 1992; 657:397–404.

175. Manley H, Haynes L. Eosinophil chemotactic responses to rat CGRP-1 increased after exposure to trypsin or guinea-pig lung particulate fraction. Neuropeptides 1989; 13:29–34.

176. Davies D, Medeiros MS, Keen J, Turner AJ, Haynes LW. Endopeptidase-24.11 cleaves a chemotactic factor from a-calcitonin gene related peptide. Biochem Pharmacol 1992; 43:1753–1756.

177. Haegerstrand A, Dalsgaard C-J, Jonzon B, Larsson O, Nilsson J. Calcitonin gene-related peptide stimulates proliferation of human endothelial cells. Proc Natl Acad Sci (USA) 1990; 87:3299–3303.

178. Raud J, Lundeberg T, Broddajansen G, Theodorsson E, Hedqvist P. Potent anti-inflammatory action of calcitonin gene-related peptide. Biochem Biophys Res Commun 1991; 180:1429–1435.

179. Umeda Y, Takamiya M, Yoshizaki H, Arisawa M. Inhibition of mitogen-stimulated T-lymphocyte proliferation by calcitonin gene-related peptide. Biophys Biochem Res Commun 1992; 154:227.

180. Nong YH, Titus RG, Riberio JM, Remold HG. Peptides encoded by the calcitonin gene inhibit macrophage function. J Immunol 1989; 143:45.

181. Barnes PJ, Baraniuk JN, Belvisi MG. Neuropeptides in the respiratory tract. Part II. Am Rev Respir Dis 1991; 144:1391–1399.

182. Kalubi B, Yamano M, Ohhata K, Matsunaga T, Tohyama M. Presence of VIP fibers of sensory origin in the rat trachea. Brain Res 1990; 522:107–111.

183. Moller K, Zhang YZ, Hakanson R, Luts A, Sjolund B, Uddman R, Sundler F. Pituitary adenylate cyclase activating peptide is a sensory neuropeptide—immunocytochemical and immunochemical evidence. Neuroscience 1993; 57:725–732.

184. Uddman R, Luts A, Arimura A, Sundler F. Pituitary adenylate cyclase-activating peptide (PACAP), a new vasoactive intestinal peptide (VIP)-like peptide in the respiratory tract. Cell Tissue Res 1991; 265:197–201.

185. Bowden JJ, Gibbins IL. Vasoactive intestinal peptide and neuropeptide Y coexist in non-noradrenergic sympathetic neurons to guinea pig trachea. J Auton Nerv Syst 1992; 38:1–19.

186. Matran R, Martling CR, Lundberg JM. Inhibition of cholinergic and non-adrenergic, non-cholinergic bronchoconstriction in the guinea pig mediated by neuropeptide Y and a 2-adrenoceptors and opiate receptors. Eur J Pharmacol 1989; 12:15–23.

187. Grundemar L, Grundstrom N, Johansson IGM, Andersson RGG, Håkanson R. Suppression by neuropeptide-Y of capsaicin-sensitive sensory nerve-mediated contraction in guinea-pig airways. Br J Pharmacol 1990; 99:473–476.

188. Shimosegawa T, Foda HD, Said SI. Immunohistochemical demonstration of enkephalin-containing nerve fibers in guinea pig and rat lungs. Am Rev Respir Dis 1989; 140:441–448.

189. Shimosegawa T, Foda HD, Said SI. [Met]enkephalin-Arg6-Gly7-Leu8-immunoreactive nerves in guinea-pig and rat lungs: distribution, origin, and co-existence with vasoactive intestinal polypeptide immunoreactivity. Neuroscience 1990; 36:737–750.

190. Belvisi MG, Rogers DF, Barnes PJ. Neurogenic plasma extravasation: inhibition by morphine in guinea pig airways in vivo. J Appl Physiol 1989; 66:268–272.

191. Cheung A, Polak JA, Bauer FE, Christofides ND, Cadieux A, Springall DR, Bloom SR. The distribution of galanin immunoreactivity in the respiratory tract of pig, guinea-pig, rat and dog. Thorax 1985; 40:889–896.

192. Dey RD, Zhu WM. Origin of galanin in the nerves of cat airways and colocalization with vasoactive intestinal peptide. Cell Tissue Res 1993; 273:193–200.

193. Lundberg JM. Tachykinins, sensory nerves, and asthma—an overview. Can J Physiol Pharmacol 1995;73:908–914.

194. Maggi CA: Pharmacology of the efferent function of primary sensory neurons. In: Geppetti P, Holzer P, eds. Neurogenic Inflammation. Boca Raton, FL: CRC Press, 1996: 81–90.

195. Szolcsányi J. Capsaicin and neurogenic inflammation: history and early findings. In: Chahl LA, Szolcsányi J, Lembeck F, eds. Antidromic Vasodilatation and Neurogenic Inflammation. Budapest: Akadémiai Kiadó, 1984:7–25.

196. Bevan S, Szolcsányi J. Sensory neuron-specific actions of capsaicin: mechanisms and applications. Trends Pharmacol Sci 1990;11:330–333.

197. Holzer P. Capsaicin: cellular targets, mechanisms of action, and selectivity for thin sensory neurons. Pharmacol Rev 1991; 43:143–201.

198. Dray A. Neuropharmacological mechanisms of capsaicin and related substances. Biochem Pharmacol 1992; 44:611–615.

199. Bevan S, Hothi S, Hughes G, James IF, Rang HP, Shah K, Walpole CSJ, Yeats JC. Capsazepine—a competitive antagonist of the sensory neurone excitant capsaicin. Br J Pharmacol 1992; 107:544–552.

200. Amann R, Donnerer J, Lembeck F. Ruthenium red selectively inhibits capsaicin-induced release of calcitonin gene-related peptide from the isolated perfused guinea pig lung. Neurosci Lett 1989; 101:311–315.

201. Lou YP, Karlsson JA, Franco-Cereceda A, Lundberg JM. Selectivity of ruthenium red in inhibiting bronchoconstriction and CGRP release induced by afferent C-fibre activation in the guinea-pig lung. Acta Physiol Scand 1991; 142:191–199.

202. Jessell TM, Iversen LL, Cuello AC. Capsaicin-induced depletion of substance P from primary sensory neurons. Brain Res 1978; 152:183–188.

203. Joó F, Szolcsányi J, Jancsó-Gábor A. Mitochondrial alterations in spinal ganglion cells of the rat accompanying the long-lasting sensory disturbance induced by capsaicin. Life Sci 1969; 621–626.

204. Jancsó G, Ferencsik M. Such G, Király E, Nagy A, Bujdosó M. Morphological effects of capsaicin and its analogues in newborn and adult mammals. In: Håkanson R, Sundler F, eds. Tachykinin Antagonists. New York: Elsevier North-Holland, 1985:35–44.

205. Lawson SN. The morphological consequences of neonatal treatment with capsaicin on primary afferent neurons in adult rats. Acta Physiol Hung 1987; 69:315–321.

206. Maggi CA, Meli A. The sensory-efferent function of capsaicin-sensitive sensory neurons. Gen Pharmacol 1988; 19:1–43.

207. Bowden JJ, Baluk P, Lefevre PM, Schoeb TR, Lindsey JR, McDonald DM. Sensory denervation by neonatal capsaicin treatment exacerbates *Mycoplasma pulmonis* infection in rat airways. Am J Physiol 1996; 270:L393–L403.

208. Jancsó G. Intracisternal capsaicin: selective degeneration of chemosensitive primary sensory afferents in the adult rat. Neurosci Lett 1981; 27:41–45.

209. Alving K, Matran R, Lundberg JM. Capsaicin-induced local effector responses, autonomic reflexes and sensory neuropeptide depletion in the pig. Naunyn Schmiedebergs Arch Pharmacol 1991; 343:37–45.

210. Coleridge HM, Coleridge JCG, Green JF, Parsons GH. Pulmonary C-fiber stimulation by capsaicin evokes reflex cholinergic bronchial vasodilation in sheep. J Appl Physiol 1992; 72:770–778.

211. Pisarri TE, Coleridge JCG, Coleridge HM. Capsaicin-Induced bronchial vasodilation in dogs—central and peripheral neural mechanisms. J Appl Physiol 1993; 74:259–266.

212. Forsberg K, Karlsson JA, Theodorsson E, Lundberg JM, Persson CG. Cough and bronchoconstriction mediated by capsaicin-sensitive sensory neurons in the guinea-pig. Pulmon Pharmacol 1988; 1:33–39.

213. Alving K, Matran R, Lacroix JS, Lundberg JM. Allergen challenge induces vasodilatation in pig bronchial circulation via a capsaicin-sensitive mechanism. Acta Physiol Scand 1988; 134:571–572.

214. Matsuse T, Thomson RJ, Chen XR, Salari H, Schellenberg RR. Capsaicin inhibits airway hyperresponsiveness but not lipoxygenase activity or eosinophilia after repeated aerosolized antigen in guinea pigs. Am Rev Respir Dis 1991; 144:368–372.

215. Riccio MM, Manzini S, Page CP. The effect of neonatal capsaicin on the development of bronchial hyperresponsiveness in allergic rabbits. Eur J Pharmacol 1993; 232:89–97.

216. Ladenius ARC, Nijkamp FP. Capsaicin pretreatment of guinea pigs in vivo prevents ovalbumin-induced tracheal hyperreactivity in vitro. Eur J Pharmacol 1993; 235:127–131.

217. Jancsó G, Király E, Such G, Joó F, Nagy A. Neurotoxic effect of capsaicin in mammals. Acta Physiol Hung 1987; 69:295–313.

218. Glinsukon T, Stitmunnaithum V, Toskulkao C, Buranawuti T, Tangkrisanavinont V. Acute toxicity of capsaicin in several animal species. Toxicon 1980; 18:215–220.

219. Spina D, McKenniff MG, Coyle AJ, Seeds EAM, Tramontana M, Perretti F, Manzini S, Page CP. Effect of capsaicin on PAF-induced bronchial hyperresponsiveness and pulmonary cell accumulation in the rabbit. Br J Pharmacol 1991; 103:1268–1274.

220. Midgren B, Hansson L, Karlsson JA, Simonsson BG, Persson CGA. Capsaicin-induced cough in humans. Am Rev Respir Dis 1992; 146:347–351.

221. Fuller RW. The human pharmacology of capsaicin. Arch Int Pharmacodynam Ther 1990; 303:147–156.

222. Szallasi A. The vanilloid receptor. In: Geppetti P, Holzer P, eds. Neurogenic Inflammation. Boca Raton, FL: CRC Press, 1996:43–52.

223. Szallasi A. The vanilloid (capsaicin) receptor—receptor types and species differences. Gen Pharmacol 1994; 25:223–243.

224. Szallasi A, Blumberg PM, Nilsson S, Hökfelt T, Lundberg JM. Visualization by [H-3]resiniferatoxin autoradiography of capsaicin-sensitive neurons in the rat, pig and man. Eur J Pharmacol 1994; 264:217–221.

225. Hua XY, Wong S, Jinno S, Yaksh TL: Pharmacology of calcitonin gene related peptide release from sensory terminals in the rat trachea. Can J Physiol Pharmacol 1995; 73: 999–1006.

226. Jänig W, Lisney SJW. Small diameter myelinated afferents produce vasodilatation but not plasma extravasation in rat skin. J Physiol 1989; 415:477–486.

227. Proud D, Kaplan AP. Kinin formation: Mechanisms and role in inflammatory disorders. Annu Rev Immunol 1988; 6:49–83.

228. Saria A, Martling CR, Yan Z, Theodorsson-Norheim E, Gamse R, Lundberg JM. Release of multiple tachykinins from capsaicin-sensitive sensory nerves in the lung by bradykinin, histamine, dimethylphenyl piperazinium, and vagal nerve stimulation. Am Rev Respir Dis 1988; 137:1330–1335.

229. Geppetti P. Sensory neuropeptide release by bradykinin—mechanisms and pathophysiological implications. Regulatory Peptides 1993; 47:1–23.

230. Regoli D, Jukic D, Gobeil F, Rhaleb NE. Receptors for bradykinin and related kinins: a critical analysis. Can J Physiol Pharmacol 1993; 71:556–567.

231. Yamawaki I, Geppetti P, Bertrand C, Chan B, Nadel JA. Airway vasodilation by bradykinin is mediated via B-2 receptors and modulated by peptidase inhibitors. Am J Physiol 1994; 266:L156–L162.

232. Martling CR, Lundberg JM. Capsaicin sensitive afferents contribute to acute airway edema following tracheal instillation of hydrochloric acid or gastric juice in the rat. Anesthesiology 1988; 68:350–356.

233. Bevan S, Geppetti P. Protons: small stimulants of capsaicin-sensitive sensory nerves. Trends Neurosci 1994; 17:509–512.

234. Ray DW, Hernandez C, Leff AR, Drazen JM, Solway J. Tachykinins mediate bronchoconstriction elicited by isocapnic hyperpnea in guinea pigs. J Appl Physiol p1989; 66:1108–1112.

235. Garland A, Ray DW, Doerschuk CM, Alger L, Eappon S, Hernandez C, Jackson M, Solway J. Role of tachykinins in hyperpnea-induced bronchovascular hyperpermeability in guinea pigs. J Appl Physiol 1991; 70:27–35.

236. Freed AN, Omori C, Hubbard WC, Adkinson NF. Dry air-induced and hypertonic aerosol-induced bronchoconstriction and cellular responses in the canine lung periphery. Eur Respir J 1994; 7:1308–1316.

237. Yoshihara S, Geppetti P, Hara M, Linden A, Ricciardolo FLM, Chan B, Nadel JA. Cold air-induced bronchoconstriction is mediated by tachykinin and kinin release in guinea pigs. Eur J Pharmacol 1996; 296:291–296.

238. Fox AJ, Barnes PJ, Dray A. Stimulation of guinea-pig tracheal afferent fibres by non-isosmotic and low-chloride stimuli and the effect of frusemide. J Physiol 1995; 482:179–187.

239. Matran R, Alving K, Lundberg JM. Cigarette smoke, nicotine and capsaicin aerosol-induced vasodilatation in pig respiratory mucosa. Br J Pharmacol 1990; 100:535–541.

240. Jinno SH, Hua XY, Yaksh TL. Nicotine and acetylcholine induce release of calcitonin gene-related peptide from rat trachea. J Appl Physiol 1994; 76:1651–1656.

241. Saria A, Lundberg JM, Skofitsch G, Lembeck F. Vascular protein leakage in various tissues induced by substance P, capsaicin, bradykinin, serotonin, histamine and by antigen challenge. Naunyn-Schmiedeberg's Arch Pharmacol 1983; 324:212–218.

242. Majno G, Palade GE. Studies on inflammation. I. The effect of histamine and serotonin on vascular permeability: An electron microscopic study. J Biophys Biochem Cytol 1961; 11:571–605.

243. Saria A, Lundberg JM, Skofitsch G, Lembeck F Vascular protein linkage in various tissues induced by substance P, capsaicin, bradykinin, serotonin, histamine and by antigen challenge. Naunyn-Schmiedeberg's Arch Pharmacol 1983; 324:212–218.

244. Manzini S, Maggi CA, Geppetti P, Bacciarelli C. Capsaicin desensitization protects from antigen-induced bronchospasm in conscious guinea-pigs. Eur J Pharmacol 1987; 138:307–308.

245. Bertrand C, Geppetti P, Baker J, Yamawaki I, Nadel JA. Role of neurogenic inflammation in antigen-induced vascular extravasation in guinea pig trachea. J Immunol 1993; 150:1479–1485.

246. Bertrand C, Geppetti P, Graf PD, Foresi A, Nadel JA. Involvement of neurogenic inflammation in antigen-induced bronchoconstriction in guinea pigs. Am J Physiol 1993; 265:L507–L511.

247. Buckley TL, Nijkamp FP. Airways hyperreactivity and cellular accumulation in a delayed-type hypersensitivity reaction in the mouse—modulation by capsaicin-sensitive nerves. Am J Respir Crit Care Med 1994; 149:400–407.

248. Lundberg JM, Martling CR, Saria A, Folkers K, Rosell S. Cigarette smoke-induced airway oedema due to activation of capsaicin-sensitive vagal afferents and substance P release. Neuroscience 1983; 10:1361–1368.

249. Geppetti P, Bertrand C, Baker J, Yamawaki I, Piedimonte G, Nadel JA. Ruthenium red, but not capsazepine reduces plasma extravasation by cigarette smoke in rat airways. Br J Pharmacol 1993; 108:646–650.

250. Wei ET, Kiang JG. Inhibition of protein exudation from the trachea by corticotropin-releasing factor. Eur J Pharmacol 1987; 140:63–67.

251. Thompson JE, Scypinski LA, Gordon T, Sheppard D. Tachykinins mediate the acute increase in airway responsiveness caused by toluene diisocyanate in guinea pigs. Am Rev Respir Dis 1987; 136:43–49.

252. Sakamoto T, Elwood W, Barnes PJ, Chung KF. Pharmacological modulation in inhaled sodium metabisulphite-induced airway microvascular leakage and bronchoconstriction in the guinea-pig. Br J Pharmacol 1992; 107:481–487.

253. Bannenberg G, Atzori L, Xue J, Auberson S, Kimland M, Ryrfeldt Å, Lundberg JM, Moldéus P. Sulfur dioxide and sodium metabisulfite induce bronchoconstriction in the isolated perfused and ventilated guinea pig lung via stimulation of capsaicin-sensitive sensory nerves. Respiration 1994; 61:130–137.

254. Patacchini R, Maggi CA. Tachykinin receptors and receptor subtypes. Arch Int Pharmacodynam Ther 1995; 329:161–184.

255. Hershey AD, Dykema PE, Krause JE. Organization, structure, and expression of the gene encoding the rat substance P receptor. J Biol Chem 1991; 266:4366–4374.

256. Nakanishi S. Mammalian tachykinin receptors. Annu Rev Neurosci 1991; 14:123–136.

257. Burcher E, Mussap CJ, Stephenson JA. Autoradiographic localization of receptors in peripheral tissues. In: Burk SH, ed. The Tachykinin Receptors. Totawa, NJ: Humana Press 1994:125–163.

258. Burcher E, Zeng XP, Strigas J, Geraghty DP, Lavielle S. Tachykinin receptors in guinea-pig airways: characterization using selective ligands. Can J Physiol Pharmacol 1995; 73:915–922.

259. Carstairs J, Barnes P. Autoradiographic mapping of substance P receptors in lung. Eur J Pharmacol 1986; 127:295–296.

260. Hoover D, Hancock J. Autoradiographic localization of substance P binding sites in guinea-pig airways. J Auton Nerv Syst 1987; 19:171–174.

261. Black JL, Diment LM, Alouan LA, Johnson PRA, Armour CL, Badgery-Parker T, Burcher E. Tachykinin receptors in rabbit airways—characterization by functional, autoradiographic and binding studies. Br J Pharmacol 1992; 107:429–436.

262. Meini S, Mak JCW, Rohde JAL, Rogers DF. Tachykinin control of ferret airways: mucus secretion, bronchoconstriction and receptor mapping. Neuropeptides 1993; 24: 81–89.

263. Sertl K, Wiedermann CJ, Kowalski ML, Hurtado S, Plutchok J, Linnoila I, Pert CB, Kaliner MA. Substance P: the relationship between receptor distribution in rat lung and the capacity of substance P to stimulate vascular permeability. Am Rev Respir Dis 1988; 138:151–159.

264. Walsh DA, Salmon M, Featherstone R, Wharton J, Church MK, Polak JM. Differences in the distribution and characteristics of tachykinin NK1 binding sites between human and guinea pig lung. Br J Pharmacol 1994; 113:1407–1415.

265. Miyayasu K, Mak JC, Nishikawa M, Barnes PJ. Characterization of guinea pig pulmonary neurokinin type 1 receptors using a novel antagonist ligand, [^{3}H]FK888. Mol Pharmacol 1993; 44:539–544.

266. Zhang XL, Mak JCW, Barnes PJ. Characterization and autoradiographic mapping of [H-3]CP96,345, a nonpeptide selective NK1 receptor antagonist in guinea pig lung. Peptides 1995; 16:867–872.

267. Kummer W, Fischer A, Preissler U, Couraud J-Y, Heym C. Immunohistochemistry of the guinea-pig trachea using an anti-idiotypic antibody recognizing substance P receptors. Histochemistry 1990; 93:541–546.

268. Kummer W, Fischer A, Couraud J-Y, Heym C. Immunohistochemistry of peptides (substance P and VIP) and peptide receptors in the trachea. J Auton Nerv Syst 1991; 33: 121–123.

269. Fischer A, Kummer W, Couraud J-Y, Adler D, Branscheid D, Heym C. Immunohistochemical localization of receptors for vasoactive intestinal peptide and substance P in human trachea. Lab Invest 1992; 67:387–393.

270. Grady EF, Garland AM, Gamp PD, Lovett M, Payan DG, Bunnett NW. Delineation of the endocytic pathway of substance P and its seven-transmembrane domain NK1 receptor. Mol Biol Cell 1995; 6:509–524.

271. Kwatra MM, Schwinn DA, Schreurs J, Blank JL, Kim CM, Benovic JL, Krause JE, Caron MG, Lefkowitz RJ. The substance-P receptor, which couples to gq/11, is a substrate of β-adrenergic receptor kinase-1 and kinase-2. J Biol Chem 1993; 268:9161–9164.

272. Garland AM, Grady EF, Lovett M, Vigna SR, Frucht MM, Krause JE, Bunnett NW. Mechanisms of desensitization and resensitization of G protein-coupled neurokinin(1) and neurokinin(2) receptors. Mol Pharmacol 1996; 49:438–446.

273. Grady EF, Baluk P, Böhm S, Gamp PD, Wong H, Payan DG, Ansel J, Portbury AL, Furness JB, McDonald DM, Bunnett NW. Characterization of antisera specific to NK1, NK2, and NK3 neurokinin receptors and their utilization to localize receptors in the rat gastrointestinal tract. J Neurosci 1996; 16;6975–6986.

274. Baluk P, Bunnett NW, McDonald DM. Localization of tachykinin NK-1, NK-2, and NK-3 receptors in airways by immunohistochemistry. Am J Respir Crit Care Med 1996; 153:A161.

275. Dennis T, Fournier A, Cadieux A, Pomerleau F, Jolicoeur FB, St. Pierre S, Quirion R. hCGRP$_{8-37}$, a calcitonin gene-receptor peptide antagonist revealing receptor heterogeneity in brain and periphery. J Pharmacol Exp Ther 1990; 254:123–128.

276. Poyner D. Pharmacology of receptors for calcitonin gene-related peptide and amylin. Trends Pharmacol Sci 1995; 16:424–428.

277. Aiyar N, Rand K, Elshourbagy NA, Zeng Z, Adamou JE, Bergsma DJ, Li Y. A cDNA encoding the calcitonin gene-related peptide type 1 receptor. J Biol Chem 1996; 271: 11325–11329.

278. Carstairs JR. Distribution of calcitonin gene-related peptide receptors in the lung. Eur J Pharmacol 1987; 140:357–358.

279. Krause JE, Bu JY, Takeda Y, Blount P, Raddatz R, Sachais BS, Chou KB, Takeda J, McCarson K, DiMaggio D. Structure, expression and 2nd messenger-mediated regulation of the human and rat substance-P receptors and their genes. Regulatory Peptides 1993; 46:59–66.

280. Krause JE, DiMaggio DA, McCarson KE. Alterations in neurokinin 1 receptor gene expression in models of pain and inflammation. Can J Physiol Pharmacol 1995; 73: 854–859.

281. Lindsay RM. Role of neurotrophins and trk receptors in the development and maintenance of sensory neurons: an overview. Philos Trans R Soc Lond B 1996; 351:365–373.

282. Friedin M, Kessler JA. Cytokine regulation of substance P expression in sympathetic neurons. Proc Natl Acad Sci (USA) 1991; 88:3200.

283. Noguchi K, Morita Y, Kiyama H, Ono K, Tohyama M. A noxious stimulus induces the preprotachykinin A gene expression in the rat dorsal root ganglion: a quantitative study using in situ hybridization histochemistry. Mol Brain Res 1988; 4:31–35.

284. McCarson KE, Krause JE. The formalin-induced expression of tachykinin peptide and neurokinin receptor messenger RNAs in rat sensory ganglia and spinal cord is modulated by opiate preadministration. Neuroscience 1995; 64:729–739.

285. Karlsson JA, Zackrisson C, Lundberg JM. Hyperresponsiveness to tussive stimuli in cigarette smoke-exposed guinea-pigs: a role of capsaicin-sensitive, calcitonin gene-related peptide-containing nerves. Acta Physiol Scand 1991; 141:445–454.

286. Killingsworth CR, Paulauskis JD, Shore SA. Substance P content and preprotachykinin gene-1 mRNA expression in a rat model of chronic bronchitis. Am J Respir Cell Mol Biol 1996; 14:334–340.

287. Lindsay RM, Harmar AJ. Nerve growth factor regulates expression of neuropeptide genes in adult sensory neurons. Nature 1989; 337:362.

288. Choudry NB, Fuller RW, Pride NB. Sensitivity of the human cough reflex: effect of inflammatory mediators prostaglandin E_2, bradykinin, and histamine. Am Rev Respir Dis 1989; 140:137–141.

289. McDonald DM, Schoeb TR, Lindsey JR. *Mycoplasma pulmonis* infections cause long-lasting potentiation of neurogenic inflammation in the respiratory tract of the rat. J Clin Invest 1991; 87:787–799.

290. Baluk P, Bowden JJ, Lefevre PL, McDonald DM. Increased expression of substance P (NK$_1$) receptors on airway blood vessels with *Mycoplasma pulmonis* infection. Am J Respir Crit Care Med 1995; 151:A719.

291. Lilly CM, Drazen JM, Shore SA. Peptidase modulation of airway effects of neuropeptides. Proc Soc Exp Biol Med 1993; 203:388–404.

292. Nadel JA. Peptidase modulation of neurogenic inflammation. In: Geppetti P, Holzer P, eds. Neurogenic Inflammation. Boca Raton, FL: CRC Press, 1996:115–127.

293. Jacoby D, Tamaoki J, Borson D, Nadel J. Influenza infection causes airway hyperresponsiveness by decreasing enkephalinase. J Appl Physiol 1988; 54:2653–2658.

294. Dusser DJ, Jacoby DB, Djokic TD, Rubinstein I, Borson DB, Nadel JA. Virus induces

airway hyperresponsiveness to tachykinins—role of neutral endopeptidase. J Appl Physiol 1989; 67:1504–1511.

295. Piedimonte G, Nadel JA, Umeno E, McDonald DM. Sendai virus infection potentiates neurogenic inflammation in the rat trachea. J Appl Physiol 1990; 68:754–760.
296. Brokaw JJ, White GW. Differential effects of phosphoramidon and captopril on NK1 receptor-mediated plasma extravasation in the rat trachea. Agents Actions 1994; 42:34–39.
297. Borson DB, Brokaw JJ, Sekizawa K, McDonald DM, Nadel JA. Neutral endopeptidase and neurogenic inflammation in rats with respiratory infections. J Appl Physiol 1989; 66:2653–2658.
298. Barnes PJ, Belvisi MG, Rogers DF. Modulation of neurogenic inflammation: novel approaches to inflammatory disease. Trends Pharmacol Sci 1990; 11:185–189.
299. Brokaw JJ, Hillenbrand CM, White GW, McDonald DM. Mechanism of tachyphylaxis associated with neurogenic plasma extravasation in the rat trachea. Am Rev Respir Dis 1990; 141:1434–1440.
300. Belvisi MG, Miura M, Stretton D, Barnes PJ. Capsazepine as a selective antagonist of capsaicin-induced activation of C-fibres in guinea-pig bronchi. Eur J Pharmacol 1992; 215:341–344.
301. Buckley TL, Brain SD, Williams TJ. Ruthenium red selectively inhibits oedema formation and increased blood flow induced by capsaicin in rabbit skin. Br J Pharmacol 1990; 99:7–8.
302. Brokaw JJ, White GW. Characterization of ruthenium red as an inhibitor of neurogenic inflammation in the rat trachea. Gen Pharmacol 1995; 26:327–331.
303. Barnes PJ. Modulation of neurotransmission in airways. Physiol Rev 1992; 72:699–729.
304. Belvisi MG. Effects of neuropeptides on neurotransmission in the airways. In: Kaliner MA, et al., eds. Neuropeptides in Respiratory Medicine. New York: Marcel Dekker 1994:477–500.
305. Barthó L, Szolcsányi J. Opiate agonists inhibit neurogenic plasma extravasation in the rat. Eur J Pharmacol 1981; 73:101–104.
306. Grundstrom N, Anderson RGG. In vivo demonstration of α_2-adrenoceptor mediated inhibition of the excitatory noncholinergic neurotransmission in guinea pig airways. Naunyn-Schmiedeberg's Arch Pharmacol 1985; 328:236–240.
307. Ray NJ, Jones AJ, Keen P. GABAB receptor modulation of the release of substance P from capsaicin-sensitive neurones in the rat trachea in vitro. Br J Pharmacol 1991; 102: 801–804.
308. Ichinose M, Belvisi MG, Barnes PJ. Histamine H3-receptors inhibit neurogenic microvascular leakage in airways. J Appl Physiol 1990; 68:21–25.
309. Ichinose M, Barnes PJ. Histamine H3 receptors modulate antigen-induced bronchoconstriction in guinea pigs. J Allergy Clin Immunol 1990:491–495.
310. Wei E, Thomas H. Anti-inflammatory peptide agonists. Annu Rev Pharmacol 1993; 33: 91–108.
311. Stretton D, Miura M, Belvisi MG, Barnes PJ. Calcium-activated potassium channels mediate prejunctional inhibition of peripheral sensory nerves. Proc Natl Acad Sci (USA) 1992; 89:1325–1329.
312. Lei YH, Barnes PJ, Rogers DF. Inhibition of neurogenic plasma exudation and bronchoconstriction by a K+ channel activator, BRL-38227, in guinea pig airways in vivo. Eur J Pharmacol 1993; 239:257–259.
313. Joos GF, Kips JC, Peleman RA, Pauwels RA. Tachykinin antagonists and the airways. Arch Int Pharmacodynam Ther 1995; 329:205–219.

314. Lowe JAI, Snider RM, MacLean DB. Nonpeptide NK_1 antagonists: from discovery to the clinic. In: Geppetti P, Holzer P, eds. Neurogenic Inflammation. Boca Raton, FL: CRC Press, 1996:299–309.

315. Adcock IM, Peters M, Gelder C, Shirasaki H, Brown CR, Barnes PJ. Increased tachykinin receptor gene expression in asthmatic lung and its modulation by steroids. J Mol Endocrinol 1993; 11:1–7.

316. Bowden JJ, Schoeb TR, Lindsey JR, McDonald DM. Dexamethasone and oxytetracycline reverse the potentiation of neurogenic inflammation in airways of rats with *Mycoplasma pulmonis* infection. Am J Respir Crit Care Med 1994; 150:1391–1401.

317. Kohrogi H, Nadel JA, Malfroy B, Gorman C, Bridenbaugh R, Patton JS, Borson DB. Recombinant human enkephalinase (neutral endopeptidase) prevents cough induced by tachykinins in awake guinea pigs. J Clin Invest 1989; 84:781–786.

318. Barnes PJ. Asthma as an axon reflex. Lancet 1986; 1:242–245.

319. Ollerenshaw SL, Jarvis D, Woolcock AJ, Sullivan CE, Scheibner T. Absence of immunoreactive vasoactive intestinal polypeptide in tissues from the lungs of patients with asthma. N Engl J Med 1989; 320:1244–1248.

320. Lilly CM, Bai TR, Shore SA, Hall AE, Drazen JM. Neuropeptide content of lungs from asthmatic and nonasthmatic patients. Am J Respir Crit Care Med 1995; 151:548–553.

321. Bai TR, Zhou DY, Weir T, Walker B, Hegele R, Hayashi S, McKay K, Bondy GP, Fong T. Substance P (NK1)- and neurokinin A (NK2)-receptor gene expression in inflammatory airway diseases. Am J Physiol 1995; 13:L309–L317.

322. Nieber K, Baumgarten CR, Rathsack R. Substance P and β-endorphin-like immunoreactivity in lavage fluids of subjects with and without allergic asthma. J Allergy Clin Immunol 1992; 90:646–662.

323. Moismann BL, White MV, Hohman RJ, Goldrich MS, Kaulbach HC, Kaliner MA. Substance P, calcitonin gene-related peptide, and vasoactive intestinal peptide increase in nasal secretions after allergen challenge in atopic patients. J Allergy Clin Immunol 1993; 92:95.

324. Tomaki M, Ichinose M, Miura M, Hirayama Y, Yamauchi H, Nakajima N, Shirato K. Elevated substance P content in induced sputum from patients with asthma and patients with chronic bronchitis. Am J Respir Crit Care Med 1995; 151:613–617.

325. Joos GF. Clinical studies on the airway effects of sensory neuropeptides. In: Geppetti P, Holzer P, eds. Neurogenic Inflammation. Boca Raton, FL: CRC Press, 1996:277–288.

326. Braunstein G, Fajac I, Lacronique J, Frossard N. Clinical and inflammatory responses to exogenous tachykinins in allergic rhinitis. Am Rev Respir Dis 1991; 144:630–635.

327. Geppetti P, Fusco BM, Marabini S, Maggi CA, Fanciullacci M, Sicuteri F. Secretion, pain and sneezing induced by the application of capsaicin to the nasal mucosa in man. Br J Pharmacol 1988; 93:509–514.

328. Ichinose M, Nakajima N, Takahashi T, Yamauchi H, Inoue H, Takishima T. Protection against bradykinin-induced bronchoconstriction in asthmatic patients by neurokinin receptor antagonist. Lancet 1992; 340:1248–1251.

329. Ichinose M, Miura M, Yamauchi H, Kageyama N, Tomaki M, Oyake T, Ohuchi Y, Hida W, Miki H, Tamura G, Shirato K. A neurokinin 1-receptor antagonist improves exercise-induced airway narrowing in asthmatic patients. Am J Respir Crit Care Med 1996; 153:936–941.

330. Fahy JV, Wong HH, Geppetti P, Reis JM, Harris SC, MacLean DB, Nadel JA, Boushey HA. Effect of an NK1 receptor antagonist (CP-99,994) on hypertonic saline-induced

bronchoconstriction and cough in male asthmatic subjects. Am J Respir Crit Care Med 1995; 152:879–884.

331. Turner CR, Andresen CJ, Patterson DK, Keir RF, Obach S, Lee P, Watson JW. Dual antagonism of NK1 and NK2 receptors by CP-99,994 and SR 48968 prevents airway hyperresponsiveness in primates. Am J Respir Crit Care Med 1996; 153:A160.

332. Cheung D, Bel EH, Den Hartigh J, Dijkman JH, Sterk PJ. The effect of an inhaled neutral endopeptidase inhibitor, thiorphan, on airway responses to neurokinin A in normal humans *in vivo*. Am Rev Respir Dis 1992; 145:1275–1280.

333. Di Maria GU, Bellofiore S, Pennisi A, Calgagirone F, Caincio N, Mistretta A. Neutral endopeptidase inhibition increases airway response to inhaled metabisulfite in normal subjects. Am Rev Respir Dis 1993; 147:A838.

334. Crimi N, Polosa R, Pulvirenti G, Magri S, Santonocito G, Prosperini G, Mastruzzo C, Mistretta A. Effect of an inhaled neutral endopeptidase inhibitor, phosphoramidon, on baseline airway calibre and bronchial responsiveness to bradykinin in asthma. Thorax 1995; 50:505–510.

335. Diamant Z, Timmers MC, Van der Veen H, Booms P, Sont JK, Sterk PJ. Effect of an inhaled neutral endopeptidase inhibitor, thiorphan, on airway responsiveness to leuko-triene-D(4) in normal and asthmatic subjects. Eur Respir J 1994; 7:459–466.

336. Nichol GM, O'Connor BJ, Lecomte JM, Chung KF, Barnes PJ. Effect of neutral endopeptidase inhibitor on airway function and bronchial responsiveness in asthmatic subjects. Eur J Clin Pharmacol 1992; 42:491–494.

337. Diamant Z, Vanderveen H, Kuijpers EAP, Bakker PF, Sterk PJ. The effect of inhaled thiorphan on allergen-induced airway responses in asthmatic subjects. Clin Exp Allergy 1996; 26:525–532.

338. Baluk P, McDonald DM. The β_2-adrenergic receptor agonist formoterol reduces micro-vascular leakage by inhibiting endothelial gap formation. Am J Physiol 1994; 266: L461–L468.

5

Sensory Neuropeptides and Bronchial Hyperresponsiveness

D. SPINA, C. P. PAGE, and J. MORLEY

The Sackler Institute of Pulmonary Pharmacology
and King's College School of Medicine and Dentistry
London, England

I. Airway Hyperresponsiveness in Asthma

A striking characteristic of asthma is a heightened responsiveness of asthmatic airways to a range of stimuli that ordinarily are without effect in normal subjects. A need to describe this characteristic prompted introduction of the adjective "hyper-reactive" and, because increased sensitivity is evidenced to diverse stimuli, use of the prefix "nonselective" in conjunction with asthmatic airways. Common usage and widespread acceptance of this nomenclature antedated experimental investigation and thereby has led to considerable confusion, since neither "hyperreactive" nor "nonselective" accurately describes the altered behaviour that distinguishes asthmatic from normal airways.

Measurement of changed airflow in asthmatic airways following inhalation of test spasmogens (e.g., methacholine, histamine) provides a quantitative technique that can be used to define increased responsiveness. Cumulative concentration–effect curves can be constructed and interpolation used to determine a spasmogen concentration that would have been required to reduce airway caliber proportionately (e.g., 20% reduction or PC_{20}). Although it is common for investigators to cite PC_{20} as an index of reactivity, it will be apparent that such measurements reflect changes in sensitivity (Fig. 1). Upon extending the concentration–effect curve to include more

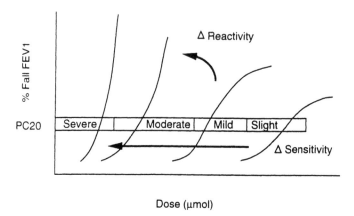

Figure 1 Cumulative dose–response curves obtained using either histamine or methacholine in asthmatics with increasing disease severity using any standardized method. The response is measured as the change in forced expiratory volume in 1 sec (FEV$_1$). Usually, airway sensitivity is obtained by interpolating the provocative concentration (PC) that causes a 20% fall in FEV$_1$. Airway hyperresponsiveness is characterized by a leftward shift in the dose–response curve (change sensitivity) and an increase in slope (change in reactivity).

pronounced obstruction, it became apparent that whereas normal subjects exhibit response limitation as a plateau of the dose–effect relationships, asthmatics have lost, or cannot effect, such limitation. Concentration–effect curves with a steeper slope and/or higher maxima than in normal subjects justify description of asthmatic airways as hyperreactive. Although hypersensitivity and hyperreactivity are distinct processes, it is usual to refer to both as hyperresponsiveness. Failure to distinguish hyperreactivity from the hypersensitivity can lead to invalid expectations. For instance, guinea pig airways become hyperreactive to a range of stimuli without any change in sensitivity. It follows that it is not prudent to anticipate that agents which increase or suppress airway reactivity in the guinea pig will necessarily have corresponding effects when assessed by measurement of sensitivity in asthmatics. This caveat is rarely heeded: It is as exceptional for experimental investigators to define changed sensitivity as it is for clinical investigators to measure changed reactivity. It should be recognized, however, that such polarization of experimental procedures is somewhat inevitable, given that changed sensitivity is exceptional in the guinea pig and that, being potentially dangerous, increased reactivity is rarely defined in asthmatics.

Comparable reservations apply to the use of "nonselective" in conjunction with hyperresponsiveness. Given that asthmatics exhibit increased sensitivity to a range of exogenous and endogenous stimuli, "nonselective" afforded a convenient, if imprecise, qualitative description of this constellation of properties. Unfortunately, "nonselective" acquired a more precise meaning following detection of very close correlation ($r = .96$) between sensitivity to inhaled methacholine and sensitivity to inhaled histamine in a group of asthmatic patients (1). Reference to hypersensitivity and hyperreactivity of asthmatic airway as being nonselective has become axiomatic, and this dogma has had disproportionate influence on concepts which seek to account for the hyperresponsiveness observed in asthmatic airway. For instance, hypertrophied airway smooth muscle (2), facilitated excitation of adjacent myocytes (3), loss of homeostatic bronchodilator mechanisms (4), loss of epithelial integrity (5), and reduction of luminal diameter secondary to increased volume of airway smooth muscle and mucosal edema (6) are artifices which have been suggested to account for nonselective increased responsiveness of asthmatic airway.

However, it is now apparent that there is a considerable body of clinical evidence at variance with the concept of nonselective hyperresponsiveness in asthmatic airways. The selective nature of hyperresponsiveness in asthmatic airways became apparent after a succession of inflammatory mediators had been considered as pivotal in the pathogenesis of asthma. It has been claimed that asthmatic airways are extraordinarily sensitive to $PGF2\alpha$. However, although asthmatics had been reported to exhibit disproportionate sensitivity to LTE_4, responsiveness to other, more recently defined spasmogens, was less than had been anticipated (7). Similarly, LTC_4 exhibited an impressive bronchoconstrictor potency in normals, yet asthmatics are only slightly more sensitive (8), and the concept of nonselective responsiveness was compounded by observations with platelet-activating factor (PAF), to which normal and asthmatic subjects experience comparable sensitivities (9,10). Such anomalies imply that hyperresponsiveness must be selective and warrant direct experimental investigation to this end. Thus, when seeking to define more precisely the capacity of salbutamol to induce hyperreactivity in guinea pig airways (11), a panel of test spasmogens was used since it was known that anaphylactic reactions in the guinea pig were exaggerated following sustained infusion of salbutamol, even though responses to histamine and acetylcholine continue to be suppressed (12). Use of a range of test spasmogens also revealed that disproportionate reactivity to leukotrienes could account for exaggerated allergic bronchospasm (13) and provided a sufficient body of data to establish that allergic reactivity is heterogeneous in this species.

A lack of uniformity of airway hyperreactivity is not restricted to the guinea pig, for comparable heterogeneity is evident in asthmatics. Thus, following regular use of terbutaline or salbutamol, there was proportionately greater increased sensitivity to pseudo-allergic (cyclic AMP) (14) or allergic (15) stimuli than to methacholine and, more recently, this finding has been extended to include bradykinin and LTC_4 (16). It should not be presumed that heterogeneity is restricted to the changed responsiveness that follows regular use of sympathomimetics. Thus, there is poor correlation between sensitivity to prostaglandin $F2\alpha$ and sensitivity to methacholine (17), between pharmacological stimuli and exercise (18,19), or hyperosmotic stimuli (20). Furthermore, survey of published data which defines responsiveness of asthmatics to a range of spasmogens has revealed that correlation between PC_{20} estimates for different spasmogens is relatively infrequent (16). The occurrence of heterogeneity is important when considering the role of sensory neuropeptides in hyperresponsiveness, since the possibility has to be considered that increased sensitivity to particular stimuli might be determined by such a mediator without necessarily extending to all manifestations of hyperresponsiveness.

II. Dale's Criteria for Mediator Identification

The proposal that nerves might activate effector cells by the release of specific chemicals is almost 100 years old. Largely because of the uncertain identity and labile nature of neurotransmitters, collection of evidence in support of this hypothesis was inordinately difficult. Moreover, such technical problems were compounded by a reluctance by many influential physiologists to acknowledge this possibility, because of a dogma that transmission between cells must be an extension of the electrical processes that determine propagation along nerve axons and into nerve terminals. In such an atmosphere, mere demonstration that a proposed constituent of nerves tissue was spasmogenic did not allow designation as a neurotransmitter, and such claims induced considerable controversy. This situation prompted Dale to enunciate a series of conditions that should be fulfilled in order that a substance might be accepted as a neurotransmitter (21). The trenchant opposition of detractors necessitated such a rigorous approach, but the outcome of this discord was productive, since application of "Dale's criteria" can with equal facility be extended to indicate endocrine secretions, neurosecretions, and lymphokines (22) and have continued to provide an appropriate way to evaluate any proposed mediator of physiological or pathological processes. Consequently, evidence implicating neuropeptides as effectors of hyperresponsiveness has been assembled under headings appropriate to these criteria.

III. Sensory Neuropeptide Distribution in the Airway

The first requirement of Dale's criteria is that any proposed mediator must occur naturally and, it might be added, that it should be present at the relevant site, in this case human airways. Dale made this condition since he was aware that synthetic substances with high potency had been shown to mimic physiological processes without necessarily being natural constituents of mammalian tissues. With more sensitive and specific assays, there is less cause for such concern. Nevertheless, it cannot be contested that any neuropeptide proposed as a mediator of hyperresponsiveness should be a constituent of asthmatic airways.

In mammals, the preprotachykinin-I (PPT-I) gene encodes substance P and neurokinin A, while PPT-II encodes neurokinin B (23–26). Alternate splicing of the PPT-I gene results in the formation of three mRNA's, designated α-, β-, and γ-PPT. Posttranslational processing of α-, β-, and γ-PPT mRNA yields substance P and, from the two latter forms of mRNA, neurokinin A. Furthermore, neuropeptide K and neuropeptide-γ are (N-terminally) extended forms of neurokinin A produced by β- and γ-PPT mRNA, respectively, while neurokinin A(3-10) is produced from β- and γ- PPT mRNA. Alpha- and beta-calcitonin gene-related peptide (α-CGRP and β-CGRP) are products of two distinct calcitonin genes (27,28), and expression of mRNA for these sensory neuropeptides has been demonstrated in primary afferent neurones (29–33).

The majority of lung and tracheobronchial afferent fibers are vagal in origin, although there is evidence that some lung afferents arise from thoracic spinal ganglia in many mammalian species. By use of specific antiserum and immunohistochemical methods, substance P and CGRP have been shown to co-localize in airway sensory nerves. For these neuropeptides, immunoreactivity was confined to primary sensory neurones in the nodose and jugular ganglia, in and around blood vessels, airway epithelium, ganglion cells, and in tracheobronchial smooth muscle of the guinea pig, rat, and cat. The conclusion that these neuropeptides were localized within sensory nerves was strengthened by the finding that immunoreactivity was markedly diminished following systemic administration of capsaicin (34–36). More recently, studies employing retrograde techniques have revealed that almost all afferent nerves within guinea pig trachea arise from cell bodies within the nodose ganglion, yet most of the nerves which contain neuropeptides have cell bodies that arise from the jugular ganglion (37). This latter finding is consistent with an earlier study in which vagal section above, but not below, the nodose ganglion resulted in significant loss in substance P-like immunoreactivity in guinea pig airway (38). Sensory neuropeptides have been shown to co-localize in large, dense-cored vesicles in the peripheral endings of afferent nerves in the guinea pig (39), but distribution of sensory neuropeptides is less well documented in other species. In the rabbit, substance P immunoreactivity has been observed in the submucosa and airway smooth muscle layer of central airways (40) and, in the pig, immunoreactivity to substance P, and CGRP has been localized in and around blood vessels, ganglion cells, exocrine glands, airway smooth muscle and in close proximity to, and within, respiratory epithelium (41).

In human airways, a number of studies have detected neuropeptides, including substance P, CGRP, neurokinin A, neuropeptide Y, and vasointestinal peptide (VIP) (34,36,42–46). Immunohistochemical techniques reveal that fibers containing substance P are sparsely distributed within the bronchial epithelium, around blood vessels, bronchial smooth muscle, and local tracheobronchial ganglia (34,45,46). It has even been suggested that there is an absence of substance P-like immunoreactivity in the lung, despite demonstrable substance P in other tissues (40). However, it is unlikely that substance P is absent from human lung, since substantial amounts have been extracted from human lung using a high-performance liquid chromatography technique (47).

It may be concluded that a number of neuropeptides, including substance P, neurokinin A, and CGRP, are constituents of human lungs, so this criterion of Dale's is adequately fulfilled.

IV. Modulation of Airway Smooth Muscle Contraction by Sensory Neuropeptides In Vitro

The second of Dale's criteria requires that a potential mediator must mimic endogenous processes both qualitatively and quantitatively. For airway smooth muscle, this may be investigated using airway smooth muscle in vitro or in situ. For the most part, airway smooth muscle contraction occurs following exposure to sensory neuropeptides, although species differences do exist. There is a greater disparity between species with regard to the effect on airway smooth muscle function of endogenously released neuorpeptides. It is therefore necessary to consider each species separately.

A. Guinea Pig

In the guinea pig, airway smooth muscle is contracted by neurokinin A or substance P (48–50). Guinea pig airway smooth muscle is also contracted by electrical field stimulation, a response that is atropine-resistant (51–53), and by capsaicin (51,54). These responses have been attributed to neuropeptide release, which has been documented in these preparations following exposure to capsaicin (36,50,55), vagal stimulation (50), and electrical field stimulation (56), an interpretation that is reinforced by the finding that a range of stimuli (including low pH, low Cl⁻, bradykinin, distilled water, and hypertonic solutions) evoke discharges from vagal C fibers of the guinea pig in vitro (57–59).

In addition to direct contractile effects, sensory neuropeptides can also modulate cholinergic neurotransmission in this species. Thus, endogenously released (60,61) and exogenously administered (60,62,63) sensory neuropeptides augment contraction of guinea pig trachea or bronchus due to stimulation of cholinergic nerves and induce depolarization of cholinergic postganglionic nerves (53).

Using an isolated airway/esophageal preparation, it has been documented that stimulation of afferent nerves by capsaicin leads to the reflex activation of inhibitory nonadrenergic, noncholinergic (iNANC) nerves (64). This suggests that endogenously released sensory neuropeptides can mediate both excitatory and inhibitory effects on airway smooth muscle in this species. The receptor subtype responsible for this facilitation of cholinergic neurotransmission appears to involve both neurokinin receptors (NK_1 and NK_2) in tracheal (61,62) and NK_1 receptors in bronchial preparations (63).

B. Rat/Mouse

The effect of sensory neuropeptides on airway smooth muscle function in the rat varies between strains. For instance, exogenously administered neuropeptides (substance P, neurokinin A, and CGRP) relax tracheal preparations. Furthermore, capsaicin and electrical field stimulation have been shown to induce the release of sensory neuropeptides and cause relaxation of tracheal preparations from Sprague-Dawley and Wistar rats (65–67). In contrast, tracheal preparations from Fischer 344 rats contracted in response to substance P and neurokinin A (67,68). Interestingly, relaxation induced by sensory neuropeptides was mediated by NK_1 receptors, required an intact epithelium, and was inhibited by cyclooxygenase inhibitors, thereby implicating prostanoids (65–67). The ability of endogenously released sensory neuropeptides to mediate relaxation is consistent with the finding that spasmolytic stimuli induced release of substance P and CGRP from perfused tracheal preparations (69–71) and with the autoradiographic localization of neuropeptide binding sites over airway epithelium, rather than smooth muscle, in Sprague Dawley rats (72). In mice also, sensory neuropeptides mediate relaxation of tracheal preparations, a process that was dependent on prostanoid formation within the epithelium (73).

C. Rabbit

Substance P and neurokinin A contract isolated bronchus in the rabbit (74). However, there was no evidence of an atropine-resistant response to electrical field stimulation, and the excitatory response to capsaicin was modest (74). The contractile response to substance P may therefore be indirect, and secondary to the stimulation of parasympathetic nerves (75,76). In accord with this interpretation, contraction of isolated rabbit trachea by substance P and by the NK_1-selective agonist septide has been shown to be atropine-sensitive. Hence the capacity of these compounds to augment contraction of isolated rabbit trachea by cholinergic stimuli may also be secondary to activation of parasympathetic nerves (77,78). Functional studies indicate an involvement of NK_1 and NK_2 receptors in the sensory neuropeptide-induced augmentation of contraction to cholinergic nerve stimulation in isolated bronchi (63). Direct confirmation of these findings has been obtained by demonstration that stimulation of NK_1 receptors on postganglionic neurons leads to an increased efflux of radiolabeled acetylcholine from parasympathetic nerves of isolated trachea (76).

D. Ferret/Pig

Neurokinin A is a more potent contractile agonist than substance P in ferret trachea (79,80), and there is evidence that substance P augments cholinergic neurotransmission (79). By way of contrast, substance P was more potent than neurokinin A in mediating contraction of porcine airway (41). However, porcine airway failed to contract in response to capsaicin, and an excitatory atropine-resistant contractile response has not been reported in this species.

E. Human

Both substance P and neurokinin A contract human bronchi (48,49), yet capsaicin elicits only a modest contractile response and is at least two to three orders of magnitude less potent than in guinea pig airway (48,81–83). Capsaicin can also mediate an inhibitory response that is not dependent on the release of sensory neuropeptides (82). To date, no study has demonstrated convincingly that there are contractile responses in human airway tissue which are secondary to release of sensory neuropeptides from excitatory nonadrenergic, noncholinergic (eNANC) nerves (48,84). Unlike the guinea pig and rabbit, substance P failed to augment release of acetylcholine from human airway (63).

F. Airway Smooth Muscle Responsiveness In Vitro

Spasmogenic and spasmolytic effects of sensory neuropeptides are now extensively documented, but relatively few studies have investigated the possibility that these peptides may increase airway responsiveness in vitro. Nevertheless, there is clear evidence of such a modulatory effect of neuropeptides. Thus, exposure of isolated perfused guinea pig lung to antigen (ovalbumin 1% for 1 min) led to an increased tracheal pressure that was largely unaffected by atropine and associated with increased release of histamine, eicosanoids (6-keto-PGF1α and TXB$_2$) and PAF (85, 86). Simultaneous bilateral vagal stimulation during exposure to ovalbumin exaggerated the increase of tracheal pressure and increased the amount of inflammatory mediators released into the superfusate (86), an affect that (for increased tracheal pressure) was mimicked by intraarterial injection of neurokinin A but not substance P.

The inference can be drawn that, if released, sensory neuropeptides could exacerbate allergic bronchospasm. No mechanism has been established to account for this effect. It may simply reflect summation of two unrelated contractile processes due to allergic mediators and sensory neuropeptides, respectively. Additionally or alternatively, it may be a consequence of changes in the permeability of epithelial and endothelial cell barriers secondary to the release of sensory neuropeptides (86), or to augmented mediator release from resident lung cells, as demonstrated for mast cells (87) and macrophages (88,89) in rats and guinea pigs, respectively. Single (90) or repeated (91) exposure to allergen augmented the eNANC contractile response in isolated guinea pig bronchus. When animals were exposed repeatedly to antigen, this effect was associated with a twofold increase in substance P-like immunoreactivity and might therefore be a consequence of enhanced peptide release (91). The threshold for activation of sensory nerves is lowered following exposure of sensitized lung to allergen. As nerves are necessarily exposed to allergic mediators derived from mast cells and/or other cells (90,92,93), a process unrelated to enhanced release and impaired generation might determine the lowered threshold.

In very many instances, the increased responsiveness that can be observed in intact animals (e.g., after an allergic response) or in humans (e.g., asthmatic tissue) is not manifest in vitro, hence, any capacity of neuropeptides to effect increased responsiveness in vitro is not definitive evidence. Nonetheless, it is apparent that there are examples of neuropeptides (including substance P and neurokinin A) that intensify contractile responses in vitro.

V. Modulation of Airway Smooth Muscle Excitation by Sensory Neuropeptides In Vivo

A. Bronchoconstriction

Guinea Pig

Bronchoconstriction follows intravenous administration of neuropeptides, including substance P and neurokinin A, in the guinea pig (35,52,54,94,95). In normal guinea pigs, bronchoconstriction to substance P was not influenced by suppression of cyclooxygenase enzymes, or by pretreatment with leukotriene or histamine receptor antagonists, and was not blocked by atropine, thereby ruling out an involvement of mast cell mediator release and of acetylcholine release from parasympathetic nerves (52,95). On the otherhand, bronchoconstriction in response to a single does of [β-A1a^8]-neurokinin A(4-10), the NK$_2$-selective agonist, was attenuated by atropine and was augmented by physostigmine in guinea pigs (96). Consistent with this latter observation were the findings that high concentrations of substance P caused depolarization of cholinergic postganglionic neurons in vitro (53), and that stimulation of NK$_2$ receptors in tracheal (61,62) and NK$_1$ receptors in bronchial (63) preparations augmented release of acetylcholine.

This interaction between sensory neuropeptides and the parasympathetic nervous system has been conclusively defined in an elegant study in vivo. Electrical stimulation of regions within the dorsal medulla oblongata induced a cholinergic bronchoconstriction which was augmented during intravenous infusion of neurokinin A, acting via the stimulation of NK$_2$ receptors (97). As neurokinin A had no effect on bronchoconstriction induced by methacholine, it may be concluded that stimulation of NK$_2$ receptors located on cholinergic nerves facilitates cholinergic neurotransmission in the guinea pig.

There is evidence which indicates that endogenous release of sensory neuropeptides mediates bronchoconstriction in the guinea pig. Thus, intravenous administration of capsaicin caused an atropine-resistant bronchoconstriction in anesthetized guinea pigs that was not evident in animals chronically exposed to capsaicin (38,52). This response to capsaicin was attenuated by the substance P antagonist, [D-Arg^1, D-Pro^2, D-Trp^{7,9}, Leu^{11}] substance P (spantide) (54), but was not altered by bilateral vagotomy (98). These findings imply that sensory neuropeptides activate airway smooth muscle directly following release in response to capsaicin, yet administration of capsaicin via the inhaled route induced bronchospasm that was significantly attenuated both by atropine as well as by the substance P antagonist, spantide, in spontaneously breathing guinea pigs (98). It has also been reported that bronchoconstriction in response to low doses of intravenously administered capsaicin was attenuated by atropine, augmented by physostigmine (96); also, the capsaicin-induced release of sensory neuropeptides from guinea pig isolated and perfused lung was attenuated by the sodium channel blocker, tetrodotoxin (99). Taken together, these findings suggest that capsaicin can induce bronchoconstriction as a response to neuropeptides released directly from sensory nerves by axon reflexes, as well as in response to activation of parasympathetic pathways via neurokinin receptors located on pre- and postjunctional sites.

In accord with these conclusions, electrical stimulation of vagal nerves induced an atropine-resistant bronchoconstriction that was not evident in animals pretreated with capsaicin (38,52,54,100) or spantide (52,54). Furthermore, the atropine-sensitive bronchoconstrictor response to vagal stimulation was partially attenuated in guinea pigs exposed chronically to capsaicin, a finding which suggests that endogenously released sensory neuropeptides may modulate cholinergic bronchomotor tone (38,100,101). However, release of acetylcholine from parasympathetic nerves per se does not induce the release of sensory neuropeptides, since selective stimulation of the parasympathetic nervous system results in a bronchoconstrictor response that is not diminished in animals chronically treated with capsaicin (97). This fully supports the view that sensory neuropeptides are not spontaneously released and so do not modulate parasympathetic nerve function, except in circumstances whereby the endogenous release of neuropeptides is elicited.

Rat/Rabbit/Pig

The effect of sensory neuropeptides on bronchomotor tone varies among different strains of rat. Intravenous administration of substance P and neurokinin A induced a bronchoconstrictor response in Fischer 344 rats that was inhibited by atropine and methysergide, implying involvement of parasympathetic nerves and mast cells (87,102,103). BDE rats, on the other hand, were less responsive to intravenously administered substance P and neurokinin A, and it might be suggested that bronchoconstriction in this strain is due to direct activation of airway smooth muscle (103). Furthermore, intravenous injection of either capsaicin or substance P caused bronchodilation in Sprague-Dawley rats, in which infusion of a muscarinic agonist has been used to increase airway resistance. This effect is not evident following inhibition of the cyclooxygenase pathway and thus represents an indirect effect (67).

Substance P evokes an atropine-sensitive bronchoconstriction in rabbits (104), an observation which is consistent with other studies that document the ability of sensory neuropeptides to facilitate cholinergic neurotransmission in the rabbit (63,76–78). Similarly, substance P has been shown to produce increased airway resistance (105) and to augment tracheal tension (106) in swine, effects that are secondary to the activation of parasympathetic nerves.

Human

Neuropeptides

Intravenous (107,108) or aerosolized (109) substance P produced marginal changes in lung function in healthy individuals, while neurokinin A produced a small bronchoconstrictor response (108). Bronchoconstriction to inhaled neurokinin A in healthy individuals (110) was augmented by prior inhalation of the neutral endopeptidase inhibitor thiorphan, which may explain the propensity of asthmatics to bronchoconstrict in response to inhaled neurokinin A (109). However, the neutral endopeptidase inhibitors thiorphan (111) and phosphoramidon (112) augmented the bronchoconstrictor response to neurokinin A in mild asthmatics to a similar degree to that observed in healthy individuals (111). It seems likely, therefore, that asthmatics are intrinsically more responsive to neurokinin A as a manifestation of the mechanism that determines differential responsivity between asthmatics and normals for other spasmogens [e.g., adenosine (113) and bradykinin (114–116)].

Anticholinergic drugs have a modest inhibitory effect on the bronchoconstrictor responses to both substance P (117) and neurokinin A (118) in asthmatics, which suggests that sensory neuropeptides may activate cholinergic nerves. On the other hand, H_1-receptor antagonists did not influence bronchoconstrictor response to substance P- (117) and neurokinin A (118), making it unlikely that mast cell activation is a consequence of exposure to these peptides. Nedocromil sodium and cromoglycate sodium attenuated bronchoconstrictor responses of mild asthmatics to substance P- (119,120) and neurokinin A (121). Similarly, furosemide attenuated bronchoconstrictor response to neurokinin A in mild asthmatics (122). Since neither nedocromil sodium (123) nor furosemide (124,125) attenuated either histamine- or methacholine-induced bronchoconstriction, nonselective suppression of bronchial responsiveness can be excluded as a mechanism of action for these drugs in humans. From the available evidence, it seems likely that neuropeptides induce bronchoconstriction indirectly, as a result of the activation of cholinergic and/or sensory nerves, in asthmatics.

Capsaicin

Excitatory Effects. No significant fall in FEV_1 was observed in response to capsaicin inhalation in several studies involving healthy individuals (126–128). However, in one study, capsaicin induced a transient fall in specific conductance (sGaw) that peaked at 10 sec and resolved after 1 min in both healthy individuals and mild asthmatics (129). However, no difference was evident between healthy individuals and mild asthmatics in whom bronchial hyperresponsiveness to histamine was demonstrable (129). By way of contrast, capsaicin-induced bronchoconstriction was observed in 7 of 17 mild asthmatics (128). When observed, bronchoconstrictor responses were modest, so the different outcome of these studies may reflect variation in investigators' techniques.

Inhibitory Effects. During adrenergic and/or cholinergic blockade, capsaicin induced a bronchodilator response in healthy individuals in whom baseline airway resistance had been increased with inhaled spasmogen (130,131). However, in the absence of ipratropium bromide or following treatment with lignocaine, a bronchoconstrictor response has been observed (131). This suggests that reflex bronchodilation and bronchoconstriction may be determined by activation of distinct nerve populations and raises the possibility that the bronchodilator response may result from reflex activation of the nonadrenergic, noncholinergic inhibitory system (130,131).

In mild asthmatics in whom baseline airway resistance has been elevated with leukotriene D_4, inhalation of capsaicin induced a bronchodilator response that was comparable with that observed in healthy individuals (132), suggesting that the nonadrenergic, noncholinergic inhibitory system was not impaired in mild asthma. Yet it has been suggested that there is a lack of vaso-intestinal peptide (VIP)-like immunofluorescence in asthmatic lung at autopsy and in lung resections from asthmatic patients (133). A more recent study disputes this conclusion, since the amount of extractable VIP from asthmatic lung was not different to that obtained from nonasthmatic controls (47). Whether VIP plays a significant role in the bronchodilator response to a capsaicin is debatable, since it has been demonstrated that the major inhibitory neurotransmitter of airway smooth muscle in humans is nitric oxide and not VIP (134).

A loss of vagal afferent innervation and vagal efferent reinnervation with retention of cholinergic and VIP-containing nerves is characteristic of lung obtained from patients with heart-lung transplants in whom retransplantation has been required (135,136). In 8 of 15 patients who received heart-lung transplants, capsaicin induced a bronchodilator response that might be attributable to the activation of nonadrenergic, noncholinergic inhibitory nerves (128). In guinea pig airway, capsaicin-induced relaxation is dependent on the presence of an intact capsaicin-sensitive afferent innervation to the esophagus which leads to the activation of nonadrenergic, noncholinergic inhibitory nerves (64). No compatible process can be operative in heart-lung transplant patients. It should be noted that there is a possibility that capsaicin may have induced relaxation directly, and independently of sensory nerves. In this context, it may be noted that capsaicin did elicit inhibitory responses in human isolated bronchi that was not subject to tachyphylaxis, indicating a lack of involvement of sensory nerves in this response, and/or to the release of mediators from sensory nerves without inducing desensitization (82).

B. Bronchial hyperresponsiveness

It is noteworthy that a range of sensory neuropeptides can augment airway responsiveness to spasmogens in vivo. Thus, administration of substance P induced an increased responsiveness of the airway to acetylcholine or histamine in the guinea pig (89,137,138). Similarly, the repetitive administration of substance P has been shown to evoke increased airway responsiveness to substance P itself and to capsaicin, a process that was inhibited by indomethacin (139). Interestingly, the increased airway responsiveness to acetylcholine that results from exposure to substance P was abrogated by vagotomy as well as by indomethacin (137). Taken together, these studies indicate that sensory neuropeptides, which are known to sensitize afferent nerves in the skin (140,141), can increase airway responsiveness via direct and indirect activation of airway sensory nerves.

In sheep, a 1.9-fold increase in airway sensitivity to carbachol was observed 30 min after inhalation of neurokinin A, but not at 24 hr (142). In Wistar rats, neither neurokinin A nor substance P altered airway responsiveness to acetylcholine (143). However, prior treatment with neutral endopeptidase inhibitors (thiorphan and phosphoramidon) significantly augmented airway responsiveness to acetylcholine that resulted from exposure to substance P and neurokinin A (143). This form of hyperresponsiveness was partially attenuated by vagotomy and by the nonselective neurokinin antagonist, [D-Arg^2, D-Trp^{7,9}]-substance P (143). Neurokinin A augmented methacholine responsiveness in monkeys for up to 4 weeks after treatment (144), and an increase in airway responsiveness to inhaled acetylcholine was observed 1 hr following aerosol administration of substance P and neurokinin A in the rabbit (145).

In contradistinction to the numerous studies in animals which reveal that sensory neuropeptides may cause increased responsivitity of the airway in vivo, there is only a single report to suggest that, following inhalation of substance P, an increase in airway narrowing (but not changed sensitivity) was evident in asthmatics (146). Thus, although the available evidence does not conflict with Dale's second criterion, it must be acknowledged that the available evidence is only qualitative and, for humans, insubstantial.

VI. Inhibition of Sensory Neuropeptide Inactivation

Dale recognized that impaired inactivation of a mediator should intensify the response to that mediator. For sensory neuropeptides, this criterion is amenable to investigation, because it is well documented that sensory neuropeptides are cleaved hydrolytically by neutral endopeptidase, a membrane-bound enzyme which effectively terminates the biological activity of these peptides (147). Various histochemical, immunohistochemical, and biochemical studies have revealed that neutral endopeptidase is found in the lungs of guinea pigs (148,149), ferrets (80), and humans (150). Neutral endopeptidase is localized predominantly to smooth muscle, to the epithelium of ferret airway (80), and to the epithelium of guinea pig airway (149). Study of the role of neutral endopeptidase in regulating the biological half-life of endogenously released sensory neuropeptides has been facilitated considerably by the emergence of neutral endopeptidase inhibitors (e.g., thiorphan and phosphoramidon).

A. Contractile Studies

The contractile potency of substance P, and to a lesser extent neurokinin A, is significantly augmented by inhibitors of neutral endopeptidase in guinea pig (151–153), ferret (79,80), rabbit (76), and human airway (81,154). From these studies, it can be concluded that much of the neutral endopeptidase activity resides within the epithelium, since removal of this structure potentiated the contractile response to sensory neuropeptides in vitro. In some species, phosphoramidon augmented acetylcholine release following electrical stimulation of the airway, consistent with the finding that sensory neuropeptides can facilitate the release of acetylcholine from parasympathetic nerves (61,76).

The nonadrenergic, noncholinergic, nerve-mediated and capsaicin-induced contraction of airway smooth muscle is enhanced in the presence of phosphoramidon or thiorphan (86,148,155). The capsaicin-induced release of substance P-like immunoreactivity was also increased following removal of the epithelium or in the presence of neutral endopeptidase inhibitors (155,156). These findings suggest that the magnitude of contractile responses to endogenous neuropeptides is determined by the activity of neutral endopeptidase in the tissue. In some studies, the contractile response to exogenous or endogenously released sensory neuropeptides following epithelium removal was further enhanced by addition of neutral endopeptidase inhibitors (148,152,155). This procedure increased further the capsaicin-induced release of substance P-like immunoreactivity in epithelium-denuded tissues (155). The supplementary effect of neutral endopeptidase inhibition may be due to inhibition of neutral endopeptidase in sites such as the epithelium of submucosal glands, fibroblasts, and chondrocytes (149). The possibility exists that sensory neuropeptides may release inhibitory factors from the epithelium which functionally antagonize the release of sensory neuropeptides and/or the contractile response to sensory neuropeptides. In guinea pig preparations, capsaicin elicited a contractile response that was significantly augmented by phosphoramidon (148); a comparable effect has been observed using human bronchial tissue, although in this circumstance capsaicin was some two to three orders of magnitude less potent than that in the guinea pig (81).

A number of other studies have employed methods other than physical destruction of the epithelium to investigate the effect of a loss of neutral endopeptidase activity in modulating airway smooth muscle responsiveness to sensory neuropeptides. Thus, viral infection (157,158) or exposure to ozone (159) result in an increased sensitivity of airway smooth muscle to substance P in vitro. This increased responsiveness was associated with the loss of neutral endopeptidase activity in the absence of any structural damage to the airway epithelium. Similarly, the allergen-induced contractile response in bronchial preparations from sensitized guinea pigs was increased by phosphoramidon and attenuated by prior treatment with capsaicin, suggesting that sensory neuropeptides are released during an immune response in vitro (160). Even so, it is apparent that repeated allergen challenge of sensitized animals is not associated with a reduction in neutral endopeptidase activity in lung tissue (91).

B. Bronchoconstriction

Neutral endopeptidase inhibitors failed to increase baseline airway resistance in control and virally infected guinea pigs (161–163) and did not augment bronchoconstriction induced by the selective stimulation of the parasympathetic nervous system (97), findings which support the notion that airway sensory nerves do not release sensory neuropeptides spontaneously.

Neutral endopeptidase inhibitors increased airway responsiveness to the bronchoconstriction mediated by intravenously administered substance P and neurokinin A in guinea pigs (94,95,161–163) and sheep (142). It is therefore likely that neutral endopeptidase regulate responses to sensory neuropeptides, since bronchoconstriction induced by endogenously released sensory neuropeptides was also increased by inhibition of neutral endopeptidase (161,162).

The role of angiotensin-converting enzyme (ACE) in augmenting the response to intravenously administered neuropeptides is controversial. Captopril has been reported to augment (94) or be without effect (95) on intravenously administered substance P-induced bronchoconstriction in the guinea pig.

Phosphoramidon failed to augment the bronchoconstrictor response to intravenous or aerosolized administration of muscarinic agonists. This finding is consistent with the view that sensory neuropeptides were not released spontaneously, nor was release stimulated by spontaneously released acetylcholine (94,95,97,161). The finding that phosphoramidon augmented bronchoconstriction to low, but not high, doses of inhaled ovalbumin in sensitized guinea pigs (164,165) suggests that, in this species, sensory neuropeptides may contribute to allergic bronchospasm.

Following inhibition of neutral endopeptidase, both substance P and neuro-kinin A-induced bronchoconstriction were enhanced both in normals and in asthmatic subjects (110–112). Similarly, bradykinin, but not histamine-induced broncho-constriction, has been shown to be significantly augmented by phosphoramidon in humans (166).

C. Bronchial Hyperresponsiveness

Airway responsiveness to intravenously administered substance P (167,168) and capsaicin (163) was significantly augmented in guinea pigs following exposure to toluene di-isocyanate (TDI) (167), ozone (168), or following viral infection (163,169). In these instances, changed airway responsiveness was associated with a reduction in tracheal neutral endopeptidase activity, raising the possibility that significant loss of neutral endopeptidase activity can occur without loss of integrity of the epithelium. Such agencies do not influence responses to all spasmogens since, although airway responsiveness to histamine and nicotine (170) were enhanced, responses to acetyl-choline (163) were unchanged in virally infected guinea pigs. The increased bron-choconstrictor response to histamine and nicotine following viral infection may involve both afferent nerve and efferent parasympathetic activity (170). Both hista-mine and nicotine can stimulate airway sensory nerves (50), and sensory neuropep-tides can facilitate cholinergic neurotransmission in the guinea pig (60–63), so bronchoconstriction to these spasmogens would be expected to be enhanced by a reduction in neutral endopeptidase activity; on the other hand, acetylcholine does not induce bronchoconstriction via the activation of sensory nerves in the guinea pig (171), and therefore responses would not be expected to be modified.

In Wistar rats, the neutral endopeptidase inhibitor phosphoramidon augmented the airway responsiveness to inhaled acetylcholine and potentiated the ability of exogenously administered neurokinin A and substance P to augment bronchocon-strictor responses to acetylcholine (143). The ability of phosphoramidon to augment airway responsiveness to acetylcholine was abrogated by vagotomy (143), implying that an enhanced vagal reflex contributes to this response and suggesting that acetylcholine may stimulate sensory nerve endings to release neuropeptides that sensitize vagal afferents in the rat. In accord with this interpretation, acetylcholine appears to stimulate afferent nerves in the Wistar rat (172) but not in the guinea pig in vivo (171). In sensitized rats, the ability of phosphoramidon to augment airway responsiveness to acetylcholine was compromised; indeed, neutral endopeptidase activity was reduced in these animals (173).

Whereas in animal experiments there is clear evidence that inhibition of enzymatic destruction of sensory neuropeptides enhances contractile effects of these substance on the airway and intensifies expression of airway hyperresponsiveness, no such unequivocal information is yet available for human airway in situ. In humans, neutral endopeptidase inhibitors augmented airway responsiveness to sensory neuropeptides but not to muscarinic agonists (110–112), but there was no evidence of altered neutral endopeptidase activity in clinically stable asthma (111).

VII. Depletion of Sensory Neuropeptides and Airway Hyperresponsiveness

The counterpart to impaired destruction of a proposed mediator is that accelerated destruction or elimination of the mediator should abrogate endogenous responses which are attributed to the release of this substance. In the case of sensory neuropeptides, this is amenable to investigation since capsaicin, the principal pungent from peppers, is a selective stimulant of sensory nerves and causes release (acutely) and depletion (chronically) of sensory neuropeptides. Furthermore, capsaicin has been widely used as a tool to investigate the role of sensory neuropeptides in various biological processes (174–177).

Functional, biochemical, and electrophysiological studies have elucidated the mechanism by which capsaicin activates sensory nerves. Thus, capsaicin is known to stimulate an ion channel subsequent to binding to the capsaicin receptor, leading to an influx of sodium and calcium ions and to the release of sensory neuropeptides (175, 178,179). It would appear that capsaicin activates sensory nerves by a similar mechanism in the airway. Thus, release of sensory neuropeptides by capsaicin in the airway was not blocked by the N-type calcium channel blocker ω-conotoxin (180), although it was suppressed by ruthenium red (181,182) and by the competitive capsaicin receptor antagonist capsazepine (183–185). In contrast, electrical depolarization of nerves or stimulation with bradykinin and prostanoids was inhibited by ω-conotoxin but not by ruthenium red or capsazepine (174).

Following chronic treatment with capsaicin, a substantial loss of substance P and CGRP immunoreactivity was evident in the lung (35,36), and there was a diminished bronchoconstrictor response (54) and reduced plasma protein extravasation (186) to vagal stimulation or capsaicin. Depletion of sensory neuropeptides by capsaicin leads to a loss in sensory nerve function, yet it is clear that a loss of sensory nerve function can antecede loss of neuropeptides from sensory nerves (187). Thus, in the rabbit, chronic treatment with capsaicin did not result in a loss in substance P-like immunoreactivity in the eye (188,189) and airway (74), despite a loss in sensory nerve function. Interestingly, the level of substance P in dorsal root ganglion was increased after capsaicin treatment, and this change was considered to coincide with the elimination of capsaicin from the body. A compensatory increased biosynthesis of sensory neuropeptide following the capsaicin-induced loss of neuropeptide in sensory nerves is thought to account for these findings (187). Similarly, CGRP-like immunoreactivity in the spinal cord dorsal horn of rats treated at birth with capsaicin was similar to that observed in vehicle-treated animals, despite a reduction in the levels of CGRP-like immunoreactivity released in the plasma (190). It can be concluded, therefore, that loss of sensory nerve function is not always dependent on the depletion of sensory neuropeptides. Clearly, species differences are evident with regard to the efficacy of capsaicin in depleting sensory nerves of their stores of neuropeptides. Unlike the guinea pig, and to a lesser extent the rat, the rabbit is partially resistant to the excitatory actions of capsaicin, as demonstrated by a lesser effect of capsaicin on conduction in cutaneous nerves (191), an inability to induce degeneration of primary afferents (192), and poor efficacy in producing desensitization (193).

Capsaicin-induced loss in sensory nerve function is a complex phenomenon that occurs via different mechanisms, a function of the dose of capsaicin employed, the duration of exposure, the interval taken before studies are performed following capsaicin treatment, and the species studied. It is evident that acute exposure to capsaicin can mediate release of sensory neuropeptides from releasable pools of neuropeptides in sensory nerves (194), which leads to a transient loss of function to various stimuli. During chronic exposure, capsaicin-induced accumulation of calcium within mitochondria of sensory nerves and osmotic swelling leads to the loss in sensory nerve function, the destruction of sensory nerve terminals, and the depletion of the total pool of sensory neuropeptides (175).

A. Acute Studies

A number of studies have relied on capsaicin to determine the relative role of sensory nerves in airway responsiveness to various stimuli. For instance, dry gas hyperpnea-induced bronchoconstriction does not appear to involve the reflex activation of parasympathetic nerves, as it was not attenuated by atropine or vagotomy (195). This response is augmented by prior treatment with phosphoramidon and abrogated when animals have been chronically exposed to capsaicin (196), thereby confirming a role for airway sensory nerves in this response. However, administration of aerosolized capsaicin to guinea pigs, which increased airway responsiveness to acetylcholine and neurokinin A, did not modify the bronchoconstrictor response to dry gas hyperpnea or capsaicin (197).

Inhalation of capsaicin or substance P increased airway responsiveness to intravenously administered histamine in guinea pigs (138). The mechanism by which endogenously released neuropeptides increase airway responsiveness to histamine may be a consequence of an increase in both afferent and efferent nervous pathways. In the guinea pig, histamine-induced bronchoconstriction is mediated both by direct stimulation of airway smooth muscle and by reflex activation of parasympathetic nerves, since histamine-induced bronchoconstriction is attenuated in capsaicin-treated animals (171) and histamine stimulates the release of neuropeptides from guinea pig lung (50). The sensory neuropeptides that are released by capsaicin may induce hyperalgesia of afferent nerves (140,141), and increase acetylcholine release from parasympathetic nerves (60–63,76) to evoke increased bronchoconstrictor responses to histamine.

B. Chronic Studies

Bronchoconstriction

In the guinea pig, chronic treatment with capsaicin reduces the lung tissue content of substance P and neurokinin A by over 90% (198). This is consistent with a loss in substance P and CGRP immunoreactivity in the lung (35,36), together with a loss in the bronchoconstrictor response (54) and plasma protein extravasation (186) to vagal stimulation or capsaicin. This effect of capsaicin is specific for sensory nerves, since bronchoconstriction induced by acetylcholine and by exogenously administered neuropeptides remained unaltered. While chronic treatment with capsaicin may be considered to have a singular action, it should be noted that a number of studies have revealed that capsaicin has other effects, including antagonism of nicotine receptor channels (199), inhibition of platelet aggregation in vitro (200), and inhibition of lipid peroxidation of membranes by various irritants (201). It is unclear to what extent these actions contribute to the effect of capsaicin treatment on bronchial hyperresponsiveness in vivo, but some caution is warranted if capsaicin alone is relied upon to indicate the role of sensory nerves in mediating bronchial hyperresponsiveness.

Chronic treatment with capsaicin has been shown to produce no protection (198), partial protection (202), or complete protection (203) against antigen-induced bronchospasm in guinea pigs. Furthermore, capsaicin treatment did not inhibit ovalbumin-induced bronchoconstriction, yet did abolish the ability of phosphoramidon to potentiate this response in sensitized guinea pigs (165). These discrepancies remain unresolved, although it can be suggested that the contribution of neuropeptides to allergen-induced bronchoconstriction may be determined by the dose of allergen employed (164) and by the species, since allergen-induced bronchoconstriction was not abrogated by capsaicin treatment of allergic rabbits (204).

Cigarette smoke-induced bronchoconstriction was partially attenuated by vagotomy and by atropine, which implies an involvement of both cholinergic and noncholinergic pathways in this response (205,206). Consistent with this interpretation are the findings that cigarette smoke-induced bronchoconstriction was abolished by capsaicin treatment (206) and augmented by phosphoramidon (206), and that cigarette smoke stimulates the release of substance P and neurokinin A-like immunoreactivity from guinea pig lung (206). Together these data support a role for sensory neuropeptides in acute bronchoconstriction due to inhaled cigarette smoke. While stimulation of nicotinic receptors on the surface of afferent nerve endings can induce the release of sensory neuropeptides (50), it is clear that removal of nicotine and tar from cigarette smoke led, nonetheless, to edema (207,208), demonstrating that irritants in the vapor phase can stimulate sensory nerves.

Bronchial Hyperresponsiveness

Repeated antigen challenge in guinea pigs augments airway responsiveness to neurokinin A and to vagal stimulation but not to histamine or acetylcholine 7 days following the last exposure to allergen (209). Similarly, pretreatment of neonatal rabbits with capsaicin attenuated the development of bronchial hyperresponsiveness to histamine at 3 months (210), and the chronic treatment of allergic guinea pigs (211) or rabbits with capsaicin significantly inhibited the allergen-induced bronchial hyper-responsiveness to histamine (Fig. 2) (204). These studies reveal the importance of sensory nerves in the expression of bronchial hyperresponsiveness. This effect of capsaicin was selective for hyperresponsiveness, since the pulmonary accumulation of eosinophils was not inhibited in these rabbits (204) or in ovalbumin-sensitized guinea pigs treated chronically with capsaicin (211). Importantly, capsaicin treat-ment did not alter airway responses to histamine in the allergic rabbit (204) or acetylcholine in the allergic guinea pig (211), ruling out a nonspecific action of capsaicin on airway smooth muscle function.

A variety of nonallergic models have confirmed that sensory neuropeptides participate in bronchial hyperresponsiveness. Thus, TDI (145,212), lipopolysac-charide (213), ozone (214,215), PAF (74,216) (Fig. 2), and 15-HPETE (217) induce bronchial hyperresponsiveness to spasmogens that is inhibited by chronic treatment with capsaicin. This effect of capsaicin is not a consequence of a nonspecific action on airway smooth muscle function, since airway responsiveness to spasmogens is unaltered in capsaicin-treated animals prior to the administration of the agents inducing hyperresponsiveness. The ability of capsaicin to inhibit bronchial hyper-responsiveness without influencing pulmonary cell recruitment is consistent with other evidence that disassociates these two processes (218).

Acute inhalation of cigarette smoke increases airway responsiveness to acetyl-choline and histamine; these effects were not attenuated by vagotomy but were prevented by acute administration of capsaicin (219,220), suggesting a role for afferent nerves in this response. Airway responsiveness to substance P was also enhanced following acute exposure to cigarette smoke and was associated with loss of neutral endopeptidase activity (221,222). Similarly, chronic exposure to cigarette smoke during early life augmented airway responsiveness to substance P, but not to acetylcholine in guinea pigs (223) or to methacholine in Sprague-Dawley rats (224). In the guinea pig, changed airway responsiveness to substance P may also be a consequence of a reduction in neutral endopeptidase activity (221). Interestingly, neonatal treatment with capsaicin abolished the cigarette smoke-induced increase in airway responsiveness to substance P in guinea pigs (223). However, capsaicin treatment also significantly reduced responsiveness to substance P in control animals, so the role of sensory nerves in the cigarette smoke-induced bronchial hyperrespon-siveness in the rat remains uncertain.

Airway responsiveness to ovalbumin and to serotonin in hybrid (BN × Wi/Fu) sensitized rats was increased following capsaicin treatment of neonates (225), whereas airway responsiveness to ovalbumin and serotonin was reduced in ovalbumin-sensitized adult rats treated with capsaicin following, but not prior to, sensitization (226). Thus, the role of sensory nerves in mediating hyperresponsiveness in the rat appears to be dependent on the strain used, the age of the animals, the timing of

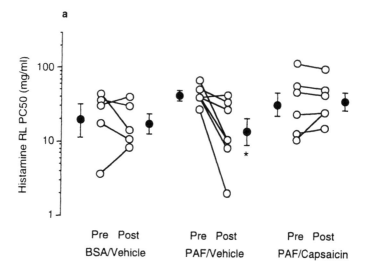

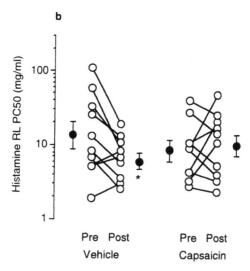

Figure 2 Airway responsiveness to histamine, defined as the dose of histamine required to increase baseline resistance by 50% (RL PC_{50}), before (pre) and 24 hr following (post) exposure to (a) PAF in naive rabbits and (b) *Alternaria tenius* in allergen-immunized rabbits. Animals were treated with either vehicle or capsaicin. Airway responsiveness to histamine was increased by threefold (a) and twofold (b) in naive rabbits exposed to PAF and in sensitized rabbits exposed to allergen, respectively. The increased airway responsiveness observed to these stimuli was abrogated in animals chronically treated with capsaicin. BSA, bovine serum albumin; PAF, platelet-activating factor. (Modified from Refs. 74 and 204.)

capsaicin usage, and the timing of the sensitization. Such strain difference was also revealed when ozone was used to induce bronchial hyperresponsiveness. Thus, ozone induced bronchial hyperresponsiveness in adult Long-Evans (227) but *not* in adult Sprague-Dawley rats (228), even though exposure to ozone treatment *induces* bronchial hyperresponsiveness in Sprague-Dawley rats that have been treated with capsaicin at birth (228). This difference may reflect the finding that endogenously released sensory neuropeptides mediate airway smooth muscle relaxation in the Sprague-Dawley strain of rats (67). In one study, inhalation of ozone induced an increase in breathing frequency and a fall in tidal volume that was exacerbated in capsaicin-treated animals (214), a finding which suggests that a population of sensory nerves might play a protective role in mucosal defense against environmental irritants in the rat. In this context, it is known that stimulation of capsaicin-sensitive afferent nerves can lead to reflex activation of iNANC nerves in the guinea pig (64), and that stimulation of afferent nerves can lead to cough, which can be considered an important reflex against the deposition of foreign agents onto the mucosal surface (229). Loss of sensory neuropeptides which mediate relaxation would leave the airway unprotected, thereby facilitating hyperresponsiveness in neonatal Sprague-Dawley rats after exposure to capsaicin.

In a further study, airway responsiveness to methacholine was increased following tracheal instillation of the cationic polypeptide, poly-L-lysine, in Sprague-Dawley rats (230). However, it appears that endogenous release of sensory neuropeptides is involved in this phenomenon, since poly-L-lysine stimulated release of CGRP from bronchial preparations and hyperresponsiveness to methacholine was abrogated by neonatal treatment with capsaicin or NK_1-receptor antagonists (230). It is instructive to note that a number of inflammatory mediators (e.g., prostaglandin E_2) can reduce the time taken for paw withdrawal in response to heat (thermal nociception) in Sprague-Dawley rats. This latter response was attenuated by capsaicin treatment (231,232), and it may be suggested that the development of hyperalgesia in capsaicin-sensitive afferents increases the pain threshold to thermal stimuli in the skin. Thus, sensory neuropeptides mediate both inhibitory and excitatory responses in Sprague-Dawley rats that appear to be dependent not only on the organ studied, but also on the stimulus used to activate sensory nerves. Capsaicin treatment in neonatal or adult Wistar rats resulted in a significant reduction in airway responsiveness to methacholine, an observation which is consistent with the view that muscarinic agonists can stimulate airway sensory nerves in this species in vivo (172).

In animal experiments, depletion of sensory neuropeptides by pretreatment with capsaicin substantially reduces allergic and other forms of airway hyperresponsiveness. It is not known whether capsaicin influences established airway hyperreactivity in asthmatic subjects, nor is it known whether capsaicin can abrogate development of allergic or other forms of hyperreactivity in the manner demonstrated in laboratory mammals. This possibility is not excluded, and study of severe, chronic, nonallergic rhinitis (233) suggests that such an inhibition may be demonstrable if investigated.

VIII. Antagonism of Sensory Neuropeptides and Airway Hyperresponsiveness

The most obvious test for the importance of a mediator is that selective antagonists in amounts sufficient to suppress responses to the mediator entirely should substantially influence the physiological response that the mediator is considered to effect. For sensory neuropeptides, a number of selective antagonists have become available in recent years, and studies have been completed on the effects of these antagonists both on responses to neuropeptides and on precesses which are known to induce airway hyperreactivity.

Three distinct types of tachykinin receptors exist, based on comparisons of the potencies of tachykinin agonists in the ileum, airway, and cardiovascular system in various animal species (174,211,234,235). The rank order of potency of mammalian tachykinins is: substance P > neurokinin A > neurokinin B for NK_1 receptors, neurokinin A > neurokinin B > substance P for NK_2 receptors, and neurokinin B > neurokinin A > substance P for NK_3 receptors.

Functional and binding studies have revealed the possible existence of neuro-kinin receptor subtypes within each family of tachykinin receptors. Thus, the NK_1-receptor antagonist [(2s,3s)-*cis*-2-(diphenylmethyl)-*N*-[(2-methoxyphenyl)-methyl]-1-aza-bicyclo[2.2.2]octan-3-amine] (CP96345) displaces radiolabeled substance P from binding sites at bovine, guinea pig, rabbit, and human NK_1-receptor sites more readily than from corresponding sites in the rat and mouse (236–238). As a corollary, the nonpeptide NK_1-receptor antagonist 7,7-diphenyl-2-[1-imino-2-(2-methoxy-phenyl)-ethyl]-perhydroisoindol-4-one (3 aR, 7 aR) (RP67580) has a relatively low affinity in the rat (237). In contrast, the potency of the nonpeptide NK_1-receptor antagonist ((S)-1-(2-[3(3,4-dichlorophenyl)-1-(3-isopropoxyphenylacetyl)piperidin-3-yl]ethyl)-4-phenyl-1-azoniabicyclo[2.2.2]octone, chloride) (SR140333) does not appear to demonstrate species specificity (239).

Similarly, it is clear from in-vitro studies that there is heterogeneity of NK_2 receptors among different tissues from different species, as revealed by divergent affinities of NK_2-receptor antagonists including (S)-*N*-methyl-N[4-(4-acetylamino-4-phenyl piperidino)-2-(3,4-dichlorophenyl)butyl)benzamide (SR48968) and [Tyr^5, D-Trp^{6,8,9}, Lys^{10}]-neurokinin A(4-10) (MEN10376) (240–244). Variation of the receptor antagonist affinities observed may reflect a species-related alteration in receptor structure (245), although this may not account for the differences observed between tissues within species and the existence of receptor subtypes has yet to be excluded (246–248).

With the introduction of neurokinin-selective antagonists, the role of sensory neuropeptides as modulators of airway function is being investigated with increasing urgency, given the possibility that these agents may prove to be of therapeutic value in asthma (234). A number of studies have used neurokinin-selective antagonists to probe for the involvement of sensory nerves in mediating bronchial hyperrespon-siveness.

A. Contractile Studies

It has been shown that the excitatory nonadrenergic, noncholinergic contraction of guinea pig isolated airway is sensitive to peptide tachykinin receptor antagonists (54, 249). Thus, the nonselective NK_1- and NK_2-receptor antagonist L-serine, N-[N^2-[N-[N-[N-[α,β-didehydro-N-methyl-N-[N-[1-oxo-3-(2-penthylphenyl)propyl]-L-threonyl]tyrosyl]-L-leucyl]-D-phenylalanyl]-L-allothreonyl]-L-asparaginyl]-ν-lactone (FK224) attenuated contraction to electrical stimulation of eNANC nerves in the guinea pig (250). With the introduction of more selective neurokinin receptor antagonists, it has become evident that contraction induced by electrical stimulation of eNANC nerves occurs predominantly via the activation of postjunctional NK_2 receptors (251–253). Thus, NK_2-receptor antagonists Asp-Tyr-D-Tyr-Val-D-Trp-D-Trp-Arg-NH_2 (MEN10207), MEN 10376, and SR48968 (252–255), unlike the NK_1-receptor antagonist cyclo(Gln-D-Trp-(NMe)-Phe(R)-Gly[ANC-2]-Leu-Met)$_2$ (L668169) (254) or CP96345 (253), attenuated the contractile response to electrical stimulation of eNANC nerves to a significant extent. However, the NK_1-receptor antagonist L668169 was more effective when the contractile response was elicited in the presence of the peptidase inhibitors bestatin, captopril, or thiorphan (254). These data have been interpreted as suggesting that, under normal conditions, NK_2 receptors are responsible for mediating the contractile response to endogenously released neuropeptides, but that, by inhibition of enzymatic degradation, the increased concentration of neuropeptides suffices to activate NK_1 receptors. Vagal stimulation in vivo (256) and in vitro (86) resulted in an atropine-resistant excitatory response due to the stimulation of eNANC nerves in the guinea pig that is mediated similarly by NK_2 receptors. However, whereas the NK_1 receptor antagonist CP96345 was without effect on this response, combined use of CP96345 with the NK_2-receptor antagonist SR48968 produced a considerable degree of inhibition (86,255).

Capsaicin induces increased tracheal pressure in guinea pig isolated perfused lung, an effect that was abolished by the NK_2-receptor antagonist SR48968 but augmented by the NK_1-receptor antagonist [(2s,3s)-3-(2-methoxybenzylamino)-2-phenylpiperidine] (CP99994) (257). In combination, SR48968 and CP99994 abolished capsaicin-induced bronchoconstriction. Why capsaicin should induce increased tracheal pressure in the presence of CP99994 is not clear, but this effect was also observed when using the NK_2-selective agonist [Nle^{10}]-neurokinin A(4-10) (257). Interestingly, NK_1 receptors located on the presynaptic nerve terminal can inhibit the release of sensory neuropeptides from rat spinal cord (258). However, such a mechanism is unlikely to account for exacerbation of the capsaicin-induced increase in tracheal pressure, since the contractile response following electrical stimulation of eNANC nerves was not augmented by NK_1-receptor antagonists (252).

Autoradiographic studies have been employed to determine the distribution of neurokinin receptor-binding sites in the airway. Binding sites for radiolabeled substance P have been detected over airway smooth muscle, mucus secreting glands, and vasculature in the guinea pig (259–262). Similarly, binding studies with radiolabeled NK_1-receptor antagonists, N^2-[(4R)-4-hydroxy-1-[(1-methyl-1H-indol-3-yl) carbonyl]-L-prolyl]-N-methyl-N-(phenyl-methyl)-3-(2-naphthyl)-L-alaninamide (FK888) (263), CP96345 (264), and competition studies (262) have confirmed the existence of NK_1 receptors over bronchial smooth muscle. The introduction of radiolabeled NK_2-receptor antagonists should facilitate detection of NK_2 receptors, particularly in airway smooth muscle of the main bronchi, thereby confirming the existence of receptors anticipated from functional studies.

Autoradiographic studies have detected binding sites for substance P over airway smooth muscle in the rabbit (265). However, little, if any, binding for substance P was detected over airway smooth muscle in Sprague-Dawley rats (72) or in humans (261,262), even though binding sites for substance P have been reported in one study of human airway smooth muscle (259). Whether methodological differences can account for these discrepancies is unclear but would seem unlikely, given that substance P-binding sites were clearly detected in the microvasculature and submucosal glands of human airway (261,262). Autoradiographic studies with radiolabeled NK_2-receptor antagonists will help resolve this issue.

B. Bronchoconstriction

FK224, a nonselective NK_1- and NK_2-receptor antagonist, attenuated substance P, neurokinin A, and capsaicin-induced bronchoconstriction in the guinea pig (250). Similarly, the NK_2-receptor antagonists MEN10367 (96,252), cyclo(Met-Asp-Trp-Phe-Dap-Leu)cyclo(2-beta-5β) (MEN10627) (266), but not the NK_1-receptor antagonist CP96345 (at doses which effectively antagonized NK_2- and NK_1-receptor agonist-induced bronchoconstriction, respectively), inhibited vagally induced atropine-resistant bronchoconstriction (96,252). In a further study, the NK_2-receptor antagonist SR48968 (255,267) and the dual NK_1-/NK_2-receptor antagonist FK224, but not the NK_1-receptor antagonist FK888, antagonized vagal-induced bronchoconstriction (256). These data suggest that NK_2 receptors play a major role in mediating the atropine-resistant response to vagal stimulation in the guinea pig. Neither CP96345 nor SR48968 reduced bronchoconstriction due to selective stimulation of parasympathetic nerves in the guinea pig (97), confirming that acetylcholine released from nerve terminals does not cause bronchoconstriction via the release of sensory neuropeptides.

Similarly, atropine-resistant bronchoconstriction produced by capsaicin in guinea pigs was reduced by the NK_2 antagonists MEN10376 (252), MEN10627 (266), and SR48968 (255,268), but not by the NK_1-receptor antagonist CP96345 (252,266,268). However, although the capsaicin-induced bronchoconstriction was only attenuated by SR48968 (269) and MEN10627 (266), the bronchoconstrictor response was inhibited totally when SR48968 was combined with the NK_1-receptor antagonist CP96345 (268). The ability of CP96345 to influence vagal bronchoconstriction is not a consequence of calcium channel blockade (270,271), since the enantiomer CP96344 (which is also a calcium antagonist) (271) was without effect on the atropine- and SR48968-resistant bronchoconstriction to vagal stimulation (268).

However, CP96345 and CP96344 lack specificity; thus, carrageenin-induced paw edema and hyperalgesia are inhibited both by CP96345 and by the inactive isomer CP96344 (272).

Allergen-induced bronchoconstriction was not attenuated by the NK_1-receptor antagonist FK888 or by the dual NK_1-/NK_2-receptor antagonist FK224 (273). Furthermore, NK_1- and NK_2-receptor antagonists (CP99994 and SR48968, respectively) failed to inhibit allergen-induced bronchoconstriction (95). This is consistent with a lack of effect of capsaicin treatment on allergen-induced bronchoconstriction (198,274). There is evidence that SR48968 (164) or MEN10627 (275) can inhibit bronchoconstriction if low doses of allergen have been used to elicit bronchospasm. CP96345 was partially effective as an inhibitor of allergic bronchospasm (275) and more effective when used in combination with SR48968 (164). Thus, it would appear that sensory neuropeptides are released during allergen challenge but contribute to only a limited extent to the bronchoconstrictor response.

Citric acid-induced bronchoconstriction in the guinea pig is mediated via the activation of sensory nerves, since this response can be abolished by capsaicin pretreatment and by the capsaicin-receptor antagonist capsazepine (267,276). This response was abrogated by the NK_2-receptor antagonist SR48968 (267,277) but not by the NK_1-receptor antagonist FK888 (277). These findings are consistent with the view that citric acid stimulates sensory C fibers and mediates bronchoconstriction via a pathway involving NK_2 receptors.

C. Bronchial Hyperresponsiveness

Studies using capsaicin have implicated sensory nerves as a common pathway by which various stimuli induce bronchial hyperresponsiveness and those of neurokinin receptor antagonists have been consistent with this proposition, being able to abrogate bronchial hyperresponsiveness induced by several stimuli. Although the importance of NK_2 receptors is usually explained in this context, it is evident that, in hyperalgesia, NK_1 receptors may also be important (278). Consequently, combined use of NK_1- and NK_2-receptor antagonists may afford better protection against hyperalgesia and hyperresponsiveness than use of selective NK_2-receptor antagonists.

Sensory Neuropeptides

Administration of sensory neuropeptides can induce changes in airway responsiveness to spasmogens, as has been demonstrated in guinea pigs (89,137,138), sheep (142), rats (143), monkeys (144), rabbits (145), and in asthmatics (146). The NK_2-receptor antagonist SR48968, but not the NK_1-receptor antagonist SR140333, suppressed increased airway responsiveness to acetylcholine in guinea pigs following infusion of substance P (279). Similarly, in the rabbit, the NK_1-receptor antagonist [Arg[6], D-Trp[7,9], MePhe[8]]-substance P(6-11) and the NK_2-receptor antagonist MEN10376 suppressed the increased airway responsiveness to acetylcholine induced by substance P and neurokinin A, respectively (145). These neuropeptide antagonists did not modify the bronchoconstrictor response to acetylcholine, which excludes the possibility that acetylcholine activates sensory nerves in vivo.

Antigen

The bronchial hyperresponsiveness to muscarinic agonists that can be induced by antigen in the guinea pig was attenuated by NK_2- (SR48968) but not by NK_1-receptor (SR140333) antagonists (273,279,280). Furthermore, the capsaicin-induced hyper-responsiveness to inhaled histamine was attenuated by FK224 but not FK888 (273), confirming the involvement of NK_2 receptors in this response. The effect of FK224 on capsaicin-induced bronchial hyperresponsiveness may be selective for allergic hyperreactivity, since hyperresponsiveness to histamine following infusion with interleukin-8 was not inhibited by FK224 of FK888 (281).

Ovalbumin (1% for 1 min) increased tracheal pressure in guinea pig isolated perfused lungs, an effect that was significantly augmented by simultaneous bilateral vagal stimulation (86). An NK_2- but not an NK_1-receptor antagonist inhibited potentiation by vagal stimulation of the ovalbumin-induced increase in tracheal pressure (86). This demonstrates the ability of endogenously released sensory neuropeptides to augment ovalbumin-induced changes in airway smooth muscle tone in the guinea pig.

Citric Acid

Inhalation of citric acid by guinea pigs pretreated with the neutral endopeptidase inhibitor thiorphan increased airway responsiveness to acetylcholine 24 hr later, an effect that was abrogated in animals that had been treated chronically with capsaicin (282). This finding is consistent with an involvement of capsaicin-sensitive sensory nerves in the response to citric acid. Furthermore, increased airway responsiveness to acetylcholine that results from exposure of thiorphan-treated guinea pigs to citric acid was antagonized by the NK_2-receptor antagonist SR48968, but not by the NK_1-receptor antagonist SR140333 (282). Interestingly, the μ-opioid agonist codeine failed to attenuate this form of bronchial hyperresponsiveness, although, the response to cough was attenuated to a similar degree to that observed in the presence of SR48968 (282). These divergent results are consistent with the suggestion that different populations of sensory nerves mediate cough and bronchoconstriction and perhaps bronchial hyperresponsiveness (229,283).

PAF

The ether phospholipid PAF increases airway responsiveness to spasmogens in a number of animal species. The ability of PAF to induce bronchial hyperresponsiveness was abrogated by chronic treatment with capsaicin in the rabbit (74) and guinea pig (216). Furthermore, both an NK_1-receptor antagonist, CP96345, and an NK_2-receptor antagonist, MEN10627, inhibited PAF-induced bronchial hyperresponsiveness to histamine in guinea pigs (275), although the effect of CP96345 on PAF-induced bronchial hyperresponsiveness has been attributed to blockade of calcium channels (275).

TDI/Ozone

Acute inhalation with TDI increased airway responsiveness to acetylcholine in guinea pigs (212). This response was abrogated if the nonselective peptide antagonist spantide had been administered before, but not after, exposure to TDI (284). Both an NK_1-receptor antagonist, [Arg^6, $D\text{-}Trp^{7,9}$,$MePhe^8$]-substance P(6-11), and an NK_2-receptor antagonist, MEN10376, antagonized TDI-induced bronchial hyperresponsiveness to acetylcholine in the rabbit (145). In a preliminary report, the bronchial hyperresponsiveness to histamine that followed exposure to ozone was significantly inhibited by prior treatment with either CP99994 and SR48968, NK_1- and NK_2-receptor antagonists, respectively (285). Together, these studies support a role for the involvement of sensory nerves in the bronchial hyperresponsiveness mediated by these noxious agents.

Dry Gas Hyperpnea

Dry gas hyperpnea caused bronchoconstriction and increased vascular permeability, processes that can be abrogated by capsaicin pretreatment (196,286). This finding is consistent with the observation that dry gas hyperpnea or hyperosmolarity, but not hypothermia, stimulated release of tachykinins from neonatal rat dorsal root ganglion cells in culture (287). Both the NK_1-receptor antagonist CP96345 and the NK_2-receptor antagonist SR48968 attenuated the increase in pulmonary resistance that follows inhalation of dry gas, although inhibition was much greater after SR48968 (288). Neither antagonist afforded significant protection against the vascular permeability that accompanies dry gas hyperpnea (288), which suggests that altered vascular permeability is not the determinant of hyperreactivity. There is considerable evidence to implicate CGRP as a mediator of neurogenic vasodilation (289), which can be expected to contribute to neurogenic plasma protein extravasation. Although this latter effect is attributed to activation of NK_1 receptors (290,291), it has recently been reported that NK_2 receptors can also mediate plasma protein extravasation in the lower airway of the guinea pig (292).

It is therefore of interest that exposure to cold air augmented the ovalbumin-induced bronchoconstriction in sensitized guinea pigs, an effect that was abrogated by the NK_2-receptor antagonist SR48968 (293). This outcome is consistent with the ability of endogenously released sensory neuropeptides to augment antigen-induced increase in tracheal pressure in vitro (86). The inability of hypothermia to stimulate the release of neuropeptides from sensory nerves in culture (287) and the ability of cold air to augment allergen-induced bronchoconstriction (293) might be accounted for by use of species and methodological differences. The effect of hypothermia on the release of sensory neuropeptides from guinea pig sensory neurons in culture has not been investigated.

Dry gas hyperpnea and cold air have often been used in experimental animals to model exercise-induced bronchoconstriction in asthma. Interestingly, recovery from exercise-induced airway narrowing in asthmatics was significantly improved following inhalation with the NK_1-receptor antagonist FK888 20 min prior to exercise (294). Since, NK_2 receptors predominate on airway smooth muscle and NK_1 receptors in the vasculature, it could be argued that the effect of FK888 on the recovery from exercise-induced bronchoconstriction is secondary to an effect on the vasculature.

Cigarette Smoke

Acute exposure to inhaled cigarette smoke induced plasma protein extravasation in the nose (208) and tracheobronchial tree via a capsaicin-sensitive mechanism that could be attenuated by the peptide neurokinin receptor antagonist spantide (186,207), additionally, cigarette smoke-induced bronchoconstriction was attenuated by the combined use of the NK_1- and NK_2-receptor antagonists CP99994 and SR48968, respectively (205).

D. Cough

Citric acid-induced cough in the guinea pig was attenuated by the NK_2-receptor antagonist SR48968 (277,295,296), but not by the NK_1-receptor antagonists SR140333 (296) or FK888 (277). The ability of NK_2-receptor antagonists to protect from cough is not secondary to bronchodilation, since baseline tone was unaltered. Furthermore, the inhibitory effect of SR48968 on citric acid-induced cough was not modified by salbutamol (296). Interestingly, combination of NK_1- with NK_2-receptor antagonists proved to be more effective than use of an NK_2-receptor antagonist alone (296). Similarly, capsaicin-induced cough was attenuated following inhalation of the NK_2-receptor antagonist SR48968 but not following the NK_1-receptor antagonist CP99994 (297), and it has been reported that the combined use of FK888 and FK224 attenuated phosphoramidon-induced cough (298). Together, these studies confirm that sensory neuropeptides mediate citric acid-induced cough in the guinea pig. As the locus of action of these antagonists has not been defined, the possibility of a central site of action should be considered.

The NK_1-receptor antagonist FK888 inhibited cough induced by cigarette smoke (298), suggesting a role for sensory neuropeptides in this response, a finding that is consistent with the observation that cough induced by citric acid was partially suppressed in animals given prior inhalation of cigarette smoke (220). These findings suggests that acute inhalation of cigarette smoke may induce tachyphylaxis to further stimulation of sensory nerves. However, this effect was inhibited by atropine, suggesting that increased secretions from bronchial glands might be responsible for loss of responsiveness.

Chronic exposure to inhaled cigarette smoke led to an enhancement of citric acid- and capsaicin- but not cigarette smoke-induced cough (299), yet bronchoconstriction induced by these stimuli was not augmented following chronic exposure to cigarette smoke. A selective increase in the function of sensory nerves that mediate cough and not bronchoconstriction is implied, as is consistent with a twofold increase in CGRP levels in lung tissue from guinea pigs chronically treated with cigarette smoke (299,300).

In animal experiments, the antagonism of neurokinin receptors by selective antagonists substantially reduced airway hyperresponsiveness due to antigen, PAF, TDI, ozone, cigarette smoke, and citric acid. However, in asthmatic subjects there was no overt improvement in pulmonary function upon acute administration. Whether processes such as allergic and other forms of hyperresponsiveness are modified clinically by neurokinin antagonists remains to be established.

IX. Overview of Evidence Implicating Neuropeptides in Asthma

Animal models have provided a wealth of information concerning the possible mechanisms by which sensory nerves might alter airway responsiveness. There are some analogous studies which demonstrate that these mechanisms may also operate in human airways. For instance, it has been claimed that substance P-containing nerves are more abundant in lungs obtained at autopsy from asthmatics as compared with healthy individuals (301), although this observation was not confirmed in a subsequent study (302). Using high-performance liquid chromatography, a reduction in substance P-like immunoreactivity was observed in the airways of individuals who died of asthma or who were undergoing thoracotomy, compared with age-matched nondiseased subjects (47). Similar changes have been reported in other diseases. In rheumatoid arthritis, there appears to be a loss of substance P- and CGRP-like immunoreactivity in sensory nerves in synovial tissue (303), while individuals with idiopathic cough who have increased sensitivity to capsaicin have increased levels of CGRP and, to a lesser extent, substance P immunoreactivity in nerves within bronchial biopsies are compared with healthy subjects (304).

This circumstantial evidence implicating release of sensory neuropeptides in asthma is consistent with the detection of increased substance P-like immunoreactivity in bronchoalveolar lavage fluid in atopic asthmatics as compared with healthy individuals (305). Furthermore, concentrations of substance P-like immunoreactivity in bronchoalveolar lavage was further increased in atopic asthmatics who had experienced an acute reaction to inhaled allergen (305). Elevated levels of substance P-like immunoreactivity has also been detected in the sputum of patients with asthma or chronic bronchitis as compared with healthy individuals following hypertonic saline inhalation (306). These findings are complimented by the observation that chronic treatment with capsaicin reduced symptoms and vascular reactivity in patients with severe, chronic, nonallergic rhinitis (233), suggesting that sensory neuropeptides are involved in the increased responsiveness of the upper respiratory tract.

Other studies have documented possible changes in neurokinin receptor expression in asthma. An increase in mRNA transcripts for neurokinin-1 (307) and neurokinin-2 (308) receptors was demonstrated in lung tissue from asthmatics as compared with nonasthmatics. This may be in response to local release of neuropeptides and consequent neuropeptide receptor tachyphylaxis. However, evidence of an increase in the expression of neuropeptide mRNA in sensory nerves and/or an increase in afferent activity awaits documentation in humans.

With the availability of neurokinin antagonists, a number of studies have begun to explore the role of sensory neuropeptides in asthma. The nonselective NK_1 and NK_2 antagonist FK224 has been reported to inhibit (309) or be marginally effective (310) against bradykinin-induced bronchoconstriction and ineffective against neurokinin A-induced bronchoconstriction in asthmatics (311). The latter outcome questions the utility of this antagonist to evaluate the role of neuropeptides in asthma. The selective NK_2 antagonist SR48968 did attenuate neurokinin A-induced bronchoconstriction in asthma (312). On the other hand, the NK_1-receptor antagonist FK888 improved the recovery from exercise-induced airway narrowing in asthmatics (294), whereas the NK_1-selective antagonist CP99994 was without effect on hypertonic saline-induced bronchoconstriction and cough in asthmatics (313).

X. Cellular and Subcellular Mechanisms

Involvement of sensory nerves in bronchial hyperresponsiveness invites analogy with hyperalgesia, and parallels have been drawn between the manifestation of hyperalgesia and bronchial hyperresponsiveness to various noxious and inflammatory insults (314). When noxious stimuli such as formalin or Freund's adjuvant are applied to the skin, there is an increased sensitivity to stimuli that would otherwise fail to elicit a pain response (allodynia) and an increase in the magnitude of the response to noxious stimuli (hyperalgesia) (278,315). When noxious stimuli are applied to the skin, hypersensitivity evident around the injection site is termed the zone of primary hyperalgesia, while that beyond this boundary is termed the zone of secondary hyperalgesia. Different mechanisms are thought to account for these changes, involving peripheral and central mechanisms, respectively. It is not clear whether similar "zones" occur in the airway and whether analogous changes contribute to heterogeneity of bronchial hyperresponsiveness.

A. Studies in Skin

A number of inflammatory mediators, including bradykinin, 15-hydroperoxyeicosatetraenoic acid (15-HPETE), prostaglandins, leukotrienes, PAF (314), substance P (140,141), neurokinin A (141), cytokines including IL-1β (316,317), and neurotrophins including nerve growth factor (NGF) (317,318) are known to induce hyperalgesia to thermal, chemical, and mechanical stimuli. The precise molecular events underlying the effect of these inflammatory mediators on sensory nerve function is not well understood; changes in the sensitivity of peripheral nociceptors may be determined by altered sodium, calcium, and potassium ion channel conductance, phosphorylation of target proteins within the nerve terminal, and recruitment of C fibers that were previously silent (278,315).

A variety of stimuli can alter sensory nerve conductivity and neuropeptide content within sensory nerves, thereby contributing to hyperalgesia. For instance, subcutaneous injection of formalin into the rat hind paw has been associated with an increase in the electrical activity of sensory nerves and an increase in substance P levels in the spinal cord within 1 hr (319–321), an effect that was blocked by tetrodotoxin and colchicine (320) and naloxone (319). Similarly, subcutaneous injection of formalin or Freund's adjuvant resulted in an increase in the levels of spinal PPT mRNA between 3 and 6 hr (322,323), and between 2 and 8 days later (324). These observations suggest that noxious stimuli induce the depolarization of sensory nerves and propagation of action potentials, which leads to increased synthesis of neuropeptides, a process that is regulated by endogenous opiates. That these processes might lead to release of neuropeptides is revealed by the effect of carrageenin in the rat hindpaw, for hyperalgesia was associated with a reduction in substance P and CGRP levels in the spinal cord within 3 hr and a concomitant increase in the release of these neuropeptides from spinal cord (325). Similarly, increased levels of substance P and CGRP were detected in the sciatic nerve, dorsal root ganglion, and spinal cord between 12 hr and 15 days following injection of Freund's adjuvant into the hindlimbs of rats (318,326–328). In a recent study, the plasticity of sensory neuropeptide expression following injection of Freund's adjuvant was examined in detail. Thus, 6 to 48 hr after injection of Freund's adjuvant into the rat hindpaw, the attendant thermal hyperalgesia was associated with a reduction of CGRP- and substance P-like immunoreactivity from the spinal cord, and there was a significant reduction in the ability of capsaicin to evoke release of substance P at 6 hr (324), most likely as a reflection of the increased release of neuropeptides during the development of inflammation and hyperalgesia. On day 8, thermal hyperalgesia had resolved. At this time the levels of CGRP and substance P had returned to basal values, whereas levels of PPT mRNA remained elevated (324). The increased level of neuropeptides in sciatic nerves which parallels a reduction or in changed levels of CGRP in the rat paw (326) may represent a balance between release of neuropeptides from peripheral nerve terminals and the centripetal transport of substance P toward peripheral terminals of sensory axons. Interestingly, 28 days after repeated intradermal injections with Freund's adjuvant in the rat knee, a loss of CGRP and substance P-like immunoreactivity was observed in the dorsal horn, and this was interpreted as a mechanism whereby sensory nerves might participate in the resolution of the inflammatory response (327).

In addition to an increase in PPT mRNA in rat spinal cord following intradermal injection of formalin or Freund's adjuvant at 6 hr and 4 days, respectively, it has been reported that there was an increased expression of NK_1- and NK_3-receptor mRNA (322). Such changed neurokinin receptor expression in the spinal cord would be consistent with the increased release of sensory neuropeptides following an inflammatory stimulus, and could represent turnover to replenish the loss of cell surface receptors or increased synthesis of receptors secondary to persistent stimulation of the sensory nerve by the inflammatory stimulus (322).

Taken together, these studies show that activation of sensory nerves by noxious stimuli leads to a transient loss of sensory neuropeptides, increased expression of genes which determine neuropeptide synthesis, and the transportion of neuropeptides to peripheral and central projections of the nerve. Sensory neuropeptides released in the spinal cord may contribute toward hyperalgesia by facilitating neurotransmission of central neurones, acting either to increase release of glutamate from afferent nerves (329,330) and/or to enhance discharge of spinal neurones (331,332). In addition, sensory neuropeptides released peripherally would contribute toward edema, vasodilation, and stimulation of inflammatory cells (174).

It appears that the activation of protein kinase C leads to the phosphorylation of target proteins, thereby increasing the open probability of the N-methyl-D-aspartate (NMDA) receptor, which reduces the ability of Mg^{2+} ions to inhibit NMDA-receptor function. This would lead to prolonged depolarization of spinal neurons and, consequently, central sensitization (333). It is therefore of interest that stimulation of NK_1 and NK_2 receptors, which leads to the facilitation of NMDA-induced depolarization of spinal cord neurons, is abolished by protein kinase C inhibition (334) and that formalin-induced hyperalgesia was also associated with increased protein kinase C activity in the spinal cord (335). The 15-HPETE-induced hyperalgesia and increase in electrical response of C fibers to mustard oil in the skin appear to be sensitive to the protein kinase inhibitor H-7 (336), suggesting that a similar mechanism may operate in this situation.

It has been noted that during an inflammatory stimulus, there is an increase in the expression of preprodynorphin and preproenkephalin mRNA (337), and that there was an increase in the transport of opiate receptors in sensory nerves (338). Opiates attenuated the formalin-induced increase in the level of NK_1- and NK_2-receptor mRNA transcripts (339) and suppressed the release of neuropeptides from sensory nerves, including the airway (71,319,339). Together these studies support the notion that an upregulation in the biosynthesis of opioids may be a mechanism to regulate the function of sensory nerves during inflammation.

Noxious stimuli can stimulate the release of sensory neuropeptides, expression of neuropeptide, and neurokinin receptors. The role of NGF in this regard has received particular attention. NGF is an important factor for survival of sensory neurons during development; although it is not essential for the survival of adult sensory neurons, it nonetheless regulates the expression of sensory neuropeptides in these cells. Thus, withdrawal of NGF from sensory nerves in culture resulted in a significant loss of substance P and CGRP content (340,341) and a decline in the levels of PPT and preproCGRP mRNA (341). Furthermore, the ability of capsaicin and low pH to stimulate membrane currents in these cells was also compromized in the absence of NGF (342,343). These studies suggest that NGF has important actions on neuropeptide content of sensory nerves and thus could have a profound effect on sensory nerve function during inflammation. Indeed, an increase in the level of NGF at sites of inflammation (317,318,326) and within sensory nerves has been reported

(326) to follow the administration of noxious stimuli in the skin. As a counterpart to these findings, intradermal injection or systemic administration of NGF can be shown to induce hyperalgesia (317,344,345) that is paralleled by raised levels of substance P and CGRP in sensory nerves (326,345). Furthermore, antibodies directed against NGF can attenuate hyperalgesia and the increased levels of substance P and CGRP in sensory nerves to noxious stimuli, yet do not prevent associated inflammation (i.e., edema and erythrema) (317,318,326,344). It has also been demonstrated that, following the administration of Freund's adjuvant into the rat paw, there is an increase in the level of the transcription factor protein, c-Fos, within the spinal cord, and that this change is preempted by prior treatment with antibodies against NGF (318).

NGF can induce hyperalgesia and increased neuropeptide synthesis within sensory neuropeptides, but hyperalgesia can precede the increase in neuropeptide levels (345). Thus, the capacity of NGF to sensitize afferents might be secondary to other processes, such as degranulation of mast cells (346), as well as to a direct effect of NGF on sensory nerves (326). Following contact with a noxious stimulus, the concentration of NGF at these sites is raised; NGF then binds to *trkA* tyrosine kinase receptors present on peripheral terminals of sensory nerves. This could account for increased sensitivity of the nociceptor at peripheral terminals, presumably as a consequence of tyrosine kinase-mediated phosphorylation of relevant target proteins within the terminal (318). Transport of NGF to the cell body leads to the increased synthesis of sensory neuropeptides (341), and the increased transmitter release at central processes may account for central sensitization (344), both of which may participate in hyperalgesia.

Production of NGF is regulated by the cytokine IL-1β, and activated macrophages are a potential source of this cytokine. IL-1β can in turn stimulate synthesis of NGF from a number of cells, including macrophages, monocytes, fibroblasts, and smooth muscle (278). A number of studies have revealed that IL-1β can induce hyperalgesia in the skin (316,317) and can induce elevated levels of substance P and of PPT mRNA in sensory nerves (347). These effects are a direct consequence of synthesis and release of NGF from target cells, since neutralizing antibodies to IL-1β reduce Freund's adjuvant-induced hyperalgesia and also abrogated the increase concentration of NGF (317).

B. Studies in the Airway

Many of the inflammatory mediators that induce hyperalgesia in the skin might be expected to sensitize airway sensory nerves, thereby contributing toward bronchial hyperresponsiveness in asthma (314). Indeed, various inflammatory mediators, including PAF (74,216), 15-HPETE (217), and sensory neuropeptides (86,89,137–139,142–144,146,197), can augment airway responsiveness to spasmogens in a variety of animal species including humans. Intratracheal administration of IL-1β augmented bronchial responsiveness to bradykinin but not acetylcholine (348), an effect which could in part be attributable to an alteration in sensory nerve function. Interestingly, the airway responsiveness induced by PAF (74,274) and 15-HPETE

Table 1 Representative Studies Demonstrating a Role for Sensory Nerves in Bronchial Hyperresponsiveness to Various Stimuli

Stimulus	Spasmogen	Effect of capsaicin	References
Antigen	Acetylcholine, histamine	Inhibited	(204,210,211)
	5-hydroxytryptamine	Augmented	(225)
TDI	Methacholine, acetylcholine	Inhibited	(145,212)
Cold air		Not determined	
LPS	Histamine	Inhibited	(213)
Ozone	Histamine	Inhibited	(214,215)
	Methacholine	Augmented	(228)
Citric acid	Acetylcholine	Inhibited	(282)
Cigarette smoke	Acetylcholine, histamine	Inhibited	(219,220)
PAF	Histamine	Inhibited	(74,216)
15-HPETE	Histamine	Inhibited	(217)
Poly-L-lysine	Methacholine	Inhibited	(230)
Neuropeptides		Not determined	

(217) was abrogated in animals chronically treated with capsaicin, and bronchial hyperresponsiveness due to PAF (275) and substance P (279) was attenuated by NK_1- and NK_2-receptor antagonists, implying a role for sensory nerves in bronchial hyperresponsiveness. There is extensive evidence which implicates sensory nerves in the process of hyperresponsiveness to ozone, TDI, LPS, and allergen (Tables 1 and

Table 2 Representative Studies Demonstrating a Role for Sensory Nerves in Bronchial Hyperresponsiveness to Various Stimuli

Stimulus	Spasmogen	Effect of NK-receptor antagonist	References
Antigen	Acetylcholine, histamine	Inhibited (SR48968)	(279,280)
	Methacholine	Inhibited (FK224)	(273)
TDI	Acetylcholine	Inhibited (spantide)	(284)
	Acetylcholine	Inhibited (MEN10327)	(145)
Cold air	Ovalbumin	Inhibited (SR48968)	(293)
LPS		Not determined	
Ozone		Inhibited (CP99994 and SR48968)	(285)
Citric acid	Acetylcholine	Inhibited (SR48968)	(282)
Cigarette smoke	Histamine	Not determined	
PAF	Histamine	Inhibited (MEN10627)	(275)
15-HPETE		Not determined	
Poly-L-lysine	Methacholine	Inhibited (CP96345 and RP67580)	(230)
Substance P	Histamine	Inhibited (SR48968)	(279)

2). A number of these stimuli have been shown to affect sensory nerves in the airway, in a manner that is analogous to those that have been described for stimuli (e.g., formalin, Freund's adjuvant) in the skin. Thus, 12 hr following inhalation of antigen there was an increase (20%) in PPT mRNA in the nodose ganglion, and at 24 hr there was an increase (300–300%) in the levels of substance P, neurokinin A, and CGRP in the trachea (349). Retrograde tracing studies confirmed that 10% of the nodose airway sensory nerves contained freshly synthesized neuropeptides in sensitized animals that had been exposed to antigen, whereas none was detected in nonsensitized animals exposed to either vehicle or antigen (349). It can be anticipated that, following exposure of sensitized animals to antigen, increased neuropeptide content in sensory nerves might be indicative of an alteration in sensory nerve function. This interpretation is consistent with the finding that allergen-induced bronchial hyperresponsiveness was attenuated by capsaicin pretreatment or by NK_2-selective antagonists (211,280). Similarly, an increase (200%) in substance P-like immunoreactivity was observed in lung tissue following repeated exposure of sensitized guinea pigs to inhaled antigen, an effect which was accompanied by augmented eNANC responses in vitro (91). It is unclear whether the change in neuropeptide content described by these studies involves NGF. A preliminary study has documented that tracheal instillation of NGF in guinea pigs resulted in an increase (200%) in the levels of substance P-like immunoreactivity in the lung, main bronchus, and nodose ganglion (350). The available evidence suggests that, as in the skin, airway sensory nerve function can be altered by NGF.

Chronic exposure of guinea pigs to TDI produced behavioral changes upon exposure to TDI which were associated with a significant increase in the intensity of substance P- and CGRP-like immunoreactivity in the nasal mucosa, features that were not observed when naive animals were exposed to TDI acutely (351). These changes in peripheral nerves were mirrored centrally, in that there was a slight increase in the number and intensity of neuropeptide-containing fibers in the spinal trigeminal nucleus (351). The cell bodies of these nasal nerves are located in the trigeminal ganglion, where there was a demonstrable loss in substance P- and CGRP-like immunoreactivity, and increased concentrations of PPT and preproCGRP mRNA. These changes were consistent with increased synthesis and transport of neuropeptides to central and peripheral endings of the sensory nerves. The change in neuropeptide levels observed centrally might result in pain hypersensitivity and might lead to secondary hyperalgesia, so sensory neuropeptides in the respiratory system, as in the skin, may play a role in central sensitisation. Recently, a preliminary study has shown that guinea pigs sensitized to TDI experienced a loss in immunoreactivity to CGRP and tachykinins upon exposure to TDI, if compared with nonsensitized animals exposed to TDI (352). Similarly, acute exposure of guinea pigs to nitrogen dioxide resulted in a reduction in CGRP and tachykinin immunoreactivity in the airways (353). Airway responsiveness was not reported, so the significance of these changes to this process cannot be assessed; nevertheless, these experiments demonstrate that various noxious stimuli can alter sensory neuropeptide levels in the airway. In other studies, daily exposure to cigarette smoke resulted in an increase in

Table 3 Summary of Data Supporting a Role for Sensory Neuropeptides as Agents in Bronchial Hyperresponsiveness in Guinea Pigs and Humans

Dale's criteria	Guinea pigs	References	Humans	References
Distribution within the lung	Present	(34,35,37)	Present	(34,36,42,44,46,47)
Neuropeptides induce BHR in vitro	Observed	(86)	Not determined	
Neuropeptides induce BHR in vivo	Observed	(89,137,138)	Observed	(146)
Inhibition of neuropeptide inactivation	Causes BHR	(167,168)	Not determined	
Depletion of sensory neuropeptides[a]	Inhibits BHR	a (211), b (212), c (213) d (298,357) e (216)	Not determined	
Antagonism of neuropeptide receptors[b]	Inhibits BHR	a (279), b (273,279,280), c (282), d (275)	Not determined	

[a]Studies showing the ability of capsaicin treatment to abrogate allergen (a)-, TDI (b)-, lipopolysaccharide (c)-, ozone (d)-, and PAF (e)-induced bronchial hyperresponsiveness in guinea pigs.

[b]Studies showing the ability of neurokinin-2 receptor antagonists to inhibit bronchial hyperresponsiveness induced by substance P (a), antigen (b), citric acid (c), and PAF (d).

the level of CGRP in guinea pig lung, which was associated with changes in airway responsiveness (299,300). Overall, a number of stimuli that cause hyperresponsiveness have been shown to induce increased neuropeptide gene transcription and to elicit levels of neuropeptide in airway sensory nerves, which would lead to increased levels of neuropeptides at central and peripheral processes of the sensory nerve and thereby facilitate central sensitization and local inflammatory changes, respectively.

In guinea pigs infected with parainfluenza virus, onset of dyspnea to inhaled methacholine and substance P was accelerated by comparison with uninfected animals and was associated with increased contractile responses of isolated trachea in vitro and increased plasma protein extravasation in vivo to substance P (354), a 30% reduction in substance P-like immunoreactivity in the trachea, and a loss in binding affinity to, and density of, NK_1 receptors (354). Loss of NK_1-receptor-binding sites on endothelial cells was also observed following infusion of substance P and appears to be secondary to internalization of NK_1 receptors into endosomes (355,356). Thus, following acute viral infection, the increased release of sensory neuropeptides could lead to desensitization of NK_1 receptors, even though this does not compromise the functional changes that are observed in vivo and in vitro. These studies illustrate that a number of immunological and nonimmunological stimuli increase airway responsiveness to various spasmogens and is associated with increased neuropeptide levels in the airway. However, it is yet to be established whether any change of neurokinin receptors occurs centrally and whether central sensitization as has been shown in studies of the skin.

XI. Conclusion

There can be no doubt that neuropeptides are constituents of animal and human airways, and there has been adequate demonstration that acute exposure of animal airways to sensory neuropeptides results in increased responsiveness to spasmogens. That release of sensory neuropeptides might account for endogenous changes in airway responsiveness is indicated by a loss of preformed material from sensory nerves and release into tissue fluids that parallels hyperresponsiveness. As a corollary, suppression of neuropeptide destruction intensifies, while depletion of neuropeptides impairs expression of hyperresponsiveness. As might be anticipated from these findings, selective antagonists of neuropeptides can prevent not only development of airway hyperreactivity by exogenous neuropeptides, but also endogenous forms airway hyperresponsiveness in laboratory animals.

By reference to Dale's criteria, it is therefore reasonable to conclude that actions of locally released sensory neuropeptides contribute to airway hyperresponsiveness during and following an acute allergic reaction in laboratory animals. Nevertheless, it is apparent that, in asthmatics, selective antagonists inhibit bronchoconstrictor responses to inhaled sensory neuropeptides without modifying extant airway hyperresponsiveness. In making this generalization, it must be acknowledged that only limited clinical data are available at present. Even so, no clinical data preclude a role for sensory neuropeptides as agencies of airway hyperresponsiveness in asthma, so it is surprising that no firm opinion as to the role of sensory neuropeptides in asthma has yet emerged.

It is evident that implication of neuropeptides in asthmatic airway hyperresponsiveness is plausible, but not proven. The capacity of these endogenous substances to produce protracted hyperresponsiveness makes them attractive, especially as they offer the prospect of a common mechanism for hyperresponsiveness due to factors as diverse as antigens, environmental chemicals, and pharmaceuticals. As studies focus on persisting effects of neuropeptides and consolidate the present fragmentary evidence that supports the general scheme proposed, it seems likely that investigation of sensory neuropeptides will provide greater attention to the involvement of neuronal events in asthma than has previously been considered.

References

1. Juniper EF, Frith PA, Dunnett C, Cockcroft DW, Hargreave FE. Reproducibility and comparison of response to inhaled histamine and methacholine. Thorax 1978;33: 705–710.
2. Dunnill MS, Masserella GA, Anderson J. A comparison of the quantitative anatomy of the bronchi in normal subjects, in status asthmaticus, in chronic bronchitis, and in emphysema. Thorax 1969;24:174–179.
3. Daniels EE. Control of airway smooth muscle. In: Kaliner MA, Barnes PJ, eds. The Airways: Neural Control in Health and Disease. New York: Marcel Dekker, 1988.
4. Woolcock AJ, Salome CM, Yan K. The shape of the dose-response curve to histamine in asthmatic and normal subjects. Am Rev Respir Dis 1984;130:71–75.
5. Barnes PJ. Asthma as an axon reflex. Lancet 1986;1:242–245.
6. Moreno RH, Hogg JC, Pare PD. Mechanics of airway narrowing. Am Rev Respir Dis 1986;133:1171–1180.
7. Arm JP, O'Hickey SP, Hawksworth RJ, Fong CY, Crea AE, Spur BW, et al. Asthmatic airways have a disproportionate hyperresponsiveness to LTE4, as compared with normal airways, but not to LTC4, LTD4, methacholine, and histamine. Am Rev Respir Dis 1990;142:1112–1118.
8. Adelroth E, Morris MM, Hargreave FE, O'Byrne PM. Airway responsiveness to leukotrienes C4 and D4 and to methacholine in patients with asthma and normal controls. N Engl J Med 1986;315:480–484.
9. Chung KF, Dent G, Barnes PJ. Effects of salbutamol on bronchoconstriction, bronchial hyperresponsiveness, and leucocyte responses induced by platelet activating factor in man. Thorax 1989;44:102–107.
10. Chung KF, Barnes PJ. Effects of platelet activating factor on airway calibre, airway responsiveness, and circulating cells in asthmatic subjects. Thorax 1989;44:108–115.
11. Hoshiko K, Morley J. Exacerbation of airway hyperreactivity by (+/−)salbutamol in sensitized guinea pig. Jpn J Pharmacol 1993;63:159–163.
12. Morley J. Anomalous effects of albuterol and other sympathomimetics in the guinea-pig. Clin Rev Allergy Immunol 1996;14:65–89.
13. Hoshiko K, Morley J. Allergic bronchospasm and airway hyperreactivity in the guinea pig. Jpn J Pharmacol 1993;63:151–157.

14. O'Connor BJ, Aikman SL, Barnes PJ. Tolerance to the nonbronchodilator effects of inhaled beta 2-agonists in asthma. N Engl J Med 1992;327:1204–1208.
15. Cockcroft DW, McParland CP, Britto SA, Swystun VA, Rutherford BC. Regular inhaled salbutamol and airway responsiveness to allergen. Lancet 1993;342:833–837.
16. Crowther SD, Chapman ID, Morley J. Heterogeneity of airway hyperresponsiveness. Clin Exp Allergy Clin Immunol. In press.
17. Thomson NC, Roberts R, Bandouvakis J, Newball H, Hargreave FE. Comparison of bronchial responses to prostaglandin F2alpha and methacholine. J Allergy Clin Immunol 1981;68:392–398.
18. Eggleston PA. A comparison of the asthmatic responses to methacholine and exercise. J Allergy Clin Immunol 1979;63:104–110.
19. Anderton RC, Cuff MT, Frith PA, Cockcroft DW, Morse JLC, Jones NL, et al. Bronchial responsiveness to inhaled histamine and exercise. J Allergy Clin Immunol 1979;63:315–320.
20. Boulet LP, Legris C, Thibault L, Turcotte H. Comparative bronchial responses to hyperosmolar saline and methacholine in asthma. Thorax 1987;42:953–958.
21. Dale HH. Progress in autopharmacology: a survey of present knowledge of the chemical regulation of certain functions by natural constituents of tissue. Johns Hopkins Med J 1933;53:297–347.
22. Morley J, Wolsencroft RA, Dumonde DC. The measurement of lymphokines. In: Weir DM, ed. Handbook of Experimental Immunology. Oxford: Blackwell, 1978:2711–2728.
23. Nawa H, Hirose T, Takashima H, Inayama S, Nakanishi S. Nucleotide sequences of cloned cDNAs for two types of bovine brain substance P precursor. Nature 1983;306:32–36.
24. Nawa H, Kotani H, Nakanishi S. Tissue-specific generation of two preprotachykinin mRNAs from one gene by alternative RNA splicing. Nature 1984;312:729–734.
25. Kotani H, Hoshimaru M, Nawa H, Nakanishi S. Structure and gene organization of bovine neuromedin K precursor. Proc Natl Acad Sci (USA) 1986;83:7074–7078.
26. Krause JE, Chirgwin JM, Carter MS, Xu ZS, Hershey AD. Three rat preprotachykinin mRNAs encode the neuropeptides substance P and neurokinin A. Proc Natl Acad Sci (USA) 1987;84:881–885.
27. Amara SG, Jonas V, Rosenfeld MG, Ong ES, Evans RM. Alternative RNA processing in calcitonin gene expression generates mRNAs encoding different polypeptide products. Nature 1982;298:240–244.
28. Amara SG, Arriza JL, Leff SE, Swanson LW, Evans RM, Rosenfeld MG. Expression in brain of a messenger RNA encoding a novel neuropeptide homologous to calcitonin gene-related peptide. Science 1985;229:1094–1097.
29. Gibson SJ, Polak JM, Giaid A, Hamid QA, Kar S, Jones PM, et al. Calcitonin gene-related peptide messenger RNA is expressed in sensory neurones of the dorsal root ganglia and also in spinal motoneurones in man and rat. Neurosci Lett 1988;91:283–288.
30. Henken DB, Tessler A, Chesselet MF, Hudson A, Baldino F, Jr., Murray M. In situ hybridization of mRNA for beta-preprotachykinin and preprosomatostatin in adult rat dorsal root ganglia: comparison with immunocytochemical localization. J Neurocytol 1988;17:671–681.
31. Boehmer CG, Norman J, Catton M, Fine LG, Mantyh PW. High levels of mRNA coding for substance P, somatostatin and alpha-tubulin are expressed by rat and rabbit dorsal root ganglia neurons. Peptides 1989;10:1179–1194.

32. Minami M, Kuraishi Y, Kawamura M, Yamaguchi T, Masu Y, Nakanishi S, et al. Enhancement of preprotachykinin A gene expression by adjuvant-induced inflammation in the rat spinal cord: possible involvement of substance P-containing spinal neurons in nociception. Neurosci Lett 1989;98:105–110.

33. Rethelyi M, Metz CB, Lund PK. Distribution of neurons expressing calcitonin gene-related peptide mRNAs in the brain stem, spinal cord and dorsal root ganglia of rat and guinea-pig. Neuroscience 1989;29:225–239.

34. Lundberg JM, Hokfelt T, Martling CR, Saria A, Cuello C. Substance P-immunoreactive sensory nerves in the lower respiratory tract of various mammals including man. Cell Tissue Res 1984;235:251–261.

35. Lundberg JM, Franco Cereceda A, Hua X, Hokfelt T, Fischer JA. Co-existence of substance P and calcitonin gene-related peptide-like immunoreactivities in sensory nerves in relation to cardiovascular and bronchoconstrictor effects of capsaicin. Eur J Pharmacol 1985;108:315–319.

36. Martling C-R, Saria A, Fischer JA, Hokfelt T, Lundberg JM. Calcitonin gene-related peptide and the lung: neuronal coexistence with substance P, release by capsaicin and vasodilatory effect. Regulatory Peptides 1988;20:125–139.

37. Kummer W, Fischer A, Kurkowski R, Heym C. The sensory and sympathetic innervation of guinea-pig lung and trachea as studied by retrograde neuronal tracing and double-labelling immunohistochemistry. Neurosci 1992;49:715–737.

38. Lundberg JM, Brodin E, Saria A. Effects and distribution of vagal capsaicin-sensitive substance P neurons with special reference to the trachea and lungs. Acta Physiol Scand 1983;119:243–252.

39. Kummer W, Fischer A, Heym C. Ultrastructure of calcitonin gene-related peptide- and substance P-like immunoreactive nerve fibres in the carotid body and carotid sinus of the guinea pig. Histochemistry 1989;92:433–439.

40. Laitinen LA, Laitinen A, Panula PA, Partanen M, Tervo K, Tervo T. Immunohistochemical demonstration of substance P in the lower respiratory tract of the rabbit and not of man. Thorax 1983;38:531–536.

41. Martling C-R, Matran R, Alving K, Hokfelt T, Lundberg JM. Innervation of lower airways and neuropeptide effects on bronchial and vascular tone in the pig. Cell Tissue Res 1990;260:223–233.

42. Polak JM, Bloom SR. Regulatory peptides and neuron-specific enolase in the respiratory tract of man and other mammals. Exp Lung Res 1982;3:313–328.

43. Uddman R, Sundler F, Emson P. Occurrence and distribution of neuropeptide-Y-immunoreactive nerves in the respiratory tract and middle ear. Cell Tissue Res 1984; 237:321–327.

44. Martling CR, Theodorsson Norheim E, Lundberg JM. Occurrence and effects of multiple tachykinins; substance P, neurokinin A and neuropeptide K in human lower airways. Life Sci 1987;40:1633–1643.

45. Hislop AA, Wharton J, Allen KM, Polak JM, Haworth SG. Immunohistochemical localization of peptide-containing nerves in human airways: age-related changes. Am J Respir Cell Mol Biol 1990;3:191–198.

46. Komatsu T, Yamamoto M, Shimokata K, Nagura H. Distribution of substance P-immunoreactive and calcitonin gene-related peptide-immunoreactive nerves in normal human lungs. Int Arch Allergy Appl Immunol 1991;95:23–28.

47. Lilly CM, Bai TR, Shore SA, Hall AE, Drazen JM. Neuropeptide content of lungs from asthmatic and nonasthmatic patients. Am J Respir Crit Care Med 1995;151:548–553.

48. Lundberg JM, Martling C-R, Saria A. Substance P and capsaicin-induced contraction of human bronchi. Acta Physiol Scand 1983;119:49–53.
49. Advenier C, Naline E, Drapeau G, Regoli D. Relative potencies of neurokinins in guinea-pig trachea an human bronchus. Eur J Pharmacol 1987;139:133–137.
50. Saria A, Martling C-R, Yan Z, Theodorsson-Norheim E, Gamse R, Lundberg JM. Release of multiple tachykinins from capsaicin-sensitive sensory nerves in the lung by bradykinin, histamine, dimethylphenyl piperazinium, and vagal nerve stimulation. Am Rev Respir Dis 1988;137:1330–1335.
51. Grundstrom N, Andersson RGG, Wikberg JES. Pharmacological characterization of the autonomous innervation of the guinea-pig tracheobronchial smooth muscle. Acta Pharmacol Toxicol 1981;49:150–157.
52. Lundberg JM, Saria A. Bronchial smooth muscle contraction induced by stimulation of capsaicin-sensitive sensory neurons. Acta Physiol Scand 1982;116:473–476.
53. Undem BJ, Myers AC, Barthlow H, Weinreich D. Vagal innervation of guinea pig bronchial smooth muscle. J Appl Physiol 1990;69:1336–1346.
54. Lundberg JM, Saria A, Brodin E, Rosell S, Folkers K. A substance P antagonist inhibits vagally induced increase in vascular permeability and bronchial smooth muscle contraction in the guinea pig. Proc Natl Acad Sci (USA) 1983;80:1120–1124.
55. Lou YP, Franco Cereceda A, Lundberg JM. Different ion channel mechanisms between low concentrations of capsaicin and high concentrations of capsaicin and nicotine regarding peptide release from pulmonary afferents. Acta Physiol Scand 1992;146:119–127.
56. Shah S, Spina D, Page CP. The release of substance P-like immunoreactivity from guinea-pig isolated bronchus. Am J Respir Crit Care Med 1996;153:A848.
57. Fox AJ, Barnes PJ, Urban L, Dray A. An in vitro study of the properties of single vagal afferents innervating guinea-pig airways. J Physiol (Lond) 1993;469:21–35.
58. Fox AJ, Urban L, Barnes PJ, Dray A. Effects of capsazepine against capsaicin- and proton-evoked excitation of single airway C-fibres and vagus nerve from the guinea-pig. Neuroscience 1995;67:741–752.
59. Fox AJ, Barnes PJ, Dray A. Stimulation of guinea-pig tracheal afferent fibres by non-isosmotic and low-chloride stimuli and the effect of frusemide. J Physiol (Lond) 1995; 482:179–187.
60. Myers AC, Undem BJ. Functional interactions between capsaicin-sensitive and cholinergic nerves in the guinea-pig bronchus. J Pharmacol Exp Ther 1991;259(1):104–109.
61. Watson N, Maclagan J, Barnes PJ. Endogenous tachykinins facilitate transmission through parasympathetic ganglia in guinea-pig trachea. Br J Pharmacol 1993;109:751–759.
62. Hall AK, Barnes PJ, Meldrum LA, Maclagan J. Facilitation by tachykinins of neurotransmission in the guinea pig pulmonary parasympathetic nerves. Br J Pharmacol 1989;97:274–280.
63. Belvisi MG, Patacchini R, Barnes PJ, Maggi CA. Facilitatory effects of selective agonists for tachykinin receptors on cholinergic neurotransmission: evidence for species differences. Br J Pharmacol 1994;111:103–110.
64. Canning BJ, Undem BJ. Relaxant innervation of the guinea pig trachealis: Demonstration of capsaicin sensitive and insensitive vagal pathways. J Physiol 1993;460:719–739.
65. Frossard N, Muller F. Epithelial modulation of tracheal smooth muscle response to antigenic stimulation. J Appl Physiol 1986;61:1449–1456.
66. Devillier P, Acker GM, Advenier C, Marsac J, Regoli D, Frossard N. Activation of an epithelial neurokinin NK-1 receptor induces relaxation of rat trachea through release of prostaglandin E2. J Pharmacol Exp Ther 1992;263:767–772.

67. Szarek JL, Stewart NL, Spurlock B, Schneider C. Sensory nerve- and neuropeptide-mediated relaxation responses in airways of Sprague-Dawley rats. J Appl Physiol 1995; 78:1679–1687.
68. Joos GF, Lefebvre RA, Kips JC, Pauwels RA. Tachykinins contract trachea from Fischer 344 rats by interaction with a tachykinin NK1 receptor. Eur J Pharmacol 1994; 271:47–54.
69. Hua XY, Jinno S, Back SM, Tam EK, Yaksh TL. Multiple mechanisms for the effects of capsaicin, bradykinin and nicotine on CGRP release from tracheal afferent nerves: role of prostaglandins, sympathetic nerves and mast cells. Neuropharmacology 1994;33:1147–1154.
70. Hua XY, Yaksh TL. Pharmacology of the effects of bradykinin, serotonin, and histamine on the release of calcitonin gene-related peptide from C-fiber terminals in the rat trachea. J Neurosci 1993;13:1947–1953.
71. Ray NJ, Jones AJ, Keen P. Morphine, but not sodium cromoglycate, modulates the release of substance P from capsaicin-sensitive neurones in the rat trachea in vitro. Br J Pharmacol 1991;102:797–800.
72. Sertl K, Kowalski ML, Slater J, Kaliner MA. Passive sensitization and antigen challenge increase vascular permeability in rat airways. Am Rev Respir Dis 1988;138:1295–1299.
73. Manzini S. Bronchodilatation by tachykinins and capsaicin in the mouse main bronchus. Br J Pharmacol 1992;105:968–972.
74. Spina D, McKenniff MG, Coyle AJ, Seeds EA, Tramontana M, Perretti F, et al. Effect of capsaicin on PAF-induced bronchial hyperresponsiveness and pulmonary cell accumulation in the rabbit. Br J Pharmacol 1991;103:1268–1274.
75. Tanaka DT, Grunstein MM. Mechanisms of substance P-induced contraction of rabbit airway smooth muscle. J Appl Physiol 1984;57:1551–1557.
76. Inoue K, Sakai Y, Homma I. An ubiquitous modulating function of rabbit tracheal epithelium: degradation of tachykinins. Br J Pharmacol 1992;105:393–399.
77. Tanaka DT, Grunstein MM. Effect of substance P on neurally mediated contraction of rabbit airway smooth muscle. J Appl Physiol 1986;60:458–463.
78. John C, Brunner S, Tanaka DT. Neuromodulation by neurokinin-1 subtype receptors in adult rabbit airways. Am J Physiol 1993;265:L228–L233.
79. Sekizawa K, Tamaoki J, Nadel JA, Borson DB. Enkephalinase inhibitor potentiates substance P- and electrically induced contraction in ferret trachea. J Appl Physiol 1987;63:1401–1405.
80. Sekizawa K, Tamaoki J, Graf PD, Basbaum CB, Borson DB, Nadel JA. Enkephalinase inhibitor potentiates mammalian tachykinin-induced contraction in ferret trachea. J Pharmacol Exp Ther 1987;243:1211–1217.
81. Honda I, Kohrogi H, Yamaguchi T, Ando M, Araki S. Enkephalinase inhibitor potentiates substance P- and capsaicin-induced bronchial smooth muscle contractions in humans. Am Rev Respir Dis 1991;143:1416–1418.
82. Chitano P, Di Blasi P, Lucchini RE, Calabro F, Saetta M, Maestrelli P, et al. The effects of toluene diisocyanate and of capsaicin on human bronchial smooth muscle in vitro. Eur J Pharmacol 1994;270:167–173.
83. Molimard M, Martin CA, Naline E, Hirsch A, Advenier C. Contractile effects of bradykinin on the isolated human small bronchus. Am J Respir Crit Care Med 1994; 149:123–127.

84. De Jongste JC, Mons H, Bonta IL, Kerrebijn KF. Nonneural components in the response of fresh human airways to electric field stimulation. J Appl Physiol 1987;63:1558–1566.

85. Selig WM, Tocker JE, Tannu SA, Cerasoli Jr F, Durham SK. Pharmacologic modulation of antigen-induced pulmonary responses in the perfused guinea pig lung. Am Rev Respir Dis 1993;147:262–269.

86. Tocker JE, Gertner SB, Welton AF, Selig WM. Vagal stimulation augments pulmonary anaphylaxis in the guinea pig lung. Am J Respir Crit Care Med 1995;151:461–469.

87. Joos GF, Pauwels RA, van der Streaten ME. The mechanism of tachykinin-induced bronchoconstriction in the rat. Am Rev Respir Dis 1988;137:1038–1044.

88. Brunelleschi S, Vanni L, Ledda F, Giotti A, Maggi CA, Fantozzi R. Tachykinins activate guinea-pig alveolar macrophages: involvement of NK2 and NK1 receptors. Br J Pharmacol 1990;100:417–420.

89. Boichot E, Lagente V, Paubert Braquet M, Frossard N. Inhaled substance P induces activation of alveolar macrophages and increases airway responses in the guinea-pig. Neuropeptides 1993;25:307–313.

90. Ellis JL, Undem BJ. Antigen-induced enhancement of noncholinergic contractile responses to vagus nerve and electrical field stimulation in guinea pig isolated trachea. J Pharmacol Exp Ther 1992;262:646–653.

91. Kageyama N, Ichinose M, Igarashi A, Miura M, Yamauchi H, Sasaki Y, et al. Repeated allergen exposure enhances excitatory nonadrenergic noncholinergic nerve-mediated bronchoconstriction in sensitised guinea-pigs. Eur Respir J 1996;9:1439–1444.

92. Myers AC, Undem BJ, Weinreich D. Influence of antigen on membrane properties of guinea-pig bronchial ganglion neurons. J Appl Physiol 1991;71:970–976.

93. Riccio MM, Myers AC, Undem BJ. Immunomodulation of afferent neurones in guinea-pig isolated airways. J Physiol 1996;491:409–509.

94. Shore SA, Stimler-Gerard NP, Coats SR, Drazen JM. Substance P-induced bronchoconstriction in the guinea pig. Enhancement by inhibitors of neutral metallo-endopeptidase and angiotensin-converting enzyme. Am Rev Respir Dis 1988;137:331–336.

95. Foulon DM, Champion E, Masson P, Rodger IW, Jones TR. NK1 and NK2 receptors mediate tachykinin and resiniferatoxin-induced bronchospasm in guinea pigs. Am Rev Respir Dis 1993;148:915–921.

96. Ballati L, Evangelista S, Maggi CA, Manzini S. Effects of selective tachykinin receptor antagonists on capsaicin- and tachykinin-induced bronchospasm in anaesthetized guinea-pigs. Eur J Pharmacol 1992;214:215–221.

97. Hey JA, Danko G, del Prado M, Chapman RW. Augmentation of neurally evoked cholinergic bronchoconstrictor responses by prejunctional NK2 receptors in the guinea pig. J Auton Pharmacol 1996;16:41–48.

98. Bergren DR. Capsaicin challenge, reflex bronchoconstriction, and local action of substance P. Am J Physiol 1988;254:R845–R852.

99. Kroll F, Karlsson J-A, Lundberg JM, Persson CGA. Capsaicin-induced bronchoconstriction and neuropeptide release in guinea pig perfused lungs. J Appl Physiol 1990;68:1679–1687.

100. Martling CR, Saria A, Andersson P, Lundberg JM. Capsaicin pretreatment inhibits vagal cholinergic and non-cholinergic control of pulmonary mechanics in the guinea pig. Naunyn-Schmiedeberg's Arch Pharmacol 1984;325:343–348.

101. Stretton D, Belvisi MG, Barnes PJ. The effect of sensory nerve depletion on cholinergic neurotransmission in guinea pig airways. J Pharmacol Exp Ther 1992;260:1073–1080.

102. Joos GF, Pauwels RA. The in vivo effect of tachykinins on airway mast cells of the rat. Am Rev Respir Dis 1993;148:922–926.

103. Joos GF, Kips JC, Pauwels RA. In vivo characterization of the tachykinin receptors involved in the direct and indirect bronchoconstrictor effect of tachykinins in two inbred rat strains. Am J Respir Crit Care Med 1994;149:1160–1166.

104. Grunstein MM, Tanaka DT, Grunstein JS. Mechanisms of substance P-induced bronchoconstriction in maturing rabbit. J Appl Physiol 1984;57:1238–1246.

105. Dreshaj IA, Martin RJ, Miller MJ, Haxhiu MA. Responses of lung parenchyma and airways to tachykinin peptides in piglets. J Appl Physiol 1994;77:147–151.

106. Haxhiu Poskurica B, Haxhiu MA, Kumar GK, Miller MJ, Martin RJ. Tracheal smooth muscle responses to substance P and neurokinin A in the piglet. J Appl Physiol 1992; 72:1090–1095.

107. Fuller RW, Maxwell DL, Dixon CM, McGregor GP, Barnes VF, Bloom SR, et al. Effect of substance P on cardiovascular and respiratory function in subjects. J Appl Physiol 1987;62:1473–1479.

108. Evans TW, Dixon CM, Clarke B, Conradson TB, Barnes PJ. Comparison of neurokinin A and substance P on cardiovascular and airway function in man. Br J Clin Pharmacol 1988;25:273–275.

109. Joos G, Pauwels R, Van Der Straeten M. Effect of inhaled substance P and neurokinin A on the airways of normal and asthmatic subjects. Thorax 1987;42:779–783.

110. Cheung D, Bel EH, den Hartigh J, Dijkman JH, Sterk PJ. The effect of an inhaled neutral endopeptidase inhibitor, thiorphan, on airway responses to neurokinin A in normal humans in vivo. Am Rev Respir Dis 1992;145:1275–1280.

111. Cheung D, Timmers MC, Zwinderman AH, den Hartigh J, Dijkman JH, Sterk PJ. Neutral endopeptidase activity and airway hyperresponsiveness to neurokinin A in asthmatic subjects in vivo. Am Rev Respir Dis 1993;148:1467–1473.

112. Crimi N, Palermo F, Oliveri R, Polosa R, Magri S, Mistretta A. Inhibition of neutral endopeptidase potentiates bronchoconstriction induced by neurokinin A in asthmatic patients. Clin Exp Allergy 1994;24:115–120.

113. Cushley MJ, Tattersfield AE, Holgate ST. Adenosine-induced bronchoconstriction in asthma. Antagonism by inhaled theophylline. Am Rev Respir Dis 1984;129:380–384.

114. Simonsson BG, Skoogh BE, Bergh NP, Andersson R, Svedmyr N. In vivo and in vitro effect of bradykinin on bronchial motor tone in normal subjects and patients with airways obstruction. Respiration 1973;30:378–388.

115. Fuller RW, Dixon CM, Cuss FM, Barnes PJ. Bradykinin-induced bronchoconstriction in humans. Mode of action. Am Rev Respir Dis 1987;135:176–180.

116. Polosa R, Holgate ST. Comparative airway response to inhaled bradykinin, kallidin, and [des-Arg9]bradykinin in normal and asthmatic subjects. Am Rev Respir Dis 1990;142: 1367–1371.

117. Crimi N, Palermo F, Oliveri R, Palermo B, Vancheri C, Polosa R, et al. Influence of antihistamine (astemizole) and anticholinergic drugs (ipratropium bromide) on bronchoconstriction induced by substance P. Ann Allergy 1990;65:115–120.

118. Joos G, Pauwels R, Van Der Straeten M. The effect of oxitropium bromide on neurokinin A-induced bronchoconstriction in asthmatic subjects. Pulmon Pharmacol 1988;1: 41–45.

119. Crimi N, Palermo F, Oliveri R, Vancheri C, Palermo B, Polosa R, et al. Bronchospasm induced by inhalation of substance P: effect of sodium cromoglycate. Respiration 1988; 54(suppl 1):95–99.

120. Crimi N, Palermo F, Oliveri R, Palermo B, Vancheri C, Polosa R, et al. Effect of nedocromil on bronchospasm induced by inhalation of substance P in asthmatic subjects. Clin Allergy 1988;18:375–382.

121. Crimi N, Palermo F, Oliveri R, Palermo B, Polosa R, Mistretta A. Protection of nedocromil sodium on bronchoconstriction induced by inhaled neurokinin A (NKA) in asthmatic patients. Clin Exp Allergy 1992;22:75–81.

122. Crimi N, Polosa R, Prosperini G, Mastruzzo C, Magri S, Santonocito G, et al. Effect of inhaled frusemide on neurokinin A-induced bronchoconstriction in asthma. Am J Respir Crit Care Med 1995;151:A108.

123. Crimi E, Brusasco V, Brancatisano M, Losurdo E, Crimi P. Effect of nedocromil sodium on adenosine- and methacholine-induced bronchospasm in asthma. Clin Allergy 1987;17:135–141.

124. Nichol GM, Alton EW, Nix A, Geddes DM, Chung KF, Barnes PJ. Effect of inhaled furosemide on metabisulfite- and methacholine-induced bronchoconstriction and nasal potential difference in asthmatic subjects. Am Rev Respir Dis 1990;142:576–580.

125. O'Connor BJ, Chung KF, Chen Worsdell YM, Fuller RW, Barnes PJ. Effect of inhaled furosemide and bumetanide on adenosine 5'-monophosphate- and sodium metabisulfite-induced bronchoconstriction in asthmatic subjects. Am Rev Respir Dis 1991;143: 1329–1333.

126. Collier JG, Fuller RW. Capsaicin inhalation in man and the effects of sodium cromoglycate. Br J Pharmacol 1984;81:113–117.

127. Hansson L, Wollmer P, Dahlback M, Karlsson JA. Regional sensitivity of human airways to capsaicin-induced cough. Am Rev Respir Dis 1992;145:1191–1195.

128. Hathaway TJ, Higenbottam TW, Morrison JF, Clelland CA, Wallwork J. Effects of inhaled capsaicin in heart-lung transplant patients and asthmatic subjects. Am Rev Respir Dis 1993;148:1233–1237.

129. Fuller RW, Dixon CM, Barnes PJ. Bronchoconstrictor response to inhaled capsaicin in humans. J Appl Physiol 1985;58:1080–1084.

130. Ichinose M, Inoue H, Miura M, Takishima T. Nonadrenergic bronchodilation in normal subjects. Am Rev Respir Dis 1988;138:31–34.

131. Lammers JW, Minette P, McCusker MT, Chung KF, Barnes PJ. Nonadrenergic bronchodilator mechanisms in normal human subjects in vivo. J Appl Physiol 1988;64: 1817–1822.

132. Lammers JW, Minette P, McCusker MT, Chung KF, Barnes PJ. Capsaicin-induced bronchodilation in mild asthmatic subjects: possible role of nonadrenergic inhibitory system. J Appl Physiol 1989;67:856–861.

133. Ollerenshaw S, Jarvis D, Woolcock A, Sullivan C, Scheibner T. Absence of immunoreactive vasoactive intestinal polypeptide in tissue from the lungs of patients with asthma. N Engl J Med 1989;320:1244–1248.

134. Ward JK, Belvisi MG, Fox AJ, Miura M, Tadjkarimi S, Yacoub MH, et al. Modulation of cholinergic neural bronchoconstriction by endogenous nitric oxide and vasoactive intestinal peptide in human airways in vitro. J Clin Invest 1993;92:736–742.

135. Springall DR, Polak JM, Howard L, Power RF, Krausz T, Manickam S, et al. Persistence of intrinsic neurones and possible phenotypic changes after extrinsic denervation of

human respiratory tract by heart-lung transplantation. Am Rev Respir Dis 1990;141: 1538–1546.

136. Stretton CD, Mak JC, Belvisi MG, Yacoub MH, Barnes PJ. Cholinergic control of human airways in vitro following extrinsic denervation of the human respiratory tract by heart-lung transplantation. Am Rev Respir Dis 1990;142:1030–1033.

137. Omini C, Brunelli G, Hernandez A, Daffonchio L. Bradykinin and substance P potentiate acetylcholine-induced bronchospasm in guinea-pig. Eur J Pharmacol 1989;163: 195–197.

138. Umeno E, Hirose T, Nishima S. Pretreatment with aerosolized capsaicin potentiates histamine-induced bronchoconstriction in guinea pigs. Am Rev Respir Dis 1992;146: 159–162.

139. Shore SA, Drazen JM. Enhanced airway responses to substance P after repeated challenge in guinea pigs. J Appl Physiol 1989;66:955–961.

140. Nakamura-Craig M, Smith TW. Substance P and peripheral inflammatory hyperalgesia. Pain 1989;38:91–98.

141. Nakamura-Craig M, Gill BK. Effect of neurokinin A, substance P and calcitonin gene related peptide in peripheral hyperalgesia in the rat paw. Neurosci Lett 1991;124:49–51.

142. Abraham WM, Ahmed A, Cortes A, Soler M, Farmer SG, Baugh LE, et al. Airway effects of inhaled bradykinin, substance P, and neurokinin A in sheep. J Allergy Clin Immunol 1991;87:557–564.

143. Chiba Y, Misawa M. Inhibition of neutral endopeptidase increases airway responsiveness to ACh in nonsensitized normal rats. J Appl Physiol 1995;78:394–402.

144. Tamura G, Sakai K, Taniguchi Y, Iijima H, Honma M, Katsumata U, et al. Neurokinin A-induced bronchial hyperresponsiveness to methacholine in Japanese monkeys. Tohoku J Exp Med 1989;159:69–73.

145. Marek W, Potthast JJW, Marczynski B, Baur X. Role of substance P and neurokinin A in toluene diisocyanate-induced increased airway responsiveness in rabbits. Lung 1996; 174:83–97.

146. Cheung D, van der Veen H, den Hartigh J, Dijkman, Sterk PJ. Effects of inhaled substance P on airway responsiveness to methacholine in asthmatic subjects in vivo. J Appl Physiol 1994;77:1325–1332.

147. Nadel JA. Neutral endopeptidase modulates neurogenic inflammation. Eur Respir J 1991;4:745–754.

148. Djokic TD, Nadel JA, Dusser DJ, Sekizawa K, Graf PD, Borson DB. Inhibitors of neutral endopeptidase potentiate electrically and capsaicin-induced non-cholinergic contraction in guinea pig bronchi. J Pharmacol Exp Ther 1989;248:7–11.

149. Kummer W, Fischer A. Tissue distribution of neutral endopeptidase 24.11 ("enkephalinase") activity in guinea pig trachea. Neuropeptides 1991;18:181–186.

150. Johnson AR, Ashton J, Schulz WW, Erdos EG. Neutral metalloendopeptidase in human lung tissue and cultured cells. Am Rev Respir Dis 1985;132:564–568.

151. Devillier P, Advenier C, Drapeau G, Marsac J, Regoli D. Comparison of the effects of epithelium removal and of an enkephalinase inhibitor on the neurokinin-induced contractions of guinea-pig isolated trachea. Br J Pharmacol 1988;94:675–684.

152. Fine JM, Gordon T, Sheppard D. Epithelium removal alters responsiveness of guinea pig trachea to substance P. J Appl Physiol 1989;66:232–237.

153. Frossard N, Rhoden KJ, Barnes PJ. Influence of epithelium on guinea pig airway responses to tachykinins: role of endopeptidase and cyclooxygenase. J Pharmacol Exp Ther 1989;248:292–298.

154. Naline E, Devillier P, Drapeau G, Toty L, Bakdach H, Regoli D, et al. Characterization of neurokinin effects and receptor selectivity in human isolated bronchi. Am Rev Respir Dis 1989;140:679–686.

155. Maggi CA, Patacchini R, Perretti F, Meini S, Manzini S, Santicioli P, et al. The effect of thiorphan and epithelium removal on contractions and tachykinin release produced by activation of capsaicin-sensitive afferents in the guinea-pig isolated bronchus. Naunyn-Schmiedeberg's Arch Pharmacol 1990;341:74–79.

156. Martins MA, Shore SA, Gerard NP, Gerard C, Drazen JM. Peptidase modulation of the pulmonary effects of tachykinins in tracheal superfused guinea pig lungs. J Clin Invest 1990;85:170–176.

157. Saban R, Dick EC, Fishleder RI, Buckner CK. Enhancement of parainfluenza 3 infection of contractile responses to substance P and capsaicin in airway smooth muscle from the guinea pig. Am Rev Respir Dis 1987;136:586–591.

158. Jacoby DB, Tamaoki J, Borson DB, Nadel JA. Influenza infection causes airway hyperresponsiveness by decreasing enkephalinase. J Appl Physiol 1988;64:2653–2658.

159. Murlas CG, Murphy TP, Chodimella V. O3-induced mucosa-linked airway muscle hyperresponsiveness in the guinea pig. J Appl Physiol 1990;69:7–13.

160. Kohrogi H, Yamaguchi T, Kawano O, Honda I, Ando M, Araki S. Inhibition of neutral endopeptidase potentiates bronchial contraction induced by immune response in guinea pigs in vitro. Am Rev Respir Dis 1991;144:636–641.

161. Thompson JE, Sheppard D. Phosphoramidon potentiates the increase in lung resistance mediated by tachykinins in guinea pigs. Am Rev Respir Dis 1988;137:337–340.

162. Dusser DJ, Umeno E, Graf PD, Djokic T, Borson DB, Nadel JA. Airway neutral endopeptidase-like enzyme modulates tachykinin-induced bronchoconstriction in vivo. J Appl Physiol 1988;65:2585–2591.

163. Dusser DJ, Jacoby DB, Djokic TD, Rubinstein I, Borson DB, Nadel JA. Virus induces airway hyperresponsiveness to tachykinins: role of neutral endopeptidase. J Appl Physiol 1989;67:1504–1511.

164. Bertrand C, Geppetti P, Graf PD, Foresi A, Nadel JA. Involvement of neurogenic inflammation in antigen-induced bronchoconstriction in guinea pigs. Am J Physiol 1993;265:L507–L511.

165. Kawano O, Kohrogi H, Yamaguchi T, Araki S, Ando M. Neutral endopeptidase inhibitor potentiates allergic bronchoconstriction in guinea pigs in vivo. J Appl Physiol 1993; 75:185–190.

166. Crimi N, Polosa R, Pulvirenti G, Magri S, Santonocito G, Prosperini G, et al. Effect of an inhaled neutral endopeptidase inhibitor, phosphoramidon, on baseline airway calibre and bronchial responsiveness to bradykinin in asthma. Thorax 1995;50:505–510.

167. Sheppard D, Thompson JE, Scypinski L, Dusser D, Nadel JA, Borson DB. Toluene diisocyanate increases airway responsiveness to substance P and decreases airway neutral endopeptidase. J Clin Invest 1988;81:1111–1115.

168. Murlas CG, Lang Z, Williams GJ, Chodimella V. Aerosolized neutral endopeptidase reverses ozone-induced airway hyperreactivity to substance P. J Appl Physiol 1992;72: 1133–1141.

169. Borson DB, Brokaw JJ, Sekizawa K, McDonald DM, Nadel JA. Neutral endopeptidase and neurogenic inflammation in rats with respiratory infections. J Appl Physiol 1989; 66:2653–2658.

170. Buckner CK, Songsiridej V, Dick EC, Busse WW. In vivo and in vitro studies on the

use of the guinea pig as a model for virus-provoked airway hyperreactivity. Am Rev Respir Dis 1985;132:305–310.

171. Biggs DF, Ladenius RC. Capsaicin selectively reduces airway responses to histamine, substance P and vagal stimulation. Eur J Pharmacol 1990;175:29–33.

172. Sakae RS, Leme AS, Dolhnikoff M, Pereira PM, do Patrocinio M, Warth TN, et al. Neonatal capsaicin treatment decreases airway and pulmonary tissue responsiveness to methacholine. Am J Physiol 1994;266:L23–L29.

173. Chiba Y, Misawa M. Antigen-induced airway hyperresponsiveness is associated with airway tissue NEP hypoactivity in rats. Life Sci 1994;55:1919–1928.

174. Maggi CA. Tachykinins and calcitonin gene-related peptide (CGRP) as co-transmitters released from peripheral endings of sensory nerves. Prog Neurobiol 1995;45:1–98.

175. Bevan S, Szolcsznyi J. Sensory neuron-specific actions of capsaicin: mechanisms and applications. Trends Pharmacol Sci 1990;11:330–333.

176. Maggi CA, Meli A. The sensory-efferent function of capsaicin-sensitive sensory neurons. Gen Pharmacol 1988;19:1–43.

177. Holzer P. Capsaicin: cellular targets, mechanisms of action, and selectivity for thin sensory neurons. Pharmacol Rev 1991;43:143–201.

178. Marsh SJ, Stansfield CE, Brown DA, Davey R, McCarthy D. The mechanism of action of capsaicin on sensory C-type neurons and their axons in vitro. Neurosci 1987;23:275–289.

179. Wood JN, Winter J, James IF, Rang HP, Yeats J, Bevan S. Capsaicin-induced ion fluxes in dorsal root ganglion cells in culture. J Neurosci 1988;8:3208–3220.

180. Maggi CA, Patacchini R, Giuliani S, Santicioli P, Meli A. Evidence for two independent modes of activation of the "efferent" function of capsaicin-sensitive nerves. Eur J Pharmacol 1988;156:367–373.

181. Amann R, Maggi CA. Ruthenium red as a capsaicin antagonist. Life Sci 1991;49:849–856.

182. Lou Y-P, Karlsson J-A, Franco-Cereceda A, Lundberg JM. Selectivity of ruthenium red in inhibiting bronchoconstriction and CGRP release induced by afferent C-fibre activation in the guinea pig lung. Acta Physiol Scand 1991;142:191–199.

183. Bevan S, Hothi S, Hughes G, James IF, Rang HP, Shah K, et al. Capsazepine: a competitive antagonist of the sensory neurone excitant capsaicin. Br J Pharmacol 1992; 107:544–552.

184. Belvisi MG, Miura M, Stretton D, Barnes PJ. Capsazepine as a selective antagonist of capsaicin-induced activation of C-fibres in guinea-pig bronchi. Eur J Pharmacol 1992; 215:341–344.

185. Lou YP, Lundberg JM. Inhibition of low pH evoked activation of airway sensory nerves by capsazepine, a novel capsaicin-receptor antagonist. Biochem Biophys Res Commun 1992;189:537–544.

186. Lundberg JM, Saria A. Capsaicin induced desensitization of airway mucosa to cigarette smoke, mechanical and chemical irritants. Nature 1983;302:251–253.

187. Lembeck F, Donnerer J. Time course of capsaicin-induced functional impairments in comparison with changes in neuronal substance P content. Naunyn-Schmiedeberg's Arch Pharmacol 1981;316:240–243.

188. Tervo K. Effect of prolonged and neonatal capsaicin treatments on the substance P immunoreactive nerves in the rabbit eye and spinal cord. Acta Ophthalmol 1981;59:737–746.

189. Butler JM, Hammond BR. The effects of sensory denervation on the responses of the rabbit eye to prostaglandin E1, bradykinin and substance P. Br J Pharmacol 1980;69:495–502.

190. Diez Guerra FJ, Zaidi M, Bevis P, MacIntyre I, Emson PC. Evidence for the release of calcitonin gene-related peptide and neurokinin A from sensory nerve endings in vivo. Neuroscience 1988;25(3):839–846.

191. Baranowski R, Lynn B, Pini A. The effects of locally applied capsaicin on conduction in cutaneous nerves in four mammalian species. Br J Pharmacol 1986;89:267–276.

192. Lynn B, Shakhanbeh J. Substance P content of skin, neurogenic inflammation and numbers of C-fibres following capsaicin application to a cutaneous nerve in the rabbit. Neuroscience 1988;24:769–775.

193. Amann R, Lembeck F. Capsaicin-induced desensitization in rat and rabbit. Ann NY Acad Sci 1991;632:363–365.

194. Hakanson R, Beding B, Ekman R, Heilig M, Wahlestedt C, Sundler F. Multiple tachykinin pools in sensory nerve fibres in the rabbit iris. Neuroscience 1987;4:943–950.

195. Ray DW, Hernandez C, Munoz N, Leff AR, Solway J. Bronchoconstriction elicited by isocapnic hyperpnea in guinea pigs. J Appl Physiol 1988;65:934–939.

196. Ray DW, Hernandez C, Leff AR, Drazen JM, Solway J. Tachykinins mediate bronchoconstriction elicited by isocapnic hyperpnea in guinea pigs. J Appl Physiol 1989;66:1108–1112.

197. Hsiue TR, Garland A, Ray DW, Hershenson MB, Leff AR, Solway J. Endogenous sensory neuropeptide release enhances nonspecific airway responsiveness in guinea pigs. Am Rev Respir Dis 1992;146:148–153.

198. Ingenito EP, Pliss LB, Martins MA, Ingram RH, Jr. Effects of capsaicin on mechanical, cellular, and mediator responses to antigen in sensitized guinea pigs. Am Rev Respir Dis 1991;143:572–577.

199. Nakazawa K, Inoue K, Koizumi S, Ikeda M. Inhibitory effects of capsaicin on acetylcholine-evoked responses in rat phaeochromocytoma cells. Br J Pharmacol 1994;113:296–302.

200. Hogaboam CM, Wallace JL. Inhibition of platelet aggregation by capsaicin. An effect unrelated to actions on sensory afferent neurons. Eur J Pharmacol 1991;202:129–131.

201. De AK, Ghosh JJ. Capsaicin action modulates lipid peroxidation induced by different irritants. Phytotherapy Res 1993;7:273–277.

202. Saria A, Lundberg JM, Skofitsch G, Lembeck F. Vascular protein linkage in various tissue induced by substance P, capsaicin, bradykinin, serotonin, histamine and by antigen challenge. Naunyn-Schmiedeberg's Arch Pharmacol 1983;324:212–218.

203. Manzini S, Maggi CA, Geppetti P, Bacciarelli C. Capsaicin desensitization protects from antigen-induced bronchospasm in conscious guinea pigs. Eur J Pharmacol 1987;138:307–308.

204. Herd CM, Gozzard N, Page CP. Capsaicin pre-treatment prevents the development of antigen-induced airway hyperresponsiveness in neonatally immunised rabbits. Eur J Pharmacol 1995;282:111–119.

205. Hong JL, Rodger IW, Lee LY. Cigarette smoke-induced bronchoconstriction: cholinergic mechanisms, tachykinins, and cyclooxygenase products. J Appl Physiol 1995;78:2260–2266.

206. Lee LY, Lou YP, Hong JL, Lundberg JM. Cigarette smoke-induced bronchoconstriction and release of tachykinins in guinea pig lungs. Respir Physiol 1995;99:173–181.

207. Lundberg JM, Martling CR, Saria A, Folkers K, Rosell S. Cigarette smoke-induced airway oedema due to activation of capsaicin-sensitive vagal afferents and substance P release. Neuroscience 1983;10:1361–1368.

208. Lundberg JM, Lundblad L, Saria A, Anggard A. Inhibition of cigarette smoke-induced

oedema in the nasal mucosa by capsaicin pretreatment and a substance P antagonist. Naunyn-Schmiedeberg's Arch Pharmacol 1984;326:181–185.

209. Ballati L, Evangelista S, Manzini S. Repeated antigen challenge induced airway hyper-responsiveness to neurokinin A and vagal non-adrenergic, non-cholinergic (NANC) stimulation in guinea pigs. Life Sci 1992;51:PL119–PL124.

210. Riccio MM, Manzini S, Page CP. The effect of neonatal capsaicin on development of bronchial hyperresponsiveness in allergic rabbits. Eur J Pharmacol 1993;232:89–97.

211. Matsuse T, Thomson RJ, Chen XR, Salari H, Schellenberg RR. Capsaicin inhibits airway hyperresponsiveness but not lipoxygenase activity or eosinophilia after repeated aerosolized antigen in guinea pigs. Am Rev Respir Dis 1991;144:368–372.

212. Thompson JE, Scypinski LA, Gordon T, Sheppard D. Tachykinins mediate the acute increase in airway responsiveness caused by toluene diisocyanate in guinea pigs. Am Rev Respir Dis 1987;136:43–49.

213. Jarreau PH, D'Ortho MP, Boyer V, Harf A, Macquin Mavier I. Effects of capsaicin on the airway responses to inhaled endotoxin in the guinea pig. Am J Respir Crit Care Med 1994;149:128–133.

214. Tepper JS, Costa DL, Fitzgerald S, Doerfler DL, Bromberg PA. Role of tachykinins in ozone-induced acute lung injury in guinea pigs. J Appl Physiol 1993;75:1404–1411.

215. Koto H, Aizawa H, Takata S, Inoue H, Hara N. An important role of tachykinins in ozone-induced airway hyperresponsiveness. Am J Respir Crit Care Med 1995;151:1763–1769.

216. Perretti F, Manzini S. Activation of capsaicin-sensitive sensory fibers modulates PAF-induced bronchial hyperresponsiveness in anesthetized guinea pigs. Am Rev Respir Dis 1993;148:927–931.

217. Riccio MM, Matsumoto T, Adcock JJ, Douglas GJ, Spina D, Page CP. The effect of 15-HPETE on airway responsiveness and pulmonary cell recruitment. Br J Pharmacol. In press.

218. Chapman ID, Foster A, Morley J. The relationship between inflammation and hyper-reactivity of the airways in asthma. Clin Exp Allergy 1993;23:168–171.

219. Daffonchio L, Hernandez A, Gallico L, Omini C. Airway hyperreactivity induced by active cigarette smoke exposure in guinea-pigs: possible role of sensory neuropeptides. Pulmon Pharmacol 1990;3:161–166.

220. Karlsson JA, Zackrisson C, Sjolin C, Forsberg K. Cigarette smoke-induced changes in guinea-pig airway responsiveness to histamine and citric acid. Acta Physiol Scand 1991;142:119–125.

221. Dusser DJ, Djokic TD, Borson DB, Nadel JA. Cigarette smoke induces bronchocon-strictor hyperresponsiveness to substance P and inactivates airway neutral endopep-tidase in the guinea pig. Possible role of free radicals. J Clin Invest 1989;84:900–906.

222. Dusser DJ, Lacroix H, Desmazes Dufeu N, Mordelet Dambrine M, Roisman GL. Nedocromil sodium reduces cigarette smoke-induced bronchoconstrictor hyperrespon-siveness to substance P in the guinea-pig. Eur Respir J 1995;8:47–52.

223. Lai YL, Thacker A, Gairola CG. Sidestream cigarette smoke exposure and airway reactivity during early life. J Appl Physiol 1994;77:1868–1874.

224. Joad JP, Ji C, Kott KS, Bric JM, Pinkerton KE. In utero and postnatal effects of sidestream cigarette smoke exposure on lung function, hyperresponsiveness, and neuro-endocrine cells in rats. Toxicol Appl Pharmacol 1995;132:63–71.

225. Ahlstedt S, Alving K, Hesselmar B, Olaisson E. Enhancement of the bronchial reac-

tivity in immunized rats by neonatal treatment with capsaicin. Int Arch Allergy Appl Immunol 1986;80:262–266.

226. Alving K, Ulfgren A, K., Lundberg J, M., Ashlstedt S. Effect of capsaicin on bronchial reactivity and inflammation in sensitized adult rats. Int Arch Allergy Appl Immunol 1987;82:377–379.

227. Evans TW, Brokaw JJ, Chung KF, Nadel JA, McDonald DM. Ozone-induced bronchial hyperresponsiveness in the rat is not accompanied by neutrophil influx or increased vascular permeability in the trachea. Am Rev Respir Dis 1988;138:140–144.

228. Jimba M, Skornik WA, Killingsworth CR, Long NC, Brain JD, Shore SA. Role of C fibers in physiological responses to ozone in rats. J Appl Physiol 1995;78:1757–1763.

229. Karlsson JA, Sant'Ambrogio G, Widdicombe J. Afferent neural pathways in cough and reflex bronchoconstriction. J Appl Physiol 1988;65:1007–1023.

230. Coyle AJ, Perretti F, Manzini S, Irvin CG. Cationic protein-induced sensory nerve activation: role of substance P in airway hyperresponsiveness and plasma protein extravasation. J Clin Invest 1994;94:2301–2306.

231. Schuligoi R, Donnerer J, Amann R. Bradykinin-induced sensitization of afferent neurons in the rat paw. Neuroscience 1994;59:211–215.

232. Amann R, Schuligoi R, Holzer P, Donnerer J. The non-peptide NK1 receptor antagonist SR140333 produces long-lasting inhibition of neurogenic inflammation, but does not influence acute chemo- or thermonociception in rats. Naunyn-Schmiedeberg's Arch Pharmacol 1995;352:201–205.

233. Lacroix JS, Buvelot JM, Polla BS, Lundberg JM. Improvement of symptoms of non-allergic chronic rhinitis by local treatment with capsaicin. Clin Exp Allergy 1991;21:595–600.

234. Bertrand C, Geppetti P. Tachykinin and kinin receptor antagonists: therapeutic perspectives in allergic airway disease. Trends Pharmacol Sci 1996;17:255–259.

235. Maggi CA, Giachetti A, Dey RD, Said SI. Neuropeptides as regulators of airway function: vasoactive intestinal peptide and the tachykinins. Pharmacol Rev 1995;75:277–322.

236. Snider RM, Constantine JW, Lowe JA 3d, Longo KP, Lebel WS, Woody HA, et al. A potent nonpeptide antagonist of the substance P (NK1) receptor. Science 1991;251:435–437.

237. Garret C, Carruette A, Fardin V, Moussaoui S, Peyronel JF, Blanchard JC, et al. Pharmacological properties of a potent and selective nonpeptide substance P antagonist. Proc Natl Acad Sci (USA) 1991;88:10208–10212.

238. Gitter BD, Waters DC, Bruns RF, Mason NR, Nixon JA, Howbert JJ. Species differences in affinities of non-peptide antagonists for substance P receptors. Eur J Pharmacol 1991;197:237–238.

239. Emonds Alt X, Doutremepuich JD, Heaulme M, Neliat G, Santucci V, Steinberg R, et al. In vitro and in vivo biological activities of SR140333, a novel potent non-peptide tachykinin NK1 receptor antagonist. Eur J Pharmacol 1993;250:403–413.

240. Maggi CA, Patacchini R, Giuliani S, Rovero P, Dion S, Regoli D, et al. Competitive antagonists discriminate between NK2 tachykinin receptor subtypes. Br J Pharmacol 1990;100:589–592.

241. Dion S, Rouissi N, Nantel F, Jukic D, Rhaleb NE, Tousignant C, et al. Structure-activity study of neurokinins: antagonists for the neurokinin-2 receptor. Pharmacology 1990;41:184–194.

242. Emonds Alt X, Vilain P, Goulaouic P, Proietto V, Van Broeck D, Advenier C, et al. A potent and selective non-peptide antagonist of the neurokinin A (NK2) receptor. Life Sci 1992;50:PL101–PL106.

243. Maggi CA, Patacchini R, Rovero P, Giachetti A. Tachykinin receptors and tachykinin receptor antagonists. J Auton Pharmacol 1993;13:23–93.

244. Wiesenfeld-Hallin Z, Luo L, Xu XJ, Maggi CA. Differential effects of selective tachykinin NK2 receptor antagonists in rat spinal cord. Eur J Pharmacol 1994;251: 99–102.

245. Fong TM, Yu H, Strader CD. Molecular basis for the species selectivity of the neurokinin-1 receptor antagonists CP-96,345 and RP67580. J Biol Chem 1992;267:25668–25671.

246. Xu XJ, Maggi CA, Wiesenfeld-Hallin Z. On the role of NK-2 tachykinin receptors in the mediation of spinal reflex excitability in the rat. Neuroscience 1991;44:483–490.

247. Brunelleschi S, Ceni E, Fantozzi R, Maggi CA. Evidence for tachykinin NK-2B-like receptors in guinea-pig alveolar macrophages. Life Sci 1992;51:PL177–PL181.

248. Croci T, Emonds Alt X, Le Fur G, Manara L. In vitro characterization of the non-peptide tachykinin NK1 and NK2-receptor antagonists, SR140333 and SR48968 in different rat and guinea-pig intestinal segments. Life Sci 1994;56:267–275.

249. Karlsson JA, Finney MJ, Persson CG, Post C. Substance P antagonists and the role of tachykinins in non-cholinergic bronchoconstriction. Life Sci 1984;35:2681–2691.

250. Murai M, Morimoto H, Maeda Y, Kiyotoh S, Nishikawa M, Fujii T. Effects of FK244, a novel compound NK1 and NK2 receptor antagonist, on airway constriction and airway edema induced by neurokinins and sensory nerve stimulation in guinea pigs. J Pharmacol Exp Ther 1992;262:403–408.

251. Martin CA, Naline E, Emonds Alt X, Advenier C. Influence of (+/−)-CP-96,345 and SR48968 on electrical field stimulation of the isolated guinea-pig main bronchus. Eur J Pharmacol 1992;224:137–143.

252. Maggi CA, Giuliani S, Ballati L, Lecci A, Manzini S, Patacchini R, et al. In vivo evidence for tachykininergic transmission using a new NK-2 receptor-selective antagonist, MEN 10,376. J Pharmacol Exp Ther 1991;257:1172–1178.

253. Renzetti LM, Shenvi A, Buckner CK. Nonadrenergic, noncholinergic contractile responses of the guinea pig hilar bronchus involve the preferential activation of tachykinin neurokinin2 receptors. J Pharmacol Exp Ther 1992;262:957–963.

254. Maggi CA, Patacchini R, Rovero P, Santicioli P. Tachykinin receptors and noncholinergic bronchoconstriction in the guinea-pig isolated bronchi. Am Rev Respir Dis 1991;144:363–367.

255. Lou YP, Lee LY, Satoh H, Lundberg JM. Postjunctional inhibitory effect of the NK2 receptor antagonist, SR 48968, on sensory NANC bronchoconstriction in the guinea-pig. Br J Pharmacol 1993;109:765–773.

256. Hirayama Y, Lei YH, Barnes PJ, Rogers DF. Effects of two novel tachykinin antagonists, FK224 and FK888, on neurogenic airway plasma exudation, bronchoconstriction and systemic hypotension in guinea-pigs in vivo. Br J Pharmacol 1993;108:844–851.

257. Lilly CM, Besson G, Israel E, Rodger IW, Drazen JM. Capsaicin-induced airway obstruction in tracheally perfused guinea pig lungs. Am J Respir Crit Care Med 1994; 149:1175–1179.

258. Malcangio M, Bowery NG. Effect of the tachykinin NK1 receptor antagonists, RP 67580 and SR 140333, on electrically-evoked substance P release from rat spinal cord. Br J Pharmacol 1994;113:635–641.

259. Castairs JF, Barnes PJ. Autoradiographic mapping of substance P receptors in lung. Eur J Pharmacol 1986;127:295–296.
260. Hoover DB, Hancock JC. Autoradiographic localization of substance P binding sites in guinea-pig airways. J Auton Nervous Syst 1987;19:171–174.
261. Goldie RG. Receptors in asthmatic airways. Am Rev Respir Dis 1990;141:S151–S156.
262. Walsh DA, Salmon M, Featherstone R, Wharton J, Church MK, Polak JM. Differences in the distribution and characteristics of tachykinin NK1 binding sites between human and guinea pig lung. Br J Pharmacol 1994;113:1407–1415.
263. Miyayasu K, Mak JC, Nishikawa M, Barnes PJ. Characterization of guinea pig pulmonary neurokinin type 1 receptors using a novel antagonist ligand, [3H]FK888. Mol Pharmacol 1993;44:539–544.
264. Zhang XL, Mak JC, Barnes PJ. Characterization and autoradiographic mapping of [3H]CP96,345, a nonpeptide selective NK1 receptor antagonist in guinea pig lung. Peptides 1995;16:867–872.
265. Black J, Diment L, Armour C, Alouan L, Johnson P. Distribution of substance P receptors in rabbit airways, functional and autoradiographic studies. J Pharmacol Exp Ther 1990;253:381–386.
266. Boni P, Ballati L, Evangelista S. Tachykinin NK1 and NK2 receptors mediate the non-cholinergic bronchospastic response to capsaicin and vagal stimulation in guinea-pigs. J Auton Pharmacol 1995;15:49–54.
267. Satoh H, Lou YP, Lee LY, Lundberg JM. Inhibitory effects of capsazepine and the NK2 antagonist SR 48968 on bronchoconstriction evoked by sensory nerve stimulation in guinea-pigs. Acta Physiol Scand 1992;146:535–536.
268. Bertrand C, Nadel JA, Graf PD, Geppetti P. Capsaicin increases airflow resistance in guinea pigs in vivo by activating both NK2 and NK1 tachykinin receptors. Am Rev Respir Dis 1993;148:909–914.
269. Berman AR, Togias AG, Skloot G, Proud D. Allergen-induced hyperresponsiveness to bradykinin is more pronounced than that to methacholine. J Appl Physiol 1995;78:1844–1852.
270. Guard S, Boyle SJ, Tang KW, Watling KJ, McKnight AT, Woodruff GN. The interaction of the NK1 receptor antagonist CP-96,345 with L-type calcium channels and its functional consequences. Br J Pharmacol 1993;110:385–391.
271. Schmidt AW, McLean S, Heym J. The substance P receptor antagonist CP-96,345 interacts with Ca2+ channels. Eur J Pharmacol 1992;219:491–492.
272. Nagahisa A, Asai R, Kanai Y, Murase A, Tsuchiya Nakagaki M, Nakagaki T, et al. Non-specific activity of (+/−)-CP-96,345 in models of pain and inflammation. Br J Pharmacol 1992;107:273–275.
273. Mizuguchi M, Fujimura M, Amemiya T, Nishi K, Ohka T, Matsuda T. Involvement of NK2 receptors rather than NK1 receptors in bronchial hyperresponsiveness induced by allergic reaction in guinea-pigs. Br J Pharmacol 1996;117:443–448.
274. Lai YL. Endogenous tachykinins in antigen-induced acute bronchial responses of guinea pigs. Exp Lung Res 1991;17:1047–1060.
275. Perretti F, Ballati L, Manzini S, Maggi CA, Evangelista S. Antibronchospastic activity of MEN10,627, a novel tachykinin NK2 receptor antagonist, in guinea-pig airways. Eur J Pharmacol 1995;273:129–135.
276. Forsberg K, Karlsson JA, Theodorsson E, Lundberg JM. Persson CG. Cough and bronchoconstriction mediated by capsaicin-sensitive sensory neurons in the guinea-pig. Pulmon Pharmacol 1988;1:33–39.

277. Yasumistu R, Hirayama Y, Imai T, Miyaysu K, Hiroi J. Effects of specific tachykinin receptor antagonists on citric acid-induced cough and bronchoconstriction in unanesthetised guinea-pigs. Eur J Pharmacol 1996;300:215–219.

278. Dray A, Urban L, Dickenson A. Pharmacology of chronic pain. Trends Pharmacol Sci 1994;15:190–197.

279. Boichot E, Biyah K, Germain N, Emonds-Alt X, Lagente V, Advenier C. Involvement of tachykinin NK1 and NK2 receptors in substance P-induced microvascular leakage hypersensitivity and airway hyperresponsiveness in guinea-pigs. Eur Respir J 1996;9: 1445–1450.

280. Boichot E, Germain N, Lagente V, Advenier C. Prevention by the tachykinin NK2 receptor antagonist, SR 48968, of antigen-induced airway hyperresponsiveness in sensitized guinea-pigs. Br J Pharmacol 1995;114:259–261.

281. Fujimura M, Tsujiura M, Nomura M, Mizuguchi M, Matsuda T, Matsushima K. Sensory neuropeptides are not directly involved in bronchial hyperresponsiveness induced by interleukin-8 in guinea-pigs *in vivo*. Clin Exp Allergy 1996;26:357–362.

282. Girard V, Yavo JC, Emonds-Alt X, Advenier C. The tachykinin NK2 receptor antagonist SR 48968 inhibits citric acid-induced airway hyperresponsiveness in guinea-pigs. Am J Respir Crit Care Med 1996;153:1496–1502.

283. Widdicombe JG. Neurophysiology of the cough reflex. Eur Respir J 1995;8:1193–1202.

284. Sheppard D, Scypinski L. A tachykinin receptor antagonist inhibits and an inhibitor of tachykinin metabolism potentiates toluene diisocyanate-induced airway hyperresponsiveness in guinea pigs. Am Rev Respir Dis 1988;138:547–551.

285. Masson P, Fluckiger JP, Rodger IW. Ozone induced airways hyperresponsiveness in conscious unrestrained guinea pigs: effects of PDE IV inhibitors and neurokinin antagonists. Am J Respir Crit Care Med 1996;153:A627.

286. Garland A, Ray DW, Doerschuk CM, Alger L, Eappon S, Hernandez C, et al. Role of tachykinins in hyperpnea-induced bronchovascular hyperpermeability in guinea pigs. J Appl Physiol 1991;70:27–35.

287. Garland A, Jordan JE, Necheles J, Alger LE, Scully MM, Miller RJ, et al. Hypertonicity, but not hypothermia, elicits substance P release from rat C-fiber neurons in primary culture. J Clin Invest 1995;95:2359–2366.

288. Solway J, Kao BM, Jordan JE, Gitter B, Rodger IW, Howbert JJ, et al. Tachykinin receptor antagonists inhibit hyperpnea-induced bronchoconstriction in guinea pigs. J Clin Invest 1993;92:315–323.

289. Hughes SR, Brain SD. A calcitonin gene-related peptide (CGRP) antagonist (CGRP8-37) inhibits microvascular responses induced by CGRP and capsaicin in skin. Br J Pharmacol 1991;104:738–742.

290. Eglezos A, Giuliani S, Viti G, Maggi CA. Direct evidence that capsaicin-induced plasma protein extravasation is mediated through tachykinin NK1 receptors. Eur J Pharmacol 1991;209:277–279.

291. Xu XJ, Dalsgaard CJ, Maggi CA, Wiesenfeld-Hallin Z. NK-1, but not NK-2, tachykinin receptors mediate plasma extravasation induced by antidromic C-fiber stimulation in rat hindpaw: demonstrated with the NK-1 antagonist CP-96,345 and the NK-2 antagonist Men 10207. Neurosci Lett 1992;139:249–252.

292. Tousignant C, Chan CC, Guevremont D, Brideau C, Hale JJ, MacCoss M, et al. NK2 receptors mediate plasma extravasation in guinea-pig lower airways. Br J Pharmacol 1993;108:383–386.

293. Yoshihara S, Geppetti P, Linden A, Hara M, Chan B, Nadel JA. Tachykinins mediate the potentiation of antigen-induced bronchoconstriction by cold air in guinea-pigs. J Allergy Clin Immunol 1996;97:756–760.

294. Ichinose M, Miura M, Yamauchi H, Kageyama N, Tomaki M, Oyake T, et al. A neurokinin 1-receptor antagonist improves exercise-induced airway narrowing in asthmatic patients. Am J Respir Crit Care Med 1996;153:936–941.

295. Advenier C, Girard V, Naline E, Vilain P, Emonds Alt X. Antitussive effect of SR 48968, a non-peptide tachykinin NK2 receptor antagonist. Eur J Pharmacol 1993;250: 169–171.

296. Girard V, Naline E, Vilain P, Emonds Alt X, Advenier C. Effect of the two tachykinin antagonists, SR 48968 and SR 140333, on cough induced by citric acid in the unanaesthetised guinea-pig. Eur J Pharmacol 1995;8:1110–1114.

297. Robineau P, Petit C, Staczek J, Peglion J, Brion J, Canet E. NK1 and NK2 receptors involvement in capsaicin-induced cough in guinea-pigs. Am J Respir Crit Care Med 1994;149:A186.

298. Ujiie Y, Sekizawa K, Aikawa T, Sasaki H. Evidence for substance P as an endogenous substance causing cough in guinea pigs. Am Rev Respir Dis 1993;148:1628–1632.

299. Karlsson JA, Zackrisson C, Lundberg JM. Hyperresponsiveness to tussive stimuli in cigarette smoke-exposed guinea-pigs: a role for capsaicin-sensitive, calcitonin generelated peptide-containing nerves. Acta Physiol Scand 1991;141:445–454.

300. Lundberg JM, Alving K, Karlsson JA, Matran R, Nilsson G. Sensory neuropeptide involvement in animal models of airway irritation and of allergen-evoked asthma. Am Rev Respir Dis 1991;143:1429–1430.

301. Ollerenshaw SL, Jarvis D, Sullivan CE, Woolcock AJ. Substance P immunoreactive nerves in airways from asthmatics and nonasthmatics. Eur Respir J 1991;4:673–682.

302. Howarth PH, Djukanovic R, Wilson JW, Holgate ST, Springall DR, Polak JM. Mucosal nerves in endobronchial biopsies in asthma and non-asthma. Int Arch Allergy Appl Immunol 1991;94:330–333.

303. Mapp PI, Kidd BL, Gibson SJ, Terry JM, Revell PA, Ibrahim NB, et al. Substance P-, calcitonin gene-related peptide- and C-flanking peptide of neuropeptide Y-immunoreactive fibres are present in normal synovium but depleted in patients with rheumatoid arthritis. Neuroscience 1990;37:143–153.

304. O'Connell F, Springall DR, Moradoghli-Haftvani A, Krausz T, Price D, Fuller RW, et al. Abnormal intraepithelial airway nerves in persistent unexplained cough? Am J Respir Crit Care Med 1995;152:2068–2075.

305. Nieber K, Baumgarten CR, Rathsack R, Furkert J, Oehme P, Kunkel G. Substance P and beta-endorphin-like immunoreactivity in lavage fluids of subjects with and without allergic asthma. J Allergy Clin Immunol 1992;90:646–652.

306. Tomaki M, Ichinose M, Miura M, Hirayama Y, Yamauchi H, Nakajima N, et al. Elevated substance P content in induced sputum from patients with asthma and patients with chronic bronchitis. Am J Respir Crit Care Med 1995;151:613–617.

307. Adcock IM, Peters M, Gelder C, Shirasaki H, Brown CR, Barnes PJ. Increased tachykinin receptor gene expression in asthmatic lung and its modulation by steroids. J Mol Endocrinol 1993;11:1–7.

308. Bai TR, Zhou D, Weir T, Walker B, Hegele R, Hayashi S, et al. Substance P (NK1)- and neurokinin A (NK2)-receptor gene expression in inflammatory airway diseases. Am J Physiol 1995;269:L309–L317.

309. Ichinose M, Nakajima N, Takahashi T, Yamauchi H, Inoue H, Takishima T. Protection against bradykinin-induced bronchoconstriction in asthmatic patients by neurokinin receptor antagonist. Lancet 1992;340:1248–1251.
310. Schmidt D, Jorres RA, Rabe KF, Magnussen H. Reproducibility of airway response to inhaled bradykinin and effect of neurokinin receptor antagonist FK-224 in asthmatic subjects. Eur J Clin Pharmacol 1996;50:269–273.
311. Joos GF, Van Schoor J, Kips JC, Pauwels RA. The effect of inhaled FK224, a tachykinin NK-1 and NK-2 receptor antagonist, on neurokinin A-induced bronchoconstriction in asthmatics. Am J Respir Crit Care Med 1996;153:1781–1784.
312. Van Schoor J, Joos G, Chasson B, Brouard R, Pauwels R. The effect of the oral nonpeptide NK2 receptor antagonist SR48968 on neurokinin A-induced bronchoconstriction in asthmatics. Eur Respir J 1996;9:289s.
313. Fahy JV, Wong HH, Geppetti P, Reis JM, Harris SC, Maclean DB, et al. Effect of an NK1 receptor antagonist (CP-99,994) on hypertonic saline-induced bronchoconstriction and cough in male asthmatic subjects. Am J Respir Crit Care Med 1995;152:879–884.
314. Adcock JJ, Garland LG. The contribution of sensory reflexes and "hyperalgesia" to airway hyperresponsiveness. In: Page CP, Gardiner PJ, eds. Airway Hyperresponsiveness: Is it Really Important for Asthma. Oxford: Blackwell, 1993:234–255.
315. Woolf CJ. An overview of the mechanisms of hyperalgesia. Pulmon Pharmacol 1995;8:161–167.
316. Ferreira SH, Lorenzetti BB, Bristow AF, Poole S. Interleukin-1 beta as a potent hyperalgesic agent antagonized by a tripeptide analogue. Nature 1988;334:698–700.
317. Safieh-Garabedian B, Poole S, Allchorne A, Winter J, Woolf CJ. Contribution of interleukin-1 beta to the inflammation-induced increase in nerve growth factor levels and inflammatory hyperalgesia. Br J Pharmacol 1995;115:1265–1275.
318. Woolf CJ, Safieh Garabedian B, Ma QP, Crilly P, Winter J. Nerve growth factor contributes to the generation of inflammatory sensory hypersensitivity. Neuroscience 1994;62:327–331.
319. McCarson KE, Goldstein BD. Time course of the alteration in dorsal horn substance P levels following formalin: blockade by naloxone. Pain 1990;41:95–100.
320. Kantner RM, Goldstein BD, Kirby ML. Regulatory mechanisms for substance P in the dorsal horn during a nociceptive stimulus: axoplasmic transport vs electrical activity. Brain Res 1986;385:282–290.
321. Kantner RM, Kirby ML, Goldstein BD. Increase in substance P in the dorsal horn during a chemogenic nociceptive stimulus. Brain Res 1985;338:196–199.
322. McCarson KE, Krause JE. NK-1 and NK-3 type tachykinin receptor mRNA expression in the rat spinal cord dorsal horn is increased during adjuvant or formalin-induced nociception. J Neurosci 1994;14:712–720.
323. Noguchi K, Morita Y, Kiyama H, Ono K, Tohyama M. A noxious stimulus induces the preprotachykinin-A gene expression in the rat dorsal root ganglion: a quantitative study using in situ hybridization histochemistry. Brain Res 1988;464:31–35.
324. Galeazza MT, Garry MG, Yost HJ, Strait KA, Hargreaves KM, Seybold VS. Plasticity in the synthesis and storage of substance P and calcitonin gene-related peptide in primary afferent neurons during peripheral inflammation. Neurosci 1995;66:443–458.
325. Garry MG, Hargreaves KM. Enhanced release of immunoreactive CGRP and substance P from spinal dorsal horn slices occurs during carrageenan inflammation. Brain Res 1992;582:139–142.

326. Donnerer J, Schuligoi R, Stein C. Increased content and transport of substance P and calcitonin gene-related peptide in sensory nerves innervating inflamed tissue: evidence for a regulatory function of nerve growth factor in vivo. Neuroscience 1992;49:693–698.

327. Mapp PI, Terenghi G, Walsh DA, Chen ST, Cruwys SC, Garret N, et al. Monoarthritis in the rat knee induces bilateral and time-dependent changes in substance P and calcitonin gene-related peptide immunoreactivity in the spinal cord. Neuroscience 1993;57:1091–1096.

328. Smith GD, Harmar AJ, McQueen DS, Seckl JR. Increase in substance P and CGRP, but not somatostatin content of innervating dorsal root ganglia in adjuvant monoarthritis in the rat. Neurosci Lett 1992;137:257–260.

329. Kangrga I, Randic M. Tachykinins and calcitonin gene-related peptide enhance release of endogenous glutamate and aspartate from the rat spinal dorsal horn slice. J Neurosci 1990;10:2026–2038.

330. Kangrga I, Larew JS, Randic M. The effects of substance P and calcitonin gene-related peptide on the efflux of endogenous glutamate and aspartate from the rat spinal dorsal horn in vitro. Neurosci Lett 1990;108:155–160.

331. Urban L, Naeem S, Patel IA, Dray A. Tachykinin induced regulation of excitatory amino acid responses in the rat spinal cord in vitro. Neurosci Lett 1994;168:185–188.

332. Rusin KI, Ryu PD, Randic M. Modulation of excitatory amino acid responses in rat dorsal horn neurons by tachykinins. J Neurophysiol 1992;68:265–286.

333. Chen L, Huang LY. Protein kinase C reduces $Mg2+$ block of NMDA-receptor channels as a mechanism of modulation. Nature 1992;356:521–523.

334. Kudlacz EM, Logan DE, Shatzer SA, Farrell AM, Baugh LE. Tachykinin-mediated respiratory effects in conscious guinea pigs: modulation by NK1 and NK2 receptor antagonists. Eur J Pharmacol 1993;241:17–25.

335. Yashpal K, Pitcher GM, Parent A, Quirion R, Coderre TJ. Noxious thermal and chemical stimulation induce increases in 3H-phorbol 12,13-dibutyrate binding in spinal cord dorsal horn as well as persistent pain and hyperalgesia, which is reduced by inhibition of protein kinase C. J Neurosci 1995;15:3263–3272.

336. Follenfant RL, Nakamura-Craig M, Garland LG. Sustained hyperalgesia in rats evoked by 15-hydroperoxyeicosatetranoic acid is attenuated by the protein kinase inhibitor H-7. Br J Pharmacol 1990;99:289P.

337. Hylden JL, Noguchi K, Ruda MA. Neonatal capsaicin treatment attenuates spinal Fos activation and dynorphin gene expression following peripheral tissue inflammation and hyperalgesia. J Neurosci 1992;12:1716–1725.

338. Jeanjean AP, Maloteaux JM, Laduron PM. IL-1 beta-like Freund's adjuvant enhances axonal transport of opiate receptors in sensory neurons. Neurosci Lett 1994;177:75–78.

339. McCarson KE, Krause JE. The formalin-induced expression of tachykinin peptide and neurokinin receptor messenger RNAs in rat sensory ganglia and spinal cord is modulated by opiate preadministration. Neuroscience 1995;64:729–739.

340. Linsay RM, Lockett C, Sternberg J, Winder J. Neuropeptide expression in cultures of adult sensory neurons: modulation of substance P and calcitonin gene-related peptide levels by nerve growth factor. Neuroscience 1989;33:53–65.

341. Linsay RM, Harmar AJ. Nerve growth factor regulates expression of neuropeptide genes in adult sensory neurons. Nature 1989;337:362–364.

342. Winter J, Forbes CA, Sternberg J, Lindsay RM. Nerve growth factor (NGF) regulates adult rat cultured dorsal root ganglion neuron responses to the excitotoxin capsaicin. Neuron 1988;1:973–981.

343. Bevan S, Winter J. Nerve growth factor (NGF) differentially regulates the chemosensitivity of adult rat cultured sensory neurons. J Neurosci 1995;15:4918–4926.

344. Lewin GR, Rueff A, Mendell LM. Peripheral and central mechanisms of NGF-induced hyperalgesia. Eur J Neurosci 1994;6:1903–1912.

345. Amann R, Schuligoi R, Herzeg G, Donnerer J. Intraplantar injection of nerve growth factor into the rat hind paw: local edema and effects on thermal nociceptive threshold. Pain 1996;64:323–329.

346. Lewin GR, Ritter AM, Mendell LM. Nerve growth factor-induced hyperalgesia in the neonatal and adult rat. J Neurosci 1993;13:2136–2148.

347. Jonakait GM, Schotland S, Hart RP. Effects of lymphokines on substance P in injured ganglia of the peripheral nervous system. Ann NY Acad Sci 1991;632:19–30.

348. Tsukagoshi H, Sakamoto T, Xu W, Barnes PJ, Chung KF. Effect of interleukin-1 beta on airway hyperresponsiveness and inflammation in sensitized and nonsensitized Brown-Norway rats. J Allergy Clin Immunol 1994;93:464–469.

349. Kummer W, Fischer A. Plasticity of the afferent innervation of the airways. Pulmon Pharmacol 1995;8:169–172.

350. Saito H, Tsukiji J, Ikeda H, Okubo T. Nerve growth factor increases the substance P contents of the adult guinea pig airway and nodose ganglion. Am J Respir Crit Care Med 1996;153:A163.

351. Kalubi B, Takeda N, Irifune M, Ogino S, Abe Y, Hong SL, et al. Nasal mucosa sensitization with toluene diisocyanate (TDI) increases preprotachykinin A (PPTA) and preproCGRP mRNAs in guinea pig trigeminal ganglion neurons. Brain Res 1992;576: 287–296.

352. Lucchini RE, Novelli A, Springall DR, Saetta M, Maestrelli P, Fabbri LM, et al. Immunization and challenge with toluene diisocyanate (TDI) decrease airway sensory neuropeptides in guinea pigs. Am J Respir Crit Care Med 1996;153:A160.

353. Lucchini RE, Springall DR, Chitano P, Fabbri LM, Polak JM, Mapp CE. In vivo exposure to nitrogen dioxide (NO2) induces a decrease in calcitonin gene-related peptide (CGRP) and tachykinin immunoreactivity in guinea-pig peripheral airways. Eur Respir J 1996;9:1847–1851.

354. Kudlacz EM, Shatzer SA, Farrell AM, Baugh LE. Parainfluenza virus type 3 induced alterations in tachykinin NK1 receptors, substance P levels and respiratory functions in guinea pig airways. Eur J Pharmacol 1994;270:291–300.

355. Bowden JJ, Garland AM, Baluk P, Lefevre P, Grady EF, Vigna SR, et al. Direct observation of substance P-induced internalization of neurokinin 1 (NK1) receptors at sites of inflammation. Proc Natl Acad Sci (USA) 1994;91:8964–8968.

356. Bowden JJ, Baluk P, Lefevre PM, Vigna SR, McDonald DM. Substance P (NK$_1$) receptor immunoreactivity on endothelial cells of the rat tracheal mucosa. Am J Physiol Lung Cell Mol Physiol 1996;270:L404–L414.

357. Jung M, Calassi R, Maruani J, Barnouin MC, Souilhac J, Poncelet M, et al. Neuropharmacological characterization of SR 140333, a non peptide antagonist of NK1 receptors. Neuropharmacology 1994;33:167–179.

6

The Tachykinin Family of Peptides and Their Receptors

DOMENICO REGOLI, PIERANGELO GEPPETTI, and GIROLAMO CALÓ

University of Ferrara
Ferrara, Italy

I. The Tachykinin System

The tachykinins and their receptors constitute a biologically important system, which emerged very early in life (see examples of tachykinins in Table 1) and has evolved to the three mammalian tachykinins (also referred to as neurokinins), substance P (SP), neurokinin A (NKA), and neurokinin B (NKB). As for other naturally occurring agents (particularly peptides), the tachykinins [including the peptides of nonmammalian origin (Table 1)] activate three specific targets, the NK-1, NK-2, and NK-3 receptors, despite the fact that SP shows higher affinity for NK-1 than for the others, NKA for NK-2 and NKB for the NK-3 (1).

II. Tachykinin Peptides

The first tachykinin to be discovered, in 1931, was substance P, whose structure, however, was elucidated only in 1971 (2); several other tachykinins were identified (and their structures elucidated) in lower species (especially anphibia) by Erspamer and his co-workers from 1962 to very recently (Table 1). Tachykinins and neurokinins are short peptides consisting of 10 to 13 residues and characterized by the

Table 1 The Tachykinin Family of Peptides

Name	Primary structure	Source	Reference
Substance P	Arg-Pro-Lys-Pro-Gln-Gln-**Phe**-Phe-**Gly-Leu-Met**-NH$_2$	Rabbit	(58)
Neurokinin A	His-Lys-Thr-Asp-Ser-**Phe**-Val-**Gly-Leu-Met**-NH$_2$	Pig	(59)
Neurokinin B	Asp-Met-His-Asp-Phe-**Phe**-Val-**Gly-Leu-Met**-NH$_2$	Pig	(59)
Eledoisin	pGlu-Pro-Ser-Lys-Asp-Ala-**Phe**-Ile-**Gly-Leu-Met**-NH$_2$	Octopus	(60)
Physalaemin	pGlu-Ala-Asp-Pro-Asn-Lys-**Phe**-Tyr-**Gly-Leu-Met**-NH$_2$	Frog	(61)
Phyllomedusin	pGlu-Asn-Pro-Asn-Arg-**Phe**-Ile-**Gly-Leu-Met**-NH$_2$	Frog	(62)
Uperulein	pGlu-Pro-Asp-Pro-Asn-Ala-**Phe**-Tyr-**Gly-Leu-Met**-NH$_2$	Frog	(63)
Kassinin	Asp-Val-Pro-Lys-Ser-Asp-Gly-**Phe**-Val-**Gly-Leu-Met**-NH$_2$	Frog	(64)
PGK II	pGlu-Pro-Asn-Pro-Asp-Glu-**Phe**-Val-**Gly-Leu-Met**-NH$_2$	Frog	(65)

presence of a common core (Phe-X-Gly-Leu-Met-NH$_2$) at the C-terminal end and by different ionic charges at the N-terminal end. In fact, SP is basic (Arg^1, Lys^3), NKA is neutral (Lys^2, Asp^4), and NKB is rather acidic (Asp^1, Asp^4). N-terminal charges appear to be determinant for receptor selectivity, especially the acidic charges for the NK-3 receptor (see apparent affinities of SP and NKB on the NK-3 receptor in Table 2). Two other neurokinins, neuropeptide K (36 amino acids) and neuropeptide γ (24 residues) containing the NKA sequence at the C-terminal end (3), as well as a fragment of NKA [NKA(4-10)], have been identified in normal or pathological tissues (4).

Work performed by molecular biologists in the last 10 years has led to identification of the genes containing the neuropeptides (5) and their receptors (6). The genetic sequences have been coded and expressed in appropriate cell systems. The precursors and their maturation processes have been well defined and various peripheral mechanisms regulating tachykinin release from sensory nerves have been investigated (7) (see also Chapter 9). The biosynthesis of neurokinins is presented schematically in Fig. 1. Mammalian neurokinins originate from two preprotachykinin (PTT) genes, which evolved from the same ancestor gene by duplication (8): the PPT-A gene, which generates only SP (α-PPT-A mRNA), SP, NKA and NPK (β-PPT-A mRNA) or SP, NKA and NPγ (γ-PPT-A mRNA), and the PPT-B gene which encodes NKB (5). The example presented in Fig. 1 shows that one of the gene sections coding

Table 2 Pharmacological Characterization of Neurokinin Receptors with Agonists (Bioassay and Binding Data)

Techniques: Preparations:	NK-1		NK-2		NK-3	
	Bioassay rbVC	Binding gpB	Bioassay rbPA	Binding rDSM	Bioassay rPV	Binding gpB
Natural nonselective agonists (pEC$_{50}$ or pK$_i$)						
SP	8.6	9.4	6.1	6.7	5.8	6.9
NKA	7.3	6.1	8.2	8.1	6.4	7.2
NKB	7.2	6.8	6.4	7.5	7.7	8.1
Synthetic selective agonists (pEC$_{50}$or pK$_i$)						
[Sar^9,Met(O)$_2^{11}$]SP	8.6	ND	Inactive	ND	Inactive	ND
Ac[Arg^6,Sar^9,Met(O)$_2^{11}$]- SP(6–11)	8.7	8.7	Inactive	ND	Inactive	4.7
[βAla^8]NKA(4–10)	<5.0	3.8	8.6	ND	6.1	4.9
[Nle^{10}]NKA(4–11)	5.4	ND	8.0	ND	6.5	ND
[MePhe^7]NKB	5.6	5.2	5.2	ND	8.3	9.0
Senktide	5.5	6.2	5.5	ND	7.6	8.5

rbVC, rabbit vena cava; gpB, guinea pig brain membranes; rbPA, rabbit pulmonary artery; rDSM, rat duodenum smooth muscle membranes; rPV, rat portal vein. pEC$_{50}$, −log of the concentration of agonist producing 50% of the maximal response. pK, −log of the affinity constant of the ligand. ND, not determined.

for neurokinins is approximately 8.4 kb long and consists of seven exons separated by six introns (5). Exon 2-7 contains the information for the β-PPT-A. The large sequence of the precursor includes the 5′-untranslated regions, consisting of a signal peptide, a spacer sequence, the neurokinin sequences, and the carboxy-terminal portion (Fig. 1). The signal sequence starts with Met-Lys and is followed by a chain of 11 hydrophobic residues and a hydrophilic processing site. Neurokinins are released from their precursors by the actions of proteases called convertases (9), followed by C-terminal amidation (10). Cleavage points for the convertases in the PPT are doublets of cationic residues such as Lys-Arg, Arg-Arg, and Arg-Lys. The C-terminal amide is obtained by the action of peptidyl-Gly-α-amidating monoxygenase using Gly as amide donor (11). Mature (biologically active) neurokinins are stored in granules in central and peripheral nerve terminals as well as in some intestinal cells and in endothelia (see Ref. 5 for a review), from which they are released by a variety of stimuli (see Chapter 9). A mixture of peptides, including calcitonin gene-related peptide (CGRP), is released from the neurokinin-producing neurons following cellular stimulation. The major metabolic pathways of mammalian neurokinins have been analyzed in recent reviews (1,12).

Biosynthesis of neurokinins

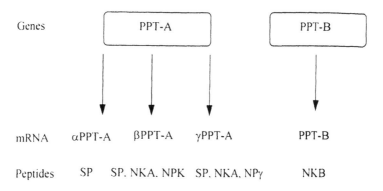

The β PPT-A gene encodes substance P (exon 3), NKA (exon 6) and neuropeptide K (exon 4, 5 and 6)

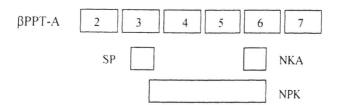

Figure 1 Schematical representation of neurokinin biosynthesis in mammals.

III. Tachykinin Receptor Types

The existence of multiple receptors for neurokinins was proposed at the beginning of the 1980s, based on pharmacological and biochemical findings (13–15). The initial proposal emerging from data obtained in the guinea pig ileum (a multiple-receptor preparation) and the rat vas deferens (an NK-2 monoreceptor system) (13) showed important discrepancies between pharmacological and biochemical findings (see Ref. 1 for an extensive discussion of this topic). However, the discovery and characterization of three monoreceptor systems (16) was instrumental in demonstrating that the functional site preferentially activated by substance P [dog carotid artery (17), rabbit vena cava (18)] is analogous to the binding site detected by [^{125}I]Bolton Hunter SP (BH-SP) (19) in rat brain homogenates. The characteristics of the binding site of

[^{125}I]BH-ELE in the rate brain (20) is similar to the contractile receptor preferentially activated by NKB in the rat portal vein (21). A consensus on neurokinin receptor nomenclature was reached at the Montreal Symposium (22), where three functional and binding sites, namely, NK-1, NK-2, and NK-3, were well defined.

Functional and binding data obtained with agonists in pharmacological and biochemical assays performed on the three receptors are presented in Table 2. These results, expressed in terms of apparent affinities of agonists (pEC$_{50}$), indicate that the most active naturally occurring agonist on the NK-1 receptor of the rabbit vena cava (rbVC) is SP, while the most active natural stimulant of the NK-2 in the rabbit pulmonary artery (rbPA) is NKA, and NKB is the most active on the NK-3 receptor of the rat portal vein (rPV). Results shown in Table 2 also indicate that the three neurokinins are potent stimulants of the NK-1 receptor and therefore they show poor selectivity (see also NKA and NKB on the NK-2 and NK-3 receptors). To improve selectivity, a series of synthetic compounds was prepared (23) and tested in a variety of biological assays (16,24). Data presented in Table 2 indicate the high selectivity of certain compounds for the NK-1, NK-2, and NK-3 receptors. Compounds reported in Table 2 were tested as competitors of selective receptor ligands, namely, [^{125}I]BH-SP for NK-1, [^{3}H]NKA for NK-2, and [^{125}I]BH-ELE for the NK-3 receptors. The comparison between pharmacological and biochemical findings reported in Table 2 shows a good correlation between the two set of data, suggesting that pharmacological and biochemical assays evaluate identical biological entities. NK-1, NK-2, and NK-3 receptors were then classified using the order of potency of agonists as follows:

NK-1: SP > NKA > NKB   [Sar^9,Met(O)$_2$^{11}]SP > [MePhe^7]NKB > [βAla^8]NKA(4-10)

NK-2: NKA > NKB > SP   [βAla^8]NKA(4-10) > [MePhe^7]NKB > [Sar^9,Met(O)$_2$^{11}]SP

NK-3: NKB > NKA > SP   [MePhe^7]NKB > [βAla^8]NKA(4-10) > [Sar^9,Met(O)$_2$^{11}]SP

Neurokinin receptors were further characterized with antagonists, initially by using peptide sequences modified in various positions such as spantide 1 or spantide 2 (25, 26). The discovery of antagonists of small molecular size such as modified hexa (e.g., R486), trypeptides (e.g., R820), or nonpeptide compounds (e.g., CP99994, SR140333) has provided adequate tools for classifying the receptors with selective and specific antagonists. Some of the compounds presented in Table 3 have been used in in-vitro or in-vivo pharmacological assays in laboratory animals. Results obtained in vitro in the three monoreceptor systems, already utilized in studies with agonists (see Table 2), are presented in Table 3. In each receptor system, nonpeptide antagonists, especially the Sanofi compounds, SR140333 (anti NK-1) (27), SR48968 (anti NK-2) (28), and SR142801 (anti NK-3) (29) have been found to posses high affinities (see pA$_2$ values in Table 3), long duration of action, and remarkable selectivity (compare affinities in the three functional assays in Table 3). Specificity of each compound for its receptor has been established by a series of comparative binding assays on receptors for a large variety of neurotransmitters, hormones or autacoids (27–29). The existence of

Table 3 Pharmacological Characterization of
Neurokinin Receptors with Antagonists (Bioassay Data)

Preparations: Agonist:	NK-1 rbVC SP	NK-2 rbPA NKA	NK-3 rPV [MePhe^7]NKB
Antagonists (pA$_2$)			
CP 99994	8.9	Inactive	Inactive
SR 140333	9.8	5.8	5.8
LY 303870	9.4	4.7	4.7
RP 67580	7.2	<5.0	4.5
SR 48968	6.1	9.8	<5.0
GR 94800	6.9	8.2	6.0
MEN 10376	Inactive	7.2	<5.0
SR 142801	6.4	7.1	9.0
R 486	Inactive	Inactive	7.5
R 820	5.4	<5.0	7.6

rbVC, rabbit vena cava; rbPA, rabbit pulmonary artery; rPV, rat
portal vein. R 486, Asp-Ser-Phe-Trp-βAla-Leu-Met-NH$_2$; R 820,
3-indolylcarbonyl-Hyp-Phg-N(Me)-Bzl; for chemical structure of
the other antagonists see Ref. 1. pA$_2$, −log of the concentration of
antagonist that reduce the effect of a double dose of agonist to that
of a single dose.

three different functional sites mediating the biological actions of neurokinins is thus confirmed unequivocally by results obtained with antagonists, especially the new, nonpeptide, orally active compounds, which discriminate among the three receptors (Table 3) by showing apparent affinities that differ by at least two log units.

The existence of the three neurokinin receptors has been definitely proved by the identification, isolation, and transfection (in CHO or COS cells) of the cDNA coding for each of the three receptors, either of rat (see Ref. 6 for a review) or human (see Ref. 30 for a review) origin. The three receptors belong to rhodopsin-like membrane structures, consisting of seven hydrophobic transmembrane domains, connected by extra- and intracellular loops and coupled to G proteins. The secondary structure of the human NK-1 receptor is illustrated in Fig. 2. When characterized by binding to transfected CHO cells [either intact cells or plasma membrane preparations (31,32)] using classical neurokinin receptor ligands, three different pharprofiles were observed (Table 4), which are similar to those obtained in classic pharmacological or binding assays (Table 2). pK$_i$ values reported in Table 4 should be considered as accurate measurements of actual affinities for the naturally occurring agonists and for the nonpeptide antagonists. Interesting enough, all values for agonists are between 9.5 and 9.8, and the same is true for the antagonists (9.6–9.7). Again, the neurokinins show less selectivity than the antagonists, as already indicated by the data obtained in biological assays. Taken together, the data of pharmacological, biochemical (bind-

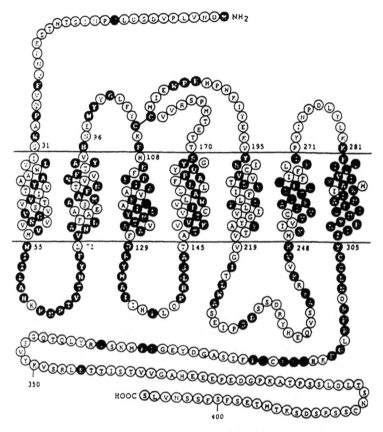

Figure 2 Molecular structure of the human NK-1 receptor. The filled circles indicate the amino acid residues which are conserved between human NK-1, NK-2, and NK-3 receptors, the open circles the residues which are divergent.

ing), and molecular biological assays are consistent with the idea that the three neurokinins and the other naturally occurring peptides (e.g., NPK, NPγ) exert their biological effects by activating three distinct receptor types, NK-1, NK-2, and NK-3. These rhodopsin-like structures are expected to mediate definite biological effects in each living organism whose genome contains the three specific sequences and whose neurons or peripheral target cells express such sequences as functional membrane receptors.

IV. Species-Related Tachykinin Receptor Subtypes

Work with antagonists has revealed differences between species and has led to the identification of receptor subtypes. In fact, each neurokinin receptor can be subdi-

Table 4 Affinities of Neurokinin
Receptor Agonists and Antagonists
Evaluated by Molecular Biology Approach

	NK-1	NK-2	NK-3
Agonists (pK_i)			
SP	9.8	7.0	7.0
NKA	7.7	9.6	7.7
NKB	7.3	8.3	9.5
Antagonists (pK_i)			
SR 140333	9.6	7.0	7.5
SR 48968	6.1	9.6	6.4
SR 142801	6.0	6.0	9.7

pK_i, $-$log of the affinity constant of the ligand.
These data, obtained in Chinese hamster ovary
cells (CHO) or in CV1-origin SV40 cells, are from
Refs. 1 and 46.

vided into at least two subtypes when comparing affinities of ligands, especially antagonists. Results obtained with neurokinin receptor agonists and antagonists in six in-vitro preparations (two for each receptor) from different animals are summarized in Table 5. The data have been taken from recent publications from our laboratory, and the reader is referred to these papers (1,33,34) for details about the experimental protocols or the specific treatments applied to the various preparations to select a single functional site.

Two NK-1 receptor preparations, the rbVC and the rat urinary bladder (rUB), two NK-2 receptor preparations, the rbPA and the hamster urinary bladder (hsUB), as well as two NK-3 receptor preparations, the rPV and the guinea pig ileum (gpI), have been investigated (Table 5) using agonists and antagonists. Despite some remarkable differences in the apparent affinities of agonists, the order of potency of agonists is similar in all the species-related subtypes; therefore, agonists do not discriminate between different species-related functional sites. Antagonists, on the other hand, not only show marked (generally greater than agonists) differences in affinity values between species, they also show opposite order of potencies (Table 5). For instance, CP 99994 is 2.9 log units more active in the rabbit compared to the rat NK-1 receptor, while RP 67580 is 1.0 log unit more active in the rat than the rabbit NK-1 receptor site. A similar difference was observed with the two NK-2 antagonists on the NK-2 receptor preparations. These observations support the hypothesis that the receptor proteins mediating the same biological effects in various species may differ in amino acid composition or disposition at the active site (where small, nonpeptide antagonists are expected to interact) and bind antagonists more or less efficiently. An illustrative example is provided by the interaction of CP 96345 with the human (pIC_{50} 9.4) and the rat (pIC_{50} 7.6) NK-1 receptor (35,36). The affinity of this compound for

Table 5 Neurokinin Receptor Subtypes in Different Species (Bioassay Data)

Preparations:	NK-1		NK-2		NK-3	
	rbVC	rUB^t	rbPA	hsUB	rPV	gpI^t
Agonists (pEC$_{50}$)						
SP	8.6	7.1	6.1	5.6	5.8	7.4
NKA	7.3	5.8	8.2	7.4	6.4	7.0
NKB	7.2	5.7	7.4	7.2	7.7	8.4
[Sar^9,Met(O)$_2$^{11}] SP	8.6	8.0	Inactive	Inactive	Inactive	7.2
Ac[Arg^{6}Sar^9,Met(O)$_2$^{11}]- SP(6–11)	8.7	ND	Inactive	Inactive	Inactive	6.6
[βAla^8]NKA(4–10)	<5.0	<5.0	8.6	7.2	6.1	5.8
[Nle^{10}]NKA(4–10)	5.4	5.2	8.0	7.1	6.5	5.8
[MePhe^7]NKB	5.6	Inactive	5.2	6.1	8.3	9.1
Senktide	Inactive	6.3	5.4	5.7	7.6	9.2
Antagonists (pA$_2$)						
CP 99994	8.9	6.0	Inactive	Inactive	Inactive	5.0
RP 67580	7.2	8.2	<5.0	<5.0	<5.0	5.3
SR 48968	6.1	5.9	9.8	7.6	<5.0	5.8
R 396	Inactive	<5.0	5.6	7.5	<5.0	Inactive
R 820	5.4	5.4	<5.0	ND	7.6	Inactive
R 486	Inactive	5.5	Inactive	PA	7.5	6.1

rbVC, rabbit vena cava; rUB^t, rat urinary bladder treated with SR 48968, 1.7 μM; rbPA, rabbit pulmonary artery; haUB, hamster urinary bladder; rPV, rat portal vein; gpI^t, guinea pig ileum treated with CP 99994, 1 μM, and SR 48968, 1.7 μM. ND, not determined; PA partial agonist. pEC$_{50}$, −log of the concentration of agonist producing 50% of the maximal response. pA$_2$, −log of the concentration of antagonist that reduce the effect of a double dose of agonist to that of a single dose.

the human receptor is increased by the presence of a His residue in position 172 (close to the active site), which is replaced by a Ile in the rat receptor (36).

V. Mechanisms of Tachykinin Actions

It is assumed that all the biological effects of neurokinins are mediated by receptors, with the possible exception of the release of histamine from rat peritoneal mast cells (37), which apparently derives from a direct interaction of the cationic peptides (especially SP) with a mastocyte G protein (see Ref. 38 for a review of this topic). NK-1, NK-2, and NK-3 receptors are coupled to G proteins, probably at the level of the third intracellular loop (1,39), which contains several cationic residues (Arg, Lys) and may therefore interact with the anionic groups of the G protein. Such a mechanism has already been suggested to interpret results obtained with other receptors (40,41). Neurokinin receptors, especially NK-1 receptors, have been shown to interact with different G proteins and activate more than one second messenger. Indeed, recent studies have shown that neurokinin-related peptides interacting with the NK-1 receptor stimulate (a) phosphatidyl inositol hydrolysis in a variety of target cells (42–44); (b) cAMP formation in other cells (42,43), and (c) activation of phospholipase A_2 with the consequent release of prostaglandins in other cells. Activation of more than one second messengers is induced by NK-1 receptors even in the same cell, namely, the CHO cells transfected with the human NK-1 receptor (45). From these facts it can be concluded that the second messenger cannot be taken as a valid criterion for neurokinin receptor classification, since the three receptors (NK-1, NK-2, and NK-3) are able to utilize the same second messengers (e.g., IP_3, cAMP, prostaglandins), and each receptor can interact with different G proteins depending on target cells.

VI. Receptors for Tachykinins in Target Cells

Previous review articles (1,46,47) have reported in detail the distribution of tachykinin receptors in different effector cells. Cells that classically express tachykinin receptors include smooth muscle, exocrine gland, epithelial and endothelial cells, as well as peripheral and central nervous system neurons. Inflammation may be profoundly affected by activation of tachykinin receptors on these cells. For example, NK-1 receptors on arteriole endothelial cells increase blood flow (48) and on postcapillary venule endothelial cells promote plasma leakage (49). In addition to these kinds of classical proinflammatory actions, tachykinins via activation of specific receptors may stimulate directly the function of inflammatory and immunocompetent cells. These effects of tachykinins comprise release of oxygen radicals, cytokines, eicosanoids, proteases, and other mediators. Activation of these pathways may lead to inflammatory mechanisms that are more prolonged than those usually promoted by tachykinins when they act on effector cells, such as smooth muscle cells or endothelia. Participation of tachykinins and tachykinin receptors to the development of delayed forms of inflammation is regarded as a possible new target for the therapeutic application of tachykinin receptor antagonists.

Table 6 summarizes information regarding the presence of tachykinin receptors on inflammatory and immunocompetent cells. Most of the early studies were performed before the discovery of selective nonpeptide antagonists; thus, attribution of effects produced by SP or other tachykinins to the activation of one of the three tachykinin receptors is sometimes not feasible. In these cases the acronym SP-R (substance-P receptor) has been adopted, in order to underline uncertainty regarding the identity of the receptor that mediates the response. The existence of a "nonclassical" neurokinin receptor has been proposed on human monocytes (50). This proposal was based on the ability of radiolabeled SP to bind with low affinity to monocyte membrane preparations, the observation that SP binding did not show any pharmacological characteristics of the NK-1, NK-2, or NK-3 receptors, and an SP-induced activation of MAP kinase. Definitive proof of the existence of such a

Table 6 Tachykinin Receptors on Inflammatory or Immunocompetent Cells

Cell type	Receptor	Effect (release, binding, expression, activation)	Species	Reference
Mast cell	NK-1	TNF-α/mRNA	Mouse	(66)
Mast cell	NK-1	5-HT	Rat (F-344)	(51)
Mast cell	G protein	Histamine	Rat, human	(38,67)
Monocyte	SP-R	TNF-α	Human	(68)
Monocyte/macrophage	NK-1	TNF-α	Human	(69)
Macrophage	NK-1	Prostanoids	Guinea pig	(70)
Macrophage	NK-2	Gelatinase	Guinea pig	(71)
Macrophage	NK-2	O_2^-	Guinea pig	(72)
Macrophage	SP-R (nonclassical)	MAP kinase	Human	(50)
IM-9 Lymphoblast	SP-R	^{125}I-SP	Human	(54)
Lymphocyte T (cell line)	NK-1	Il-2	Mouse	(73)
Lymphocyte T (cell line)	SP-R	IL-2	Mouse	(74)
Lymphocyte B (cell line)	SP-R	IgA, IgM	Mouse	(75)
Eosinophil	NK-1	IFN-γ	Mouse	(76)
Microglia cell	SP-R (nonclassical)	TNF-γ, IL-1	Rat	(77)
Neuroglial cell	SP-R	TNF-α	Rat	(78)

SP-R, SP receptor noncharacterized with selective antagonists; SP-R (nonclassical), effect produced by SP and probably not mediated by NK-1, NK-2, or NK-3 receptor activation.

receptor is lacking, and the physiological relevance, if any, of these findings is unknown. Mast cells from different tissues present diverse phenotypes. Peptides, including SP, containing positively charged amino acids release mediators (histamine) via a nonreceptor mechanism that is probably due to direct activation of G proteins (38). This observation has been documented in skin, peritoneal mast cells, and other mast cells. However, tachykinins may release histamine from lung mast cells via NK-1 receptor activation in Fisher 344 rats (51).

Much evidence has been reported on the presence of tachykinin receptors on macrophages (see Table 6). Either mouse or human cells of the monocyte/macrophage lineage respond to tachykinin challenge with release of different mediators, including O_2^-, TNF-α, and gelatinase, by activation of both NK-1 and NK-2 receptors. The pathophysiological significance of these observations is at present unclear. However, recent findings showing that macrophages may produce SP (52), and that macrophagic SP participates into inflammation induced by deposition of antigen and antibodies complexes (53) suggest that tachykinins released from inflammatory cells may be important in the regulation of inflammation, possibly through autocrine mechanisms. Certain lymphoblastic cell lines, including human IM-9 lymphoblasts, express NK-1 receptors (54). Various reports claimed the presence of tachykinin receptors on polymorphonuclear leukocytes or lymphocytes, whereas other findings did not confirm these observations. Nevertheless, effects of tachykinins on B and T lymphocytes have been documented (55), and endogenous tachykinins seem to play a role on antibodie production in different models of disease (56,57).

VII. Conclusions

The tachykinin system appeared very early in evolution and remains an important functional entity in mammals. The system consists of three neurokinins, SP, NKA, and NKB, which show high affinities for NK-1, NK-2, and NK-3 receptors, respectively. Peptides and receptors have been cloned, and their precursors as well as their molecular structure are known. Receptors are rhodopsin-like membrane proteins (with seven hydrophobic transmembrane domains) coupled to G proteins. NK-1, NK-2 and NK-3 receptor types, when characterized pharmacologically in various species, show differences in their pharmacological spectrum, especially with antagonists, and have been subdivided into species-related subtypes. NK-1, NK-2, and NK-3 receptors have been characterized and classified by the use of selective agonists and antagonists; they are able to activate more than one second messenger (IP$_3$, cAMP, prostaglandins), depending on the target cell. The three neurokinins receptors are found in endothelia, vascular smooth muscles, other smooth muscles, epithelia, neurons, blood cells—where they mediate a large variety of biological effects, in particular nervous transmission both motor and sensory (pain), cardiovascular reflexes, vasodilatation, increase of vascular permeability, venoconstriction, central regulation of blood pressure, release of several neurotransmitters both in the central nervous system and in peripheral tissues, and modulation of the immunological responses though actions on inflammatory and immunocompetent cells. The various biological actions confer to neurokinins and their receptors a very important role as a proinflammatory system, especially in the airways, in the skin, and in the intestinal and urogenital tracts. Selective as well as ambivalent (anti-NK-1 and -NK-2— antagonists of neurokinin receptors are now available for further characterization of the proinflammatory role of these peptides. Several orally active antagonists possessing favorable pharmacodynamic and pharmacokinetic features for human applications are at present in clinical trial.

References

1. Regoli D, Boudon A, Fauchere JL. Receptors and antagonists for substance P and related peptides. Pharmacol Rev 1994;46:551–599.
2. Chang MM, Leeman SE, Niall HD. Amino acid sequence of substance P. Nature 1971;232:86–87.
3. Tatemoto K, Lundberg JM, Jornvall H, Mutt V. Neuropeptide K: isolation, structure and biological activities of a novel brain tachykinin. Biochem Biophys Res Commun 1985; 128:947–953.

4. Theodorsson-Norheim E, Jornvall H, Andersson M, Norheim I, Oberg K, Jacobsson G. Isolation and characterization of neurokinin A, neurokinin A (3-10) and neurokinin A (4-10) from a neutral water extract of a metastatic ileal carcinoid tumor. Eur J Biochem 1987;166:693–698.

5. Nakanishi S. Substance P precursor and kininogen: their structures, genes organization and regulation. Physiol Rev 1987;67:1117–1142.

6. Nakanishi S. Mammalian tachykinin receptors. Annu Rev Neurosci 1991;14:123–136.

7. Bertrand C, Geppetti C. Tachykinin and kinin receptor antagonists: therapeutic perspectives in allergic airway disease. Trends Pharmacol Sci 1996;17:255–259.

8. Kotani H, Hoshimaru M, Nawa H, Nakanishi S. Structure and gene organization of bovine neuromedin K precursor. Proc Natl Acad Sci (USA) 1986;83:7074–7078.

9. Chretien M, Sikstrom R, Lazure C, et al. Expression of the diversity of neural and hormonal peptides via the cleavage of precursor molecules. In: Martinez J, ed. Peptide hormones as prohormones. London: Ellis Horwood, 1989:1–24.

10. Harris RB. Processing of prohormone precursor proteins. Arch Biochem Biophys 1989;275:315–333.

11. Kream RM, Schenfeld TA, Mancuso R, Clancy AN, el-Bermani W, Macrides F. Precursor forms of SP in nervous tissue: detection with antisera to SP, SP-Gly and SP-Gly-Lys. Proc Natl Acad Sci (USA) 1985;84:4832–4836.

12. Mussap CJ, Geraghty DP, Burcher E. Tachykinin receptors: a radioligand binding perspective. J Neurochem 1993;60:1987–2009.

13. Lee CM, Iversen CL, Hanley MR, Sandberg BE. The possible existence of multiple receptors for substance P. Naunyn Schmiedebergs Arch Pharmacol 1982;318:281–287.

14. Buck SH, Burcher E, Shults CW, Lovenberg W, O'Donohue TL. Novel pharmacology of substance K-binding sites: a third type of tachykinin receptor. Science 1984;226:987–989.

15. Regoli D, D'Orleans-Juste P, Drapeau G, Dion S, Escher E. Pharmacological characterization of substance P antagonists. In: Hakanson R, Sundler F, eds. Tachykinin antagonists. Amsterdam: Elsevier Scientific, 1985.

16. Regoli D, Drapeau G, Dion S, Couture R. New selective agonists for neurokinin receptors: pharmacological tools for receptor characterization. Trends Pharmacol Sci 1988;9:290–295.

17. Couture R, Regoli D. Mini review: smooth muscle pharmacology of substance P. Pharmacology 1982;24:1–25.

18. Nantel F, Rouissi N, Rhaleb NE, Jukic D, Regoli D. Pharmacological evaluation of the angiotensin, kinin and neurokinin receptors on the rabbit vena cava. J Cardiovasc Pharmacol 1991;18:398–405.

19. Cascieri MA, Liang T. Characterization of the substance P receptor in rat brain cortex membranes and the inhibition of radioligand binding by guanine nucleotides. J Biol Chem 1983;258:5158–5164.

20. Buck SH, Burcher E. The tachykinins: a family of peptides with a brood of "receptors." Trends Pharmacol Sci 1986;7:65–68.

21. Mastrangelo D, Mathison R, Huggel HJ, et al. The rat isolated portal vein: a preparation sensitive to neurokinins, particularly neurokinin B. Eur J Pharmacol 1987;134:321–326.

22. Henry JL, Couture R, Cuello AC, Pelletier G, Quirion R, Regoli D. Montreal Symposium. Substance P and neurokinins. New York: Springer-Verlag, 1987.

23. Drapeau G, D'Orleans-Juste P, Dion S, Rhaleb NE, Rouissi N, Regoli D. Selective agonists for substance P and neurokinin receptors. Neuropeptides 1987;10:43–54.

24. Dion S, D'Orleans-Juste P, Drapeau G, et al. Characterization of neurokinin receptors in various isolated organs by the use of selective agonists. Life Sci 1987;41:2269–2278.

25. Folkers K, Hakanson R, Horig J, Xu JC, Leander S. Biological evaluation of substance P antagonists. Br J Pharmacol 1984;83:449–456.

26. Folkers K, Feng DM, Asano N, Hakanson R, Weisenfeld-Hallin Z, Leander S. Spantide II, an effective tachykinin antagonist having high potency and negligible neurotoxicity. Proc Natl Acad Sci (USA) 1990;87:4833–4835.

27. Emonds-Alt X, Doutremepuich JD, Heaulme M, et al. In vitro and in vivo biological activities of SR 140333, a novel potent non-peptide tachykinin NK-1antagonist. Eur J Pharmacol 1993;250:403–413.

28. Emonds-Alt X, Vilain P, Goulaouic P, et al. A potent and selective non-peptide antagonist of the neurokinin A (NK-2) receptor. Life Sci 1992;50:PL101–PL106.

29. Emonds-Alt X, Bichon D, Ducoux JP, et al. SR 142801, the first potent non-peptide antagonist of the tachykinin NK3 receptor. Life Sci 1995;56:PL27–PL32.

30. Gerard NP, Bao L, Ping HX, Gerard C. Molecular aspects of the tachykinin receptors. Regulatory Peptides 1993;43:21–35.

31. Yokota Y, Sasai Y, K.T., et al. Molecular characterization of a functional cDNA for rat substance P receptor. J Biol Chem 1989;264:17649–17652.

32. Shigemoto R, Yokota Y, Tsuchida K, Nakanishi S. Cloning and expression of a rat neuromedin K receptor cDNA. J Biol Chem 1990;265:623–628.

33. Nguyen-Le XK, Nguyen QT, Gobeil F, Jukic D, Chretien L, Regoli D. Neurokinin receptors in the guinea pig ileum. Pharmacology 1996;52:35–45.

34. Nguyen-Le XK, Nguyen QT, Emonds-Alt X, et al. Pharmacological characterization of SR142801, a new non peptide antagonist of the neurokinin NK-3 receptor. Pharmacology 1996;52:283–291.

35. Fong TM, Yu H, Strader CD. Molecular basis for the species selectivity of the neuro-kinin-1 receptor antagonists CP 96345 and RP 67580. J Biol Chem 1992;267:25668–25671.

36. Fong TM, Cascieri MA, Yu H, Bansal A, Swain C, Strader CD. Amino aromatic interaction between histidine 197 of the neurokinin-1 receptor and CP 96345. Nature 1993;362:350–353.

37. Devillier P, Drapeau G, Renoux M, Regoli D. Role of the N-terminal arginine in the histamine releasing activity of substance P, bradykinin and related peptides. Eur J Pharmacol 1989;168:53–60.

38. Mousli M, Bueb JL, Bronner C, Rouot B, Landry Y. G protein activation: a receptor-independent mode of action for cationic amphiphilic neuropeptides and venom peptides. Trends Pharmacol Sci 1990;11:358–362.

39. Regoli D, Nantel F. Direct activation of G-proteins. Trends Pharmacol Sci 1990;11:400–401.

40. Kasugi S, Okajima F, Ban T, Hidaka A, Shenker A, Kohn LD. Mutation of alanine 623 in the third cytoplasmatic loop of the rat thyrotropin (TSH) receptor results in a loss in the phosphoinositide but not cAMP signal induced by TSH and receptor autoantibodies. J Biol Chem 1992;267:24153–24156.

41. Spengler D, Woeber C, Pantaloni C, et al. Differential signal transduction by five splice varians of the PACAP receptor. Nature 1993;365:170–175.

42. Guard S, Watson SP. Tachykinin receptor types: classification and membrane signalling mechanisms. Neurochem Int 1991;18:149–165.

43. Nakajima Y, Tsuchida K, Negishi M, Ito S, Nakanishi S. Direct linkage of three tachy-

kinin receptors to stimulation of both phosphatidyl-inositol hydrolysis and cyclic AMP cascade in transfected chinese hamster ovary cells. J Biol Chem 1992;267:2437–2442.

44. Otsuka M, Yoshioka K. Neurotransmitter functions of mammalian tachykinins. Physiol Rev 1993;73:229–308.

45. Sagan S, Chassaing G, Pradier L, Lavielle S. Tachykinin peptides affect differently the second messenger pathways after binding to CHO-expressed human NK-1 receptors. J Pharmacol Exp Ther 1996;276:1039–1048.

46. Longmore J, Swain CJ, Fill RG. Neurokinin receptors. DN&P 1995;8:5–23.

47. Maggi CA. The mammalian tachykinin receptors. Gen Pharmacol 1995;26:911–944.

48. Piedimonte G, Hoffman JE, Husseini WK, Hiser WL, Nadel JA. Effect of neuropeptides released from sensory nerves on blood flow in the rat airway microcirculation. J Appl Physiol 1992;72:1563–1570.

49. Bowden JJ, Garland AM, Baluk P, et al. Direct observation of substance P-induced internalization of neurokinin1 (NK$_1$) receptors at sites of inflammation. Proc Natl Acad Sci (USA) 1994;91:8964–8968.

50. Jeurissen F, Kavelaars A, Korstjens M, et al. Monocytes express a non-neurokinin substance P receptor that is functionally coupled to a MAP kinase. J Immunol 1994;152:2987–2993.

51. Joos GF, Kips JC, Pauwels RA. In vivo characterization of the tachykinin receptor involved in the direct and indirect bronchoconstrictor effect of tachykinins in two inbred rat strains. Am J Respir Crit Care Med 1994;149:1160–1166.

52. Bost KL, Breeding SA, Pascual DW. Modulation of the mRNAs encoding substance P and its receptor in rat macrophages by LPS. Regional Immunol 1992;4:1988–1992.

53. Bozic CR, Lu B, Hopken UE, Gerard C, Gerard NP. Neurogenic amplification of immune complex inflammation. Science 1996;273:1722–1725.

54. Payan DG, Brewster DR, Goetzl EJ. Stereospecific receptors for substance P on cultured human IM-9 lymphoblasts. J Immunol 1984;133:3260–3265.

55. Bost KL, Pascual DW. Substance P: a late-acting B lymphocyte differentiation cofactor. Am J Physiol 1992;262:C537–C545.

56. Helme RD, Eglezos A, Dandie GW, Andrews PV, Boyd RL. The effect of substance P on the regional lymph node antibody response to antigenic stimulation in capsaicin-pretreated rats. J Immunol 1987;139:3470–4373.

57. Scicchitano R, Bienenstock J, Stanitz AM. In vivo immunomodulation by the neuropeptide substance P. Immunology 1988;63:733–735.

58. Euler US, Gaddum JH. An unidentified depressor substance in certain tissue extracts. J Physiol 1931;72:74–87.

59. Kimura S, Okada M, Sugita Y, Kanazawa I, Munekata E. Novel neuropeptides, neurokinin α and β, isolated from porcine spinal cord. Proc Jpn Acad Sci 1983;(B)59:101–104.

60. Erspamer V, Anastasi A. Structure and pharmacological actions of eledoisin, the active endecapeptide of the posterior salivary glands of *Eledone*. Experientia 1962;18:58–59.

61. Erspamer V, Anastasi A, Bertaccini G, Cei JM. Structure and pharmacological actions of physalaemin, the main active polypeptide of the skin of *Physalaemus fuscumaculatus*. Experientia 1964;20:489–490.

62. Anastasi A, Falconieri-Erspamer G. Occurrence of phyllomedusin, a physalaemin-like decapeptide in the skin of *Phyllomedusa bicolor*. Experientia 1970;26:866–867.

63. Anastasi A, Erspamer V. Endean R. Structure of uperolein, a physalaemin-like endecapeptide occurring in the skin of *Uperoleia rugosa* and *Uperoleia marmorata*. Experientia 1978;31:393–395.

64. Anastasi A, Montecucchi PC, Erspamer V, Visser J. Aminoacid composition and sequence of kassinin, a tachykinin from the skin of the african frog *Kassina senegalensis*. Experientia 1977;33.

65. Simmaco M, Severini C, DeBiase D, et al. Six novel tachykinin and bombesin related peptides from the skin of the australian frog, *Pseudophryne guntheri*. Peptides 1990;11: 299–304.

66. Ansel JC, Brown JR, Payan DG, Brown MA. Substance P selectively activates TNF-a gene expression in murine mast cells. J Immunol 1993;150:4478–4485.

67. Heaney LG, Cross LJM, Stanford CF, Ennis M. Substance P induces histamine release from human pulmonary mast cells. Clin Exp Allergy 1993;25:179–186.

68. Lotz M, Vaughan JH, Carson DA. Effect of neuropeptides on production of inflammatory cytokines by human monocytes. Science 1988;241:1218–1221.

69. Lee HR, Ho WZ, Douglas SD. Substance P augments tumor necrosis factor release in human-derived macrophages. Clin Diagnostic Lab Immunol 1994;1:419–423.

70. Murris-Espin M, Pinelli E, Pipy B, Leophonte P, Didier A. Substance P and alveolar macrophages: effects on oxidative metabolism and eiocsanoid production. Allergy 1995; 50:334–339.

71. D'Ortho M-P, Jarreau P-H, Delacourt C, et al. Tachykinins induce gelatinase production by guinea pig alveolar macrophages. Am J Physiol 1995;269:L631–L636.

72. Brunelleschi S, Vanni L, Ledda F, Giotti A, Maggi, CA, Fantozzi R. Tachykinins activate guinea-pig alveolar macrophages: involvement of NK1 and Nk2 receptors. Br J Pharmacol 1990;100:417–420.

73. Rameshwar P, Gascon P, Ganea D. Stimulation of IL-2 production in murine lymphocytes by substance P and related tachykinins. J Immunol 1993;151:2484–2496.

74. Nio DA, Moylan RN, Roche JK. Modulation of T lymphocyte function by neuropeptides. J Immunol 1993;150:5281–5288.

75. Pascual DW, Xu-Amano J, Kiyono H, McGhee JR, Bost KL. Substance P acts directly upon cloned B lymphoma cells to enhance IgA and IgM production. J Immunol 1991;146:2130–2136.

76. Blum AM, Metwali A, Cook G, Mathew RC, Elliott D, Weinstock JV. Substance P modulates antigen-induced, INF-γ production in murine *Schistosomiasis mansoni*. J Immunol 1993;151:225–233.

77. Martin FC, Anton PA, Gornbein JA, Shanahan F, Merrill JE. Production of interleukin-1 by microglia in response to substance P: role for a non-classical NK-i receptor. J Neuroimmunol 1993;42:53–60.

78. Luber-Narod J, Kage R, Leeman SE. Substance P enhances the secretion of tumor necrosis factor-α from neuroglia cells stimulated with lipopolysaccharide. J Immunol 1994;152:819–824.

7

Endothelins in Inflammation

GILES A. RAE

Universidade Federal de Santa Catarina
Florianópolis, Brazil

MARIA G. M. O. HENRIQUES

Institute of Drug Technology
Fundação Oswaldo Cruz
Rio de Janeiro, Brazil

I. Introduction

Endothelins (ETs) are peptides that produce potent and widespread effects in multiple tissues and systems (1–3). Besides ET-1, its most prominent representative, this peptide family also includes ET-2 and ET-3, all of which display 21 amino acids and two disulfide bonds linking the two pairs of cysteine residues (4,5). The ETs are closely related in structure to the sarafotoxins (SRTXs), found in the venom of the burrowing asp *Atractaspis engaddensis* (6). Since their discovery, a formidable amount of evidence has accumulated concerning the mechanisms of action of and potential roles played by the ETs in many systems. The present chapter attempts to provide a comprehensive overview of the evidence accumulated to date which suggests that ETs participate in inflammatory phenomena. Nevertheless, before launching ourselves into this effort, we will very briefly summarize the basic mechanisms involved in the synthesis and cellular modes of action of these peptides.

A. Synthesis of Endothelins

The sequence of each ET is determined by a distinct gene which encodes the corresponding prepro-ET (5). These precursors are cleaved, possibly by furin, to yield propeptides named big-ETs (7), which are, in turn, converted to the mature peptides by action of an "endothelin-converting enzyme" (ECE). Two such ECEs have been cloned and consistently identified as metallopeptidases: ECE-1, which displays optimal activity at neutral pH, is processed via the secretory pathway and is found in both cytosol and on the plasma membrane as an ectoenzyme (8,9); and ECE-2, which is more active at acidic pH and seems to be restricted to the trans-Golgi complex (10). Both enzymes display greater efficacy for processing big-ET-1 over big-ET-2 or big-ET-3 and are inhibited by phosphoramidon. There is also functional and biochemical evidence for occurrence of other alternative ECEs (11,12).

B. Endothelin Receptors

ETs produce their various effects via stimulation of specific 7-transmembrane-helix G-protein-coupled receptors. Three ET receptors have been cloned: the ET_A receptor, which shows higher affinity for ET-1 and ET-2 than for ET-3 (13); the ET_B receptor, which is nonisopeptide selective (14); and the ET_C receptor, which has been cloned only from an amphibian genome and displays higher affinity for ET-3 than for ET-1 (15). Subtypes of both ET_A (ET_{A1} and ET_{A2}) and ET_B (ET_{B1} and ET_{B2}) receptors, as well as alternative receptors not conforming to the ET_A or ET_B receptor profiles, have been proposed on the basis of functional evidence (16,17). Effects mediated via ET_A or ET_B receptors can be discriminated by use of selective agonists or antagonists (18–22). Thus, ET_A receptors are selectively blocked by peptidic (BQ-123 and FR 139317, among others) or nonpeptidic (BMS-182874 or A-127722) antagonists. On the other hand, ET_B receptors are selectively stimulated by SRTX S6c, [Ala[1,3,11,15]]ET-1, BQ-3020, and IRL 1620, and are blocked by peptidic antagonists such as BQ-788 and RES-701-1 or the nonpeptidic compound Ro 46-8443. There are also several mixed (nonselective) ET_A/ET_B receptor antagonists, such as bosentan, SB 217242, and TAK-044.

C. Signal Transduction Pathways

Both ET_A and ET_B receptors can be coupled to various signal transduction pathways, including activation of phospholipases C, A_2, and D, as well as of Na^+/H^+ exchange and modulation of adenylate cyclase or guanylate cyclase activities (23). Through activation of phospholipase C, ETs stimulate the formation of both inositol 1,4,5-trisphosphate (IP_3), which by mobilizing intracellular Ca^{2+} stores acts to rapidly increase intracellular calcium ($[Ca^{2+}]_i$), and diacylglycerol (DAG), which activates protein kinase C. The initial IP_3-mediated transient rise in $[Ca^{2+}]_i$ is followed closely by a smaller, more prolonged increase, due to Ca^{2+} influx through voltage-operated and, possibly, receptor-operated channels. Elevation of $[Ca^{2+}]_i$ can activate Ca^{2+}-dependent protein kinases and also phospholipase A_2, thus providing arachidonic acid for synthesis of eicosanoids (24).

Eicosanoid formation triggered by ETs can also be mediated via direct phospholipase A_2 activation, or through the phospholipase C pathway, whereby DAG is converted by action of lipases to arachidonic acid (23). Acting through ET_A receptors, ETs can also stimulate phospholipase D in fibroblasts and glioma cells, either directly or via a protein kinase C-dependent mechanism, leading to formation of phosphatidic acid and DAG. On the other hand, ETs can cause cellular alkalinization by promoting Na^+/H^+ exchange, which also increases $[Ca^{+2}]_i$ indirectly and sensitizes Ca^{+2}-dependent enzymes to the cation. ETs have also been shown to either inhibit or, in a few instances, activate adenylyl cyclase-mediated cAMP production. Though the former action may be direct, the latter is possibly dependent on production of eicosanoids. Finally, the guanylyl cyclase pathway appears to be important for certain depressor actions of the ETs and SRTXs, but the rise in cGMP levels in these cases seems to be mediated via synthesis of either nitric oxide or carbon monoxide (25,26).

II. Endothelin Production in Response to Inflammatory Stimuli

Expression and/or production of ETs can be stimulated either in vitro or in vivo by bacterial lipopolysaccharide (LPS), as well as by several cytokines, growth factors, humoral inflammatory mediators, and autacoids.

A. In-Vitro Findings

LPS enhances the production of ET in cultured endothelial cells from bovine aorta (27) or pulmonary artery (28), epithelial cells from guinea pig bronchi (29) or human trachea (30), human amniotic cells (31), macrophages (32), and monocytes (33,34). Although ET production in human umbilical vein endothelial cells is insensitive to LPS (35) or tumor necrosis factor-α (TNF-α) (36), these cells display a robust response to infection with *Trypanosoma cruzi* (37). TNF-α increases ET secretion in bovine (38–40) or porcine endothelial cells (41), bovine (40) or rat mesangial cells (41), porcine renal epithelial cells (42), and in human bronchial epithelial cells (30), amniotic cells (31,43), and keratinocytes (44). Also, HIV-1 glycoprotein 120 is a powerful stimulus for production of both ET-1 and TNF-α in human circulating monocytes (33), but it remains unclear to what extent the production of ET-1 is dependent on TNF-α. Interleukin-1 (IL-1) is another cytokine with potent ET-releasing properties in vitro in porcine (45,46) and human endothelial cells (47), guinea pig (29) and human respiratory epithelial cells (30), rabbit (41) and human mesangial cells (48), and in human amniotic (31,43) or endometrial stromal cells (49) and keratinocytes (44). Other interleukins, such as IL-2, IL-6, and IL-8, appear to be less effective than TNF-α or IL-1 at stimulating ET production, and indeed do not modify production by bovine endothelial cells (38). Nevertheless, they all increase ET levels in guinea pig tracheal epithelial cells (50). IL-2 treatment in vivo (for 7 days) in the rat enhances ET release from the perfused mesentery (51), and IL-6 is effective in human amniotic cells (43) and in a breast cancer cell line (52). Finally, interferon-β (IFN-β) stimulates ET production in human skin fibroblasts (53), whereas IFN-γ is an effective stimulus in bovine pulmonary, but not aortic or retinal, endothelial cells (38,39), and actually inhibits ET production in cells of the inner medullar duct of the rat kidney (54).

On the other hand, several growth factors also stimulate ET production. Significant enhancement is observed with epidermal growth factor (EGF) in human amniotic (55) and omental vascular smooth muscle cells (56), with platelet-derived growth factor (PDGF) in human and rat vascular smooth muscle cells (56,57), and with fibroblast growth factor (FGF) and insulin-like growth factor (IGF) in rat adenohypophiseal cells (58). IGF is also active on porcine endothelial cells (59), whereas transforming growth factor-β (TGF-β) is a potent stimulus for ET release in bovine (28,60–62), porcine (57,63,64), and guinea pig endothelial cells (65), as well as in rabbit gastric epithelial cells (66) and human endothelial (67), mesangial (48), vascular smooth muscle (56), and amniotic cells (55).

Other inflammatory factors which can also trigger ET production include bradykinin (BK), platelet-activating factor (PAF), thromboxane A_2 (TXA$_2$), and thrombin. BK and PAF are both effective in bovine glomerular endothelial cells (68), whereas the TXA$_2$-mimetic U 46619 stimulates production by human mesangial cells (48). Thrombin is active in bovine (28,61,62,68,69), porcine (4,57,64), guinea pig (65), and human endothelial cells (70–72), in addition to guinea pig lungs (73), rabbit gastric epithelial cells (66), and human mesangial cells (48). In most cases of stimulation of ET production in cultured cells, the time course is quite slow, with peak effects occurring after at least a few hours and, sometimes, up to 1 or 2 days of incubation.

B. In-Vivo Findings

Intravenous bolus or infusion of LPS elevates plasma levels of ET-1 in the rat (27, 74–79), pig (80–83), dog (84), and sheep (85,86), and of circulating big-ET-1 in baboons (87). Such effects of LPS are most likely mediated via production of cytokines such as IL-1, IL-2, and TNF-α, because each of these agents can increase plasma ET-1 levels in the rat (51,74,75,79), and levels of both ET-1 and ET-3 are increased in dogs given TNF-α (88). Furthermore, pretreatment with anti-TNF-α antibodies reduces the capability and LPS to raise plasma levels of ET-1 in pigs (83) and of big-ET-1 in baboons (87) and that of *Escherichia coli*-induced bacteremia to increase plasma ET-1 in rats (89). Anti-TNF-α antibodies also block the ET production triggered by IL-2 in the rat (79) and attenuate the life-threatening cardiopulmonary sequelae of bacteremia in baboons (87). Allergens are another important stimulus for ET production in vivo. In guinea pigs sensitized to ovalbumin or to *Ascaris sum*, systemic (ovalbumin) or aerosol (*A. sum*) challenge with the antigen triggered significant increases in ET levels in plasma or in bronchoalveolar lavage (BAL) fluid, respectively (90,91). The time course for stimulation of ET production in vivo is generally faster than that observed in vitro, in that significant effects and sometimes even peak effects are detected within 15 min to 1 hr of administration of the stimulus.

Humans with septicemia display significant increases in circulating ET-1 and ET-3 levels (81,92–95), and ET-1 levels are positively correlated to both the severity of the illness (96) and plasma TNF-α levels (97,98). The increased plasma levels of ET-1 in septicemic patients are also correlated to levels of creatinine, type II phospholipase A_2, C3a complement, and atrial natriuretic peptide in plasma, but not to those of PAF, leukotriene B_4 (LTB4), or lactate (96–98). Nevertheless, it remains unclear if, in addition to TNF-α, any of these factors actually contributes to increase plasma ET-1 levels during sepsis or vice versa. On the other hand, circulating levels of ET-1 are increased in patients displaying HIV infection with retinal microangiopathy (99), and monocytes taken from HIV-infected patients, unlike those from healthy controls, display a pronounced expression of ET-1 mRNA, which correlates well with disease progression (33). Moreover, high levels of ET-1 were detected in brain macrophages of patients who died of HIV-encephalopathy (33,100). Patients with complicated *Plasmodium falciparum* infection (malaria) display 10-fold higher plasma big-ET-1 levels than healthy controls, and these increased levels are correlated to those of plasma TNF-α, but not to parasitemia, fever, or other signs of severe infection (101).

Most of the in vivo studies concerning ET production have focused on alterations in plasma levels of these peptides. It is important to bear in mind that ETs most likely act primarily, in either autocrine or paracrine fashion, as local hormones or autacoids. Circulating ET-1 concentrations are normally far below those needed to trigger most of the peptide's biological effects, because it is cleared rapidly and most effectively from the blood upon passage through the lungs (102,103). Also, monolayers of cultured endothelial cells (104,105) or renal epithelial cells (54) preferentially release ET-1 across the abluminal (basal) membrane. Thus, luminal concentrations of the peptide in vivo possibly represent only spillover of peptide into the blood or tubular fluid. The findings of the in vivo studies on ET production frequently agree well with the upregulation of ET mRNA expression or production detected in vitro, particularly with endothelial cells. Nevertheless, the elevation in ET plasma levels observed following intravenous injection of an inflammatory stimulus may well, in some cases, reflect an increased spillover of ET-1 from damaged endothelium and/or decreased clearance of circulating peptide, rather than true stimulation of production.

III. Endothelin-Induced Release of Inflammatory Mediators

As expected from substances acting as local hormones or autacoids, ETs can also trigger the release of several other locally acting mediators, many of which also stimulate ET production. Such actions of the ETs are summarized, together with other actions, in Fig. 1.

A. Cytokines and Growth Factors

ET-1 is an effective stimulus for production of TNF-α, IL-1, IL-6, IL-8, monocyte chemotactic protein-1, and granulocyte-macrophage colony-stimulating factor in human cultured monocytes/macrophages (106–109), of IL-6 from rat stromal cells of

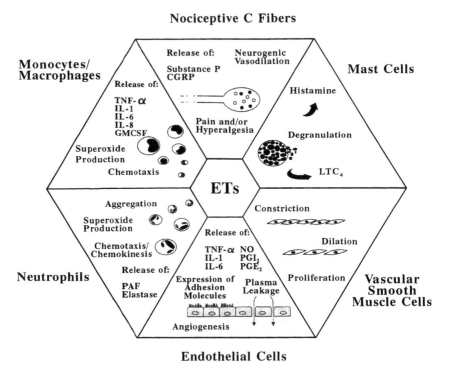

Figure 1   Diagram summarizing the main actions underlying the proinflammatory ⌐irects of endothelins (ETs).

the bone marrow (110) or endothelial cells (111), and of TNF-α, IL-1, and IL-6 from human endothelial cells (112).

B.   PAF and Eicosanoids

As ETs can activate phospholipase A_2 in many cells, they can trigger release of PAF, as well as products of arachidonic acid processed via the cyclooxygenase or lipooxygenase pathways. Thus, ET-1 and ET-3 can stimulate the release of PAF from cultured hepatic Kupffer cells (113). PAF receptor antagonists inhibit the proaggregatory effects of ET-1 in isolated human neutrophils (114) and its constrictor effects in guinea pig airways in vitro (115–117) and in vivo (118). Also, leukotriene receptor antagonists blunt ET-1-induced constriction of guinea pig airways in vitro (117,119), whereas lipooxygenase inhibitors inhibit the early phase of ET-1-induced bronchoconstriction in dogs (120). ET-1 also stimulates release of LTC_4 from mouse bone marrow-derived mast cells (121), and the production of 15-HETE in rat distal lung (122) and of 5-HETE and 12-HETE in cat cultured airway epithelium (123). ET-1 is a potent releaser of cyclooxygenase-derived eicosanoids (prostanoids) in guinea pig or rat isolated perfused lungs, among many different tissues and cells (124–126). It releases TXA_2 from guinea pig airways (119), PGD_2, PGE_2, PGI_2, $PGF_{2\alpha}$, and TXA_2 from human bronchus (127), PGE_2 and $PGF_{2\alpha}$ from cat airway epithelial cells (123), TXA_2 and PGD_2 from dog airway macrophages (128), and PGE_2, PGI_2, and TXA_2 from rabbit perfused spleen and kidney (129). Thus, it is not surprising that the effects of ETs in vivo or in vitro are influenced in many instances by inhibitors of cyclooxygenase, blockers of prostaglandin isomerases, or antagonists of the various prostanoid receptors. Furthermore, there is evidence that, at least in rat lungs and human bronchus, ET-1 induces eicosanoid release via stimulation of ET_A receptors (126, 127). Finally, a recent study has found that ET-1 stimulates the expression of inducible cyclooxygenase-2, but not of constitutive cyclooxygenase-1, in cultured rat mesangial cells, and that the increased production of PGE_2 induced by ET-1 in these cells is only correlated to levels of the inducible isoform (130). Since upregulation of cyclooxygenase-2 expression and of prostanoid synthesis are both key features of inflammatory states, these may be important targets of action of the ETs during inflammation.

C. Histamine

On the other hand, ET-1 releases histamine from guinea pig pulmonary (but not peritoneal) mast cells, through activation of what appear to be ET_B-like receptors (117,120). Furthermore, ET-1 can potently and directly trigger BQ-123-sensitive (ET_A receptor-mediated) degranulation of mouse freshly harvested peritoneal mast cells and cultured bone marrow mast cells, leading to release of both histamine and leukotriene C_4 (121,131–133). Interestingly, primary bone marrow mast cells of the mouse produce ET-1 and express ET_A receptor-like binding sites, but challenge with the peptide only evokes release of histamine and serotonin if the cells are primed (differentiated) with IL-4 (133,134). In sharp contrast, ET-1 fails to degranulate human skin mast cells directly, but can activate such cells indirectly via stimulation of sensory nerves (135). Thus, the mechanisms underlying the mast cell-degranulating effect of ET-1 display considerable species and/or tissue variability.

D. Oxygen Free Radicals and Enzymes

Another aspect of ET-1 action, which could be potentially important in inflammation, is its ability to modulate oxygen free-radical formation by leukocytes. ET-1 increases oxygen radical and 15-lipooxygenase product formation in rat BAL fluid and lung homogenates (122) and stimulates superoxide production in human alveolar macrophages (136), but does not directly affect superoxide production by human circulating monocytes (137). Superoxide production by guinea pig alveolar macrophages in response to ET-1 has also been reported (138), but others have failed to confirm this observation, even though the peptide did enhance the production stimulated by either FMLP or PAF, apparently via ET_A receptors (139). Also, ET-1 does not seem to precipitate superoxide production in human neutrophils directly, but it can enhance the generation stimulated by the chemotactic peptide FMLP (140,141) or by monocyte supernatants (142), but not by zymosan (143,144). However, ET-1 can directly and very effectively trigger elastase release from human neutrophils (145), but, unlike FMLP, it does not stimulate release of lysozyme from rabbit neutrophils (146) or of myeloperoxidase from guinea pig neutrophils (139).

IV. Effects of Endothelins on Leukocyte Functions

A. Neutrophils

As pointed out above, ET-1 can induce neutrophils to aggregate via PAF release (114), as well as to secrete elastase (145) and greater amounts of superoxide in response to FMLP (140,141). Though some studies have found human cultured neutrophils to be unresponsive to ET-1 (147), all three mammalian ETs can display chemotactic activity toward human neutrophils (148). Surprisingly, this activity seemed to be endowed by the N-terminal portion of the peptides, rather than the C-terminal portion which is usually more important for interaction with ET_A or ET_B receptors. Perhaps this finding is related to the fact that human neutrophils, which cannot produce ETs of its own, not only can promptly form ET-1 from big-ET-1, but can also destroy this peptide via action of either a serine protease or elastase (149,150).

On the other hand, ET-1 potently induces an ET_A receptor-mediated chemokinetic effect in rabbit neutrophils, acting through mechanisms which are clearly distinct from those implicated in the chemotactic effect of FMLP (146). In contrast, ET-2 induces what seems to be a truly chemotactic effect in these same cells, which is susceptible to blockade of either ET_A or ET_B receptors (151). At higher concentrations ($\geqslant 30$ nM), ET-1 and ET-2 actually display antichemokinetic and antichemotactic activity in rabbit neutrophils. Intravenous infusion of ET-1 in rats induces, within 2 hr, pronounced adherence of neutrophils to the endothelium of pulmonary vessels, and a sixfold accumulation of this cell type in the alveolar wall (152). A similar study conducted in guinea pigs found that intravenous bolus injection of ET-1 rapidly induces an ET_A receptor-mediated neutropenia (139). Since this effect is not associated with significant neutrophil migration into the alveolar space, at least up to 20 min after administration, the neutropenia is likely to reflect intravascular sequestration. Marked ET-1-induced adherence of neutrophils to endothelial cells has also been observed in the perfused heart (153), and in the cat isquemic coronary vessels following reperfusion (154). It may also be worth mentioning that human neutrophils can upregulate ET-1 synthesis in endothelial cells (155).

B. Monocytes and Macrophages

Unlike neutrophils, both monocytes and macrophages are able to generate their own ET-1, and synthesis is stimulated by inflammatory stimuli such as LPS (32,34). Although ET-1 does not seem to affect human monocyte adhesiveness to endothelium or superoxide production directly (137), it can act as a chemotactic factor for these cells (156). Furthermore, it stimulates the release of superoxide in human alveolar macrophages (136) and upregulates superoxide production by guinea pig alveolar macrophages challenged with either FMLP or PAF (139). It also stimulates human monocytes and/or monocyte-derived macrophages to release TNF-α, IL-1, IL-6, IL-8, monocyte chemotactic protein-1, and granulocyte-mediated colony-stimulating factor (106–109), as well as PGE_2 (157). Thus, monocytes and macrophages may be important targets for autocrine or paracrine actions of ET-1, through which it can orchestrate the recruitment to and modulate the activity of other leukocytes during inflammation.

C. Other Leukocytes

To our knowledge, there is virtually no information on the possible effects of ETs on the functions of other leukocytes. Intravenous bolus injection of ET-1 causes significant lymphocytosis in the guinea pig, yet this may be simply the consequence of the severe hemoconcentration induced by the peptide (139). On the other hand, ETs may be important in the recruitment of eosinophils, but not of neutrophils, to the pleural cavity of OVA-sensitized mice challenged with the antigen, since eosinophil accumulation was reduced by intrapleural pretreatment with an ET_A receptor antagonist (158).

V. Effects of Endothelins on Endothelial Cell Functions

The vascular endothelium plays a prominent role in inflammatory processes, by producing several mediators which control the tone of underlying vascular smooth muscle, the susceptibility of platelets to aggregation, blood coagulation, the expression of adhesion molecules and thus trafficking of leukocytes across the vessel wall, vascular permeability, and angiogenesis. The ETs, particularly ET-1, can influence each of these aspects of endothelial cell function, as summarized in Fig. 1.

A. Release of Inflammatory Mediators

As mentioned above, ET-1 can stimulate endothelial cells to release the cytokines TNF-α, IL-1, and IL-6, all of which are important for leukocyte activation during inflammation (111,112). ETs also induce marked increases in plasma prostanoid levels when injected intravenously in vivo (24,159,160), or in the perfusate when infused into isolated perfused organs or vascular beds, particularly of PGI_2 and PGE_2 (24,124). Indeed, production of these vasodilator prostanoids appears to be important in counteracting the vasoconstrictor actions of ET-1, as these can be markedly potentiated following inhibition of prostanoid synthesis in several species, including the human (2,24,129,161). It appears most likely that endothelial cells contribute importantly to prostanoid output in such cases. Indeed, ETs release PGI_2 from cultured bovine endothelial cells (116,162) and PGD_2 and $PGF_{2\alpha}$ from cultured endothelial cells of human brain microvessels, where PGD_2 can further enhance production of $PGF_{2\alpha}$, PGE_2, TXA_2, and suffer conversion to a potent vasoconstrictor metabolite (163,164). Many studies have found that ETs control prostanoid production exclusively via ET_A receptors (126,164,165), but others suggest that ET_B receptors may also contribute significantly (166,167). Nevertheless, the important point in the context of this chapter is that endothelium-derived prostanoids contribute importantly to inflammation, controlling local blood flow, the threshold of activation of nociceptive nerve terminals, and leukocyte functions. ETs also stimulate the release of nitric oxide from endothelial cells, via activation of constitutive nitric oxide synthase (24,124). However, more relevant to the context of inflammation, ET-1 can inhibit, via ET_A receptors, the expression of the inducible nitric oxide synthase isoform and production of nitric oxide triggered in mesangial cells by TNF-α and IL-1 (168).

B. Release of Mediators Affecting Platelet Function and Blood Coagulation

ETs inhibit platelet aggregation ex vivo in rabbits (160,169) and in vivo in dogs (166). This antiaggregatory effect, which is not observed in vitro, is most likely the result of ET_B receptor-mediated release of PGI_2 or PGD_2 from endothelial cells (166,167). Indeed, ET-1 actually potentiates ADP-induced aggregation of platelets isolated from rabbits or dogs, but not from humans (170,124). ET-1 and ET-3 also display important fibrinolytic activity ex vivo in rabbit arterial blood, which is mediated by stimulation of tissue plasminogen activator (t-PA) release from endothelial cells (171). Release of t-PA has also been detected in the rat perfused hindleg challenged with either ET-1 or ET-3 (172). In human cultured endothelial cells, however, ET-1 enhances the inhibition of t-PA release induced by TNF-α or IL-1, whereas it inhibits the release of PA-inhibitor-1 stimulated by these same cytokines (173). On the other hand, ETs also stimulate the release of von Willebrand factor in the rat perfused hindleg, acting through cellular mechanisms clearly distinct from those involved in t-PA release (172). Although the endothelium appears to play an important antithrombotic role under normal conditions, and ETs may participate in this process, such a role is likely to be compromised in conditions in which there is extensive endothelial cell damage and inflammation. Patients with disseminated intravascular coagulation associated with malignancy exhibit raised plasma levels of ET-1 and big-ET-1, which are closely related to the onset and progression of this condition (174,175). It is at present unclear if ET-1 actively participates in bringing about this condition, by favouring thrombus formation due to its vasoconstrictor and/or fibrinolytic effects, or is solely a good marker of ongoing endothelial cell damage. Perhaps it is also pertinent to mention that red blood cells can release massive amounts of ET-1 and its precursor upon lysis (176).

C. Expression of Adhesion Molecules

ET-1 markedly increases the expression of the adhesion molecules ICAM-1, VCAM, and E-selectin on human cultured brain microvascular endothelial cells (177). The effects of ET-1 are relatively rapid in onset and equivalent to those produced by TNF-α. Furthermore, as ET-2 and ET-3 are as potent as ET-1 in promoting expression of all three adhesion molecules, the receptors involved are likely to be of the ET_B

type, but it remains to be seen if these effects are mediated directly or via cytokine production. Since adhesion molecules play a critical role in the trafficking of leukocytes across the vessel wall during inflammation (178), it will also be important to assess if these findings with the ETs can be extended to endothelial cells of other vascular beds. Indeed, ET-1 enhances adhesion of leukocytes to endothelial cells in the pulmonary (139,152) and coronary circulations (153,154), and of neutrophils to human cultured endothelial cells (179), whereas a low concentration of ET-3 was ineffective in the mesenteric bed, at least within the first 30 min (180).

D. Vascular Permeability

Several inflammatory mediators enhance the permeability of the venular endothelium by inducing endothelial cell contraction and hence formation of interendothelial cell gaps, and, at later stages, emigration of recruited leukocytes across the vessel wall (180–182). Both ET-1 and ET-3 can increase the formation of interendothelial cell gaps in cultured monolayers of bovine endothelial cells (183). Given intravenously, ET-1 markedly enhances extravasation of plasma proteins from the microvasculature in rat stomach, duodenum, spleen, heart, diaphragm, trachea, and bronchus (116,184–186). In several, but not all, of these vascular beds the effect is mediated indirectly via release of PAF and TXA_2. The use of selective ET_A receptor antagonists (BQ-123 or FR 139317) or a mixed ET_A/ET_B receptor antagonist (bosentan) has revealed that plasma protein leakage triggered by intravenous ET-1 in the rat appears to be mediated exclusively via ET_A receptors in the stomach and duodenum, whereas both ET_A and ET_B receptors control extravasation in the heart, bronchus, spleen, and kidney (139,184,187,189). Confirming this view, the selective ET_B receptor agonist IRL 1620 is active only in those tissues in which bosentan blocks ET-1 induced plasma extravasation to a greater extent than the selective ET_A receptor antagonist FR 139317 (139,187,188). Also, superfusion of the rat mesentery in vivo with ET-3, in the presence of the ET_A receptor antagonist BQ-123, enhances permeability of the venules to albumin by 10-fold (189). Infusion of ET-1 into the brachial artery also induces edema in the human forearm (190), but the receptors responsible for this effect remain to be characterized. As expected of an agent that causes widespread plasma leakage, intravenous ET-1 also causes pronounced hemoconcentration in rats (116), guinea pigs (139), and mice (191).

The potentially edematogenic actions of the ETs, detected in the studies mentioned above are most likely due to interendothelial gap formation, as none of them found any relationship between ET-induced plasma extravasation and leukocyte emigration into extravascular spaces. In addition, pretreatment of rats with antineutrophil serum failed to inhibit plasma leakage induced by close arterial injection of ET-1 in the stomach (186). Nevertheless, the apparent lack of contribution of leukocytes toward plasma extravasation triggered by ETs may simply be because all these studies focused on rather short-term changes in vascular permeability, occurring 10–30 min after administration. Clearly, a more systematic analysis over longer periods should be conducted in vivo to clarify this aspect, as leukocytes seem to be essential in mediating the increase in microvascular permeability induced by ET-1 in rat isolated and perfused lungs (108). Another aspect which should merit consideration is that, given intravenously, ETs act simultaneously in all vascular beds, inducing a shocklike state which could be quite different from what might happen when these peptides act only in a bed supplying a single tissue or organ, particularly with regard to the possible participation of leukocytes. However, endogenous ETs seem to contribute to paw edema induced by local injection of ovalbumin in sensitized mice or by carrageenan, as these responses, unlike those induced by zymosan or histamine, are inhibited by local injection of selective ET_A receptor antagonists (192).

Notwithstanding the proposed stimulatory influence of ETs on plasma extravasation, various studies have found that ETs, injected extravascularly (usually intradermally), inhibit protein leakage induced by several edematogenic agents in different animal models (193–197). Also, intradermal injection of ET-1 into the human forearm has been found to cause localized palor due to vasoconstriction, but this area is surrounded by a flare response (194,198), even though others detected only a wheal-and-flare-like effect (199). The flare response to ET-1 reflects neurogenic vasodilation which is mediated, at least in part, via mast cell degranulation secondary to direct activation of nociceptive C fibers (200). The apparent paradox concerning the dual effects of ET-related peptides on plasma extravasation (and therefore edema) may be explained by their differential effects on smooth muscle of the arterial and venous vasculatures (201), as outlined in Fig. 2. In the rat perfused mesentery, for example, ETs cause endothelium-dependent (nitric oxide-mediated) arterial dilation and intense venoconstriction when injected intravascularly (202), hemodynamic changes which enhance hydrostatic pressure in the microvasculature and potentiate plasma extravasation through the interendothelial cell gaps. However, when ETs are superfused over the mesenteric bed, they constrict both arterial and venous sides of this vascular bed (180). In this condition, the increase in microvascular hydrostatic pressure due to venoconstriction can be counterbalanced, or even fully abolished, by the sharp reduction in blood flow determined by arterial constriction.

E. Angiogenesis and Tissue Repair

The proliferation and migration of endothelial cells play a crucial role in the vascular remodeling which takes place during the healing of injured tissue. In this regard, ETs have been found to stimulate proliferation of many cells types, including endothelial cells, as recently reviewed elsewhere (203). ETs can act as true mitogens in some instances, but in most cases they are potent primers or co-mitogens. In the later case, they synergize with various growth factors by stimulating, via Ras protein activation, the serine-threonine and tyrosine kinase cascades which control the expression of immediate early genes such as c-*myc* and c-*fos* (204). ET-1 and ET-3 are equipotent in stimulating proliferation and DNA synthesis in either bovine or human cultured endothelial cells, with efficacies similar to that of 10% fetal bovine serum, but lower than that of fibroblast growth factor (205). In addition and like fibroblast growth factor, ET-1 and ET-3 also induced migration of both kinds of endothelial cells in

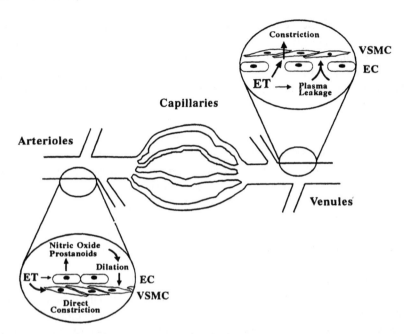

Figure 2 Diagram illustrating the effects of endothelins (ET) on endothelial cells (EC) and on vascular smooth muscle cells (VSMC) of arterioles and venules. Note that the venular actions promote plasma extravasation, and thus edema. However, by acting on the arterial side of the vascular bed, ET can either potentiate edema by increasing blood flow via arteriolar endothelium-dependent dilation, or limit edema via direct contraction of arteriolar smooth muscle.

microchemotaxis chambers. The proliferative and migratory effects of the ETs were unaffected by an ET_A receptor antagonist, but virtually abolished by an ET_B receptor antagonist. This contrasts with findings in vascular smooth muscle cells, in which the proliferative effects of ET-1 involve activation of ET_A receptors (206).

One of the well-known instances in which neovascularization appears to be an outstanding feature is in atherosclerosis, an inflammatory-fibroproliferative response of the vascular wall to several stimuli (207). Atherosclerotic plaques constitute agglomerations of lipid-laden macrophages, lymphocytes, endothelial cells, smooth muscle cells, connective tissue, and debris in the vessel intima. Particularly pronounced levels of ET-1-binding sites and immunoreactive ET-1 associated to areas of neovascularization and vasa vasorum have been observed in human atheromatous saphenous veins, or in pig femoral arteries with intimal hyperplastic lesions (208), and in the neointima of coronary atherosclerotic vessels of the human heart (209). Furthermore, the ET-1-binding sites observed in the regions of neovascularization in human atherosclerotic epicardial coronary arteries possibly represent ET_B receptors associated to endothelial cells as, unlike the binding sites present in the tunica media, they show low affinity to a selective ET_A receptor antagonist (210). It also seems pertinent to point out that ET-1 plays a major role in vascular restenosis, a phenomenon in which lesions to the endothelial layer (such as those occurring during angioplasty) induce extensive neointimal formation due to smooth muscle cell proliferation (211,212).

A major aspect of tissue repair during an inflammatory response is that cells are stimulated to change from a quiescent phenotype to a synthetic dividing one. In the process of atherosclerosis, the phenotype transition of vascular smooth muscle cells is controlled by various cytokines released from macrophages or endothelial cells, such as IL-1, TNF-α, and TGF (207). In this context, it has been observed that IL-1 dramatically inhibits ET-1-induced DNA synthesis and proliferation of human cultured aortic smooth muscle, whereas IL-8, TNF-α, and TGF are ineffective (213). It appears plausible to anticipate that similar interactions between ETs, cytokines, and growth factors possibly occur with regard to proliferation, migration, and differentiation of other cell types during inflammation.

Wound repair involves the formation of granulation tissue, that is comprised of leukocytes, fibroblastic cells, and new blood vessels, all implanted in a loose collagenous extracellular matrix, which ultimately contracts to help close the wound. Indeed, ET-1 has been found to contract granulation tissue (214). Another study has since shown that ET-1 in fact actively mediates cross-talk between endothelial cells and fibroblasts (215). Rat aortic endothelial cells grown between sheets of collagen rapidly reorganize into a dense network of microvessels which degenerate within 3–4 days of culture. However, if the endothelial cells are co-cultured with tendon fibroblasts, the microvessels stabilize for up to 21 days and exhibit increased deposition of subendothelial extracellular matrix. On the other hand, endothelial cells and ET-1 were both capable of stimulating the expression of smooth muscle α-actin in the cultured fibroblasts, as well as fibroblast-mediated collagen contraction. These exciting findings suggest that paracrine interactions between fibroblasts and endothelial cells are important determinants of granulation tissue formation, whereby the fibroblasts stimulate (or amplify) angiogenesis and stabilize the neovascular endothelium which, by releasing ET-1, transforms the fibroblasts into myofibroblasts to resolve the granuloma.

VI. Effects of Endothelins on Pain and Fever

A. Pain

Given intraperitonealy to mice, ET-1, ET-2, and ET-3 each elicit dose-dependent visceral pain, measured as an increased number of abdominal constrictions or writhes (199,216–218). All three peptides are roughly equipotent in evoking this rapid and short-lasting effect, but pretreatment with a selective ET_A receptor antagonist attenuates the responses to ET-1 and ET-2, but not to ET-3, suggesting that both ET_A and ET_B (or even ET_C) receptors are implicated (218). One study reported full suppression of ET-1-induced writhing following pretreatment with indomethacin (199), whereas another failed to detect any influence of this or another two cyclooxygenase inhibitors (216). Therefore, it is unclear if hyperalgesic prostanoids, such as PGE_2 and PGI_2, mediate ET-1-induced writhing. However, ET-1 injected into the rat hind paw induces a long-lasting hyperalgesic effect in a modified version of the Randal-Selitto paw pressure test which, unlike that caused by carrageenan, is not inhibited by indomethacin or the sympathetic neuron blocker guanethidine (199). In the dog knee joint, ET-1 evokes a relatively transient articular incapacitation when injected alone, but can very markedly sensitize the joint to the hyperalgesic effects of PGE_2, in such a way that a low and usually ineffective dose of the prostanoid evokes long-lasting incapacitation (199). We have also found that injections of ET-1 into the mouse hind paw, at doses which per se do not cause nociceptive responses, can potentiate formalin-induced nociception and paw edema (219). Interestingly, ET-3 is more potent than ET-1 at potentiating the nociceptive response to formalin, but does not influence paw edema.

Humans receiving intradermal injections of ET-1 in the forearm report intense itching, increased sensitivity to pinching, and tenderness to pressure throughout the area developing the wheal-and-flare reaction (199). Moreover, volunteers given infusions of ET-1 into the brachial artery also report deep muscular forearm pain, which is enhanced by touch or muscle contraction (1990).

There is some evidence suggesting that ETs may modulate nociception by acting directly on nociceptive C-fiber neurons. ET-1 evokes a substance P-mediated ventral root depolarization in spinal cord isolated from the newborn rat (220). This peptide also enhances the basal release of substance P from rat cultured sensory neurons, and potentiates the release of both substance P and CGRP triggered by capsaicin in these cells (221). Furthermore, ET-1 itself is found in both porcine (220) and human spinal cord, especially in the more superficial laminae of the dorsal horn, as well as in human dorsal root ganglia, where it coexists with substance P and CGRP in several sensory neurons (222). Accordingly, binding sites for ET-1 have been demonstrated in the spinal-cord gray matter of rats and humans (223,224). Somewhat surprisingly, ET-1 has been found to display very potent antinociceptive activity when injected intrathecally to mice submitted to a radiant heat tail-flick test, which is blocked by opioid receptor antagonists (225). However, this apparent analgesia-induced by ET-1 may well be due to activation of the opioid system involved in heterosegmental diffuse noxious inhibitory control, which is so important for localized pain perception (226). On the other hand, intrathecal injection of ET-1 to lumbar segments of the spinal cord induced an ET_A receptor-mediated analgesic effect against formalin-induced hind paw nociception in rats (227). Interestingly, unlike ET-1, ET-3 and the selective ET_B receptor agonist SRTX S6c both potentiated nociception in this model, but none of the three peptiedes altered sensitivity to thermal stimulation. Clearly, further studies are needed to better elucidate the mechanisms and receptors involved in modulation of nociception by the ETs, as well as to establish if these peptides actually play any physiological or pathological roles in the processing of nociceptive input at the spinal-cord level.

B. Fever

Fever is a well-established systemic sign of infection and inflammation, which is controlled by an interplay among several endogenous pyrogens and cryogens (228). In this regard, ET-1 induces fever when injected intracerebroventricularly in rabbits (229). This effect, which is partially prevented by treatment with indomethacin, is mediated at least in part through ET_B receptor activation, as it can also be produced by a selective ET_B receptor agonist. It will be interesting to determine which other mechanisms, besides prostanoid generation, are involved in the pyrogenic effects of the ETs.

VII. Possible Involvement of Endothelins in Inflammatory Diseases

At this stage, we hope that those readers who have been patient enough to reflect on all the information contained in the previous sections will be convinced that ETs are endowed with many properties which are pertinent to inflammatory conditions (Fig. 1). However, to what extent does the evidence accumulated thus far actually implicate endogenous ETs in the deflagration and/or maintenance of different chronic inflammatory diseases? We will now attempt to outline briefly some of the evidence which constitutes a partial—and no doubt fragmentary—answer to this question.

A. Asthma

The chronic inflammatory disease asthma is associated with episodes of nonspecific airway hyperresponsiveness to spasmogens, airway edema, leukocyte accumulation, smooth muscle and submucous gland hyperplasia, overproduction of mucus, and epithelial cell dysfunction and desquamation. The synthesis of ET-1 is upregulated in bronchial epithelial cells collected from asthmatic patients, in comparison to cells taken from either chronic bronchitic patients or healthy donors (230–232), and this overproduction can be stimulated by histamine or IL-1 (232) and inhibited by hydrocortisone (231). Likewise, both dexamethasone and triamcinolone inhibit ET-1 production triggered by IL-2 and fetal bovine serum in a human pulmonary epithelial cell line (233). Increased amounts of ET-1 have also been observed in BAL fluid of patients with symptomatic asthma, and these levels were normalized following antiasthmatic treatment with glucocorticoids (234–237). Another study has failed to confirm these findings, and actually detected lower ET-1 levels in BAL fluid of patients with nocturnal asthma, as compared to levels seen in either diurnal asthmatics or healthy subjects (238). Nevertheless, a greater expression of ET-1 is also seen in alveolar macrophages obtained by BAL in asthmatic subjects, even though release of the peptide into the supernatant was similar to that of nonasthmatic control cells (239). Interestingly, the alveolar macrophages isolated from stable asthma patients, and especially those with unstable asthma, spontaneously released more TNF-α than healthy control cells, and production of this cytokine was remarkably increased by ET-1 in cells from patients with stable, but not unstable, asthma. Asthmatic children show raised plasma levels of ET-1, and monocytes isolated from these patients produce more ET-1, but these alterations are offset by specific allergen immunotherapy (240).

ETs are extremely potent bronchospastic peptides both in vivo and in airways isolated from many species, including humans, through actions mediated via ET_A and/or ET_B receptors (3,127). Guinea pigs sensitized to ovalbumin display exaggerated bronchospastic responses to aerosolized ET-1 18–24 hr after antigen challenge (241). Moreover, phosphoramidon and indomethacin, which potentiate and inhibit responses to ET-1 in unchallenged animals, respectively, fail to do so following ovalbumin challenge. Thus, in this animal model of asthma, responsiveness to ET-1 is enhanced, in part due to diminished catabolism of ET-1 by a phosphoramidon-sensitive metalloprotease (most likely neutral endopeptidase), and involves different cellular mechanisms. On the other hand, allergic humans sneeze more and exhibit greater nasal mucosal secretion and rhinorrhea in response to intranasal instilation of ET-1 than do nonallergic subjects (242). Nevertheless, human nonasthmatic and asthmatic airways show similar sensitivities to ET-1 and densities of smooth muscle ET-1 binding sites, which in both cases consist mainly of ET_B receptors (243,244). In addition, asthmatic human peripheral airway smooth muscle and alveolar tissue both fail to display altered densities of ET_A or ET_B receptors (245). However, this does not rule out ET-1 as a putative mediator of the excessive bronchospasm associated with asthma, especially if its synthesis is increased and its catabolism is decreased in this condition. Indeed, there is a high correlation between ET-1 levels in plasma or in BAL fluid and the degree of airflow obstruction in asthmatic patients who are not medicated with glucocorticoids (237,246). On the other hand, there is controversy as to whether ETs can cause airway smooth muscle hyperresponsiveness. ET-1 does not directly sensitize guinea pig trachea (247) or conscious guinea pigs to the bronchoconstrictor effects of acetylcholine (248), and BAL fluid levels of ET-1 do not correlate with cholinergic bronchoreactivity in asthmatic patients (237). Nevertheless, it does sensitize guinea pig trachea to histamine (249), and bronchial biopsies taken from asthmatic patients with the most exaggerated bronchoresponsiveness to nebulized water show the highest levels of mast cells, eosinophils, and ET-1-positive epithelial cells (250). In sheep sensitized to *Ascaris sum*, treatment with an ET_A receptor antagonist significantly reduced (~50%) the late bronchoconstriction induced by antigen, as well as airway hyperresponsiveness to inhaled carbachol (251). Also, we have observed that an antagonist for ET_A receptors inhibits late ovalbumin-induced leukocyte recruitment (especially eosinophils, which appear to be most relevant to asthma) to the pleural cavity of sensitized mice, whereas an ET_B receptor antagonist is ineffective (192). Significant correlations have been detected between eosinophil influx and increased levels of ET-1 in BAL fluid collected 24 hr after (but not 1 week after) antigen challenge in guinea pigs sensitized to *Ascaris sum* (91), and between ET-1 content and interstitial edema in lungs of rats following intratracheal instillation of Sephadex beads, an eosinophilic model of alveolitis (252).

As mentioned in the previous section, ETs are potent inducers of ET_A and ET_B receptor-mediated airway edema, which can be achieved via formation of interendothelial cell gaps or transcapillary fluid transfer following elevation in capillary hydrostatic pressure. Although guinea pigs fail to display gross signs of lung inflammation following acute ET-1 administration (253), they can display an increased number of total leukocytes in BAL fluid (254). Also, ET-1 may, either directly or via a priming effect, activate various pulmonary leukocytes to release oxygenated radicals, eicosanoids, and 15-lipooxygenase products (117,122,136,140). On the other hand, ETs display mitogenic/proliferative properties in pulmonary artery fibroblasts (255) and in airway smooth muscle (256–259). Thus, these peptides may well contribute, alongside cytokines released from infiltrating leukocytes and growth factors, to both the thickening of the airways and the hyperplasia of submucosal glands seen in asthma. Moreover, ET-1 directly stimulates mucus secretion from feline cultured tracheal submucosal glands, but inhibits secretion from tracheal explants with epithelium (260), as well as from submucosal glands of ferret trachea (with epithelium) stimulated by methacoline or phenylephrine (261). As asthma is associated with airway epithelium dysfunctions and desquamation (262), the direct stimulatory effects of ET-1 on submucosal glands may well prevail over its epithelium-dependent inhibitory effects during this condition. Thus, by inhibiting the widespread actions of ETs in the airways, ET_A and ET_B receptor antagonists may constitute a promising therapeutic alternative for the treatment of asthma.

B. Systemic Sclerosis

Systemic sclerosis is a connective tissue disease of unknown etiology, which is characterized by microvascular lesions and disturbances and varying degrees of fibrosis in different tissues and organs including, quite frequently, the lungs. Airway epithelium from patients with interstitial lung fibrosis display increased production of ET-1 which correlates well with the intensity of histological lesions (263). Plasma ET-1 levels are markedly elevated in patients with systemic sclerosis, especially in those with large cutaneous involvement or the diffuse type of the disease (264,265). Dermal fibroblasts from patients with systemic sclerosis produce far more ET-1 than normal cells, particularly those taken from patients with the diffuse type (266). Furthermore, it is important to mention that the enhancement of plasma ET levels during systemic sclerosis apparently fails to correlate specifically with the severity of lung fibrosis in this condition (267,268), and may reflect spillover of these peptides due to endothelial cell damage. Enhanced levels of ET-1 are also observed in BAL fluid of patients with idiopathic lung fibrosis and, to a lesser extent, with interstitial lung disease (236), as well as in that of patients with systemic sclerosis already manifesting pulmonary fibrosis (269). Interestingly, however, BAL fluid ET-1 levels were even greater in systemic sclerosis patients not yet displaying clear signs of pulmonary fibrosis, suggesting that this could be a potentially useful marker for the early stages of lung disease associated with this condition.

C. Lupus Erythematosus and Rheumatic Diseases

Remarkably high levels of ET-1 have been detected in patients with systemic lupus erythematosus (270,271), Raynaud's disease (271–274), mixed connective tissue disease associated with the presence of circulating antiendothelial antibodies (275), and rheumatic arthritis (particularly in patients where it is active), but not osteo-arthritis or gout (276). As most of these conditions are associated with extensive endothelial cell damage, it appears likely that the increase in ET-1 levels is merely a consequence, rather than a causative factor. Indeed, the sera of patients with systemic lupus erythematosus stimulated greater secretion of ET-1 from human cultured endothelial cells than that of normal controls, and this effect was correlated to serum titers of IgM antiendothelial cell antibodies and immunocomplex (277).

Nevertheless, in a strain of mice that develops spontaneous lupus nephritis, prolonged treatment with a selective ET_A receptor antagonist enhanced survival dramatically (278). This treatment also substantially impaired the development of renal histological lesions, proteinuria, hypertension, and the accumulation of colla-gens I, III, and IV, laminin, metalloproteases, extracellular matrix proteins, and heparan sulfate proteoglycan in the renal cortex. Interestingly, in a rat model of glomerular nephritis, there is also marked upregulation of ET-1 synthesis and of ET_B (but not ET_A) receptor expression in the glomeruli, as well as an increased urinary excretion of ET-1, during the mesangial proliferative phase of the disease (279). The upregulation of ET_B receptors plays a critical role in the stimulation of ET-1 synthesis by rat mesangial cells (280). However, it may well be that, once produced, ET-1 acts via ET_A receptors to trigger mesangial cell proliferation and matrix expansion in this model, as an ET_A receptor antagonist ameliorated proteinuria, glomerular injury, and other renal function disturbances in a closely related rat model of progressive glomerulosclerosis (281). Therefore, at least in animal models, both ET_A and ET_B receptors appear to be involved in the mediation of different stages of nephritis. As increased urinary excretion of ET-1 also occurs in humans with different progressive nephropathies, including lupus nephritis (282,283), perhaps antagonists for ET_A and ET_B receptors may prove useful in the treatment of such conditions.

On the other hand, ET-1 levels are also increased in plasma of rats with adjuvant-induced arthritis (284) and in synovial fluid of inflamed joints in humans (285), where it seems to be produced mainly by macrophage-like type A macro-phages, at least in joints affected by rheumatoid arthritis (276). Interestingly, the peptide can act in autocrine fashion to exert a potent proliferative effect on synovio-cytes and may thus contribute to proliferation and outgrowth of the synovial mem-brane. Nevertheless, it is unknown if ET-1 plays any role in the erosion of articular cartilage and subchondral bone in rheumatoid arthritic joints. Indeed, ET-1 has been found in the hypertrophic zone of cartilage and in bone associated with osteoclasts and osteoblasts (but not chondrocytes), where it can stimulate proliferation of chondrocytes and osteoblasts (286–288). However, there are contradictory reports on the influence of ET-1 on osteoclast activity and bone resorption, one finding inhibition (289) and another stimulation of prostanoid-mediated bone resorption (290).

D. Gastric Ulcers and Inflammatory Bowel Disease

ETs are potent ulcerogenic agents when injected either into the gastric artery (291,292), gastric submucous layer (293), or intravenously to rats (294–296), and synergizes with other ulcerogenic agents such as indomethacin and ethanol (294–298). Mucosal damage induced by ET-1 is characterized by areas of vasocongestion, hemorrhage, necrosis, and sloughing of a mucoid cap overlying the lesioned sites (291,299). Importantly, as rats treated with an anti-ET-1 antibody become less sensitive to gastric ulcers induced by ethanol (298) or indomethacin (300), ET-1 may be a key endogenous mediator of gastric injury. PGI_2 and PGE_2 are important cytoprotectors if the gastric mucosa, by increasing local blood flow and inhibiting acid secretion. Thus, as gastric chief cells and neuroendocrine cells contain ET-1 (301) indomethacin may not only prevent the synthesis of cytoprotective prostanoids, leaving the damaging actions of locally produced ET-1 unopposed, but may actually increase ET-1 secretion from these cells. As mentioned earlier, intravenous ET-1 increases vascular permeability in the stomach and jejunum of rats, via activation of ET_A receptors coupled to release of PAF and TXA_2 (116,185–187), though some studies have failed to demonstrate the involvement of both mediators in this effect (185,295). It also seems that the vascular leakage triggered by ET-1 in rat stomach does not depend on neutrophil migration (185).

The ulcerogenic effects of ET-1 in the stomach have been attributed to its potent constrictor action on the gastric microvasculature, which limits washout of back-diffusing acid from and/or oxygen and nutrient supply to the mucosa (291,293,295–298). Indeed, a recent study found that an ET_A receptor antagonist fully abolished the reduction in gastric mucosa blood flow and substantially reduced mucosal damage triggered by high-frequency electrical stimulation of the rat stomach in vivo (302). However, although ET-1 is a more effective constrictor of the rat gastric microvasculature than ET-3, both peptides are equipotent in causing gastric ulcers in anesthetized rats (296). Also, the PGI_2-mimetic iloprost and sensory nerve stimulation with capsaicin both protect the gastric mucosa from the damaging effects of ET-1, without antagonizing its local vasoconstrictor effect (297,303). Furthermore, CGRP, which is also released from sensory nerves, can either potentiate or prevent ET-1-induced microvascular leakage in the gastric mucosa (304). Finally, the anti-ulcer drug cetraxate counteracts ET-1-induced mucosal hypoxia and ulcers, without modifying the effects of ET-1 on gastric mucosal blood flow (305). Therefore, additional mechanisms may well be involved in the gastric ulcerogenic effects of ETs. The relevance of these findings to the human situation has yet to be established, particularly as ulcers in this species are located almost exclusively in the duodenum (306).

ET-1-like immunoreactivity can be elevated 10-fold in the supernatants of intestinal samples from patients with Crohn's disease or with ulcerative colitis (307), although others have failed to substantiate this finding (308). In the Crohn's disease samples, ET-1 appeared to be localized mainly in the perivasculature of the submucosa, in aggregated macrophage-like cells, and to a lesser extent in the lamina propria (307). No perivascular changes were noted in ulcerative colitis samples, which showed more immunoreactive ET-1 in the lamina propria than in the submucosa. It is hoped that these findings will soon be corroborated by studies in animal models of inflammatory bowel disease, which should then enable the unraveling of the putative roles of ET-1 and other ETs in these conditions. In the meantime, it seems pertinent to state that PAF antagonists can block ET-1-induced vasoconstriction and necrosis in rat intestine (309), and that ETs are potent stimulators of prostanoid-dependent electrolyte secretion in both rat and rabbit colonic mucosa (310–313) and could thus underlie the diarrhea associated with inflammatory bowel diseases.

E. Systemic Inflammatory Response Syndromes

Due to space limitations, we will not review the evidence relating ETs to the pathogenesis of systemic inflammatory response syndromes, such as sepsis, septicemia, endotoxemia, septic shock, hemorrhagic shock, anaphylactic shock, adult respiratory distress syndrome, multiple organ failure, and disseminated intravascular coagulation. Nevertheless, it is important to mention that the significant and widespread roles played by the ETs in each of these life-threatening conditions have recently been outlined in an excellent review which deals specifically with this subject (314).

VIII. Perspectives for Future Research

The amount of data accumulated to date on the different influences of ETs in the various aspects of inflammation is truly amazing, particularly for substances which are more known for their potent and important actions on the cardiovascular system. It is now evident that ETs can induce, either alone or in concert with several other mediators, each of the cardinal signs of inflammation, that several inflammatory disorders are associated with altered local or circulating levels of ETs, and that these peptides may play key roles in at least certain inflammatory diseases. Nevertheless, there is still clearly a lot to be done in order to determine the true implications of ETs in inflammation. One of the important aspects to be tackled in future studies is to assess to what extent the altered local or circulating levels of ETs, detected in inflammatory diseases, actually contribute to development or maintenance of specific characteristics of the disease. Perhaps one of the more feasible approaches to achieve this is to employ selective nonpeptidic antagonists for either ET_A or ET_B receptors, and attempts in this direction with peptidic antagonists have already led to important findings. Selective inhibitors of ET synthesis (ECE inhibitors) may also prove useful in this respect, but no such drugs are yet available. However, even in the cases where

ET receptor antagonists may come to display significant antiinflammatory effects, the results should be carefully analyzed to rule out bias due to actions that are not linked directly to inflammation, such as hemodynamic changes related to antagonism of ET receptors on vascular smooth muscle or endothelial cells. On the other hand, we have outlined several examples in which ETs can affect the production of cytokines or mediators and vice versa. There is no inflammatory condition which can be ascribed to the actions of a single inflammatory mediator. Thus, and considering that inflammation is normally a process activated to restore the function of damaged tissue, it would seem extremely important for future studies to elucidate the roles of ETs in the complex interplay of different mediators and cells at the various stages of the inflammatory process. Finally, it will be necessary to assess to what extent the findings obtained in the various animal models of acute and chronic inflammation correlate with the human condition. In this context, new animal models and human studies yielding data which go beyond merely the measurement of ET levels will be more than welcome. Collectively, all these approaches may lead to a better understanding of inflammation and, perhaps, to the development of new antiinflammatory strategies based on selective downregulation of production or action of the ETs.

Acknowledgments

The authors wish to thank the support of the Brazilian National Research Council (CNPq), and André Sampaio for helping in preparing the illustrations.

References

1. Masaki T, Yanagisawa M, Goto K. Physiology and pharmacology of endothelins. Med Res Rev 1992; 12:391–421.
2. Haynes WG, Webb DJ. The endothelin family of peptides: local hormones with diverse roles in health and disease? Clin Sci 1993; 84:485–500.
3. Rae GA, Calixto JB, D'Orléans-Juste P. Effects and mechanisms of action of endothelins on non-vascular smooth muscle of the respiratory, gastrointestinal and urogenital tracts. Regulatory Peptides 1995; 55:1–46.
4. Yanagisawa M, Kurihara H, Kimura S, Tomobe Y, Kobayashi M, Mitsui Y, Yasaki Y, Goto K, Masaki T. A novel potent vasoconstrictor peptide produced by vascular endothelial cells. Nature 1988; 332:441–445.
5. Inoue A, Yanagisawa M, Kimura S, Kasuya Y, Miyauchi T, Goto K, Masaki T. The human endothelin family: three structurally and pharmacologically distinct isopeptides predicted by three separate genes. Proc Natl Acad Sci (USA) 1989; 86:2863–2867.
6. Kochva E, Bdolah A, Wollberg Z. Sarafotoxins and endothelins: evolution, structure and function. Toxicon 1993; 31:541–568.
7. Denault JB, Claing A, D'Orléans-Juste P, Sawamura T, Kido T, Masaki T, Leduc R. Processing of proendothelin-1 by human furin convertase. FEBS Lett 1995; 362: 276–280.
8. Xu D, Emoto N, Giaid A, Slaughter C, Kaw S, deWit D, Yanagisawa M. ECE-1: a

membrane-bound metalloprotease that catalyses the proteolytic activation of endothelin-1. Cell 1994 78:473–485.

9. Takahashi M, Fukuda L, Shimada L, Barnes L, Turner AJ, Ikeda M, Koike H, Yamamoto Y, Tanzawa K. Localization of rat endothelin-converting enzyme to vascular endothelial cells and some secretory cells. Biochem J 1995; 311:657–665.

10. Emoto N, Yanagisawa M. Endothelin converting enzyme 2 is a membrane bound, phosphoramidon sensitive metalloprotease with acidic pH optimum. J Biol Chem 1995; 270:15262–15268.

11. Opgenorth TJ, Kimura S, Wu-Wong JR. Characterization of endothelin-converting enzymes. Meth Neurosci 1995; 23:251–265.

12. Turner AJ, Murphy LJ. Molecular pharmacology of endothelin converting enzymes. Biochem Pharmacol 1996; 51:91–102.

13. Arai H, Hori S, Aramori I, Ohkubo H, Nakanishi S. Cloning and expression of a cDNA encoding an endothelin receptor. Nature 1990; 348:730–732.

14. Sakurai T, Yanagisawa M, Takuwa Y, Miyazaki H, Kimura S, Goto K, Masaki T. Cloning of a cDNA encoding a non-isopeptide-selective subtype of the endothelin receptor. Nature 1990; 348:732–735.

15. Karne S, Jayawickreme CK, Lerner MR. Cloning and characterization of an endothelin-3 specific receptor (ET_C receptor) from *Xenopus laevis* derman melanophores. J Biol Chem 1993; 268:19126–19133.

16. Bax, WA, Saxena PR. The current endothelin receptor classification: time for reconsideration? Trends Pharmacol Sci 1994; 15:379–386.

17. Yoneyama T, Hori M, Makatani M, Yamamura T, Tanaka T, Matsuda Y, Karaki H. Subtypes of endothelin ET_A and ET_B receptors mediating tracheal smooth muscle contraction. Biochem Biophys Res Commun 1995; 207:668–674.

18. Masaki T, Vane JR, Vanhoutte PM. International Union of Pharmacology nomenclature of endothelin receptors. Pharmacol Rev 1994; 46:137–142.

19. Hiley CR. Endothelin receptor ligands. Neurotransmissions 1995; 11:1–6.

20. Webb ML, Bird JE, Liu EC, Rose PM, Serafino R, Stein PD, Moreland S. BMS 182874 is a selective, nonpeptide endothelin ETA receptor antagonist. J Pharmacol Exp Ther 1995; 272:1124–1134.

21. Breu V, Clozel M, Burri K, Hirth G, Neidhart W, Ramuz H. In vitro characterisation of Ro 46-8443, the first non-peptide antagonist selective for endothelin ET(B) receptor. FEBS Lett 1996; 383:37–41.

22. Opgenorth TJ, Adler AL, Calzadilla SV, Chiou WJ, Dayton BD, Dixon DB, Gehrke LJ, Hernandez L, Magnuson SR, Marsh KC, Novosad EI. Pharmacological characterization of A-127722: an orally active and highly potent ET(A)-selective receptor antagonist. J Pharmacol Exp Ther 1996; 276:473–481.

23. Sokolovsky M. Endothelin receptor subtypes and their role in transmembrane signaling mechanisms. Pharmacol Ther 1995; 68:435–471.

24. Hyslop S, De Nucci G. Vasoactive mediators released by endothelins. Pharmacol Res 1992; 26:223–242.

25. Shagra-Levine Z, Galron R, Sokolowsky M. Cyclic GMP formation in rat cerebellar slices is stimulated by endothelins via nitric oxide formation and by sarofotoxins via formation of carbon monoxide. Biochemistry 1994; 33:14656–14659.

26. Moritoki H, Miyano H, Takeuchi S, Yamaguchi M, Hisayama T, Kondoh . Endothelin-3-induced relaxation of rat thoracic aorta: a role for nitric oxide formation. Br J Pharmacol 1993; 108:1125–1130.

27. Sugiura M, Inagami T, Kon V. Endotoxin stimulated endothelin-release in vivo and in vitro as determined by radioimmunoassay. Biochem Biophys Res Commun 1989; 161: 1220–1227.

28. Ohlstein EH, Storer BL, Butcher JA, Debouck C, Feuerstein G. Platelets stimulate expression of endothelin mRNA and endothelin biosynthesis in cultured endothelial cells. Circ Res 1991; 69:832–841.

29. Ninomiya H, Uchida Y, Ishii Y, Nomura A, Kameyama M, Saotome M, Endo T, Hasegawa S. Endotoxin stimulates endothelin-1 release from cultured epithelial cells of guinea pig trachea. Eur J Pharmacol 1991; 203:299–302.

30. Nakano J, Takizawa H, Ohtoshi T, Shoji S, Yamaguchi M, Ishii A, Yanagisawa M, Ito K. Endotoxin and pro inflammatory cytokines stimulate endothelin 1 expression and release by airway epithelial cells. Clin Exp Allergy 1994; 24:330–336.

31. Casey ML, Word RA, MacDonald PC. Endothelin-1 gene expression and regulation of endothelin mRNA and protein biosynthesis in avascular human amnion. J Biol Chem 1991; 266:5762–5767.

32. Ehrenreich H, Anderson RW, Fox CH, Rieckmann P, Hoffman GS, Travis WD, Coligan JE, Kehrl JH, Fauci AS. Endothelins, peptides with potent vasoactive properties, are produced by human macrophages. J Exp Med 1990; 173:1741–1748.

33. Ehrenreich H, Rieckmann P, Sinowatz F, Weih KA, Arthur LO, Goebel FD, Burd PR, Coligan JE, Clouse KA. Potent stimulation of monocytic endothelin-1 production by HIV-1 glycoprotein 120. J Immunol 1993; 150:4601–4609.

34. Cunningham ME, Huribal M, McMillen MA. Endotoxin-stimulated monocytes produce endothelin. FASEB J 1991; 8:A214.

35. Moldovan F. The effect of endotoxin on human umbilical vein endothelial cells in culture. FASEB J 1994; 122:346A.

36. Clozel M, Fischli W. Human cultured endothelial cells do secrete ET-1. J Cardiovasc Pharmacol 1989; 13(suppl 5):S229–S231.

37. Wittner M, Christ GJ, Huang J, Weiss LM, Hatcher VB, Morris SA, Orr GA, Berman JW, Zeballos GA, Douglas SA, Tanowitz HB. Trypanosoma cruzi induces endothelin release from endothelial cells. J Infect Dis 1995; 171:493–497.

38. Kanse SM, Takahashi L, Lam H-C, Rees A, Warren JB, Porta M, Molinatti P, Ghatei M, Bloom SR. Cytokine stimulated endothelin release from endothelial cells. Life Sci 1991; 48:1379–1384.

39. Lamas S, Michael T, Collins T, Brenner BM, Marsden PA. Effects of interferon-gamma on nitric oxide synthase activity and endothelin-1 production by vascular endothelial cells. J Clin Invest 1992; 90:879–887.

40. Marsden PA, Brenner BM. Transcriptional regulation of the endothelin-1 gene by TNF-alpha. Am J Physiol 1992; 262:C854–C861.

41. Kohan DE. Production of ET-1 by rat mesangial cells: Regulation by tumor necrosis factor. J Lab Clin Med 1992; 119:477–484.

42. Ohta K, Hirata Y, Imai T, Kanno K, Emori T, Shichiri M, Marumo F. Cytokine-induced release of endothelin-1 from porcine renal epithelial cell line. Biochem Biophys Res Commun 1990; 169:578–584.

43. Mitchell MD, Lundin-Schiller S, Edwin SS. Endothelin production by amnion and its regulation by cytokines. Am J Obstet Gynecol 1991; 165:120–124.

44. Tsuboi R, Sato C, Shi CM, Nakamura T, Sakurai T, Ogawa H. Endothelin 1 acts as an autocrine growth factor for normal human keratinocytes. J Cell Physiol 1994; 159: 213–220.

45. Yoshizumi M, Kurihara H, Morita T, Yamashita T, Oh-hashi Y, Sugiyama T, Takaku F, Yanagisawa M, Masaki T, Yazaki Y. Interleukin 1 increases the production of endothelin-1 by cultured endothelial cells. Biochem Biophys Res Commun 1990; 166:324–329.

46. Maemura K, Kurihara H, Morita T, Oh-hashi Y, Yazaki Y. Production of endothelin-1 in vascular endothelial cells is regulated by factor associated with vascular injury. Gerontology 1992; 38:29–35.

47. Katabami T, Shimizu M, Okano K, Yano Y, Nemoto K, Ogura M, Tsukamoto T, Suzuki S, Ohira K, Yamada Y. Intracellular signal transduction for interleukin-1 beta-induced endothelin production in human umbilical vein endothelial cells. Biochem Biophys Res Commun 1992; 188:565–570.

48. Zoja C, Osirio S, Perico N, Benigni A, Morigi M, Benatti L, Rambaldi A, Remuzzi G. Constitutive expression of endothelin gene in cultured human mesangial cells and its modulation by transforming growth factor-β, thrombin, and a thromboxane A_2 analogue. Lab Invest 1991; 64:16–20.

49. Economos K, MacDonald PC, Casey ML. Endothelin-1 gene expression and protein biosynthesis in human endometrium: potential modulator of endometrial blood flow. J Clin Endocrinol Metabol 1992; 74:14–19.

50. Endo T, Uchida Y, Matsumoto H, Suzuki N, Nomura A, Hirata F, Hasegawa S. Regulation of endothelin-1 synthesis in cultured guinea pig airway epithelial cells by various cytokines. Biochem Biophys Res Commun 1992; 186:1594–1599.

51. Miyamori I, Takeda Y, Yoneda T, Iki K, Takeda R. Interleukin-2 enhances the release of endothelin-1 from the rat mesenteric artery. Life Sci 1991; 49:1295–1300.

52. Yamashita J, Ogawa M, Nomura K, Matsuo S, Inada K, Yamashita S, Nakashima Y, Saishoji T, Takano S, Fujita S. Interleukin 6 stimulates the production of immunoreactive endothelin 1 in human breast cancer cells. Cancer Res 1993; 53:464–467.

53. Zebalos GA, An S, Wu JM. ET-1 secretion by human fibroblasts in culture: effects of cell density and IFN-β. Biochem Int 1991; 25:845–852.

54. Kohan DE, Padilla E. Endothelin-1 is an autocrine factor in rat inner medullary collecting ducts. Am J Physiol 1992; 263:F607–F612.

55. Sunnergren KP, Word RA, Sambrook JF, McDonald PC, Casey ML. Expression and regulation of endothelin precursor mRNA in avascular human amnion. Mol Cell Endocrinol 1990; 68:R7–R14.

56. Resink TJ, Hahn AWA, Scott-Burden T, Powell J, Weber E, Buhler FR. Inducible endothelin mRNA expression and peptide secretion in cultured human vascular smooth muscle cells. Biochem Biophys Res Commun 1990; 168:1303–1310.

57. Hahn AWA, Resink T, Scott-Burden T, Powell J, Dohi Y, Buhler FR. Stimulation of endothelin mRNA and secretion in rat vascular smooth muscle cells, a novel autocrine function. Cell Regul 1990; 1:649–659.

58. Matsumoto H, Suzuki N, Shiota K, Inoue K, Tsuda M, Fujino M. Insulin-like growth factor-1 stimulates endothelin-3 secretion from rat anterior pituitary cells in primary culture. Biochem Biophys Res Commun 1990; 172:661–668.

59. Hattori Y, Kasai K, Nakamura T, Emoto T, Shimoda SI. Effect of glucose and insulin on immunoreactive endothelin-1 release from cultured porcine aortic endothelial cells. Metabolism 1991; 40:165–169.

60. Hexum TD, Hoeger C, Riever JE, Baird A, Brown MR. Characterization of endothelin secretion by bascular endothelial cells. Biochem Biophys Res Commun 1990; 167:294–300.

61. Brown MJ, Vaughan J, Walsh J, Jimenez L, Hexum TD, Baird A, Vale W. Endothelin

releasing activity in calf serum and porcine follicular fluid. Biochem Biophys Res Commun 1990; 173:807–815.

62. Ohlstein EH, Arleth A, Ezekiel M, Horohonich S, Ator MA, Calbatiano MM, Sung CP. Biosynthesis and modulation of endothelin from bovine pulmonary arterial endothelial cells. Life Sci 1990; 46:181–188.

63. Suzuki N, Matsumoto H, Kitada C, Kumura S, Fujino M. Production of ET-1 and big ET-1 by tumor cells with epithelial-like morphology. J Biochem 1989; 106.

64. Kurihara H, Yoshizumi M, Sugiyama T, Takaku F, Yanagisawa M, Masaki T, Hamaoki M, Kato H, Yazaki Y. Transforming growth factor-β stimulates the expression of endothelin mRNA by vascular endothelial cells. Biochem Biophys Res Commun 1989; 159:1435–1440.

65. Rieder H, Ramadori G, Meyer zum Buschenfelde KH. Sinusoidal endothelial liver cells in vitro release endothelin—augmentation by transforming growth factor-β and Kupffer cell-conditioned medium. Klin Wochnschr 1991; 69:387–391.

66. Ota S, Hirata Y, Sugimoto T, Kohmoto O, Hata Y, Yoshiura K, Nakada R, Terano A, Sugimoto T. Endothelin-1 secretion from cultured rabbit gastric epithelial cells. J Cardiovasc Pharmacol 1991; 17(suppl 7):S406–S407.

67. MacCumber MW, Ross CA, Glaser BM, Snyder SH. Endothelin: visualization of mRNAs by in situ hybridization provides evidence for local action. Proc Natl Acad Sci (USA) 1989; 86:7285–7289.

68. Marsden PA, Dorfman DM, Collins T, Brenner BM, Orkin SH, Ballermann BJ. Regulated expression of endothelin 1 in glomerular capillary endotehlial cells. Am J Physiol 1991; 261:F117–F125.

69. Emori T, Hirata Y, Marumo F. Specific receptors for endothelin-3 in cultured bovine endothelial cells and its cellular mechanism of action. FEBS Lett 1990; 263:261–264.

70. Kohno M, Yashunari K, Yokokawa K, Murakawa K, Takeda T. Thrombin stimulates the production of immunoreactive endothelin-1 in cultured human umbilical vein endothelial cells. Metabolism 1990; 39:1003–1005.

71. Saijonmaa O, Ristimaki A, Fyrquist F. Atrial natriuretic peptide, nitroglycerine, and nitroprusside reduce basal and stimulated endothelin production from cultured endothelial cells. Biochem Biophys Res Commun 1990; 173:514–520.

72. Takagi Y, Fukase M, Takata S, Yoshimi H, Tokonaga O, Fujita T. Autocrine effect of endothelin on DNA synthesis in human vascular endothelial cells. Biochem Biophys Res Commun 1990; 168:537–543.

73. Moon DG, Horgan ML, Anderson TT, Krytek SR Jr, Fenton JW II, Malik AB. Endothelin-like pulmonary vasoconstrictor peptide release by α-thrombin. Proc Natl Acad Sci (USA) 1989; 86:9529–9533.

74. Vemulapalli S, Chiu PJS, Rivelli M, Foster CJ, Sybertz EJ. Modulation of circulating endothelin levels in hypertension and endotoxemia in rats. J Cardiovasc Pharmacol 1991; 18:895–903.

75. Vemulapalli S, Chiu PJ, Griscti K, Brown A, Kurowski S, Sybertz EJ. Phosphoramidon does not inhibit endogenous endothelin 1 release stimulated by hemorrhage, cytokines and hypoxia in rats. Eur J Pharmacol 1994; 257:95–102.

76. Pollock DM, Divish BJ, Opgenorth TJ. Stimulation of endogenous endothelin release in the anesthetized rats. J Cardiovasc Pharmacol 1993; 22(suppl 8):S295–S298.

77. Morise Z, Ueda M, Aiura K, Endo M, Kitajima M. Pathophysiologic role of endothelin-1 in renal function in rats with endotoxin shock. Surgery 1994; 115:199–204.

78. Nambi P, Pullen M, Slivjak MJ, Ohlstein EH, Storer B, Smith EF. Endotoxin-mediated changes in plasma endothelin concentrations, renal endothelin receptor and renal function. Pharmacology 1994; 48:147–156.

79. Klemm P, Warner TD, Hohlfeld T, Corder R, Vane JR. Endothelin-1 mediates ex vivo coronary vasoconstriction caused by exogenous and endogenous cytokines. Proc Natl Acad Sci (USA) 1995; 92:2691–2695.

80. Pernow J, Hemsen A, Hallen A, Lundberg JM. Release of endothelin-like immunoreactivity in relation to neuropeptide Y and catecholamines during endotoxin shock and asphyxia in the pig. Acta Physiol Scand 1990; 140:311–322.

81. Lundberg JM; Ahlborg G, Hemsen A, Nisell H, Lunell NO, Pernow J, Rudehill A, Weitzberg E. Evidence for release of endothelin-1 in pigs and humans. J Cardiovasc Pharmacol 1991; 17(suppl 7):S350–S353.

82. Myhre U, Pettersen JT, Risoe C, Giercksky KE. Endothelin-1 and endotoxemia. J Cardiovasc Pharmacol 1993; 22(suppl 8):S291–S294.

83. Han JJ, Windsor A, Drenning DH. Release of endothelin-1 in relation to tumour necrosis factor-alpha in porcine *Pseudomonas aeruginosa*-induced septic shock. Shock 1994; 1:342–346.

84. Nakamura T, Kasai K, Sekiguchi Y, Banba N, Takahashi K, Emoto T, Hattori Y, Shimoda S. Elevation of plasma endothelin concentrations during endotoxin shock in dogs. Eur J Pharmacol 1991; 205:277–282.

85. Morel DR, Lacroix JS, Hemsen A, Steinig DA, Pittet JF, Lundberg JM. Increased plasma and pulmonary lymph levels of endothelin during endotoxin shock. Eur J Pharmacol 1989; 167:427–428.

86. Morel DR, Pittet JF, Gunning K, Hemsen A, Lacroix JS, Lundberg JM. Time course of plasma and pulmonary lymph endothelin-like immunoreactivity during sustained endotoxemia in chronically instrumented sheep. Clin Sci 1991; 81:357–365.

87. Redl H, Schlag G, Baharami S, Kargl R, Hartter W, Woloszczuk W, Davies J. Big-endothelin release in baboon bacteremia is partially TNF dependent. J Lab Clin Med 1994; 124:796–801.

88. Mitaka C, Hirata Y, Ichikawa K, Yokoyama K, Emori T, Kanno K, Amaha K. Effects of TNF alpha on hemodynamic changes and circulating endothelium derived vasoactive factors in dogs. Am J Physiol 1994; 267:H1530–H1536.

89. Takahashi K, Ghatei MA, Lam HC, O'Halloran DJ, Bloom SR. Elevated plasma endothelin in patients with diabetes mellitus. Diabetologia 1990; 33:306–310.

90. Filep JG, Télémaque S, Battistini B, Sirois P, D'Orleans-Juste P. Increased plasma levels of endothelin during anaphylactic shock in the guinea-pig. Eur J Pharmacol 1993; 239:231–236.

91. Andersson SE, Zackrisson C, Behrens K, Hemsén A, Forsberg K, Linden M, Lundberg JM. Effect of allergen provocation on inflammatory cell profile and endothelin-like immunoreactivity in guinea-pig airways. Allergy 1995; 50:349–358.

92. Weitzberg E, Lundberg JM, Rudehill A. Elevated plasma levels of endothelin in patients with sepsis syndrome. Circ Shock 1991; 33:222–227.

93. Voerman HJ, Stehouwer CD, van-Kamp GJ, Strack-van-Schijndel RJ, Groeneveld AB, Thijs LG. Plasma endothelin levels are increased during septic shock. Crit Care Med 1992; 20:1097–1101.

94. Hirata Y, Mitaka C, Emori T, Amaha K, Marumo F. Plasma endothelins in sepsis syndrome. JAMA 1993; 270:2182.

95. Mitaka C, Hirata Y, Makita K, Nagura T, Tsunoda Y, Amaha K. Endothelin-1 and atrial natriuretic peptide in septic shock. Am Heart J 1993; 126:466–468.

96. Pittet JF, Morel DR, Mensen A, Gunning K. Elevated plasma endothelin-1 concentrations are associated with the severity of illness in patients with sepsis. Ann Surg 1991; 213:261–264.

97. Takakuwa T, Endo S, Nakae H, Suzuki T, Inada K, Yoshida M, Ogawa M, Uchida K. Relationships between plasma levels of type II phospholipase A2, PAF acetylhydrolase, leukotriene B4, complements, endothelin-1, and thrombomodulin in patients with sepsis. Res Commun Chem Pathol Pharmacol 1994; 84:271–281.

98. Takakuwa T, Endo S, Nakae H, Kikichi M, Suzuki T, Inada K, Yoshida M. Plasma levels of TNF alpha, endothelin-1 and thrombomodulin in patients with sepsis. Res Commun Chem Pathol Pharmacol 1994; 84:261–269.

99. Rolinski B, Geier SA, Sadri I, Klauss V, Bogner JR, Ehrenreich H, Goebel FD. Endothelin-1 immunoreactivity in plasma is elevated in HIV-1 infected patients with retinal microangipathic syndrome. Clin Invest 1994; 72:288–293.

100. Rieckmann P, Albrecht M, Ehrenreich H, Weber T, Michel U. Semi-quantitative analysis of cytokine gene expression in blood and cerebrospinal fluid cells by reverse transcriptase polymerase chain reaction. Res Exp Med (Berl) 1995; 195:17–29.

101. Wenisch C, Wenisch H, Wilairatana P, Looareesuwan S, Vannaphan S, Wagner O, Graninger W, Schönthal E, Rumpold H. Big endothelin in patients with complicated *Plasmodium falciparum* malaria. J Infect Dis 1996; 173:1281–1284.

102. Battistini B, D'Orleans-Juste P, Sirois P. Endothelins: circulating plasma levels and presence in other biologic fluids. Lab Invest 1993; 68:600–628.

103. Angaard E, Galton S, Rae G, Thomas R, McLoughlin L, De Nucci G, Vane JR. The fate of radioiodinated endothelin-1 and endothelin-3 in the rat. J Cardiovasc Pharmacol 1989; 13(suppl 5):546–549.

104. Yoshimoto S, Ishizaki Y, Sasaki T, Murota S. Effect of carbon dioxide and oxygen on endothelin production by cultured porcine cerebral endothelial cells. Stroke 1991; 22: 378–383.

105. Wagner OF, Christ G, Wojta J, Vierhapper H, Parzer S, Nowotny PJ, Schneider B, Waldhausl W, Binder BR. Polar secretion of endothelin-1 by cultured endothelial cells. J Biol Chem 1992; 267:16006–16008.

106. Huribal M, Kumar R, Cunningham M, Sumpio BE, McMillen MA. Endothelin causes production of interleukin 6 but not interleukin 1 by human monocytes. FASEB J 1992; 6:A1613.

107. Cunningham ME, Huribal M, McMillen MA. Endothelin-stimulated monocytes produce IL-1β, IL-8 and GM-CSF. Mol Biol Cell 1993; 4:451A.

108. Helset E, Sildnes T, Selkelid R, Konopski ZS. Endothelin-1 stimulates human monocytes in vitro to release TNF-α, IL-β and IL-6. Mediators of Inflammation 1993; 2: 417–422.

109. Helset E, Sildnes T, Konopski ZS. Endothelin-1 stimulates monocytes in vitro to release chemotactic activity identified as interleukin-8 and monocyte chemotactic protein-1. Mediators of Inflammation 1994; 3:155–160.

110. Agui T, Xin X, Cai Y, Sakai T, Matsumoto K. Stimulation of interleukin-6 production by endothelin in rat bone marrow-derived stromal cells. Blood 1994; 84:2531–2538.

111. Xin X, Cai Y, Matsumoto K, Agui T. Endothelin induced interleukin 6 production by rat aortic endothelial cells. Endocrinology 1995; 136:132–137.

112. Stankova J, D'Orleans-Juste P, Rola-Pleszczynski M. Endothelin-1 augments produc-

tion of cytokines by human lymphocytes and endothelial cells. Peptide Receptors, Montreal, July 28–August 1, 1996.

113. Mustafa SB, Gandhi CR, Harvey SA, Olson MS. Endothelin stimulates platelet activating factor synthesis by cultured rat Kupffer cells. Hepatology 1995; 21:545–553.

114. Gomez-Garre D, Guerra M, Gonzalez E, Lopez-Farre A, Riesco A, Caramelo C, Escanero K, Egido J. Aggregation of human polymorphonuclear leukocytes by endothelin: role of platelet activating factor. Eur J Pharmacol 1992; 224:167–172.

115. Battistini B, Sirois P, Braquet P, Filep J. Endothelin-induced constriction of guinea-pig airways: role of platelet-activating factor. Eur J Pharmacol 1990. 186:307–310.

116. Filep JG, Sirois P, Rousseau A, Fournier A, Sirois P. Effects of endothelin-1 on vascular permeability in the conscious rat: interactions with platelet-activating factor. Br J Pharmacol 1991; 104:797–804.

117. Ninomiya H, Uchida Y, Saotome M, Nomura A, Ohse H, Matsumoto H, Hirata F, Hasegawa S. Endothelins constrict guinea pig trachea by multiple mechanisms. J Pharmacol Exp Ther 1992; 262:570–576.

118. Lagente V, Chabrier PE, Mencia-Huerta JM, Braquet P. Pharmacological modulation of the bronchopulmonary action of the vasoactive peptide, endothelin, administered by aerosol in the guinea-pig. Biochem Biophys Res Commun 1989; 158:625–632.

119. Filep JG, Battistini B, Sirois P. Pharmacological modulation of endothelin-induced contraction of guinea-pig isolated airways and thromboxane release. Br J Pharmacol 1991; 103:1633–1640.

120. Uchida Y, Ninomiya H, Sakamoto T, Lee Y, Endo T, Nomura A, Hasegawa S, Hirata D. ET-1 released histamine from guinea-pig pulmonary but not peritoneal mast cells. Biochem Biophys Res Commun 1992; 189:1106–1110.

121. Yamamura H, Nabe T, Kohno S, Ohata K. Endothelin-1 induces release of histamine and leukotriene C_4 from mouse bone marrow-derived mast cells. Eur J Pharmacol 1994; 257:235–242.

122. Nagase T, Fukuchi Y, Jo C, Teramoto S, Uejima Y, Ishida K, Shimuzu T, Orimo H. Endothelin-1 stimulates arachidonate 15-lipoxygenase activity and oxygen radical formation in the rat distal lung. Biochem Biophys Res Commun 1990; 168:485–489.

123. Wu T, Rieves RD, Larivee P, Logun C, Lawrence MG, Shelhamer JH. Production of eicosanoids in response to endothelin-1 and identification of specific endothelin-1 binding sites in airway epithelial cells. Am J Respir Cell Mol Biol 1993; 8:282–290.

124. De Nucci E, Thomas R, D'Orleans-Juste P, Antunes E, Walder C, Warner TD, Vane JR. Pressor effects of circulating endothelin are limited by its removal in the pulmonary circulation and by the release of prostacyclin and endothelium-derived relaxing factor. Proc Natl Acad Sci (USA) 1988; 85:9797–9800.

125. Touvay C, Vilain B, Pons F, Chabrier PE, Mencia-Huerta JM, Braquet P. Bronchopulmonary and vascular effect of endothelin in the guinea-pig. Eur J Pharmacol 1990; 176:23–33.

126. D'Orleans-Juste P, Télémaque S, Claing A, Ihara M, Yano M. Human big-endothelin-1 and endothelin-1 release prostacyclin via the activation of ET_A receptors in the rat perfused lung. Br J Pharmacol 1992; 105:773–775.

127. Hay DW, Luttmann MA, Hubbard WC, Undem BJ. Endothelin receptor subtypes in human and guinea pig pulmonary tissues. Br J Pharmacol 1993; 110:1175–1183.

128. Ninomiya H, Yu XY, Hasegawa S, Spanhake EW. Endothelin-1 induces stimulation of prostaglandin synthesis in cells obtained from canine airways by brochoalveolar lavage. Prostaglandins 1992; 43:401–411.

129. Rae GA, Trybulec M, De Nucci G, Vane JR. Endothelin-1 release eicosanoids from rabbit isolated perfused kidney and spleen. J Cardiovasc Pharmacol 1989; 13(suppl 5): S89–S92.

130. Kester M, Coroneos E, Thomas PJ, Dunn MJ. Endothelin stimulates prostaglandin endoperoxide synthase 2 mRNA expression and protein synthesis through a tyrosine kinase signaling pathway in rat mesangial cells. J Biol Chem 1994; 269:22574–22580.

131. Yamamura H, Nabe T, Kohno S, Ohata K. Endothelin-1, one of the most potent histamine releasers in mouse peritoneal mast cells. Eur J Pharmacol 1994; 265:9–15.

132. Yamamura H, Nabe T, Kohno S, Ohata K. Mechansim of histamine release by endothelin 1 distinct from that by antigen in mouse bone marrow derived mast cells. Eur J Pharmacol 1995; 288:269–275.

133. Egger F, Geuenich S, Denzlinger C, Schmitt E, Mailhammer R, Ehrenreich H, Dörmer P, Hültner L. IL-4 Renders mast cells functionally responsive to endothelin-1. J Immunol 1995; 154:1830–1837.

134. Ehrenreich H, Burd PR, Rottem M, Hültner L, Hylton JB, Garfield M, Colligan JE, Metcalfe DD, Fauci AS. Endothelins belong to the assortment of mast cell-derived and mast-cell bound cytokines. New Biol 1992; 4:147–154.

135. Brain SD, Thomas E, Crossman DC, Fuller R, Church MK. Endothelin-1 induces a histamine-dependent flare in vivo, but does not activate human skin mast cells in vitro. Br J Pharmacol 1992; 33:117–120.

136. Haller H, Schaberg T, Lindschau C, Lode H, Histler A. Endothelin increases $[Ca^{2+}]_i$, protein phosphorylation and O^{2-} production in human alveolar macrophages. Am J Physiol 1991; 261:L478–L484.

137. Bath PM, Mayston SA, Martin JF. Endothelin and PDGF do not stimulate peripheral blood monocyte chemotaxis, adhesion to endothelium and superoxide production. Exp Cell Res 1990; 187:339–342.

138. Millul V, Lagente V, Gillardeaux O, Boichot E, Dugas B, Mencia-Huerta JM, Bereziat G, Braquet P, Masliah J. Activation of guinea pig alveolar macrophages by endothelin-1. J Cardiovasc Pharmacol 1991; 17(suppl 7):S233–S235.

139. Filep JG, Fournier A, Földes-Filep É. Acute pro-inflammatory actions of endothelin-1 in the guinea-pig lung: involvement of ET_A and ET_B receptors. Br J Pharmacol 1995; 115:227–236.

140. Ishida K, Takeshige K, Minakami S. Endothelin-1 enhances superoxide generation of human neutrophils stimulated by the chemotactic peptide N-formyl-methionyl-leucyl-phenyl-alanine. Biochem Biophys Res Commun 1990; 173:496–500.

141. Hafström I, Ringertz B, Lundberg T, Palmblad J. The effect of endothelin, neuropeptide Y, calcitonin gene-related peptide and substance P on neutrophil functions. Acta Physiol Scand 1993; 148:341–346.

142. Huribal M, Kumar R, Cunningham ME, Sampio BE, McMillen MA. Endothelin-stimulated monocyte supernatants enhance neutrophil superoxide production. Shock 1994; 1:184–187.

143. Prasad S, Lee P, Kalra J. Influence of endothelin on cardiovascular function, oxygen free radicals and blood chemistry. Am Heart J 1991; 121:178–187.

144. Kopprasch S, Gatzweiler A, Kohl M, Schröder H-E. Endothelin-1 does not prime polymorphonuclear leukocytes for enhanced production of reactive oxygen metabolites. Inflammation 1995; 19:679–687.

145. Halim A, Kanayama N, El Maradny E, Maehara K, Terao T. Activated neutrophil by

endothelin 1 caused tissue damage in human umbilical cord. Thromb Res 1995; 77: 321–327.

146. Elferink JGR, Koster BM. Endothelin-induced activation of neutrophil migration. Biochem Pharmacol 1994; 48:865–871.

147. Gallois A, Bueb JL, Tschirhart E. Endothelin-1 does not modulate O_2 release and $[Ca^{2+}]_i$ variations in resting or differentiated HL-60 cells. Fundam Clin Pharmacol 1996; 20:28–32.

148. Wright CD, Cody WL, Dunbar Jr JB, Doherty AM, Hingorani GP, Rapundalo ST. Characterization of endothelins as chemoattractants for human neutrophils. Life Sci 1994; 21:1633–1641.

149. Sessa WC, Kaw S, Hecker M, Vane JR. The biosynthesis of endothelin-1 by human polymorphonuclear leukocytes. Biochem Biophys Res Commun 1991; 174:613–618.

150. Kaw S, Hecker M, Vane JR. The two-step conversion of big endothelin 1 to endothelin 1 and degradation of endothelin 1 by subcellular fractions from human polymorphonuclear leukocytes. Proc Natl Acad Sci (USA) 1992; 89:6886–6890.

151. Elferink JGR, Koster MB. The effect of endothelin-2 (ET-2) on migration and changes in cytosolic free calcium of neutrophils. Naunyn-Schmiedeberg's Arch Pharmacol 1996; 353:130–135.

152. Helset E, Ytrehus K, Tveita T, Kjæve T, Jørgensen L. Endothelin-1 causes accumulation of leukocytes in the pulmonary circulation. Circ Shock 1994; 44:201–209.

153. Lopes-Farré A, Riesco A, Espinosa G, Digiuni E, Cernadas MR, Alvarez V, Monton M. Effect of endothelin-1 on neutrophil adhesion to endothelial cells and perfused heart. Circulation 1993; 88:1166–1171.

154. Lefer AM, Albertine KH, Weyrich AS, Ma XL. Polymorphonuclear (PMN) leucocytes accumate intravascularly but not migrate to the myocardium following ischemia/reperfusion in the cat. FASEB J 1993; 7:A344.

155. Morita T, Kurihara H, Yoshizumi M, Maemura K, Sugiyama T, Nagai R, Yazaki Y. Human polymorphonuclear leukocytes have dual effects on endothelin-1: the induction of endothelin-1 mRNA expression in vascular endothelial cells and modification of the endothelin-1 molecule. Heart Vessels 1993; 8:1–6.

156. Achmad TH, Rao GS. Chemotaxis of human blood monocytes toward endothelin 1 and the influence of calcium channel blockers. Biochem Biophys Res Commun 1992; 189: 994–1000.

157. McMillen MA, Huribal M, Kumar R, Sumpio BE. Endothelin-stimulated human monocytes produce prostaglandin F_2 but not leukotriene B_4. J Surg Res 1993; 54:331–335.

158. Sampaio ALF, Rae GA, D'Orléans-Juste P, Henriques MGMO. Effects of endothelin ET_A and ET_B receptor antagonists on allergic- or zymozan-induced pleurisy in mice. Peptide Receptors, Montreal, July 28–August 1, 1996.

159. Miura K, Yukimura T, Yamashita Y, Shimmen T, Okumura M, Imanishi M, Yamamoto K. Endothelin stimulates the renal production of prostaglandin E_2 and I_2 in anesthetized dogs. Eur J Pharmacol 1989; 170:91–93.

160. Thiemermann C, Lidbury P, Thomas G, Vane J. Endothelins inhibits ex vivo platelet aggregation in the rabbit. Eur J Pharmacol 1988; 158:181–182.

161. Granstam E, Wang L, Bill A. Vascular effects of endothelin-1 in the cat: modification by indomethacin and L-NAME. Acta Physiol Scand 1993; 148:165–176.

162. Emori T, Hirata Y, Marumo F. Endothelin-3 stimulates prostacyclin production in cultured bovine endothelial cells. J Cardiovasc Pharmacol 1991; 17(suppl 7):S140–S144.

163. Spatz M, Stanimirovic DB, Uematsu S, McCarron RM. Vasoactive peptides and prostaglandin D_2 in human cerebromicrovascular endothelium. J Auton Nervous Syst 1994; 49:S123–S127.

164. Stanimirovic DB, Yamamoto T, Uematsu S, Spatz M. Endothelin-1 receptor binding and cellular signal transduction in cultured human brain endothelial cells. J Neurochem 1994; 62:592–601.

165. Télémaque S, Gratton JP, Claing A, D'Orléans-Juste P. Endothelin-1 induces vasoconstriction and prostacyclin release via the activation of endothelin ET_A receptors in the perfused rabbit kidney. Eur J Pharmacol 1993; 237:275–281.

166. Hérman F, Yano M, Filep JG. The in vivo antiaggreatory action of endothelin-1 is not mediated through the endothelin ET_A receptor. Eur J Pharmacol 1993; 236:143–146.

167. McMurdo L, Thiemermann C, Vane JR. The endothelin ET_B receptor agonist, IRL 1620, causes vasodilatation and inhibits ex vivo platelet aggregation in the anaesthetised rabbit. Eur J Pharmacol 1994; 259:51–55.

168. Beck KF, Mohaupt MG, Sterzel RB. Endothelin-1 inhibits cytokine-stimulated transcription of inducible nitric oxide synthase in glomerular mesangial cells. Kidney Int 1995; 48:1893–1899.

169. Lidbury PS, Thiemermann C, Thomas GR, Vane JR. Endothelin-3: selectivity as an anti-aggregatory peptide in vivo. Eur J Pharmacol 1989; 166:335–338.

170. Ohlstein EH, Storer BL, Nambi P, Given M, Lippton H. Endothelin and platelet function. Thromb Res 1990; 57:967–974.

171. Lidbury PS, Thiemermann C, Korbut R, Vane JR. Endothelins release tissue plasminogen activator and prostanoids. Eur J Pharmacol 1990; 186:205–212.

172. Pruis J, Emeis JJ. Endothelin-1 and -3 induce the release of tissue-type plasminogen activator and Von Willebrand factor from endothelial cells. Eur J Pharmacol 1990; 187:105–112.

173. Yamamoto C, Kaji T, Sakamoto M, Kozuka H. Modulation by endothelin-1 of tissue plasminogen activator and plasmogen activator inhibitor-1 release from culture human vascular endothelial cells: Interaction of endothelin-1 with cytokines. Biol Pharm Bull 1993; 16:714–715.

174. Ishibashi M, Saito K, Watanabe K, Eusugi S, Furue H, Yamaji T. Plasma endothelin-1 levels in patients with disseminated intravascular coagulation. N Engl J Med 1991; 324:1516–1517.

175. Shibashi M, Ito N, Fujita M, Furue H, Yamaji T. Endothelin-1 as an aggravating factor of disseminated intravascular coagulation associated with malignant neoplasms. Cancer 1994; 1:191–195.

176. Tippler B, Herbst C, Simmet T. Evidence for the formation of endothelin by lysed red blood cells from endogenous precursor. Eur J Pharmacol 1994; 271:131–139.

177. McCarron RM, Wang L, Stanimirovic DB, Spatz M. Endothelin induction of adhesion molecule expression on human brain microvascular endothelial cells. Neurosci Lett 1993; 156:31–34.

178. Duperray A, Mantovani A, Introna M, Dejana E. Endothelial cell regulation of leukocyte infiltration in inflammatory tissues. Mediators of Inflammation 1995; 4:322–330.

179. McGregor PE, Agerawal DK, Edwards JD. Technique for assessment of leukocyte adherence to human umbilical vein endothelial cells monolayers. J Pharmacol Toxicol Meth 1994; 32:73–77.

180. Kurose I, Kubes P, Wolf R, Anderson DC, Paulson J, Miyasaka M, Granger DN.

Inhibition of nitric oxide production. Mechanisms of vascular albumin leakage. Circ Res 1993; 73:164.

181. Grega JG, Adamski SW, Dobbins DE. Physiological and pharmacological evidence for the regulation of permeability. Fed Proc 1986; 45:96–100.

182. Lum H, Malik AB. Mechanisms of increased endothelial permeability. Can J Physiol Pharmacol 1996; 74:787–800.

183. Farmer P, Kawagushi T, Sirois MG, D'Orleans-Juste P, Sirois P. Endothelin-1 and -3 increase albumin permeability in monolayers of bovine aortic endothelial cells. Meeting on Cells and Cytokines in Pulmonary Inflammation, Paris, June 22–24, 1993.

184. Filep JG, Fournier A, Földes-Filep É. Endothelin-1-induced myocardial ischaemia and oedema in the rat: involvement of the ET_A receptor, platelet-activating factor and thromboxane A_2. Br J Pharmacol 1994; 112:963–971.

185. Sirois MG, Filep JG, Rousseau A, Fournier A, Plante GE, Sirois P. Endothelin-1 increases protein extravasation in conscious rats: role of thromboxane A_2. Eur J Pharmacol 1992; 214:119–125.

186. Lopez-Belmonte J, Whittle BRJ. Endothelin-1 induced neutrophil-independent vascular injury in the rat gastric microcirculation. Eur J Pharmacol 1995; 278:R7–R9.

187. Filep JG, Clozel M, Fournier A, Földes-Filep É. Characterization of receptors mediating vascular responses to endothelin-1 in the conscious rat. Br J Pharmacol 1994; 112: 845–852.

188. Filep JG, Fournier A, Földes-Filep É. Effects of the ET_A/ET_B receptor antagonist, bosentan on endothelin-1-induced myocardial ischaemia and oedema in the rat. Br J Pharmacol 1995; 116:1745–1750.

189. Kurose I, Fukumura D, Miura S, Sekizuka E, Nagata H, Suematsu M, Tsuchiya M. Nitric oxide mediates vasoactive effects of endothelin 3 on rat mesenteric microvascular beds in vivo. Angiology 1993; 44:483–490.

190. Dalhoff B, Gustafsson D, Hedner D, Jern S, Hansson L. Regional hemodynamic effects of endothelin-1 in rat and man: unexpected adverse reactions. J Hypertension 1990; 8: 811–818.

191. Okumura H, Ashizawa N, Aotsuka T, Asakura R, Kobayashi F, Matsuura A. Possible mechanisms of sudden death and hemoconcentration induced by endothelin-1 and big-endothelin-1 in mice. Biol Pharm Bull 1994; 17:645–650.

192. Sampaio ALF, Rae GA, D'Orleans-Juste P, Henriques MGMO. ET_A receptor antagonists inhibit allergic inflammation in the mouse. J Cardiovasc Pharmacol 1995; 26(suppl 3):S416–S418.

193. Chander CL, Moore AR, Desa FM, Howat DW, Willoughby DA. Anti-inflammatory effects of endothelin-1. J Cardiovasc Pharmacol 1989; 13(suppl 5):S218–S219.

194. Brain SD, Crossman DC, Buckley TL, Williams TJ. Endothelin-1: demonstration of potent effects on the microcirculation of humans and other species. J Cardiovasc Pharmacol 1989; 13(suppl 5):S147–S149.

195. Henriques MGMO, Rae GA, Cordeiro RSB, Williams TJ. Endothelin-1 inhibits PAF-induced paw oedema and pleurisy in the mouse. Br J Pharmacol 1992; 106:579–582.

196. Lawrence E, Brain SD. Responses to endothelins in the rat cutaneous microvasculature: a modulatory role of locally produced nitric oxide. Br J Pharmacol 1992; 106:733–738.

197. Lawrence E, Siney L, Wilsoncroft P, Knock GA, Terenghi E, Polak JM, Brain SD. Evidence for ET_A and ET_B receptors in rat skin and an investigation of their function in the cutaneous microvasculature. Br J Pharmacol 1995; 115:840–844.

198. Wenzel RR, Noll G, Luscher TF. Endothelin receptor antagonists inhibit endothelin in human skin microcirculation. Hypertension 1994; 23:581–586.

199. Ferreira SH, Romitelli M, De Nucci G. Endothelin-1 participation in overt and inflammatory pain. J Cardiovasc Pharmacol 1989; 13(suppl 5):S220–S222.

200. Brain SD, Thomas G, Crossman DC, Fuller R, Church MK. Endothelin-1 induces a histamine-dependent flare in vivo, but does not activate human skin mast cells in vitro. Br J Pharmacol 1992; 33:117–120.

201. D'Orléans-Juste P, Claing A, Regoli D, Sirois P, Plante GE. Endothelial and smooth muscle pharmacology of pre- and postcapillary microcirculation: correlation with plasma extravasation. Prostaglandins Leukotrines Essential Fatty Acids 1996; 54: 31–37.

202. D'Orleans-Juste P, Claing A, Warner TD, Yano M, Télémaque S. Characterization of receptors for endothelins in perfused arterial and venous mesenteric vasculatures of the rat. Br J Pharmacol 1993; 110:687–692.

203. Battistini B, Chailler P, D'Orleans-Juste P, Brière N, Sirois P. Growth regulatory properties of endothelins. Peptides 1993; 14:385–399.

204. Herman WH, Simonson MS. Nuclear signaling by endothelin-1. A Ras pathway for activation of the c fos serum response element. J Biol Chem 1995; 270:11654–11661.

205. Morbidelli L, Orlando C, Maggi CA, Ledda F, Ziche M. Proliferation and migration of endothelial cells is promoted by endothelins via activation of ETB receptors. Am J Physiol 1995; 269:H686–H695.

206. Ohlstein EH, Arleth A, Bryan H, Elliott JD, Sung CP. The selective endothelin ETA receptor antagonist BQ123 antagonizes endothelin-1-mediated mitogenesis. Eur J Pharmacol 1992; 225:347–350.

207. Ross R. The pathogenesis of artherosclerosis: a perspective for the 1990s. Nature 1993; 362:801–809.

208. Dashwood MR, Barker SGE, Muddle JR, Yacoub MH, Martin JF. [^{125}I]-endothelin-1 binding to vasa vasorum and regions of neovascularization in human and porcine blood vessels: a possible role for endothelin in intimal hyperplasia and atherosclerosis. J Cardiovasc Pharmacol 1993; 22(suppl 8):S343–S347.

209. Watschinger B, Sayegh MH, Hancock WW, Russel ME. Up-regulation of endothelin-1 mRNA and peptide expression in rat cardiac allografts with rejection and arteriosclerosis. Am J Pathol 1995; 146:1065–1072.

210. Dashwood MR, Allen SP, Luu TN, Muddle JR. The effect of the ET$_A$ receptor antagonist, FR 139317, on [^{125}I]-ET-1 binding to the atherosclerotic human coronary artery. Br J Pharmacol 1994; 112:386–389.

211. Douglas SA, Ohlstein EH. Endothelin-1 promotes neointima formation after balloon angioplasty in the rat. J Cardiovasc Pharmacol 1993; 22(suppl 8):S371–S373.

212. Douglas SA, Louden C, Vickery-Clarck LM, Storer BL, Hart T, Feuerstein GZ, Elliott JD, Ohlstein EH. A role for endogenous endothelin-1 in neointimal formation after rat carotid artery balloon angioplasty. Protective effects of the novel nonpeptide endothelin receptor antagonist SB 209670. Circ Res 1994; 75:190–197.

213. Fujitani Y, Ninomiya H, Okada T, Urade Y, Masaki T. Suppression of endothelin 1 induced mitogenic responses to human aortic smooth muscle cells by interleukin 1β. J Clin Invest 1995; 95:2474–2482.

214. Appleton I, Tomlinson A, Chander CL, Willoughby DA. Effect of endothelin-1 on croton oil induced granulation tissue in the rat. A pharmacologic and immuno-histochemical study. Lab Invest 1992; 67:703–710.

215. Villaschi S, Nicosia RF. Paracrine interactions between fibroblasts and endothelial cells in a serum-free coculture model. Lab Invest 1994; 71:291–299.
216. Raffa RB, Jacoby HI. Endothelin-1, -2 and -3 directly and big-endothelin-1 indirectly elicit an abdominal constriction response in mice. Life Sci 1991; 48:PL85–PL90.
217. Raffa RB, Schupsky JJ, Martinez RP, Jacoby HI. Endothelin-1-induced nociception. Life Sci 1991; 49:PL61–PL65.
218. Raffa RB, Schupsky JJ, Jacoby HI. Endothelin-induced nociception in mice: mediation by ET_A and ET_B receptors. J Pharmacol Exp Ther 1996; 276:647–651.
219. Piovezan AP, D'Orléans-Juste P, Tonussi CR, Rae GA. Endothelins potentiate formalin-induced pain in mice. Peptide Receptors, Montreal, July 28–August 1, 1996.
220. Yoshizawa T, Kimura S, Kanazawa I. Uchiyama Y, Yanagisawa M, Masaki T. Endothelin localizes in the dorsal horn and acts on the spinal neurones: possible involvement of dihydropyridine-sensitive calcium channels and substance P release. Neurosci Lett 1989; 102:179–184.
221. Dymshitz J, Vasko MR. Endothelin-1 enhances capsaicin-induced peptide release and cGMP accumulation in cultures of rat sensory neurons. Neurosci Let 1994; 167:128–132.
222. Giaid A, Gibson SJ, Ibrahim NBN, Legon S, Bloom SR, Yanagisawa M, Masaki T, Varndell IM, Polak JM. Endothelin 1 an endothelium-derived peptide, is expressed in neurons of human spinal cord and dorsal root ganglia. Proc Natl Acad Sci (USA) 1989; 86:7634–7638.
223. Kar S, Chabot J-G, Quirion R. Quantitative autoradiographic localization of [^{125}I]endothelin binding sites in spinal cord and dorsal root ganglia of the rat. Neurosci Lett 1991; 133:117–120.
224. Jones CR, Hiley CR, Pelton JT, Mohr M. Autoradiographic visualization of the binding sites for [^{125}I]endothelin in rat and human brain. Neurosci Let 1989; 97:276–279.
225. Kamei J, Hitosugi H, Kawashima N, Misawa M, Kasuya Y. Antinociceptive effects of intrathecally administered endothelin-1 in mice. Neurosci Lett 1993; 153:69–72.
226. Le Bars D, Bourgoin S, Clot AM, Hamon M, Cesselin F. Noxious mechanical stimuli increase the release of Met-enkephalin-like material heterosegmentally in the rat spinal cord. Brain Res 1987; 402:188–192.
227. Yamamoto T, Shimoyama N, Asano H, Mizuguchi T. Analysis of the role of endothelin-A and endothelin-B receptors on nociceptive information transmission in the spinal cord with FR139317, and endothelin-A receptor antagonist, and sarafotoxin S6c, an endothelin-B receptor agonist. J Pharmacol Exp Ther 1994; 271:156–163.
228. Kluger MJ. Fever: role of pyrogens and cryogens. Physiol Rev 1991; 71:93–127.
229. Koshi T, Edano T, Arai K, Chiyoka S, Ehara Y, Hirata M, Ohkuchi M, Okabe T. Pyrogenic action of endothelin in conscious rabbit. Biochem Biophys Res Commun 1992; 186:1322–1326.
230. Springall DR, Howarth PH, Counihan H, Djukanovic R, Holgate ST, Polak JM. Endothelin immunoreactivity of airway epithelium in asthmatic patients. Lancet 1992; 337:697–701.
231. Vittori E, Marini M, Fasoli A, De Franchis R, Mattoli S. Increased expression of endothelin in bronchial epithelial cells of asthmatic patients and effect of corticosteroids. Am Rev Respir Dis 1992; 146:1320–1325.
232. Ackerman V, Carpi S, Bellini A, Vassalli G, Marini M, Mattoli S. Constitutive expression of endothelin in bronchial epithelial cells of patients with symptomatic and asymptomatic asthma and modulation by histamine and interleukin-1. J Allergy Clin Immunol 1995; 95:618–627.

233. Calderón E, Gómez-Sánchez CE, Cozza EN, Zhou M, Coffey RG, Lockey RF, Prockop LD, Szentivanyi A. Modulation of endothelin-1 production by a pulmonary epithelial cell line. Biochem Pharmacol 1994; 48:2065–2071.

234. Nomura A, Uchida Y, Kameyama M, Saotome M, Oki K, Hasegawa S. Endothelin and bronchial asthma. Lancet 1989; II:747–748.

235. Mattoli S, Soloperto M, Marinni M, Fasoli A. Levels of endothelin-1 in the broncho-alveolar lavage fluid of patients with symptomatic asthma and reversible airflow obstruction. J Allergy Clin Immunol 1991; 88:376–384.

236. Sofia M, Mormile M, Faraone S, Alifano M, Zofra S, Romano L, Carratu L. Increased endothelin-like immunoreactive material on bronchoalveolar lavage fluid from patients with bronchial asthma and patients with interstitial lung disease. Respiration 1993; 60: 89–95.

237. Redington AE, Springall DR, Ghatei MA, Lau LC, Bloom SR, Holgate ST, Polak JM, Howarth PH. Endothelin in bronchoalveolar lavage fluid and its relation to airflow obstruction in asthma. Am J Respir Crit Care Med 1995; 151:1034–1039.

238. Kraft M, Beam WR, Wenzel SE, Zamora MR, O'Brien RF, Martin RJ. Blood and bronchoalveolar lavage endothelin 1 levels in nocturnal asthma. Am J Respir Crit Care Med 1994; 149:946–952.

239. Chanez P, Vignola AM, Albat B, Spingall DR, Polak JM, Godard P, Bousquet J. Involvement of endothelin in mononuclear phagocyte inflammation in asthma. J Allergy Clin Immunol 1996; 98:412–420.

240. Chen WY, Yu J, Wang JY. Decreased production of endothelin-1 in asthmatic children after immunotherapy. J Asthma 1995; 32:29–35.

241. Boichot E, Lagente V, Mencia-Huerta JM, Braquet P. Bronchopulmonary response to endothelin-1 in sensitized and challenged guinea-pigs: role of cyclooxygenase metabolites and platelet-activating factor. Fundam Clin Pharmacol 1993; 7:281–291.

242. Riccio MM, Reynolds CJ, Hay DW, Proud D. Effects of intranasal administration of endothelin-1 to allergic and nonallergic individuals. Am J Respir Crit Care Med 1995; 152:1757–1764.

243. Goldie RG, Henry PJ, Paterson JW, Preuss JM, Rigby PJ. Contractile effects and receptor distribution for endothelin-1 (ET-1) in human and animal airways. Agents Actions 1990; 31:229–232.

244. Goldie RG, Henry PJ, Knott PG, Self GJ, Luttmann MA, Hay DW. Endothelin-1 receptor density, distribution, and function in human isolated asthmatic airways. Am J Respir Crit Care Med 1995; 152:1653–1658.

245. Knott PG, D'Aprile AC, Henry PJ, Hay DW, Goldie RG. Receptors for endothelin 1 in asthmatic human peripheral lung. Br J Pharmacol 1995; 114:1–3.

246. Aoki T, Kojima T, Ono A, Unishi G, Yoshijima S, Kameda-Hayashi N, Yamamoto C, Hirata Y, Kobayashi Y. Circulating endothelin-1 levels in patients with bronchial asthma. Ann Allergy 1994; 73:365–369.

247. White SR, Hathaway DP, Umans JG, Leff AR. Direct effects on airway smooth muscle contractile response caused by endothelin 1 in guinea pig trachealis. Am Rev Respir Dis 1992; 145:491–493.

248. Lagente V, Boichot E, Mencia-Huerta JM, Braquet P. Failure of aerosolized endothelin (ET-1) to induce bronchial hyperreactivity in the guinea pig. Fundam Clin Pharmacol 1990; 4:275–280.

249. Kanasawa H, Kurihara N, Hirata K, Fujiwara H, Matsushita H, Takeda T. Low concen-

tration endothelin-1 enhanced histamine-mediated bronchial contractions of guinea pigs in vivo. Biochem Biophys Res Commun 1992; 187:717–721.

250. Carpi S, Marini M, Vittori E, Vassalli G, Mattoli S. Bronchoconstrictive responses to inhaled ultrasonically nebulized distilled water end airway inflammation in asthma. Chest 1993; 104:1346–1351.

251. Noguchi K, Ishikawa K, Yano M, Ahmed A, Cortes A, Abraham WM. Endothelin-1 contributes to antigen-induced airway hyperresponsiveness. J Appl Physiol 1995; 79: 700–705.

252. Andersson SE, Eirefelt S, Zackrisson C, Hemsén A, Dahlbäck M, Lundberg JM. Regulatory effects of aerosolized budesonide and adrenalectomy on the lung content of endothelin-1 in the rat. Respiration 1995; 64:34–39.

253. Macquin-Mavier I, Levame M, Istin N, Harf A. Mechanisms of endothelin-mediated bronchoconstriction in the guinea pig. J Pharmacol Exp Ther 1989; 250:740–745.

254. Lueddeckens E, Bigl H, Sperling J, Becker K, Braquet P, Foster W. Importance of secondary TXA_2 release in mediating of endothelin-1 induced bronchoconstriction and vasopressin in the guinea-pig. Prostaglandins Leukotrienes Essential Fatty Acids 1993; 48:261–263.

255. Peacock AJ, Dawes KE, Shock A, Gray AJ, Reeves JT, Laurent GJ. Endothelin 1 and endothelin 3 induce chemotaxis and replication of pulmonary artery fibroblasts. Am J Respir Cell Mol Biol 1992; 7:492–499.

256. Noveral JP, Rosenberg SM, Anbar RA, Pawlowski NA, Grunstein MM. Role of endothelin-1 in regulating proliferation of cultured rabbit airway smooth muscle cells. Am J Physiol 1992; 263:L317–L324.

257. Glassberg MK, Ergul A, Wanner A, Puett D. Endothelin-1 promotes mitogenesis in airway smooth muscle cells. Am J Respir Cell Mol Biol 1994; 10:316–321.

258. Stewart AG, Grigoriadis G, Harris T. Mitogenic actions of endothelin-1 and epidermal growth factor in cultured airway smooth muscle. Clin Exp Pharmacol Physiol 1994; 21:277–285.

259. Tomlinson PR, Wilson JW, Stewart AG. Inhibition by salbutamol of the proliferation of human airway smooth muscle cells grown in culture. Br J Pharmacol 1994; 111: 641–647.

260. Shimura S, Ishihara H, Satoh M, Masuda T, Nagaki N, Sasaki H, Takashima T. Endothelin regulation of mucus glycoprotein secretion from feline tracheal submucosal glands. Am J Physiol 1992; 262:L208–L213.

261. Yurdakos E, Webber SE. Endothelin-1 inhibits pre-stimulated tracheal submucosal gland secretion and epithelial albumin transport. Br J Pharmacol 1991; 104:1050–1056.

262. Goldie RG, Fernandes LB, Farmer SG, Hay DW. Airway epithelium-derived inhibitory factor. Trends Pharmacol Sci 1990; 11:67–70.

263. Giaid A, Michel RP, Stewart DJ, Sheppard M, Corrin B, Hamid Q. Expression of endothelin-1 in lungs of patients with cryptogenic fibrosin alveolitis. Lancet 1993; 341: 1550–1554.

264. Zachariae R, Heickendorff L, Bjerring P, Halkier-Sørensen L, Søndergaard K. Plasma endothelin and the aminoterminal propeptide of type III procollagen (PIIINP) in systemic sclerosis. Acta Derm Venereol 1994; 74:368–370.

265. Kadono T, Kikuchi K, Sato S, Soma Y, Tamaki K, Takehara K. Elevated plasma endothelin levels in systemic sclerosis. Arch Dermatol Res 1995; 287:439–442.

266. Kawaguchi Y, Suzuki K, Hara M, Hidaka T, Ishizuka T, Kawagoe M, Nakamura H.

Increased endothelin-1 production in fibroblasts derived from patients with systemic sclerosis. Ann Rheum Dis 1994; 53:506–510.

267. Morelli S, Ferri C, Polettini E, Bellini C, Gualdi GF, Pittoni V, Valesini G, Santucci A. Plasma endothelin-1 levels, pulmonary hypertension, and lung fibrosis in patients with systemic sclerosis. Am J Med 1995; 99:255–260.

268. Uguccioni M, Pulsatelli L, Grigolo B, Facchini A, Fasano L, Cinti C, Fabbri M, Gasbarrini G, Meliconi R. Endothelin-1 in idiopathic pulmonary fibrosis. J Clin Pathol 1995; 48:330–334.

269. Cambrey AD, Harrison NK, Dawes KE, Southcott AM, Black CM, du Bois RM, Laurent GJ, McAnulty RJ. Increased levels of endothelin-1 in bronchoalveolar lavage fluid from patients with systemic sclerosis contribute to fibroblast mitogenic activity in vitro. Am J Respir Cell Mol Biol 1994; 11:439–445.

270. Julkunen H, Saijonmaa O, Gröhnagen-Riska C, Teppo AM, Fyhrquist F. Raised plasma concentrations of endothelin-1 in systemic lupus erythematosus. Ann Rheum Dis 1991; 50:526–527.

271. Ferri C, Latorraca A, Catapano G, Greco F, Mazzoni A, Clerico A, Pedrinelli R. Increased plasma endothelin-1 immunoreactive levels in vasculitis: a clue to the use of endothelin-1 as a marker of vascular damage? J Hypertension 1993; 11(suppl 5):S142–S143.

272. Zamorra MR, O'Brien RF, Rutherford RB, Weil JV. Serum endothelin-1 concentrations and cold provocation in primary Raynaud's phenomenon. Lancet 1990; 2:1144–1147.

273. Biondi ML, Marasini B, Bassani C, Agastoni A. Increased plasma endothelin levels in patients with Raynaud's phenomenon. N Engl J Med 1991; 324:1139–1140.

274. Cimminiello C, Milani M, Uberti T, Arpaia G, Perolini S, Bonfardeci G. Endothelin vasoconstriction and endothelial damage in Raynaud's phenomenon. Lancet 1991; 1: 114–115.

275. Filep JG, Fournier A, Földes-Filep É. Acute pro-inflammatory actions of endothelin-1 in the guinea-pig lung: involvement of ET_A and ET_B receptors. Circulation 1995; 92: 2969–2974.

276. Miyasaka N, Hirata Y, Ando K, Sato K, Morita H, Shichiri M, Kanno K, Tomita K, Marumo F. Increased production of endothelin-1 in patients with inflammatory arthritides. Arthr Rheum 1992; 35:397–400.

277. Yoshio T, Masuyama J, Mimori A, Takeda A, Minota S, Kano S. Endothelin-1 release from cultured endothelial cells induced by sera from patients with systemic lupus erythematosus. Ann Rheum Dis 1995; 54:361–365.

278. Nakamura T, Ebihara I, Tomino Y, Koide H. Effect of a specific endothelin A receptor antagonist on murine lupus nephritis. Kidney Int 1995; 47:481–489.

279. Yoshimura A, Iwasaki S, Inui K, Ideura T, Koshikawa S, Yanagisawa M, Masaki T. Endothelin-1 and endothelin B type receptor are induced in mesangial proliferative nephritis in the rat. Kidney Int 1995; 48:1290–1297.

280. Iwasaki S, Homma T, Matsuda Y, Kon V. Endothelin receptor subtype B mediates auto-induction of endothelin-1 in rat mesangial cells. J Biol Chem 1995; 270:6997–7003.

281. Benigni A, Zoja C, Corna D, Orisio S, Longaretti L, Bertani T, Remuzzi G. A specific endothelin subtype A receptor antagonist protects against injury in renal disease progression. Kidney Int 1993; 44:440–444.

282. Ohta K, Hirata Y, Shichiri M, Kanno K, Emori T, Tomita K, Marumo F. Urinary excretion of endothelin-1 in normal subjects and patients with renal disease. Kidney Int 1991; 39:307–311.

283. Simonson MS. Endothelins: multifunctional renal peptides. Physiol Rev 1993; 73:375–411.

284. Klemm P, Warner TD, Corder R, Vane JR. Endothelin-1 mediates coronary vaso-constriction caused by exogenous and endogenous cytokines. J Cardiovasc Pharmacol 1995; 26(suppl 3):S419–S421.

285. Nahir AM, Hoffman A, Lorber M, Keiser HR. Presence of immunoreactive endothelin in synovial fluid: analysis of 22 cases. J Rheumatol 1991; 18:678–680.

286. Sasaki T, Hong M-H. Endothelin-1 localization in bone cells and vascular endothelial cells in rat bone marrow. Anatom Rec 1993; 237:332–337.

287. Stern PH, Tatrai A, Semler DE, Lee SK, Lakatos P, Strieleman PJ, Tarjan G, Sanders JL. Endothelin receptors, second messengers, and actions in bone. J Nutr 1995; 125(suppl 7):2028S–2032S.

288. Sasaki T, Hong M-H. Localization of endothelin-1 in the osteoclast. J Electron Microsc 1993; 42:193–196.

289. Towhidul Alam T, Gallaguer ASM, Shankar V, Ghatei MA, Datta HK, Huang CL-H, Moonga BS, Chambers TJ, Bloom SR, Zaidi M. Endothelin inhibits osteoclastic bone resorption by a direct effect on cell motility: implications for the vascular control of bone resorption. Endocrinology 1992; 130:3617–3624.

290. Tatrai A, Lakatos P, Thompson S, Stern PH. Effects of endothelin-1 on signal transduc-tion in UMR 106 osteoblastic cells. J Bone Miner Res 1992; 7:1201–1209.

291. Whittle BJR, Esplugues JV. Induction of rat gastric damage by the endothelium-derived peptide, endothelin. Br J Pharmacol 1988; 95:1011–1013.

292. Whittle BJR, Payne AN, Espluges JV. Cardiopulmonary and gastric ulcerogenic actions of endothelin-1 in the guinea pig and rat. J Cardiovasc Pharmacol 1989; 13(suppl 5): S103–S107.

293. Lazaratos S, Kashimura H, Nakahara A, Fukutomi H, Osuga T, Urushidani T, Miyauchi T, Goto K. Gastric ulcer induced by submucosal injection of ET-1—role of potent vasoconstriction and intraluminal acid. Am J Physiol 1993; 265:G491–G498.

294. MacNaughton W, Keenan CM, McKnight WG, Wallace JL. The modulation of gastric mucosal integrity by endothelin-1 and prostacyclin. J Cardiovasc Pharmacol 1989; 13 (suppl 5):S118–S122.

295. Wallace JL, Cirino E, de Nucci G, McKnight W, MacNaughton WK. Endothelin has potent ulcerogenic and vasoconstrictor actions in the stomach. Am J Physiol 1989; 256: G661–G666.

296. Wallace JL, Keenan CM, MacNaughton WK, McKnight W. Comparison of the effects of endothelin-1 and endothelin-3 on the rat stomach. Eur J Pharmacol 1989; 167:41–47.

297. Peskar BM, Nowak P, Lambrecht N. Effect of prostaglandins and capsaicin on gastric vascular flow and mucosal injury in endothelin-1 treated rats. Agents Actions 1992; 37: 85–91.

298. Masuda E, Kawano S, Nagano K, Tsuji S, Takei Y, Hayashi N, Tsujii M, Oshita M, Michida T, Kobayashi I. Role of endogenous endothelin in pathogenesis of ethanol-induced gastric mucosal injury in rats. Am J Physiol 1993; 265:G474–G481.

299. Wallace H, McNight GW. Mucoid cap over superficial gastric damage in the rat—a high pH microenvironment dissipated by nonsteroidal anti-inflammatory drugs and endo-thelin. Gastroenterology 1990; 99:295–304.

300. Kitajima T, Yamaguchi T, Tani K, Kubota Y, Okuhira M, Inoue K, Yamada H. Role of endothelin and platelet-activating factor in indomethacin-induced gastric mucosal in-jury in rats. Digestion 1993; 54:156–159.

301. Saeki Y, Higuchi K, Nakamura S, Arakawa T, Nagura H, Kobayashi K. Location of endothelin detected immunohistochemically in gastric mucosa. Third International Conference on Endothelin, Houston, TX, February 15–17, 1993.

302. Fukumura D, Kurose I, Miura S, Serizawa H, Sekizuka E, Nagata H, Tsuchiya M, Ishii H. Role of endothelin-1 in repeated electrical stimulation-induced microcirculatory disturbance and mucosal damage in rat stomach. J Gastroenterol Hepatol 1996; 11: 279–285.

303. Whittle BJR, Lopez-Belmonte J. Interactions between the vascular peptide endothelin-1 and sensory neuropeptides in gastric mucosal injury. Br J Pharmacol 1989; 102: 950–954.

304. Lopez-Belmonte J, Whittle BJR. Calcitocin-gene related peptide can augment or prevent endothelin-1 induced gastric microvascular leakage. Eur J Pharmacol 1994; 271: R15–R17.

305. Lazaratos S, Kashimura H, Nakahara A, Fukutomi H, Osuga T, Miyauchi T, Goto K. Endothelin-1-induced gastric ulcer is attenuated by cetraxate. Life Sci 1993; 53:PL123–PL128.

306. Wolfe MM, Soll AH. The physiology of gastric acid secretion. N Engl J Med 1988; 319: 1707–1715.

307. Murch SH, Braegger CP, Sessa WC, MacDonald TT. High endothelin-1 immunoreactivity in Crohn's disease and ulcerative colitis. Lancet 1992; 339:381–385.

308. Rachmilewitz D, Eliakim R, Ackerman Z, Karmeli F. Colonic endothelin-1 immunoreactivity in active ulcerative colitis. Lancet 1992; 399:1062.

309. Miura S, Kurose I, Fukumura D, Suematsu M, Sekizuka E, Tashiro H, Serizawa H, Asako H, Tsuchiya M. Ischemic bowel necrosis induced by endothelin-1: an experimental model in rats. Digestion 1991; 48:163–172.

310. Brown MA, Smith PL. Endothelin: a potent stimulator of intestinal ion secretion in vitro. Regulatory Peptides 1991; 36:1–19.

311. Roden M, Plass H, Vierhapper H, Turnheim K. Endothelin-1 stimulates chloride and potassium secretion in rabbit descending colon. Eur J Pharmacol 1992; 421:163–167.

312. Moummi C, Xie Y, Kachur JF, Gaginella TS. Endothelin-1 stimulates contraction and ion transport in the rat colon: different mechanisms of action. J Pharmacol Exp Ther 1992; 262:409–414.

313. Kiyohara T, Okuno M, Nakanishi T, Shinomura Y, Matsuzawa Y. Effect of endothelin 1 on ion transport in isolated rat colon. Gastroenterology 1993; 104:1328–1336.

314. Battistini B, Forget M-A, Laight D. Potential roles for endothelins in systemic inflammatory response syndrome with a particular relationship to cytokines. Shock 1996; 5: 167–183.

8

Endothelins in the Lung

PETER J. HENRY
and ROY G. GOLDIE

University of Western Australia
Perth, Western Australia, Australia

DOUGLAS W. P. HAY

SmithKline Beecham Pharmaceuticals
King of Prussia, Pennsylvania

I. Introduction

The initial pharmacological characterization of a potent vasoconstrictor peptide called endothelin (ET), released from porcine aortic endothelial cells, was reported by Yanagisawa and colleagues in 1988 (1). This peptide was shown subsequently to be one member of a family of three vasoconstrictor peptides, designated ET-1, ET-2, and ET-3, which bear a close structural and functional homology to the sarafotoxins (Stxs), a group of snake venom toxins. As shown in Fig. 1, ET-2 differs by two amino acids from ET-1, and ET-3 differs from ET-1 and ET-2 by six amino acids (2–4).

Although early research interest focused on the activity of the ETs in the cardiovascular system, it is now clear that these peptides induce an array of effects in the respiratory tract. Many of these effects, including smooth muscle contraction, mitogenesis and migration, microvascular leakage and edema, mucous gland hypersecretion, and neuromodulation are thought to be important in the pathophysiology of various lung disorders. This review brings together much of the current information relating to the synthesis, release, and mechanisms of actions of ET in the airways and peripheral lung, and discusses the evidence for a role of the ETs in inflammatory lung diseases, particularly asthma.

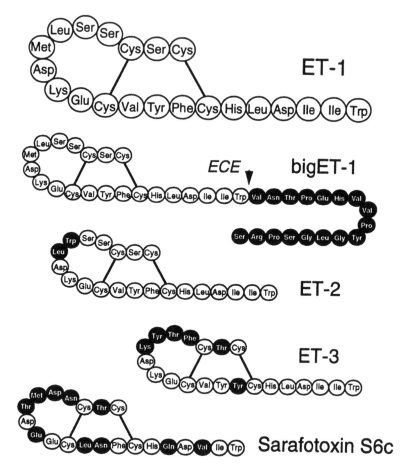

Figure 1 Diagramatic representations of the chemical structures of ET-1 and related peptides.

II. Endothelin Synthesis

Levels of immunoreactive ET (ir-ET) in the lung are among the highest of any tissue (5–8). In human airways, immunoreactivity for ET-like peptides was detected in the vascular endothelium, airway epithelium and submucosal glands (9), and in macrophages (10). The initial gene product in the synthesis of human ET-1 is preproET-1, a 212-amino acid protein. This precursor peptide is then cleaved sequentially at pairs of basic amino acids to generate big ET-1, an intermediate 38-amino acid peptide the biologic activity of which is approximately 100-fold less than that of ET-1 (11,12). It has been proposed that the processing of preproET-1 involves the action of furin, a mammalian convertase (13). The final step in the synthesis of ET-1 occurs when big ET-1 is cleaved between Trp^{21} and Val^{22}, to form ET-1 via a putative endopeptidase called ET-converting enzyme (ECE). The most recent evidence indicates the existence of two distinct forms of ECE, called ECE-1 (14) and ECE-2 (15). These enzymes in bovine aortic endothelial cells are both phosphoramidon-sensitive metalloproteases that are not sensitive to the neutral endopeptidase (NEP) inhibitor thiorphan or the angiotensin-converting enzyme inhibitor captopril. ECE-1 is a cell membrane-bound enzyme with optimal activity at neutral pH. In contrast, ECE-2 appears to be associated with the trans Golgi apparatus, has an activity optimum at pH 5.5, and is about 250 times more sensitive to phosphoramidon than ECE-1 (15). Thus, whereas ECE-2 is probably responsible for intracellular conversion of big ET-1 to ET-1, ECE-1 may be able to convert both intracellular and exogenously supplied big ET-1 to ET-1. The proteolytic conversion of big ET-1 to ET-1 also occurred via a phosphoramidon-sensitive pathway in rabbit (16) and guinea pig lung (17–19), as well as in guinea pig upper bronchus (18) and airway epithelial Clara cells (20,21). These latter in-vitro findings are consistent with in vivo observations that a phosphoramidon-sensitive pathway was essential for bronchoconstrictor responses following intravenous administration of big ET-1 to guinea pigs (22). Inhibition of ECE may provide an important therapeutic target in diseases associated with elevated ET-1 levels. The development of selective, potent, nonpeptide inhibitors of ECE and their application to the investigation of pulmonary disease is of particular interest.

III. Sites of Endothelin Synthesis and Release in the Lung

As indicated above, ET is synthesized and thus could be released within the bronchial wall from the airway epithelium, some inflammatory cells, as well as from vascular endothelial cells. ET-like immunoreactivity and/or mRNA for preproET-1 is present in epithelial cells in the conducting airways of the rat, mouse (23,24), rabbit (25), guinea pig (20,26), and human (9,27,28). In the rat and mouse epithelium, there was little or no ir-ET in basal cells or ciliated epithelial cells, whereas mucous, serous, and Clara cells stained intensely (23). Recent studies of guinea pig tracheal epithelial cell monolayers have reported that the ET-1 content of the basal side of these cells was over 30 times higher than that of the apical side (29), suggesting that upon formation, ET-1 is released from airway epithelial cells toward the submucosal surface and into the airway wall. Thus, the airway epithelium is likely to be an important source of ET-1 in the airway wall. Additional studies, which have assessed the topographical location of vascular endothelial cells and the ability of these cells to generate ET-1, suggest that endothelial cell-derived ET-1 might also contribute to airway smooth muscle contraction (30). In addition to the above locations, the expression of mRNA for ET-1 in human lung macrophages has been demonstrated (10).

Consistent with the reports of significant levels of ET-1-like immunoreactivity and mRNAs for preproET-1 in epithelial cells, the basal release of ET-1 has been detected in culture from bronchial epithelial cells from the pig, dog (31), and human (32), as well as from tracheal epithelial cells from rabbit (25) and guinea pig (33,34). ET-3 has also been detected in supernatants from canine and porcine cultured tracheal epithelial cells (31). As indicated below, the synthesis and release of the ETs from airway cells is stimulated by inflammatory mediators including endotoxin and thrombin (26,34–36).

IV. Elevated Levels of ET-1 in Asthma

It is well established that asthma is associated with increased expression and/or release of the ETs. Furthermore, the severity of asthma may be related to the airway levels of ET-1, both of which were attenuated by glucocorticoid therapy (37). A role for ET-1 in asthma was first suggested by a report concerning a patient with status asthmaticus in whom elevated levels of ir-ET were detected in BAL fluid (38). Larger studies have subsequently shown that symptomatic asthma is associated with higher levels of ET in BAL fluid than is asymptomatic asthma or chronic bronchitis (37,39). The increased amounts of ET in BAL fluid from asthmatic subjects were disproportionately high in relation to BAL fluid albumin levels, indicating that this was most likely caused by augmented local production of ET, e.g., by epithelial cells (40–42), rather than by leakage from the circulation (37). This is also supported by evidence for positive staining for ir-ET in bronchial epithelium in most asthmatics examined, but in relatively few of control subjects (40) or in asymptomatic asthmatic patients (42). Other studies using epithelial cells isolated from asthmatic patients have also demonstrated that the expression of preproET mRNA and release high amounts of ir-ET (43).

It is important to note that inflammatory cytokines and mediators previously implicated in asthma are released from inflammatory cells in response to ET-1. For example, ET-1 caused mast cell activation in the presence of IL-4 (44) and potently stimulated human monocytes to release IL-6 (45). However, several proinflammatory mediators are also known to stimulate the release of ET from airway cells, consistent with the notion that ET is a mediator in asthma. For example, in primary cultures of human airway epithelial cells derived from individuals without airway inflammation, Il-1, Il-2, Il-6, and IGF-1 elevated levels of the ET-1 precursor bigET-1 (36). Furthermore, Il-1α, Il-1β, and TNF-α stimulated the expression of mRNA for preproET-1 and the release of ET-1 (36). These cytokines also elevated mRNA levels for preproET-1 in human bronchial epithelial cells (28). In addition, bronchial epithelial cells from asymptomatic asthmatics synthesized and released ET-1 in response to Il-1β and histamine (4). Il-8, TNF-α, and TFG-β may also stimulate ET-1 release from cultured airway epithelial cells (33). The influence of these cytokines on ET-1 synthesis and release is likely to be significant in inflammatory respiratory diseases such as asthma. Interestingly, CD23-positive (low-affinity receptor for IgE) epithelial cells from a group of allergic asthmatics subjects responded to IgE by releasing ET-1 (46).

V.   Endothelin Catabolism in the Airways

The ETs are good substrates for the catabolic enzyme neutral endopeptidase (NEP), which has similar affinity for each of the isopeptides (47,48). This is consistent with the observation that in guinea pig isolated trachea, the presence of an intact epithelium, which is a rich source of NEP, attenuated contractile responses induced by ET-1 via a mechanism which was sensitive to the NEP inhibitor phosphoramidon (49–51). Studies in vivo have also demonstrated that phosphoramidon-sensitive NEP modulated ET-1-induced bronchoconstriction in the guinea pig (52). Interestingly, in rabbit bronchus, phosphoramidon potentiated contractile responses to ET-3, but not to ET-1 (53). Conflicting data have been reported in human isolated bronchus, where phosphoramidon potentiated (54) or had no effect (53) on ET-1-induced contraction.

Pathways for the degradation of ET-1 other than that involving NEP also exist. For example, in activated human polymorphonuclear neutrophils, phosphoramidon failed to inhibit ET-1 catabolism, whereas soybean trypsin inhibitor abolished degradation (55). This suggests that cathepsin G, which is 300 to 3600 times more abundant than NEP in these cells, is involved in ET-1 metabolism at this site.

VI.   Endothelin Receptors

Two ET receptor subtypes, known as ET_A and ET_B, are responsible for mediating the actions of the ETs in mammalian systems. Both receptor subtypes have been cloned in human cells and in a variety of cells and tissues from mammalian species (56–57). With the aid of recently developed ET receptor subtype-selective ligands, ET_A and ET_B receptors have been detected throughout the lung (68–73). Functional and radioligand-binding studies have suggested the existence of an ET_C receptor which exhibits selectivity for ET-3 over ET-1 (74), although this subtype has not been detected in mammalian tissues.

A.   Endothelin Receptor Subtype-Selective Ligands

Stx S6b and Stx S6c have significant sequence homology with the ETs (75). As with ET-1, Stx S6b is an agonist which does not discriminate between ET_A and ET_B receptor subtypes, whereas Stx S6c is highly selective for the ET_B receptor subtype (76). The synthetic linear peptides BQ-3020 (77) and IRL-1620 (78) are also selective ET_B receptor agonists (Table 1). Of the available ET_A receptor-selective antagonists, the cyclic pentapeptide BQ-123 (79) is the most commonly used and the best characterized. Several peptide ET_B receptor-selective antagonists have also been developed, with the tripeptide BQ-788 being the current "gold standard" in this group (80). Other ET_B receptor-selective antagonists include the 16-amino acid peptide RES-701-1, purified from the culture broth of *Streptomyces* sp. (81,82).

ET-1 binds to ET_A and ET_B receptors in a pseudo-irreversible manner (83–86). Similarly, binding of the ET_A receptor-selective antagonist $[^{125}I]$-PD151242 was

Table 1 Some Commonly Used ET Receptor
Subtype-Selective Ligands

Drug	Action	ET receptor selectivity
BQ-123	Antagonist	ET_A
FR-139317	Antagonist	ET_A
PD-151242	Antagonist	ET_A
BQ-788	Antagonist	ET_B
Sarafotoxin S6c	Agonist	ET_B
BQ-3020	Agonist	ET_B
IRL-1620	Agonist	ET_B

only slowly reversible (dissociation half-life of approximately 300 min) (87), as was the binding of the ET_B receptor-selective agonist $[^{125}I]$-BQ-3020 (77). In contrast, the ET_B receptor-selective agonist $[^{125}I]$-IRL-1620 (78) and the ET_A receptor-selective antagonists BQ-123 (86) and FR139317 (88) bound reversibly. Interestingly, the binding of $[^{125}I]$-IRL-1620 to ET_B receptors was reversible in canine and human tissues, but was pseudo-irreversible in the rat (83).

B. Atypical Endothelin Receptors

Recently a number of groups have reported that the ability of ET receptor antagonists, such as BQ-123, to inhibit functional responses mediated by ET receptors is dependent on the agonist used (89–91). In some cases this has been used as evidence for the existence of novel ET_A and ET_B receptor subtypes (92,93). For example, atypical ET_A and ET_B receptors have been suggested to mediate the contractile actions of ET-1 in guinea pig hilar bronchus (94) and rabbit trachea (92). However, due to the pitfalls associated with the classification of receptor subtypes based solely on functional data (95), any extension to the current ET receptor subclassification requires additional support from structural, operational, and signal transduction data (96). Indeed, recent studies of fibroblasts expressing cloned ET_A receptors indicate that the differential sensitivities of ET-1 and ET-3 to blockade by BQ-123 is an intrinsic property of activated ET_A receptors and that there is no need to postulate the existence of new ET_A receptor isoforms (97). For a general discussion of atypical endothelin receptors, the reader is directed to a recent review by Bax and Saxena (98).

VII. Endothelin Receptors in the Lung

A. Airway Wall

Airway Smooth Muscle

Marked species differences have been identified with respect to the density and proportions of airway smooth muscle ET_A and ET_B receptors (Table 2). Two recent studies indicate that between 60% and 90% of specific $[^{125}I]$-ET-1 binding in human

Table 2 ET Receptor Subtype Proportions in Pulmonary Tissue

Tissue cell	Species	Ratio $ET_A : ET_B$	Reference
Tracheal airway smooth muscle	Human (cultured cells)	35 : 65	(200)
	Sheep	100 : 0	(68)
	Sheep (cultured cells)	85 : 15	(139)
	Rat	50 : 50	(101)
	Mouse	40 : 60	(100)
	Pig	30 : 70	(72)
Bronchial airway smooth muscle	Human	10 : 90	(70)
		40 : 60	(99)
	Pig	70 : 30	(72)
Pulmonary artery	Human	90 : 10	(73)
		93 : 7	(111)
	Rabbit	20 : 80	(292)
		23 : 77	(111)
		40 : 60	(112)
	Pig	80 : 20	(293)
Peripheral lung	Human	30 : 70	(69)
	Guinea pig	15 : 85	(71)
	Rat	55 : 45	(71)
	Rat (alveolar epithelial cell line, L2)	100 : 0	(102)
	Pig		(71)
		30 : 70	

bronchial airway smooth muscle was to the ET_B receptor subtype (70,99), with the remainder identified as BQ-123-sensitive, ET_A-receptor binding. At the other end of the spectrum, specific binding of $[^{125}I]$-ET-1 to sheep tracheal airway smooth muscle was completely abolished by the ET_A receptor-selective ligand BQ-123 (68), suggesting that this tissue possessed a homogeneous population of ET_A receptors. Mixtures of ET_A and ET_B receptors were present in approximately equal proportions in both mouse and rat tracheal airway smooth muscle (100,101). However, regional variation in the proportion of airway smooth muscle ET_A and ET_B receptors complicates this picture. For example, pig tracheal smooth muscle contained approximately 70% ET_B receptors and 30% ET_A receptors. In contrast, in pig bronchial airway smooth muscle these percentages were reversed (72). Thus, although both ET_A and ET_B receptors were present in the upper and lower airways of the pig, the ET_B receptor predominated in tracheal smooth muscle, whereas the ET_A receptor was the major subtype in bronchial airway smooth muscle. Similar regional differences may also occur in the respiratory tracts of sheep and humans, where apparently homogeneous receptor populations have been identified.

Other Sites

Both ET_A and ET_B receptors have been detected in peripheral lung tissues from human and several animal species. This has been confirmed in functional (102,103), molecular biological (64,104), and radioligand-binding (69,71,83,103,105,106) experiments. In autoradiographic studies, ET_A receptors were found in submucosal glands (68,69,106), whereas ET_B receptors were associated with pulmonary parasympathetic ganglia and submucosal nerve plexuses (107). Binding sites for [^{125}I]-ET-1 have also been demonstrated in parasympathetic ganglia, paravascular nerves, and nerves in the connective tissues in other mammalian lung (108). In human peripheral lung, ET_A sites were identified in nerves associated with small bronchi (69). Epithelial cell binding is present but generally sparse compared with that seen in airway smooth muscle in guinea pig (109), ovine (68), or human bronchi (69,70).

Blood vessels within the lung also contain ET receptors. Messenger RNA encoding both ET_A and ET_B receptors has been detected in human pulmonary artery (110), although the majority of ET receptors were of the ET_A subtype (110,111). ET_A receptors were also found in interpulmonary blood vessels in several mammalian species (68,69,71,106). In contrast, about 80% of the receptors in rabbit pulmonary artery were of the ET_B subtype (112,113).

B. Alveolar Wall

In the rat, the lung takes up approximately 80% of the bolus dose of [^{125}I]-ET-1 given via a jugular vein, with binding primarily to alveolar capillary endothelial cells (105). A similar amount was removed after one passage in perfused rat lung (114), suggesting that some of these binding sites may play an important role in the metabolic clearance of ET-1 (115). The lung also appears to be a primary site for ET-1 clearance in humans (116,117). Certainly, both ET_A and ET_B sites have been detected within the alveolar wall in rat, guinea pig, ovine, porcine, and human lung (68–71). Given that ET-1 causes contraction of ovine peripheral lung parenchyma strip preparations (68), some alveolar wall ET receptors may be associated with contractile alveolar interstitial myofibroblasts (118). Recent studies in rat and rabbit lung have also shown that ET_B receptors were present on alveolar type II pneumocytes, suggesting a role in either the synthesis or secretion of surfactant (119). Rat cultured alveolar epithelial cells also express functional ET_A receptors which mediated ET-1 induced increases in prostaglandin E_2 and adenosine 3'5'-cyclic monophosphate (c-AMP) production (102).

VIII. Endothelin Receptor-Mediated Effects in the Lung

Functional studies in vitro and in vivo strongly suggest that the specific ET_A- and ET_B-binding sites for ET-1 are indeed active receptors that mediate a myriad of actions within the airways (Fig. 2). Importantly, cells containing these receptors are potential target sites for epithelium- and/or endothelium-derived ET-1 (120).

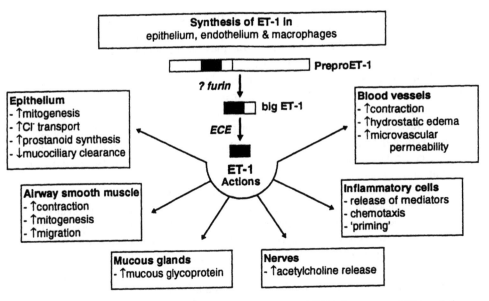

Figure 2 Diagramatic representation of the synthesis of ET-1 from preproET-1 and the proposed actions of ET-1 in the airway wall.

A. Airway Smooth Muscle

Contraction

In-Vitro Studies

It is now well recognized that the ETs and related peptides induce potent contraction in tracheal and bronchial airway smooth muscle from the human (70,99,121–129) and many animal species (68,72,124,130–138). However, the extent to which ET_A and ET_B receptors contribute to ET-1-induced contraction differs markedly between species. In ovine isolated airway smooth muscle, where ET_A sites are overwhelmingly the majority subtype (68,139), ET-1-induced contraction is abolished by the ET_A receptor-selective antagonist BQ-123, and the ET_B receptor selective agonist Stx S6c is inactive (68). However, in human bronchial preparations, Stx S6c caused powerful contraction, and the contractile response to ET-1 was not inhibited by the ET_A receptor-selective antagonists BQ-123 and FR139317, strongly suggesting that ET_B receptors were linked to contraction in this tissue (70,127,129). Nevertheless, additional studies using ET receptor subtype-selective antagonists indicate that ET_A receptors are also linked to contraction in human isolated bronchus (99). ET-1-induced contractions were mediated predominantly by ET_B receptors in rabbit airways, whereas the ET_A subtype mediated contraction in canine bronchus (138). In rat and mouse tracheal smooth muscle which contained approximately equal proportions of ET_A and ET_B receptors, ET-1-induced contraction could be triggered via either receptor subtype (100,101). However, as reported recently in human bronchus (99), blockade of just one subtype had little effect on either the potency or efficacy of ET-1. Effective blockade of ET-1-induced contractions occurred only following their simultaneous blockade (92,100,101,137).

Marked regional differences in the function of ET_A and ET_B receptors also exist, which presumably reflect differences in receptor density between these regions. For example, Stx S6c induced greater contraction in guinea pig primary bronchus than in trachea (122), suggesting a higher proportion of ET_B receptors in primary bronchus than in trachea. In contrast, ET_B receptor density and function were greater in porcine trachea than in bronchus (72). Such regional variations may also exist in the human respiratory tract.

In some species, ET-1-induced airway smooth muscle contraction may also involve the release of secondary mediators in addition to its direct effects on airway smooth muscle, although the evidence is somewhat conflicting. For example, cyclooxygenase inhibitors have been reported to inhibit (140–142), to potentiate (124,130, 143,144), or to have no effect (124,145) on contractions induced by ET-1. In human bronchus, ET-1 evoked the release of an array of prostanoids (121), but there is no evidence that cyclooxygenase products modulated ET-1-induced contractions in this tissue (121,123,126,146). The evidence for a role for histamine, thromboxane A_2, platelet-activating factor, and the leukotrienes as secondary mediators of ET-1-induced contraction in guinea pig isolated airways is also inconsistent, with some studies demonstrating inhibition of ET-1-induced contractions in the presence of receptor antagonists for these mediators (141,147,148) and no significant effect reported in others (130,141,145,149).

Signal Transduction

In the species examined so far, it is generally agreed that the generation of inositol 1,4,5-trisphosphate [Ins(1,4,5)P3] and diacylglycerol plays a part in ET-1-induced contraction of airway smooth muscle. InsP accumulation in response to ET-1 occurs in human and animal airway smooth muscle (101,130,136,146,150–154) and is associated with increased levels of cytosolic Ca^{2+} released from intracellular stores (150,154,155). An exception to this is ovine tracheal smooth muscle, where little evidence for InsP generation was observed (68). Preliminary studies suggest that diacylglycerol, though activation of protein kinase C (PKC), might also contribute to ET-1-induced contraction of airway smooth muscle in rat (156) and rabbit (136) airways, but not in guinea pig (130) or bovine airways (151). In human airway smooth muscle, the data are conflicting, with one study indicating a role for PKC in ET-1-induced contraction(138) and another finding no evidence for this (91).

The role of extracellular Ca^{2+} in ET-1-induced contraction of airway smooth muscle is even less well understood. Studies in human isolated bronchus (126) and in guinea pig trachea (130) have reported that the influx of extracellular Ca^{2+} did not contribute significantly to the contractile response to ET-1. Conversely, numerous studies in airway smooth muscle have demonstrated that at least a part of the ET-1-induced contraction was dependent on the presence of extracellular Ca^{2+} (68,101,

123,132,142,144,157,158). For example, in ferret (159), rabbit (158), and sheep (68) tracheal smooth muscle, ET-1-induced contractions were significantly inhibited by verapamil, nifedipine, and nicardipine, respectively, suggesting that Ca^{2+} influx through L-type, voltage-dependent Ca^{2+} channels contributed significantly to this response (159). Interestingly, several other studies have shown that the extracellular Ca^{2+}-dependent component of ET-1-induced contraction was insensitive to inhibition by L-type Ca^{2+} channel blockers (101,132,142,146), although they were blocked by cadmium and lanthanum ions (144,157).

In rat trachea, ET-1-induced contractions utilized both intracellular and extracellular sources of Ca^{2+} and this was dependent on which receptor subtype was activated (101). Thus, InsP accumulation and intracellular Ca^{2+} release resulted from ET_A receptor stimulation, whereas ET_B receptor stimulation facilitated the influx of extracellular Ca^{2+} through non-L-type Ca^{2+} channels (101). It is important to note that ET_B receptors are not always linked to mobilization of extracellular Ca^{2+}. Preliminary data in human bronchus indicate that ET_B receptors activated by Stx S6c did not appear to be linked to the influx of extracellular Ca^{2+} (160). Furthermore, in ovine trachea, ET_A receptors were not strongly linked to the mobilization of intracellular Ca^{2+} (68). Additional mechanisms, including activation of Na^+/H^+ exchange, may also contribute to ET-1-induced contraction of guinea pig airway smooth muscle (161).

In-Vivo Studies

Although it has not yet been established that ET-1 increases airway tone in the human, many studies have established the bronchoconstrictor effect of ET-1 in the guinea pig (147,162–175), rat (176,177), cat (178–181), dog (182), and sheep (183). In guinea pig airways, these ET-1-induced bronchoconstrictor responses were mediated via the activation of both ET_A (175) and ET_B receptors (173,175). It is also clear that the ET_B receptor-mediated response in the guinea pig often involved the actions of cyclooxygenase products, such as thromboxane A_2, since they were significantly attenuated by indomethacin, meclofenamate (162–169,174) and thromboxane A_2 synthesis inhibitors (171,174,184) or receptor antagonists (173,174). Similarly, thromboxane A_2 generation was reported in cats (179,180) and rats (185). Platelet-activating factor may also play a role as a secondary spasmogenic mediator in the guinea pig (162,164,167), although this was not seen in all studies (186).

Some evidence suggests that bronchoconstrictor responses to ET-1 in allergen-sensitized guinea pigs are enhanced over those observed in unsensitized animals, perhaps as a result of a reduction in the proteolytic activity of the airway epithelium which usually metabolizes ET-1 (52,187). In contrast, there have been various reports that neither the acute nor prolonged administration of ET-1 induced any significant enhancement in the responsiveness of guinea pig airways to the bronchoconstrictor actions of acetylcholine, histamine or 5-hydroxytryptamine (162,165,188). However, it has been reported that ET-1 can induce bronchial hyperreactivity in guinea pigs, since a very low concentration of ET-1 markedly increased the bronchoconstrictor response to aerosolized histamine (172). Similarly, in sheep, inhaled ET-1 caused airway hyperresponsiveness to carbachol (189). Currently, there is no direct evidence in humans linking ET-1 to the induction of bronchial hyperresponsiveness to other spasmogens. Unfortunately, the few animal studies provide conflicting data.

Relaxation

Low concentrations of ET-1 have been reported to cause airway smooth muscle relaxation (89,158,190–193). In guinea pig isolated tracheal preparations, this response to ET-1 was mediated at least in part via the release of epithelial nitric oxide (NO) (191,192), an effect linked to ET_A receptors (193). In addition, the opening of charybdotoxin-sensitive K^+ channels has been implicated in this relaxant response (192) as has the production of inhibitory cyclooxygenase products and the generation of intracellular c-AMP (194). The cycloxygenase metabolites PGE_2 and PGI_2 (158) have also been suggested as playing a minor role in ET-1-induced airway smooth muscle relaxation in guinea pig airways (190,191,195). The simultaneous production of spasmogenic cyclooxygenase products, such as thromboxane A_2, may functionally antagonize ET-1-induced relaxations (191). ET-1-induced relaxations might also be mediated by lipoxygenase-generated hydroperoxides of arachidonic acid (195).

Mitogenesis

ET-1 can act as a mitogen in human and animal airway smooth muscle cells, since it stimulated [^{3}H]-thymidine incorporation into DNA, stimulated protein synthesis, enhanced the transient expression of protoocogenes such as c-*fos*, and increased cell number (196–200). This effect is relatively weak compared with those of other mitogens such as epidermal growth factor (EGF) (198). However, recent evidence indicates that ET-1 acts as a potent co-mitogen (200), an effect also described in other cell systems (201). In human airway smooth muscle cells, this action was mediated via ET_A receptors, since mitogenic activity was inhibited by BQ-123 (200) and not induced by ET-3 (202). This is consistent with the presence of ET_A receptors detected on human isolated airway smooth muscle (70,99). In rabbit cultured airway smooth muscle cells, ET-1-induced mitogenesis was associated with the activation of a pertussis-sensitive G protein coupled to the stimulation of phospholipase A_2, and the generation of thromboxane A_2 (196). Pertussis toxin-sensitive mitogen-activated protein (MAP) kinase has been linked to ET-1-induced DNA synthesis in bovine airway smooth muscle cells (203).

In addition to mitogenic effects in airway smooth muscle, ET-1 acts as a chemotactic factor in guinea pig tracheal airway smooth muscle cells, i.e., it promotes the migration of these cells (204) and stimulates fibroblast proliferation (205) and collagen metabolism (206). ET-1 has also been shown to increase fibronectin production from human bronchial epithelial cells via activation of ET_A receptors (207) and to enhance the production of cytoskeletal protein markers of human lung myofibroblasts (208). These effects are consistent with a role for endogenous ET-1 release in promoting airway wall remodeling, including regulating the extent and distribution of fibrosis in lung disease (206,207).

B. Vascular Smooth Muscle

Contraction

ET_A receptors mediate sustained, concentration-dependent contractions to ET-1 in pulmonary arterial preparations from the human (111,122,125,209,210) and from several animal species (122,211,212) with the exception of the rabbit, in which ET_B receptors contribute primarily to the response (111–113). Interestingly, although the major pulmonary arteries are often insensitive to the ET_B receptor agonists Stx S6c and BQ-3020, arterioles and veins contract strongly in response to these agents (211–213), indicating the presence of functional ET_B receptors at these sites. ET-1 is usually more potent in pulmonary veins than in pulmonary arteries (70,214–216).

Relaxation

ET-1 can also induce ET_B receptor-mediated relaxation in precontracted arteries and veins that were at least in part the result of the generation of endothelium-derived NO and prostacyclin (212,217).

Vascular Responses In Vivo

Studies in vivo demonstrate that ET-1 induces both vasodilator and vasoconstrictor effects (178,181,215,218–227).ET_A receptors mediate ET-1-induced pulmonary vasoconstriction in sheep (226), where cyclooxygenase metabolites including thromboxane A_2 have been implicated as secondary mediator (227), as in the guinea pig (228), but not the rat (225) or rabbit (215). ET_B receptors directly mediate ET-1-induced vasoconstriction in the cat (181,223). ET-1-induced pulmonary vasodilatation seems to be mediated via ET_B receptors, with the generation of NO and the activation of ATP-dependent K^+ channels (227). In the few studies conducted in the human, intravenous ET-1 produced no change in one instance (117), but a 67% increase in pulmonary vascular resistance in another (116).

Permeability and Edema

Hydrostatic pulmonary edema occurs in response to ET-1 in rat and guinea pig isolated perfused lungs following increases in microvascular pressure (213,228–232), rather than as a result of an increase in microvascular permeability and plasma extravasation (233). Although a non-ET_A receptor-mediated process was reportedly involved in this response in rat isolated perfused lung (213), other studies have shown that activation of both ET_A and ET_B receptors mediated hydrostatic pulmonary edema in this preparation in response to ET-1 (231,232).

In contrast, in blood perfused rat lung (234) and in the conscious rat (235–237), ET-1 caused edema via an increase in microvascular permeability. This response was apparently due to increases in endothelial cell gap formation caused directly or via the release of secondary mediators including platelet-activating factor (236) and thromboxane A_2 (238), following activation of ET_A receptors (235). ET-1 has also been implicated in neurogenic pulmonary edema (239).

C. Nerves

Recent data show that ET-1 and ET-3 significantly potentiated nerve function. This was first described in rabbit isolated bronchial rings, where ET-3 enhanced contractile response to postganglionic cholinergic nerve activation (240). It was subsequently shown that ET-1 and related peptides enhanced cholinergic nerve-mediated contraction in mouse trachea via activation of prejunctional ET_B receptors (241), consistent with the enhanced release of acetylcholine during electrical field stimulation (EFS). Most recently, activation of both prejunctional ET_A and ET_B receptors has been shown to modulate cholinergic nerve-mediated contraction in rabbit (242), mouse (243), and rat trachea (244). Importantly, prejunctional ET_B receptor-mediated potentiation of similar responses in human isolated bronchus has also recently been demonstrated (245), suggesting yet another mechanism through which the elevated levels of endogenous ET-1 in asthma could exacerbate bronchial obstruction.

However, enhanced cholinergic nerve function is not universally observed. For example, ET-1 virtually abolished contractions in sheep tracheal smooth muscle induced by EFS of cholinergic, postganglionic nerves (246,247). This prejunctional ET_B receptor-mediated effect was similar to the neuro-inhibitory effects of ET-1 observed in the gastrointestinal tract (248–250). The influence of ET-1 receptor activation on nonadrenergic, noncholinergic excitatory (e-NANC) and inhibitory (i-NANC) neurotransmission in human airways is an important area for future investigation.

D. Mucous Glands

ET-1 is known to increase airway mucus production. For example, in human cultured nasal mucosal explants, ET-1 increased secretion from both mucous and serous cells (251). In contrast, ET-1 had no effect on basal mucus production and inhibited methacholine-stimulated production of mucus in ferret trachea (252). The net effect of differing actions in various airway cell types might explain these apparently conflicting data. This is consistent with a report that the increase in mucous glycoprotein production from isolated submucosal glands induced by ET-1 was abolished in the presence of epithelial cells, perhaps because of the production of an inhibitory mediator from the epithelium (253).

E. Epithelium

It is now clear that the airway epithelium is a major cellular source of ET-1 (9,23,24,31–33,40,43,254). Target cells for released ET-1 include adjacent airway smooth muscle cells, submucosal glands, nerves, inflammatory cells, and vascular smooth muscle cells. However, various epithelial cell functions can also be influenced by ET-1, suggesting that epithelial ET-1 might act as an autocrine, as well as a paracrine hormone within the airways (255–261). For example, ET-1 caused BQ-123-sensitive increases in [^{3}H]thymidine incorporation and cell numbers in cultured porcine tracheal epithelial cells, indicating that ET-1, via ET_A receptors, is mitogenic for epithelial cells (261). Furthermore, in epithelial cell cultures, ET-1 caused increases in the negativity of the transepithelial potential difference (258,260) and in the short-circuit current (255,257,260) caused by stimulation of Cl^- transport across the epithelium (255,257,260). Dose-dependent increases in cilia beating frequency (257) and in the mucociliary activity in both nasal sinus and tracheal mucosae in the rabbit (259) have also been demonstrated in response to ET-1. The generation of cyclooxygenase products (255,257,259,260), including PGE_2 (102,255,26), was responsible for directly mediating these effects. Of particular interest is the recent report that ET-1, via ET_A receptors, depresses mucociliary clearance in ovine airways, independently of the involvement of prostanoids or peptidoleukotrienes (262). Epithelium-derived oxidative products of arachidonic acid metabolism released in response to ET-1 (256) are likely to modulate a range of physiological and pathophysiological processes within the lung (263).

Removal of the airway epithelium in airway preparations from some species has been shown to enhance the contractile potency of ET-1 (49,140,143), presumably because of the removal of degradative enzymes such as epithelial NEP (49,50,109). However, a contribution from the additional loss of an epithelium-derived relaxant mediator cannot be ruled out (109,143).

F. Inflammatory Cells

The few studies that have been conducted concerning the direct effects of ET-1 on the function of different inflammatory cells have sometimes provided conflicting conclusions. Thus, ET-1 has been reported to be without effect on human peripheral blood monocyte chemotaxis, adhesion, or superoxide production (264,265), whereas another study concluded that it provoked human blood monocyte chemotaxis (266). Nevertheless, there is little doubt that ET-1 can induce the release of a range of mediators from various inflammatory cells. For example, ET-1 was a potent agonist in human monocytes, stimulating the release of TNF-α, IL-1β, and IL-6 (45). Increased arachidonic acid and thromboxane release in response to ET-1 was observed in guinea pig alveolar macrophages (267), and high concentrations of ET-1 increased superoxide production in human alveolar macrophages (268). The release of PGD_2 but not of histamine or peptidoleukotrienes from guinea pig and human bronchus (121,149) and from canine airways (269) in the presence of ET-1 has been reported, as has thromboxane A_2 release from perfused guinea pig lung (270), although the cellular sources of these mediators are unknown. ET-1 also enhanced the generation of oxygen radicals in BAL cells from rats (271) and from alveolar macrophages from the rabbit (272). Other effects of ET-1 include histamine release from guinea pig pulmonary mast cells but not from peritoneal mast cells (273), and the adhesion of leukocytes to the endothelium in pulmonary capillaries in perfused rat lung (274), although in guinea pigs, intravenous or aerosolized ET-1 did not produce inflammatory cell influx into the lung (165,188,275). Interestingly, ET-1 potentiated superoxide production from guinea pig alveolar macrophages induced by the bacterial peptide formylmethionylleucylphenylalanine (FMLP) or platelet-activating factor (276) and from rabbit alveolar macrophages by phorbol ester, via ET_A receptor-mediated mechanisms (272). The "priming" effects of ET-1 on responses produced by other mediators is worthy of further study.

G. Alveolar Wall

ET-1 may have a role in the regulation of alveolar epithelial cell function (102,277). A cloned rat alveolar epithelial cell line was shown to synthesize ET-1, express ET_A receptors, and generate PGE_2 and cAMP in response to ET_A receptor activation (102). Furthermore, the binding of ET-1 to alveolar type II cells was coupled to activation of protein kinase C and increases in Ca^{2+} influx through L-type calcium channels. This resulted in the enhanced secretion of lung surfactant (277), suggesting that alveolar binding sites for ET-1 were functional receptors.

IX. Endothelins and Lung Disease

A. Asthma

Asthma is a chronic inflammatory lung disease characterized by bronchial obstruction and airflow limitation. Bronchial obstruction in asthma can result from the combined effects of airway smooth muscle contraction, hypersecretion of mucus, epithelial damage, and desquamation contributing to mucous plug formation, airway wall thickening caused by microvascular leakage and edema formation, airway smooth muscle hyperplasia/hypertrophy, and thickening of the extracellular matrix, including epithelial basement membrane. ET-1 has been implicated in this disease because it mimics many of these features of asthma.

Actions of Endothelin-1 Relevant to Asthma

Activation by ET-1 of bronchial smooth muscle ET receptors induces a marked constrictor response. Interestingly, asthma was not associated with any significant change in the proportions of ET_A and ET_B receptors in airway smooth muscle, and the contractile sensitivity to ET-1 was not significantly enhanced (70). The proportions of ET_A and ET_B receptors were also similar in nonasthmatic and asthmatic peripheral lung structures (69). Nevertheless, a significant increase in ET-1-induced tone in asthmatic airways is likely to result from the markedly elevated tissue levels and release of ET-1 that have been recently described in asthmatic subjects.

Chronic asthma also involves significant changes to airway architecture, including proliferation of airway smooth muscle. ET-1 is a potent mitogen, but it produces a relatively weak (about twofold) rise in human cultured bronchial smooth muscle cells (199). However, it markedly potentiated the effects of the powerful mitogen, epidermal growth factor (200). ET_A receptors mediated both the direct and potentiating effects of ET-1 (199,200). These findings, together with reports that ET is a mitogen in Swiss 3T3 cultured fibroblasts (278) and that increased numbers of fibroblasts have been noted in asthmatic airways (279), are indicative of a role for endothelin-1 in airway wall remodeling.

As previously indicated, several animal studies have clearly demonstrated that relatively low concentrations of ET-1 and related peptides, including ET-3 and Stx S6c, cause significant potentiation of cholinergic nerve mediated airway smooth muscle contraction (240,241). New evidence indicates that such potentiation also occurs in human bronchial preparations and that this is mediated via prejunctional ET_B receptors (245). This raises the possibility that in asthma, elevated levels of epithelium-derived ET-1 might overflow to modulate cholinergic nerve function and that this might represent another pathological action of ET-1 in this disease.

Epithelial damage and desquamation coupled with mucous hypersecretion and impaired mucociliary clearance may contribute to the formation of mucus plugs, a characteristic feature in asthmatic airways. Although not yet studied in human airways, the findings that ET-1 significantly inhibited tracheal mucus velocity (an index of mucociliary clearance) in sheep (262) and caused mucus hypersecretion in feline airways (253) indicate another possible mechanism for ET-1 to contribute to airway obstruction.

Asthma is known to involve airway wall edema which is an accompaniment to inflammation and increased microvascular permeability. Although no direct evidence for ET-1-induced microvascular leakage in human lung has been obtained, there are data from animal studies in support of this postulate.

Endothelin Levels and Asthma Severity

Increased plasma ET levels have also been linked to the severity of asthma, and ET levels were correlated with the extent of airway obstruction as assessed by FEV_1/ FVC ratio (280). Furthermore, this study reported a positive correlation between the plasma levels of ET after methacholine challenge and the degree of bronchial hyperresponsiveness (280). These data are consistent with a significant negative correlation between BAL fluid ET levels and predicted FEV_1 in non-steroid-treated patients with asthma (39) and with increased numbers of epithelial cells expressing ET-like immunoreactivity in patients with more severe asthma and bronchial hyper-reactivity (41). Conversely, Springall and co-workers did not detect a positive correlation between staining for ET-1 in bronchial epithelial cells and the degree of airflow obstruction or the level of bronchial responsiveness (40), raising doubts about the reported links between elevated tissue and BAL levels of ET and asthma severity. Further doubts concerning the validity of such relationships derive from a study of nocturnal asthma, in which a significant decrease in BAL fluid ET-1 levels was detected compared with those measured in the daytime asthmatic group or night-time control group (281). Furthermore, the fall in FEV_1 overnight was significantly but negatively correlated with BAL fluid ET-1 levels. It is possible that ET release was enhanced overnight, contributing to a worsening of asthma, and also that reduced

BAL fluid ET-1 levels were explained by binding of this ET-1 to airway smooth muscle and epithelium. However, the precise link between levels of ET and severity of disease in asthma remains to be clarified in future studies.

Perhaps a key indicator of a link between elevated airway ET levels and the expression of asthma symptoms is the fact that some current asthma therapies reduce elevated airway ET levels. For example, attenuation of ir-ET-1 levels in BAL fluid to control levels, concomitant with improvement in lung function, were observed in asthmatics receiving inhaled β-agonist bronchodilator and oral corticosteroid therapy for 15 days (37). Continuous theophylline infusion controlled asthmatic attacks and deceased plasma ET-1 levels (280). Studies in vitro also support a link between asthma and ET-1 production in the airways. Thus, allergen immunotherapy for asthma was shown to significantly blunt the production of ET-1 by mononuclear cells taken from these patients (282), and hydrocortisone markedly attenuated ET-1 release from bronchial epithelial cells from six patients with symptomatic asthma (43).

B. Allergic Sensitization

Early and late "asthmatic" responses to antigen challenge have been modeled in the ovalbumin-sensitized guinea pig. The ET_B receptor-selective antagonist BQ-788 blocked the immediate, but not the late "asthmatic" response, whereas the ET_A receptor-selective antagonist BQ-123 blocked only the late response (283). Taken together, these observations suggest that ET(s) influence pulmonary functions by constricting airway smooth muscle via ET_B receptors during the immediate response and by modulating pulmonary inflammation via ET_A receptors during the late response, respectively. Ovalbumin challenge in sensitized guinea pigs was also followed by an acute rise in plasma ET levels, suggesting an antigen-stimulated, IgE-mediated release of ET (284). The late response to allergen and the associated hyperresponsiveness to inhaled carbachol in sheep was also attenuated by ET_A receptor blockade (189). Collectively, these data indicate that ET receptor activation played a pivotal role in allergic sensitization in these species.

C. Allergic Rhinitis

ET-1 mRNA and ir-ET-1 have been detected in human nasal mucosa (251,285). Although studies in vitro have demonstrated binding sites of ET-1 submucosal glands, venous sinusoids, and small muscular arterioles (251), and that ET-1 stimulated serous and mucous cell secretions (251) and prostanoid release (286), there has been only one report describing the effects of intranasal administration of ET-1 to humans (287). In that study, involving both symptomatic allergic and nonallergic individuals, ET-1 caused concentration-related increases in secretion weights, lysozyme secretion, symptoms of rhinorrhea, and symptoms of itch and sneezing. Several of these effects were more pronounced in allergic rhinitics than in nonallergic subjects, suggesting that allergic inflammation enhanced the responsiveness of the nasal mucosa to ET-1. The levels of mRNAs for ET-1 and ECE were significantly increased in chronic rhinitis, further supporting the concept of a role for ET-1 in this disease (288).

D. Cryptogenic Fibrosing Alveolitis (CFA) and Pulmonary Fibrosis

CFA is another lung disease characterized by inflammation, as well as proliferation of type II pneumocytes and fibroblasts with collagen deposition. The levels of ir-ET in lung specimens from patients with CFA correlated with the extent of type II cell proliferation. Levels of ir-ET in BAL fluid were also increased in systemic sclerosis, where pulmonary fibrosis is a major pathology (289). This study also showed that lavage fluid from these patients contained a factor that induced fibroblast proliferation in vitro, a response that was blocked by ET-1 antisera and the ET_A receptor antagonist BQ-123.

In patients with idiopathic pulmonary fibrosis, both ir-ET- and ir-ET-1-converting enzyme (ir-ECE-1) expression were increased and these were co-localized to sites in the airway epithelium, proliferating type II pneumocytes, endothelial cells, and inflammatory cells (290). In a bleomycin model of pulmonary fibrosis, the morphological changes associated with this disease were attenuated by chronic administration of BQ-123 prior to and over the study period (291), perhaps pointing to a major role for ET-1 in the pathogenesis of this disease in humans.

X. Summary

The ETs have wide-ranging effects in the respiratory tract. Importantly, ET-1 is synthesized, stored, released, and metabolized in the lung, and these activities may be relevant to both physiological function and to pathophysiological processes in this organ. The actions of the ETs are mediated by ET_A and/or ET_B receptors, although the possibility that further receptor subtypes might be subsequently defined cannot be dismissed.

Substantial but as yet largely indirect evidence suggests that the ETs play a role in the pathogenesis of airway diseases such as asthma and CFA. With respect to asthma, future studies targeting the effects for the ETs in nerves, inflammatory cells, and airway smooth muscle are likely to provide important new information concerning this relationship. However, data concerning the influence of potent and selective receptor antagonists for the various ET receptor subtypes in asthma and other lung diseases is essential to a definitive evaluation of potential the pathophysiological roles of the ETs.

Acknowledgments

Peter J. Henry and Roy G. Goldie are funded by grants from the National Health & Medical Research Council of Australia.

References

1. Yanagisawa M, Kurihara H, Kimura S, Tomobe Y, Kobayashi M, Mitsui Y, Yazaki Y, Goto K, Masaki T. A novel potent vasoconstrictor peptide produced by vascular endothelial cells. Nature 1988; 332:411–415.

2. Yanagiasawa M, Masaki T. Endothelin, a novel endothelium-derived peptide. Pharmacological activities, regulation and possible roles in cardiovascular control. Biochem Pharmacol 1989; 38:1877–1883.

3. Yanagisawa M, Masaki T. Molecular biology and biochemistry of the endothelins. Trends Pharmacol Sci 1989; 10:374–378.

4. Masaki T, Yanagisawa M, Goto K. Physiology and pharmacology of endothelins. Med Res Rev 1992; 12:391–421.

5. Matsumoto H, Suzuki N, Onda H, Fujino M. Abundance of endothelin-3 in rat intestine, pituitary gland and brain. Biochem Biophys Res Commun 1989; 164:74–80.

6. Kitamura K, Tanaka T, Kato J, Eto T, Tanaka K. Regional distribution of immunoreactive endothelin in porcine tissue: abundance in inner medulla of kidney. Biochem Biophys Res Commun 1989; 161:348–352.

7. Pernow J, Hemsen A, Lundberg JM. Tissue specific distribution, clearance and vascular effects of endothelin in the pig. Biochem Biophys Res Commun 1989 161:647–653.

8. Yoshimi H, Hirata Y, Fukuda Y, Kawano Y, Emori T, Kuramochi M, Omae T, Marumo F. Regional distribution of immunoreactive endothelin in rats, Peptides 1989; 10:805–808.

9. Giaid A, Polak JM, Gaitonde V, Hamid QA, Moscoso G, Legon S, Uwanogho D, Roncalli M, Shinmi O, Sawamura T, Kimura S, Yanagisawa M, Masaki T, Springall DR. Distribution of endothelin-like immunoreactivity and mRNA in the developing and adult human lung. Am J Respir Cell Mol Biol 1991; 4:50–58.

10. Ehrenreich H, Anderson RW, Fox CH, Rieckmann P, Hoffman GS, Travis WD, Coligan JE, Kehrl JH, Fauci AS. Endothelins, peptides with potent vasoactive properties, are produced by human macrophages. J Exp Med 1990; 172:1741–1748.

11. Kashiwabara T, Inagaki Y, Ohta H, Iwamatsu A, Nomizu M, Morita A, Nishikori K. Putative precursors of endothelin have less vasoconstrictor activity *in vitro* but a potent pressor effect *in vivo* FEBS Lett 1989; 247:73–76.

12. Kimura S, Kasuya Y, Sawamura T, Shinimi O, Sugita Y, Yanagisawa M, Goto K, Masaki T. Conversion of big endothelin-1 to 21-residue endothelin-1 is essential for expression of full vasoconstrictor activity: structure-activity relationships of big endothelin-1. J Cardiovasc Pharmacol 1989; 13(suppl 5):S5–S7.

13. Denault JB, Claing A, D'Orleans-Juste P, Sawamura T, Kido T, Masaki T, Leduc R. Processing of proendothelin-1 by human furin convertase. FEBS Lett 1995; 362: 276–280.

14. Xu D, Emoto N, Giaid A, Slaughter C, Kaw S, deWit D, Yanagisawa M. ECE-1: a membrane-bound metalloprotease that catalyzes the proteolytic activation of big endothelin-1. Cell 1994; 78:473–485.

15. Emoto N, Yanagisawa M. Endothelin-converting enzyme-2 is a membrane-bound, phosphoramidon-sensitive metalloprotease with acidic pH optimum. J Biol Chem 1995; 270:15262–15268.

16. Ishikawa S, Tsukada H, Yuasa H, Fukue M, Wei S, Onizuka M, Miyauchi T, Ishikawa T, Mitsui K, Goto K. Effects of endothelin-1 and conversion of big endothelin-1 in the isolated perfused rabbit lung. J Appl Physiol 1992; 72:2387–2392.

17. Vemulapalli S, Rivelli M, Chiu PJ, del Prado M, Hey JA. Phosphoramidon abolishes the increases in endothelin-1 release induced by ischemia-hypoxia in isolated perfused guinea pig lungs. J Pharmacol Exp Ther 1992; 262:1062–1069.

18. Lebel N, D'Orleans-Juste P, Fournier A, Sirois P. Characterization of the endothelin-converting enzyme in guinea pig upper bronchus. J Cardiovasc Pharmacol 1995; 26 (suppl 3):S81–S83.

19. Lebel N, Dorleansjuste P, Fournier A, Sirois P. Role of neutral endopeptidase 24.11 in the conversion of big endothelins in guinea-pig lung parenchyma. Br J Pharmacol 1996; 117:184–188.

20. Laporte J, D'Orleans-Juste P, Sirois P. Guinea pig Clara cells secrete endothelin-1 through a phosphoramidon-sensitive pathway. Am J Respir Cell Mol Biol 1996; 14: 356–362.

21. Laporte J, D'Orleans-Juste P, Singh G, Sirois P. Dexamethasone and phosphoramidon inhibit endothelin release by cultured nonciliated bronchiolar epithelial (Clara) cells. J Cardiovasc Pharmacol 1995; 26(suppl 3):S53–S55.

22. Pons F, Touvay C, Lagente V, Mencia-Huerta JM, Braquet P. Involvement of a phosphoramidon-sensitive endopeptidase in the processing of big endothelin-1 in the guinea-pig. Eur J Pharmacol 1992; 217:65–70.

23. Rozengurt N, Springall DR, Polak JM. Localization of endothelin-like immunoreactivity in airway epithelium of rats and mice. J Pathol 1990; 160:5–8.

24. MacCumber MW, Ross CA, Glaser BM, Snyder SH. Endothelin: visualization of mRNAs by in situ hybridization provides evidence for local action. Proc Natl Acad Sci (USA) 1989; 86:7285–7289.

25. Rennick RE, Loesch A, Burnstock G. Endothelin, vasopressin, and substance P like immunoreactivity in cultured and intact epithelium from rabbit trachea. Thorax 1992; 47:1044–1049.

26. Shima H, Yamanouchi M, Omori K, Sugiura M, Kawashima K, Sato T. Endothelin-1 production and endothelin converting enzyme expression by guinea pig airway epithelial cells. Biochem Mol Biol Int 1995; 37:1001–1010.

27. Sun G, De Angelis G, Nucci F, Ackerman V, Bellini A, Mattoli S. Functional analysis of the preproendothelin-1 gene promoter in pulmonary epithelial cells and monocytes. Biochem Biophys Res Commun 1996; 221:647–652.

28. Furukawa K, Saleh D, Tsao MS, Giaid A. Cytokine-mediated increase of endothelin-1 and endothelin converting enzyme-1 in cultured human bronchial epithelial cells. Am J Respir Crit Care Med 1996; 153:A728.

29. Noguchi Y, Uchida Y, Endo T, Ninomiya H, Nomura A, Sakamoto T, Goto Y, Haraoka S, Shimokama T, Watanabe T, et al. The induction of cell differentiation and polarity of tracheal epithelium cultured on the amniotic membrane. Biochem Biophys Res Commun 1995; 210:302–309.

30. Mariassy AT, Glassberg MK, Salathe M, Maguire F, Wanner A. Endothelial and epithelial sources of endothelin-1 in sheep bronchi. Am J Physiol 1996; 14:L54–L61.

31. Black PN, Ghatei MA, Takahashi K, Bretherton-Watt D, Krausz T, Dollery CT, Bloom SR. Formation of endothelin by cultured airway epithelial cells. FEBS Lett 1989; 255: 129–132.

32. Mattoli S, Mezzetti M, Riva G, Allegra L, Fasoli A. Specific binding of endothelin on human bronchial smooth muscle cells in culture and secretion of endothelin-like material from bronchial epithelial cells. Am J Respir Cell Mol Biol 1990; 3:145–151.

33. Endo T, Uchida Y, Matsumoto H, Suzuki N, Nomura A, Hirata F, Hasegawa S. Regulation of endothelin-1 synthesis in cultured guinea pig airway epithelial cells by various cytokines. Biochem Biophys Res Commun 1992; 186:1594–1599.

34. Ninomiya H, Uchida Y, Ishii Y, Nomua A, Kameyama M, Saotome M, Endo T, Hasegawa S. Endotoxin stimulates endothelin release from cultured epithelial cells of guinea-pig trachea. Eur J Pharmacol 1991; 203:299–302.

35. Rennick RE, Milner P, Burnstock G. Thrombin stimulates release of endothelin and vasopressin, but not substance P, from isolated rabbit tracheal epithelial cells. Eur J Pharmacol 1993; 230:367–370.

36. Nakano J, Takizawa H, Ohtoshi T, Shoji S, Yamaguchi M, Ishii A, Yanagisawa M, Ito K. Endotoxin and pro-inflammatory cytokines stimulate endothelin-1 expression and release by airway epithelial cells. Clin Exp Allergy 1994; 24:330–336.

37. Mattoli S, Soloperto M, Marini M, Fasoli A. Levels of endothelin in the bronchoalveolar lavage fluid of patients with symptomatic asthma and reversible airflow obstruction. J Allergy Clin Immunol 1991; 88:376–384.

38. Nomura A, Uchida Y, Kameyana M, Saotome M, Oki K, Hasegawa S. Endothelin and bronchial asthma. Lancet 1989; 2:747–748.

39. Redington AE, Springall DR, Ghatei MA, Lau LC, Bloom SR, Holgate ST, Polak JM, Howarth PH. Endothelin in bronchoalveolar lavage fluid and its relation to airflow obstruction in asthma. Am J Respir Crit Care Med 1995; 151:1034–1039.

40. Springall DR, Howarth PH, Counihan H, Djukanovic R, Holgate ST, Polak JM. Endothelin immunoreactivity of airway epithelium in asthmatic patient. Lancet 1991; 337:697–701.

41. Carpi S, Marini M, Vittori E, Vassalli G, Mattoli S. Bronchoconstrictive responses to inhaled ultrasonically nebulized distilled water and airway inflammation in asthma. Chest 1993; 104:1346–1351.

42. Ackerman V, Carpi S, Bellini A, Vassalli G, Marini M, Mattoli S. Constitutive expression of endothelin in bronchial epithelial cells of patients with symptomatic and asymptomatic asthma and modulation by histamine and interleukin-1. J Allergy Clin Immunol 1995; 96:618–627.

43. Vittori E, Marini M, Fasoli A, De Franchis R, Mattoli S. Increased expression of endothelin in bronchial epithelial cells of asthmatic patients and effect of corticosteroids. Am J Respir Dis 1992; 146:1320–1325.

44. Egger D, Geuenich S, Denzlinger C, Schmitt E, Mailhammer R, Ehrenreich H, Dormer P, Hultner L. IL-4 renders mast cells functionally responsive to endothelin-1. J Immunol 1995; 154:1830–1837.

45. McMillen MA, Huribal M, Cunningham ME, Kumar R, Sumpio BE. Endothelin-1 increases intracellular calcium in human monocytes and causes production of interleukin-6. Crit Care Med 1995; 23:34–40.

46. Campbell AM, Vignola AM, Chanez P, Godard P, Bousquet J. Low-affinity receptor for IgE on human bronchial epithelial cells in asthma. Immunology 1994; 82:506–508.

47. Vijayaraghaven J, Scicli AG, Carretero OA, Slaughter C, Moomaw C, Hersh LB. The hydrolysis of endothelins by neutral endopeptidase 24.11 (enkephalinase). J Biol Chem 1990; 265:14150–14155.

48. Fagny C, Michel A, Leonard I, Berkenboom G, Fontaine J, Deschodt-Lanckman M. *In vitro* degradation of endothelin-1 by endopeptidase 24.11 (enkephalinase). Peptides 1991; 12:773–778.

49. Hay DW. Guinea-pig tracheal epithelium and endothelin. Eur J Pharmacol 1989; 171: 241–245.

50. Di Maria GU, Katayama M, Borson DB, Nadel JA. Neutral endopeptidase modulates endothelin-1-induced airway smooth muscle contraction in guinea-pig trachea. Regulatory Peptides 1992; 39:137–145.

51. Noguchi K, Fukuroda T, Ikeno Y, Hirose H, Tsukada Y, Nishikibe M, Ikemoto F, Matsuyama K, Yano M. Local formation and degradation of endothelin-1 in guinea pig airway tissues. Biochem Biophys Res Commun 1991; 179:830–835.

52. Boichot E, Pons F, Lagente V, Touvay C, Mencia-Huerta JM, Braquet P. Phosphoramidon potentiates the endothelin-1-induced bronchopulmonary response in guinea-pigs. Neurochem Int 1991; 18:477–479.

53. McKay KO, Black JL, Armour CL. Phosphoramidon potentiates the contractile response to endothelin-3, but not endothelin-1 in isolated airway tissue. Br J Pharmacol 1992; 105:929–932.

54. Yamaguchi T, Kohrogi H, Kawano O, Ando M, Araki S. Neutral endopeptidase inhibitor potentiates endothelin-1-induced airway smooth muscle contraction. J Appl Physiol 1992; 73:1108–1113.

55. Fagny C, Michel A, Nortier J, Deschodt-Lanckman M. Enzymatic degradation of endothelin-1 by activated human polymorphonuclear neutrophils. Regulatory Peptides 1992; 42:27–37.

56. Cyr C, Huebner K, Druck T, Kris R. Cloning and chromosomal localization of a human endothelin ET_A receptor. Biochem Biophys Res Commun 1991; 181:184–190.

57. Hosoda K, Nakao K, Hiroshi-Arai, Suga S, Ogawa Y, Mukoyama M, Shirakami G, Saito Y, Nakanishi S, Imura H. Cloning and expression of human endothelin-1 receptor cDNA. FEBS Lett 1991; 287:23–26.

58. Sakurai T, Yanagisawa M, Takuwa Y, Miyazaki H, Kimura S, Goto K, Masaki T. Cloning of a cDNA encoding a non-isopeptide-selective subtype of the endothelin receptor. Nature 1990; 348:732–735.

59. Arai H, Hori S, Aramori I, Ohkubo H, Nakanishi S. Cloning and expression of a cDNA encoding an endothelin receptor. Nature 1990; 348:730–732.

60. Saito Y, Mizuno T, Itakura M, Suzuki Y, Ito T, Hagiwara H, Hirose S. Primary structure of bovine endothelin ET_B receptor and identification of signal peptidase and metal proteinase cleavage sites. J Biol Chem 1991; 266:23433–23437.

61. Sakamoto A, Yanagisawa M, Sakurai T, Takuwa Y, Yanagisawa H, Masaki T. Cloning and functional expression of human cDNA for the ET_B endothelin receptor. Biochem Biophys Res Commun 1991; 178:656–663.

62. Ogawa Y, Nakao K, Arai H, Nakagawa O, Hosoda K, Suga S, Nakanishi S, Imura H. Molecular cloning of a non-isopeptide-selective human endothelin receptor. Biochem Biophys Res Commun 1991; 178:248–255.

63. Nakamuta M, Takayanagi R, Sakai Y, Sakamoto S, Hagiwara H, Mizuno T, Saito Y, Hirose S, Yamamoto M, Nawata H. Cloning and sequence analysis of a cDNA encoding human non-selective type of endothelin receptor. Biochem Biophys Res Commun 1991; 177:34–39.

64. Hori S, Komatsu Y, Shigemoto R, Mizuno N, Nakanishi S. Distinct tissue distribution and cellular localization of two messenger ribonucleic acids encoding different subtypes of rat endothelin receptors. Endocrinology 1992; 130:1885–1895.

65. Lin HY, Kaji EH, Winkel GK, Ives HE, Lodish HF. Cloning and functional expression

of a vascular smooth muscle endothelin-1 receptor. Proc Natl Acad Sci (USA) 1991; 88:3185–3189.

66. Elshourbagy NA, Lee JA, Korman DR, Nuthalaganti P, Sylvester DR, Dilella AG, Sutiphong JA, Kumar CS. Molecular cloning and characterization of the major endothelin receptor subtype in porcine cerebellum. Mol Pharmacol 1992; 41:465–473.

67. Adachi M, Yang YY, Furuichi Y, Miyamoto C. Cloning and characterization of cDNA encoding human A-type endothelin receptor. Biochem Biophys Res Commun 1991; 180: 1265–1272.

68. Goldie RG, Grayson PS, Knott PG, Self GJ, Henry PJ. Predominance of endothelinA (ET$_A$) receptors in ovine airway smooth muscle and their mediation of ET-1-induced contraction. Br J Pharmacol 1994; 112:749–756.

69. Knott PG, Daprile AC, Henry PJ, Hay DWP, Goldie RG. Receptors for endothelin-1 in asthmatic human peripheral lung. Br J Pharmacol 1995; 114:1–3.

70. Goldie RG, Henry PJ, Knott PG, Self GJ, Luttmann MA, Hay DWP. Endothelin-1 receptor density, distribution and function in human isolated asthmatic airways. Am J Respir Crit Care Med 1995; 152;1653–1658.

71. Goldie RG, D'Aprile AC, Self GJ, Rigby PJ, Henry PJ. The distribution and density of receptor subtypes for endothelin-1 in the peripheral lung of the rat, guinea-pig and pig. Br J Pharmacol 1996; 117:729–735.

72. Goldie RG, D'Aprile AC, Cvetkovski R, Rigby PJ, Henry PJ. Influence of regional differences in ET$_A$ and ET$_B$ receptor subtype proportions on endothelin-1-induced contractions in porcine isolated trachea and bronchus. Br J Pharmacol 1996; 117:736–742.

73. Russell FD, Davenport AP. Characterization of endothelin receptors in the human pulmonary vasculature using bosentan, SB209670, and 97-139. J Cardiovasc Pharmacol 1995; 26:S346–S347.

74. Masaki T, Vane JR, Vanhoutte PM. International Union of Pharmacology nomenclature of endothelin receptors. Pharmacol Rev 1994; 46:137–142.

75. Bdolah A, Wollberg Z, Fleminger G, Kochva E. SRTX-d, a new native peptide of the endothelin/sarafotoxin family. FEBS Lett 1989; 256:1–3.

76. Williams DL, Jr., Jones KL, Pettibone DJ, Lis EV, Clineschmidt BV. Sarafotoxin S6c: an agonist which distinguishes between endothelin receptor subtypes. Biochem Biophys Res Commun 1991; 175:556–561.

77. Ihara M, Saeki T, Fukuroda T, Kimura S, Ozaki S, Patel AC, Yano M. A novel radioligand [^{125}I]BQ-3020 selective for endothelin (ET$_B$) receptors. Life Sci 1992; 51:PL47–PL52.

78. Watakabe T, Urade Y, Takai M, Umemura I, Okada T. A reversible radioligand specific for the ET$_B$ receptor: [^{122}I]Tyr^{13}-Suc-[Glu^9,Ala^{11,15}]-endothelin-1(8-21), [^{125}I]IRL 1620. Biochem Biophys Res Commun 1992; 185:867–873.

79. Ihara M, Fukuroda T, Saeki T, Nishikibe M, Kojiri K, Suda H, Yano M. An endothelin receptor (ET$_A$) antagonist isolated from *Streptomyces misakiensis*. Biochem Biophys Res Commun 1991; 178:132–137.

80. Ishikawa K, Ihara M, Noguchi K, Mase T, Mino N, Saeki T, Fukuroda T, Fukami T, Ozaki S, Nagase T. Biochemical and pharmacological profile of a potent and selective endothelin B-receptor antagonist, BQ-788. Proc Natl Acad Sci (USA) 1994; 91:4892–4896.

81. Karaki H, Sudjarwo SA, Hori M, Tanaka T, Matsuda Y. Endothelin ET$_B$ receptor antagonist, RES-701-1: effects on isolated blood vessels and small intestine. Eur J Pharmacol 1994; 262:255–259.

82. Tanaka T, Tsukuda E, Nozawa M, Nonaka H, Ohno T, Kase H, Yamada K, Matsuda Y. RES-701-1, a novel, potent, endothelin type B receptor-selective antagonist of microbial origin. Mol Pharmacol 1994; 45:724–730.

83. Nambi P, Pullen M, Spielman W. Species differences in the binding characteristics of [^{125}I]IRL-1620, a potent agonist specific for endothelin-B receptors. J Pharmacol Exp Ther 1994; 268:202–207.

84. Waggoner WG, Genova SL, Rash VA, Kinetic analyses demonstrate that the equilibrium assumption does not apply to [^{125}I]endothelin-1 binding data. Life Sci 1992; 51: 1869–1876.

85. Wu-Wong JR, Chiou WJ, Magnuson SR, Opgenorth TJ. Endothelin receptor agonists and antagonists exhibit different dissociation characteristics. Biochim Biophys Acta 1994; 1224:288–294.

86. Ihara M, Yamanaka R, Ohwaki K, Ozaki S, Fukami T, Ishikawa K, Towers P, Yano M. [^{3}H]BQ-123, a highly specific and reversible radioligand for the endothelin ET_A receptor subtype. Eur J Pharmacol 1995; 274:1–6.

87. Peter MG, Davenport AP. Selectivity of [^{125}I]-PD151242 for human, rat and porcine endothelin ET_A receptors in the heart. Br J Pharmacol 1995; 114:297–302.

88. Wu-Wong JR, Chiou WJ, Dixon D, Opgenorth J. Dissociation characteristics of endothelin ET_A receptor agonist and antagonists. J Cardiovasc Pharmacol 1995; suppl 26: S380–S384.

89. Battistini B, Warner TD, Fournier A, Vane JR. Characterization of ET_B receptors mediating contractions induced by endothelin-1 or IRL 1620 in guinea-pig isolated airways: effects of BQ-123, FR139317 or PD 145065. Br J Pharmacol 1994; 111:1009–1016.

90. Battistini B, Warner TD, Fournier A, Vane JR. Comparison of PD 145065 and Ro 46-2005 as antagonists of contractions of guinea-pig airways induced by endothelin-1 or IRL 1620. Eur J Pharmacol 1994; 252:341–345.

91. Nally JE, McCall R, Young LC, Wakelam MJO, Thomson NC, McGrath JC. Mechanical and biochemical responses to endothelin-1 and endothelin-1 in human bronchi. Eur J Pharmacol 1994; 288:53–60.

92. Yoneyama T, Hori M, Makatani M, Yamamura T, Tanaka T, Matsuda Y, Karaki H. Subtypes of endothelin ET_A and ET_B receptors mediating tracheal smooth muscle contraction. Biochem Biophys Res Commun 1995; 207:668–674.

93. Sudjarwo SA, Hori M, Tanaka T, Matsuda Y, Okada T, Karaki H. Subtypes of endothelin ET_A and ET_B receptors mediating venous smooth muscle contraction. Biochem Biophys Res Commun 1994; 200:627–633.

94. Nakajima Y, Kizawa Y, Nakano J, Kotake H, Inami T, Kusama T, Murakami H. Effects of ET antagonists (PD143296 and PD145065) on contractions in guinea-pig hilar bronchus induced by endothelin-1 and its related peptide. Receptor 1995; 5:177–183.

95. Kenakin TP, Bond RB, Bonner TI. Definition of pharmacological receptors. Pharmacol Rev 1992; 44:351–362.

96. Hoyer D, Clarke DE, Fozard JR, Hartig PR, Martin GR, Mylecharane EJ, Saxena PR, Humphrey PP. International Union of Pharmacology classification of receptors for 5-hydroxytryptamine (Serotonin). Pharmacol Rev 1994; 46:157–203.

97. Gresser O, Chayard D, Herbert D, Cousin MA, Lemoullec JM, Bouattane F, Guedin D, Frelin C. ET-1 and ET-3 actions mediated by cloned ET_A endothelin receptors exhibit different sensitivities to BQ-123. Biochem Biophys Res Commun 1996; 224:169–171.

98. Bax WA, Saxena PR. The current endothelin receptor classification: time for reconsideration. Trends Pharmacol Sci 1994; 15:378–386.

99. Fukuroda T, Ozaki S, Ihara M, Ishikawa K, Yano M, Miyauchi T, Ishikawa S, Onizuka M, Goto K, Nishikibe M. Necessity of dual blockade of endothelin ET(A) and ET(B) receptor subtypes for antagonism of endothelin-1-induced contraction in human bronchi. Br J Pharmacol 1996; 117:995–999.

100. Henry PJ, Goldie RG. ET_B but not ET_A receptor-mediated contractions to endothelin-1 attenuated by respiratory tract viral infection in mouse airways. Br J Pharmacol 1994; 112:1188–1194.

101. Henry PJ. Endothelin-1 (ET-1)-induced contraction in rat isolated trachea: involvement of ET_A and ET_B receptors and multiple signal transduction systems. Br J Pharmacol 1993; 110:435–441.

102. Markewitz BA, Kohan DE, Michael JR. Endothelin-1 synthesis, receptors, and signal transduction in alveolar epithelium: evidence for an autocrine role. Am J Physiol 1995; 268:L192–L200.

103. Cioffi CL, Neale RF Jr, Jackson RH, Sills MA. Characterization of rat lung endothelin receptor subtypes which are coupled to phosphoinositide hydrolysis. J Pharmacol Exp Ther 1992; 262:611–618.

104. Li H, Elton TS, Chen YF, Oparil S. Increased endothelin receptor gene expression in hypoxic rat lung. Am J Physiol 1994; 266:L553–L560.

105. Furuya S, Naruse S, Nakayama T, Nokihara K. Effect and distribution of in rat kidney and lung examined by electron microscopic radioautography. Anat Embryol 1992; 185: 87–96.

106. Nakamichi K, Ihara M, Kobayashi M, Saeki T, Ishikawa K, Yano M. Different distribution of endothelin receptor subtypes in pulmonary tissues revealed by the novel selective ligands BQ-123 and [Ala^{1,3,11,15}]ET-1. Biochem Biophys Res Commun 1992; 182: 144–150.

107. Kobayashi M, Ihara M, Sato N, Saeki T, Ozaki S, Ikemoto F, Yano M. A novel ligand, [^{125}I]BQ-3020, reveals the localization of endothelin ET_B receptors. Eur J Pharmacol 1993; 235:95–100.

108. Power RF, Wharton J, Zhao Y, Bloom SR, Polak JM. Autoradiographic localization of endothelin-1 binding sites in the cardiovascular and respiratory systems. J Cardiovasc Pharmacol 1989; 13:Suppl 5:S50–S56.

109. Tschirhart EJ, Drijfhout JW, Pelton JT, Miller RC, Jones CR. Endothelins: functional and autoradiographic studies in guinea pig trachea. J Pharmacol Exp Ther 1991; 258: 381–387.

110. Davenport AP, O'Reilly G, Kuc RE. Endothelin ET_A and ET_B mRNA and receptors expressed by smooth muscle in the human vasculature: majority of the ET_A sub-type. Br J Pharmacol 1995; 114:1110–1116.

111. Fukuroda T, Kobayashi M, Ozaki S, Yano M, Miyauchi T, Onizuka M, Sugishita Y, Goto K, Nishikibe M. Endothelin receptor subtypes in human versus rabbit pulmonary arteries. J Appl Physiol 1994; 76:1976–1982.

112. LaDouceur DM, Flynn MA, Keiser JA, Reynolds E, Haleen SJ. ET_A and ET_B receptors coexist on rabbit pulmonary artery vascular smooth muscle mediating contraction. Biochem Biophys Res Commun 1993; 196:209–215.

113. Fukuroda T, Ozaki S, Ihara M, Ishikawa K, Yano M, Nishikibe M. Synergistic inhibition by BQ-123 and BQ-788 of endothelin-1-induced contractions of the rabbit pulmonary artery. Br J Pharmacol 1994; 113:336–338.

114. Fukuroda T, Fujikawa T, Ozaki S, Ishikawa K, Yano M, Nishikibe M. Clearance of

circulating endothelin-1 by ET_B receptors in rats. Biochem Biophys Res Commun 1994; 199:1461–1465.

115. Sirvio ML, Metsarinne K, Saijonmaa O, Fyhrquist F. Tissue distribution and half-life of ^{125}I-endothelin in the rat: importance of pulmonary clearance. Biochem Biophys Res Commun 1990; 167:1191–1195.

116. Weitzberg E, Ahlborg G, Lundberg JM, Differences in vascular effects and removal of endothelin-1 in human lung, brain, and skeletal muscle. Clin Physiol 1993; 13:653–662.

117. Wagner OF, Vierhapper H, Gasic S, Nowotny P, Waldhausl W. Regional effects and clearance of endothelin-1 across pulmonary and splanchnic circulation. Eur J Clin Invest 1992; 22:277–282.

118. Kapanci Y, Assimacopoulos A, Irle C, Zwahlen A, Gabbiani G. "Contractile interstitial cells" in pulmonary alveolar septa: a possible regulator of ventilation-perfusion ratio? Ultrastructural, immunofluorescence, and *in vitro* studies. J Cell Biol 1974; 60:375–392.

119. Durham SK, Goller NL, Lynch JS, Fisher SM, Rose PM. Endothelin receptor B expression in the rat and rabbit lung as determined by in situ hybridization using nonisotopic probes. J Cardiovasc Pharmacol 1993; 22(suppl 8):S1–S3.

120. Hay DW, Henry PJ, Goldie RG. Endothelin and the respiratory system. Trends Pharmacol Sci 1993; 14:29–32.

121. Hay DW, Hubbard WC, Undem BJ. Endothelin-induced contraction and mediator release in human bronchus. Br J Pharmacol 1993; 110:392–398.

122. Hay DW, Luttmann MA, Hubbard WC, Undem BJ. Endothelin receptor subtypes in human and guinea-pig pulmonary tissues. Br J Pharmacol 193; 110:1175–1183.

123. Advenier C, Sarria B, Naline E, Puybasset L, Lagente V. Contractile activity of three endothelins (ET-1, ET-2 and ET-3) on the human isolated bronchus. Br J Pharmacol 1990; 100:168–172.

124. Henry PJ, Rigby PJ, Self GJ, Preuss JM, Goldie RG. Relationship between endothelin-1 binding site densities and constrictor activities in human and animal airway smooth muscle. Br J Pharmacol 1990; 100:786–792.

125. Hemsen A, Franco-Cereceda A, Matran R, Rudehill A, Lundberg JM. Occurrence, specific binding sites and functional effects of endothelin in human cardiopulmonary tissue. Eur J Pharmacol 1990; 191:319–328.

126. McKay KO, Black JL, Armour CL. The mechanism of action of endothelin in human lung. Br J Pharmacol 1991; 102:422–428.

127. Adner M, Cardell LO, Sjoberg T, Ottosson A, Edvinsson L. Contractile endothelin-B (ET_B) receptors in human small bronchi. Eur Respir J 1996; 9:351–355.

128. Bertrand C, Naline E, Biyah K, Fujitani Y, Okada T, Sakaki J, Advenier C. Influence of ET_A receptor blockade on the ET_B receptor antagonism following ET-1 and IRL1620-induced contraction of the human bronchi. Am J Respir Crit Care Med 1996; 153:A645.

129. Gater PR, Wasserman MA, Renzetti LM. Effects of Ro-0203 on endothelin-1 and sarafotoxin S6c-induced contractions of human bronchus and guinea-pig trachea. Eur J Pharmacol 1996; 304:123–128.

130. Hay DW. Mechanism of endothelin-induced contraction in guinea-pig trachea: comparison with rat aorta. Br J Pharmacol 1990; 100:383–392.

131. Sakata K, Ozaki H, Kwon SC, Karaki H. Effects of endothelin on the mechanical activity and cytosolic calcium level of various types of smooth muscle. Br J Pharmacol 1989; 98:483–492.

132. Turner NC, Power RF, Polak JM, Bloom SR, Dollery CT. Endothelin-induced contrac-

tions of tracheal smooth muscle and identification of specific endothelin binding sites in the trachea of the rat. Br J Pharmacol 1989; 98:361–366.

133.   Maggi CA, Giuliani S, Patacchini R, Santicioli P, Rovero P, Giachetti A, Meli A. The C-terminal hexapeptide, endothelin-(16-21), discriminates between different endothelin receptors. Eur J Pharmacol 1989; 166:121–122.

134.   Uchida Y, Ninomiya H, Saotome M, Nomura A, Ohtsuka M, Yanagisawa M, Goto K, Masaki T, Hasegawa S. Endothelin, a novel vasoconstrictor peptide, as potent bronchoconstrictor. Eur J Pharmacol 1988; 154:227–228.

135.   Lee HK, Sperelakis N. Electromechanical effects of endothelin in ferret bronchial and tracheal smooth muscle. Prog Clin Biol Res 1990; 327:683–685.

136.   Grunstein MM, Rosenberg SM, Schramm CM, Pawlowski NA. Mechanisms of action of endothelin 1 in maturing rabbit airway smooth muscle. Am J Physiol 1991; 260(pt 1): L434–L443.

137.   O'Donnell SR, Kay CS. Effects of endothelin receptor selective antagonists, BQ-123 and BQ-788, on IRL 1620 and endothelin-1 responses of airway and vascular preparations from rats. Pulmon Pharmacol 1995; 8:11–19.

138.   McKay KO, Armour CL, Black JL. Endothelin receptors and activity differ in human, dog, and rabbit lung. Am J Physiol 1996; 14:L37–L43.

139.   Ergul A, Glassberg MK, Wanner A, Puett D. Characterization of endothelin receptor subtypes on airway smooth muscle cells. Exp Lung Res 1995; 21:453–468.

140.   Maggi CA, Patacchini R, Giuliani S, Meli A. Potent contractile effect of endothelin in isolated guinea-pig airways. Eur J Pharmacol 1989; 160:179–182.

141.   Filep JG, Battistini B, Sirois P. Pharmacological modulation of endothelin-induced contraction of guinea-pig isolated airways and thromboxane release. Br J Pharmacol 1991; 103:1633–1640.

142.   Ninomiya H, Uchida Y, Saotome M, Nomura A, Ohse H, Matsumoto H, Hirata F, Hasegawa S. Endothelins constrict guinea pig tracheas by multiple mechanisms. J Pharmacol Exp Ther 1992; 262:570–576.

143.   Maggi CA, Giuliani S, Patacchini R, Santicioli P, Giachetti A, Meli A. Further studies on the response of the guinea-pig isolated bronchus to endothelins and sarafotoxin S6b. Eur J Pharmacol 1990; 176:1–9.

144.   Chand N, Diamantis W, Sofia RD. Pharmacologic modulation of endothelin-induced contraction in isolated rat tracheal segments. Res Commun Chem Pathol Pharmacol 1990; 70:173–181.

145.   Schumacher WA, Steinbacher TE, Allen GT, Ogletree ML. Role of thromboxane receptor activation in the bronchospastic response to endothelin. Prostaglandins 1990; 40:71–79.

146.   Nally JE, McCall R, Young LC, Wakelam MJ, Thomson NC, McGrath JC. Mechanical and biochemical responses to endothelin-1 and endothelin-3 in human bronchi. Eur J Pharmacol 1994; 288:53–60.

147.   Touvay C, Vilain B, Pons F, Chabrier PE, Mencia-Huerta JM, Braquet P. Bronchopulmonary and vascular effect of endothelin in the guinea-pig. Eur J Pharmacol 1990; 176:23–33.

148.   Filep JG, Battistini B, Sirois P. Endothelin induces thromboxane release and contraction of isolated guinea-pig airways. Life Sci 1990; 47:1845–1850.

149.   Hay DW, Hubbard WC, Undem BJ. Relative contributions of direct and indirect mechanisms mediating endothelin-induced contraction of guinea-pig trachea. Br J Pharmacol 1993; 110:955–962.

150.   Mattoli S, Soloperto M, Mezzetti M, Fasoli A. Mechanisms of calcium mobilization and

phosphinositide hydrolysis in human bronchial smooth muscle cells by endothelin-1. Am J Respir Cell Mol Biol 1991; 5:424–430.

151. Nally JE, McCall R, Young LC, Wakelam MJ, Thomson NC, McGrath JC. Mechanical and biochemical responses to endothelin-1 and endothelin-3 in bovine bronchial smooth muscle. Br J Pharmacol 1994; 111:1163–1169.

152. Henry PJ, Rigby PJ, Self GJ, Preuss JM, Goldie RG. Endothelin-1-induced [^{3}H]-inositol phosphate accumulation in rat trachea. Br J Pharmacol 1992; 105:135–141.

153. Yang CM, Yo YL, Ong R, Hsieh JT. Endothelin- and sarafotoxin-induced phospho-inositide hydrolysis in cultured canine tracheal smooth muscle cells. J Neurochem 1994; 62:1440–1448.

154. Yang CM, Ong R, Chen YC, Hsieh JT, Tsao HL, Tsai CT. Effect of phorbol ester on phosphoinositide hydrolysis and calcium mobilization induced by endothelin-1 in cultured canine tracheal smooth muscle cells. Cell Calcium 1995; 17:129–140.

155. Yang CM, Yo YL, Ong R, Hsieh JT, Tsao HL. Calcium mobilization induced by endothelins and sarafotoxin in cultured canine tracheal smooth muscle cells. Naunyn Schmiedeberg's Arch Pharmacol 1994; 350:68–76.

156. Henry PJ. Inhibitory effects of nordihydroguaiaretic acid on ET_A-receptor-mediated contractions to endothelin-1 in rat trachea. Br J Pharmacol 1994; 111:561–569.

157. Sarria B, Naline E, Morcillo E, Cortijo J, Esplugues J, Advenier C. Calcium dependence of the contraction produced by endothelin (ET-1) in isolated guinea-pig trachea. Eur J Pharmacol 1990; 187:445–453.

158. Grunstein MM, Chuang ST, Schramm CM, Pawlowski NA. Role of endothelin-1 in regulating rabbit airway contractility. Am J Physiol 1991; 260:L75–L82.

159. Lee HK, Leikauf GD, Sperelakis N. Electromechanical effects of endothelin on ferret bronchial and tracheal smooth muscle. J Appl Physiol 1990; 68:417–420.

160. Hay DWP, Luttmann MA, Goldie RG. Calcium (CA^{2+}) translocation mechanisms mediating endothelin-1 (ET-1)- and sarafotoxin S6c (S6c)-induced contractions in isolated human bronchus. Am J Respir Crit Care Med 1994; 149:A1083.

161. Battistini B, Filep JG, Cragoe EJ Jr, Fournier A, Sirois P. A role for Na^+/H^+ exchange in contraction of guinea pig airways by endothelin-1 *in vitro*. Biochem Biophys Res Commun 1991; 175:583–588.

162. Lagente V, Chabrier PE, Mencia-Huerta JM, Braquet P. Pharmacological modulation of the bronchopulmonary action of the vasoactive peptide, endothelin, administered by aerosol in the guinea-pig. Biochem Biophys Res Commun 1989; 158:625–632.

163. Payne AN, Whittle BJ. Potent cyclo-oxygenase-mediated bronchoconstrictor effects of endothelin in the guinea-pig *in vivo*. Eur J Pharmacol 1988; 158:303–304.

164. Braquet P, Touvay C, Vilain B, Pons F, Hosford D, Chabrier PE, Mencia-Huerta JM. Effect of endothelin-1 on blood pressure and bronchopulmonary system of the guinea pig. J Cardiovasc Pharmacol 1989; 13(suppl 5):S143–S146.

165. Macquin-Mavier I, Levame M, Istin N, Harf A. Mechanisms of endothelin-mediated bronchoconstriction in the guinea pig. J Pharmacol Exp Ther 1989; 250:740–745.

166. Pons F, Loquet I, Touvay C, Roubert P, Chabrier PE, Mencia-Huerta JM, Braquet P. Comparison of the bronchopulmonary and pressor activities of endothelin isoforms ET-1, ET-2, and ET-3 and characterization of their binding sites in guinea pig lung. Am J Respir Dis 1991; 143:294–300.

167. Boichot E, Lagente V, Mencia-Huerta JM, Braquet P. Bronchopulmonary responses to endothelin-1 in sensitized and challenge guinea-pigs: role of cyclooxygenase metabolites and platelet-activating factor. Fundam Clin Pharmacol 1993; 7:281–291.

168. Pons F, Touvay C, Lagente V, Mencia-Huerta JM, Braquet P. Bronchopulmonary and pressor activities of endothelin-1 (ET-1), ET-2, ET-3, and big ET-1 in the guinea pig. J Cardiovas Pharmacol 1991; 17(suppl 7):S326–S328.

169. Nagai H, Suda H, Kitagaki K, Koda A. Effect of tranilast on endothelin-induced bronchoconstriction in guinea pigs. J Pharmacobiol Dyn 191; 14:309–314.

170. Gratton JP, Rae GA, Claing A, Telemaque S, Dorleansjuste P. Different pressor and bronchoconstrictor properties of human big-endothelin-1, 2 (1-38) and 3 in ketamine xylazine-anaesthetized guinea-pigs. Br J Pharmacol 1995; 114:720–726.

171. Nambu F, Yube N, Omawari N, Sawada M, Okegawa T, Kawasaki A, Ikeda S. Inhibition of endothelin-induced bronchoconstriction by OKY-046, a selective thromboxane A_2 synthetase inhibitor, in guinea pigs. Adv Prostaglandin Thromboxane Leukotriene Res 1991; 21A:453–456.

172. Kanazawa H, Kurihara N, Hirata K, Fujiwara H, Matsushita H, Takeda T. Low concentration endothelin-1 enhanced histamine-mediated bronchial contractions of guinea pigs *in vivo* Biochem Biophys Res Commun 1992; 187:717–721.

173. Noguchi K, Noguchi Y, Hirose H, Nishikibe M, Ihara M, Ishikawa K, Yano M. Role of endothelin ET_B receptors in bronchoconstrictor and vasoconstrictor responses in guinea-pigs. Eur J Pharmacol 1993; 233:47–51.

174. Leuddeckens G, Bigl H, Sperling J, Becker K, Braquet P, Forster W. Importance of secondary TXA_2 release in mediating of endothelin-1 induced bronchoconstriction and vasopressin in the guinea-pig. Prostaglandins Leukotrienes Essential Fatty Acids 1993; 4:261–263.

175. Nagase T, Fukuchi Y, Matsui H, Aoki T, Matsuse T, Orimo H. *In vivo* effects of endothelin A- and B-receptor antagonists in guinea pigs. Am J Physiol 1995; 268:L846–L850.

176. Matsuse T, Fukuchi Y, Suruda T, Nagase T, Ouchi Y, Orimo H. Effect of endothelin-1 on pulmonary resistance in rats. J Appl Physiol 1990; 68:2391–2393.

177. Di Maria GU, Malatino LS, Bellofiore S, Mistretta A. Endothelin potently constricts rat airways *in vivo*. Agents Actions Suppl 1991; 34:439–443.

178. Minkes RK, Bellan JA, Saroyan RM, Kerstein MD, Coy DH, Murphy WA, Nossaman BD, McNamara DB, Kadowitz PJ. Analysis of cardiovascular and pulmonary responses to endothelin-1 and endothelin-3 in the anesthetized cat. J Pharmacol Exp Ther 1990; 253:1118–1125.

179. Dyson MC, Kadowitz PJ. Influence of SK&F 96148 on thromboxane-mediated responses in the airways of the cat. Eur J Pharmacol 1991; 197:17–25.

180. Dyson MC, Kadowitz PJ. Analysis of responses to endothelins 1, 2, and 3 and sarafotoxin 6b in airways of the cat. J Appl Physiol 1991; 71:243–251.

181. Kadowitz PJ. McMahon TJ, Hood JS, Feng CJ, Minkes RK, Dyson MC. Pulmonary vascular and airway responses to endothelin-1 are mediated by different mechanisms in the cat. J Cardiovasc Pharmacol 1991; 17(suppl 7):S374–S377.

182. Uchida Y, Hamada M, Kameyama M, Ohse H, Nomura A, Hasegawa S, Hirata F. ET-1 induced bronchoconstriction in the early phase but not late phase of anesthetized dogs is inhibited by indomethacin and ICI 198615. Biochem Biophys Res Commun 1992; 183:1197–1202.

183. Abraham WM, Ahmed A, Cortes A, Spinella MJ. Malik AB, Andersen TT. A specific endothelin-1 antagonist blocks inhaled endothelin-1-induced bronchoconstriction in sheep. J Appl Physiol 1993; 74:2537–2542.

184. del Basso P, Argiolas L. Cardiopulmonary effects of endothelin-1 in the guinea pig: role of thromboxane A_2. J Cardiovasc Pharmacol 1995; 26(suppl 3):S120–S122.

185. Uhlig S, Vonbethmann A. Endothelin-induced bronchoconstriction in perfused rat lungs is partly mediated by thromboxane. J Cardiovasc Pharmacol 1995; 26(suppl 3):S111–S114.

186. Lueddeckens G, Becker K, Rappold R, Forster W. Influence of aminophylline and ketotifen in comparison to the lipoxygenase inhibitors NDGA and esculetin and the PAF antagonists WEB 2170 and BN 52021 on endothelin-1 induced vaso- and bronchoconstriction. Prostaglandins Leukotrienes Essential Fatty Acids 1991; 44:155–158.

187. Boichot E, Lagente V, Mencia-Huerta JM, Braquet P. Effect of phosphoramidon and indomethacin on the endothelin-1 (ET-1) induced bronchopulmonary response in aerosol sensitized guinea pigs. J Vasc Med Biol 1990; 2:206.

188. Boichot E, Carre C, Lagente V, Pons F, Mencia-Huerta JM, Braquet P. Endothelin-1 and bronchial hyperresponsiveness in the guinea pig. J Cardiovasc Pharmacol 1991; 17(suppl 7):S329–S331.

189. Nogushi K, Ishikawa K, Yano M, Ahmed A, Cortes A, Abraham WM. Endothelin-1 contributes to antigen-induced airway hyperresponsiveness. J Appl Physiol 1995; 79:700–705.

190. White SR, Hathaway DP, Umans JG, Tallet J, Abrahams C, Leff AR. Epithelial modulation of airway smooth muscle response to endothelin-1. Am J Respir Dis 1991; 144:373–378.

191. Filep, JG, Battistini B, Sirois P. Induction by endothelin-1 of epithelium-dependent relaxation of guinea-pig trachea *in vitro*: role for nitric oxide. Br J Pharmacol 1993; 109:637–644.

192. Hadjkaddour K, Michel A, Chevillard C. Different mechanisms involved in relaxation of guinea-pig trachea by endothelin-1 and -3. Eur J Pharmacol 1996; 298:145–148.

193. Naline E, Bertrand C, Biyah K, Okada T, Fujitani Y, Sakaki J, Advenier C. Characterization of endothelin-ET_A receptors in epithelium of human isolated airways responsible for a relaxant component of smooth muscle through NO release. Am J Respir Crit Care Med 1996; 153:A644.

194. Elmowafy AM, Aboumohamed GA. Endothelins-induce cyclicAMP formation in the guinea-pig trachea through an ET(A) receptor- and cyclooxygenase-dependent mechanism. Br J Pharmacol 1996; 118:531–536.

195. Uchida Y, Saotome M, Nomura A, Ninomiya H, Ohse H, Hirata F, Hasegawa S. Endothelin-1-induced relaxation of guinea-pig trachealis muscles. J Cardiovasc Pharmacol 1991; 17(suppl 7):S210–S212.

196. Noveral JP, Rosenberg SM, Anbar RA, Pawlowski NA, Grunstein MM. Role of endothelin-1 in regulating proliferation of cultured rabbit airway smooth muscle cells. Am J Physiol 1992; 263:L317–L324.

197. Glassberg MK, Ergul A, Wanner A, Puett D. Endothelin-1 promotes mitogenesis in airway smooth muscle cells. Am J Respir Cell Mol Biol 1994; 10:316–321.

198. Stewart AG, Grigoriadis G, Harris T. Mitogenic actions of endothelin-1 and epidermal growth factor in cultured airway smooth muscle. Clin Exp Pharmacol Physiol 1994; 21:277–285.

199. Tomlinson PR, Wilson JW, Stewart AG. Inhibition by salbutamol of the proliferation of human airway smooth muscle cells grown in culture. Br J Pharmacol 1994; 111:641–647.

200. Panettieri RA, Goldie RG, Rigby PJ, Eszterhas AJ, Hay DWP. Endothelin-1-induced potentiation of human airway smooth muscle proliferation: an ET_A receptor-mediated phenomenon. Br J Pharmacol 1996; 118:191–197.

201. Battistini B, Chailler P, D'Orleans-Juste P, Briere N, Sirois P. Growth regulatory properties of endothelins. Peptides 1993; 14:385–399.

202. Fujitani Y, Bertrand C. Differential role of endothelin A and B receptors in cultured human airway smooth muscle cells. Am J Respir Crit Care Med 1996; 153:A843.

203. Malarkey K, Chilvers ER, Lawson MF, Plevin R. Stimulation by endothelin-1 of mitogen-activated protein kinases and DNA synthesis in bovine tracheal smooth muscle cells. Br J Pharmacol 1995; 116:2267–2273.

204. Kashihara Y, Noguchi S. Endothelin-1 (ET-1) augments platelet-derived growth factor-BB (PDGF-BB)-induced migration on cultured airway smooth muscle cells in guinea-pigs. Am J Respir Crit Care Med 1996; 153:A743.

205. Peacock AJ, Dawes KE, Shock A, Gray AJ, Reeves JT, Laurent GJ. Endothelin-1 and endothelin-3 induce chemotaxis and replication of pulmonary artery fibroblasts. Am J Respir Cell Mol Biol 1992; 7:492–499.

206. Dawes KE, Cambrey AD, Campa JS, Bishop JE, McAnulty RJ, Peacock AJ, Laurent GJ. Changes in collagen metabolism in response to endothelin-1—evidence for fibroblast heterogeneity. Int J Biochem Cell Biol 1996; 28:229–238.

207. Marini M, Carpi S, Bellini A, Patalano F, Mattoli S. Endothelin-1 induces increased fibronectin expression in human bronchial epithelial cells. Biochem Biophys Res Commun 1996; 220:896–899.

208. Shahar I, Fireman E, Topilsky M, Grief J, Kivity S, Swcharz Y, Mann A, Spirer Z, Ben-Efraim S. Effect of endothelin-1 (ET-1) on cytoskeltal protein markers expression of myofibroblasts in interstitial lung diseases (ILD). Am J Respir Crit Care Med 1996; 153:A307.

209. Buchan KW, Magnusson H, Rabe KF, Sumner MJ, Watts IS, Characterisation of the endothelin receptor mediating contraction of human pulmonary artery using BQ123 and Ro 46-2005. Eur J Pharmacol 1994; 260:221–226.

210. McKay KO, Black JL, Diment LM, Armour CL. Functional and autoradiographic studies of endothelin-1 and endothelin-2 in human bronchi, pulmonary arteries, and airway parasympathetic ganglia. J Cardiovasc Pharmacol 1991; 17(suppl 7):S206–S209.

211. MacLean MR, McCulloch KM, Baird M. Endothelin ET_A- and ET_B-receptor-mediated vasoconstriction in rat pulmonary arteries and arterioles. J Cardiovasc Pharmacol 1994; 23:838–845.

212. Sudjarwo SA, Hori M, Takai M, Urade Y, Okada T, Karaki H. A novel subtype of endothelin B receptor mediating contraction in swine pulmonary vein. Life Sci 1993; 53:431–437.

213. Bonvallet ST, Oka M, Yano M, Zamora MR, McMurtry IF, Stelzner TJ. BQ123, an ET_A receptor antagonist, attenuates endothelin-1-induced vasoconstriction in rat pulmonary circulation. J Cardiovasc Pharmacol 1993; 22:39–43.

214. Wang Y, Coceani F. Isolated pulmonary resistance vessels from fetal lambs. Contractile behavior and responses to indomethacin and endothelin-1. Circ Res 1992; 71:320–330.

215. Lippton HL, Ohlstein EH, Summer WR, Hyman AL. Analysis of responses to endothelins in the rabbit pulmonary and systemic vascular beds. J Apply Physiol 1991; 70:331–341.

216. Toga H, Ibe BO, Raj JU. *In vitro* responses of ovine intrapulmonary arteries and veins to endothelin-1. Am J Physiol 1992; 263:L15–L21.

217. Zellers TM, McCormick J, Wu Y. Interaction among ET-1, endothelium-derived nitric oxide, and prostacyclin in pulmonary arteries and veins. Am J Physiol 1994; 267:H139–H147.

218. Bradley LM, Czaja JF, Goldstein RE. Circulatory effects of endothelin in newborn piglets. Am J Physiol 1990; 259:H1613–H1617.

219. Hasunuma K, Rodman DM, O'Brien RF, McMurtry IF. Endothelin 1 causes pulmonary vasodilation in rats. Am J Physiol 1990; 259:H48–H54.

220. Lippton HL, Hauth TA, Summer WR, Hyman AL. Endothelin produces pulmonary vasoconstriction and systemic vasodilation. J Apply Physiol 1989; 66:1008–1012.

221. Minkes RK, Bellan JA, Higuera TR, Kadowitz PJ. Comparison of response to sarafotoxins 6a and 6c in pulmonary and systemic vascular beds. Am J Physiol 1992; 262:H852–H861.

222. Lippton HL, Hauth TA, Cohen GA, Hyman AL. Functional evidence for different endothelin receptors in the lung. J Appl Physiol 1993; 75:38–48.

223. McMahon TJ, Hood JS, Kadowitz PJ. Analysis of responses to sarafotoxin 6a and sarafotoxin 6c in the pulmonary vascular bed of the cat. J Appl Physiol 1991; 71:2019–2025.

224. Cassin S, Kristova V, Davis T, Kadowitz P, Gause G. Tone-dependent responses to endothelin in the isolated perfused fetal sheep pulmonary circulation in situ. J Appl Physiol 1991; 70:1228–1234.

225. Raffestin B, Adnot S, Eddahibi S, Macquin-Mavier I, Braquet P, Chabrier PE. Pulmonary vascular response to endothelin in rats. J Appl Physiol 1991; 70:567–574.

226. Ivy DD, Kinsella JP, Abman SH. Physiologic characterization of endothelin A and B receptor activity in the ovine fetal pulmonary circulation. J Clin Invest 1994; 93:2141–2148.

227. Wong J, Vanderford PA, Fineman JR, Soifer SJ. Developmental effects of endothelin-1 on the pulmonary circulation in sheep. Pediatr Res 1994; 36:394–401.

228. Horgan MJ, Pinheiro JM, Malik AB. Mechanism of endothelin-1-induced pulmonary vasoconstriction. Circ Res 1991; 69:157–164.

229. Ercan ZS, Kilinc M, Yazar O, Korkusuz P, Turker RK. Endothelin-1-induced oedema in rat and guinea-pig isolated perfused lungs. Arch Int Pharmacodyn Ther 1993; 323:74–84.

230. Rodman DM, Stelzner TJ, Zamora MR, Bonvallet ST, Oka M, Sato K, O'Brien RF, McMurtry IF. Endothelin-1 increases the pulmonary microvascular pressure and causes pulmonary edema in salt solution but not blood-perfused rat lungs. J Cardiovasc Pharmacol 1992; 20:658–663.

231. Lal H, Woodward B, Williams KI. Actions of endothelins and sarafotoxin 6c in the rat isolated perfused lung. Br J Pharmacol 1995; 115:653–659.

232. Sato K, Oka M, Hasunuma K, Ohnishi M, Kira S. Effects of separate and combined ET(A) and ET(B) blockade on ET-1-induced constriction in perfused rat lungs. Am J Physiol 1995; 13:L668–L672.

233. Waypa GB, Vincent PA, Morton CA, Minnear FL. Thrombin increases fluid flux in isolated rat lungs by a hemodynamic and not a permeability mechanism. J Appl Physiol 1996; 80:1197–1204.

234. Helset E, Kjaeve J, Hauge A. Endothelin-1-induced increases in microvascular permeability in isolated, perfused rat lungs requires leukocytes and plasma. Circ Shock 1993; 39:15–20.

235. Filep JG, Sirois MG, Foldes-Filep E, Rousseau A, Plante GE, Fournier A, Yano M,

Sirois P. Enhancement by endothelin-1 of microvascular permeability via the activation of ET_A receptors. Br J Pharmacol 1993; 109:880–886.

236. Filep JG, Sirois MG, Rousseau A, Fournier A, Sirois P. Effects of endothelin-1 on vascular permeability in the conscious rat: interactions with platelet-activating factor. Br J Pharmacol 1991; 104:797–804.

237. Filep JG, Fournier A, Foldes-Filep E. Effects of the ET_A/ET_B receptor antagonist, bosentan on endothelin-1-induced myocardial ischaemia and oedema in the rat. Br J Pharmacol 1995; 116:1745–1750.

238. Sirois MG, Filep JG, Rousseau A, Fournier A, Plante GE, Sirois P. Endothelin-1 enhances vascular permeability in conscious rats: role of thromboxane A_2. Eur J Pharmacol 1992; 214:119–125.

239. Herbst C, Tippler B, Shams H, Simmet T. A role for endothelin in bicuculline-induced neurogenic pulmonary oedema in rats. Br J Pharmacol 1995; 115:753–760.

240. McKay KO, Armour CL, Black JL. Endothelin-3 increases transmission in the rabbit pulmonary parasympathetic nervous system. J Cardiovasc Pharmacol 1993; 22(suppl 8):S181–S184.

241. Henry PJ, Goldie RG. Potentiation by endothelin-1 of cholinergic nerve-mediated contractions in mouse trachea via activation of ET_B receptors. Br J Pharmacol 1995; 114:563–569.

242. Yoneyama T, Hori M, Tanaka T, Matsuda Y, Karaki H. Endothelin ET(A) and ET(B) receptors facilitating parasympathetic neurotransmission in the rabbit trachea. J Pharmacol Exp Ther 1995; 275:1084–1089.

243. Carr MJ, Goldie RG, Henry PJ. Influence of respiratory tract viral infection on endothelin-1-induced potentiation of cholinergic nerve-mediated contraction in mouse trachea. Br J Pharmacol 1996; 119:891–898.

244. Knott PG, Fernandes LB, Henry PJ, Goldie RG. Influence of endothelin-1 on cholinergic nerve-mediated contractions and acetylcholine release in rat isolated tracheal smooth muscle. J Pharmacol Exp Ther 1996; 279:1142–1147.

245. Fernandes LB, Henry PJ, Rigby PJ, Goldie RG. Endothelin$_B$ (ET_B) receptor-activated potentiation of cholinergic nerve-mediated contraction in human bronchus. Br J Pharmacol 1996; 118:1873–1874.

246. Henry PJ, Goldie RG. Endothelin-1, via activation of an ET_B receptor, inhibits cholinergic nerve-mediated contractions in sheep trachea. J Cardiovasc Pharmacol 1995; 26(suppl 3):S117–S119.

247. Henry PJ, Shen A, Mitchelson F, Goldie RG. Inhibition by endothelin-1 of cholinergic nerve-mediated acetylcholine release and contraction in sheep isolated trachea. Br J Pharmacol 1996; 118:762–768.

248. Warner TD, Allcock GH, Mickley EJ, Vane JR. Characterization of endothelin receptors mediating the effects of the endothelin/sarafotoxin peptides on autonomic neurotransmission in the rat vas deferens and guinea-pig ileum. Br J Pharmacol 1993; 110:783–789.

249. Guimaraes CL, Rae GA. Dual effects of endothelins-1, -2 and-3 on guinea pig field-stimulated ileum: possible mediation by two receptors coupled to pertussis toxin-insensitive mechanisms. J Pharmacol Exp Ther 1992; 261:1253–1259.

250. Wiklund NP, Wiklund CU, Ohlen A, Gustafsson LE. Cholinergic neuromodulation by endothelin in guinea pig ileum. Neurosci Lett 1989; 101:342–346.

251. Mullol J, Chowdhury BA, White MV, Ohkubo K, Rieves RD, Baraniuk J, Hausfeld JN,

Shelhamer JH, Kaliner MA. Endothelin in human nasal mucosa. Am J Respir Cell Mol Biol 1993; 8:393–402.

252. Yurdakos E, Webber SE. Endothelin-1 inhibits pre-stimulated tracheal submucosal gland secretion and epithelial albumin transport. Br J Pharmacol 1991; 104:1050–1056.

253. Shimura S, Ishihara H, Satoh M, Masuda T, Nagaki N, Sasaki H, Takishima T. Endothelin regulation of mucus glycoprotein secretion from feline tracheal submucosal glands. Am J Physiol 1992; 262:L208–L213.

254. Giaid A, Michel RP, Stewart DJ, Sheppard M, Corrin B, Hamid Q. Expression of endothelin-1 in lungs of patients with cryptogenic fibrosing alveolitis. Lancet 1993; 341: 1550–1554.

255. Plews PI, Abdel-Malek ZA, Doupnik CA, Leikauf GD. Endothelin stimulates chloride secretion across canine tracheal epithelium. Am J Physiol 1991; 261:L188–L194.

256. Wu T, Rieves RD, Larivee P, Logun C, Lawrence MG, Shelhamer JH. Production of eicosanoids in response to endothelin-1 and identification of specific endothelin-1 biding sites in airway epithelial cells. Am J Respir Cell Mol Biol 1993; 8:282–290.

257. Tamaoki J, Kanemura T, Sakai N, Isono K, Kobayashi K, Takizawa T. Endothelin stimulates ciliary beat frequency and chloride secretion in canine cultured tracheal epithelium. Am J Respir Cell Mol Biol 1991; 4:426–431.

258. Webber SE, Yurdakos E, Woods AJ, Widdicombe JG. Effects of endothelin-1 on tracheal submucosal gland secretion and epithelial function in the ferret. Chest 1992; 101:63S–67S.

259. Amble FR, Lindberg SO, McCaffrey TV, Runer T. Mucociliary function and endothelins 1, 2, and 3. Otolaryngol Head Neck Surg 1993; 109:634–645.

260. Satoh M, Shimura S, Ishihara H, Nagaki M, Sasaki H, Takishima T. Endothelin-1 stimulates chloride secretion across canine tracheal epithelium. Respiration 1992; 59: 145–150.

261. Murlas CG, Gulati A, Singh G, Najmabadi F. Endothelin-1 stimulates proliferation of normal airway epithelial cells. Biochem Biophys Res Commun 1995; 212: 953–959.

262. Sabater JR, Otero R, Abraham WM, Wanner A, O'Riordan TG, Endothelin-1 depresses tracheal mucus velocity in ovine airways via ET-A receptors. Am J Respir Crit Care Med 1996; 154:341–345.

263. Holtzman MJ. Arachidonic acid metabolism Implications of biological chemistry for lung function and disease. Am J Respir Dis 1991; 143:188–203.

264. Bath PM, Mayston SA, Martin FJ. Endothelin and PDGF do not stimulate peripheral blood monocyte chemotaxis, adhesion to endothelium, and superoxide production. Exp Cell Res 1990; 187:339–342.

265. Hafstrom I, Ringertz B, Lundeberg T, Palmblad J. The effect of endothelin, neuropeptide Y, calcitonin gene-related peptide and substance P on neutrophil functions. Acta Physiol Scand 1993; 148:341–346.

266. Achmad TH, Rao GS. Chemotaxis of human blood monocytes toward endothelin-1 and the influence of calcium channel blockers. Biochem Biophys Res Commun 1992; 189: 994–1000.

267. Millul V, Lagente V, Gillardeaux O, Boichot E, Dugas B, Mencia-Huerta JM, Bereziat G, Braquet P, Masliah J. Activation of guinea pig alveolar macrophages by endothelin-1. J Cardiovasc Pharmacol 1991; 17(suppl 7):S233–S235.

268. Haller H, Schaberg T, Lindschau C, Lode H, Distler A. Endothelin increases [Ca^{2+}]i,

protein phosphorylation, and O_2^- production in human alveolar macrophages. Am J Physiol 1991; 261:L478–L484.

269. Ninomiya H, Yu XY, Hasegawa S, Spannhake EW. Endothelin-1 induces stimulation of prostaglandin synthesis in cells obtained from canine airways by bronchoalveolar lavage. Prostaglandins 1992; 43:401–411.

270. de Nucci G, Thomas R, D'Orleans-Juste P, Antunes E, Walder C, Warner TD, Vane JR. Pressor effects of circulating endothelin are limited by its removal in the pulmonary circulation and by the release of prostacyclin and endothelium-derived relaxing factor. Proc Natl Acad Sci (USA) 1988; 85:9797–9800.

271. Nagase T, Fukuchi Y, Jo C, Teramoto S, Uejima Y, Ishida K, Shimizu T, Orimo H. Endothelin-1 stimulates arachidonate 15-lipoxygenase activity and oxygen radical formation in the rat distal lung. Biochem Biophys Res Commun 1990; 168:485–489.

272. Kojima T, Hattori K, Hirata Y, Aoki T, Sasaitakedatsu M, Kino M, Kobayashi Y. Endothelin-1 has a priming effects on production of superoxide anion by alveolar macrophases—its possible correlation with bronchopulmonary dysplasia. Pediatr Res 1996; 39:112–116.

273. Uchida Y, Ninomiya H, Sakamoto T, Lee JY, Endo T, Nomura A, Hasegawa S, Hirata F. ET-1 released histamine from guinea pig pulmonary but not peritoneal mast cells. Biochem Biophys Res Commun 1992; 189:1196–1201.

274. Helset E, Ytrehus K, Tveita T, Kjaeve J, Jorgensen L. Endothelin-1 causes accumulation of leukocytes in the pulmonary circulation. Circ Shock 1994; 44:201–209.

275. Pons F, Boichot E, Lagente V, Touvay C, Mencia-Huerta JM, Braquet P. Role of endothelin in pulmonary function. Pulmon Pharmacol 1992; 5:213–219.

276. Filep JG, Fournier A, Foldes-Filep E. Acute pro-inflammatory actions of endothelin-1 in the guinea-pig lung: involvement of ET_A and ET_B receptors. Br J Pharmacol 1995; 115:227–236.

277. Sen N, Grunstein MM, Chander A. Stimulation of lung surfactant secretion by endothelin-1 from rat alveolar type II cells. Am J Physiol 1994; 266:L255–L262.

278. Takuwa N, Takuwa Y, Yanagisawa M, Yamashita K, Masaki T. A novel vasoactive peptide endothelin stimulates mitogenesis through inositol lipid turnover in Swiss 3T3 fibroblasts. J Biol Chem 1989; 264:7856–7861.

279. Brewster CE, Howarth PH, Djukanovic R, Wilson J, Holgate ST, Roche WR. Myofibroblasts and subepithelial fibrosis in bronchial asthma. Am J Respir Cell Mol Biol 1990; 3:507–511.

280. Aoki T, Kojima T, Ono A, Unishi G, Yoshijima S, Kameda-Hayashi N, Yamamoto C, Hirata Y, Kobayashi Y. Circulating endothelin-1 levels in patients with bronchial asthma. Ann Allergy 1994; 73:365–369.

281. Kraft M, Beam WR, Wenzel SE, Zamora MR, O'Brien RF, Martin RJ. Blood and bronchoalveolar lavage endothelin-1 levels in nocturnal asthma. Am J Respir Crit Care Med 1994; 149:946–952.

282. Chen WY, Yu J, Wang JY. Decreased production of endothelin-1 in asthmatic children after immunotherapy. J Asthma 1995; 32:29–35.

283. Uchida Y, Jun T, Ninomiya H, Ohse H, Hasegawa S, Nomura A, Sakamoto T, Sardessai MS, Hirata F. Involvement of endothelins in immediate and late asthmatic responses of guinea pigs. J Pharmacol Exp Ther 1996; 277:1622–1629.

284. Filep JG, Telemaque S, Battistini B, Sirois P, D'Orleans-Juste P. Increased plasma levels of endothelin during anaphylactic shock in the guinea-pig. Eur J Pharmacol 1993; 239: 231–236.

285. Casasco A, Benazzo M, Casasco M, Cornaglia AI, Springall DR, Calligaro A, Mira E, Polak JM. Occurrence, distribution and possible role of the regulatory peptide endothelin in the nasal mucosa. Cell Tissue Res 1993; 274:241–247.

286. Wu T, Mullol J, Rieves RD, Logun C, Hausfield J, Kaliner MA, Shelhamer JH. Endothelin-1 stimulates eicosanoid production in cultured human nasal mucosa. Am J Respir Cell Mol Biol 1992; 6:168–174.

287. Riccio MM, Reynolds CJ, Hay DWP, Proud D. Effects of intranasal administration of endothelin-1 to allergic and nonallergic individuals. Am J Respir Crit Care Med 1995; 152:1757–1764.

288. Furukawa K, Saleh D, Bayan F, Emoto N, Kaw S, Yanagisawa M, Giaid A. Co-expression of endothelin-1 and endothelin-converting enzyme-1 in patients with chronic rhinitis. Am J Respir Cell Mol Biol 1996; 14:248–253.

289. Cambrey AD, Harrison NK, Dawes KE, Southcott AM, Black CM, du Boi RM, Laurent GJ, McAnulty RJ. Increased levels of endothelin-1 in bronchoalveolar lavage fluid from patients with systemic sclerosis contribute to fibroblast mitogenic activity in vitro. Am J Respir Cell Mol Biol 1994; 11:439–445.

290. Saleh D, Yanagisawa M, Corrin B, Giaid A. Co-expression of endothelin and endothelin converting enzyme-1 in the lungs of normal humans and patients with idiopathic pulmonary fibrosis. Am J Respir Crit Care Med 1995; 151:A51.

291. Seino S, Kato S, Takahashi H, Iwabuchi K, Yuki H, Nakamura H, Tomoike H. BQ123, endothelin type A receptor antagonist, attenuates bleomycin-induced pulmonary fibrosis. Am J Respir Crit Care Med 1995; 151:A63.

292. Warner TD, Allcock GH, Corder R, Vane JR. Use of the endothelin antagonists BQ-123 and PD 142893 to reveal three endothelin receptors mediating smooth muscle contraction and the release of EDRF. Br J Pharmacol 1993; 110:777–782.

293. Hislop AA, Zhao YD, Springall DR, Polak JM, Haworth SG. Postnatal changes in endothelin-1 binding in porcine pulmonary vessels and airways. Am J Respir Cell Mol Biol 1995; 12:557–566.

9

Proinflammatory Peptides in Relation to Other Inflammatory Mediators

PIERANGELO GEPPETTI, COSTANZA EMANUELI, MICHELA FIGINI, and DOMENICO REGOLI

University of Ferrara
Ferrara, Italy

I. Introduction

Mediators of inflammations include molecules that belongs to different chemical species, including amines, lipid derivatives, gases, peptides, and others. Mediators of different chemical nature may often interact, sharing a final common inflammatory pathway. Thus, one mediator that in certain experimental conditions produces one or more inflammatory responses, may in different models of inflammation cause inflammatory responses indirectly, via release of different messengers. The complexity of these interactions usually reflects the complexity of the system under investigation. In vivo experiments offer the possibility to fully appreciate interactions between different inflammatory systems and mediators. The aim of the present chapter is to focus on the interactions between proinflammatory peptides and other inflammatory mediators.

Various peptides are powerful mediators of inflammation. Some inflammatory peptides appear to act directly on effector cells, whereas other peptides preferentially act indirectly, by releasing a variety of inflammatory mediators. On the other hand, inflammatory peptides may be released by diverse inflammatory mediators. A variety of compounds may also inhibit the action or reduce the release of inflammatory peptides, thus attenuating their inflammatory actions. In describing some examples of

241

the interactions between inflammatory peptides and other inflammatory mediators, major attention is paid to tachykinins and kinins. These two classes of mediators are, on a molar basis, among the most powerful inflammatory peptides described so far. They can also be considered as paradigms of inflammatory mediators that act mainly by direct activation of effector cells (tachykinins) or by the release of intermediate mediators (kinins). Examples reported in this chapter that illustrate the various interactions between kinins, tachykinins, and other mediators derive mainly from studies in airways, although in a few instances, observations obtained in tissues other than airways are included.

II. One Stimulus May Release Tachykinins Directly and Indirectly

The reader is referred to other chapters of this book or previous review articles (1–3) that cover the biochemical and pharmacological aspects of tachykinins, including their localization in sensory nerves and mechanisms of their release from peripheral endings of these nerves. A variety of agents regulate the release of calcitonin gene-related peptide (CGRP) and the tachykinins substance P (SP) and neurokinin A (NKA). Many of these stimuli appear to act directly via activation of channels or receptors located on sensory nerves, although, depending on the experimental setting, the same agents may stimulate neurogenic inflammatory mechanisms indirectly, through release of different mediators.

A. Protons

Low-pH media are often encountered in the body, either in physiological (gastric juice, or urine) or pathophysiological (sites of injury or inflammation, venous blood from infarcted cardiac tissue) conditions. Protons, depending on their concentration in the medium, may initiate either protective inflammatory responses or, if the pH is markedly low and not buffered efficiently, may lead to tissue damage and cell death. Low-pH media (pH 6.5–5) may also excite capsaicin-sensitive sensory neurons and activate their secretory function, thereby causing neurogenic inflammation (4). Protons can be regarded as a typical example of agents that may stimulate sensory neurons via direct or indirect mechanisms. Exposure to low-pH media of slices of rat or guinea pig tissues was found to evoke a Ca^{2+}-dependent release of CGRP (5,6), as well as inward current to cations in dorsal root ganglion (7) or trigeminal ganglion neurons in culture (8). These effects are blocked by capsazepine, a compound that antagonizes the effect of capsaicin (9,10), presumably by inhibiting the capsaicin-activated "receptor" (11), and by ruthenium red, an inorganic dye which blocks Ca^{2+} influx through the capsaicin-activated receptor/channel (12). These observations suggested the hypothesis that protons produce excitation of sensory nerves directly, by stimulation of the channel/receptor activated by capsaicin. The fact that capsazepine abolished the cough response to citric in guinea pigs supports the hypothesis that protons could also activate the capsaicin "receptor" on sensory nerves also in in-vivo conditions (13).

Other in-vivo studies, however, did not favor the view that sensory nerve excitation and sensory neuropeptide release following exposure to low-pH media involves the stimulation of the capsaicin-activated receptor/channel. Plasma extrava-sation in the guinea pig conjunctiva, caused by instillation of a low-pH medium, was abolished by a tachykinin NK_1 receptor antagonist and by a bradykinin B_2 receptor antagonist (14). These findings suggest that not only tachykinins but also kinins are involved in the response to protons in the conjunctiva. They also indicate that kinins and tachykinins share a final common pathway to mediate acid-induced plasma extravasation in the guinea pig conjunctiva. Previous studies have clearly demon-strated the ability of bradykinin to release sensory neuropeptides (reviewed in Ref. 15), whereas there is no evidence to support the possibility that tachykinin-induced inflammation is mediated by kinin release. Therefore, the data in the conjunctiva suggest that kinins released by low-pH media stimulate the release of tachykinins that increase conjunctival plasma extravasation (14). A similar mechanism is involved in the increase in blood flow induced by acid back-diffusion in the mucosa of the rat stomach (16). This effect of protons, which is mediated by CGRP release from sensory nerves, was partially reduced by the bradykinin B_2 receptor antagonist icatibant (HOE 140) (16). Neither sensory neuron depolarization nor sensory neuro-peptide release by bradykinin are attenuated by ruthenium red or capsazepine (6,15). These studies indicate that protons stimulate sensory nerves by different mechanisms that appear to be dependent on the experimental conditions and the tissue under investigation. Although the final inflammatory mediators are tachykinins and CGRP, their release may be stimulated by two different mechanisms: a direct action of protons on the capsaicin-activated receptor/channel or activation of kinins whose release is brought about by acidified extracellular fluid (Fig. 1).

III. Prejunctional Modulation of Sensory Neuropeptide Release

Inflammatory mediators, originating from different sources and of different chemical nature, have been shown to release sensory neuropeptides from peripheral endings of capsaicin-sensitive sensory nerves. Histamine (17), serotonin (5-HT) (18), pros-taglandins (19–21), sulfidopeptide-leukotrienes (22), eosinophil cationic protein (23), nitric oxide (24), and others cause the release of sensory neuropeptides or produce effects that, being inhibited by capsaicin pretreatment or by tachykinin or CGRP receptor antagonists, are considered to be due to sensory neuropeptide release. For many of these mediators the presence of specific receptors on cell bodies or terminals of sensory neurons has been documented. For instance, histamine H_1 receptors and 5-HT_3 receptors are considered to excite sensory nerves and cause sensory neuropeptide release directly. However, little evidence has been obtained so far for the involvement of these mediators in the stimulation of sensory nerves in

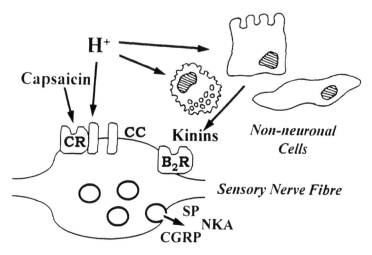

Figure 1 Schematic representation of the direct and indirect actions of protons (H^+) on capsaicin-sensitive sensory nerve fibers. CR, capsaicin "receptor"; CC, capsaicin-activated channel; B_2R, bradykinin receptor.

pathophysiological models of disease. Recently, evidence has been represented that kinins, released following different types of insult, produce their proinflammatory effects mainly by releasing tachykinins from sensory nerves. Section V of this chapter describes in more detail the experimental proof that led to this proposal.

A. Prostaglandins and Kinins

The role of prostaglandins in the activation of sensory nerves induced by kinins merits specific attention. Three issues appear of particular relevance: first, bradykinin releases large amounts of prostaglandins from a variety of cells; second, prostaglandins are known to sensitize sensory nerves to the excitatory action of different mediators including kinins (25); and third, sensory neuropeptide release induced by bradykinin is markedly inhibited by indomethacin (26,27). Therefore, it appears that kinins and prostaglandins act synergistically to activate the dual afferent (pain transmission) and efferent functions (neuropeptide release) of primary sensory neurons. The ionic basis and the mechanism of release of sensory neuropeptides by prostaglandins and bradykinin has been reviewed recently (15). Kinins and prostaglandins are involved in the pathogenesis of painful and inflammatory diseases, including migraine and arthritis. In these pathologies, cyclooxygenase inhibitors have a major therapeutic impact. The therapeutic potential of the new peptide and nonpeptide antagonists of bradykinin B_2 receptors will reflect the pathogenetic role played by kinins, but also how much of this role is independent from prostaglandin release.

B. Serotonin Receptors

A large variety of transmitters and mediators has been shown to prevent the release of neuropeptides from peripheral endings of primary sensory neurons. Also in this case, none of these inhibitory mechanisms has been proven convincingly to play a major role in human disease. For examples, neurogenic inflammation or, more specifically, neurogenic plasma extravasation in the meninges, has been proposed as a major pathophysiological mechanism of migraine headache. The inhibitory action of selective agonists for 5-hydroxytryptamine 5-HT$_{1B/D}$ receptors, including sumatriptan, on neurogenic plasma extravasation in rodent dura mater (28) has been proposed as the mechanism by which this class of drugs interrupts migraine attack (28). Indeed, the presence of 5-HT$_{1D\alpha}$ receptors has been documented in guinea pig and human dorsal root ganglion neurons (29). However, 5-HT$_{1D\beta}$ receptors have been found on smooth muscle cells of pial vessels (30), and the question remains unresolved as to whether sumatriptan, and sumatriptan-like drugs, act via presynaptic inhibition of SP, NKA, and CGRP release from sensory nerves in the meninges (28), or if they constrict arterial vessels of the cephalic circulation, which are abnormally dilated during the migraine attack (31).

Tachykinin NK$_1$ receptors mediate neurogenic plasma extravasation in various tissues, including the meninges (28). SP release from central terminals of sensory nerves may play a role in pain transmission (32). These observations represent the rationale for ongoing clinical studies with nonpeptide tachykinin NK$_1$ receptor antagonists for treatment of migraine attack. These studies may contribute to determining the role of neurogenic inflammatory mechanisms in migraine headache and, indirectly, may shed light on the site of action of sumatriptan-like drugs. Activation of receptors for mediators, such as histamine (33) and purines (34), which are released in large amounts during injury, inflammation, and anaphylaxis, may reduce sensory neuropeptide release. Although there is no evidence that these receptors are activated by their respective endogenous agonists, they might represent interesting targets for the clinical development of drugs that limit neurogenic inflammation.

IV. Are Tachykinins the Final Mediators or Their Proinflammatory Actions?

A. Endothelial Cells

Tachykinins cause a series of inflammatory responses in the airways, which include arteriolar vasodilatation, plasma extravasation, leukocyte adhesion to the vascular endothelium, secretion from seromucous glands, and bronchoconstriction. All these effects are believed to occur upon stimulation of specific receptors of the NK$_1$ or NK$_2$ type on effector cells. This conclusion applies to most of the tachykinin-mediated responses in different animal species. Thus, bronchoconstriction mediated by NK$_2$ receptor stimulation is most likely an effect due to direct stimulation of receptors on airway smooth muscle cells. Immunocytochemical demonstration of NK_1 receptors at sites of inflammation (postcapillary venules) of the rat trachea suggests that tachykinins act on these endothelial receptors to increase vascular permeability (35). However, when other effects of tachykinins are studied, the ability of tachykinins to stimulate effector cells directly cannot be held as a general rule. For instance, the mechanism by which SP and NKA cause arteriolar vasodilatation is unknown. In various large arteries, neurogenic inflammatory vasodilatation is mediated by release of CGRP, which relaxes smooth muscle cells (36). Less clear is the mechanism of neurogenic-mediated vasodilatation in the microcirculation. In the mucosa of the rat stomach, CGRP is the mediator of neurogenic vasodilatation, but in rat airways this effect appears to be mediated by tachykinins and NK_1 receptor activation (37). In large arteries SP causes an endothelium-dependent NK_1 receptor-mediated vaso-dilatation that is due to nitric oxide (NO) release. This indirect dilatory mechanism does not seem to be responsible for the dilatation of arterioles, because SP-induced increase in blood flow in rat airways was not affected by pretreatment with NO-synthase inhibitors, which, however, markedly decreased baseline blood flow (I. Yamawaki, J. A. Nadel, and P. Geppetti, unpublished observation). The mechanism activated by tachykinins to increase blood flow in the airway microcirculation remains unknown. A direct relaxation by NK_1 receptors of arteriolar smooth muscle cells or different epithelial-dependent (hyperpolarizing) factors could be taken into consideration.

Leukocyte adhesion following administration of SP or capsaicin was found to be blocked by an NK_1 receptor antagonist (38). There is also evidence for adhesion of leukocytes to bronchial epithelial cells via NK_1 receptor activation (39). However, the precise location of the NK_1 receptors that mediate this response has not been determined, and the presence of NK_1 receptors on leukocytes have not been unequiv-ocally confirmed.

B. Epithelial Cells

Epithelium-dependent inhibition of tachykinin-induced contraction of strips of guinea pig trachea was found to be blocked by pretreatment with indomethacin (40). The order of potency of naturally occurring tachykinins suggests that the epithelial receptor involved in prostaglandin-mediated tracheal relaxation is of the NK_1 type (40). Similar findings have been obtained in the mouse trachea (41). More recently, it has been shown that NK_1 receptor activation in guinea pig tracheal tube preparations causes epithelium-dependent relaxations that are mediated by release of NO (42). Various studies have reported consistently that NKA and SP stimulate human large to mid-size bronchial preparation in vitro by direct activation of NK_2 receptors in the smooth muscle. The contribution of epithelium-derived mediators to tachykinin-induced contractions in medium to small-size bronchi, a tissue where bradykinin has been shown to release contractile tromboxane A_2 from the epithelium (43), has not been fully investigated yet.

C. Mast Cells

SP and NKA may stimulate airway epithelial cells, macrophages, and mast cells. These effects of tachykinins cause the release of different mediators, often endowed with proinflammatory potential. The presence of the positively charged amino acids, arginine and lysine, confer on SP the ability to release histamine from mast cells by direct G-protein activation, without the involvement of receptor activation (44). This effect is produced only by elevated concentrations of SP, which are unlikely to be achieved in vivo and therefore might not have physiological relevance. However, in certain strains of rats, release of 5-HT via activation of NK_1 receptors on mast cells appears to be the mechanisms of tachykinin-induced bronchoconstriction (45). Indirect mechanisms have been proposed to mediate the bronchoconstriction induced by inhalation of NKA and SP in humans, because these effects were inhibited by the mast cell stabilizer, nedocromil sodium (46). SP has been reported to release tumor necrosis factor-α (TNF-α) from murine mast cells (47). Thus, mast cells from different tissues and various species may express tachykinin receptors, and tachykinins may release inflammatory mediators from these cells.

D. Other Inflammatory and Immunocompetent Cells

The presence of tachykinin receptors on resident or circulating inflammatory cells have been proposed. Although conflicting results have been reported in the past regarding the presence of tachykinin receptors in leukocytes and particularly lymphocytes, more consistent findings indicate that cells of the monocytic lineage in the lung and outside the lung cause the release of diverse inflammatory mediators via activation of specific tachykinin receptors. Mononuclear phagocytes, either as monocytes circulating in the bloodstream or as tissue macrophages, modulate the host defense response through their ability to present antigens and the release of soluble mediators. Interleukin-1 (IL-1), TNF-α, and IL-6 were released from human blood monocytes by SP, an effect that was blocked by a SP receptor antagonist (48–50). Stimulation by naturally occurring tachykinins and by selective NK_2 receptor agonists caused a respiratory burst-dependent release of superoxide anion (51) and gelantinase production (52) from rat pulmonary macrophages, thus suggesting that NK_2 receptors were involved. "Nonclassical" tachykinin receptors have been proposed to be involved in cytokine release from microglia cells and macrophages (53,54).

In the airways, tachykinins acting directly on effector cells produce short-lived inflammatory responses. They may also amplify and extend their inflammatory actions over time by releasing polypeptide cytokines or other inflammatory mediators from proinflammatory or immunocompetent cells. Indirect proinflammatory effects of tachykinins, for instance, via cytokine release have not been explored carefully so far. Future research should be directed to reconsider more carefully the possible involvement of tachykinins in delayed inflammatory responses in human disease (Fig. 2).

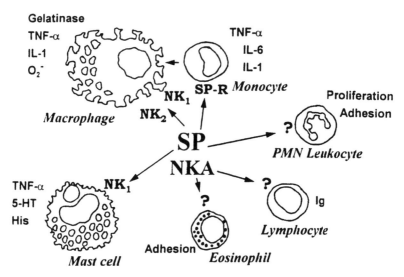

Figure 2 Schematic representation of the actions of tachykinins on inflammatory and immunocompetent cells. IL-1, interleukin-1; IL-6, interleukin-6; TNF-α, tumor necrosis factor-α; 5-HT, 5-hydroxytryptamine; His, histamine; Ig, immunoglobulins; SP-R, nontachykinin substance P receptor.

V. The Kinin/Tachykinin Inflammatory Pathway

The kinins, bradykinin, kallidin, and des Arg^9-bradykinin, are formed through enzymatic cleavage of their precursors, the kininogens, by various proteases, termed kallikreins (55). Kinins are activated by different stimuli, including injury, inflammation, and low pH. Kinins mediate their biological effects by activating B_1 and B_2 receptors (56). Chapters 10 and 20 in this book describe in detail the pharmacological and pathophysiological aspects of kinins. Kinins are powerful algesic agents and cause all the main signs of inflammation. In the airways, kinins increase vascular permeability and blood flow, and cause secretion from seromucous glands and bronchoconstriction. There is evidence that kinins produce these effects by activating B_2 receptors on effector cells directly. Kinins may also stimulate a variety of cells to release diverse mediators that markedly enhance, or sometimes limit, their own inflammatory action. One of the main proinflammatory pathways by which kinins cause inflammation is stimulation of sensory nerves and release of sensory neuropeptides. Pharmacological studies in which bradykinin was applied locally to the airways showed that plasma extravasation and bronchoconstriction were mediated by tachykinin release from sensory nerves (57,58).

A. Endogenous Kinins Release Tachykinins

The importance of the ability of kinins to cause inflammation by releasing tachykinins is better appreciated from studies that show the involvement of this pathway in various pathophysiological models of inflammation. These studies suggested the hypothesis that endogenous kinins, released during a first phase of inflammation, cause the release of tachykinins, which markedly potentiate inflammatory responses to kinins; plasma extravasation produced by antigen (ovalbumin) challenge in sensitized guinea pigs increased over time, reaching a plateau at 15–20 min. Neither NK_1 receptor or B_2 receptor antagonists produced any inhibitory effect in the early 5 min of extravasation after ovalbumin challenge (58,59). However, 10 min after ovalbumin, both antagonists caused marked inhibition, and the combination of both inhibitors did not inhibit further ovalbumin-induced plasma extravasation (58,59). As proposed previously for acid-induced plasma extravasation in the conjunctiva, kinins and tachykinins may share a final common pathway to increase vascular permeability after antigen challenge in guinea pig airways. It is possible that, during the anaphylactic reaction, kinins accumulate in the tissue and release the potent vasoactive neuropeptide SP, which increases plasma extravasation. Consistent with this hypothesis is the observation that 10 min after ovalbumin challenge there is a 10-fold increase in the levels of immunoreactive bradykinin in the bronchoalveolar lavage in guinea pigs (60).

The possibility that a neurogenic pathway potentiates the inflammatory effects of endogenous kinins has been investigated in various pathophysiological models of inflammation in the airways and outside the airways: two of these models are reported on Section II.A of this chapter. An additional example is offered by experiments with cold air inhalation. Inflammation induced by cold air in guinea pig airways consists of increased vascular permeability and bronchoconstriction, two effects mediated by activation of tachykinin NK_1 and NK_2 receptor, respectively. Inhibition of these responses by the B_2-recetor antagonist icatibant suggested that also in the case of cold air kinins mediate at least part of the activation of sensory nerves that leads to tachykinin release and to plasma extravasation and bronchoconstriction (61). Similarly, endogenous kinins participate in the activation of neurogenic plasma extravasation in the guinea pig urinary bladder that follows cyclophosphamide administration (62). However, there is no conclusive proof that endogenous kinins cause inflammation in human airways by activation of a neurogenic inflammatory pathway and the activation of tachykinin NK_1 and NK receptors.

B. The Role of Peptidases in the Control of the Kinin-Tachykinin Inflammatory Pathway

Interaction between kinins and tachykinins seems to play an important role in different models of inflammation. The use of peptidase inhibitors has been particularly useful for the discovery of this proinflammatory cascade. Thus, the inhibition of ovalbumin-induced bronchoconstriction or plasma extravasation was more pronounced when experiments were performed in the presence of peptidase inhibitors (58,59). Neutral endopeptidase (NEP) and angiotensin-converting enzyme (ACE, kininase II) are the most important peptidases for the catabolism of tachykinins and kinins. Chapters 4, 5, 20, and 21 of this book offer a broad overview of the physiological and pathophysiological aspects related to the ability of NEP and ACE to metabolize tachykinins and kinins. The hypothesis that in certain pathophysiological conditions decreased expression and/or function of NEP may exaggerate the effect of neurogenic inflammation has been proposed (63). NEP or ACE inhibition could also exaggerate the kinin/tachykinin inflammatory pathway by reducing the cleavage of either kinins, tachykinins, or both groups of peptides. A large body of evidence shows that when appropriate stimuli are provided, kinins and tachykinins are released from their precursors and from sensory nerves, respectively. In contrast, there is no evidence for tachykinins and little evidence for kinins that these peptides are continuously released under baseline conditions.

C. Are Kinins Released Continuously?

In the case of the kinin/tachykinin pathway stimuli such as lowering of the pH, antigenic stimulation or chemical injury are required in order to promote tissue kinin accumulation, which causes tachykinin release that results in increased vascular permeability and bronchoconstriction. Recent studies (64) in mice in which the gene encoding for NEP was disrupted by homologous recombination [NEP($-/-$)] indicate that activation of the kinin-tachykinin inflammatory pathway may occur even in the absence of any obvious proinflammatory stimulus. NEP($-/-$) mice showed baseline levels of plasma extravasation in the trachea, urinary bladder, pancreas, and various areas of the gut that were higher than those measured in their wild-type controls (64). Studies with tachykinin and kinin receptor antagonists provided information regarding the mechanism that causes increased inflammation in NEP($-/-$) mice. Thus, icatibant, a B_2 receptor antagonist abolished and SR 140333, an NK_1 receptor antagonist, markedly reduced the increased baseline levels of plasma extravasation in NEP($-/-$) mice (64). Similar findings were obtained if plasma extravasation was measured 10–15 min after the pharmacological blockade of NEP with phosphoramidon (64). These observations suggested that in mice, (a) kinins are continuously formed from their precursors, possibly by constitutively activated kallikreins; (b) NEP is a major factor controlling kinin levels, and the absence of NEP leads to uncontrolled inflammation mediated by kinins; and (c) the inflammatory response caused by uncontrolled kinin levels is mediated mainly by neurogenic inflammatory mechanisms. The study of the role of peptidases in the regulation of inflammation under baseline conditions might provide new insights into the mechanisms of initiation and termination of inflammatory responses, particularly at the vascular level.

VI. Multiple Mediators Released by Kinins in the Airways

A. Bradykinin-Induced Bronchoconstriction in Guinea Pigs

Bradykinin is a pleiotropic molecule that interacts with a variety of cells, producing a broad series of biological responses. Inflammatory responses to bradykinin may be mediated by direct activation of bradykinin receptors, usually of the B_2 type, on effector cells. However, in-vivo conditions and particularly when bradykinin is administered locally, direct contribution of kinin receptor activation on effector cells is minor, whereas the ability of bradykinin to release different mediators that orchestrate a complex biological response is more evident. Recent studies have also shown that bradykinin may release mediators that exert antiinflammatory action. Thus, the final response to bradykinin reflects the contrasting actions of the different mediators released by this autacoid. One example of these possible interactions is the bronchoconstriction induced by bradykinin in guinea pigs. Bradykinin, if injected intravenously, causes a bronchoconstrictor response that is mediated by prostanoids, and in minor part by an atropine-sensitive cholinergic reflex pathway. If is administered by aerosolization, the constrictor effect of bradykinin is mediated mainly by tachykinin release from sensory nerve endings (57), although in this case also, a minor contribution of cholinergic nerves has been described (57).

B. The Role of Nitric Oxide in Guinea Pigs and Humans

In atropinized guinea pig, bradykinin aerosol is more potent in causing plasma extravasation than bronchoconstriction (65). Aerosolized capsaicin, which causes plasma extravasation and bronchoconstriction by a mechanism identical to that of bradykinin (tachykinin release from sensory nerves), is more potent, however, in causing bronchoconstriction than plasma extravasation (65). One possible explanation for this apparent discrepancy is that bradykinin, but not capsaicin, releases, along with bronchoconstrictor tachykinins, a bronchorelaxant molecule (65). Bradykinin releases prostanoids from airway epithelial cells in guinea pigs, which may cause relaxant (66) and contractile responses in airway smooth muscle cells. Despite these findings, bronchoconstriction induced by bradykinin in vivo does not seem to involve prostanoid release (57). More recently, it has been shown that various agents and autacoids, including histamine (67), endothelins (68), and high K^+ (69), relax guinea pig airways by releasing NO from the airway epithelium. Bradykinin also causes an epithelium-dependent relaxation of guinea pig tracheal tube preparations, an effect that is changed into a contraction by inhibitors of NO-synthase, thus indicating NO as the relaxant mediator released by bradykinin (70). Indeed, the observation that in guinea pigs NO-synthase inhibition markedly potentiated bronchoconstriction induced by aerosolized or intravenous (Yoshihara, Nadel, and Geppetti, unpublished observation) bradykinin, whereas it did not affect bronchoconstriction induced by aerosolized capsaicin, suggests that NO released by bradykinin, possibly from the airway epithelium, markedly inhibits the indirect bronchoconstrictor mechanisms activated by bradykinin (Fig. 3).

These pharmacological data indicate that a major inflammatory effect of bradykinin, such as bronchoconstriction, results from the release of different mediators that, with their opposing forces, modulate the final response. In this regard, two major questions arise. First, is this interaction between kinins, tachykinins, and NO relevant in pathophysiological models of inflammation? Second, is this interaction present in human airways? Experiments in which cold air was administered to the airways of atropinized guinea pigs may answer the first question. Cold air inhalation for more than 10 min caused bronchoconstriction, which, being completely abolished by an NK_2 receptor antagonist, is considered to be mediated by tachykinin release from sensory nerves (71). The observation that icatibant, a B_2 receptor antagonist, markedly reduces cold air-induced bronchoconstriction suggests that kinins, released following exposure to cold air, increase airway tone by releasing tachykinins (71). Cold air inhalation for 5 min does not affect bronchoconstriction. However, pretreatment with NO-synthase inhibitors unmasks a bronchoconstrictor response to cold air that is mediated by kinins and tachykinins (Yoshihara, Nadel, & Geppetti, unpublished observations). Therefore, the hypothesis was advanced that kinins released by cold air not only release contractile tachykinins, but also release relaxant NO, probably from the airway epithelium.

Bradykinin does not substantially affect lung resistance in healthy human subjects, and it causes a slight bronchoconstrictor effect in moderately asthmatic subjects (72). The mechanism of bradykinin-induced bronchoconstriction in humans is indirect: pharmacological evidence suggesting that, as already shown in guinea

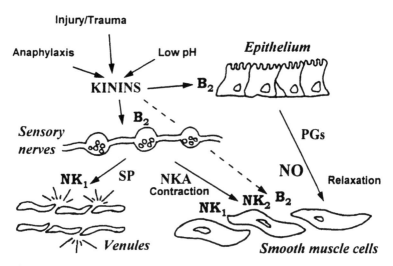

Figure 3   Schematic representation of the interactions between kinins and other mediators in airway inflammatory response.

pigs, a cholinergic reflex pathway (72) and release of tachykinins from sensory nerves (73) are involved. A recent study showed that bradykinin-induced bronchoconstriction in moderately asthmatic patients is markedly increased after pretreatment with an NO-synthase inhibitor (74). These findings suggested that NO released by bradykinin plays a major role in inhibiting the bronchoconstriction induced by bradykinin in asthma. In severe asthmatics the potentiation by NO-synthase inhibitors of bradykinin-induced bronchoconstriction was markedly reduced (F. L. M. Ricciardolo and P. Geppetti, personal communication). Therefore in severe asthma the protective effect mediated by NO seems to be absent, thus leaving unopposed the action of bronchoconstrictor mediators released by bradykinin. Thus, hyperreactivity to bradykinin may result from the decreased ability of bradykinin to release of bronchorelaxant NO from the airway epithelium.

VII. Conclusions

The number of mediators involved in the regulation of inflammation most likely reflects the need to modulate this important biological response carefully. Interactions between various mediators magnify the possibilities of fine-tuning the inflammatory response. Examples reported in this chapter focused on mediators involved in the early phase of inflammation, namely, kinins and tachykinins. Even at this early stage, kinins and tachykinins can interact in order to potentiate or reduce the inflammatory response. Inflammation brought about by sensory nerves (neurogenic inflammation) may be modulated by an individual stimulus that acts on sensory nerves via activation of different receptors with opposing actions. Evidence has been also provided that an individual stimulus may activate neurogenic inflammation by direct or indirect mechanisms.

Tachykinins, released from sensory nerve endings, can be regarded as final mediators of various short-lived inflammatory responses. Their potential as initiators of prolonged inflammatory reactions, including those mediated by cytokine release, remains to be fully investigated. Kinins are generally considered as indirect stimulants of the release of more potent mediators. Kinins may also activate the release of mediators that limit their proinflammatory activity. Our understanding of these complex processes and interactions is only at its beginning. Definition of the cascades of mediators involved in the enhancement/limitation of the inflammatory process may be a critical step in the development of more selective receptor agonists and antagonists that ultimately may result in more effective therapeutic activity and significant reduction of drug-related side effects.

References

1.   Holzer P. Local effector functions of capsaicin-sensitive sensory nerves endings: involvement of tachykinins, calcitonin gene-related peptide and other neuropeptides. Neuroscience 1988; 24:739–768.

2. Holzer P. Capsaicin: cellular targets, mechanisms of actions, and selectivity for thin sensory neurons. Pharmacol Rev 1991; 43:143–201.

3. Maggi CA. The pharmacology of the efferent function of sensory nerves. J Auton Pharmacol 1991; 11:173–208.

4. Bevan S, Geppetti P. Protons, small stimulants of capsaicin-sensitive sensory nerves. Trends Neurosci 1994; 26:509–512.

5. Geppetti P, Tramontana M, Patacchini R, Del Bianco E, Santicioli P, Maggi CA. Neurochemical evidence for the activation of the "efferent" function of capsaicin-sensitive nerves by lowering of the pH in the guinea-pig urinary bladder. Neurosci Lett 1990; 114:101–106.

6. Geppetti P, Del Bianco E, Patacchini R, Santicioli P, Maggi CA, Tramontana M. Low pH-induced release of calcitonin gene-related peptide from capsaicin-sensitive sensory nerves: mechanism of action and biological response. Neuroscience 1991; 41:295–301.

7. Bevan S, Yeats JC. Protons activate a cation conductance in a sub-population of rat dorsal ganglion neurons. J Physiol 1991; 433:145–161.

8. Liu L, Simon SA. A rapid capsaicin-activated current in rat trigeminal ganglion neurons. Proc Natl Acad Sci (USA) 1994; 91:738–741.

9. Bevan S, Hothi S, Hughes G, et al. Capsazepine: a competitive antagonist of the sensory neurone excitant capsaicin. Br J Pharmacol 1992; 107:544–552.

10. Santicioli P, Del Bianco E, Figini M, Bevan S, Maggi CA. Effect of capsazepine on the release of calcitonin gene-related peptide-like immunoreactivity (CGRP-LI) induced by low pH, capsaicin and potassium in the rat soleus muscle. Br J Pharmacol 1993; 2:609–612.

11. Szallasi A, Goso C, Blumberg PM, Manzini S. Competitive antagonism by capsazepine of [^{3}H]resineferatoxin binding to central (spinal cord and dorsal root ganglia) and peripheral (urinary bladder and airways) vanilloid (capsaicin) receptors in the rat. J Pharmacol Exp Ther 1993; 2:728–733.

12. Maggi CA, Bevan S, Walpole CSJ, Rang HP, Giuliani S. A comparison of capsazepine and ruthenium red as capsaicin antagonists in the rat isolated urinary bladder and vas deferens. Br J Pharmacol 1993; 108:801–805.

13. Lalloo UG, Fox AJ, Belvisi MG, Chung KF, Barnes PJ. Capsazepine inhibits cough induced by capsaicin and citric acid but not hypertonic saline in guinea pigs. J Appl Physiol 1995; 79:1082–1087.

14. Figini F, Javdan P, Cioncolini F, Geppetti P. Involvement of tachykinins in plasma extravasation induced by bradykinin and low pH medium in the guinea-pig conjunctiva. Br J Pharmacol 1995; 115:128–132.

15. Geppetti P. Sensory neuropeptide release by bradykinin: mechanism and pathophysiological implications. Regulatory Peptides 1993; 47:1–23.

16. Petho G, Jocic M, Holzer P. The role of bradykinin in the hyperaemic following acid challenge of the rat gastric mucosa. Br J Pharmacol 1994; 113:1036–1042.

17. Saria A, Martling C-R, Yan Z, Theodorsson-Norheim E, Gamse R, Lundberg JM. Release of multiple tachykinins from capsaicin-sensitive nerves in the lung by bradykinins histamine, dimethylphenyl piperazinum, and vagal nerve stimulation. Am Rev Respir Dis 1988; 137:1330–1335.

18. Tramontana M, Giuliani S, Del Bianco E, et al. Effects of capsaicin and 5HT$_3$ receptor antagonist on 5-HT-evoked release of calcitonin gene-related peptide in the guinea-pig heart. Br J Pharmacol 1993; 108:431–435.

19. Ueda N, Maramatsu I, Fujiwara M. Prostaglandins enhance trigeminal substance P-ergic responses in the rabbit iris sphincter muscles. Brain Res 1985; 337:347–351.

20. Geppetti P, Del Bianco E, Tramontana M, et al. Arachidonic acid and bradykinin share a common pathway to release neuropeptide from capsaicin-sensitive sensory nerve fibers of the guinea pig heart. J Pharmacol Exp Ther 1991; 259:759–765.

21. Vasko MR. Prostaglandin-induced neuropeptide release from spinal cord. Prog Brain Res 1995; 104:367–380.

22. Ellis J, Undem B. Role of peptidoleukotrienes in capsaicin-sensitive sensory fibre-mediated responses in guinea-pig airways. J Physiol 1991; 436:469–484.

23. Coyle AJ, Perretti F, Manzini S, Irvin CG. Cationic protein induced sensory nerve activation: role of substance P in airway hyperresponsiveness and plasma extravasation. J Clin Invest 1994; 94:2301–2306.

24. Hughes SR, Brain SD. Nitric oxide-dependent release of vasodilator quantities of calcitonin gene-related peptide from capsaicin-sensitive nerves in rabbit skin. Br J Pharmacol 1994; 111:425–430.

25. Weinreich D. Bradykinin inhibits a slow spike after hyperpolarization in visceral sensory neurons. Eur J Pharmacol 1986; 132:61–63.

26. Geppetti P, Tramontana M, Santicioli P, Del Bianco E, Giuliani S, Maggi CA. Bradykinin-induced release of calcitonin gene-related peptide from capsaicin-sensitive nerves in guinea-pig atria: mechanism of action and calcium requirements. Neuroscience 1990; 38:687–692.

27. Nicol G, Klinberg D, Vasko M. Prostaglandin E_2 increases calcium conductance and stimulates release of substance P in avian sensory neurons. J Neurosci 1992; 12:1917–1927.

28. Moskowitz MA. Neurogenic versus vascular mechanisms of action of sumatriptan and ergot alkaloids in migraine. Trends Pharmacol Sci 1992; 13:307–311.

29. Rebeck GW, Maynard KI, Hyman B, Moskowitz MA. Selective 5HT1Dα receptor gene expression in trigeminal ganglia: implications for anti-migraine drug development. Proc Natl Acad Sci (USA) 1994; 91:3666–3670.

30. Hamel E, Fan E, Linville D, Ting V, Villemure JG, Chia LS. Expression of mRNA for the serotonin 5-hydroxytryptamine $1D_{beta}$ receptor subtype in human and bovine cerebral arteries. Mol Pharmacol 1993; 44:242–246.

31. Ferrari MD, Saxena PR. Clinical and experimental effects of sumatriptan in humans. Trends Pharmacol Sci 1993; 14:129–133.

32. Otsuka M. Yoshioka K. Neurotransmitter functions of mammalian tachykinins. Physiol Rev 1993; 73:229–308.

33. Imamura M, Smith NC, Garbarg M, Levi R. Histamine H_3-receptor-mediated inhibition of calcitonin gene-related peptide release from cardiac C fibers. A regulatory negative-feedback loop. Circ Res 1996; 78:863–869.

34. Chen C-C, Akopian AN, Silviotti L, Colquoun D, Burnstock G, Wood JN. A P2X purinoceptor expressed by a subset of sensory neurons. Nature 1995; 377:428–431.

35. Bowden JJ, Garland AM, Baluk P, et al. Direct observation of substance P-induced internalization of neurokinin1 (NK_1) receptors at sites of inflammation. Proc Natl Acad Sci (USA) 1994; 91:8964–8968.

36. Hall JM, Brain SD. Pharmacology of calcitonin gene-related peptide. In: Geppetti P, Holzer P, eds. Neurogenic Inflammation. Boca Raton, FL: CRC Press, 1996:101–114.

37. Piedimonte G, Hoffman JE, Husseini WK, Hiser WL, Nadel JA. Effect of neuropeptides released from sensory nerves on blood flow in the rat airway microcirculation. J Appl Physiol 1992; 72:1563–1570.

38. Baluk P, Bertrand C, Geppetti P, McDonald MD, Nadel JA. NK_1 receptors mediate

leukocyte adhesion in neurogenic inflammation in the rat trachea. Am J Physiol 1994; 12:L263–L269.

39. De Rose V, Robbins RA, Snider RM, et al. Substance P increases neutrophil adhesion to bronchial epithelial cells. J Immunol 1994; 152:1339–1346.

40. Frossard N, Rhoden KJ, Barnes PJ. Influence of epithelium on guinea pig airway responses to tachykinins: role of endopeptidase and cyclooxygenase. J Pharmacol Exp Ther 1989; 248:292–298.

41. Manzini S. Bronchodilatation by tachykinins and capsaicin in the mouse main bronchus. Br J Pharmacol 1992; 105:968–972.

42. Figini M, Emanuelli C, Bertrand P, Javdan P, Geppetti P. Evidence that tachykinins relax the guinea-pig trachea via nitric oxided release and by stimulation of a septide-insensitive NK_1 receptor. Br J Pharmacol 1996; 117:1270–1276.

43. Molinard M, Martin CA, Naline E, Hirsch A, Advenier C. Role of thromboxane A_2 in bradykinin-induced human isolated small bronchi contraction. Eur J Pharmacol 1995; 278:49–54.

44. Mousli M, Bueb J-L, Bronner C, Rout B, Landry Y. G protein activation: a receptor-independent mode of action for cationinc amphiphilic neuropeptides and venom peptides. Trend Pharmacol Sci 1990; 11:358–362.

45. Joos GF, Kps JC, Pauwels RA. In vivo characterization of the tachykinin receptor involved in the direct and indirect bronchoconstrictor effect of tachykinins in two inbred rat strains. Am J Respir Crit Care Med 1994; 149:1160–1166.

46. Joos G, Pauwels R, Van Straeten M. The effect of nedocromil sodium on the bronchoconstrictor effect of neurokinin A in subjects with asthma. J Allergy Clin Immunol 1989; 83:663–668.

47. Ansel JC, Brown JR, Payan DG, Brown MA. Substance P selectively activates TNF-a gene expression in murine mast cells. J Immunol 1993; 150:4478–4485.

48. Lotz M, Vaughan JH, Carson DA. Effect of neuropeptides on production of inflammatory cytokines by human monocytes. Science 1988; 241:1218–1221.

49. Lee HR, Ho WZ, Douglas S. Substance P augments tumor necrosis factor release in human-derived macrophages. Clin Diag Lab Immunol 1994; 1:419–423.

50. Bahl AK, Foreman JC. Stimulation and release of interleukin-1 from peritoneal macrophages. Agents Actions 1994; 42:154–158.

51. Brunelleschi S, Vanni L, Ledda F, Giotti A, Maggi CA, Fantozzi R. Tachykinins activate guinea-pig alveolar macrophages: involvement of NK1 and Nk2 receptors. Br J Pharmacol 1990; 100:417–420.

52. D'Ortho MP, Jareau PH, Delacourt C, Pezet S, Lafuma C, Harf A. Tachykinins induce gelatinase production by guinea pig alveolar macrophages: involvement of NK2 receptors. Am J Physiol 1995; 269:L361–L366.

53. Martin FC, Anton PA, Gornbein JA, Shanahan F, Merrill JE. Production of interleukin-1 by microglia in response to substance P: role for a non-classical NK-1 receptor. J Neuroimmunol 193; 42:53–60.

54. Jeurissen F, Kavelaars A, Korstjens M, et al. Monocytes express a non-neurokinin substance P receptor that is functionally coupled to a MAP kinase. J Immunol 1994; 152:2987–2993.

55. Bhoola K, Figueroa C, Worthy K. Bioregulation of kinins: kallikreins, kininogens, and kininases. Pharmacol Rev 1992; 44:1–80.

56. Regoli D, Barabe J. Pharmacology of bradykinin and related kinins. Pharmacol Rev 1980; 32:1–46.

57. Ichinose M, Belvisi MG, Barnes PJ. Bradykinin-induced bronchoconstriction in guinea pig in vivo: role of neural mechanisms. J Pharmacol Exp Ther 1990; 253:594–599.

58. Bertrand C, Yamawaki I, Nadel JA, Geppetti P. Role of kinins in the vascular extravasation evoked by antigen and mediated by tachykinins in guinea pig trachea. J Immunol 1993; 151:4902–4907.

59. Bertrand C, Geppetti P, Baker J, Yamawaki I, Nadel JA. The role of neurogenic inflammation in antigen-induced vascular extravasation in guinea pig trachea. J Immunol 1993; 150:1479–1485.

60. Erjfält I, Greiff L, Alkner U, Persson CG. Allergen-induced biphasic plasma exudation responses in guinea pig large airways. Am Rev Respir Dis 1993; 148:695–701.

61. Yoshihara S, Chan B, Yamwaki I, et al. Plasma extravasation in the rat trachea induced by cold air is mediated by tachykinin release from sensory nerves. Am J Respir Crit Care Med 1995; 151:1011–1017.

62. Ahluwalia A, Maggi CA, Santicioli P, Lecci A, Giuliani S. Characterization of the capsaicin-sensitive component of cyclophosphamide-induced inflammation in the rat urinary. Br J Pharmacol 1994; 111:1017–1022.

63. Nadel JA. Neutral endopeptidase modulates neurogenic inflammation. Eur Respir J 1991; 4:745–749.

64. Lu B, Figini M, Emanueli C, et al. The control of microvascular permeability and blood pressure by neutral endopeptidase. Nature Med 1997; 3:904–907.

65. Ricciardolo FLM, Nadel JA, Yoshihara S, Geppetti P. Evidence for reduction of bradykinin-induced bronchoconstriction in guinea-pigs by release of nitric oxide. Br J Pharmacol 1994; 113:1147–1152.

66. Frossard N, Stretton CD, Barnes PJ. Mechanism of epithelial modulation of bradykinin responses in airway smooth muscle. Am Rev Respir Dis 1989; 139:A351.

67. Nijkamp FP, van Der Linde HJ, Folkerts G. Nitric oxide synthesis inhibitors induce airway hyperresponsiveness in the guinea pig in vivo and in vitro. Role of the epithelium. Am Rev Respir Dis 1993; 148:727–734.

68. Filep J, Battistini B, Sirois P. Induction by endothelin-1 of epithelium-dependent relaxation of guinea-pig trachea in vitro: role of nitric oxide. Br J Pharmacol 1993; 109:637–644.

69. Folkerts G, van der Linde HJ, Verheyen AK, Nijkamp FP. Endogenous nitric oxide modulates potassium-induced changes in guinea pig airway tone. Br J Pharmacol 1995; 115:1194–1198.

70. Figini M, Ricciardolo FLM, Javdan P, et al. Evidence that epithelium-derived relaxing factor released by bradykinin in guinea pig trachea is nitric oxide. Am J Respir Crit Care Med 1995; 153:918–923.

71. Yoshihara S, Geppetti P, Hara M, et al. Cold air-induced bronchoconstriction is mediated by tachykinin and kinin release in guinea pigs. Eur J Pharmacol 1996; 296:291–296.

72. Fuller RW, Dixon CMS, Cuss FCM, Barnes PJ. Bradykinin-induced bronchoconstriction in humans: mode of action. Am Rev Respir Dis 1987; 135:176–180.

73. Ichinose M, Nakajima N, Takahashi T, Yamauchi H, Inoue H, Takashima T. Protection against bradykinin-induced bronchoconstriction in asthmatic patients by neurokinin receptor antagonist. Lancet 1992; 340:1248–1251.

74. Ricciardolo FLM, Geppetti P, Mistretta A, et al. Randomised double-blind placebo-controlled study of the effect of inhibition of nitric-oxide synthesis in bradykinin-induced asthma. Lancet 1996; 348:374–377.

10

Cellular Regulation of Kinin Receptors
Potential Role in Inflammation

ADELBERT A. ROSCHER

Children's Hospital
University of Munich
Munich, Germany

ALEXANDER FAUSSNER

Johns Hopkins Asthma and Allergy Center
Baltimore, Maryland

I. Introduction

In this communication we review the regulatory mechanisms that may contribute significantly to the pronounced local inflammatory action of bradykinin. In particular, we will focus on the role of the cellular kinin receptors and their relationship to acute inflammatory airway conditions and chronic repair mechanisms. We describe the mechanisms involved in transcriptional regulation, expression and up- and down-regulation of the two known subtypes of B_2 and B_1 kinin receptors. Furthermore, we discuss the influence of G-protein coupling and signal transduction events and the variety of cross-interactions occurring at different levels between kinin receptors and other mediators or other inflammatory agents. Finally, we discuss the possible therapeutic interventions that aim to influence respiratory diseases by virtue of kinin receptor antagonism.

A. Kinins, Kinin Receptors, and Inflammation

The nonapeptide bradykinin (BK) and related kinins such as kallidin (Lys-BK,KD) are considered to be key inflammatory mediators, causing vasodilation, increased blood flow, pain, and increased vascular permeability (1). Kinins are released by limited proteolysis from kininogen precursors in the circulation, in interstitial tissue fluids, or from the surface of kininogen-binding cells including leukocytes. The large array of physiological activities is mediated mainly by specific receptors, classified pharmacologically as B_2 receptors, which exhibit high affinity for KD and BK (2–4). In contrast, the expression of B_1 kinin receptors is possibly enhanced in pathological situations, particularly in response to inflammatory conditions (5). Recently, the human B_2 and B_1 receptors have been cloned, and their genomic organization was subsequently characterized (6–9). By homology search, both receptor subtypes can be grouped to the superfamily of the G-protein-coupled seven-transmembrane-domain receptors. Depending on the different cell types and tissues, B_2 receptors couple to G proteins, thereby triggering activation of phospholipase C and/or phospholipase A_2, accompanied by increased intracellular levels of Ca^{2+}, NO, cGMP, and/or cAMP (10–13).

Kinins are rapidly broken down by a variety of enzymes, particularly, ACE (angiotensin-converting enzyme), NEP (neutral endopeptidase), and the carboxypeptidases (CP). While the products of ACE and NEP are inactive, those produced by the CPs, des-Arg^9-BK and des-Arg^{10}-KD, can serve as selective B_1-receptor agonists. For example, interconversion of B_2 and B_1 agonists has been found to be very pronounced in the nasal discharge fluid of patients with allergic rhinitis (14). Due to their rapid degradation in tissues and in the circulation, kinins are regarded as "autacoids" or local hormones, since action on remote organs is not favored.

B. Kinins in Human Airway Diseases

The contribution of kinins and kinin receptors to the pathogenesis of human airway diseases occurs at different cells and sites within the respiratory tract. In the upper airways, kinins can stimulate glandular secretion, increase vascular permeability, and stimulate sensory nerves to produce symptoms of nasal obstruction, rhinorrhea, and nasal and throat irritation (15,16). BK is generated during natural rhinovirus colds (17) and as a result of experimental infection with rhinovirus and attenuated influenza virus. Its levels in infected persons are shown to correlate with clinical symptoms and mucous secretion (18). Kallikrein activities and kinin levels are also increased in nasal or bronchial lavage fluids after antigen challenge in subjects with allergic rhinitis (19) or asthma (20). This points to a contribution of kinins not only to viral infections but also in immediate-type hypersensitivity events. In the lower airways of asthmatics, kinins are potent inducers of edema and cause bronchoconstriction by mechanisms that appear to involve neural reflexes. Recently, a role of kinins in interstitial lung disease, such as sarcoidosis and progressive systemic sclerosis, has been implicated because of the demonstration of B_1 receptors in transbronchial tissues of such patients (21).

C. Interactions Between Inflammatory Mediators and Components of the Kinin System

Among tissue-specific activation of second-messenger pathways, kinins also possess the ability to release neurotransmitters or prostaglandins (PGs) and to stimulate the synthesis of cytokines such as interleukin-1 (IL-1) or tumor necrosis factor (22). On the other hand, inflammatory mediators or BK itself, once released, are capable of modifying the expression of components of the kallikrein-kininogen-kinin system. IL-1 is reported to amplify B_2 receptor-induced PGE_2 production (23) or to upregulate kinin B_1 receptor-mediated responses during tissue injury (5). Kininogene gene expression has also been shown to be induced in response to cAMP, PGE_2, and tumor necrosis factor (24), and kininogen-binding sites of endothelial cells appear to be induced by BK itself (25).

In general, these regulatory events, occurring during inflammation, may modify the BK action at three different levels:

1. By "prereceptor" mechanisms due to the action of kinin-degrading enzymes that influence the bioavailability of active kinin receptor agonists.
2. At the "receptor level," by the mechanisms that control the expression, characteristics, and subtype of cell-surface kinin receptors.
3. At the "postreceptor level," by the second-messenger pathways that mediate and modify BK-induced signal transduction.

In this limited report we will focus mainly on mechanisms involved in the cellular expression and regulation of kinin receptors.

II. Heterologous Regulation of Kinin Receptor Expression and Post-Receptor Events

A. Kinin B_2 Receptor

Induction of B_2 Kinin Receptors by IL-1 and Agents That Increase Intracellular cAMP Levels

Despite the fact that B_2 kinin receptors are constitutively expressed in a wide variety of tissues and cultured cell lines, several studies have demonstrated that extracellular hormones and mediators can modulate the level of receptor expression and thereby the amplitude of the BK-stimulated signal.

The inflammatory cytokine IL-1, which is involved in the pathogenesis of a variety of diseases (26), has been shown to double the number B_2 receptors in human synovial fibroblasts and to increase drastically the synthesis of PGs in response to BK (27). The appearance of new B_2 receptors on the cell surface was prevented by the protein synthesis inhibitor cycloheximide, indicating that neosynthesis and not activation of spare receptors was responsible for the observed effect. IL-1 also has been reported to induce phospholipase A_2 and prostaglandin H synthase in synovial fibroblasts (28). Therefore a combination of increased B_2 receptors and induction of PG-synthesizing enzymes may be responsible for the dramatically enhanced susceptibility of synovial cells upon BK stimulation after IL-1 pretreatment. The release of arachidonic acid in IL-1-pretreated synovial cells in response to BK can selectively be abolished by downregulation of protein kinase C with the phorbolester PMA, while the formation of diacylglycerol or elevation of intercellular (Ca^{2+}) is not affected by PKC downregulation (29).

In previous studies (30), is was reported that a variety of manipulations which activate intracellular cAMP pathways (forskolin, cholera toxin-modified G-alpha proteins, inactivation of G-alpha-i, cell membrane-permeable cAMP analogs), are able to increase B_2 receptor numbers. This effect was not observed in a mutant PC12 cell line that lacks functional cAMP-dependent protein kinase type II.

In our own studies utilizing human foreskin fibroblasts, a direct stimulation of adenylate cyclase with forskolin—in the presence of the phosphodiesterase inhibitor IBMX—promoted a two- to fivefold increase of B_2 receptors after 24 hr of incubation (Fig. 1). The specific receptor ligand BK, which also induces cAMP accumulation in these cells, had no effect. A similar increment of B_2 receptors was also seen with IL-1, but not with IL-4 or IL-6. The effect of both IL-1 and forskolin (not shown) was blocked when they were added to the cells along with increasing concentrations of PKC-activating phorbol esters such as PDBu, PMA, or PDD-beta (Fig. 2). No inhibition was observed in the presence of the inactive phorbol ester PDD-alpha. The moderate specific inhibitor staurosporine increased [³H]BK binding, either alone or in combination with IL-1. Particularly in non-foreskin fibroblasts, which have a

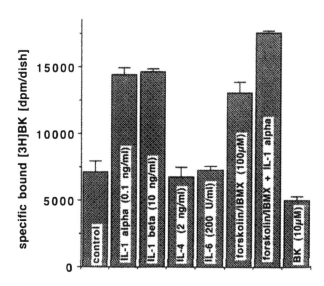

Figure 1 Influence of cytokines and forskolin/IBMX on B_2 kinin receptor expression. Human foreskin fibroblasts in 6-well trays were incubated with the indicated reagents in DMEM medium containing 0.05% BSA. After 24 hr, specific [³H]BK-binding activity was determined at 0°C in triplicate.

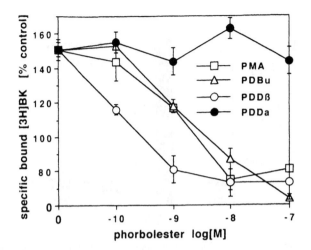

Figure 2 Influence of phorbol esters on IL-1-induced increase of BK binding. Human foreskin fibroblasts were incubated with the indicated concentrations of phorbol ester in DMEM medium. After 24 hr, specific [^{3}H]BK-binding activity was determined at 0°C in triplicate.

[^{3}H]BK-binding activity much lower than that of foreskin fibroblasts (M. Bidling-maier, unpublished data) the combination of IL-1 and staurosporine increased the binding activity by up to 10-fold, reaching the receptor level of foreskin fibroblasts (not shown). Interestingly, staurosporin alone promoted B$_2$ receptor number in non-foreskin fibroblasts about sevenfold and in foreskin fibroblasts about fourfold (Fig. 3). However, neither of the much more specific PKC inhibitors bisindolylmaleimide or chelerythrin (not shown) was capable of reproducing the effect of staurosporine. Therefore, other qualities of staurosporine than inhibition of PKC may be responsible for its B$_2$ receptor-inducing effect. Another possibility is that the PKC inhibitors used display different specificities for the various isoenzymes of protein kinase C. Similarly, inhibition of protein kinase A by staurosporine (IC$_{50}$ = 7 nM) should not be responsible, since the protein kinase A antagonist (cAMP-S, Rp isomer) also was not able to reproduce the effect of staurosporine.

After pretreatment with IL-1, we also observed a markedly increased cAMP response to forskolin stimulation in these cells as well as increased basal cAMP levels (A. Faussner, unpublished). Therefore one might speculate that IL-1 induces expression of at least part of the new B$_2$ receptors via promotion of adenylate cyclase activity and elevation of intracellular basal cAMP levels. In vascular smooth muscle cells, incubation with cholera toxin, forskolin, or isobutylmethylxanthine led to an increase in B$_2$ receptor number and BK-stimulated phosphoinositide hydrolysis. This effect appears to be specific for BK, since the same treatment decreased the responsiveness of these cells to angiotensin II and vasopressin (31). Interestingly, in these cells isoproterenol and PGE2 were not capable of inducing B$_2$ receptor synthesis,

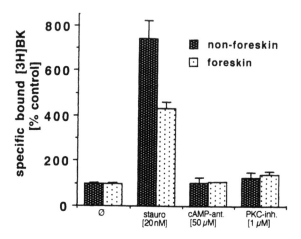

Figure 3 Binding of [³H]BK to foreskin and non-foreskin fibroblasts after treatment with stauroporine, bisindoylmaleimide (BIM), and cAMP-antagonist (cAMP-S, Rp isomer). Human foreskin fibroblasts in 6-well trays were incubated with the indicated reagents in DMEM medium containing 0.05% BSA. After 24 hr, specific [³H]BK-binding activity was determined at 0°C in triplicate.

although both agents clearly elevated cAMP. This may indicate that formation of cAMP alone is not entirely sufficient to increase B_2 receptors. Similarly, we found in human fibroblasts that stimulation with BK, despite increasing intracellular cAMP levels via release of PGs and subsequent activation of their receptors, does not lead to synthesis of its own receptor (Fig. 1). However, this failure to detect a ligand-specific upregulation of B_2 receptors may be explained by the BK-induced concommitant activation of protein kinase C, which is counteracting the cAMP-effect, as described above.

Overexpression of p21ras Induces B_2 Kinin Receptors

Several studies have reported that transformation of cultured cells with the activated *ras* protein leads to increased responsiveness of these cells to BK (32,33). Rat-1 fibroblasts, normally not sensitive to BK, showed an inositolphosphate response to the peptide after a transfection that led to the overexpression of normal or constitutive active p21*ras* protein (34). This increased responsiveness was due to a strong increase in B_2 receptor number, which correlated well with BK-stimulated phosphatidylinositol turnover, but not with the amount of overexpressed p21*ras*. This may indicate that the number of receptors is important for the amplitude of the response. In addition, this effect was found to be specific for BK and could not be attributed to a general transformation phenomenon, since *src*- or N-*myc*-transformed cells showed no (increased) response to BK.

Glucocorticosteroids Reduce B_2 Kinin Receptor Number

Glucocorticosteroids are well known as endogenous antiinflammatory agents and are frequently and successfully used in the therapy of patients with asthma (35). We observed that in-vitro treatment of cultured fibroblasts with glucocorticosteroids for 24–48 hr is capable of reducing B_2 receptor number (36) in a dose-dependent fashion. Slight effects were seen with concentrations as low as 1 nM dexamethasone and reached a maximal reduction in B_2 receptor number by almost 50% at 1 µM. Paralleling their antiinflammatory potencies, steroids reduced B_2 kinin receptors with the following rank order: fluocinolone acetonide > dexamethasone > hydrocortisone > corticosterone > deoxycorticosterone > estradiol.

B. Cytokines Stimulate de-Novo Synthesis of Kinin B_1-Receptors

In contrast to the constitutive expressed kinin B_2 receptor, the B_1 receptor is believed to be synthesized de novo under pathological conditions. Many smooth muscle preparations, which under normal circumstances did not respond to B_1 agonists at all, became responsive after prolonged in-vitro incubation or after treatment with cytokines such as IL-1, IL-2, or IL-8 (see Ref. 5 and references cited therein). This sensitization of B_1 agonists can be prevented by a permanent addition of the protein synthesis inhibitor cycloheximide. Interestingly, short-term (pulse of 1 hr) addition of cycloheximide had an opposite effect. It led to a subsequent increase in the responsiveness to B_1 agonists that was even stronger than in preparations without cycloheximide treatment. IMR 90 or WI 38 cell lines (fetal lung fibroblasts) represent two of the few examples of human cells that constitutively express B_1 receptors, even if only in low numbers. Cell culture, however, can be regarded as a kind of a pathological situation, inducing B_1 receptor expression, since these cells are removed from their natural environment and are in continuous contact with fetal calf serum ("wound fluid") present in the medium. In IMR 90 fibroblasts, pretreatment with IL-1 increased B_1 receptors about sevenfold (8). mRNA from such IL-1-stimulated IMR 90 cells was used to clone the human B_1 receptor. In cell culture the signal transduction pathways operating on B_1 receptor stimulation very much resemble the ones observed with B_2 receptor activation, such as the formation of inositolphosphates or the release of PGs. For example, in smooth muscle cells a sevenfold increase in B_1 receptor-mediated secretion of prostacyclin was observed after priming with IL-1 (37).

III. Ligand-Induced Internalization and Desensitization of Kinin Receptors

It is well established that incubation of human foreskin fibroblasts with BK leads to a strong reduction in BK B_2 receptor number (about 80% within 10 min), due to the loss of active receptors at the cell surface as indicated by a large change in B_{max} but only a small change in K_d. This effect is paralleled by a reduction in the responsiveness of the cells to BK regarding prostacyclin synthesis (38). The desensitization is homologous, since the responses to isoproterenol or exogenous PGE_2 were not changed. When the first stimulation lasted no longer than 15 min, almost all of the initial binding activity and responsiveness was completely recovered after about 30 min incubation at 37°C in the absence of BK. Longer stimulation, however, led to a permanent loss of about 40% binding activity, which was not reversed even several hours after removal of BK. However, it is not clear yet if this long-term reduction in BK-binding activity is due to a change in receptor number or to a permanent change in receptor affinity as described in another study using rat-13 fibroblasts (39). In this report, stimulation with BK for 5 hr led to an affinity shift of the B_2 receptor from 2 nM to 40 nM. This shift was observed only in intact cells, but not in assays utilizing membrane preparations (no guanine nucleotides were added). Neither publication clarified whether the fraction of B_2 receptors that remains detectable after long-term stimulation with BK is still coupled to a biological effect.

Several studies thereafter revealed that the primary loss of binding activity from the cell surface is due to ligand-induced internalization of the B_2 receptor (40–42). As a consequence of BK binding to its receptor, the B_2 receptors (and bound BK) are removed from the cell surface within minutes, presumably via clathrin-coated vesicles (C. Klier, unpublished data). As described for other receptor systems (43), these vesicles may then fuse with early endosomes, where the ligand dissociates rapidly from its receptor due to the more acid environment. The receptors—presumably in conjunction with most of the (degraded) ligand—are then recycled back to the cell surface and are available for a new cycle of activation. This cycle takes less than 1 hr. The use of the anticancer drug taxol is limited by significant cardiovascular side effects that might possibly be explained by interference with intracellular BK receptor cycling. In studying BK receptor desensitization, which reflects receptor recycling, it was demonstrated that taxol inhibits BK-evoked Ca^{2+} transients by 50% within endothelial cells (44).

One question still to be answered relates to the physiological importance and need for BK-induced receptor internalization in the case of such a small ligand as BK. It is certainly not a mechanism to remove BK from the extracellular environment via intracellular degradation, since there are potent peptidases at the cell surface that can exert this function effectively. Furthermore, the small amount of internalized ligand can hardly change the extracellular concentration of BK. Receptor-mediated endocytosis of BK might be either a way to protect the cell from overstimulation or, as proposed for the beta-adrenergic receptor, a process necessary to desensitize the receptor. Beta-adrenergic receptors, after being desensitized at the cell surface by phosphorylation, become dephosphorylated and are reactivated through phosphatases located in an intracellular compartment. The amino acid sequence of the human B_2 receptor displays several consensus sites for phosphorylation through protein kinases A and C; however, phosphorylation of the B_2 BK-receptor protein in response to BK binding has not yet been reported. Several reports show that activation of protein kinase C strongly reduces BK-stimulated phosphatidylinositol hydrolysis (29,45). In contrast, PG synthesis upon BK stimulation is increased after short-term

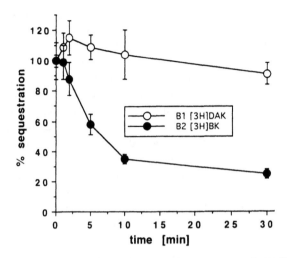

Figure 4 Sequestration of B_1 and B_2 kinin receptor in human IMR-90 fibroblasts. Monolayers were incubated at 37°C for the indicated time with 1 μM des-Arg[10]-Kallidin (DAK) for B_1 receptors or 1 μM bradykinin (BK) for B_2 receptors. After removing free and bound unlabeled ligand using treatment with 0.2 M acetic acid/0.5 M NaCl (10 min, 0°C), the remaining surface biding was determined at 0°C with the appropriate [³H] ligand in physiological buffer. Data are expressed in percent of the binding activity of cells not treated with unlabeled ligand. The experiment was done in triplicate.

phorbolester treatment (29,46). Up to now no information is available as to whether G-protein-coupled receptor kinases (47) or arrestins (48) are involved in B_2 kinin receptor desensitization and internalization, as reported for other G-protein-coupled receptors.

In contrast to the B_2 kinin receptor, the human B_1 receptor showed no ligand induced-receptor internalization or sequestration in human IMR 90 fibroblasts (Fig. 4). The inositolphosphate formation, induced by the B_1 agonist desArg[10]-KD in chinese hamster ovary cells stably expressing B_1 receptors, displayed no sign of fast desensitization (not shown). Furthermore, there was no difference in the dissociation constants K_d between cells incubated for 2 hr with [³H]desArg[10]-KD at 0°C, and cells incubated for 30 min at 37°C (A. Faussner, unpublished data). This is again in contrast to the B_2 receptor, for which agonist binding at 37°C results in a permanent shift to lower affinity as a result of the activation of G proteins (see below).

IV. The B_2 Kinin Receptor Is a G-Protein-Coupled Receptor; Is the B_1 Receptor Too?

The genes of both human kinin receptors, the B_2 and B_1 receptors, exhibit a sequence homology at the amino acid level of only 36%. Nonetheless, hydrophobicity analysis of their amino acid sequence revealed in both receptors seven transmembrane domains, indicating that they belong to the superfamily of the seven-transmembrane receptors, also known as G-protein-coupled receptors. The later term was created because, for almost all members of this family, involvement of guanine nucleotides in the stimulation of signal cascades or in the regulation of receptor affinity could be demonstrated (49,50). Indeed, years before the cloning of the first kinin receptor, a rat B_2 subtype (51), several studies in cultured cells had described effects of guanine nucleotides on inositolphosphate formation (45,52) or Ca^{2+} mobilization (53) upon BK stimulation. Other investigators reported sensitivity of B_2 receptor binding in membranes to guanine nucleotides (54–57). Up to now, for the B_1 subtype no effect of guanine nucleotides on either receptor affinity or biological effects has been described. However, B_1 receptors, artificially expressed in CHO cells, stimulate formation of inositolphosphates (A. Faussner, unpublished data), a response that commonly is considered to be G-protein-mediated.

The B_2 Subtype Couples to G-alpha-q (and Others)

Even though there are no doubts that the B_2 kinin receptor is coupled to G proteins, it is still not completely clear which G protein(s) is (are) involved. There is a strong body of evidence that G-alpha-q is involved in the formation of inositolphosphates. Antibodies directed against the C-terminal end of G-alpha-q/11 were able to attenuate the inositolphosphate formation in response to BK in membranes derived from NG 108-15 cells (58). In membrane preparations made from bovine aortic endothelial cells, a prior incubation with similar antibodies against G-alpha-q reduced BK-induced GTPase activity and shifted the affinity of a part of the BK receptors from high to low affinity (59). Additional pretreatment of the intact cells with pertussis toxin or additional incubation of the membranes with antibodies against G-alpha-i2 and G-alpha-i3 lead to an almost complete reduction of BK-induced GTPase activity and shifted the full complement of receptors from the high- to the low-affinity state. This clearly indicates that in these cells the B_2 receptor is coupled to pertussis toxin-insensitive G-alpha-q/11 and to pertussis toxin-sensitive G-alpha-i2 and G-alpha-i3 (59). In Swiss 3T3 fibroblasts, stimulation with BK induced inositolphosphate formation and release of arachidonic acid that subsequently served as substrate for PG synthesis (45). Both actions were influenced by GTP analogs and therefore were presumably G protein-mediated. This indicates that in these cells the BK receptor is coupled (directly or indirectly) via G proteins to phospholipase C and phospholipase A_2. Phorbol ester activation of protein kinase C on the one hand reduced the BK-induced inositol phosphate response, but on the other hand also strongly increased the PG formation in response to BK. Preincubation of the cells with pertussis toxin did not interfere with either response, indicating that G proteins of G-alpha-q/11 are involved. G-alpha-q has been found to activate specifically the beta-1 isoenzyme of phospholipase C (60), which on the other hand acts as a GTPase-activating protein for Gq (61), thereby limiting its action. Therefore, to summarize the literature presented, it appears that a signal pathway BK $\rightarrow$ B_2 receptor $\rightarrow$ G-alpha-q $\rightarrow$

phospholipase C beta-1 $\rightarrow$ inositolphosphate $\rightarrow$ Ca^{2+} is common to all cells expressing the B_2 kinin receptor. In addition, depending on the cell type, other G-alpha subunits or other signaling pathways, directly or indirectly activated through G-alpha-q, might be involved in BK signaling via the B_2 receptor subtype. No data are available so far on the role of the G-beta/gamma subunits in B_2 receptor signal transduction. It is also not clear whether there is a preferred combination of G-protein subunits that interact specifically with the B_2 kinin receptor. Similarly, future experiments have to show which G proteins are involved in the signal transmission of the B_1 receptor.

V. Transcriptional Regulation of the Kinin Receptors and B_2 Kinin Receptor Polymorphisms

A. The Regulatory Region of the B_2 Receptor Gene

Our own studies on the genomic structure of the gene revealed that it consists of three exons (7). The first exon is not translated, whereas the second and third ones comprise the entire open reading frame and a large 3'-not-translated region. A region upstream of exon 1 was identified by luciferase reporter gene assays as the promoter/enhancer region. In accordance with data on the rat B_2 receptor gene (62), there are no typical TATA or CCAAT boxes in the putative promoter region of the human B_2 receptor gene. A comparison of the 5' upstream region in these two species revealed a homology of more than 80% in the first 85 bp upstream of the transcription start site, determined for the rat B_2 receptor gene. Therefore, it is very likely that this region represents the core promoter of the human B_2 receptor gene. In this region a cAMP-responsive element-like sequence was defined (63). It has been shown that activated *ras* protein specifically increases the B_2 receptor number on the cell surface (33,34). A recent study by Pesquero et al. (63) appears to indicate that this effect is also caused by transcriptional activation of the B_2 receptor gene.

B. B_2 Receptor Polymorphisms

The B_2 receptor gene has been proposed as one of the candidate genes involved in the complex genetic underpinnings of common chronic disorders such as allergic asthma. Knowledge of the genomic structure enabled us to search systematically for polymorphic markers in the human B_2 receptor gene that are suitable to be used in association and/or linkage studies in patient cohorts. So far we have identified four polymorphic sites (64,65), one in the promoter region and three others in each of the three exons. Although they are speculative at the present stage, a variety of hypotheses may be discussed with respect to the biological consequences of these B_2 receptor variants.

The polymorphism of the promoter region may influence the transcription rate of the gene. A similar effect may also result from a tandem repeat polymorphism, consisting of three common alleles, since this site is located in exon 1 only 12 bp downstream of the putative transcription start site. By combining the promoter region with various alleles of exon 1[BE1-2G(C), BE1-3G(C), BE1-3T(C), and BE1-3T(T)] in all of the combinations, a reduction of the transcriptional activity was observed in luciferase reporter gene assays (66). These findings are consistent with the hypothesis of allele-specific different biological action of the B_2 receptor. A polymorphism in exon 2 (C-to-T transition) at nucleotide position 181 of the cDNA is characterized by an amino acid substitution from arginine to cysteine in the receptor protein at position 14 (R14C). This results in two different allelic polypeptides that could alter either the affinity of the receptor for BK or the efficiency of the signal transduction pathway. It may also influence the biological half-time of the receptor. The 3′-untranslated region of genes has been shown to participate in the regulation of RNA stability by a variety of mechanisms. Therefore the more complex repeat polymorphism of the BK receptor gene, located in the 3′ untranslated region of exon 3, could theoretically affect the processing and transport of the mRNA from the nucleus to the ribosomes, or the stability of the mRNA. In a population study, the various genotypes have been shown to be in equilibrium according to the Hardy-Weinberg principles (64).

Therefore these new genetic markers provide valuable tools to elucidate a potential role of a hereditary dysfunction of the B_2 receptor gene in allergic asthma and other diseases.

C. The Regulatory Region of the B_1 Receptor Gene

Very recently, the genomic organization of the B_1 receptor gene has been unraveled. The gene is present as a single copy and is comprised of three exons interrupted by two introns. Sequence analysis of the 5′-flanking region revealed the presence of a consensus TATA box and numerous candidate transcription factor binding sequences (9). Other data appear to indicate that two distinct functional promoters may exist in the human B_1 receptor gene (67). Based on the pharmacological evidence, the hypothesis of cytokine-driven upregulation of B_1 receptors can be considered to be generally verified, although in some tissues, such as the rat aorta, this never happens (5). It is not yet known whether cytokines directly influence the transcriptional rate of B_1 receptors in general or whether other tissue-specific (transcriptional) factors are required for the expression of B_1 receptors. The presence of multiple transcription initiation sites in the 5′ region may point to tissue-specific expression and/or alternative splicing of the mRNA of the B_1 receptor gene (9). More detailed studies of the putative promoter regions of the B_1 receptor gene should provide an avenue to further elucidate the role of this receptor in inflammatory conditions.

VI. Discussion

The nonapeptide BK is considered to be a key mediator in provoking acute inflammatory signs of vasodilation, increased blood flow, and increased vascular permeability. Another important action is the activation of sensory nerves, thereby contributing to the mechanisms involved in inflammatory pain and hyperalgesia. The potential role of this peptide in chronic inflammation is not yet defined in detail. However, because of their mitogenic properties, kinins are proposed to influence repair mechanisms and to be involved in the complex underpinnings of cell multiplication events that occur as a consequence of inflammatory tissue damage.

There is also compelling evidence linking BK with the pathophysiological processes that accompany a variety of inflammatory or allergic airway diseases, such as allergic rhinitis, viral rhinitis, and asthma (15,16). The local release of kinins in the airways involves activation of the kinin-forming system by tissue kallikrein from airway cells or from invaded neutrophils. In neutrophils the coexistence of tissue kallikrein and kininogens has been demonstrated, which might provide an important additional source for kinin generation during inflammatory conditions (68). This activation of the kinin-forming system can be mediated via viral or bacterial inflammation inducing local trauma, activation of coagulation pathways (Hageman factor and thrombin), or immune reactions. Excessive release of kinins can then activate specific B_1 and B_2 receptors that are widely distributed in the lung. High densities of B_2 receptors have been identified mainly in larger airways occurring on bronchial and pulmonary blood vessels of all sizes and in the lamina propria subjacent to the basal epithelial cell layer (69). In smaller airways, B_2 receptors are also detectable on airway smooth muscle, submucosal glands, and nerve fibers of intrapulmonary bronchi. B_1 receptors have recently been shown to occur in pathological states of the lung within the fibrous tissues and basement membrane of alveoli and capillaries (21).

A. Acute Kinin Effects in Airway Diseases

The loss of functions due to activation of B_2 kinin receptors may relate to increased vaculatur permeability, mucosal edema, and activation of sensory nerves. These mechanisms are important determinants of airway inflammation and airway hyperreactivity, thereby contributing to the symptomatology of allergic and viral rhinitis as well as of chronic asthma (15,70–72). Activation of kinin receptors can also cause release of other powerful noncytokine and cytokine mediators of inflammation, e.g., PGE_2, PGI_2, leukotrienes, histamine, PAF, IL-1, TNF, or eosinophil-granule major basic protein, derived mainly from leukocytes, macrophages, endothelial cells, and mucosal glands (3,73). This provides a great variety of interactions between BK and other inflammatory mediators and sets the stage to demonstrate potential positive and negative controlling mechanisms that enhance or inhibit BK action during airway inflammation.

B. Positive B_2 Receptor-Controlling Mechanisms

At the *prereceptor level*, inhibition of kinin degradation, thereby increasing local kinin concentration, may lead to an enhanced response to BK.

It has been proposed that coughing, a common side effect during treatment with angiotensin-converting enzyme (ACE) inhibitors, is linked to the action of kinins, since ACE is able to degrade kinins, and since the effects of ACE inhibitors (kininase II inhibitor) are reduced by kinin antagonists (15). This side effect might be explained by the activation of sensory nerves in airways, leading to reflex bronchoconstriction and coughing through the release of neuropeptides from sensory nerves (72). It is also tempting to speculate that the recent identification of polymorphisms in the B_2 receptor gene (64) may allow definition of a subset of persons carrying a "gain of function" variant of the B_2 receptors and thereby making them more susceptible to develop side effects upon ACE inhibitor treatment. ACE inhibitors also cause increases in the kinin levels in nasal secretions during an allergic response (74). With chronic therapy, it is possible that they could exacerbate allergic symptoms during repeated seasonal exposure.

Functional effects of BK, mediated via *B_2 receptors*, are shown to be potentiated by co-administration of PGE_2, PGI_2, calcitonin gene-related peptide, 5-hydroxytryptamine, substance P, IL-1, or platelet-activating factor (27,75). IL-1 and PAF are both able to increase B_2 receptor synthesis and expression, leading to enhanced responsiveness to BK (27,76 and data in this report). In many cell types, PGE_2 and PGI_2 elicit a marked accumulation of cAMP, which in turn might directly promote B_2 receptor expression via the cAMP-responsive elements, recently identified in the B_2 receptor gene promoter (63). TNF, PGE_2, and cAMP have also been shown to upregulate the expression of kininogens on cell surface receptors (24), which points to an additional mechanism that may facilitate the delivery of BK at sites of inflammation.

Postreceptor mechanisms also appear to be responsible for potentiation of BK action. Recent studies in tracheal smooth muscle suggest that there may be a causal link between phospholipase D-derived phosphatidate and a positive modulation of adenylate cyclase (13), and that the synergistic action of IL-1 and BK on prostanoid production is dependent on the activation of the phospholipase D pathway (23).

C. Negative B_2 Receptor-Controlling Mechanisms

On the other hand, the immediate effects of kinins on B_2 receptors are probably self-limiting due to the rapid homologous desensitization and B_2 receptor internalization observed upon receptor occupancy (38). Agonist occupancy of B_2 receptors also promotes a functional uncoupling of the receptors from signal transduction pathways (42). In addition, the activity of the kinins generated is only of short duration, due to the degrading capacity of potent kininases associated with cells or present in plasma.

Regulation of tissue kallikrein release may provide a "prereceptor" mechanism to reduce kinin generation. Tissue kallikrein is the major kininogenase detected in bronchoalveolar lavage fluids from asthmatics (77) and may play a particularly important role in kinin generation during asthma. In the human trachea, immunoreactive tissue kallikrein was localized in submucosal glands of the lamina propria (78).

In experimental model systems, BK-induced plasma extravasation can be inhibited by stimulation of C-fiber afferents (79). This indicates that in vivo a negative feedback inhibition on B_2 receptor-mediated effects may also be operative at some sites of inflammation and that it is dependent on an intact neuroaxis.

The variety of signal transduction events activated by BK and the G proteins coupled to the B_2 receptor allow for amplification and diversification of the BK signal in different tissues (10,13). One can also visualize that the overall effect of BK will be regulated by the interplay between these second messengers in lung tissues and the cross-talk at the "postreceptor" level. For instance, the BK-induced activation of protein kinase C can be counteracted by a Gs-mediated stimulation of the cAMP pathway (12). This may potentially be elicited by a variety of other mediators involved in the inflammatory reaction. On the other hand, in human fibroblasts the BK-induced activation of protein kinase C may counteract and slow down in a negative feedback loop the simultaneously occurring cAMP-accumulation that is elicited by BK-induced prostanoid release. This may explain the failure to detect a ligand-specific upregulation of B_2 receptors that was otherwise observed by activation of the cAMP pathway. We have also seen that upregulation of B_2 receptors by IL-1 is prevented by simultaneous treatment with phorbol esters (Fig. 2).

The very recent identification of several polymorphisms in the human B_2 receptor gene (64) supports the idea that genetic background may also influence BK susceptibility and BK effects. Theoretically, some of these B_2 receptor gene variants may alter transcription rates, receptor-binding characteristics, or mRNA stability, thereby representing susceptibility loci for altered responsiveness to BK. The demonstration that the combination of the promoter region with various alleles of exon 1, located only 12 b downstream of the putative transcription start site, can result in a reduction of the transcriptional activity in luciferase reporter gene assays is consistent with the hypothesis of allele-specific different biological action of the B_2 receptor (66). More studies are needed to confirm this possibility.

It is of course very difficult to extrapolate in-vitro concepts on B_2 receptor regulatory mechanisms to the in-vivo situation in pulmonary inflammation. However, the variety of processes that can affect kinin responsiveness at the level of the B_2 receptor may include positive controlling mechanisms that inhibit the detrimental consequences due to airway inflammation. The lack of such controlling mechanisms may be involved in the development of the airway hyperresponsiveness (70).

D. Chronic Inflammation, Repair Mechanisms, and Cell Multiplication

Several lines of evidence suggest that when inflammation is prolonged, B_1 receptors, which are not expressed in healthy tissues to a significant degree, as well as B_2 receptors, play an important role in the coordinated response to inflammatory injury (5). B_1 receptor stimulation has been demonstrated to increase collagen synthesis and cell multiplication in lung fibroblasts (80). In arterial smooth muscle cells, platelet-derived growth factor specifically upregulates cell surface B_2 receptors that may enhance mitogenesis (76). These mitogenic effects, however, are strongly opposed by BK acting via a B_1 receptor. The interaction between growth factors and kinins acting on both of their receptors may well be a factor in the link between chronic inflammation and fibrosis (21). It is also known that overexpression of the oncogenic forms of *ras* lead to increased expression of cell surface B_2 receptors (33,34). These observations support the hypothesis of a functional interaction between growth factors and kinins, since growth factors are known to activate *ras* (76).

Recently, BK was shown to be the most potent peptide among the neuropeptides that function as autocrine growth factors for human small cell lung cancer cell lines (81). Consequently, novel BK receptor antagonists were designed that induce growth inhibition and apoptosis in these cells and thus may have clinical potential for the treatment of these cancers (82).

In addition, B_1 receptors appear to play a prominent role in chronic inflammatory hyperalgesia (83), and they are likely to be involved in inflammatory vascular remodeling. In conjunction with cytokines, substance P and BK have been shown to produce an intense angiogenic response in a model of chronic inflammation (84). It has also been speculated that the antiproliferative effects of ACE inhibitors (85) may be due to increased levels of B_1 agonist that are generated by a preferred action of carboxypeptidases on BK metabolism (76).

The understanding of the cellular regulation of the B_1 receptor has lagged behind that of the B_2 receptor. It is well established that epidermal growth factor (EGF), IL-1, or lipopolysaccharide (LPS) can drive upregulation of B_1 receptors in most tissues (5). IL-1 most likely exerts this long-term upregulation on the transcriptional level via some regulatory elements recently identified on the B_1 receptor gene promoter (9). The synthesis of new, functionally active B_1 receptors may also depend on accurate processing of the receptor protein through the endoplasmic reticulum-Golgi apparatus pathway, since the protein translocation inhibitor brefeldin A and the glycosylation inhibitor tunicamycin both depress B_1 receptor upregulation (5). In contrast to IL-1, EGF can immediately enhance B_1 receptor responses (86). This interaction is shared only by thrombin and is prevented by the tyrosine kinase inhibitor genistein. This may suggest an interaction between tyrosine kinase signals and the function of B_1 receptors. From the still limited information available, one can delineate that in general B_1 receptors appear to utilize the same signal transduction pathways as described for the B_2 receptor (5). Neither the cross-regulations that might occur within this network nor the nature of the G proteins involved have yet been investigated. Interestingly, B_1 receptors appear not to be susceptible to homologous desensitization or internalization (Fig. 4), which is in contrast to the behavior of the B_2 receptors. At present the pathophysiological impact of the upregulation of B_1 kinin receptors is thus not fully elucidated. It may, however, be one of a number of important adaptive mechanisms that occur following the prolonged activation of the network of inflammatory mediators.

The understanding of the cellular regulation of BK receptors is now increasing very rapidly with the event of the molecular characterization of their genes. The identification of the regulatory regions of the B_2 and B_1 receptors will greatly facilitate the further identification of the signaling pathways, the cis-acting DNA elements, and the trans-acting protein factors or of cytokine or glucocorticoid responsive elements. This will allow an in-depth understanding of the effects described here of interleukins, adenylate cyclase activation, or activated *ras* protein on human kinin receptor expression. The availability of novel and specific B_2 and B_1 receptor antagonists provides another avenue to further delineate the importance of the kinins in defined pathologies.

With respect to respiratory airway and lung diseases, several studies have already aimed to influence chronic asthma or allergic rhinitis by virtue of local kinin receptor antagonism (87) or by inhibition of kallikrein activity (19). BK receptor antagonists also appear to have clinical potential to inhibit the growth of human small cell lung cancers (82). In addition, one has to consider the complex network of neuropeptides and mediators involved in the development of the pathology of inflammatory airway diseases. Therefore, a variety of other drugs, such as kininase (ACE, CP) inhibitors, glucocorticosteroids (36,88), or interferon-gamma (89) are described to modify BK effects either by direct or indirect means.

In human inflammatory lung diseases, kinins, acting via their B_2 and B_1 receptors, contribute significantly to both the initial presence of injury and ultimately to the processes that set the stage for repair or chronic airway hyperresponsiveness. In this report we have described and discussed some of the important regulatory modes for kinin receptors that might be involved in the balance between host protection and destruction during inflammation. The further characterization of these mechanisms may in the long term provide a rational basis for designing specific interventions within the kallikrein-kinin system in lung diseases.

References

1. Stewart JM. The kinin system in inflammation. Agents Actions 1993; suppl 42:145–157.
2. Bathon JM, Proud D. Bradykinin antagonists. Annu Rev Pharmacol Toxicol 1991; 31: 129–162.
3. Bhoola KD, Figueroa CD, Worthy K. Bioregulation of kinins: kallikreins, kininogens, and kininases. Pharmacol Rev 1992; 44:1–80.
4. Hall JM. Bradykinin receptors: pharmacological properties and biological roles. Pharmacol Ther 56; 1992:131–190.
5. Marceau F. Kinin B1 receptors: a review. Immunopharmacology 1995; 30:1–26.
6. Hess JF, Borkowski JA, Young GS, Strader CD, Ransom RW. Cloning and pharmacological characterization of a human bradykinin (BK-2) receptor. Biochem Biophys Res Commun 1992; 184:260–268.
7. Kammerer S, Braun A, Arnold N, Roscher AA. The human bradykinin B_2 receptor gene: full length cDNA, genomic organization and identification of the regulatory region. Biochem Biophys Res Commun 1995; 211:226–233.
8. Menke JG, Borkowski JA, Bierilo KK, MacNeil T, Derrick AW, Schneck KA, Ransom RW, Strader CD, Linemeyer DL, Hess JF. Expression cloning of a human B_1 bradykinin receptor. J Biol Chem 1994; 269:21583–21586.
9. Bachvarov DR, Hess F, Menke JG, Larrivee JF, Marceau F. Structure and genomic organization of the human bradykinin B_1 receptor gene for kinins (BDKRB$_1$). Genomics 1996; 33:374–381.

10. Smith JA, Webb C, Holford J, Burgess GM. Signal transduction pathways for B_1 and B_2 bradykinin receptors in bovine pulmonary artery endothelial cells. Mol Pharmacol 1995; 47:525–534.

11. Field JL, Butt SK, Morton IK, Hall JM. Bradykinin B_2 receptors and coupling mechanisms in the smooth muscle of the guinea-pig taenia caeci. Br J Pharmacol 1994; 113:607–613.

12. Liebmann C, Graness A, Ludwig B, Adomeit A, Boehmer A, Boehmer FD, Nurnberg B, Wetzker R. Dual bradykinin B_2 receptor signaling in A431 human epidermoid carcinoma cells: activation of protein kinase C is counteracted by a GS-mediated stimulation of the cyclic AMP pathway. Biochem J 1996; 313:109–118.

13. Stevens PA, Pyne S, Grady M, Pyne NJ. Bradykinin-dependent activation of adenylate cyclase activity and cyclic AMP accumulation in tracheal smooth muscle occurs via protein kinase C-dependent and -independent pathways. Biochem J 1994; 297:233–239.

14. Proud D, Baumgarten CR, Naclerio RM, Ward PE. Kinin metabolism in human nasal secretions during experimentally induced allergic rhinitis. J Immunol 1987; 138: 428–434.

15. Trifilieff A, Da Silva A, Gies JP. Kinins and respiratory tract diseases. Eur Respir J 1993; 6:576–587.

16. Proud D. Kinins in the pathogenesis of human airway diseases. Braz J Med Biol Res 1994; 27:2021–2031.

17. Proud D, Naclerio RM, Gwaltney JM, Hendley JO. Kinins are generated in nasal secretions during natural rhinovirus colds. J Infect Dis 1990; 161:120–123.

18. Shibayama Y, Skoner D, Suehiro S, Konishi JE, Fireman P, Kaplan AP. Bradykinin levels during experimental nasal infection with rhinovirus and attenuated influenza virus. Immunopharmacology 1996; 33:311–313.

19. Katori M, Nishiyama K, Iguchi Y, Majima M, Yao K. Reduction by a kallikrein inhibitor of the increased nasal airway resistance of allergic patients after antigen challenge. Immunopharmacology 1996; 33:308–310.

20. Christiansen SC, Proud D, Sarnoff RB, Juergens U, Cochrane CG, Zuraw BL. Elevation of tissue kallikrein and kinin in the airways of asthmatic subjects after endobronchial allergen challenge. Am Rev Respir Dis 1992; 145:900–905.

21. Nadar R, Derrick A, Naidoo S, Naiddoo Y, Hess F, Bhoola K. Immunoreactive B_1 receptors in human transbronchial tissue. Immunopharmacology 1996; 33:317–320.

22. Paegelow I, Werner H, Vietinghoff G, Wartner U. Release of cytokines from isolated lung strips by bradykinin. Inflamm Res 1995; 44:306–311.

23. Angel J, Audubert F, Bismuth G, Fournier C. IL-1 beta amplifies bradykinin-induced prostaglandin E_2 production via a phospholipase D-linked mechanism. J Immunol 1994; 152:5032–5040.

24. Takano M, Yokoyama K, Yayama K, Okamoto H. Murine fibroblasts synthesize and secrete kininogen in response to cyclic-AMP, prostaglandin E_2 and tumor necrosis factor. Biochim Biophys Acta 195; 1265:189–195.

25. Zini JM, Schmaier AH, Cines DB. Bradykinin regulates the expression of kininogen binding sites on endothelial cells. Blood 193; 81:2936–2946.

26. Dinarello CA, Wolff SM. The role of interleukin-1 in disease. N Engl J Med 1993; 328:106–113.

27. Bathon JM, Manning DC, Goldman DW, Towns MC, Proud D. Characterization of kinin receptors on human synovial cells and upregulation of receptor number by interleukin-1. J Pharmacol Exp Ther 1992; 260:483–392.

28. Hulkower KI, Wertheimer SJ, Lein W, Coffey JW, Anderson CM, Chen T, DeWitt DL, Crowl RM, Hope WC, Morgan DW. Interleukin-1beta induces cytosolic phospholipase A_2 and prostaglandin H synthase in rheumatoid synovial fibroblasts. Arthritis Rheum 1994; 17:653–661.

29. Cisar LA, Mochan E, Schimmel R. Interleukin-1 selectively potentiates bradykinin-stimulated arachidonic acid release from human synovial fibroblasts. Cell Signal 1993; 5:463–472.

30. Etscheid BG, Ko PH, Villereal ML. Regulation of bradykinin receptor level by cholera toxin, pertussis toxin and forskolin in cultured human fibroblasts. Br J Pharmacol 1991; 103:1347–1350.

31. Dixon BS. Cyclic AMP selectively enhances bradykinin receptor synthesis and expression in cultured arterial smooth muscle. J Clin Invest 1994; 93:2535–2544.

32. Roberts RA, Gullick WJ. Bradykinin receptor number and sensitivity to ligand stimulation of mitogenesis is increased by expression of a mutant *ras* oncogene. J Cell Sci 1989; 94:527–535.

33. Parries G, Hoebel R, Racker E. Opposing effects of a *ras* oncogene on growth factor-stimulated phosphoinosite hydrolysis: desensitization of platelet-derived growth factor and enhanced sensitivity to bradykinin. J Biol Chem 1987; 84:2648–2652.

34. Downward J, De Gunzburg J, Riehl R, Weinberg RA. $_p$21*ras*-induced responsiveness of phosphatidylinositol turnover to bradykinin is a receptor number effect. Proc Natl Acad Sci (USA) 1988; 85:5774–578.

35. Barnes PJ. Inhaled glucocorticoids for asthma. N Engl J Med 1995; 332:868–875.

36. Roscher AA, Paschke E, Manganiello VC. Glucocorticoids reduce the number of specific bradykinin receptor in cultured human fibroblasts. Clin Res 1984; 32:468A.

37. Gallizi JP, Bodinier MC, Chapelain B, Ly SM, Coussy L, Gireaud S, Neliat G, Jean T. Upregulation of [³H]des-Arg¹⁰-kallidin binding to the bradykinin B_1 receptor by interleukin 1β in isolated smooth muscle cells: correlation with B_1 agonist-induced PGI_2 production. Br J Pharmacol 1994; 113:389–394.

38. Roscher AA, Manganiello VC, Jelsema LJ, Moss J. Autoregulation of bradykinin receptors and bradykinin-induced prostacyclin formation in human fibroblasts. J Clin Invest 1984; 74:552–558.

39. Roberts RA, Gullick WJ. Bradykinin receptors undergo ligand-induced desensitization. Biochemistry 1990; 29:1975–1979.

40. Roscher AA, Klier C, Dengler R, Faussner A, Müller-Esterl W. Regulation of bradykinin action at the receptor level. J Cardiovasc Pharmacol 1990; 15(suppl 6):S39–S43.

41. Munoz FM, Leeb-Lundberg LMF. Receptor-mediated internalization of bradykinin. J Biol Chem 1992; 267:303–309.

42. Munoz CM, Cotecchia S, Leeb-Lundberg LMF. B_2 kinin receptor-mediated internalization of bradykinin in DDT$_1$ MF-2 smooth muscle cells is paralleled by sequestration of the occupied receptors. Arch Biochem Biophys 1993; 301:336–344.

43. Trowbridge IS, Collawn JF. Signal-dependent membrane protein trafficking in the endocytic pathway. Annu Rev Cell Biol 1993; 9:129–161.

44. Hamm-Alvarez SF, Alayof BE, Himmel HM, Kim PY, Crews AL, Strauss HC, Sheetz MP. Coordinate depression of bradykinin receptor recycling and microtubule-dependent transport by taxol. Proc Natl Acad Sci (USA) 1994; 91:7812–7816.

45. Burch RM, Axelrod J. Dissociation of bradykinin-induced prostaglandin formation from phosphatidylinositol turnover in swiss 3T3 fibroblasts: evidence for G protein regulation of phospholipase A_2. Proc Natl Acad Sci (USA) 1987; 84:6374–6378.

46. Burch RM, Ma AL, Axelrod J. Phorbol esters and diacylglycerols amplify bradykinin-stimulated prostaglandin synthesis in swiss 3T3 fibroblasts. J Biol Chem 1988; 263: 4764–4767.

47. Premont RT, Inglese J, Lefkowitz RJ. Protein kinases that phosphorylate activated G protein-coupled receptors FASEB J 1995; 9:175–182.

48. Wilson CJ, Applebury ML. Arresting G-protein coupled receptor activity. Curr Biol 1993; 3:683–686.

49. Gudermann T, Nürnberg B, Schultz G. Receptors and G proteins as primary components of transmembrane signal transduction. Part 1. G-protein-coupled receptors: structure and function. J Mol Med 1995; 73:51–63.

50. Strader CD, Fong TM, Tota MR, Underwood D. Structure and function of G protein-coupled receptors. Annu Rev Biochem 1994; 63:101–132.

51. McEachern AE, Shelton ER, Bhakta S, Obernolte R, Bach C, Zuppan P, Fujisaki J, Aldrich RW, Jarnagin K. Expression cloning of a rat B_2-bradykinin receptor. Proc Natl Acad Sci (USA) 1991; 8:7724–7728.

52. Etscheid BG, Villereal ML. Coupling of bradykinin receptors to phospholipase C in cultured fibroblasts is mediated by a G-protein. J Cell Physiol 1989; 140:264–271.

53. Muldoon LL, Rodland KD, Magun E. Transforming growth factor beta and epidermal growth factor alter calcium influx and phosphatidylinositol turnover in rat-1 fibroblasts. J Biol Chem 1988; 263:18834–18841.

54. Leeb-Lundberg LMF, Mathis SA. Guanine nucleotide regulation of B_2 kinin receptor. J Biol Chem 1990; 265:9621–9627.

55. Mathis SA, Leeb-Lundberg LMF. Bradykinin recognizes different molecular forms of the B_2 kinin receptor in the presence and absence of guanine nucleotides. Biochem J 1991; 276:141–147.

56. Keravis TM, Nehlig H, Delacroix MF, Regoli D, Hiley CR, Stoclet JC. High-affinity bradykinin B_2 binding sites sensitive to guanine nucleotides in bovine aortic endothelial cells. Eur J Pharmacol 1991; 207:149–155.

57. Liao JK, Homcy CJ. Specific receptor-guanine nucleotide binding protein interaction mediates the release of endothelium-derived relaxing factor. Cir Res 1992; 70:1018–1026.

58. Gutowski S, Smrcka A, Nowak L, Wu D, Simon M, Sternweis PC. Antibodies to the alpha q subfamily of guanine nucleotide-binding regulatory protein alpha subunits attenuate activation of phosphatidylinositol 4.5-bisphosphate hydrolysis by hormones. J Biol Chem 1991; 266:20519–20524.

59. Liao JK, Homcy CJ. The G proteins of the G alpha i and G alpha q family couple the bradykinin receptor to the release of endothelium-derived relaxing factor. J Clin Invest 1993; 92:2168–2172.

60. Taylor SJ, Chae HZ, R SG, Exton JH. Activation of the beta1 isoenzyme of phospholipase C by alpha subunits of the Gq class of G proteins. Nature 1991; 350:516–518.

61. Berstein G, Blank JL, Jhon DJ, Exton JH, Rhee SG, Ross EM. Phospholipase C-beta1 is a GTPase-activating protein for Gq/11, its physiologic regulator. Cell 1992; 70:411–418.

62. Pesquero JB, Lindsey CJ, Zeh K, Paiva AC, Ganten D, Bader M. Molecular structure and expression of rat bradykinin B_2 receptor gene. Evidence for alternative splicing. J Biol Chem 1994; 269:26920–26925.

63. Pesquero JB, Lindsey CJ, Paiva ACM, Ganten D, Bader M. Transcriptional regulatory elements in the rat bradykinin B_2 receptor gene. Immunopharmacology 1996; 33:36–41.

64. Braun A, Kammerer S, Böhme E, Müller B, Roscher AA. Identification of polymorphic sites of the human bradykinin B$_2$ receptor gene. Biochem Biophys Res Commun 1995; 211:234–240.

65. Braun A, Maier E, Kammerer S, Müller B, Roscher AA. A novel sequence polymorphism in the promoter region of the human B$_2$-bradykinin receptor gene. Hum Genet 1996; 97:688–689.

66. Braun A, Kammerer S, Maier E, Böhme E, Roscher AA. Polymorphisms in the gene for the human B$_2$-bradykinin receptor. New tools in assessing a genetic risk for bradykinin associated diseases. Immunopharmacology 1996; 33:32–35.

67. Yang X, Polgar P. Genomic structure of the human bradykinin B$_1$ receptor gene and preliminary characterization of its regulatory regions. Biochem Biophys Res Commun 1996; 222:718–725.

68. Figueroa CD, Henderson LM, Kaufmann J, De La Cadena RA, Colman RW, Muller-Esterl W, Bhoola KD. Immunovisualization of high (HK) and low (LK) molecular weight kininogens on isolated human neutrophils. Blood 1992; 79:754–759.

69. Mak JC, Barnes PJ. Autoradiographic visualization of bradykinin receptors in human and guinea pig lung. Eur J Pharmacol 1991; 194:37–43.

70. Pauwels P, Kips J, Joos G. Airway hyperreactivity, an introduction. Agents Actions 1990; suppl 31:11–24.

71. Nakajima N, Ichinose M, Takahashi T, Yamauchi H, Igarashi A, Miura M, Inoue H, Takishima T, Shirato K. Bradykinin-induced airway inflammation. Contribution of sensory neuropeptides differs according to airway site. Am J Respir Crit Care Med 1994; 149:694–698.

72. Barnes PJ. Effect of bradykinin on airway function. Agents Actions 1992; 38(suppl 3): 432–438.

73. Coyle AJ, Ackerman SJ, Burch R, Proud D, Irvin CG. Human eosinophil-granule major basic protein and synthetic polycations induce airway hyperresponsiveness in vivo dependent on bradykinin generation. J Clin Invest 1995; 95:1735–1740.

74. Proud D, Naclerio RM, Meyers DA, Kagey-Sobotka A, Lichtenstein LM, Valentine MD. Effects of a single-dose pretreatment with captopril on the immediate response to nasal challenge with allergen. Int Arch Allergy Appl Immunol 1990; 93:165–170.

75. Campos MM, Calixto JB. Involvement of B$_1$ and B$_2$ receptors in bradykinin-induced rat paw oedema. Br J Pharmacol 1995; 114:1005–1013.

76. Dixon BS, Dennis M. Interaction between growth factors and kinins in arterial smooth muscle cells. Immunopharmacology 1996; 33:16–23.

77. Christiansen SC, Proud D, Cochrane CG. Detection of tissue kallikrein in the bronchoalveolar lavage fluid of asthmatic subjects. J Clin Invest 1987; 79:188–197.

78. Proud D, Vio CP. Localization of immunoreactive tissue kallikrein in human trachea. Am J Respir Cell Mol Biol 1993; 8:16–19.

79. Green PG, Miao FJ, Janig W, Levine JD. Negative feedback neuroendocrine control of the inflammatory response in rats. J Neurosci 1995; 15:4678–4686.

80. Goldstein RH, Wall M. Activation of protein formation and cell division by bradykinin and [des-Arg^9]bradykinin. J Biol Chem 1984; 259:9263–9268.

81. Bunn PA, Chan D, Stewart J, Gera L. Tolley R, Jewett P, Tagawa M, Alford C, Mochzuki T, Yanaihara N. Effects of neuropeptide analogues on calcium flux and proliferation in lung cancer cell lines. Cancer Res 194; 54:3602–3610.

82. Chan D, Gera L, Helfrich B, Helm K, Stewart J, Walley E, Bunn PA. Novel bradykinin

antagonist dimers for the treatment of human lung cancers. Immunopharmacology 1996; 33:201–204.

83. Davis AJ, Perkins MN. The involvement of bradykinin B_1 and B_2 receptor mechanisms in cytokine-induced mechanical hyperalgesia in the rat. Br J Pharmacol 1994; 113:63–68.

84. Hu DE, Hori Y, Fan TP. Interleukin-8 stimulates angiogenesis in rats. Inflammation 1993; 17:135–143.

85. Farhy RD, Ho KL, Carretero OA, Scicli AG. Kinins mediate the antiproliferative effect of ramipril in rat carotid artery. Biochem Biophys Res Commun 1992; 182:83–88.

86. deBlois D, Drapeau G, Petitclerc E, Marceau F. Synergism between the contractile effect of epidermal growth factor and that of des-Arg^9-bradykinin or of alpha-thrombin in rabbit aortic rings. Br J Pharmacol 1992; 105:959–967.

87. Akbary AM, Wirth K, Schölkens BA. Efficacy and tolerability of icatibant (Hoe 140) in patients with moderately severe chronic bronchial asthma. Immunopharmacology 1996; 33:238–242.

88. Roisman GL, Peiffer C, Lacronique JG, Le Cae A, Dusser DJ. Perception of bronchial obstruction in asthmatic patients. Relationship with bronchial eosinophilic inflammation and epithelial damage and effect of corticosteroid treatment. J Clin Invest 1995; 96: 12–21.

89. De Kimpe SJ, Tielemans W, Van Heuven-Nolsen D, Nijkamp FP. Reversal of bradykinin-induced relaxation to contraction after interferon-gamma in bovine isolated mesenteric arteries. Eur J Pharmacol 1994; 261:111–120.

Part Three

ANTIINFLAMMATORY PEPTIDES

11

Modulation of Inflammatory Effects of Cytokines by the Neuropeptide α-MSH

J. M. LIPTON, NILUM RAJORA, R. A. STAR, and S. TAHERZADEH

University of Texas Southwestern Medical
  Center at Dallas
Dallas, Texas

ANNA CATANIA, GIULIANA CERIANI, and G. BOCCOLI

IRCCS Ospedale Maggiore de Milano
Milan, Italy

I. Introduction

Inflammation is perhaps the most primitive of the host responses to challenge, and it is believed to promote survival (1). The inflammatory reaction involves multiple host cells, especially neutrophils and macrophages, and an ever-growing list of proinflammatory agents including certain cytokines (e.g., IL-1, TNF-α). Identification of cytokines, their receptors, antagonists, the functions of cytokines, and their cells of origin has progressed rapidly. It is now clear that whereas most cytokines have proinflammatory influences, others such as IL-10 and IL-4 have antiinflammatory effects. Cytokines are believed to participate in normal physiological events, but in infection and inflammation the influence of proinflammatory cytokines can be exaggerated. It is not surprising that endogenous modulators of inflammation have evolved, agents that tend to reduce, but perhaps not eliminate, the effects of proinflammatory mediators.

The evidence that certain cytokines are crucial to inflammatory reactions in peripheral tissue is clear, and new details of the complexity of the influence of these soluble mediators in peripheral inflammation come with successive issues of scientific journals. There is likewise increasing evidence of the importance of proinflammatory cytokines in CNS disease such as multiple sclerosis (2–5) and Alzheimer's disease or CNS infection with HIV (6). In the case of CNS inflammation, proinflammatory cytokines may be produced by glia or neurons or by peripheral monocyte/macrophages that migrate into the tissue. Whatever the origin of inflammatory cytokines within the brain, there is movement in the scientific community toward evaluation of anticytokine agents for the treatment of neurodegenerative disease and CNS reperfusion injury. It may be that the endogenous anticytokine neuropeptide α-MSH will be useful in such treatments.

II. The Antiinflammatory Neuropeptide α-MSH

α-MSH, a tridecapeptide derived from proopiomelanocortin (POMC), occurs in the pituitary, brain, skin, and other body sites (7–10). Its concentration in the circulation in humans is normally regulated within narrow limits, although plasma and local α-MSH concentrations are increased in certain inflammatory disorders (9,10). Although it is named for its influence on pigmentation in amphibians, the widespread distribution of α-MSH in tissues of higher organisms, particularly in barriers to the external environment such as the skin and gut, suggests that α-MSH mediates functions other than skin darkening. There is substantial evidence that α-MSH is important to control of host reactions, particularly inflammation.

III. α-MSH Receptors

One of the most enlightening pursuits in recent research on α-MSH and related melanocortins (α-MSH- and ACTH-like peptides) concerns identification and cloning of their receptors. The nomenclature is not complete, and other receptors will undoubtedly be identified, but at this time five G-protein-linked receptors (MC-1 through MC-5) are generally considered. Certain receptors appear to occur primarily or solely in peripheral cells. For instance, MC-1 occurs in melanocytes of mice and humans (11,12), and MC-2 (the ACTH receptor) is found only in the adrenal cortex (12). These and the other receptors transfected into carrier cells increase intracellular cAMP when stimulated with α-MSH. MC-3 was identified in a screen of a rat hypothalamus cDNA library (13). This receptor responds to: NDP-α-MSH $>$ γ_2-MSH = γ_1-MSH = α-MSH = ACTH (1-39) $> \gamma_3$-MSH $>$ des α-MSH $> >$ ACTH (4-10). It occurs in brain regions outside the hypothalamus, but not in melanocytes or adrenals. A murine MC-3 has been identified, and a gene encoding a similar receptor has been isolated from human tissue (14). A gene encoding a fourth human melanocortin receptor (MC-4) that has a characteristic profile of responses to melanocortin peptides was subsequently identified and cloned by Gantz and colleagues (15). Northern analysis and in-situ hybridization studies indicate that MC-4 is expressed mainly in brain, especially in thalamus, hypothalamus, hippocampus, and amygdala. MC-4 receptor distribution differs somewhat from that of MC-3 receptors. α- and β-MSH and ACTH all activate MC-4 receptors with equal potency, whereas γ-MSH and truncated ACTH peptides (ACTH 4-10 and 1-10) are effective only in very high concentrations. Rat MC-5 receptors have been cloned and charac-terized, and their expression found in brain (16). Unlike other melanocortin recep-tors, α-MSH was twice as potent as NDP α-MSH, 10 times as potent as ACTH, and 100 times as potent as γ-MSH in promoting intracellular cAMP. The MC-5 receptor has been cloned, expressed, and characterized in the mouse by Gantz et al. (17) and by Labbe et al. (18). In both laboratories this α-MSH receptor was expressed in brain as well as peripheral tissue, including lymphoid tissue. The most potent stimulus of MC-5 receptors was α-MSH in the studies by Gantz et al. (17); in the competitive binding studies of Labbe et al. (18), NDP-α-MSH was the most potent competitor and α-MSH was second. The combined results indicate that at least three melanocor-tin receptor subtypes exist within the brain: MC-3, MC-4, and MC-5. Stimulation of one or more of these subtypes, or other central melanocortin receptors not yet identified in the brain, is likely responsible for the antiinflammatory influence of centrally administered α-MSH. Undoubtedly, identification of the precise receptors involved will improve our understanding of control of inflammation in both the periphery and the CNS.

IV. α-MSH Inhibits Inflammatory Actions of Cytokines and Related Inflammatory Mediators

A. Antiinflammatory Action of Peripheral α-MSH

In our original observations, enhanced vascular permeability, an early hallmark of in-flammation, was induced with intradermal injections of histamine in rabbits (Lipton, 1989). Both α-MSH (1-13) and α-MSH (11-13, KPV), the tripeptide antipyretic "message sequence" of α-MSH, given intravenously, inhibited the increase in vascular permeability. Enhancement of vascular permeability induced by intradermal endogenous pyrogen (EP), which contains multiple mediators of inflammation, was likewise inhibited by α-MSH (1-13) (19). α-MSH (11-13), given intraperitoneally, inhibited picryl chloride-induced inflammation in a dose-related fashion in a murine ear edema model; the most effective dose was as effective as a very large dose of glucocorticoid (20). Antiinflammatory effects of α-MSH molecules have been dem-onstrated in acute inflammation induced by picryl chloride (20), carrageenan (21), and EP (22). α-MSH (1-13) and α-MSH (11-13) inhibited contact sensitivity reac-tions in mice (21,22). We also found that α-MSH (1-13) and the tripeptide α-MSH (11-13) inhibited edema caused by intradermal injections of IL-1β, IL-6, and TNF-α; we subsequently showed that the peptide inhibited inflammation caused by leuko-triene B_4, platelet-activating factor, and IL-8 (22). Thus, systematically administered melanocortin peptides inhibit inflammation caused by some of the most prominent local mediators of inflammatory reactions.

B. Central α-MSH Inhibits Peripheral Inflammation

In experiments on acute inflammation in mice, α-MSH (0.1–10 μg) injected into the cerebral ventricles inhibited inflammation measured 3 and 6 hr later in a dose-related fashion (23). The inhibition with the maximum dose (10 μg/mouse) was approximately 50%, greater inhibition than that seen in other experiments after peripheral administration of glucocorticoid. Systemic administration of 10 μg of α-MSH had little or no effect on this model of inflammation (20). Further, central administration of as little as 1 ng of a stable α-MSH analog (NDP-α-MSH) markedly reduced inflammation (24). The inhibitory effect of central α-MSH on inflammation could not be attributed to a peptide-induced increase in circulating glucocorticoids because α-MSH (10 μg) did not affect circulating corticosterone concentration. Subsequent studies (22,25) confirmed and extended these observations. Central administration of α-MSH (1-13) inhibited murine ear edema caused by local injection of major mediators of tissue inflammation: Il-1, Il-8, leukotriene B_4, and platelet-activating factor (22). In tests confined to IL-1β-induced inflammation, central administration of α-MSH (11-13) was effective, although the concentrations required were greater than for α-MSH (1-13). D-Val^{13}α-MSH (1-13) was also effective, although potency was not increased by this substitution (25); D-Val^{13}α-MSH(8-13) likewise inhibited inflammation caused by Il-1β, but its potency was less than that of the larger molecules.

The antiinflammatory effect of centrally administered α-MSH (1-13) and α-MSH (11-13) was further confirmed (26). In tests on inflammation induced in the mouse ear by intradermal injections of recombinant Il-1β, central administration of α-MSH (1-13) effectively reduced inflammation. The antiinflammatory effect of centrally administered α-MSH (1-13) was prevented in mice with inflammation induced in a hind paw and with spinal cord transection, indicating that intact descending inhibitory neural pathways are required for the antiinflammatory influence of the central peptide. Intraperitoneal administration of α-MSH (1-13) in mice with spinal transection had a smaller and delayed antiinflammatory effect consistent with an action of the peptide, albeit lesser, in the periphery. α-MSH (11-13) given i.p. had rapid and marked antiinflammatory activity in mice with spinal transection, suggesting that the tripeptide can act below the transection.

The combined results indicate that α-MSH can act solely within the brain to inhibit inflammation but that melanocortin receptors in the periphery also contribute when the peptide is given systemically. Perhaps a major antiinflammatory influence occurs through stimulation by the peptide of its receptors within the brain, activation of which promotes neuronal signal transduction and subsequent transmission in descending neurogenic antiinflammatory pathways, thereby inhibiting inflammatory events by modulating release of mediators at the inflammatory site. This hypothesis parallels current ideas of descending pain modulation pathways.

V. α-MSH Mechanisms of Antiinflammatory Action

The mechanisms that underlie the antiinflammatory effects of α-MSH have thus been traced to local effects on peripheral host cells and mediators and to actions of the peptide on CNS receptors that drive descending antiinflammatory neuronal pathways.

There are several hypotheses about the mechanisms of antiinflammatory action of α-MSH in the periphery mechanisms likely applicable to CNS inflammation. There is no reason to believe that the hypothesized actions are mutually exclusive; indeed, they may be interactive and/or derived from a more primary and basic common influence. To the extent these hypothesized actions do take place in vivo, it is possible that they are, in some major part, induced or sustained via actions of the peptide within the brain.

A. Inhibition of Neutrophil Migration

Migration of polymorphonuclear leukocytes (PMNs) into tissue sites of injury or invasion is one of the earliest stages of the inflammatory reaction. A number of antiinflammatory substances, including glucocorticoids, appear to inhibit inflammation by inhibiting PMN migration, thereby modulating local injury caused by the host cells themselves. There is evidence that α-MSH has such an inhibitory influence on neutrophil migration.

Mason and Van Epps (27) first showed that neutrophil migration into subcutaneous sponges treated with IL-1, TNF, or complement C_{5a} was inhibited by systemic administration of α-MSH. In a rodent model of adult respiratory distress syndrome, an often-lethal syndrome marked by migration of PMNs into the lungs, we noted that such migration is markedly inhibited by systemic α-MSH treatment (28). Our recent evidence (29) indicates that α-MSH inhibits PMN migration in IL-8 or FMLP gradients in vitro. Related evidence indicates that these cells can express specific receptors for α-MSH and that the expression is increased in the presence of TNF. The combined evidence thus supports the idea that at least one of the actions of α-MSH is to inhibit neutrophil migration. Because of the prominence of PMNs in acute and recurrent inflammation, as opposed to chronic inflammation, it may be that this influence of the peptide is greater in acute (both local and systemic) inflammatory reactions.

B. Inhibition of Production of Proinflammatory Cytokines

Because many in-vivo experiments have shown the anticytokine influences of α-MSH, the effects of the peptide on production of such cytokines in vitro has been investigated. Concentrations of α-MSH similar to those found in aqueous humor inhibited the production of interferon-γ (IFN-γ) by antigen-stimulated mouse lymph node cells (30). In related research, α-MSH inhibited IFN-γ production by lymph node cell cultures in the presence of anti-IL-4 antibodies and suppressed nitric oxide generation induced by IFN-γ (31). In experiments on cytokine production by mitogen-stimulated peripheral blood mononuclear cells from normal humans, there was inhibition of production of both TNF-α (Fig. 1) and IFN-γ (Fig. 2) (Catania et al., unpublished observations). These observations suggest that α-MSH can inhibit immune-mediated inflammation by modulating production of proinflammatory cytokines.

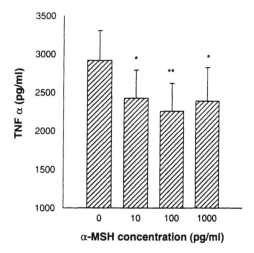

Figure 1   Concentration-related inhibition of TNF-α production (ELISA, R&D System) by α-MSH in human mononuclear cells. Parameters in this figure and Fig. 2: 10^6 cells/well; $n = 15$ normal subjects; 24-hr incubation with PHA 5%; scores are means ± SEM.

C. Inhibition of Nitric Oxide

Nitric oxide production by macrophages is believed to contribute to inflammatory reactions, perhaps particularly chronic inflammation. It is therefore important to know if α-MSH inhibits production of nitric oxide, an agent induced by multiple inflammatory cytokines. In inflammatory cells (RAW 264.7), this peptide inhibited production of NOS-II mRNA, NOS-II protein, nitrite, and peroxynitrate (32). These

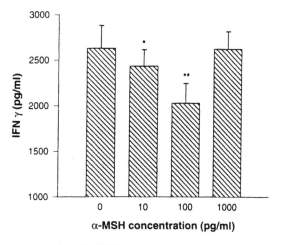

Figure 2   Concentration-related inhibition of IFN-γ production of α-MSH.

cells produced α-MSH in response to stimulation with TNF-α, and they were found to contain mRNA for α-MSH receptors (MC-1). The combined evidence suggests that α-MSH is an autocrine modulator of inflammatory responses in murine macrophages, one that acts on specialized melanocortin receptors to inhibit expression of nitric oxide synthase. If these observations represent a general phenomenon in macrophages, they may explain the increased α-MSH in synovial sites of inflammation in adult human rheumatoid arthritis (33) in terms of self-modulation by local phagic cells. The potential clinical significance of an endogenous α-MSH regulatory circuit in macrophages could be very great, particularly in terms of development of new pharmacological agents for therapeutic intervention.

D. Inhibition of Neopterin

Although cultured RAW cells share characteristics of normal human macrophages, there is controversy concerning human monocyte/macrophage production of nitric oxide. The THP-1 cell line, a human monocytic leukemia cell line, can be induced to differentiate into macrophages by the phorbol ester PMA. This cell line resembles normal human monocytes with regard to morphology, secretory products, oncogene expression, expression of membrane antigens, and expression of genes involved in lipid metabolism. To determine if the antiinflammatory effects of α-MSH obtained with murine macrophages also occur in cells of human monocyte/macrophage characteristics, a series of in-vitro tests were performed using THP-1 cells (34). These cells produce neopterin, and α-MSH inhibited production of this primate homolog of cytotoxic nitric oxide. It thus appears that α-MSH inhibits inflammatory products in human macrophages much as it does in murine macrophages. Modulation by α-MSH of specific inflammatory mediators/cytotoxic agents appears to differ depending on the importance of the mediators in myelomonocytic cells of different species.

E. Inhibition of CNS Prostaglandin Induced by LPS or IL-1

Particularly pertinent to CNS inflammation is the observation that α-MSH injected into the mouse brain inhibits production of PGE_2 synthesis in the dorsal hippocampus caused by intraperitoneal injection of LPS or IL-1α (35). These observations confirm that induction of IL-1 by LPS, or administration of IL-1, induces the inflammatory autocoid PGE_2 within the CNS. The data further indicate that α-MSH can act centrally to inhibit the induced rise in PGE_2 without altering PGE_2 production. Whether this result represents a specific direct action of α-MSH, or simply another example of antagonism of the actions of a proinflammatory cytokine, remains to be determined.

VI. Clinical Observations

Perhaps not surprisingly from the basic evidence on the peptide, and from its structural consistency over many millions of years of evolution, α-MSH is believed to be a factor in infectious and inflammatory disorders in humans. The primary evidence comes from recent observations on patients with HIV infection, endotoxemia, rheumatoid arthritis, and myocardial infarction.

A. HIV Infection

Cytokines are believed to have a role in disease progression in HIV-infected patients. IL-1, TNF-α, and IL-6 are known to potentiate HIV replication in certain cell lines (36–38). TNF-α upregulates viral expression in HIV-infected T and monocyte cell lines (36). Because cytokine production can be induced by HIV (39) and cytokines can amplify virus replication and expression, a vicious cycle can occur. It seems, therefore, that modulation of cytokine production could influence disease progression in HIV-infected individuals (40).

In early research on HIV-infected patients of the Centers for Disease Control (CDC) group IV classification, we found that plasma α-MSH was elevated (41). The data also showed that plasma α-MSH concentration was correlated with 6-month survival: 33% of patients with plasma α-MSH within the 20- to 40-pg/mL range were alive 6 months after the test, whereas virtually all patients with either greater or lesser concentrations were dead.

B. Endotoxemia

α-MSH increases in the circulation after administration of endogenous pyrogen (42) or endotoxin (43) to rabbits. To test the idea that α-MSH is also important in host reactions in humans, plasma α-MSH was measured in normal human subjects given endotoxin i.v. The stimulus caused a fever-related increase in plasma α-MSH (44). Subjects with very high temperatures had marked increases in plasma α-MSH, whereas those with lower fevers did not. These results suggest that release of α-MSH is part of the acute-phase response to infection in humans and that the molecule rapidly becomes available to modulate fever in subjects with high fever. Although the trigger for release of α-MSH in these subjects is unknown, the evidence cited above suggests that actions of circulating cytokines may be responsible.

C. Rheumatoid Arthritis

It is clear that cytokines such as IL-1 and TNF are significant in inflammation of rheumatoid arthritis (45,46). Joint inflammation and subsequent evolution to joint destruction may result from an imbalance between the production of proinflammatory cytokines and cytokine antagonists. Concentrations of α-MSH, sTNFr, and IL-1ra in plasma and synovial fluid of patients with rheumatoid arthritis were measured (33). Plasma concentrations of these cytokine antagonists were similar to those found in normal subjects. However, synovial sTNFr and IL-1ra were elevated and were correlated with parameters of disease activity such as erythrocyte sedimentation rate. α-MSH was found in the synovial fluid, and its concentration was greater in the synovium than in plasma in the majority of patients. Further, there was no correlation between plasma and synovial α-MSH. This suggests that the increased peptide in the synovium did not simply permeate from plasma but was likely produced by local cells. This was the first demonstration of the presence of α-MSH directly within a site of inflammation. Of particular interest, there was a negative correlation between α-MSH and white blood cells in the synovium. This might indicate that α-MSH has a protective influence on migration of neutrophils to inflamed synovial fluid. Such an idea is consistent with the observations in which α-MSH inhibited development of experimental arthritis (47).

D. Myocardial Infarction

It is clear that during myocardial infarction there is an increase in circulating cytokines such as TNF and IL-6 (48,49). We tested the idea that compensatory increases in cytokine antagonists likewise occur (50). We also investigated relationships between plasma concentrations of cytokine antagonists and changes in plasma concentrations of enzymes released from injured myocardium, including creatine kinase, aspartate serum transferase, and lactate dehydrogenase. Samples were obtained at presentation in the Coronary Unit, every 3 hr for 24 hr, and daily thereafter. We found elevations in the plasma cytokines IL-1 and TNF-α in a proportion of subjects, whereas cytokine antagonists α-MSH, IL-1ra, and sTNFr were elevated in all myocardial infarction patients. α-MSH was elevated at presentation but decreased progressively in successive tests, reaching normal values within 24 hr, whereas sTNFr and IL-1ra peaked at 3–6 hr or later. It thus appears that compensatory changes in the antiinflammatory peptide α-MSH occur in this form of sterile reperfusion injury and that these changes are linked to alterations in other cytokine antagonists.

VII. Conclusion

In conclusion, experimental and clinical data indicate that α-MSH is a modulator of host reactions including acute, chronic, and systemic inflammation. Although the precise mechanism of action is still under investigation, antagonism of the actions and production of cytokines is responsible for a portion of its antiinflammatory/antipyretic activity. The peptide can act solely within the brain to inhibit fever and peripheral inflammation, but α-MSH and fragments of it also have antiinflammatory influences directly on peripheral cells. The remarkable potency of the molecule together with its apparent low toxicity makes this peptide a candidate for therapeutic use in disorders in which inflammation and/or cytokines are present.

Acknowledgments

This work was supported by National Institutes of Health Grant NS 10046; NATO Collaborative Research Grant 200467; and VIII progetto AIDS, Istituto Superiore di Sanita', Italy.

References

1. Gallin JI, Goldstein IM, Snyderman R. (1992) Inflammation: Basic Principles and Clinical Correlates. New York: Raven Press.
2. Brosnan CF, Cannella B, Battistini L, Raine CS. (1995) Cytokine localization in multiple sclerosis lesions: correlation with adhesion molecule expression and reactive nitrogen species. Neurology 45:S16–S21.
3. Olsson T. (1995) Cytokine-producing cells in experimental autoimmune encephalomyelitis and multiple sclerosis. Neurology 45:S11–S15.
4. Rieckmann P, Albrecht M, Kitze B, Weber T, Tumani H, Broocks A, Luer W, Helwig A, Poser S. (1995) Tumor necrosis factor-alpha messenger RNA expression in patients with relapsing-remitting multiple sclerosis is associated with disease activity. Ann Neurol 37:82–88.
5. Huberman M, Shalit F, Roth-Deri I, Gutman B, Brodie C, Kott E, Sredni B. (1994) Correlation of cytokine secretion by mononuclear cells of Alzheimer patients and their disease stage. J Neuroimmunol 52:147–152.
6. Tyor WR, Glass JD, Griffin JW, Becker PS, McArthur JC, Bezman L, Griffin DE. (1992) Cytokine expression in the brain during the acquired immunodeficiency syndrome. Ann Neurol 31:349–360.
7. Lipton JM. (1990) Modulation of host defense by the neuropeptide α-MSH. Yale J Biol Med 63:173–182.
8. Lipton JM, Catania A. (1997) Antiinflammatory actions of the neuroimmunomodulator α-MSH. Immunol Today 18:140–145.
9. Catania A, Lipton JM. (1993) α-Melanocyte stimulating hormone in the modulation of host reactions. Endocrine Rev 14:564–576.
10. Catania A, Lipton JM. (1994) The neuropeptide alpha-melanocyte-stimulating hormone: a key component of neuroimmunomodulation. NeuroImmunoModulation 1:93–99.
11. Chhajlani V, Wikberg JES. (1992) Molecular cloning and expression of the human melanocyte stimulating hormone receptor cDNA. FEBS Lett 309:417–420.
12. Mountjoy KG, Robbins LS, Mortrud MT, Cone RD. (1992) The cloning of a family of genes that encode the melanocortin receptors. Science 257:1248–1251.
13. Roselli-Rehfuss L, Mountjoy KG, Robbins LS, Mortrud MT, Low MJ, Tatro JB, Entwistle ML, Simerly RB, Cone RD. (1993) Identification of a receptor for gamma melanotropin and other proopiomelanocortin peptides in the hypothalamus and limbic system. Proc Natl Acad Sci (USA) 90:8856–8860.
14. Gantz I, Konda Y, Tashiro T, Shimoto Y, Miwa H, Munzert G, Watson SJ, DelValle J, Yamada T. (1993) Molecular cloning a novel melanocortin receptor. J Biol Chem 268:8246–8250.
15. Gantz I, Miwa H, Konda Y, Shomoto Y, Tashiro T, Watson SJ, DelValle J, Yamada T. (1993) Molecular cloning, expression, and gene localization of a fourth melanocortin receptor. J Biol Chem 268:15174–15179.
16. Griffon N, Mignon V, Facchinetti P, Diaz J, Schwartz JC, Sokoloff P. (1994) Molecular cloning and characterization of the rat fifth melanocortin receptor. Biochem Biophys Res Commun 200:1007–1014.
17. Gantz I, Shomoto Y, Konda Y, Miwa H, Dickinson CJ, Yamada T. (1994) Molecular cloning, expression, and characterization of a fifth melanocortin receptor. Biochem Biophys Res Commun 200:1214–1220.

18. Labbe O, Desarnaud F, Eggerickx D, Vassart G, Parmentier M. (1994) Molecular cloning of a mouse melanocortin 5 receptor gene widely expressed in peripheral tissues. Biochemistry 33:4543–4549.

19. Lipton JM. (1989) The neuropeptide α-MSH in control of fever, the acute phase response and inflammation. In: Gortzl E, Spector NJ, eds. Neuroimmune Networks: Physiology and Diseases. New York: Alan R. Liss:243–250.

20. Hiltz ME, Lipton JM. (1989) Anti-inflammatory activity of a C-terminal fragment of the neuropeptide α-MSH. FASEB J 3:2282–2284.

21. Hiltz ME, Lipton JM. (1990) α-MSH peptides inhibit acute inflammation and contact hypersensitivity. Peptides 1990:979–982.

22. Ceriani G, Macaluso A, Catania A, Lipton JM. (1994b) Central neurogenic antiinflammatory action of α-MSH: modulation of peripheral inflammation induced by cytokines and other mediators of inflammation. Neuroendocrinology 59:138–143.

23. Lipton JM, Macaluso A, Hiltz ME, Catania A. (1991) Central administration of the peptide α-MSH inhibits inflammation in the skin. Peptides 12:795–798.

24. Dulaney R, Macaluso A, Woerner J, Hiltz ME, Catania A, Lipton JM. (1992) Changes in peripheral inflammation induced by CNS actions of an α-MSH analog and of endogenous pyrogen. Neuroendocrinimmunology 5:179–186.

25. Watanabe T, Hiltz ME, Catania A, Lipton JM. (1993) Inhibition of IL-1β-induced peripheral inflammation by peripheral and central administration of analogs of the neuropeptide α-MSH. Brain Res Bull 32:311–314.

26. Macaluso A, McCoy D, Ceriani G, Watanabe T, Biltz J, Catania A, Lipton JM. (1994) Antiinflammatory influences of α-MSH molecules: central neurogenic and peripheral actions. J Neurosci 14:2377–2382.

27. Mason MJ, Van Epps D. (1989) Modulation of IL-1, tumor necrosis factor, and C5a-mediated murine neutrophil migration by α-melanocyte-stimulating hormone. J Immunol 142:1646–1651.

28. Lipton JM, Ceriani G, Macaluso A, McCoy D, Carnes K, Biltz J, Catania A. (1994) Antiinflammatory effects of the neuropeptide α-MSH in acute, chronic and systemic inflammation. Ann NY Acad Sci 741:137–148.

29. Catania A, Rajora N, Capsoni F, Minonzio F, Star RA, Lipton JM. (1996) The neuropeptide α-MSH has specific receptors on neutrophils and reduces chemotaxis in vitro. Peptides 17:675–679.

30. Taylor AW, Streilein JW, Cousins SW. (1992) Identification of alpha-melanocyte stimulating hormone as a potential immunosuppressive factor in aqueous humor. Curr Eye Res 11:1199–1206.

31. Taylor AW, Streilein JW, Cousins SW. (1994) alpha-Melanocyte stimulating hormone suppresses antigen-stimulated T cell production of gamma interferon. NeuroImmunoModulation 1:188–194.

32. Star RA, Rajora N, Huang J, Chavez R, Catania A, Lipton JM. (1995) Evidence of autocrine modulation of macrophage nitric oxide synthase by α-MSH. Proc Natl Acad Sci (USA) 92:8016–8020.

33. Catania A, Gerloni V, Procaccia S, Airaghi L, Manfredi MG, Lomater C, Grosi L, Lipton JM. (1994) The anti-cytokine neuropeptide α-MSH in synovial fluids of patients with rheumatic diseases: comparisons with other anti-cytokine molecules. NeuroImmunoModulation 1:321–328.

34. Rajora N, Ceriani G, Catania A, Star RA, Murphy MT, Lipton JM. (1996) α-MSH

production, receptors, and influence on neopterin, in a human monocyte/macrophage cell line. J Leukocyte Biol.

35. Weidenfeld J, Crumeyrolle-Avas M, Haour F. (1995) Effect of bacterial endotoxin and interleukin-1 on prostaglandin biosynthesis by the hippocampus of mouse brain: role of interleukin-1 receptors and glucocorticoids. Neuroendocrinology 62:39–46.

36. Osborn L, Kunjel S, Nabel GJ. (1989) Tumor necrosis factor α and interleukin 1 stimulate the human immunodeficiency virus enhancer by activation of the nuclear factor kB. Proc Natl Acad Sci (USA) 86:2336–2340.

37. Poli G, Bressler P, Kinter A, Duh E, Timmer WC, Rabson A, Justement JS, Stanley S, Fauci AS. (1990a) Interleukin 6 induces human immunodeficiency virus expression in monocyte cells alone and in synergy with tumor necrosis factor alpha by transcriptional and postranscriptional mechanisms. J Exp Med 172:151–158.

38. Poli G, Kinter A, Justement JS, Kehrl JH, Bressler P, Stanley S, Fauci AS. (1990b) Tumor necrosis factor alpha functions in an autocrine manner in the induction of human immunodeficiency virus expression. Proc Natl Acad Sci (USA) 87:78882–78885.

39. Merril JE, Koyanagi Y, Chen ISY. (1989) Interleukin 1 and tumor necrosis factor α can be induced from mononuclear phagocytes by human immunodeficiency virus type I binding to the CD4 receptor. J Virol 63:4404–4408.

40. Benveniste EN. (1994) Cytokine circuits in the brain. Implications for AIDS dementia complex. Res Pub Assoc Res Nervous Mental Dis 72:71–88.

41. Catania A, Airaghi L, Manfredi MG, Vivirito MC, Milazzo F, Lipton JM, Zanussi C. (1993) Proopiomelanocortin-derived peptides and cytokines: relations in patients with acquired immunodeficiency syndrome. Clin Immunol Immunopathol 66:73–79.

42. Martin LW, Deeter LB, Lipton JM. (1989) Acute phase response to endogenous pyrogen in rabbit: effects of age and route of administration. Am J Physiol 257:R189–R193.

43. Martin LW, Lipton JM. (1990) Acute phase response to endotoxin: rise in plasma α-MSH and effects of α-MSH injection. Am J Physiol 259:R768–R772.

44. Catania A, Suffredini AF, Lipton JM. (1995) Endotoxin causes α-MSH release in normal human subjects. NeuroImmunoModulation 2:258–262.

45. Ridderstad A, Abedi-Valugerdi M, Moller E. (1991) Cytokines in rheumatoid arthritis. Ann Med 23:219–223.

46. Altomonte L, Zoli A, Mirone L, Scolieri IP, Magaro M. (1992) Serum levels of interleukin-1β, tumor necrosis factor-α and interleukin-2 in rheumatoid arthritis. J Clin Rheumatol 11:202–205.

47. Ceriani G, Diaz J, Murphree S, Catania A, Lipton JM. (1994a) The neuropeptide α-MSH inhibits experimental arthritis in rats. NeuroImmunoModulation 1:28–32.

48. Maury CP, Teppo AM. (1989) Circulating tumor necrosis factor-alpha (cachectin) in myocardial infarction. J Intern Med 225:333–336.

49. Sturk A, Hack CE, Aarden LA, Brouwer M, Koster RR, Sanders GT. (1992) Interleukin-6 release and the acute-phase reaction in patients with acute myocardial infarction: A pilot study. J Lab Clin Med 119:574–579.

50. Airaghi L, Lettino M, Manfredi MG, Lipton JM, Catania A. (1995) Endogenous cytokine antagonists during myocardial ischemia and thrombolytic therapy. Am Heart J 130: 204–211.

12

Regulation of VIP Gene Expression

AVIVA J. SYMES and J. STEPHEN FINK
Uniformed Services University of the Health Sciences
Bethesda, Maryland

I. Introduction

Vasoactive intestinal peptide (VIP) is an important neuropeptide with diverse roles including neurotransmission, trophic support of neurons and glia, and immuno-modulation. These many functions require the VIP gene to be regulated in a tempo-rally and spatially specific manner and to respond rapidly to a vast array of growth factors, cytokines, and hormones. The VIP gene has been cloned from several organisms, including human, rat, and mouse (1–3), and there is strong conservation not only of coding sequences, but also of important upstream regulatory regions among the different species (4,5). We will review the regulation of VIP mRNA in vivo and in vitro and proceed to detail what is known about the mechanisms through which VIP gene expression is regulated at the transcriptional and posttranscriptional levels.

II. Cell-Specific Expression

In the adult rat, VIP has specific, discrete localization throughout the central, periph-eral, and enteric nervous systems (6–8) and in the anterior pituitary (9,10). There are also reports that VIP mRNA is expressed in specific subpopulations of B and T lymphocytes (11). In the central nervous system, VIP mRNA expression is highest in the hypothalamus (particularly the paraventricular nucleus and the superchiasmatic nucleus), in the median eminence, and the cerebral cortex (12–15). VIP mRNA is predominantly expressed in the bipolar neurons of layers II–VI of the cortex (7,16). Other CNS sites of VIP mRNA expression include hippocampus, amygdala, inferior and superior colliculus, various thalmic structures, the raphé nucleus, the periaqueductal gray matter, and the dorsal horn of the spinal cord (reviewed in Ref. 8). VIP mRNA is also localized to the anterior pituitary, most probably in large stellate cells, although there have been some reports of VIP expression in lactotrophs (9,10). This is one of the only nonneuronal sites of VIP gene expression.

VIP is localized throughout the peripheral and enteric nervous systems. VIP-ergic neurons innervate blood vessels in the urinary tract, salivary glands, bowel, pancreas, and respiratory tract (6). Additionally, VIP nerve terminals innervate smooth muscle in urogenital organs, intestine and colon, and respiratory tract (8). VIP is typically found in cholinergic parasympathetic neurons, but small populations of sympathetic neurons express VIP (17,18). A small percentage of neurons in the dorsal root ganglia also contain VIP mRNA (19). Thus VIP expression is broadly distributed, reflecting its many diverse roles within and outside the nervous system. Such specific but widespread localization must be under very tight developmental, spatial, and inducible control.

III. Developmental Expression of VIP

It was originally thought that VIP was barely detectable in the CNS before birth (20–22). However, reports which proposed a role for VIP in embryonic development (23–25) led to a search for VIP expression in the developing fetus. Waschek et al. (26) recently reported finding VIP mRNA in the mouse hindbrain as early as E11. VIP mRNA expression increases from birth, reaching maximal levels in the cerebral cortex around P20 and decreasing to a fifth of maximal levels by P90 (20). The hippocampus and hypothalamus show a similar, if less dramatic, increase and decline at these stages (20).

VIP expression in the peripheral and enteric nervous systems is also detected early in development. VIP mRNA is detectable in the intestine in the mouse embryo at E19 (26), and specific peripheral ganglia strongly express VIP mRNA and immunoreactivity (IR) transiently during embryonic development. The most well characterized of these is the stellate ganglia, where VIP is expressed in one-third of cells at E14.5, being undetectable one day earlier (27). This level of expression declines, and by birth VIP is detectable in only 2% of stellate neurons. VIP also shows a transient increase at the same time in the superior cervical ganglia (SCG) (28). Neurons in the pterygopalatine ganglia express VIP mRNA at E16, but the time course of this expression has not yet been reported (26). The large transient increase in VIP expression in these ganglia supports the hypothesis of a trophic role for VIP in the developing nervous system. However, the mechanisms which regulate this expression have not yet been examined.

IV. Inducible Regulation of VIP Gene Expression

A. VIP Expression Is Elevated by Nerve Injury

VIP gene expression is markedly altered by neuronal injury. Axotomy of sensory, motor, or sympathetic neurons upregulates VIP mRNA and protein in the ganglia of the axotomized neurons (29–32). Axotomy of the neurons of the SCG is the most well characterized example of this plasticity of VIP expression. Under normal conditions only a few cell bodies in the SCG express VIP mRNA (33). However, 48 hr after axotomy of the major postganglionic efferents, VIP mRNA expression in the SCG increased 25-fold, with a similar rise in VIP-IR (33,34). There is a marked increase in the number of cells expressing detectable VIP mRNA and IR, as well as an overall increase in the levels of VIP in each cell.

Increased VIP expression is partially dependent on increased expression of the neuropoietic cytokine, leukemia inhibitory factor (LIF). In the injured ganglia, increased LIF expression precedes the rise in VIP mRNA (35–37). As LIF is known to induce VIP gene expression in sympathetic neurons (38,39), LIF may induce expression of VIP in the axotomized ganglia. Indeed, the axotomy-induced increase in VIP is significantly reduced in mice which do not express the LIF gene (37,40). However, in LIF-knockout mice, axotomy still induces a moderate increase in VIP expression (37,40), suggesting the involvement of a second factor.

Axotomy also prevents retrograde transport of target-derived neurotrophic factors. Nerve growth factor (NGF) is retrogradely transported by sensory and sympathetic neurons (41,42). Thus, the absence of NGF (or another retrogradely transported factor) after axotomy may lead to an upregulation of VIP expression, suggesting a repressive effect of NGF on basal VIP expression. In support of the existence of an inhibitory target-derived factor, vinblastine inhibition of retrograde transport (without axon damage) leads to increased VIP expression in sensory neurons (43). Further, in sympathetic neurons, reinnervation of the target after injury leads to a reduction in VIP expression to control levels (44). More direct evidence of the negative regulation of NGF on VIP gene expression was provided by examining levels of VIP in primary neonatal DRG cultures in the presence or absence of NGF (45). VIP was significantly repressed in the presence of NGF. Additionally, intrathecal injection of NGF lowered axotomy-induced expression of VIP in the DRG (19). Thus, there is increasing evidence that the injury-induced upregulation of VIP is mediated by the presence of positively acting factors together with the absence of negatively acting factors (Fig. 1).

B. Expression of VIP Is Regulated by Light

The ventrolateral part of the superchiasmatic nucleus (SCN) of the hypothalamus is a major site of VIP expression (7,13). There is much evidence suggesting that the SCN

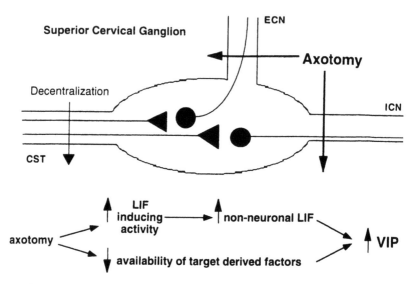

Figure 1 Axotomy of the superior cervical ganglion induces expression of VIP. Axotomy of the two major postganglionic trunks, the internal and external carotid nerves (ICN, ECN), but not decentralization of the preganglionic cervical sympathetic trunk (CST) leads to large increases in VIP gene expression. (Adapted from Ref. 29.) It has been proposed that axonal injury induces a LIF-inducing activity, which mediates increased expression of the cytokine LIF in nonneuronal cells. LIF then acts on the neurons in the ganglia to increase expression of VIP and other neuropeptides. Axotomy also reduces the availability of target-derived factors, such as NGF, to the neuronal cell body. In the intact nerve, NGF may act to suppress expression of VIP. Thus, removal of target-derived factors may contribute to the increased VIP expression (reviewed in Ref. 85).

acts as a circadian clock, regulating various physiological functions with the 24-hr light–dark cycle (46). As VIP-containing neurons of the SCN receive synapses from photic projections (47), VIP expression was examined to see if it showed any diurnal variation. In the SCN but not in the cerebral cortex, VIP mRNA expression was significantly higher in the dark phase than after the onset of the light period (48), suggesting that light may downregulate VIP expression. In support of this hypothesis, VIP mRNA levels in the SCN in rats enucleated at birth were double those of normal rats (49). However, Gozes and colleagues also showed that rats reared in complete darkness exhibited diurnal variation in VIP mRNA expression in the SCN, providing evidence that this variation may be intrinsic as well as, or instead of, a response to environmental cues (50). The mechanism of this regulation is not understood.

C. Neuroendocrine Regulation of VIP Expression

VIP is a potent prolactin-releasing factor in mammals and birds (51,52). Regulation of VIP expression therefore impacts directly on prolactin secretion from the anterior pituitary, and thus VIP expression is under strict regulation by the neuroendocrine system. VIP gene expression in the anterior pituitary is influenced by several different manipulations of the neuroendocrine axes. Adrenalectomy (53), acute or chronic estrogen treatment (54–56), and hypothyroidism (9,10,12) all lead to increased VIP expression in the anterior pituitary. However, whether there is direct regulation of the VIP gene by glucocorticoids or estrogen is not clear. Hypothyroidism increases VIP mRNA and immunoreactivity in the anterior cingulate and cerebral cortices of rats in addition to the anterior pituitary, without altering the expression in the thalamus or hypothalamus (9,10,57), suggesting that circulating thyroid hormones suppress VIP gene expression in a tissue-specific manner. A direct suppressive effect of T3 and T4 on VIP IR was demonstrated in cultured fetal cerebral cortical cells (58). Thus, in hypothyroid animals, a release of the inhibition of thyroid hormones rather than an indirect effect of their reduced levels may be responsible for elevated VIP mRNA expression.

There is sexual dimorphism in VIP expression in the anterior pituitary and hypothalamus, but not in other tissues examined (59,60). In the rat anterior pituitary, VIP mRNA expression is twice as high in males as in females (59). In the human SCN, VIP IR is also higher in males than females, implying a possible role for VIP neurons in the SCN in sexually dimorphic functions such as reproduction and sexual behavior (61). Hormonal regulation of VIP expression in the SCN is suggested by studies showing that lactating rats express twofold higher levels of VIP mRNA in the SCN than in nonlactating controls (15). Additionally, ovariectomy significantly decreases VIP mRNA in the rat SCN, an effect which is reversed with estradiol treatment (60). Thus, VIP gene expression is hormonally regulated in a tissue-specific manner. However, the promoter elements which mediate this hormonal regulation have not yet been characterized.

D. Inducible Gene Expression in Cell Culture

In cultured cells the VIP gene is induced by a variety of other stimuli and extracellular messengers. In primary culture of rat pituitary cells, insulin-like growth factor I induces VIP mRNA and protein in a dose-dependent manner (62). Much work has been reported on VIP gene expression in a variety of neuroblastoma cells. In neuroblastoma cells VIP gene expression is increased by cAMP (63,64), phorbol esters (63,64), calcium ionophore (65), KCl (65), retinoic acid (66), and the cytokines CNTF, LIF, and oncostatin M (67,68). These are all direct effects, and many have been studied at the transcriptional level to elucidate the molecular mechanisms of transcriptional regulation of the VIP gene.

V. Studies on the Transcriptional Regulation of the VIP Gene

The 5'-flanking region of the VIP gene contains many DNA sequences which mediate the response of the gene to a variety of extracellular stimuli (Fig. 2). The first

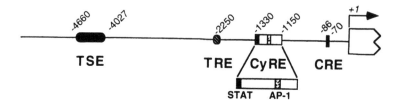

Figure 2 The VIP promoter. Binding sites for known transcription factors within the 5'-flanking region of the human VIP gene. TSE, tissue-specific element; TRE, TPA-response element; CyRE, cytokine response element; CRE, cAMP response element.

regulatory element to be characterized was the cAMP response element (CRE) (69). This sequence, located between -86 and -70 nucleotides upstream from the transcriptional start site, is necessary for cAMP responsiveness of the VIP gene and is sufficient to confer cAMP responsiveness to a gene that is not normally regulated by cAMP. The VIP CRE contains two inverted CGTCA sequences which are also found in a variety of other cAMP-induced genes (70) and confer cAMP responsiveness through the binding and subsequent phosphorylation of the transcription factor CREB (71). The VIP CRE was subsequently shown to confer responsiveness to phorbol esters on a heterologous viral promoter in HeLa cells (64). The VIP CRE can bind CREB and the AP-1 transcription factors c-fos and c-jun, through which the phorbol ester induction may be mediated (64). However, in SHEP neuroblastoma cells, mutations in the VIP CRE which abolish the response of VIP promoter-luciferase reporter constructs to forskolin, do not abolish their response to phorbol esters (72). These data suggest that, in the context of the endogenous VIP promoter in SHEP cells, an additional element, distinct from the previously characterized CRE, contributes to the response of the VIP gene to phorbol esters (72). There is a second consensus phorbol-ester response element (TRE) 2.25 kb upstream of the transcription start site. This upstream TRE may contribute to the phorbol ester responsiveness of the full-length promoter in SHEP cells (72). Additionally, the VIP CRE contributes to basal expression of the VIP gene. Mutation of the CRE within the context of the VIP promoter significantly reduced the level of expression of VIP promoter-luciferase reporters, in each of three cell lines tested (72). Therefore, although the VIP CRE mediates inducible expression of the VIP gene, it also plays a role in basal expression.

Little is known of the elements which control the tissue-specific or developmental expression of the VIP gene in vivo. Experiments with transgenic mice in which different lengths of VIP promoter sequences were placed upstream of reporter genes have provided contrasting results. In one of two published studies, 5.2 kb of human VIP promoter driving expression of the CAT gene directed detectable expression in four separate lines of mice of CAT exclusively to the ileum (73). These results indicate that 5.2 kb of VIP promoter sequence were not necessarily sufficient to direct expression of the transgene to other sites of VIP expression. However, a second study which generated transgenic mice containing 1.9 kb of human VIP promoter sequence linked to the β-galactosidase reporter gene demonstrated expression of the transgene in most central and peripheral nervous system sites in which endogenous VIP expression is expressed, with the major exception of the SCN of the hypothalamus (74). Thus, these studies indicate that 1.9 kb of promoter sequence contains within it sufficient information for the majority of cell-specific expression of the VIP gene. However, inclusion of further-upstream sequence may be able to repress transcription of this gene in vivo. Developmental expression of the transgene was not examined in either report.

Examination of the role of sequences between 2.5 and 5.2 kb demonstrated the presence of a tissue-specific enhancer (TSE) element active in neuroblastoma cell lines. This element, located between −4.66 and −4.02 kb, contributes to a high level of VIP expression of SHEP and SH-IN cells (72,73,75). Full constitutive expression of the VIP promoter in VIP-expressing neuroblastoma cells was absolutely dependent on the presence of the upstream TSE together with the more proximal CRE (72).

The response of the VIP gene to the neuropoietic cytokines ciliary neurotrophic factor (CNTF), LIF, oncostatin M, and interleukin-6 (IL-6) has been mapped to a 180-bp cytokine response element (CyRE), 1.15 kb upstream of the transcription start site (76). The VIP CyRE is necessary and sufficient to mediate the response to CNTF, LIF, or oncostatin M in NBFL neuroblastoma cells (76). Deletion analysis of the CyRE demonstrates that is comprises a complex element composed of multiple functional domains (76). A Stat-binding site is present at the extreme 5′ end of the CyRE. CNTF or LIF induce Stat1 and Stat3 to bind to this site in neuroblastoma cells and primary sympathetic neurons (76). Deletion or mutation of the Stat site reduces the cytokine-mediated inducibility of the CyRE by 80%, demonstrating the importance of the Stat site to cytokine regulation of VIP gene expression. However, one copy of the Stat site alone is not inducible by CNTF, and three multimerized copies can be induced only threefold, in comparison to the 40- to 100-fold induction for the whole CyRE. Thus, sequences distinct from the Stat site are important to mediating the cytokine inducibility of the CyRE.

A second site with the CyRE which binds proteins induced by neuropoietic cytokine treatment is a noncanonical AP-1 site. CNTF induces AP-1 proteins to bind to this site in neuroblastoma cells. The induction of AP-1 proteins contributes to the CNTF-mediated induction of VIP gene transcription, as mutations in this site reduce the CNTF induction by 50% (77). However, neither a multimerized AP-1 site nor a canonical AP-1 site is able to mediate any induction to CNTF (Fig. 2). Therefore, although the CyRE Stat site and AP-1 site are important for mediating CNTF-induced transcriptional activation of the CyRE, neither is sufficient. Other proteins binding within the CyRE contribute to a combinatorial induction by CNTF and related cytokines.

The VIP CyRE also encodes within it cell-specific information. Hepatoma HepG2 cells do not express VIP mRNA and do not induce VIP in response to IL-6 or LIF. However, HepG2 cells respond to LIF to induce the acute-phase genes. LIF stimulation of HepG2 cells induces Stat1 and Stat3 protein binding to the VIP Stat site, but does not induce transcriptional activation of the CyRE (5). Therefore the VIP CyRE responds to neuropoietic cytokine induction in a cell-specific and cytokine-specific manner. The mechanisms through which this specificity is attained are not known.

VI. Posttranscriptional Regulation of the VIP Gene

The 5'-flanking region of the VIP gene contains many different regulatory regions which control the transcriptional response of the VIP gene to different stimuli. However, transcriptional regulation is only one way in which VIP gene expression is regulated. There are now several reports suggesting that VIP mRNA stability is also regulated. Phorbol ester treatment of NBFL neuroblastoma cells increases the stability of VIP mRNA (78,79). There are two distinct VIP mRNA species in human, rat, and mouse (2,68,80), which differ with respect to the length of their 3'-untranslated region (UTR) and use of alternative polyadenylation sites (2,81). In the majority of tissues where VIP mRNA is detectable, the larger 1.7-kb VIP mRNA is the predominant species (2). However, this 1.7-kb mRNA includes a longer 3'-UTR, which in the rat anterior pituitary may be destabilized and differentially degraded, leaving the 1-kb mRNA species as the major VIP gene product (80). Such regulation has been demonstrated in the anterior pituitary in vivo after termination of an estrogen stimulus, and in explanted pituitaries (80). The 1.7-kb VIP transcript is stabilized by inhibitors of RNA and protein synthesis, supporting the involvement of a de-novo-synthesized labile factor which acts selectively to downregulate the 1.7-kb VIP mRNA (80). The VIP 3' UTR in the 1.7-kb transcript encodes multiple AUUUA motifs in an AU-rich context (2), found in several genes whose mRNA stability is regulated through interaction with RNA-binding proteins (82). This 3' region of VIP mRNA binds tissue-specific proteins in a concentration-dependent manner (83). These cytoplasmic RNA-binding proteins are competed by an AU-rich region of the c-fos 3' UTR, suggesting that the VIP and c-fos mRNA 3' UTRs bind similar proteins (83). Thus, just as the sequences in the c-fos mRNA mark the mRNA for rapid degradation (84), so too these sequences in the VIP 3' UTR may have a similar function. As the 1.7-kb mRNA is the predominant VIP transcript in the brain, posttranscriptional regulation of VIP mRNA stability may be important in determination of VIP levels.

VII. Conclusion

There is a plethora of descriptive data demonstrating precise developmental, hormonal, and environmental control of VIP gene expression in vivo and increasingly complex mechanisms of the transcriptional and posttranscriptional regulation of the VIP promoter in cultured cells in vitro. However, the connection between these two substantial bodies of work remains largely undetermined. Although all the data resulting from experiments in cell culture are suggestive of mechanisms which may operate to regulate the VIP gene in vivo, in most cases the extracellular signaling factors, signal transduction pathways, and DNA-binding sites which control VIP gene expression during development and in response to diverse stimuli in the adult have yet to be identified.

A more complete understanding of these regulatory mechanisms may lead to approaches to manipulate expression of the VIP gene in situations where increased or decreased amounts of VIP would be beneficial.

Acknowledgments

We thank Dr. Susan Lewis and Dr. Richard Zigmond for helpful discussions. This work was supported by National Institutes of Health grant NS27514 (JSF) and a Uniformed Services University of the Health Sciences intramural grant to Aviva J. Symes.

References

1. Itoh N, Obata K, Yanaihara N, Okamoto H. Human preprovasoactive intestinal polypeptide contains a novel PHI-27-like peptide, PHM-27. Nature 1983; 304:547–549.
2. Lamperti E, Rosen K, Villa-Kamaroff L. Characterization of the gene and messages for vasoactive intestinal peptide (VIP) in the rat and mouse. Mol Brain Res 1991; 9:217–231.
3. Nishizawa M, Hayakawa Y, Yanaihara N, Okamoto H. Nucleotide sequence divergence and functional constraint in VIP precursor mRNA evolution between human and rat. FEBS Lett 1985; 183:55–59.
4. Sena M, Bravo DT, Von Agoston D, Waschek JA. High conservation of upstream regulatory sequences on the human and mouse vasoactive intestinal peptide (VIP) genes. DNA Seq 1994; 5:25–29.
5. Symes AJ, Corpus L, Fink JS. Differences in nuclear signaling by LIF and IFN-γ: the role of STAT proteins in regulating VIP gene expression. J Neurochem 1995; 65:1926–1933.
6. Sundler F, Ekblad E, Grunditz T, Hakanson R, Uddman R. Vasoactive intestinal peptide in the peripheral nervous system. Ann NY Acad Sci 1988; 527:143–167.
7. Dussaillant M, Sarrieau A, Gozes I, Berod A, Rostene W. Distribution of cells expressing vasoactive intestinal peptide/peptide histidine isoleucine-amide precursor messenger RNA in the rat brain. Neuroscience 1992; 50:519–530.
8. Gozes I, Brenneman DE. VIP: molecular biology and neurobiological function. Mol Neurobiol 1989; 3:201–236.
9. Segerson TP, Lam KS, Cacicedo L, et al. Thyroid hormone regulates vasoactive intestinal peptide (VIP) mRNA levels in the rat anterior pituitary gland. Endocrinology 1989; 125:2221–2223.
10. Lam KS, Lechan RM, Minamitani N, Segerson TP, Reichlin S. Vasoactive intestinal peptide in the anterior pituitary is increased in hypothyroidism. Endocrinology 1989; 124:1077–1084.
11. Gomariz RP, Leceta J, Garrido E, Garrido T, Delgado M. Vasoactive intestinal peptide (VIP) mRNA expression in rat T and B lymphocytes. Regulatory Peptides 1994; 50: 177–184.

12. Buhl T, Georg B, Nilsson C, Mikkelsen JD, Wulff BS, Fahrenkrug J. Effect of thyroid hormones on vasoactive intestinal polypeptide gene expression in the rat cerebral cortex and anterior pituitary. Regulatory Peptides 1995; 55:237–251.

13. Card JP, Fitzpatrick-McElligott S, Gozes I, Baldino F. Localization of vasopressin, somatostain and VIP messenger RNA in the rat suprachiasmatic nucleus. Cell Tissue Res 1988; 252:307–315.

14. Ceccatelli S, Giardino L, Calza L. Response of hypothalamic peptide mRNAs to thyroidectomy. Neuroendocrinology 1992; 56:694–703.

15. Gozes I, Avidor R, Biegon A, Baldino F Jr. Lactation elevates vasoactive intestinal peptide messenger ribonucleic acid in rat suprachiasmatic nucleus. Endocrinology 1989; 124:181–186.

16. Magistretti PJ. VIP neurons in the cerebral cortex. Trends Pharmacol Sci 1990; 11: 250–254.

17. Landis SC, Fredieu JR. Coexistence of calcitonin gene-related peptide and vasoactive intestinal peptide in cholinergic sympathetic innervation of rat sweat glands. Brain Res 1986; 377:177–181.

18. Landis SC, Keefe D. Evidence for neurotransmitter plasticity in vivo: developmental changes in properties of cholinergic sympathetic neurons. Dev Biol 1983; 98:349–372.

19. Verge V, Richardson P, Wiesenfeld-Hallin Z, Hokfelt T. Differential influence of nerve growth factor on neuropeptide expression in vivo: a novel role in peptide suppression in adult sensory neurons. J Neurosci 1995; 15:2082–2096.

20. Gozes I, Shani Y, Rostene WH. Developmental expression of the VIP-gene in brain and intestine. Brain Res 1987; 388:137–148.

21. Hill JM, Agoston DV, Gressens P, McCune SK. Distribution of VIP mRNA and two distinct VIP binding sites in the developing rat brain: relation to ontogenic events. J Comp Neurol 1994; 342:186–205.

22. Fink JS, Montminy MR, Tsukada T, et al. In situ hybridization of somatostatin and vasoactive intestinal peptide mRNA in rat nervous system. In: Uhl G, ed. In Situ Hybridization in the Brain. New York: Plenum Press, 1986:181–191.

23. Gressens P, Hill JM, Gozes I, Fridkin M, Brenneman DE. Growth factor function of vasoactive intestinal peptide in whole cultured mouse embryos. Nature 1993; 362: 155–158.

24. Gressens P, Hill JM, Paindaveine B, Gozes I, Fridkin M, Brenneman DE. Severe microcephaly induced by blockade of vasoactive intestinal peptide function in the primitive neuroepithelium of the mouse. J Clin Invest 1994; 94:2020–2027.

25. Hill JM, Mervis RF, Politi J, et al. Blockade of VIP during neonatal development induces neuronal damage and increases VIP and VIP receptors in brain. Ann NY Acad Sci 1994; 739:211–225.

26. Waschek JA, Ellison J, Bravo DT, Handley V. Embryonic expression of vasoactive intestinal peptide (VIP) and VIP receptor genes. J Neurochem 1996; 66:1762–1765.

27. Tyrrell S, Landis SC. The appearance of NPY and VIP in sympathetic neuroblasts and subsequent alterations in their expression. J Neurosci 1994; 14:4529–4547.

28. Pincus DW, DiCicco-Bloom E, Black I. Trophic mechanisms regulate mitotic neuronal precursors: role of vasoactive intestinal peptide. Brain Res 1994; 663:51–60.

29. Hyatt-Sachs H, Schreiber RC, Bennett TA, Zigmond RE. Phenotypic plasticity in adult sympathetic ganglia in vivo: effects of deafferentation and axotomy on the expression of vasoactive intestinal peptide. J Neurosci 1993; 13:1642–1653.

30. Hockfelt T, Zhang X, Wiesenfeld-Hallin Z. Messenger plasticity in primary sensory

neurons following axotomy and its functional implications. Trends Neurosci 1994; 17: 22–30.

31. Noguchi K, De Leon M, Nahin RL, Senba E, Ruda MA. Quantification of axotomy-induced alteration of neuropeptide mRNAs in dorsal root ganglion neurons with special reference to neuropeptide Y mRNA and the effects of neonatal capsaicin treatment. J Neurosci Res 1993; 35:54–66.

32. Zhang X, Verge VM, Wiesenfeld-Hallin Z, Piehl F, Hokfelt T. Expression of neuropeptides and neuropeptide mRNAs in spinal cord after axotomy in the rat, with special reference to motoneurons and galanin. Exp Brain Res 1993; 93:450–461.

33. Mohney RP, Siegel RE, Zigmond RE. Galanin and vasoactive intestinal peptide messenger RNAs increase following axotomy of adult sympathetic neurons. J Neurobiol 1994; 25:108–118.

34. Hyatt-Sachs H, Bachoo M, Schreiber R, Vaccariello A, Zigmond RE. Chemical sympathectomy and postganglionic nerve transection produce similar increases in galanin and VIP mRNA but differ in their effects on peptide content. J Neurobiol 1996; 30:543–555.

35. Sun Y, Rao MS, Zigmond RE, Landis SC. Regulation of vasoactive intestinal peptide expression in sympathetic neurons in culture and after axotomy: the role of cholinergic differentiation factor/leukemia inhibitory factor. J Neurobiol 1994; 25:415–430.

36. Sun Y, Landis SC, Zigmond RE. Signals triggering the induction of leukemia inhibitory factor in sympathetic superior cervical ganglia and their nerve trunks after axonal injury. Mol Cell Neurosci 1996; 7:152–163.

37. Sun Y, Zigmond RE. Involvement of leukemia inhibitory factor in the increases in galanin and vasoactive intestinal peptide mRNA and the decreases in neuropeptide Y and tyrosine hydroxylase mRNA in sympathetic neurons after axotomy. J Neurochem 1996; 67:1751–1760.

38. Fann MJ, Patterson PH. A novel approach to screen for cytokine effects on neuronal gene expression. J Neurochem 1993; 61:1349–1355.

39. Lewis SE, Rao MS, Symes AJ, et al. Coordinate regulation of choline acetyltransferase, tyrosine hydroxylase, and neuropeptide mRNAs by CNTF and LIF in cultured sympathetic neurons. J Neurochem 1994; 63:429–438.

40. Rao MS, Sun Y, Escary JL, et al. Leukemia inhibitory factor mediates an injury response but not a target-directed developmental transmitter switch in sympathetic neurons. Neuron 1993; 11:1175–1185.

41. Stockel K, Schwab M, Thoenen H. Comparison between the retrograde axonal transport of nerve growth factor and tetanus toxin in motor, sensory and adrenergic neurons. Brain Res 1975; 99:1–16.

42. Hendry IA, Stach RA, Herrup K. Characteristics of the retrograde axonal transport system for nerve growth factor in the sympathetic nervous system. Brain Res 1974; 82: 117–128.

43. Kashiba H, Senba E, Kawai Y, Ueda Y, Tohyama M. Axonal blockade induces the expression of vasoactive intestinal polypeptide and galanin in rat dorsal root ganglion neurons. Brain Res 1992; 577:19–28.

44. Klimaschewski L, Grohmann I, Heym C. Target-dependent plasticity of galanin and vasoactive intestinal peptide in the rat superior cervical ganglion after nerve lesion and re-innervation. Neuroscience 1996; 72:265–272.

45. Mulderry PK. Neuropeptide expression by newborn and adult rat sensory neurons in culture: effects of nerve growth factor and other neurotrophic factors. Neuroscience 1994; 59:673–688.

46. Meijer JH, Rietveld WJ. Neurophysiology of the suprachiasmatic circadian pacemaker in rodents. Physiol Rev 1989; 69:671–707.

47. Ibata Y, Takahashi Y, Okamura H, et al. Vasoactive intestinal peptide (VIP)-like immuno-reactive neurons located in the rat superchiasmatic nucleus receive direct retinal projection. Neurosci Lett 1989; 97:1–5.

48. Albers HE, Stopa EG, Zoeller RT, et al. Day-night variation in prepro vasoactive intestinal peptide/peptide histidine isoleucine mRNA within the rat suprachiasmatic nucleus. Mol Brain Res 1990; 7:85–89.

49. Holtzman RL, Malach R, Gozes I. Disruption of the optic pathway during development affects vasoactive intestinal peptide mRNA expression. New Biol 1989; 1:215–221.

50. Glazer R, Gozes I. Diurnal oscillation in vasoactive intestinal peptide gene expression independent of environmental light entraining. Brain Res 1994; 644:164–167.

51. Nagy G, Mulchahey JJ, Neill JD. Autocrine control of prolactin secretion by vasoactive intestinal peptide. Endocrinology 1988; 122:364–366.

52. Rostene W. Neurobiological and neuroendocrine functions of the vasoactive intestinal peptide (VIP). Progr Neurobiol 1984; 22:103–129.

53. Lam KS, Srivastava G, Tam SP. Divergent effects of glucocorticoid on the gene expression of vasoactive intestinal peptide in the rat cerebral cortex and pituitary. Neuroendocrinology 1992; 56:32–37.

54. Kasper S, Popescu RA, Torsello A, Vrontakis ME, Ikejiani C, Friesen HG. Tissue-specific regulation of vasoactive intestinal peptide messenger ribonucleic acid levels by estrogen in the rat. Endocrinology 1992; 130:1796–1801.

55. Montagne MN, Dussaillant M, Chew LJ, et al. Estradiol induces vasoactive intestinal peptide and prolactin gene expression in the rat anterior pituitary independently of plasma prolactin levels. J Neuroendocrinol 1995; 7:225–231.

56. Lam KS, Srivastava G, Lechan RM, Lee T, Reichlin S. Estrogen regulates the gene expression of vasoactive intestinal peptide in the anterior pituitary. Neuroendocrinology 1990; 52:417–421.

57. Giardino L, Ceccatelli S, Zanni M, Hokfelt T, Calza L. Regulation of VIP expression by thyroid hormone in different brain areas of adult rat. Mol Brain Res 1994; 27:87–94.

58. Lorenzo M, Sanchez-Franco F, de los Frailes M, Tolon RM, Fernandez G, Cacicedo L. Thyroid hormones regulate release and content of vasoactive intestinal peptide in cultured fetal cerebral cortical cells. Neuroendocrinology 1992; 55:59–65.

59. Lam KS, Srivastava G. Sex-related differences and thyroid hormone regulation of vasoactive intestinal peptide gene expression in the rat brian and pituitary. Brain Res 1990; 526:135–137.

60. Gozes I, Werner H, Fawzi M, et al. Estrogen regulation of vasoactive intestinal peptide mRNA in rat hypothalamus. J Mol Neurosci 1989; 1:55–61.

61. Swaab DF, Zhou JN, Ehlhart T, Hofman MA. Development of vasoactive intestinal polypeptide neurons in the human suprachiasmatic nucleus in relation to birth and sex. Brain Res Dev Brain Res 1994; 79:249–259.

62. Lara JI, Lorenzo MJ, Cacicedo L, et al. Induction of vasoactive intestinal peptide gene expression and prolactin secretion by insulin-like growth factor I in rat pituitary cells: evidence for an autoparacrine regulatory system. Endocrinology 1994; 135:2526–2532.

63. Ohsawa K, Hayakawa Y, Nishizawa M, et al. Synergistic stimulation of VIP-PHM-27 gene expression by cyclic AMP and phorbol esters in human neuroblastoma cells. Biochem Biophys Res Commun 1985; 132:885–891.

64. Fink JS, Verhave M, Walton K, Mandel G, Goodman RH. Cyclic AMP- and phorbol

ester-induced transcriptional activation are mediated by the same enhancer element in the human vasoactive intestinal peptide gene. J Biol Chem 1991; 266:3882–3887.

65. Adler EM, Fink JS. Calcium regulation of vasoactive intestinal polypeptide mRNA abundance in SH-SY5Y human neuroblastoma cells. J Neurochem 1993; 61:727–737.

66. Georg B, Wulff BS, Fahrenkrug J. Characterization of the effects of retinoic acid on vasoactive intestinal polypeptide gene expression in neuroblastoma cells. Endocrinology 1994; 135:1455–1463.

67. Symes AJ, Rao MS, Lewis SE, Landis SC, Hyman SE, Fink JS. Ciliary neurotrophic factor coordinately activates transcription of neuropeptide genes in a neuroblastoma cell line. Proc Natl Acad Sci USA 1993; 90:572–576.

68. Rao MS, Symes A, Malik N, Shoyab M, Fink JS, Landis SC. Oncostatin M regulates VIP expression in a human neuroblastoma cell line. Neuroreport 1992; 3:865–868.

69. Tsukada T, Fink JS, Mandel G, Goodman RH. Identification of a region in the human vasoactive intestinal polypeptide gene responsible for regulation by cyclic AMP. J Biol Chem 1987; 262:8743–8747.

70. Hyman SE, Comb MC, Lin Y-S, Pearlberg J, Green M, Goodman HM. A common trans-acting factor is involved in transcriptional regulation of neurotransmitter genes by cyclic AMP. Mol Cell Biol 1988; 8:4225–4233.

71. Habener J. Cyclic AMP response element binding proteins: a cornucopia of transcription factors. Mol Endocrinol 1990; 4:1087–1094.

72. Hahm S-H, Eiden LE. Tissue-specific expression of the vasoactive intestinal peptide gene requires both an upstream tissue specific element and the 5′ proximal cyclic-AMP responsive element. J Neurochem 1996; 67:1872–1881.

73. Agoston DV, Bravo DT, Waschek JA. Expression of a chimeric VIP gene is targeted to the intesting in transgenic mice. J Neurosci Res 1990; 27:479–486.

74. Tsuruda L, Lamperti E, Lewis S, et al. Region-specific central nervous system expression and axotomy-induced regulation in sympathetic neurons of a VIP-β-galactosidase fusion gene in transgenic mice. Mol Brain Res 1996; 42:181–192.

75. Waschek J, Hsu C-M, Eiden L. Lineage-specific regulation of the vasoactive intestinal peptide gene in neuroblastoma cells is conferred by 5.2 kilobases of 5′ flanking sequence. Proc Natl Acad Sci (USA) 1988; 85:9547–9551.

76. Symes AJ, Lewis SE, Corpus L, Rajan P, Hyman SE, Fink JS. Stat proteins participate in the regulation of the VIP gene by the CNTF family of cytokines. Mol Endocrinol 1994; 8:1750–1763.

77. Symes AJ, Gearan T, Eby J, Fink JS. Integration of Jak-Stat and AP-1 signaling pathways at the VIP cytokine response element in CNTF-dependent transcription. J Biol Chem 1997; 272:9648–9654.

78. Davidson A, Moody T, Gozes I. Regulation of VIP gene expression in general. J Mol Neurosci 1996; 7:99–110.

79. Tolentino P, Villa-Komaroff L. Regulation of vasoactive intestinal polypeptide and galanin mRNA stabilities. Mol Brain Res 1996; 39:89–98.

80. Chew L-J, Murphy D, Carter D. Alternatively polyadenylated vasoactive intestinal peptide mRNAs are differentially regulated at the level of stability. Mol Endocrinol 1994; 8:603–613.

81. Chew LJ, Murphy D, Carter DA. Differential use of 3′ poly(A) addition sites in vasoactive intestinal peptide messenger ribonucleic acid of the rat anterior pituitary gland. J Neuroendocrinol 1991; 3:351–355.

82. Gillis P, Malter JS. The adenosine-uridine binding factor recognizes the AU-rich ele-

ments of cytokine, lymphokine, and oncogene mRNAs. J Biol Chem 1991; 266:3172–3177.

83. Wolford JK, Signs SA. Binding of sequence-specific proteins to the 3'-untranslated region of vasoactive intestinal peptide mRNA. Biochem Biophys Res Commun 1995; 211:819–825.

84. You Y, Chen CY, Shyu AB. U-rich sequence-binding proteins (URBPs) interacting with a 20-nucleotide U-rich sequence in the 3' untranslated region of c-fos mRNA may be involved in the first step of c-fos mRNA degradation. Mol Cell Biol 1992; 12:2931–2940.

85. Zigmond RE. LIF, NGF and the cell body response to axotomy. Neuroscientist 1997. In press.

13

Neuroregulation of Pulmonary Immune Responses by Vasoactive Intestinal Peptide and Substance P

EDWARD J. GOETZL and
SUNIL P. SREEDHARAN

University of California, San Francisco
San Francisco, California

PATRICIA K. BYRD and
H. BENFER KALTREIDER

University of California, San Francisco
and Veterans Affairs Medical Center
San Francisco, California

MENGHANG XIA

Amgen Inc.
Thousand Oaks, California

I. Introduction and Significance

Neuroendocrine–immune communication influences mammalian development, cellular differentiation, normal functions of organs in both systems, inflammatory responses to microbial invasion and tissue injury, tissue repair, and the pathogenesis of diverse diseases (1–4). Interactions between the nervous and immune systems are facilitated by anatomical connections (5–7) and mediated by many factors that are released and recognized in both systems (8–10). The initial descriptive findings suggesting such interactions included changes in circulating levels and functions of T cells after lesions were induced in certain areas of the central nervous system (CNS) and alterations in CNS electrical activity during systemic immune responses (1,3,8). Results of increasingly sophisticated studies have delineated extensive sympathetic, cholinergic, and sensory innervation of primary immune organs and lymphoid tissues of other systems.

Morphometric enumeration of nerve endings in lymphoid tissues has established proximity and connections with most T lymphocytes and mast cells, including occasional plasma membrane intercalations (11,12). Lymphatics and high endothelial venules of specialized lymphoid compartments in the gastrointestinal tract and lungs have been shown to receive many peptidergic neurons, rich in substance P (SP) and vasoactive intestinal peptide (VIP), and to contain a much higher percentage of T cells than in blood that express the receptors for these neuropeptides (7,13,14). Neural ablation and pharmacological antagonism of neuromediators in vivo have revealed functional roles for many components of lymphoid innervation in immune development and responses (15,16). Detailed studies of neuropeptide effects on T cells and other immune cells in vitro have defined distinctive patterns of regulation of mobility, adherence, surface protein expression, synthetic functions, and immune contributions. The documentation of increases in concentration of some neuropeptides at sites of normal immune and inflammatory responses, and in lesions of human autoimmune and inflammatory diseases suggest that selected neuropeptide pharmacological agonists and/or antagonists may have novel therapeutic possibilities in such diseases.

II. Generation and Release of VIP and SP in Immunity and Hypersensitivity Reactions

Lymphoid organs are heavily supplied by noradrenergic autonomic nerves and cholinergic nerves, which also contain neuropeptide Y, somatostatin, galanin, and VIP, and by the nonadrenergic and noncholinergic sensory nerves, which contain the tachykinins SP and neurokinin A as well as calcitonin gene-related peptide (CGRP) (17,18). The endings of these nerves are especially dense in T-cell corridors and near specialized effector cells, such as mast cells, where a high percentage of the immune cells are in close contact with such endings (11,12). Stimulation of nerves supplying lymphoid organs and adoptive transfer of lymphocytes after in-vitro exposure to neuromediators has revealed the involvement of regional innervation in distinctive regulation of lymphocyte traffic and distribution in tissues (19,20).

The intestinal and respiratory tracts illustrate specialized relationships between regional innervation and compartmental immune tissues. Only rare extrinsic nerves serving the intestines have neuropeptides, whereas many intrinsic neurons of the myenteric plexus and submucosal plexus are peptidergic, with a predominance of VIP, SP, and somatostatin (21). VIP- and SP-containing nerve fibers are at the basement membrane and among epithelial cells, around crypts and in villi of the lamina propria, at the margins of Peyer's patches, and in areas of greatest T-cell traffic, such as the lymphatics and high endothelial venules (7). The SP- and VIP-ergic nerves of the respiratory system of mammalian species are found predominantly in nasal mucosa and large pulmonary airways, with the highest density in tracheo-bronchial smooth muscle and fewer around submucosal serous and mucous glands, in the lamina propria, in the walls of pulmonary and bronchial arteries, and around central lymphoid follicles (22). The density of SP- and VIP-ergic nerves is lowest in the pulmonary parenchyma, but some are observed around blood vessels, small bronchi, and peribronchiolar lymphoid tissue and can release sufficient amounts of neuropeptides to attain nanomolar concentrations in respiratory secretions (23,24). Many of the VIP-ergic nerves, especially in nonvascular smooth muscle, also contain pituitary adenylyl cyclase-activating peptide (PACAP), peptide histidine isoleucine/methionine, and nitric oxide synthase (NOS) (25). Deficiencies of VIP-ergic nerves have been documented in esophageal achalasia and Hirschsprung's disease of the colon, where the absence of VIP may contribute to the abnormalities of intestinal motility, and in cystic fibrosis and some patients with asthma, where respective disorders of exocrine secretion and bronchial reactivity may result from the loss of VIP secretory and bronchodilatory effects (21,25). In contrast, overexpression of VIP in pancreatic beta cells of transgenic mice improves insulin secretion and glucose tolerance (26).

VIP, SP, and some other neuropeptides are released in functionally relevant amounts in numerous regional immune responses. After respiratory allergen challenge of allergic patients and in spontaneous attacks of allergic rhinitis and asthma, the concentrations of VIP, SP, and CGRP increased in nasal secretions and bronchoalveolar lavage fluid (BAL) (27–29). VIP and SP levels in BAL also rose with distinctive time courses after intratracheal antigen challenge of mice that had been sensitized by systemic immunization (24,30,31). The concentrations of VIP, SP, and occasionally other neuropeptides also were elevated in tissue and fluid from lesions of some patients with atopic dermatitis, urticaria, and other dermatitides, along with increases in the density of cutaneous peptidergic nerve fibers (32,33). VIP and SP thus are released in the courses of immune responses and inflammation, have physiological effects that mimic and augment the primary reactions, and both mediate and modify the recruitment and contributions of diverse cytokines and immune cells.

III. Overview of Effects of VIP in Immunity and Inflammation

The contributions of VIP to inflammation are attributable in part to physiological responses, such as vasodilatation and secretion. That VIP affects mononuclear phagocytes and mast cells in vitro in some systems (34–36) suggests that concentrations attainable in vivo may recruit and activate diverse inflammatory cells directly. Many of the principal immunological activities of VIP are T-cell-dependent (37,38). Three major effects of VIP on human blood and rodent tissue T cells, and on human cultured lines of T cells with defined VIP receptors, have been elucidated by in vitro studies. First, VIP evokes T-cell migration through basement membranes and connective tissues by increasing expression of adhesive proteins, stimulating chemotaxis, and inducing secretion of specific matrix metalloproteinases (MMPs) (39–42). VIP facilitates interactions of T cells with endothelium and connective tissues by increasing expression of P- and E-selectins (43), and enhancing T-cell β_1 integrin-dependent binding to VCAM-1 (42) and fibronectin (41,42). Chemotaxis of T cells through micropore filters is stimulated by VIP at concentrations as low as 10^{-10} M (39,42). VIP stimulation of T-cell chemotaxis through Matrigel model basement membranes on micropore filters is dependent on T-cell secretion of MMP-2 and MMP-9, which create channels in the basement membrane matrix (39,40). Stimulation of T-cell chemotaxis through a model basement membrane shows the same VIP concentration dependence as induction of T-cell MMP activity, both require T-cell β_1 integrin binding to fibronectin and/or type IV collagen, and basement membrane transmigration of T cells elicited by VIP is suppressed by specific inhibitors of MMPs (39).

Second, concentrations of VIP attained in tissues mounting immune responses inhibit T-cell production and secretion of IL-2, IL-4, IL-10, and, in some instances IFN-γ (44,45). In contrast, VIP may enhance the production of other cytokines, such as IL-5, by recently established collections of T cells as in hepatic granulomae evoked in rodents by schistosomal antigens (46). Third, T-cell dependent production of IgG and, at early times, IgM also is inhibited more than 90% by VIP (47), that concurrently enhances generation of IgA and IgE (47,48). A combination of VIP and monoclonal anti-CD40 antibody induced alpha-germ-line transcripts and IgA production by IgD+ human B cells, as well as switch circular DNA directly, indicative of isotype switching from mu to alpha (49). The integrated, usually immunosuppressive effects of VIP are apparent in some model compartmental responses to antigen, such as the anterior chamber of the eye, where VIP is released at nanomolar levels by antigen and/or specific cytokines (50). Absorption or antagonism of VIP in aqueous humor demonstrated its major role in suppressing delayed-type hypersensitivity and IFN-γ production (51). The capacity of VIP to stimulate T-cell migration through tissues and concurrently prevent immune activation appears to be mediated principally by type II VIP receptors (VIPR2), as will be described in the next section. Recent results suggest that VIP may have opposite effects on subsets of T cells expressing predominantly type I VIP receptors, such as inhibition of migration and initiation of secretion of IL-2 and IL-5, as will be discussed next and as has been observed on occasion in analyses of hybridomas and selected populations of rodent native T cells (52,53).

As for VIP, SP is a potent mediator of many physiological responses of intestinal and airway smooth muscles, arteries and arterioles, and secretory epithelial cells and glands that mimic inflammatory reactions and facilitate development of compartmental immune responses by enhancing regional delivery of immune proteins and cells. The immune effects of SP often differ from those of VIP in target cellular specificity, direction, and/or magnitude. An example of the first is the stimulation of vascular adherence and tissue migration of mononuclear phagocytes by SP and of T cells by VIP. The second is illustrated by augmentation of T-cell proliferative responses and of NK cell function by SP and suppression of both by VIP. The greater degree of stimulation of IgA production by VIP than SP in most systems is characteristic of the last type of difference. These and other immune effects of SP, and the critical regulatory roles of peptidases in determining the effective level of SP and, to a much lesser extent, VIP, have been reviewed recently (24,30,65,66).

IV. Immune Cellular Receptors for SP and VIP

VIP and PACAP are bound with nearly the same affinity by two receptors (Rs) with seven transmembrane domains, which are designated the type I VIPR (VIPR1) or type II PACAPR and the type II VIPR (VIPR2) or type III PACAPR (Table 1). VIPR1 and VIPR2 are members of a distinct family of G-protein-associated Rs that also includes Rs for other large peptides, such as secretin, calcitonin, and parathyroid hormone. Rs of the VIP/PACAP family have only 15% or less homology with other members of the superfamily of G-protein-associated Rs and lack the signature sequences of the β-adrenergic R family, but exhibit up to 50% homology between family members, which all have long extracellular amino-terminal domains with eight conserved cysteines (1,54–57). Although the amino acid (aa) sequences of the 457-aa VIPR1 and the 437-aa VIPR2 of humans are only 49% identical (Table 1), the expression of either type alone by transfectants and cultured cell lines results in VIP binding of the same specificity, as defined by competition with related peptides (57–59). The affinity of binding of VIP is approximately 10-fold higher for VIPR 1 than for VIPR2, but at these respective concentrations both VIPR1 and VIPR2 signal increases in the intracellular concentrations of cyclic AMP ($[cAMP]_i$) and Ca^{2+} ($[Ca^{2+}]_i$), which are similar in magnitude (Table 1). To date, a series of novel peptide analogs of VIP are the only ligands known to usefully distinguish between VIPR1 and VIPR2 at the levels of binding and biochemical signals.

Recent analyses of the structural determinants of binding and signal transduction by VIPR1 have demonstrated critical requirements for both the large amino-terminal domain (60) and the carboxyl-terminal cytoplasmic tail (61), which are not typical of other Rs of the G-protein-associated superfamily. The dependence of functions of the closely related secretin R on the first extracellular loop was not shared by the VIPR1 (60). Preliminary studies of G-protein coupling, which also have so far been limited to the VIPR1, have shown a dependence of high-affinity binding and cAMP responses to VIP on coupling to Gs (62). There have been no studies of the structural determinants of VIPR2 expression, signal transduction, or effects on cellular functions. SP is bound with high affinity by the G-protein-associated neurokinin receptors designated NK-1, NK-2, and NK-3, of which NK-1 has the greatest specificity for SP (63). NK-1 Rs have been identified on macro-

Table 1 Functional Properties of Human Receptors for VIP and PACAP

Receptor type	Binding of VIP (K_d)	Responses to VIP of:		AP-2	SP1	CREB	Transcriptional signaling by:[a]		
		$[cAMP]_i$	$[Ca^{2+}]_i$				NF-IL-6	STAT 1	STAT 5
		EC_{50}		Electrophoretic mobility shift assays					
VIPR1	0.2 nM[b]	0.3 nM	2.5 nM	−[c]	−	+	ND	−	−
VIPR2	19 nM	30 nM	27 nM	+	+	+	+	+	+

[a]These are the results of studies of HUT78 (VIPR1) and Tsup-1 (VIPR2) T cells.
[b]Each value is the mean of results of 3–7 studies of 293 cell transfectants.
[c]+ = increase in level of nuclear transcriptional regulatory factor (TRF) by VIP; − = decrease in stimulated level of TRF by VIP; ND = not done.

phages, B cells, and T cells that all respond functionally to SP (64,65), as do many different types of non-immune lung cells (66). The T-cell subset specificity of effects of SP and the biochemical mechanisms of signal transduction to immune cells are largely unknown.

The principal differences between VIPR1 and VIPR2 already elucidated in the immune system are cellular and tissue distribution, and regulation of specific effector functions. Recent applications of nucleic acid probes and type-specific anti-VIPR antibodies have delineated the expression of VIPR1 and VIPR2 on T cells and other immune cells (7,14,34,52,57). Platelets, some types of mast cells, some sets of B cells, and some T cells in the spleen bear VIPR1 exclusively, and some CD4[+]8[+] thymocytes and T cells of early granulomae have only VIPR2, but much more commonly T cells and other immune cells co-express VIPR1 and VIPR2. Human and mouse lines of T cells have been identified that express solely one or the other type of VIPR, as exemplified by human CD4[+]8[+]3[low] Tsup-1 cells and human PEER T cells, which have only VIPR2, and human HUT78 T cells, which have only VIPR1 (52,57). For murine splenic T cells, the fine specificity of regulation of expression of the two types of VIPRs extends to subsets of helper T cells, where VIPR1 mRNA predominates in Th1 cells and VIPR2 on Th2 cells. In some defined populations or cultured lines of T cells, VIPR2 mediates increased adherence to connective tissue proteins and endothelial cells, chemotaxis, and enhanced secretion of MMPs, which together are necessary for movement through and localization in tissues. In these T cells, VIPR2 concurrently mediates suppression of aspects of T-cell activation, such as proliferation and generation of cytokines, which could have deleterious effects on normal tissues in the path of migrating T cells. In other T cells, the predominant VIPR1 has little stimulatory effect on functions required for recruitment of T cells to tissues, but instead mediates inhibition of adherence, chemotaxis, and MMP secretion evoked by other stimuli and may augment production of some cytokines. The resultant hypothesis is that T cells responding to VIP and other regional stimuli first express VIPR2s, which mediate migration and localization without immune activation, as will be described. It is further postulated that after T cells have localized in a tissue compartment, they then express more VIPR1s than VIPR2s, and the former transduce inhibition of further migration as well as enhancement of production of some cytokines. The abundance of cytokine-regulated elements in the 5′-flanking regions of the VIPR1 and VIPR2 genes suggest that cytokines also are a major determinant of the ratio of VIPR1 to VIPR2 on T cells.

V. VIPR Type-Specific Determinants of the Immune Effects of VIP

At the concentrations delivered neurally to primary immune organs and lymphoid follicles in nonimmune organs, VIP has major effects on T-cell migration into tissues, homing to gastrointestinal and other lymphoid tissues, and production of cytokines, as well as on immunoglobulin synthesis and NK cell activity (1,19–21,25). Of these

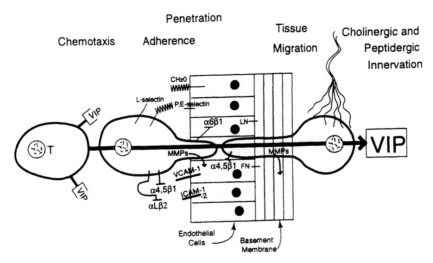

Figure 1 Enhancement of T-cell migration through vascular basement membranes by VIP. FN = fibronectin, LN = laminin, CH_2O = carbohydrate ligands for selectins, T = T cell, alpha/beta = designation of different integrins, VCAM-1 = type 1 vascular cell adhesion molecule, ICAM-1/-2 = types 1 and 2 intercellular adhesion molecules, MMP = matrix metalloprotease, thin arrow = release by T cells of MMPs that adhere to T-cell surface or enter extracellular fluid, thick arrow = path of T-cell chemotaxis.

effects of VIP, regulation of T-cell migration has been most definitively evaluated in relation to the two types of VIPRs. VIPR2 mediates enhancement of expression of selectins and of active beta-1 integrins on T cells, which adhere to VCAM-1 and then fibronectin (41–43). Such adherent T cells become more susceptible to the MMP-stimulating and chemotactic effects of VIP, which also are transduced by VIPR2 predominating on the responding T cells (Fig. 1). The migrating T cells form clusters of two to six, secrete and bind active MMP-9 and MMP-2 on their surface, degrade matrix proteins of the basement membrane, and migrate from blood vessels into the tissues with an elevated concentration of VIP (39,40). VIPR2 concurrently mediates suppression of T-cell proliferative and cytokine synthetic responses to antigens (44, 45,47,50,51), which would reduce any immune impact on tissues in the pathway of T-cell transmigration (Fig. 2). Once the T cells reside in established infiltrates and corridors of immune organs, VIPR1 predominates and mediates some effects of VIP opposite to those transduced by VIPR2, including inhibition of MMP secretion and chemotactic responses evoked by diverse stimuli (52) (Fig. 2). In this setting VIP would diminish the recruitment of lymphoid T cells to other tissues. The present tentative hypothesis thus suggests that VIP facilitates T-cell recruitment through VIPR2 and stabilization of T cells in lymphoid follicles through VIPR1.

The roles of the two types of VIPRs in regulating other immune functions have been explored only partially in preliminary investigations (Table 2). The inhibition of

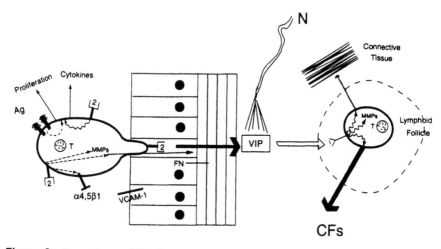

Figure 2 Dependence of T-cell responses to VIP on the predominantly expressed type of VIP receptor. T = T cell, 1 = VIPR1, 2 = VIPR2, MMPs = matrix metalloproteinases, N = cholinergic or peptidergic neuron, CFs = chemotactic factors such as interleukins or chemokines, Ag = antigen bound to T-cell receptor, alpha/beta = designation of integrins, FN = fibronectin, VCAM-1 = type 1 vascular cell adhesion molecule, narrow solid arrow = functional response or secretion of protein, wide solid arrow = chemotactic response, reversed wide open arrow = lack of chemotactic response to VIP, dashed arrow = stimulation or activation, curly arrow = inhibition.

Table 2 Receptor Mediation of Lymphocyte Responses to SP and VIP

		NK-1 R	VIPR1	VIPR2
Immune cell expression		M > B ≥ T	M = T > B	M = T > B
Tissue migration	CT	M+	T−	T+
	MMPs	ND	T−	T+
Adherence	BM	M+	ND	T+
	Ls	±	B+	ND
Proliferation (mitogen/ag)		T+		T−
Ig production		IgA+		IgG−
			IgA+	IgM−
				IgE+
Cytokine secretion		IL-1+, IL-2+, IL-6+, TNF-α+, IFN-γ+		IL-2-, IL-4-, IL-10-, IL-5+
NK cell activity		+		−

M = macrophage, B = B lymphocyte, T = T lymphocyte, CT = chemotaxis, MMP = matrix metalloproteinase, BM = basement membrane, L = lymphocyte, + = stimulation, − = inhibition, ND = not done. The VIPR subset specificity has not been determined for VIP direct enhancement of IgA synthesis by B cells.

T-cell proliferative responses to mitogens and antigens by VIP, which is most prominent in T-helper cells, appears to be mediated by VIPR2 based on results of limited studies of human blood-derived T cells. In contrast, NK-1 Rs transduce SP enhancement of T-cell proliferative responses to the same primary stimuli. The striking hibitory effects of VIP on T-cell dependent mitogen-stimulated production of IgG and, in some circumstances, IgM appear to be principally mediated by VIPR2. It is presumed that VIPR2 also transduces the enhancing effect of VIP on IL4-induced T-cell-dependent production of IgE in the same experimental system. However, it is not known which VIPR transduces the direct stimulation of B-cell production of IgA, which resembles the NK-1 R-mediated effect of SP on IgA synthesis. The effects of VIP on cytokine production by human cultured lines of T cells and murine T cells isolated from subacutely elicited hepatic granulomae are mediated predominantly by VIPR2 (Table 2). Although VIPR2 transduces profound inhibition of generation of IL-2, IL-4, and IL-10, IL-5 secretion by T cells from murine granulomas is enhanced consistently. In contrast, the NK-1 R mediates enhancement of T-cell production of IL-2, IL-6, and IFN-γ as well as macrophage generation of IL-1 and TNF-α. The inhibitory effect of VIP on NK cell activity is presumed to be mediated by the VIPR2 and also is the opposite of SP enhancement mediated by the NK-1 R.

VI. Appearance and Activities of Neuropeptides and Neuropeptide Receptors in Pulmonary Parenchymal Immune Responses

Many unmyelinated C fibers of sensory nerves in mammalian upper and lower respiratory structures, including the pulmonary parenchyma, contain SP, along with NKs and CGRP, which are released into airway secretions after physical or chemical stimulation. Efferent parasympathetic nerves of the airways and, to a lesser degree, pulmonary parenchyma contain VIP, PHI/M, and PACAP, which also are released by a range of physiological stimuli. Until very recently, determination of respiratory tissue distribution of neuropeptide receptors was dependent on the results of auto-radiographic analyses of the localization of bound ligands. SP is known to be capable of dilating and increasing permeability of microvasculature in the respiratory tract, enhancing adhesion of leukocytes to pulmonary vascular endothelial cells, evoking bronchial smooth muscle contraction and cough, and stimulating epithelial cell and glandular secretions and mucociliary clearance (66). VIP in the lungs also has the capacity to dilate and enhance permeability of respiratory microvasculature, and stimulate secretions, but mediates some effects opposite to those of SP and most neurokinins, such as being a potent bronchodilator. Some neuropeptides released in lung tissues lack the effects observed in other tissues, as for example the usually highly potent CGRP, which elicits only marginal vasodilation in lung tissues, and SP, which activates mast cells of human skin but not those from the lungs (66).

The finding of immunologically relevant concentrations of VIP and SP in broncho-alveolar lavage (BAL) fluid after intratracheal antigen challenge of sensitized mice provided an opportunity to evaluate their roles in mobilizing and regulating T-cell and some other immune responses in the pulmonary parenchymal compartment (24,30,31). The BAL fluid concentration of SP increased at 1 day after antigen challenge to a peak that was up to sixfold higher than that prior to the challenge, at the same time as the BAL fluid concentrations of IL-2 and IFN-γ were maximal. At 2–4 days after antigen challenge, the BAL fluid concentration of VIP also increased to a peak up to sixfold higher than the baseline level, concurrent with the increases in BAL fluid concentrations of IL-6 and then IL-4. The maximal increases in number of lymphocytes and alveolar macrophages in BAL fluid, which reflected the composition of pulmonary parenchymal infiltrates, were observed at 4–6 days after intratracheal antigen challenge (31). PCR radioamplification of neuropeptide receptor mRNA in BAL fluid leukocytes, normalized to the level of a constitutively expressed mRNA, revealed significant increases of up to 10-fold in those encoding NK-1 R, VIPR1, and VIPR2 in lymphocytes on days 2, 4, and 4, respectively (31). Significant increases of up to 12.5-fold in the relative levels of mRNAs encoding NK-1 Rs and VIPR1s, but not VIPR2s, were observed in alveolar macrophages during the same interval. Immunohistochemical studies of parenchymal tissues taken during the same period, after antigen challenge, revealed expression of NK-1 Rs, VIPR1s, and VIPR2s by a high percentage of mononuclear leukocytes and granulocytes (31). Each receptor was found on most airway epithelial cells, and on some alveolar macrophages and vascular smooth muscle cells prior to antigen challenge. After antigen challenge, Rs of airway epithelial cells disappeared rapidly, followed by the appear-

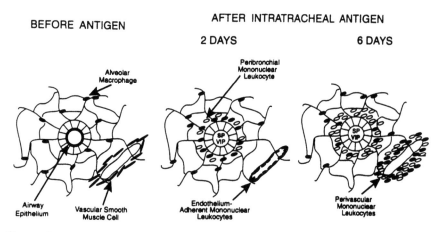

Figure 3   Course of appearance of SP and VIP, and expression of NK-1 R and VIPRs in pulmonary parenchymal immune responses. Dark cell = site of expression of one or more neuropeptide receptors. The broad dark ring on the inner surface of airways prior to antigen challenge represents a high density of VIP and NK-1 Rs, which almost completely disappear after intratracheal administration of antigen. Heavy lettering for SP at 2 days and VIP at 6 days indicate their respective higher concentrations than the other neuropeptide at each time.

ance on day 2 of R-bearing granulocytes and mononuclear leukocytes adherent to vascular endothelial cells and localized in peribronchial infiltrates (Fig. 3). By day 6, the level of expression of Rs had decreased in peribronchial infiltrates and increased in perivascular accumulations.

To begin to define the functional consequences of increases in the parenchymal and airway fluid concentrations of the neuropeptides and in the levels of expression of the respective receptors, a specific nonpeptide NK-1 R antagonist was administered intraperitoneally to mice once daily from 3 days before to 5 days after antigen challenge. The BAL fluid levels of lymphocytes and granulocytes, but not macrophages, were decreased significantly on day 5 by this regimen of NK-1 R antagonist. Thus the early peak of SP appears to contribute significantly to the antigen-induced mobilization of some types of leukocytes and possibly their compartmental immunological contributions.

VII. Conclusions and Future Directions

Pulmonary immune responses to inhaled antigens that are of sufficient magnitude to release neurostimulatory levels of cytokines evoke release of SP, VIP, and some other neuropeptides by peptidergic and cholinergic neurons. Although the significance of differences in maximal concentrations and time courses of release of each of the neuropeptides is not known, it is clear that the macrophages and T cells infiltrating parenchymal tissues after intratracheal antigen challenge express receptors for several neuropeptides at far higher frequencies than in blood and many lymphoid organs. In one murine model of pulmonary immunity and airway hyperreactivity, specific pharmacological antagonism of the NK-1 R for SP during times of maximal concentration of released SP significantly blunted the influx of T cells and eosinophils, but not of macrophages.

At least three additional types of evidence will be required to establish primary roles for SP and VIP in pulmonary immune responses. The first is more detailed delineation of the respective time courses of appearance and the relative maximal peaks of expression of the NK-1 R, VIPR1, and VIPR2 on macrophages and each subset of T cells during immune responses. This will require application of in-situ hybridization or in-situ PCR together with immune cell-defining imunohistochemistry. In parallel, the principal cytokines that regulate expression of each of the Rs will have to be identified from the results of studies of changes in T cell and macrophage R mRNA, protein, and functions after in-vitro exposure to individual interleukins and interferons. The second required line of investigation will consist of challenge of mouse lungs in vivo with a range of concentrations of stable R subtype-selective agonists of VIP, without and with antigen challenge and NK-1 R blockade, before, at, and after the times of expected maximal pulmonary concentrations of endogenous VIP. The third type of study requires separate disruption of the neuropeptide receptor genes encoding NK-1 R, VIPR1, and VIPR2, with specific targeting to different immune cells and components of lung tissues. If these additional results confirm a critical role for one or more neuropeptides, then it will be useful to begin to develop therapeutic applications for agents that alter specifically one or more distinct mechanisms of neuroregulation of pulmonary immune responses.

Acknowledgments

This work was supported by grants AI 29912 (EJG), DK 44876 (SPS), and AI 34570 (Edward J. Goetzl, H. Benfer Kaltreider, and Sunil P. Sreedharan) from the National Institutes of Health.

References

1. Goetzl EJ, Sreedharan SP. Mediators of communication and adaptation in the neuroendocrine and immune systems. FASEB J 1992; 6:2646–2652.
2. Steinman L. Connections between the immune system and the nervous system. Proc Natl Acad Sci (USA) 1993; 90:7912–7914.
3. Lotan M, Schwartz M. Cross talk between the immune system and the nervous system in response to injury: Implications for regeneration. FASEB J 1994; 8:1026–1033.
4. Merrill JE, Jonakait GM. Interactions of the nervous and immune systems in development, normal brain homeostasis, and disease. FASEB J 1995; 9:611–618.
5. Felten DL, Felten SY, Bellinger DL, Carlson SL, Ackerman KD, Madden KS, Olschowka JA, Livnat S. Noradrenergic sympathetic neural interactions with the immune system: structure and function. Immunol Rev 1987; 100:225–260.
6. Fink T, Weihe E. Multiple neuropeptides in nerves supplying mammalian lymph nodes: messenger candidates for sensory and autonomic neuroimmunomodulation. Neurosci Lett 1988; 90:39–54.
7. Ichikawa S, Sreedharan SP, Goetzl EJ, Owen RL. Immunohistochemical localization of peptidergic nerve fibers and neuropeptide receptors in Peyer's patches of the cat ileum. Regulatory Peptides 1994; 54:385–395.
8. Goetzl EJ, Adelman DC, Sreedharan SP. Neuroimmunology. Adv Immunol 1990; 48: 161–190.
9. Patterson PH. Leukemia inhibitory factor, a cytokine at the interface between neurobiology and immunology. Proc Natl Acad Sci (USA) 1994; 91:7833–7835.
10. Benveniste EN, Benos DJ. TNF-alpha- and IFN-gamma-mediated signal transduction pathways: effects on glial cell gene expression and function. FASEB J 1995; 9:1577–1584.
11. Stead RH, Tomioka M, Quinonez G, Simon GT, Felten SY, Bienenstock J. Intestinal mucosal mast cells in normal and nematode-infected rat intestines are in intimate contact with peptidergic nerves. Proc Natl Acad Sci (USA) 1987; 84:2975–2979.
12. Felten SY, Olschowka JA. Noradrenergic sympathetic innervation of the spleen. II. Tyrosine hydroxylase (TH)-positive nerve terminals form synaptic-like contacts on lymphocytes in the splenic white pulp. J Neurosci Res 1987; 18:37–48.
13. Stanisz AM, Scicchitano R, Dazin P, Bienenstock J, Payan DG. Distribution of substance P receptors on murine spleen and Peyer's patch T and B cells. J Immunol 1987; 139: 749–754.

14. Ichikawa S, Sreedharan SP, Owen RL, Goetzl EJ. Immunochemical localization of type I VIP receptor and NK-1-type substance P receptor in rat lung. Am J Physiol 1995; 268 (Lung Cell Mol Physiol 12):L584–L588.

15. Donnerer J, Eglezos A, Hemle RD. Neuroendocrine and immune function in the capsaicin-treated rat: evidence for afferent neural modulation in vivo. In: Freier S, ed. The Neuroendocrine-Immune Network. Boca Raton, FL: CRC Press, 1991:70–85.

16. Ackerman KD, Madden KS, Livnat S, Felten SY, Felten DL. Neonatal sympathetic denervation alters the development of in vitro spleen cell proliferation and differentiation. Brain, Behavior, and Immunity 1991; 5:235–261.

17. Weihe E, Hohr D, Michael S. Molecular anatomy of the neuroimmune connection. Int J Neurosci 1991; 59:1–23.

18. Gomariz RP, Delgado M, Naranjo JR, Mellstrom B, Tormo A, Mata F, Leceta J. VIP gene expression in rat thymus and spleen. Brain, Behavior, and Immunity 1993; 7:271–278.

19. Ottaway CA. In vitro alteration of receptors for vasoactive intestinal peptide changes the in vivo localization of mouse T cells. J Exp Med 1984; 160:1054–1069.

20. Ohkubo N, Miura S, Serizawa H, Yan HJ, Kimura H, Imaeda H, Tashiro H, Tsuchiya M. In vivo effect of chronic administration of vasoactive intestinal peptide on gut-associated lymphoid tissues in rats. Regulatory Peptides 1994; 50:127–135.

21. Ottaway CA. Neuroimmunomodulation in the intestinal mucosa. Gastroenterol Clin NA 1991; 20:511–529.

22. Barnes PJ, Baraniuk JN, Belvisi MG. Neuropeptides in the respiratory tract. Part I. Am Rev Respir Dis 1991; 144:1187–1198.

23. Luts A, Uddman R, Alm P, Basterra J, Sundler F. Peptide-containing nerve fibers in human airways: distribution and coexistence pattern. Int Arch Allergy Immunol 1993; 101:52–60.

24. Goetzl EJ, Sreedharan SP, Kaltreider HB. Neuroregulation of pulmonary immunity: the roles of substance P and vasoactive intestinal peptide. In: Kaliner MA, Barnes PJ, Kunkel GHH, Baraniuk JN, eds. Neuropeptides in Respiratory Medicine. New York: Marcel Dekker, 1994:607–616.

25. Said S. Vasoactive intestinal polypeptide in the respiratory tract. In: Kaliner MA, Barnes PJ, Kunkel GHH, Baraniuk JN, eds. Neuropeptides in Respiratory Medicine. New York: Marcel Dekker, 1994:143–160.

26. Kato I, Suzuki Y, Akabane A, Yonekura H, Tanaka O, Kondo H, Takasawa S, Yoshimoto T, Okamoto H. Transgenic mice overexpressing human vasoactive intestinal peptide (VIP) gene in pancreatic beta cells. Evidence for improved glucose tolerance and enhanced insulin secretion by VIP and PHM-27 in vivo. J Biol Chem 1994; 269:21223–21228.

27. Walker KB, Serwonska MH, Valone FH, Harkonen WS, Frick OL, Scriven KH, Ratnoff WD, Browning JG, Payan DG, Goetzl EJ. Distinctive patterns of release of neuroendocrine peptides after nasal challenge of allergic subjects with ryegrass antigen. J Clin Immunol 1990; 8:108–113.

28. Mosimann BL, White MV, Hohman RJ, Goldrich MS, Kaulbach HC, Kaliner MA. Substance P, calcitonin gene-related peptide, and vasoactive intestinal peptide increase in nasal secretions after allergen challenge in atopic patients. J Allergy Clin Immunol 1993; 92:95–104.

29. Nieber K, Baumgarten CR, Rathsack R, Furkert J, Oehme P, Kunkel G. Substance P and beta-endorphin-like immunoreactivity in lavage fluid of subjects with and without allergic asthma. J Allergy Clin Immunol 1992; 90:646–654.

30. Goetzl EJ, Ichikawa S, Ingram DA, Kishiyama JL, Xia M, Sreedharan SP, Byrd PK, Kaltreider HB. Neuropeptide regulation of pulmonary parenchymal immune responses. In: Basomba A, Sastre J, eds. Proceedings XVI European Congress of Allergology and Clinical Immunology. Bologna: Monduzzi Editore, 1995:161–167.

31. Kaltreider HB, Ichikawa S, Byrd PK, Ingram DA, Kishiyama JL, Sreedharan SP, Warnock ML, Beck JM, Goetzl EJ. Upregulation of neuropeptides and neuropeptide receptors in a murine model of immune inflammation in lung parenchyma. Am J Respir Cell Mol Biol 1997; 16:133–144.

32. Wallengren J, Moller H, Ekman R. Occurrence of substance P, vasoactive intestinal peptide, and calcitonin gene-related peptide in dermographism and cold urticaria. Arch Dermatol Forsch 1987; 279:512–515.

33. Tobin D, Nabarro G, Baart de la Faille H, van Vloten WA, van der Putte SC, Schuurman HJ. Increased number of immunoreactive nerve fibers in atopic dermatitis. J Allergy Clin Immunol 1992; 90:613–622.

34. Roy Choudhury A, Goetzl EJ, Xia M, Sreedharan SP, Furuta GT, Morys-Wortmann C, Galli SJ, Schmidt WE, Wershil BK. Mouse mast cells grown in stem cell factor express only type I vasoactive intestinal peptide. FASEB J. In press.

35. Litwin DK, Wilson AK, Said SI. Vasoactive intestinal polypeptide (VIP) inhibits rat alveolar macrophage phagocytosis and chemotaxis in vitro. Regulatory Peptides 1992; 40:63–74.

36. Sakakibara H, Shima K, Said SI. Characterization of vasoactive intestinal peptide receptors on rat alveolar macrophages. Am J Physiol 1994; 267:L256–L262.

37. Goetzl EJ, Sreedharan SP. Lymphocyte recognition and effects of neuropeptides: a determinant of compartmental immunity. In: Bloom FE, Campbell IL, Mucke L, eds. Discussions in Neuroscience. Amsterdam: Elsevier, 1993:IX, 97–101.

38. Goetzl EJ, Xia M, Ingram DA, Kishiyama JL, Kaltreider HB, Byrd PK, Ichikawa S, Sreedharan SP. Neuropeptide signaling of lymphocytes in immunological responses. Int Arch Allergy Immunol 1995; 107:202–204.

39. Xia M, Leppert D, Hauser SL, Sreedharan SP, Nelson PJ, Krensky AM, Goetzl EJ. Stimulus-specificity of matrix metalloproteinase-dependence of human T cell migration through a model basement membrane. J Immunol 1996; 156:160–167.

40. Goetzl EJ, Banda MJ, Leppert D. Matrix metalloproteinases in immunity. J Immunol 1996; 156:1–4.

41. Xia M, Sreedharan SP, Dazin P, Damsky CH, Goetzl EJ. Integrin-dependent induction of human T cell matrix metalloproteinase activity in chemotaxis through a model basement membrane. J Cell Biochem 1996; 61:452–458.

42. Johnston JA, Taub DD, Lloyd AR, Conlon K, Oppenheim JJ, Kevlin DJ. Human T lymphocyte chemotaxis and adhesion induced by vasoactive intestinal peptide. J Immunol 1994; 153:1762–1768.

43. Smith CH, Barker JN, Morris RW, MacDonald DM, Lee TH. Neuropeptides induce rapid expression of endothelial cell adhesion molecules and elicit granulocytic infiltration in human skin. J Immunol 1993; 151:3274–3282.

44. Ganea D, Sun L. VIP downregulates the expression of IL-2 but not of IFN-gamma from stimulated murine T lymphocytes. J Neuroimmunol 1993; 47:147–155.

45. Sun L, Ganea D. VIP inhibits IL-2 and IL-4 production through different mechanisms in T cells activated via the TCR/CD3 complex. J Neuroimmunol 1993; 48:59–66.

46. Metwali A, Elliott D, Blum AM, Li J, Sandor M, Weinstock JV. T cell vasoactive

intestinal peptide receptor subtype expression differs between granulomas and spleen of Schistosome-infected mice. J Immunol 1996; 157:265–270.

47. Hassner A, Lau MS, Goetzl EJ, Adelman DC. Isotype-specific regulation of human lymphocyte production of immunoglobulins by sustained exposure to vasoactive intestinal peptide. J Allergy Clin Immunol 1993; 92:891–901.

48. Boirivant M, Fais S, Annibale B, Agostini D, Delle Fave G, Pallone F. Vasoactive intestinal polypeptide modulates the in vitro immunoglobulin A production by intestinal lamina propria lymphocytes. Gastroenterology 1994; 106:576–582.

49. Fujieda S, Waschek JA, Zhang K, Saxon A. Vasoactive intestinal peptide induces Salpha/Smu switch circular DNA in human B cells. J Clin Invest 1996; 98:1527–1532.

50. Taylor AW, Streilein JW, Cousins SW. Immunoreactive vasoactive intestinal peptide contributes to the immunosuppressive activity of normal aqueous humor. J Immunol 1994; 153:1080–1086.

51. Ferguson TA, Fletcher S, Herndon J, Griffith TS. Neuropeptides modulate immune deviation induced via the anterior chamber of the eye. J Immunol 1956; 155:1746–1756.

52. Goetzl EJ, Gaufo G, Sreedharan SP, Xia M. Transduction of specific inhibition of HuT78 human T cell chemotaxis by type I vasoactive intestinal peptide receptors. J Immunol 1996; 157:1132–1138.

53. Mathew RC, Cook GA, Blum AM, Metwali A, Felman R, Weinstock JV. VIP stimulates T lymphocytes to release IL-5 in murine *Schistosomiasis mansoni* infection. J Immunol 1992; 148:3572–3580.

54. Ishihara T, Shigemoto R, Mori K, Takahashi K, Nagata S. Functional expression and tissue distribution of a novel receptor for vasoactive intestinal polypeptide. Neuron 1992; 8:811–819.

55. Lutz EM, Sheward WJ, West KM, Morrow JA, Fink G, Harmar AJ. The VIP2 receptor: molecular characterisation of a cDNA encoding a novel receptor for vasoactive intestinal peptide. FEBS Lett 1993; 334:3–8.

56. Inagaki N, Yoshida H, Mizuta M, Mizuno N, Fujii Y, Gonoi T, Miyazaki J, Seino S. Cloning and functional characterization of a third pituitary adenylate cyclase-activating polypeptide receptor subtype expressed in insulin-secreting cells. Proc Natl Acad Sci (USA) 1994; 91:2679–2683.

57. Xia M, Sreedharan SP, Goetzl EJ. Predominant expression of type II vasoactive intestinal peptide receptors by human T lymphoblastoma cells: Transduction of both Ca^{2+} and cAMP signals. J Clin Immunol 1996; 16:21–30.

58. Sreedharan SP, Patel DR, Huang J-X, Goetzl EJ. Cloning and functional expression of a human neuroendocrine vasoactive intestinal peptide receptor. Biochem Biophys Res Commun 1993; 193:546–553.

59. Sreedharan SP, Patel DR, Xia M, Ichikawa S, Goetzl EJ. Human vasoactive intestinal peptide 1 receptors expressed by stable transfectants couple to two distinct signaling pathways. Biochem Biophys Res Commun 1994; 203:141–148.

60. Holtmann MH, Hadac EM, Miller LJ. Critical contributions of amino-terminal extracellular domains in agonist binding and activation of secretin and vasoactive intestinal polypeptide receptors. Studies of chimeric receptors. J Biol Chem 1995; 270:14394–14398.

61. Mu Y, Xia M, Kong Y, Goetzl EJ, Sreedharan SP. Mediation of calcium signals by type I and II vasoactive intestinal peptide receptors in stable transfectants. Biochem Biophys Res Commun. In press.

62. Goetzl EJ, Kishiyama JL, Shames RS, Liu Y-F, Albert PR, An S, Birke FW, Yang J, Sreedharan SP. Specific inhibition of receptor-dependent human cellular responses by antisense mRNA depletion of individual G-proteins. Trans Assoc Am Physicians 1993; 56:69–76.
63. Nakanishi S. Mammalian tachykinin receptors. Annu Rev Neurosci 1991; 14:123–136.
64. Payan DG, Brewster DR, Goetzl EJ. Substance P recognition by a subset of human T lymphocytes. J Clin Invest 1984; 74:1532–1539.
65. Goetzl EJ, Turck CW, Sreedharan SP. Production and recognition of neuropeptides by cells of the immune system. In: Ader R, Felten DL, Cohen N, eds. Psychoneuro-immunology. 2d ed. New York: Academic Press, 1991:263–280.
66. Piedimonte G. Tachykinin peptides, receptors, and peptidases in airway disease. Exp Lung Res 1995; 21:809–834.

14

Role of Vasoactive Intestinal Peptide in Myocardial Ischemia Reperfusion Injury

**DIPAK K. DAS and
NILANJANA MAULIK**

University of Connecticut School of Medicine
Farmington, Connecticut

RICHARD M. ENGELMAN

Baystate Medical Center
Springfield, Massachusetts

I. Introduction

Vasoactive intestinal peptide (VIP) is a 28-amino acid neuropeptide with diverse physiological properties. Originally isolated from porcine upper intestine (1), it is now known to be present in neural and many other tissues including heart, lung, brain, colon, and genitourinary tract (2). It is abundantly present in neurons of the myenteric plexus that project into circular muscle and specialized longitudinal muscle (3). VIP also causes relaxation to the muscle strips (4). This peptide has been found to exist in the postganglionic nerve terminals in the heart (5). The origins of the VIP immunoreactive nerve fibers are primarily intrinsic, but to some extent also extrinsic (6).

The sequence originally proposed by Mutt and Said (1) for porcine VIP, His-Ser-Asp-Ala-Val-Phe-Thr-Asp-Asn-Tyr-Thr-Arg-Phe-Arg-Lys-Gln-Met-Ala-Val-Lys-Lys-Tyr-Leu-Asn-Ser-Ile-Leu-Asn-NH_2, was subsequently confirmed by many investigators. Its amino acid composition is considerably conserved regardless of its biological source. The gene for VIP has been sequenced and its chromosomal localization established.

Investigations in the last two decades have not only established an indisputable role of VIP in central neurotransmission, but also its diverse biological functions in many organs. Several properties of VIP, including its free radical-scavenging activities (7), Ca^{2+}-modulating properties (8,9), and its role in signal transduction (10,11) make it an ideal therapeutic agent for the prevention of ischemia reperfusion injury. The intention of this review is to elaborate the properties of VIP in the context of myocardial ischemia reperfusion injury, and discuss the results of a few experiments to demonstrate its cardioprotective role.

II. VIP and Cellular Protection

VIP has been found to protect mammalian organs including heart, lung, and intestine against injury resulting from reactive oxygen species and other toxins (12). The beneficial effects of VIP on lung preservation were attributed to many factors, including the free radical-scavenging activity of VIP. In a recent study on intestinal ischemia/reperfusion injury, release of VIP was noticed during reperfusion and was found to play a role in postischemic vasodilation (13).

In another study, mast cell activation in hemorrhagic shock was inhibited by VIP (14). The same authors also demonstrated that VIP increased the survival rate and protected renal tissue from reperfusion injury when used in conjunction with other therapeutic agents (15). The beneficial effect of VIP on hemorrhagic ischemia/reperfusion injury in renal tissue was attributed to the antioxidant properties of VIP.

In heart, VIP functions as a potent inotrope and coronary vasodilator. For example, in a study using dogs, intracoronary injections of VIP improved coronary blood flow and myocardial oxygen consumption (16). The authors further demonstrated that VIP had a greater coronary vasodilator/inotropic ratio than either isoproterenol or forskolin, agents that also activate the adenylate cyclase pathway.

III. VIP and Heart

Among neuropeptides that may be involved in the regulation of myocardial function, VIP probably possesses the most marked influence on coronary circulation in the regulation of vascular tone (17). This peptide is located in various functionally important regions of the heart as well as in the perivascular neuroplexus of the coronary vessels (18). Coronary arteries, atrial myocardium, and nodal regions of the heart contain the highest density of VIP-immunoreactive cardiac nerve fibers (19). VIP relaxes isolated coronary arteries and increases coronary blood flow in vivo (20). VIP possesses positive inotropic effects and improves left ventricular function (21). In a study using patients suffering from heart disease, serum VIP levels were found to be significantly increased in the patients with heart failure, but not in those with myocardial infarction or noncardiac shock (22), suggesting that VIP has a pathophysiological role in heart failure.

Coronary arteries isolated from cats relaxed after administration of VIP (10–100 nM). VIP (0.1 μM)-induced relaxation was of nearly the same order of magnitude as that of noradrenaline (1 μM)-induced relaxation (Fig. 1). However, in the

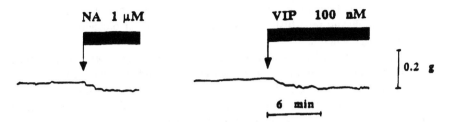

Figure 1 Mechanical activity of a coronary artery muscle strip isolated from cat heart: (left) before and after noradrenaline (NA) (1 mM); and (right) before and after VIP (0.1 mM). (Data from Ref 21, with permission.)

presence of propranolol (1 μM), noradrenaline reverted its action from inhibitory into excitatory, while VIP produced the same relaxation as that observed in the absence of propranolol. In the presence of propranolol, noradrenaline reverted its action from inhibitory into excitatory, while VIP produced the same relaxation as that observed in the absence of propranolol, suggesting that VIP induces relaxation of the cat coronary arteries in vitro, which is not mediated through β-adrenergic pathways. Thus, it appears that the presence of this vasoactive neuropeptide in the coronary arteries may have potential significance in the pathophysiology of coronary heart disease, especially diseases related to ischemia and reperfusion. It has been suggested that VIP has a role in maintaining a state of relaxation of the epicardial coronary arteries, and any change in the vasoactive intestinal polypeptidergic innervation may result in focal coronary spasm (6).

IV. Myocardial Ischemia and Reperfusion

Ischemia, by definition, represents a phenomenon in which tissue is deprived of either partial or total blood flow in conjunction with hypoxia. Clinically this can occur during thrombolytic therapy and angioplasty, in which blood flow is temporarily interrupted (23). Prolonged myocardial ischemia can occur during open-heart surgery, when the heart must be arrested in order to permit surgical manipulations (24,25). Thus, ischemic insult may occur regionally or globally, either for a short period of time or for relatively longer periods, depending on the clinical setting. While short-term ischemia (usually less than 20 min) causes reversible tissue injury which may be completely restored, prolonged myocardial ischemia (more than 30 min) can lead to irreversible cellular injury, resulting from energy depletion, lactic acidosis, and inhibition of both glucose and fatty acid utilization (26,27).

Reperfusion of previously ischemic myocardium has been shown to be associated with exacerbation of ischemic injury as judged by morphological, physiological, and biochemical criteria. Reperfusion occasionally potentiates release of intracellular enzymes, reduction of contractile functions, influx of Ca^{2+}, and disruption of cell membranes, which either alone or in combination result in ultimate cell death (28–

32). These sequela of events are known as *reperfusion injury*, because these changes specially occur during reperfusion rather than as a result of biochemical changes occurring during ischemia.

The term *myocardial stunning* is generally used to define short-term ischemia. However, myocardial stunning is characterized by a number of adverse functional events which are not immediately restored upon reperfusion. Complete recovery of myocardial contractile functions often takes hours or even days after myocardial stunning (33). Thus, the beneficial effects of reperfusion of ischemic tissue are often blunted, because revascularization while terminating ischemia simultaneously can cause additional damage to the already jeopardized cells.

V. Pathophysiology of Ischemia Reperfusion Injury

The scientific community has witnessed the development of the concept of ischemia reperfusion injury associated with the revascularization of ischemic tissue in a relatively short time. Despite the fact that the exact mechanism of reperfusion injury remains unknown, the importance of reperfusion injury in the biological system has become increasingly recognized and appreciated in recent years. It is generally believed that several interrelated factors are responsible for reperfusion injury; among them three are considered to be major: (a) free radical generation, (b) intracellular Ca^{2+} overloading, and (c) loss of membrane phospholipids.

A. Role of Free Radicals

Evidence in the literature supports that a large quantity of oxygen free radicals are generated in the postischemic heart upon reperfusion. They have been detected directly by HPLC using electrochemical detection technique or by ESR spectroscopy as well as indirectly by the formation of malonaldehyde and conjugated dienes (34). Numerous reports from many laboratories including our own have demonstrated that reperfusion of ischemic myocardium leads to the accumulation of free arachidonic acid, presumably resulting from the breakdown of membrane phospholipids (35,36). Arachidonic acid is generally broken down through the cyclooxygenase pathway to produce prostaglandins (PGs) and thromboxanes (Tx), and production of PGs can also occur in conjunction with the reperfusion of the ischemic heart (37). Arachidonic acid is a substrate for prostaglandin synthase present in heart, a critical enzyme complex serving two enzyme functions; the first, cyclooxygenase, adds molecular O_2 to the fatty acid and cyclizes it to the cyclic endoperoxides, PGG_2. This initial conversion of arachidonate into PGG_2 is associated with the production of a carbon-centered radical associated with the enzyme (38). PGG_2 is then acted upon by the second function of the enzyme, the peroxidase activity, thereby converting PGG_2 to PGH_2, with the co-production of unstable and toxic oxygen radicals which feed back and tend to deactivate cyclooxygenase (39). A number of recent studies have indicated that superoxide dismutase (SOD) stimulates the prostacyclin production, presumably by scavenging the oxygen free radical generated during the conversion of

PGG_2 into PGH_2, thereby in turn stimulating cyclooxygenase activity (40). More recently, we have demonstrated that a specific prostacyclin synthase stimulator, ONO-3144, could block the free-radical generation, simultaneously protecting the heart from reperfusion injury (41). Furthermore, prostaglandin hydroperoxide in the presence of a suitable substrate can cause the oxidation of a large number of reducing cosubstrates, which can evoke chain reactions involving the generation of free radicals (42). Superoxide could be produced by the interaction of one of these radicals with O_2, which is introduced during reperfusion of ischemic myocardium. In addition, prostaglandin hydroperoxidase can also give rise to enzyme-centered inter-mediate radical species, which can then interact with O_2 to produce superoxide, hydrogen peroxide, superoxide anion, singlet oxygen, and peroxidase (43).

Although the development of oxidative stress in the ischemic reperfused myocardium has been confirmed, the literature is full of conflicting reports regarding the source(s) of these free radicals. In addition to the sources described above, polymorphonuclear leukocytes, mitochondrial and microsomal respiratory chains, xanthine oxidase, catecholamine autooxidation, and oxygen-carrying proteins have been pointed out as potential sources of oxygen free radicals (44–48). The precise role of free radicals in the mediation of ischemic reperfusion injury also remains unknown.

B. Role of Ca^{2+}

Nearly 25 years ago, it was first documented that myocardial reperfusion after temporary coronary artery occlusion accelerated an accumulation of tissue Ca^{2+} in the canine heart (49). There is no doubt that the Ca^{2+} which is involved in the typical manifestation of Ca^{2+} overload during reperfusion is of extracellular origin. Sarco-lemmal membrane serves as a strong permeability barrier which restricts Ca^{2+} influx against 10,000-fold concentration gradients of Ca^{2+} across the membrane. Thus, disruption of the sarcolemmal membrane permits free diffusion of Ca^{2+} into the cytosol, and therefore represents the most powerful mechanism of Ca^{2+} overload. A variety of metabolic products are known to contribute to the destructive processes of the membrane during reperfusion, including phospholipid breakdown products. For example, breakdown of membrane phospholipids generates free radicals in the ischemic reperfused myocardium, which in turn lead to Ca^{2+} infiltration into the myocytes. For example, phospholipid breakdown products, arachidonic acid and prostaglandins, can function as ionophoretic agents (50), causing intracellular Ca^{2+} overloading. The lipophilic phospholipid hydroperoxides generally possess a ligand environment capable of interaction with Ca^{2+}, inducing its transport to the myocar-dial membrane (51). In addition, the degradation of phospholipid hydroperoxides leads to formation of hydroxyl derivatives and dialkylperoxides, which are capable of promoting Ca^{2+} transport (52). Myocardial reperfusion stimulates phosphodiester-atic breakdown, and the turnover of phosphoinositides is coupled with phosphoinositol-specific phospholipase C activation (35). The product of receptor-mediated hydro-lysis of inositol-1,4,5-triphosphate (IP_3), phosphatidyl-inositol 4,5-bisphosphate, can play a role as a second messenger by promoting mobilization of intracellular Ca^{2+} (53). A recent study demonstrated that IP_3 not only releases Ca^{2+} from intracellular microsomes, but also can increase cytosolic Ca^{2+} by altering flux (54).

Generally, the infiltration of Ca^{2+} into the cell during ischemia and reperfusion is detrimental to myocardial hemodynamic and metabolic functions. The mechanism of Ca^{2+}-mediated ischemic reperfusion injury is multifaceted and complex, and is affected by altered Ca^{2+} homeostasis. The Ca^{2+} slow channel blockers, $Ca^{2+}/$ calmodulin receptor antagonists, and Na^+/Ca^{2+} exchange blockers all can provide myocardial protection to some extent (30,55–57). Intracellular Ca^{2+} that gains access during the reperfusion functions as a second messenger for signal transduction (58–61).

C. Role of Signal Transduction

Recent advances in our understanding of membrane phospholipid biochemistry have led to the conclusion that α-adrenergic receptors (specifically, α_1 subtypes) control cytosolic Ca^{2+} by stimulating hydrolysis of highly phosphorylated inositol phospholipids (62). One product of this hydrolysis, inositol-1,4,5-trisphosphate, has been shown to release Ca^{2+} sequestered in intracellular stores, primarily the endoplasmic reticulum (34); α_1 receptors thus can utilize changes in intracellular free Ca^{2+} as a primary signal transduction mechanism (63). We previously reported that phosphoinositide response is mediated by α_1-adrenoceptor stimulation but is not linked with excitation–contraction coupling in cardiac muscle (64). In addition, prazosin, an α_1-adrenoceptor antagonist, and neomycin, which binds with $PI-4,5-P_2$ and thereby prevents its enzymatic degradation (65), were found to inhibit $PI-4,5-P_2$ hydrolysis, simultaneously blocking all components of inotropic responses in the left ventricular papillary muscle (54). Very recently, we have found that IP_3 releases Ca^{2+} from sarcoplasmic reticulum, a phenomenon blocked by prazosin (66). Therefore, it is likely that inhibition of α_1-adrenergic receptors during ischemia and reperfusion may lead to the preservation of membrane phospholipids, simultaneously providing protection to the myocardium from reperfusion injury.

As described above, α_1-adrenoceptor stimulation is coupled with $PI-4,5-P_2$ hydrolysis, which generates two second messengers leading to a number of intracellular events. Besides IP_3, diglyceride (DG) is also produced, which has as a most important consequence of these events probably the activation of protein kinase C. The involvement of protein kinase C activation can be supported by the work of Lindermann et al., who demonstrated that α_1-adrenoceptor-mediated slow response in the partially depolarized rat ventricles was accompanied by phosphorylation of a 15-kDa sarcolemmal protein that appears identical to the protein phosphorylated by protein kinase C in the isolated cardiac sarcolemmal vesicles (67). During the reperfusion of ischemic myocardium, when diacylglycerol (DAG) is generated in the presence of increased cytosolic Ca^{2+}, the Ca^{2+}-activated phospholipid-dependent protein kinase C is likely to be translocated to the plasmalemmal membrane from the cytosol and then activated with formation of quaternary complex with phospholipids, DAG, and

Ca^{2+} (68). Nevertheless, several investigators have demonstrated that a significant amount of protein kinase C activity and some of its substrate proteins are present in both membrane and cytosolic fractions in the heart (69). In contrast to protein kinase C activation induced by $PI-4,5-P_2$ hydrolysis, tumor-promoting phorbol ester-induced activation of protein kinase C appears to have more diverse effects on Ca^{2+} channel activity and intracellular Ca^{2+} levels. Thus, it appears that the effects of protein kinase C activation on Ca^{2+} channel activity and on cellular Ca^{2+} levels are different depending on whether there is an increase in intracellular Ca^{2+} prior to the induction of protein kinase C or not. A study from our group demonstrated that inhibition of protein kinase C activity could reduce myocardial reperfusion injury (70).

Recently, phospholipase D has been found to be involved in the intracellular signaling in ischemic heart (71). Phospholipase D signaling, like the phenomenon of phospholipase C signaling, also occurs in the ischemic heart; but unlike phospholipase C signaling, phospholipase D signaling is actually beneficial to the ischemic heart. The results of this study also suggested that the activation of phospholipase D contributes significantly to the formation of DG, because phospholipase D-produced phosphatidic acid is dephosphorylated by phosphatidate phosphatase into DG (whereas phospholipase C activation produces DG directly). It seems likely that phosphatidic acid functions as a second messenger either indirectly, through generation of DG and subsequent protein kinase C activity, or directly, through an as yet unknown mechanism (as suggested also by recent studies in neutrophils and other cell types). If this is true (this possibility has never been explored), many unanswered questions may be resolved.

VI. VIP as Free-Radical Scavenger

VIP possesses potent antioxidant property. Although VIP does not have significant O_2^-, OH·, or H_2O_2 scavenging ability, it can inhibit, in a dose-dependent manner, the $^1O_2$-dependent 2,2,6,6-tetramethylpiperidine N-oxyl (TEMPO) formation (72). To demonstrate the ability of VIP to quench $^1O_2$, the free radical was generated in photosensitizing systems using rose bengal or methylene blue as sensitizer. Electron paramagnetic resonance (EPR) spectroscopy was used to detect the TEMP-$^1O_2$ product, TEMPO. In this study, VIP inhibited the TEMPO formation in a dose-dependent manner. The amount of VIP needed to cause 50% inhibition of the rate of $^1O_2$ quenching was found to be 37 μM.

VIP also possesses free-radical scavenging properties in in-vivo systems. One of the biological sources of the reactive oxygen species is the xanthine/xanthine oxidase system. The addition of this oxygen free-radical-generating system to perfused rat lungs increased both peak airway pressure and perfusion pressure, simultaneously resulting in pulmonary edema and increased protein content in bronchoalveolar lavage fluid (73). Treatment with 1–10 mg/kg/min of VIP significantly inhibited or completely abolished all signs of injury and reduced or abolished the generation of arachidonic acid products, suggesting that VIP may function as a physiological modulator of inflammatory tissue damage resulting from toxic oxygen species. Another, related study demonstrated that VIP does not cause generation of oxygen free radicals in any form and that the vasodilatory actions of VIP is not mediated through the generation of reactive oxygen species (74).

VII.   VIP and Ca^{2+}

Many peptides including VIP exert potent positive contractile responses directly in ventricular cardiomyocytes. In a recent study, the involvement of L-type calcium channels in the contractile responses elicited by VIP was investigated using selective antagonists at L-type calcium channels, verapamil and diltiazem (75). The results of this study indicated that positive contractile responses to VIP in ventricular cardio-myocytes involve the influx of Ca^{2+} via L-type calcium channels. Crude membrane fractions prepared from rabbit gastric fundic muscle degraded VIP (76). The VIP degradation was inhibited by EDTA and enhanced by Ca^{2+} in the concentration range 0.3–1.0 mM. In another, related study, VIP stimulated cAMP production and its interaction with protein kinase C activation and elevation of intracellular Ca^{2+} in NIE-115 neuroblastoma cells (77). The results of this study indicated that VIP stimulates cAMP accumulation in NIE-115 cells, and although activation of protein kinase C inhibits the VIP-stimulated cAMP response, elevation of intracellular Ca^{2+} potentiates this signaling pathway. VIP can induce glycogenolysis through a Ca^{2+}-dependent mechanism, and the restriction of cAMP accumulation during the infusion of high concentrations of VIP is caused by Ca^{2+}-induced phosphodiesterase activation (78). In another study, Ca^{2+}-mobilizing agonists were found to potentiate VIP-stimulated cAMP production in human colonic cell line, HT29-cl.19A (79). Calcium regulation of VIP mRNA abundance in SH-SY5Y human neuroblastoma cells has been described (80).

In a recent study, a role of VIP was shown in the regulation of NO inhibition of intestinal function (81). In this study, an addition of L-NAME to the perfusate obtained from perfused canine ileal segments caused, after a delay, a concentration-dependent persistent increase in tonic and phasic activity of circular muscle with corresponding reduction in tonic VIP output. Removal of Ca^{2+} to the perfusate markedly reduced this response. VIP has been described as the primary inhibitory transmitter which stimulates the production of NO. VIP-mediated G-protein-coupled Ca^{2+} influx was found to activate the constitutive NOS in dispersed gastric muscle cells (82).

VIII.   VIP in Signal Transduction

The accumulated data indicate that the physiological action of VIP is mediated through the adenylyl cyclase system (83). The α-adrenergic agonists were shown to act synergistically with VIP in cerebral cortex (84). In type I astrocytes from rat cerebral cortex, VIP at concentrations below 1 nm evoked an increase in intracellular

Ca^{2+} concentration (85). Treatment of these astrocytes with 0.1 nM VIP together with an α-adrenergic receptor agonist, phenylephrine, at subthreshold concentrations produced a large increase in intracellular Ca^{2+} concentration and oscillations. VIP (0.1 nM) and phenylephrine were found to increase cellular levels of inositol phosphates. These observations suggested a calcium-mediated second messenger system for the high-affinity VIP receptor in astrocytes and that α-adrenergic receptors act synergistically with the VIP receptor to augment an intracellular Ca^{2+} signal.

In a recent study, Ca^{2+}-mobilizing agonists were found to potentiate VIP-stimulated cAMP production in human colonic cell line, HT29-cl.19A (82). In the same study, protein kinase C activator phorbol ester markedly inhibited the cAMP response to the receptor-mediated cAMP agonist, VIP. A related study demonstrated the involvement of cGMP in potentiating VIP-mediated increase in intracellular Ca^{2+} transients (86). The results of this study demonstrated that VIP (5–200 nM) increased the $[Ca^{2+}]_i$ in 64% of isolated individual pinealocytes. These results further showed that the dominant effect of VIP causing transient elevation of $[Ca^{2+}]_i$ was mediated through cGMP gating α_1-*cis*-diltiazem-sensitive rod-type cyclic nucleotide-gated cation channels. This observation was substantiated by the fact that VIP is the primary inhibitory transmitter, and in smooth muscle cells it stimulates production of NO, which also exerts its signaling through cGMP (87). VIPs which are co-localized with nitric oxide synthase in a subpopulation of enteric neurons serve as parallel neurotransmitters with NO. Interestingly, plasma membranes isolated from dispersed gastric muscle cells exhibited calmodulin-dependent NOS activity that was stimulated by Ca^{2+} in the range 0.1–1 mM, and VIP augmented this NOS activity in a concentration-dependent fashion above the maximally stimulated by Ca^{2+} (88). This increase in NOS activity induced by VIP was abolished by GDPβS, which had no effect on NOS activity stimulated by Ca^{2+}. The authors concluded that Ca^{2+}/calmodulin-dependent NOS present in plasma membranes of gastric muscle cells is activated by two homologous peptide transmitters, VIP and pituitary adenylate cyclase-activating peptide (PACAP), via a common receptor coupled to pertussis toxin-sensitive Gil-2.

IX. Role of VIP in Myocardial Ischemia Reperfusion Injury

As mentioned earlier, VIP relaxes isolated coronary arteries, increases coronary blood flow, possesses positive ionotropic effects and thereby improves left ventricular functions, and directly scavenges oxygen free radicals. These properties of VIP make it an ideal cardioprotective agent against myocardial ischemia reperfusion injury. In a recent study, we measured the release of VIP in the coronary effluents of rats undergoing ischemia and reperfusion. The results are summarized in Fig. 2. The basal level of VIP in the effluent was 3.19 $\pm$ 0.05 pg/mL. The amount of VIP released from the heart increased progressively with the duration of reperfusion (15-min R = 5.16 $\pm$ 0.17 pg/mL; 30-min R = 6.11 $\pm$ 0.11 pg/mL; 45-min R = 7.69 $\pm$ 0.58

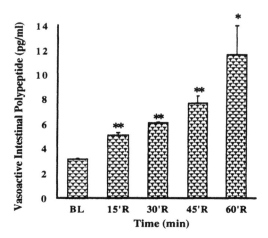

Figure 2 Quantitation of VIP in the coronary effluents collected after stabilization (baseline), and during 15, 30, 45, and 60 min of reperfusion. Mean values ± SEM of 5 observations are presented. Significant differences from the baseline levels are indicated; $*p < .01$; $**p < .001$. (Data from Ref. 21, with permission.)

pg/mL, $p < .001$ in each case). Maximal VIP level (11.66 ± 2.34 pg/mL, $p < .007$) was noted at the end of the 60-min reperfusion period.

In an attempt to examine the effects of VIP on ischemia reperfusion injury, isolated rat hearts were preperfused with three different doses of VIP for 15 min, followed by 30 min of ischemia and 60 min of reperfusion. Administration of VIP significantly increased the LV contractility before ischemia ($p < .001$). A 0.3 μM dose of VIP was found to be the optimal dose (Table 1). A 0.1 μM was much less effective, and 1 μM increased the heart rate to over 70 beats/min and showed extremely potent positive inotropic action. The left ventricular functions were lowered in the postischemic hearts as expected, but significantly better recoveries in LVDP and LVdp/dt were observed in the VIP-treated group (0.3 and 1 μM) ($p < .05$). For example, after 30 min of reperfusion, LVDP in the control group was 77 ± 1, while in the treated group the values were 107 ± 9 (0.3 μM) and 114 ± 11 (1 μM). LVdp/dt in the control and treated groups were 1637 ± 114 and 2667 ± 566, respectively ($p < .05$). At 60 min of reperfusion after the ischemic insult, the VIP-treated group again showed much better recovery of LV developed pressure (81 ± 3 [0.3 μM] and 78 ± 5 [1 μM], compared to only 54 ± 7 in the controls, $p < .01$) and its maximum first derivative LVdp/dt max (1847 ± 152 [0.3 μM] and 1675 ± 166 [1 μM], compared to 1410 ± 142 in the controls, $p < .01$). Administration of VIP in preischemic hearts increased the coronary flow ($p < .05$) compared with control. Throughout the reperfusion, recovery of coronary flow for the VIP-treated group was significantly better compared to the control group. After 5 min of reperfusion, coronary flow in the VIP-treated group was 12.38 ± 0.47 mL/min (0.3 μM) and $13.41 + 0.68$ mL/min (1 μM), compared to only 9.25 ± 0.25 mL/min ($p < .001$)

Table 1 Effects of VIP on Postischemic Myocardial Recovery

Left ventricular functions	Control				VIP			
	Pre-I	R15	R30	R60	Pre-I	R-15	R-30	R60
Coronary flow (mL/min)	10.5 ± 1.0	9.2 ± 0.2	7.0 ± 0.5	6.1 ± 0.3	10.1 ± 0.3	11.7 ± 0.5*	11.8 ± 0.3*	11.5 ± 0.4*
LVDP (mmHg)	120 ± 10.3	73 ± 7.5	63 ± 6.0	52 ± 3.5	117 ± 5.4	109 ± 8.5*	107 ± 6.2*	92 ± 5.2*
LVdp/dt (mmHg/sec)	2545 ± 141	2302 ± 114	2123 ± 136	1548 ± 103	2732 ± 207	2520 ± 122	2432 ± 97*	2080 ± 67*
CK release (IU/L)	0.8 ± 0.2	8.3 ± 1.7	11.8 ± 2.0	14.7 ± 2.3	0.7 ± 0.2	6.3 ± 1.0	7.5 ± 1.0*	8.2 ± 1.7*

$*p < .05$ compared to control.

in the control group. In the control group, coronary flow was further reduced during the reperfusion, so that at the end of reperfusion coronary flow in the treated group was 10.15 ± 0.45 mL/min (0.3 µM) and 10.0 ± 0.5 mL/min (1 µM), as compared to $7.65 \pm .075$ mL/min ($p < .05$) in the untreated group.

The myocardial CK release is a sensitive indicator of cellular injury. During the stabilization period, the release of CK in all groups was negligible. Reperfusion following 30 min of ischemia enhanced CK release. However, in the 0.3 µM VIP-treated group, CK release from postischemic hearts was significantly lower compared to the untreated group. For example, after 30 and 60 min of reperfusion, CK release from the heart in the control group was 11.79 ± 1.5 and 13.01 ± 1.9 IU/L, respectively, as compared to only 7.68 ± 1.2 and 8.8 ± 0.2 ($p < .05$ in each case), respectively, in the 0.3 µM VIP-treated group. VIP in 0.1 µM was totally ineffective, and 1 µM reduced the CK release minimally but not significantly.

It has already been mentioned that several interrelated factors including intra-

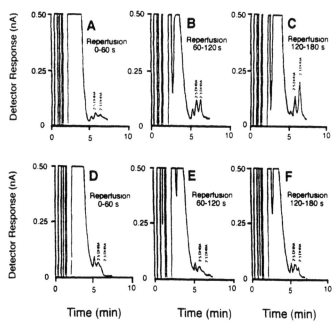

Figure 3 High-performance liquid chromatography (HPLC) detection of 2,3- and 2,5-dihydroxybenzoic acid (DHBA) production during reperfusion. Rat hearts were subjected to 30 min of global ischemia followed by reperfusion. Effluents were collected from control rat hearts at: (A) 0–60 sec, (B) 60–120 sec, and (C) 120–180 sec of reperfusion as well as from VIP-treated hearts at (D) 0–60 sec, (E) 60–120 sec, and (F) 120–180 sec. The flow rate for HPLC mobile phase was 1 mL/min, and the chromatographs were recorded at 1 cm/min. The electrochemical detector was set at a 0.5-nA sensitivity. (Data from Ref. 21, with permission.)

cellular Ca^{2+} overloading and oxygen free-radical generation play a role in the pathogenesis of myocardial ischemia reperfusion injury. We therefore examined the effects of VIP on the myocardial OH· formation and $[Ca^{2+}]_i$ during ischemia and reperfusion. The OH· formation was monitored in the heart by examining the hydroxylated benzoic acid formation after treating the hearts with salicylate. As shown in Fig. 3, both 2,3- and 2,5-dihydroxybenzoic acid (DHBA) peaks were significantly increased in the postischemic myocardium, the maximal amount forming after 2–3 min of reperfusion. Although treated (0.3 μM VIP) hearts also showed the activities of both 2,3- and 2,5-DHBA, the amount of these DHBAs was significantly lower in the VIP-treated group.

We monitored beat-to-beat change in $[Ca^{2+}]_i$ transients that occurred during each cardiac cycle (Fig. 4). Isolated and perfused rat hearts loaded with Fura-2 exhibited a reciprocal change in fluorescence at excitation wavelengths of 340 and 380 nm associated with clear $[Ca^{2+}]_i$ transients (Fig. 4A). As shown in Fig. 4B, in the control group, the amplitude of the $[Ca^{2+}]_i$ transients increased during ischemia and peaked during the initial phase of reperfusion. Sixty minutes of reperfusion of the ischemic myocardium did not restore the amplitude of $[Ca^{2+}]_i$ transients to the basal level. VIP (0.3 μM) reduced the amplitude of $[Ca^{2+}]_i$ significantly, especially in the 5- to 30-min period of reperfusion, demonstrating inhibition of intracellular Ca^{2+} overloading by VIP.

The results of this study indicated significant improvement of both biochemical and functional parameters of ischemic reperfused hearts in the VIP-treated group. In addition to positive ionotropic and vasodilatory actions, VIP also decreased the $[Ca^{2+}]_i$ transients and reduced the amount of free radicals generated in the reperfused heart.

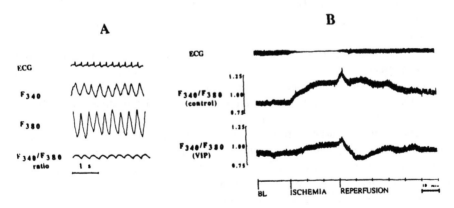

Figure 4   (A) Recordings of electrocardiogram (ECG), fluorescence at 340 (F_{340}) and 380 (F_{380}) nm, and fluorescence ratio (F_{340}/F_{380}) from an isolated and perfused rat heart containing Fura-2 AM. (B) The effect of VIP on the F_{340}/F_{380} fluorescence ratio in a rat heart containing Fura-2 AM. (Data from Ref. 21, with permission.)

X. Summary and Conclusions

VIP was initially postulated to function as a local neurotransmitter, being localized primarily in perivascular nerves surrounding coronary arteries and within the vascular wall (19). Subsequently it was found that VIP is a potent vasodilator of many vascular beds, including coronary arteries. More recent studies demonstrated the presence of VIP-like immunoreactive substance in the nerves of mammalian (including human) coronary arteries (89). VIP has also been implicated in the pathogenesis of hypertension, hemorrhagic shock, and heart failure (90).

The results of our study as well as reports from other laboratories demonstrated progressive release of VIP from the ischemic and reperfused myocardium. The amount of VIP release from the heart during reperfusion increased with the duration of reperfusion, similar to the CK release from the heart, suggesting that cellular injury induced by reperfusion could cause the loss of VIP from the heart. When supplied to the heart prior to ischemia, VIP at a concentration of 0.3 μM (1 $\times$ 10^6 pg/mL) provided excellent cardioprotection. In order to examine whether a lower dose or a higher dose was also effective, we perfused the heart with 0.1 μM and 1 μM and 1 μM. A 0.1 μM dose was much less effective and did not scavenge the free radicals, while 1 mM increased the heart rate to over 70 beats/min and showed extremely potent positive inotropic action. VIP at 0.3 μM and 1 μM reduced the ischemic reperfusion injury, as evidenced by improved left ventricular contractile functions and coronary flow. It is interesting to note that VIP only in 0.3 μM concentration reduced the CK release compared to control. It seems therefore reasonable to assume that VIP plays a role in the pathogenesis of myocardial ischemic reperfusion injury.

VIP has been shown to be a potent coronary vasodilator (22). In normal heart, intracoronary injection of VIP improved left ventricular functions and increased the coronary blood flow (20). The same study also demonstrated that VIP had a greater coronary vasodilator/inotropic ratio than either isoproterenol (β-adrenergic agonist), or forskolin, a compound that directly stimulates the catalytic subunit of adenylate cyclase. The authors postulated that VIP receptors, coupled to adenylate cyclase to a relatively greater degree in coronary resistance vessels, could allow for a greater coronary vasodilation/inotropic stimulation ratio compared to β-adrenergic receptor activation or adenylate cyclase stimulation. Our studies demonstrated that VIP can improve postischemic contractile functions and coronary flow.

As mentioned earlier, VIP has been shown to exert a positive inotropic effect on heart. However, it is unlikely that the positive inotropic action has any contribution to the improvement of postischemic myocardial functions observed in our study, because the isolated heart was preperfused with VIP for only 15 min. The half-life of VIP is only 48 sec, and the positive inotropic effect of VIP on heart is only transitory (2–5 min at 20 nmol). Therefore, it seems quite reasonable to assume that the postischemic myocardium is not affected by any positive inotropic effect derived from VIP. We strongly believe that the cardioprotective effects are derived from its other biological actions, such as free-radical scavenging and Ca^{2+} antagonism.

In a recent study, VIP was shown to prevent superoxide anion-mediated lung injury induced by xanthine/xanthine oxidase, simultaneously abolishing the generation of arachidonate products associated with the injury (16). The authors also reported that injured lungs released large amounts of VIP, suggesting that this peptide may function as a physiological modulator of injury due to toxic oxygen free radicals. Numerous evidence supports that reperfusion of ischemic myocardium generates oxygen free radicals, which play a significant role in the pathogenesis of reperfusion injury. In one study, VIP was found to reduce OH· formation during the reperfusion of ischemic heart. OH· formation was monitored by detecting the hydroxylated benzoic acid formation after perfusing the heart with salicylic acid as described previously (21). As shown in Fig. 3, both 2,3- and 2,5-dihydroxybenzoic acid formation was significantly reduced with VIP. Recently, it has been pointed out that 2,5-dihydroxybenzoic acid may also be formed via cytochrome P-450, while 2,3-dihydroxybenzoic acid is formed exclusively by the action of OH· on salicylic acid (90). This study thus supports the previous observation and further demonstrates that VIP can quench in-vivo OH· formation. A previous study indicated that VIP can directly scavenge singlet oxygen as efficiently as β-carotene (72). Evidence exists to support that singlet oxygen can also be formed in the ischemic and reperfused heart (91) and that a pathway exists in the heart to convert the singlet oxygen into hydroxyl radical (92). Taken together, the results of our study suggest that VIP may reduce the reperfusion injury by scavenging OH· formation.

Uncontrolled Ca^{2+} influx into the myocardial cell leading to the genesis of severe myocardial dysfunction has been observed during reperfusion of ischemic myocardium (60). The overload occurs as soon as ischemic hearts are exposed to calcium-containing blood during revascularization. Ca^{2+} may enter into the cell by three major routes: Ca^{2+}-selective voltage-activated channels, Na^+/Ca^{2+} exchanger, and uncontrolled entry through disrupted sarcolemma (59). Interestingly, in this study, VIP reduced the increased Ca^{2+} transients during ischemia and reperfusion. However, the VIP-mediated inhibition of Ca^{2+} transients remains to be explained.

Much has been learned during the last decade regarding the pathobiology of VIP in health and disease. VIP is now known to play various roles, such as regulation of regional blood flow and smooth muscle tone, secretion of macromolecules, immune and neuroendocrine functions, etc., in different tissues and cells, including epithelial cells, vascular as well as nonvascular smooth muscle, T lymphocytes, certain neurons, brain, and lung. VIP has also been found to be released from tissues during pathological conditions such as septic shock (93). In the GI tract, VIP functions as a neurotransmitter mediating descending relaxation (94). In rat superior cervical ganglion, VIP has been shown to increase inositol phospholipid breakdown and generate second messengers (94). VIP has been implicated as a mediator of coronary vasodilation during hypoxia (95). Free radical-injured lungs that were not treated with VIP released a large amount of this peptide in the perfusate, and the protective action of VIP was believed to be due to its ability to deal with the toxic oxygen metabolites (73). In another, related study, VIP could protect against acid-induced lung injury (96). In a recent study, VIP receptors were found in an endothelial cell line which controlled inwardly rectifying K^+ channels (97), suggesting a role of VIP in K^+ channel opening. In the cerebral cortex, VIP receptors were found to be coupled with cAMP-generating systems, and the VIP released from the neurons is believed to be mediated by a mechanism that involves Ca^{2+} (98). Involvement of Ca^{2+} was also reported in VIP-mediated fetal brain maturation.

In summary, VIP plays a cardioprotective role against ischemia and reperfusion injury. VIP, which was found to exert positive ionotropic and vasodilatory actions on the coronary artery, also possesses free-radical-scavenging ability and can reduce intracellular Ca^{2+} overloading. However, VIP has several different physiological roles on biological cells, and any of such properties could be instrumental in reducing the ischemia reperfusion injury.

Acknowledgments

These studies were in part supported by National Institutes of Health grants HL 34360 and HL 22559 and an award from the American Heart Association.

References

1. Mutt V, Said SI. Structure of the porcine vasoactive intestinal octacosapeptide. The amino acid sequence. Use of kallikrein in its determination. Eur J Biochem 1974; 42: 581–589.
2. Said SI, ed. Vasoactive Intestinal Peptide, Advances in Peptide Hormone Research Series, New York: Raven Press, 1982.
3. Costa M, Furness JB. The origins, pathways and terminations of neurons with VIP-like immunoreactivity in the guinea pig small intestine. Neuroscience 1983; 8:665–676.
4. Bitar KN, Makhlouf GM. Relaxation of isolated gastric smooth muscle cells by vasoactive intestinal peptide. Science 1982; 216:531–533.
5. Weihe E, Reinecke M, Forssmann WG. Distribution of vasoactive intestinal polypeptide-like immunoreactivity in the mammalian heart. Interrelation with neurotensin and substance P-like immunoreactive nerves. Cell Tissue Res 1984; 236:527–540.
6. Brum JM, Bove AA, Sufan Q, Reilly W, Go VLW. Action and localization of vasoactive intestinal peptide in the coronary circulation: evidence for nonadrenergic, noncholinergic coronary regulation. J Am Coll Cardiol 1986; 7:406–413.
7. Misra BR, Misra HP. The role of vasoactive intestinal peptide in scavenging singlet oxygen. FASEB J 1990; 4:630.
8. Mulderry PK. Synergistic regulation of vasoactive intestinal polypeptide expression by cyclic AMP and calcium in newborn but not adult rat sensory neurons in culture. Neuroscience 1993; 53:229–238.
9. Tatsuno I, Yada T, Vigh S, Hidaka H, Arimura A. Pituitary adenylate cyclase activating polypeptide and vasoactive intestinal peptide increase cytosolic free calcium concentration in cultured rat hippocampal neurons. Endocrinology 1992; 131:73–81.
10. deNeef P, Vertongen P, Ciccarelli E, Svoboda M, Herchuelz A, Winland J, Robberecht P. Pituitary adenylate cyclase activating polypeptide (PACAP) and vasoactive intestinal peptide stimulate two signaling pathways in CHO cells stably transfected with the selective type I PACAP receptor. Mol Cell Endocrinol 1995; 107:71–76.

11. Duner-Engstrom M, Fredholm BB. Breakdown of membrane choline-phospholipids induced by endogenous and exogenous muscarinic agonist is potentiated by VIP in rat submandibular gland. Acta Physiol Scand 1993; 149:41–49.

12. Pakbaz H, Berisha H, Foda HD, Absood A, Said SI. Vasoactive intestinal peptide (VIP) and related peptides: a new class of anti-inflammatory agents? In: Rosselin G, ed. International Symposium on Vasoactive Intestinal Peptide Pituitary Adenylate Cyclase Activating Polypeptide and Related Regulatory Peptides. Singapore: World Scientific Publishing, 1994:597–605.

13. Meleagros L, Ghatei MA, Bloom SR. Release of vasodilator, but not vasoconstrictor, neuropeptides and of enteroglucagon by intestinal ischemia/reperfusion in the rat. Gut 1994; 35:1701–1706.

14. Tikiz H, Tuncel N, Gurer F, Baycu C. Mast cell degranulation in hemorrhagic shock in rats and the effect of vasoactive intestinal peptide, aprotinin and H1 and H2-recepter blockers on degranulation. Pharmacology 1991; 43:47–52.

15. Akin MZ, Tuncel N, Gurer F, Kural N, Uslu S. Effect of vasoactive intestinal peptide and naloxone combination on urinary N-acetyl-beta-D-glucoseaminidase level and kidney histology of rats exposed to severe hemorrhage. Pharmacology 1993; 47:194–199.

16. Anderson FL, Kralios AC, Hershberger R, Bristow MR. Effect of vasoactive intestinal peptide on myocardial contractility and coronary blood flow in the dog: comparison with isoproterenol and forskolin. J Cardiovasc Pharmacol 1988; 12:365–371.

17. Smitherman TC, Sakio H, Geumei AM, Yoshida T, Oyamada M, Said SI. Coronary vasodilator action of VIP. In: Vasoactive Intestinal Peptide. New York: Raven Press. 1982:169–176.

18. Della NG, Papka RE, Furness JB, Costa M. Vasoactive intestinal peptide-like immunoreactivity in nerves associated with the cardiovascular system of the guinea pigs. Neuroscience 1983; 9:605–619.

19. Forssmann WG, Triepel J, Daffner C, Heym C. Vasoactive intestinal peptide in the heart. Ann NY Acad Sci 1988; 527:405–420.

20. Smitherman TC, Dehmer GJ. Vasoactive intestinal peptide as a coronary vasodilator. Ann NY Acad Sci 1988; 527:421–430.

21. Kalfin R, Maulik N, Engelman RM, Cordis GA, Milenov K, Kasakov L, Das DK. Protective role of intracoronary vasoactive intestinal peptide in ischemic and reperfused myocardium. J Pharmacol Exp Ther 1994; 268:952–958.

22. Clark AJL, Adrian TE, McMichael HB, Bloom SR. Vasoactive intestinal peptide in shock and heart failure. Lancet 1983; 1:539.

23. Gang ES, Lew AS, Hong M, Wang FZ, Sieber CA, Peter T. Decreased incidence of ventricular late potentials after successful thrombolytic therapy for acute myocardial infarction. N Engl J Med 1989; 321:712–716.

24. Engelman RM, Rousou JA, Flack JE, Deaton DW, Liu X, Das DK. A prospective randomized analysis of cold crystalloid, cold blood and warm blood cardioplegia for coronary revasulcarization. In: Engelman RM, Levitsky S, eds. Advanced Textbook of Difficult Clinical Problems. New York: Futura, 1992:143–150.

25. Engelman RM, Rousou JA, Flack JE, Deaton DW, Pleet AB, Das DK. Normothermic myocardial preservation, an optimal approach for myocardial protection during all forms of open-heart surgery. In: Cardiac Surgery: Current Issues II. New York: Plenum Press, 1994:25–37.

26. Das DK, Maulik N. Energy metabolism in ischemic reperfused heart. In: Karmazyn M, ed. Myocardial Ischemia: Mechanisms, Reperfusion, Protection. Basel: Birkhauser Verlag, 1996:155–173.

27. Datta S, Das DK, Engelman RM, Otani HG, Rousou JA, Breyer RH, Klar J. Enhanced myocardial preservation by nicotinic acid, an antilipolytic compound: mechanism of action. Basic Res Cardiol 1989; 84:63–76.

28. Kimura Y, Engelman RM, Rousou J, Flack J, Iyengar J, Das DK. Moderation of myocardial ischemia reperfusion injury by calcium channel and calmodulin receptor inhibition. Heart Vessels 1992; 7:189–195.

29. Otani H, Engelman RM, Rousou JA, Breyer RH, Clement T, Prasad R, Das DK. Improvement of myocardial function by trifluoperazine, a calmodulin antagonist, during experimental acute coronary artery occlusion and reperfusion. J Thoracic Cardiovasc Surg 1989; 97:267–274.

30. Das DK, Engelman RM, Prasad R, Rousou JA, Breyer RH, Jones R, Young H, Cordis GA. Improvement of ischemia reperfusion induced myocardial dysfunction by modulating calcium-overload using a novel specific calmodulin antagonist, CGS 9343B. Biochem Pharmacol 1989; 38:465–471.

31. Liu X, Engelman RM, Wei Z, Bagchi D, Rousou JA, Nath D, Das DK. Attenuation of myocardial reperfusion injury by reducing intracellular calcium overloading with dihydropyridines. Biochem Pharmacol 1993; 45:1333–1341.

32. Nayler WG, Sturrock WJ. An inhibitory effect of verapamil and diltiazem on the release of noradrenaline from ischemic and reperfused hearts. J Mol Cell Cardiol 1984; 16: 331–343.

33. Bolli R, Zhu WX, Thornby JI, O'Neil PG, Roberts R. Time course and determinants of recovery of function after reversible ischemia in conscious dogs. Am J Physiol 1988; 254:H102–H114.

34. Tosaki A, Bagchi D, Hellegouarch D, Pali T, Cordis GA, Das DK. Comparisons of ESR and HPLC methods for the detection of hydroxyl radicals in ischemic/reperfused hearts. A relationship between the genesis of oxygen-free radicals and reperfusion-induced arrhythmias. Biochem Pharmacol 1993; 45:961–969.

35. Otani H, Prasad R, Engelman RM, Otani H, Cordis GA, Das DK. Enhanced phosphodiesteratic breakdown and turnover of phosphoinositides during reperfusion of ischemic rat heart. Circ Res 1988; 63:930–936.

36. Das DK, Engelman RM, Rousou JA, Breyer RH, Otani H, Lemeshow S. Role of membrane phospholipids in myocardial injury induced by ischemia and reperfusion. Am J Physiol 1986; 251:H71–H79.

37. Otani H, Engelman RM, Rousou JA, Breyer RH, Das DK. Enhanced prostaglandin synthesis due to phospholipase breakdown in the ischemic-reperfused myocardium. Control of its production by a phospholipase inhibitor or free radical scavengers. J Mol Cell Cardiol 1986; 18:953–961.

38. Kukreja RC, Kontos HA, Hess ML, Ellis ER. PGH synthase and lipoxygenase generate superoxide in the presence of NADH or NADPH. Circ Res 1986; 59:612–619.

39. Kuehl FA, Humes JL, Egan RM, Ham EA, Beveridge GC, vanArman CG. Role of prostaglandin endoperoxide PGG2 in inflammatory process. Nature 1977; 265:170–173.

40. Bagchi D, Iyengar J, Jones R, Stockwell P, Das DK. Enhanced prostaglandin production in the ischemic reperfused myocardium by captopril linked with its free radical scavenging action. Prostaglandin Leukotrienes Essential Fatty Acids 1989; 38:145–150.

41. Kimura Y, Iyengar J, Engelman RM, Das DK. Prevention of myocardial reperfusion injury in experimental coronary revascularization following ischemic arrest by a novel anti-inflammatory drug, ONO-3144. J Cardiovasc Pharmacol 1990; 16:992–999.

42. Egan RW, Gale PH, Baptista EM, Kennicott KL, Vanden Heuvel WJA, Walker RW,

Fagerness PE, Kuehl FA. Oxidation reactions by prostaglandin cyclooxygenase hydroperoxidase. J Biol Chem 1981; 256:7352–7361.

43. Mullane KM. Free radical generation from prostaglandin hydroperoxides. Adv Inflammation Res 1988; 12:191–213.

44. Das DK, Engelman RM, Clement R, Otani H, Prasad MR, Rao PS. Role of xanthine oxidase inhibitor as free radical scavenger: a novel mechanism of action of allopurinol and oxypurinol in myocardial salvage. Biochem Biophys Res Commun 1987; 148: 314–318.

45. Das DK, George A, Liu X, Rao PS. Detection of hydroxyl radicals in the mitochondria in ischemic-reperfused myocardium by trapping with salicylate. Biochem Biophys Res Commun 1989; 165:1004–1009.

46. Bagchi D, Das DK, Engelman RM, Prasad MR, Subramanian R. Polymorphonuclear leukocytes as a potential source of free radicals in the ischemic reperfused myocardium. Eur Heart J 1990; 11:800–813.

47. Das DK, Engelman RM, Liu X, Maity S, Rousou JA, Flack J, Laksmipati J, Jones RM, Prasad MR, Deaton DW. Oxygen-derived free radicals and hemolysis during open heart surgery. J Mol Cell Biochem 1992; 111:77–86.

48. Godin DV, Bhimji S. Effects of allopurinol on myocardial ischemic injury induced by coronary artery ligation and reperfusion. Biochem Pharmacol 1987; 36:2101–2109.

49. Shen AC, Jennings RB. Myocardial calcium and magnesium in acute ischemic injury. Am J Pathol 1972; 67:417–425.

50. Janero DR, Burghardt C. Nonesterified fatty acid accumulation and release during heart muscle cell (myocyte) injury: modulation by intracellular "acceptor." J Cell Physiol 1989; 140:150–160.

51. Grover AK, Samson SE. Effect of superoxide radical on Ca^{2+} pumps of coronary artery. Am J Physiol 1988; 225:C297–C303.

52. Grinwald PM, Nayler WG. Calcium entry and calcium paradox. J Mol Cell Cardiol 1981; 13:867–880.

53. Otani H, Otani H, Das DK. α_1-Adrenoceptor-mediated phosphoinositide breakdown and inotropic response in rat left ventricular papillary muscles. Circ Res 1988; 62:8–17.

54. Nosek TM, Williams MF, Zeigler ST, Godt RE. Inositol triphosphate enhances calcium release in skinned cardiac and skeletal muscle. Am J Physiol 1986; 250:C807–C811.

55. Otani H, Engelman RM, Breyer RH, Rousou JA, Lemeshow S, Das DK. Mepacrine, a phospholipae inhibitor. A potential tool for modifying myocardial reperfusion injury. J Thoracic Cardiovasc Surg 1986; 92:247–254.

56. Otani H, Engelman RM, Rousou JA, Breyer RH, Clement R, Prasad R, Das DK. Improvement of myocardial function by trifluoperazine, a calmodulin antagonist, during experimental acute coronary artery occlusion and reperfusion. J Thoracic Cardiovasc Surg 1989; 97:267–274.

57. Liu X, Engelman RM, Iyengar J, Cordis GA, Das DK. Amiloride enhances postischemic ventricular recovery during cardioplegic arrest: a possible role of Na^+/Ca^{2+} exchange. Ann NY Acad Sci 1991; 639:471–474.

58. Otani H. Role of calcium in the pathogenesis of myocardial reperfusion injury. In: Das DK, ed. Pathophysiology of Reperfusion Injury. Boca Raton, FL: CRC Press, 1993: 181–219.

59. Nayler WG, Sturrock WJ, Panogiotopoulos S. Calcium and myocardial ischemia. In: Parratt Jr, ed. Control and Manipulation of Calcium Movement. New York: Raven Press, 1985:303–317.

60. Rasmussen H, Barrett PQ. Calcium and messenger system: an integrated view. Physiol Rev 1984; 64:938–962.

61. McCormack JG, Denton RM. Role of Ca^{2+} ions in the regulation of intramitochondrial metabolism in rat heart. Biochem J 1989; 218:235–240.

62. Poggioli J, Sulpice JC, Vassort G. Inositol phosphate production following α_1-adrenergic, muscarinic or electrical stimulation in isolated rat heart. FEBS Lett 1986; 206:292–298.

63. Osnes JB, Aass H, Skomedal T. In: Herefosum ODM, ed. α-Adrenoceptor Blockers in Cardiovascular Disease. E. London: Churchill Livingstone, 1985:69–110.

64. Otani H, Otani H, Das DK. Evidence that phosphoinositide response is mediated by α_1-adrenoceptor stimulation but not linked with excitation-contraction coupling in cardiac muscle. Biochem Biophys Res Commun 1986; 136:863–869.

65. DeMarinis RM, Wise M, Hieble JP, Ruffolo RR. In: Ruffolo RR, ed. α_1-Adrenergic Receptors. Clifton, NJ: Humana Press, 1987:1–25.

66. Moraru II, Jones RM, Popescu LM, Engelman RM, Das DK. Prazocin reduces myocardial ischemia/reperfusion-induced Ca^{2+} overloading in rat heart by inhibiting phosphoinositide signaling. Biochim Biophys Acta 1995; 1268:1–8.

67. Lindermann JP. α-Adrenergic stimulation of sarcolemmal protein phosphorylation and slow responses in intact myocardium. J Biol Chem 1986; 261:4860–4867.

68. Katoh N, Wrenn RW, Wise BC, Shoji M, Kuo JF. Substrate proteins for calmodulin-sensitive and phospholipid sensitive Ca^{2+}-dependent protein kinases in heart and inhibition of their phosphorylation of palmitoyl carnitine. Proc Natl Acad Sci (USA) 1981; 78: 4813–4817.

69. Wolf M, Levine H, May WS Jr, Cuatrecasas P, Sahyoun N. A model for intracellular translocation of protein kinase C involving synergism between Ca^{2+} and phorbol esters. Nature 1985; 317:546–549.

70. Otani H, Kato Y, Nonoyama A, Das DK. The role of protein kinase C in the pathogenesis of myocardial reperfusion injury. Seventeenth International Conference on Cyclic Nucleotides, Calcium and Protein Phosphorylation, Kobe, Japan, Oct 8–13, 1989.

71. Moraru II, Popescu LM, Maulik N, Liu X, Das DK. Phospholipase D signaling in ischemic heart. Biochim Biophys Acta 1992; 1139:148–154.

72. Misra BR, Misra HP. Vasoactive intestinal peptide, a singlet oxygen quencher. J Biol Chem 1990; 265:15371–15274.

73. Berisha H, Foda H, Sakakibara H, Trotz M, Pakbaz H, Said SI. Vasoactive intestinal peptide prevents lung injury due to xanthine/xanthine oxidase. Am J Physiol 1990; 259: L151–L155.

74. Ballon BJ, Wei EP, Kontos HA. Superoxide anion radical does not mediate vasodilation of cerebral arterioles by vasoactive intestinal polypeptide. Stroke 1986; 17:1287–1290.

75. Bell D, McDermott BJ. Inhibition by verapamil and diltiazem of agonist-stimulated contractile responses in mammalian ventricular cardiomyocytes. J Mol Cell Cardiol 1995; 27:1977–1987.

76. Kobayashi R, Chen Y, Lee TD, Davis MT, Ito O, Walsh JH. Degradation of vasoactive intestinal polypeptide by rabbit gastric smooth muscle membranes. Peptides 1994; 15: 323–332.

77. Inukai T, Chik CL, Ho AK. Vasoactive intestinal polypeptide stimulates cyclic AMP production in mouse NIE-115 neuroblastoma cells: modulation by a protein kinase C activator and ionomycin. Peptides 1994; 15:1361–1365.

78. Saito K, Yamatani K, Manaka H, Takahashi K, Tominaga M, Sasaki H. Role of Ca^{2+} on

vasoactive intestinal peptide-induced glucose and adenosine 3',5'-monophosphate production in the isolated perfused rat liver. Endocrinology 1992; 130:2267–2273.

79. Warhurst G, Fogg KE, Higgs NB, Tonge A, Grundy J. Ca^{2+}-mobilizing agonists potentiate forskolin- and VIP-stimulated cAMP production in human colonic cell line, HT29-cl.19A: role of $[Ca^{2+}]_i$ and protein kinase C. Cell Calcium 1994; 15:162–174.

80. Adler EM, Fink JS. Calcium regulation of vasoactive intestinal peptide mRNA abundance in SH-SY5Y human neuroblastoma cell. J Neurochem 1993; 61:727–737.

81. Daniel EE, Haugh C, Woskowska Z, Cipris S, Jury J, Fox-Threlkeld JE. Role of nitric oxide-related inhibition in intestinal function: relation to vasoactive intestinal polypeptide. Am J Physiol 1994; 266:G31–G39.

82. Murthy KS, Zhang KM, Jin JG, Grider JR, Makhlouf GM. VIP-mediated G protein-coupled Ca^{2+} influx activates a constitutive NOS in dispersed gastric muscle cells. Am J Physiol 1993; 265:G660–G671.

83. Christophe J, Svoboda M, Lambert M, Waelbroeck M, Winnd J, Dehaye JP, Vandermeers-Piret MC, Vandermeers A, Robberecht P. Effector mechanisms of peptides of the VIP family. Peptides 1986; 7(suppl 1):101–107.

84. Magistretti PJ, Schorderet M. VIP and noradrenaline act synergistically to increase cyclic AMP in cerebral cortex. Nature 1984; 308:280–282.

85. Fatatis A, Holtzclaw LA, Avidor R, Brenneman DE, Russell JT. Vasoactive intestinal peptide increases intracellular calcium in astroglia: synergism with alpha-adrenergic receptors. Proc Natl Acad Sci (USA) 1994; 91:2036–2040.

86. Schaad NC, Vanecek J, Rodriguez IR, Klein DC, Holtzclaw L, Russell JT. Vasoactive intestinal peptide elevates pinealocyte intracellular calcium concentrations by enhancing influx: evidence for involvement of a cyclic GMP-dependent mechanism. Mol Pharmacol 1995; 47:923–933.

87. Keef KD, Shuttleworth CW, Xue C, Bayguinov O, Publicover NG, Sanders KM. Relationship between nitric oxide and vasoactive intestinal polypeptide in enteric inhibitory neurotransmission. Neuropharmacology 1994; 33:1303–1314.

88. Murthy KS, Makhlouf GM. Vasoactive intestinal peptide/pituitary adenylate cyclase-activating peptide-dependent activation of membrane-bound NO synthase in smooth muscle mediated by pertussis toxin-sensitive Gil-2. J Biol Chem 1994; 269:15977–15980.

89. Popma JJ, Smitherman TC, Bedotto JB, Eichhorn EJ, Said SI, Dehmer GJ. Direct vasodilation induced by intracoronary vasoactive-intestinal peptide. J Cardiovasc Pharmacol 1990; 16:1000–1006.

90. Das DK, Engelman RM. Mechanism of free radical generation in ischemic and reperfused myocardium. In: Das DK, Essman WB, eds. Oxygen Radicals: System Events and Disease Processes. Basel: Karger, 1990:97–128.

91. Kukreja RC, Hess ML. The oxygen free radical systems: from equations through membrane protein interaction to cardiovascular injury and protection. Cardiovasc Res 1992; 26:641–650.

92. Bagchi D, Bagchi M, Douglas DM, Das DK. Generation of singlet oxygen and hydroxyl radical from sodium chlorite and lactic acid. Free Radical Res Commun 1992; 17:109–120.

93. Revhaug Lygren I, Jenssen TG, Giercksky KE, Burhol PG. Vasoactive intestinal peptide in sepsis and shock. Ann NY Acad Sci 1988; 527:536–545.

94. Biancani P, Walsh J, Behar J. Vasoactive intestinal peptide: a neurotransmitter for relaxation of the internal and anal sphincter. Gastroenterology 1985; 89:867–875.

95. Smitherman TC, Dehmer GJ. Vasoactive intestinal peptide as a coronary vasodilator. Ann NY Acad Sci 1988; 527:421–430.

96. Kawanaga T, Kitamura S, Hirose T, Said SI. Vasoactive intestinal peptide (VIP) protects against acid-induced acute lung injury in isolated perfused rat lungs. Nippon Kyobu Shikkan Gakkai Zasshi 1989; 27:789–795.

97. Pasyk E, Mao YK, Ahmad S, Shen SH, Daniel EE. An endothelial cell-line contains functional vasoactive intestinal polypeptide receptors: they control inwardly rectifying K^+ channels. Eur J Pharmacol 1992; 212:209–214.

98. Magistretti PJ. VIP neurons in the cerebral cortex. Trends Pharmacol Sci 1990; 11: 250–254.

15

Antiinflammatory Actions of VIP in the Lungs and Airways

SAMI I. SAID

State University of New York Health Sciences Center
Stony Brook
and Northport Veterans Affairs Medical Center
Northport, New York

I. Introduction

Over the past several years, the ability of VIP to protect the lungs against acute inflammatory injury has been documented in a variety of experimental models. For the most part, these models attempted to duplicate the type of injury seen in the adult respiratory distress syndrome (ARDS), but some represented acute immunological injury and others reproduced acute airway inflammation. The evidence from these studies, and from related studies on other organ systems, is here reviewed.

II. Protection Against Acute Lung Injury (Table 1)

A. Experiments on Isolated, Perfused, and Ventilated Lungs

In these models, the lungs were perfused through the pulmonary artery (PA), at a constant flow rate, with physiological solution containing 4% bovine serum albumin, and mechanically ventilated at a constant breath volume and rate with air or O_2 containing 5% CO_2. Criteria for grading lung injury in these experiments included: (a) increased peak airway pressure during mechanical ventilation, an index of decreased dynamic lung compliance; (b) increased PA perfusion pressure, a measure of

Table 1 Models of Acute Lung and Airway Injury That May Be Attenuated or Prevented by VIP

Method of injury[a]	Preparation
1. HCl intratracheally	
2. PAF into the PA	
3. Xanthine + xanthine oxidase into the PA	Isolated rat lung
4. Phospholipase C into the PA	
5. Prolonged perfusion ex vivo	
6. Paraquat into the PA	Isolated guinea pig lung
7. Capsaicin into the airways	Rat and guinea pig lungs, perfused via airway
8. PAF i.v.	Anesthesized dogs
9. Cobra venom factor	Anesthetized rats (in vivo)

[a]Abbreviations: PAF, platelet-activating factor; PA, pulmonary artery; i.v., intravenously.

increased pulmonary vascular resistance; (c) increased wet-to-dry lung weight ratio (W/D), a function of increased extravascular fluid or pulmonary edema; and (d) increased leakage of protein molecules, such as albumin, from the intravascular space into the air spaces and lung tissue, measured in bronchoalveolar lavage (BAL) fluid, reflecting increased permeability of the endothelial–epithelial barrier.

HCl-Induced Injury (1)

The instillation of 0.2 M HCl (2 mL/kg) intratracheally in mechanically ventilated rat lungs, perfused in situ at constant flow with Krebs-4% albumin solution, caused an increase in peak airway pressure, lasting at least 1 hr and reaching 5 times basal value at 30 min. Mean PA pressure increased by 68%; dynamic compliance decreased by 78%; and W/D increased by 74%. Infusion of VIP into the PA (1 μg/kg min), beginning 10 min before HCl and for the rest of the experiment, markedly reduced or totally prevented all abnormalities.

Injury Caused by Platelet-Activating Factor (PAF) (2)

In perfused rat lungs, injection of 30 μg/kg PAF into the PA caused an immediate rise in PA pressure, which almost doubled at 30 min. W/D increased by 56%, and BAL fluid protein content increased 15-fold. In lungs pretreated with VIP (1 μg/kg min), the rise in PA pressure was attenuated in degree and duration. Lung weight gain and protein leakage were reduced by 35% ($p < .01$).

Injury Due to Prolonged Perfusion ex Vivo

In perfused rat lungs, ventilated mechanically with 95% O_2/5% CO_2, airway pressure increased by 71% and PA pressure by 72% at the end of 2 hr; BAL protein content was markedly increased, and W/D was 81% higher than in unperfused lungs. Infusion of VIP (1 μg/kg min) attenuated these changes: Airway pressure increase was reduced by 73%, PA pressure elevation by 74%, W/D gain by 66%, and BAL protein content by 34% (3). In a parallel study (4), lung perfusion was continued until overt injury occurred, as evidenced by the appearance of foam in the upper airway. The duration of perfusion up to this point was an index of lung survival. This duration was 213.2 ± 10.5 min with Krebs-BSA perfusion; addition of prostacyclin (PGI_2, 0.3 μg/kg min) extended the survival time to 250.6 ± 13.6 min ($p < .05$), but VIP, at equimolar concentration (3 μg/kg min), prolonged the time to 349.6 ± 11.3 ($p < .01$).

In complementary studies (5,6), the preservation of the lung was compared when four different solutions were used: Krebs solution, Krebs solution with VIP, the University of Wisconsin (UW) solution, and UW solution with VIP. The lungs of male Sprague-Dawley rats were flushed and stored in these solutions for 24 hr, then examined by light, scanning, and transmission electron microscopy (EM) at regular intervals. Casts of the vasculature, made after 4 hr and viewed by a scanning electron microscope, showed that edema around the large vessels was least in lungs treated with VIP ($p < .01$). Quantitative transmission EM demonstrated that lungs stored in the VIP solutions had more normal-shaped mitochondria, with less edema and less distortion of their cristae, thinner basal lamina, and less aggregation of nuclear chromatin ($p < .01$).

Injury Due to Xanthine + Xanthine Oxidase (X/XO) (7)

The addition of X and XO to perfused rat lungs led to increases in peak airway pressure and perfusion pressure, pulmonary edema, and increased BAL protein content. Treatment with 1–10 μg/kg min of VIP markedly reduced or totally prevented all signs of injury. Simultaneously, VIP also diminished or abolished the associated generation of cyclooxygenase metabolites (7). Similar protection was provided by catalase (100 μg/mL).

Injury Due to Paraquat (8)

The pesticide paraquat (methyl viologen), known to cause tissue injury by oxidant mechanisms, induced acute injury of isolated guinea pig lungs, perfused in situ with Krebs-4% albumin, and ventilated with 95% O_2/5% CO_2. Upon infusion of paraquat (100 mg/kg) into the PA, airway pressure increased promptly by 230%, PA pressure increased moderately, and W/D and BAL protein, after 1 hr were significantly elevated. Pre- and co-treatment with VIP (3 μg/kg min) markedly attenuated or totally prevented all evidence of injury ($p < .001$).

Phospholipase C (PLC)-Induced Lung Injury (8a)

PLC mediates the degradation of cell membrane phosphoinositides, leading to activation of protein kinase C, Ca^{2+} mobilization, and other proinflammatory pathways. PLC (15 units), infused into the PA in isolated guinea pig lungs, elicited marked increases in airway pressure (from 8.5 ± 0.3 to 26 ± 3.5 cm H_2O), PA pressure (from

8 ± 0.4 to 16 ± 1.4 cm H_2O), W/D (from $5.20 \pm$ to 6.25 ± 0.22), and BAL protein content (to 1.25 ± 0.7 mg/mL). When VIP was added to the perfusate (1 µg/kg min) just before the PLC challenge, the injury was considerably attenuated.

Anaphylaxis in Guinea Pig Lungs (9)

The effectiveness of VIP against acute anaphylaxis was investigated in lungs from immunologically sensitized guinea pigs, perfused via the trachea instead of the PA, a model that more closely mimics clinical allergic reactions. Guinea pigs had been sensitized with i.p. and i.m. injections of ovalbumin (OA, 100 mg each) 4 weeks earlier. The isolated lungs were suspended in a warm, humidified chamber and perfused (2.2 mL/min) with Krebs-Ringer-phosphate buffer equilibrated with 5% CO_2 in O_2 at 37°C. Airway perfusion pressure was monitored throughout (30 min preceding OA challenge, and 30 min after). The perfusate escaped through puncture holes made in the lung. Following OA challenge (1.5 µg/mL min), intratracheally over a 15-min period, airway pressure increased from 0–2.4 to a peak of 61.7 ± 6.3 cm H_2O. VIP (10^{-7} M), added to perfusate, attenuated peak airway pressure in a dose-dependent manner (Fig. 1), and markedly decreased or prevented the release of cyclooxygenase metabolites into the perfusate. VIP thus protected guinea pig lungs against the increased airway pressure and release of eicosanoids triggered by anaphylaxis.

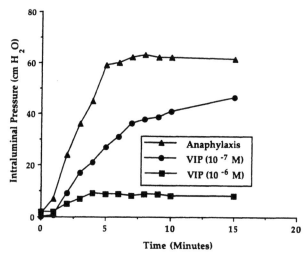

Figure 1 Marked increase in airway perfusion pressure in previously sensitized guinea pig lung challenged with ovalbumin (filled triangles). Pretreatment with VIP at 10^{-7} M (filled circles), and 10^{-6} M (filled squares) attenuated this airway response in a dose-dependent manner. The lung was perfused with physiological solution via trachea (see text).

Injury Caused by Overactivation of N-Methyl-ᴅ-Aspartate (NMDA) Receptors in the Lung (10,11)

Glutamate and related agonists, acting via glutamate NMDA receptors, form the major excitatory transmitter system in the mammalian brain (12). There is strong evidence that NMDA receptors, by promoting excessive entry of Ca^{2+} into neurons, play an important role in the neuronal damage that follows head injury, strokes, and epileptic seizures, and are associated with degenerative diseases such as Alzheimer's disease, Huntington's disease, Parkinson's disease, and amyotrophic lateral sclerosis (13). We recently presented evidence that NMDA receptors exist in the lung, and that their activation can trigger acute edematous lung injury (10,11). As in NMDA toxicity to central neurons, pulmonary excitotoxicity was associated with stimulation of nitric oxide (NO) synthesis, and could be attenuated by inhibition of this synthesis, as well as by NMDA receptor antagonists (10,11) (Fig. 2). The injury was also prevented by VIP (Fig. 2), apparently through inhibition of a key neurotoxic action of NO (see below; 11).

B. In Vivo Experiments

Injury Caused by Platelet-Activating Factor (PAF) (2)

In anesthetized and mechanically ventilated dogs, i.v. infusion of PAF (750 ng/kg/min for 2 min) caused increases in PA pressure, pulmonary vascular resistance, and

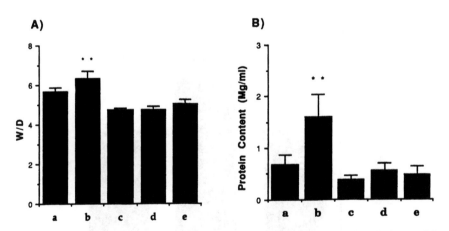

Figure 2 Induction of high-permeability edema in guinea pig lungs by NMDA and its prevention by NMDA receptor antagonist AP-5, by NOS inhibitor L-NAME, and by VIP. Wet/dry (W/D) lung weight ratios (A) and bronchoalveolar lavage (BAL) protein content (B) in several experimental groups: a = control, untreated lungs ($n = 8$); b = NMDA (1 mM) + ʟ-Arg (10 mM), $n = 16$; c = NMDA + ʟ-Arg + AP-5 (100 mM), $n = 3$; d = NMDA + ʟ-Arg + L-NAME (2 mM), $n = 4$; e = NMDA + ʟ-Arg + VIP (10 mM/kg min), $n = 5$. **$p < .01$ versus control. (From Ref. 11, with permission.)

airway pressure, systemic shock, and a sharp decrease in cardiac output. Treatment with VIP (500 ng/kg min), infused into the PA for 40 min beginning 10 min prior to the PAF infusion, attenuated all abnormalities, including the systemic hypotension.

Injury Due to Cobra Venom Factor (CVF) (14)

In this model, lung injury results from intravascular complement activation, depends on toxic oxygen metabolites generated from neutrophils, and requires P-selectin (15–18). The injury therefore is representative of that commonly seen in clinical ARDS. CVF was infused i.v. in anesthetized, pathogen-free, and mechanically ventilated rats. One group of rats received CVF only (40 U/kg, in 0.5 mL phosphate buffer, over 30 sec). A second group received VIP (10 μg/kg/min) by i.v. infusion, beginning 10 min before CVF and for the balance of the experiment. A third group received only phosphate buffer (negative control). In most experiments, the rats were also injected i.v. with ^{125}I-BSA and ^{51}Cr-red blood cells, for measurement of two indicators of lung injury known, respectively, as permeability index and hemorrhage index. The experiment was terminated 30 min after the injection of CVF, when lung injury was already in full evidence. Following the injection of CVF, systemic arterial blood pressure (BP) decreased from 110.0 ± 7.63 to a low value of 80.0 ± 8.19 at 10 min, and 88.0 ± 7.5 mm Hg at 30 min, and peak airway pressure increased from 9.4 ± 0.4 to 13.2 ± 1.2 cm H_2O. Permeability and hemorrhage indices, measured at the end of the experiment, were markedly increased, as was W/D (Fig. 3). Infusion of VIP attenuated or abolished all manifestations of lung injury (Fig. 3). As in the case of PAF

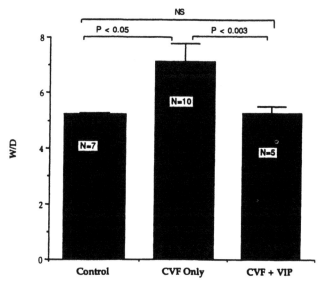

Figure 3 Prevention of cobra venom factor (CVF)-induced lung injury by VIP. W/D, wet/dry lung weight ratio, a function of pulmonary edema. Details in text.

experiments in vivo, VIP, itself a vasodilator and potentially hypotensive agent, actually minimized the decrease in BP due to CVF.

Specificity of VIP Action

Of the VIP-related peptides, only helodermin was uniformly protective against injury due to paraquat in guinea pig lungs or xanthine/xanthine oxidase in rat lungs. Secretin and glucagon were ineffective, and pituitary adenylate cyclase-activating peptide (PACAP) had an inconsistent effect (7,19). Recently synthesized, smaller helodermin analogs (19a) appear to have antiinflammatory activity equal to that of VIP or helodermin in acute lung injury models (E. Jaeger and S. I. Said, unpublished; see chapter by E. Jaeger in this volume).

III. Protection Against Airway Injury and Inflammation

A. Acute Airway Injury Due to Capsaicin (53)

Capsaicin, the active ingredient in red peppers, stimulates sensory C fibers in the lung, triggering the release of proinflammatory neuropeptides and the onset of airway inflammation. We investigated whether VIP could attenuate the acute effects of capsaicin, and whether it is released as a part of the airway inflammatory response to capsaicin.

In the same guinea pig lung model in which the lungs are perfused via the airways, as in the anaphylaxis experiments outlined above, single intratracheal additions of capsaicin, 10^{-9} M/10^{-7} M, in separate experiments, elicited concentration-dependent increases in airway pressure of up to 1200%. The pressure increase was immediate, moderated within 10–20 min, but persisted for > 2 hr. When VIP was added to the perfusate (10 µg/kg min), however, the capsaicin-induced increase in airway pressure was attenuated by $>70\%$ in both amplitude and duration (20). When VIP was not added, its concentrations in the effluent increased markedly; the increase was sustained, lasting >1 hr after a single dose, and generally paralleled the pressure increase (20). The results show that capsaicin induces a dramatic increase in airway resistance of guinea pig lungs, consistent with an acute inflammatory reaction; this increase is markedly attenuated by pre- and co-treatment with VIP; and capsaicin elicits the release of VIP, as well as substance P, from guinea pig airways. The stimulation of VIP release by capsaicin probably represents upregulation of the expression of the peptide, as a mechanism of counteracting the pro-inflammatory tachykinins.

In complementary studies, 3 µM of capsaicin (or solvent only, in control studies) was infused intratracheally (in 0.5 mL of solvent) in anesthetized, mechanically ventilated rats and guinea pigs. Ten minutes earlier, Evans blue was injected i.v. (30 mg/kg). A third group of animals received an i.v. infusion of VIP (20 µg/kg min), beginning 5 min before capsaicin, and for the rest of the experiment. Airway pressure and arterial BP were monitored throughout. At the end of the experiment, the trachea was removed, blotted clean, and extracted in formamide for spectrophotometric measurement of Evans blue at 620 nM. The dye content increased in tracheas from capsaicin-treated animals (to 22.8 $\pm$ 2.8 μg/g), but was at control value in tracheas from animals that also received VIP ($p < .05$).

B. Eosinophil-Mediated Injury of Cultured Bronchial Epithelial Cells

Bronchial epithelial injury is a characteristic early feature of bronchial asthma (21), and activated eosinophils may play an important part in mediating this injury (22). Having demonstrated VIP binding to cultured bronchial epithelial cells (23), we examined the injury of these cells caused by activated eosinophils, and its possible attenuation by VIP. Eosinophils were obtained by peritoneal lavage, at 97% purity. When activated by phorbol myristate acetate (PMA), eosinophils induced damage of branchial epithelial cells, in confluent monolayers, as evidenced by release of ^{51}Cr from cells that were radiolabeled with $Na_2[^{51}Cr]O_4$. PMA alone or unstimulated eosinophils did not increase ^{51}Cr release. VIP (10^{-9} and 10^{-6} M) plus phosphodiesterase inhibitor isobutylmethylxanthine dose-dependently decreased eosinophil-mediated cytotoxicity (24). The protection by VIP of bronchial epithelial cells against damage by activated eosinophil suggests an important mechanism by which VIP may modulate bronchial epithelial injury in asthma.

IV. Protection of Other Tissues and Organs

The tissue-protective effect of VIP is not limited to the lung. VIP may protect the *heart* against reperfusion injury, a common cause of oxidant injury of cardiac muscle. In isolated rat hearts perfused by the Langendorff technique, VIP accelerates the recovery of left ventricular function after ischemia-perfusion, enhances coronary flow, reduces myocardial tissue injury, and lowers intracellular Ca^{2+} transients and the level of hydroxyl radical (25; see chapter by Das et al. in this volume). In the normal heart, and in patients with anginal syndromes, VIP, given intravenously or into the coronary circulation, produces coronary vasodilation in concentrations that are too low to lower arterial BP (26). VIP also increases myocardial contractility (27) and favorably influences ventriculovascular coupling and transfer of mechanical energy to the circulatory bed (28).

In the *nervous system*, VIP, and a more potent lipophilic analog, [stearyl-norleucine17] VIP (29), protect cortical and hippocampal neurons in several experimental models of injury that are relevant to Alzheimer's disease (30,31). A special link to this disease is further suggested by two sets of observations: (a) VIP is richly present in neurons of the cerebral cortex and hippocampus and often localized in cholinergic neurons (32); and (b) in Alzheimer's disease, neuronal loss leads to major deficits in three neuronal systems—cholinergic, hippocampal, and cortical neurons (33). Like other neurotrophic factors, VIP also promotes the survival and differentiation of developing neuronal cells (34,35). The administration of a VIP antagonist in neonatal mice resulted in neuronal damage and retardation of behavioral development (36). VIP also induces astroglial cells to secrete a neuroprotective 14-kDa protein, called activity-dependent neurotrophic factor (ADNF) (30). A 14-amino acid-residue peptide, contained within ADNF (ADNF-14), has been reported to prevent neuronal cell death caused by glycoprotein 120, the HIV envelope protein, as well as by NMDA, by β-amyloid peptide, a neurotoxin associated with the pathogenesis of Alzheimer's disease, and by the neurotoxin tetrodotoxin—all at surprisingly low concentrations (10^{-17}–10^{-12} M) (30; see chapters by Brenneman et al. and DiCicco-Bloom in this volume).

V. Mechanisms of Protective Effect of VIP

Several mechanisms have been identified by which VIP may protect against tissue injury, including the following. Other mechanisms, still unidentified, may participate.

A. Modulation of Inflammatory Cell Function

In addition to its wide distribution in nerves supplying various structures in the lungs (37) and other organs, VIP is localized in, and binds to receptors on, several key inflammatory cells.

VIP in Inflammatory Cells

Immunoreactive VIP has been localized, either by radioimmunoassay or by immunohistochemistry or both, in three types of leukocytes, neutrophils, eosinophils, and mononuclear cells (38–41), as well as in pulmonary and peritoneal mast cells (42). Of the human leukocytes, eosinophils contain the highest levels of this peptide (38). The VIP in mast cells is present as a mixture of several structurally related but different peptides (41), and is released together with histamine by mast cell degranulators such as Ca^{2+} ionophore A23187 and compound 48/80 (42).

Receptor Sites on Inflammatory Cells

VIP binds to specific receptor sites on murine and human T lymphocytes, MOLT 4b lymphoblasts, cultured lines of T and B lymphocytes (43–46), human blood monocytes (47–49), as well as on rat alveolar macrophages (50), and rabbit and rat platelets (51,51a,52).

Inhibition of Inflammatory Cells

Lymphocytes

VIP inhibits mitogen-induced T-lymphocyte proliferation (44,53,54) and other aspects of T-lymphocyte function, including the release of certain cytokines, especially interleukin-2 (54,55) and interleukin-4 (55). VIP also inhibits immunoglobulin secretion (56), and modulates natural killer cell activity (57). Infusion of VIP into afferent lymphatics caused prompt and marked reduction of both recirculating and blast lymphocyte traffic in efferent lymph (58). On the other hand, VIP stimulates T-lymphocyte chemotaxis (59), homing to gastrointestinal lymphoid tissue (60), and interleukin-5 production (61).

Mononuclear Cells

VIP inhibits the respiratory burst in human monocytes (62) and inhibits phagocytosis and superoxide radical production by rat alveolar macrophages (63). Both of these effects are associated with stimulated cyclic AMP production.

Mast Cells

VIP moderately inhibits antigen-induced release of histamine (and possibly other mast cell mediators) from guinea pig lung (64), and inhibits mast cell degranulation induced by hemorrhagic shock in rats (65).

Platelets

The binding of VIP to receptors on platelets (52) increases cyclic AMP levels and inhibits platelet aggregation and serotonin secretion induced by PAF (51).

B. Inhibition of Release of Inflammatory Mediators

The suppression by VIP of the increased lung synthesis of cyclooxygenase metabolites induced by X/XO (7), and of the lipoxygenase-catalyzed leukotrienes provoked by bradykinin or by the chemotactic peptide fMLP (66), suggested a common mechanism of action—namely, inhibition of phospholipase A_2 (PLA_2). This suggestion was confirmed in vitro, as VIP dose-dependently inhibited pancreatic and cobra venom PLA_2 with the same potency as mepacrine (67). The related peptide helodermin was similarly effective, but secretin and PHI were inactive. PLA_2 inhibition decreases the liberation of arachidonic acid and its proinflammatory metabolites, as well as the synthesis of PAF, a powerful inflammatory mediator, from lyso-PAF (67).

C. Neutralization of Effects of Inflammatory Mediators

VIP prevents or reverses the bronchoconstrictor and pulmonary vasoconstrictor effects of histamine, prostaglandin $F_{2\alpha}$ leukotrienes C_4 and D_4, neurokinins A and B, and endothelin (68–70).

D. Antioxidant Activity

The protective effect of VIP in three models of oxidant lung injury suggested some antioxidant activity as a basis. It has now been demonstrated that VIP effectively scavenges singlet oxygen, and the hydroxyl radical, both highly reactive oxygen metabolites that cause oxidant tissue injury (25,71).

E. Prevention of NO Toxicity

In several of the lung injury models tested, excessive NO production appears to be a critical mediator of the injury (11,72,73). Because these same forms of injury are preventable by VIP (see above), the possibility arises that VIP may attenuate the injury by inhibiting the synthesis, or a key toxic effect, of NO. These questions were tested in the NMDA lung injury model. VIP (10 mg/kg min) in the perfusate totally prevented all manifestations of injury: 60 min after addition of NMDA, airway and perfusion pressures had not increased significantly from control values, and W/D and BAL protein content were normal. Treatment with VIP, however, did not prevent the NMDA-induced increase in cyclic GMP production, indicating that the increased NO synthesis persisted (11). These results suggest that the protective effect of VIP in this model is attributable to inhibition of a critical toxic action of NO rather than to inhibition of its production (11). The toxic action in question may be activation of the DNA-repair enzyme poly(ADP-ribose) polymerase, which leads to cell death through depletion of cell energy sources (11).

F. Importance of Cyclic AMP and Other Signal Transduction Pathways

Adenylyl cyclase stimulation and cAMP production is the dominant mode of action of VIP in most systems. Whether the antiinjury effect of VIP is entirely mediated by stimulation of intracellular cAMP levels is uncertain, however. Several observations suggest that this mediator relationship may not be a simple one. (a) Although the related peptide PACAP is more potent than VIP in stimulating cAMP production, it has not consistently protected the lung in the paraquat or X/XO injury models (7). (b) Of the other related peptides, all of which act via cAMP as a second messenger, only helodermin was as effective as VIP against injury in both models, while PHI, secretin, and glucagon were ineffective. Thus, the degree of lung protection does not always correlate with overall cAMP stimulation.

With some exceptions, the evidence for other second messengers of VIP action is less compelling. In the superior cervical ganglion (74) and adrenal chromaffin cells (75), relatively high (10^{-6} M) concentrations of VIP increased the breakdown of phosphoinositides to inositol phosphates, enhancing the intracellular mobilization of Ca^{2+}. In the adrenal medulla, this effect was linked to the induction of catecholamine secretion (75). In concentrations as low as 10^{-10} M, VIP increased intracellular $[Ca^{2+}]$ at least some rat cortical astrocytes (76) and, at higher concentrations, in cultured rat hippocampal neurons (77). In the latter preparation, PACAP exerted the same action at greater potency (77). The VIP-induced rise in intracellular $[Ca^{2+}]$ in astrocytes was correlated with increased secretory activity of these cells, including the production of trophic factor ADNF (see above). Similar, though more subtle, changes in intracellular $[Ca^{2+}]$ due to VIP have been described in prolactin-producing rat anterior pituitary cells (78). The significance of intracellular $[Ca^{2+}]$ and inositol phosphates in mediating the physiological or pharmacological actions of VIP, especially its antiinjury effect, remains uncertain. Finally, as with other agents, the actions of VIP on target cells are probably mediated by more than one signal transduction pathway. Possible interactions between these pathways have been described (79–83), but are still incompletely understood.

G. Role of Vasodilation

The importance of VIP-induced vasodilation as a factor in minimizing lung injury was investigated in the X/XO injury model in isolated lungs. The pulmonary vasodilator papaverine (0.15 mg/mL) actually increased pulmonary edema in this preparation, indicating that vasodilation per se does not attenuate the injury (7). It seems likely, therefore, that the protective effect of VIP, at least in the lung, is independent of its pulmonary vasodilator action.

VI. Upregulation and Modulatory Role of VIP in Acute Injury

Several observations suggest that VIP not only reduces acute lung injury when added exogenously, but is also an endogenous modulator of injury in the lung and elsewhere: (a) Unexpectedly high concentrations of the peptide were released from isolated lungs upon the induction of injury (7); (b) plasma VIP levels increased sharply in response to PAF given i.v. (2); (c) similarly high circulating concentrations are found in experimental and clinical septic shock (84,85); (d) VIP mRNA levels are increased in ferret tracheal ganglia and rat airway segments upon incubation with paraquat, NMDA, or capsaicin in vitro (Bandyopadhyay, Rattan, and Said, unpublished); and (e) VIP expression is also stimulated after axotomy of certain neurons (86).

The increased production of VIP in these models of local and systemic injury appears to represent an attempt to modulate the local or systemic inflammatory response. The sharply elevated circulating levels of VIP in experimental and clinical septic shock may also help to maintain adequate blood flow to vital organs. Because VIP suppresses or prevents injury in these models, the upregulation of its expression— an example of neurotransmitter plasticity (87)—probably represents an adaptive response designed to limit the damage and promote recovery (88).

References

1. Foda HD, Iwanaga T, Liu L-W, Said SI. Vasoactive intestinal peptide protects against HCl-induced pulmonary edema in rats. Ann NY Acad Sci 1988; 527:633–636.
2. Pakbaz H, Liu L-W, Foda HD, Berisha H, Said SI. Vasoactive intestinal peptide (VIP) as a modulator of PAF-induced lung injury. Clin Res 1988; 36:626A.
3. Pakbaz H, Foda HD, Sharaf H, Alessandrini F, Schraugnagel DE, Said SI. Vasoactive intestinal peptide (VIP) is more effective than prostacyclin in prolonging the viability of rat lungs ex vivo. Am Rev Respir Dis 1992; 145:A843.
4. Pakbaz H, Berisha H, Sharaf H, Foda HD, Said SI. VIP enhances, and nitric xynthase inhibitor reduces, survival of rat lungs perfused ex vivo. Ann NY Acad Sci 1994; 723:426–428.
5. Alessandrini F, Thakkar M, Foda HD, Pakbaz H, Said SI, Lodi R, Schraufnagel DE. Vasoactive intestinal peptide enhances lung preservation. Transplantation 1993; 56: 964–973.

6. Alessandrini F, Sasaki S, Said SI, Lodi R, LoCicero III J. Enhancement of extended lung preservation with a vasoactive intestinal peptide-enriched University of Wisconsin solution. Transplantation 1995; 59:1253–12589.

7. Berisha H, Foda H, Sakakibara H, Trotz M, Pakbaz H, Said SI. Vasoactive intestinal peptide prevents lung injury due to xanthine/xanthine oxidase. Am J Physiol 1990; 259:L151–L155.

8. Pakbaz H, Foda HD, Berisha HI, Trotz M, Said SI. Paraquat-induced lung injury: prevention by vasoactive intestinal peptide (VIP) and the related peptide helodermin. Am J Physiol 1993; 265 (Lung Cell Mol Physiol 9):L369–L373.

8a. Pakbaz H, Higuchi J, Foda HD, Said SI. Phospholipase C-induced acute lung injury and its attenuation by vasoactive intestinal peptide (VIP). Am Rev Respir Dis 1991; 143:A577.

9. Takamatsu J, Shima K, Sakakibara H, Trotz M, Said SI. Anaphylaxis in guinea pig lungs: attenuation of physiologic and biochemical consequences by vasoactive intestinal peptide. Am Rev Respir Dis 1991; 143:A44.

10. Said SI, Berisha HI, Pakbaz H. NMDA receptors outside the CNS: activation causes acute lung injury that is mediated by nitric oxide synthesis and prevented by VIP. Neuroscience 1995; 65:943–946.

11. Said SI, Berisha HI, and Pakbaz H. Excitotoxicity in lung: N-methyl-D-aspartate-induced, nitric oxide-dependent, pulmonary edema is attenuated by vasoactiveintestinal peptide and by inhibitors of poly (ADP-ribose) polymerase. Proc Natl Acad Sci (USA) 1996; 93:4688–4692.

12. Mayer ML, Westbrook GL. The physiology of excitatory amino acids in the vertebrate central nervous system. Prog Neurobiol 1987; 28:197–276.

13. Hollmann M, Heinemann S. Cloned glutamate receptors. Annu Rev Neurosci 1994; 17: 31–108.

14. Berisha H, Pakbaz H, Lyubsky S, Said SI. Acute lung injury due to cobra venom factor in vivo is prevented by vasoactive intestinal peptide (VIP). Am J Respir Crit Care Med 1995; 151:A767.

15. Till GO, Johnson KJ, Kunkel R, Ward PA. Intravascular activation of complement and acute lung injury: dependency on neutrophils and toxic oxygen metabolites. J Clin Invest 1982; 69:1126–1135.

16. Mulligan MS, Varani J, Dame MK, Lane CL, Smith CW, Anderson DC, Ward PA. Role of endothelial-leukocyte adhesion molecule 1 (ELAM-1) in neutrophil-mediated lung injury in rats. J Clin Invest 1991; 88:1396–1406.

17. Mulligan MS, Polley MJ, Bayer RJ, Nunn MF, Paulson JC, Ward PA. Neutrophil-dependent acute lung injury requirement for P-selectin (GMP-140). J Clin Invest 1992; 90:1600–1607.

18. Mulligan MS, Paulson JC, DeFrees S, Zheng Z-L, Lowe JB, Ward PA. Protective effects of oligosaccharides in P-selectin-dependent lung injury. Nature 1993; 364:149–151.

19. Pakbaz H, Berisha H, Foda HD, Absood A, Said SI. Vasoactive intestinal peptide (VIP) and related peptides: a new class of anti-inflammatory agents? In: Rosselin G, ed. VIP, PACAP, and Related Regulatory Peptides. River Edge, NJ: World Scientific; 1994: 597–605.

19a. Jaeger E, Bauer S, Joyce M, Foda HD, Berisha HI, Said SI. Structure-activity studies on VIP: IV. The synthetic agonist helodermin-fragment-(1-28)-amide is a potent VIP-agonist with prolonged duration of tracheal relaxant activity. Ann NY Acad Sci 1996; 805:499–504.

20. Pakbaz H, Berisha H, Absood A, Foda HD, Said SI. VIP in sensory nerves of the lung: capsaicin-induced release of immunoreactive vasoactive intestinal peptide (VIP) from guinea pig lungs. Am Rev Respir Dis 1993; 147:A477.

21. Laitinen LA, Heino M, Laitinen A, Kava T, Haahtela T. Damage of the airway epithelium and bronchial reactivity in patients with asthma. Am Rev Respir Dis 1985; 131:599–606.

22. Holgate ST. Mediator and cellular mechanisms in asthma. J Roy Coll Physicians (Lond) 1990; 24:304–312.

23. Sakakibara H, Shima K, Said SI. Vasoactive intestinal polypeptide (VIP) binds to cultured bronchial epithelial cells derived from normal rhesus monkey (4MBr-5). Clin Res 1990; 38:824A.

24. Sakakibara H, Takamatsu J, Said SI. Eosinophil-mediated injury of cultured bronchial epithelial cells: attenuation by vasoactive intestinal peptide (VIP). Am Rev Respir Dis 1991; 143:A44.

25. Kalfin R, Maulik N, Engelman RM, Cordis GA, Milenov K, Kasakov L, Das DK. Protective role of intracoronary vasoactive intestinal peptide in ischemic and reperfused myocardium. J Pharmacol Exp Ther 1994; 268:952–958.

26. Smitherman TC, Popma JJ, Said SI, Krejs GJ, Dehmer GJ. Coronary hemodynamic effects of intravenous vasoactive intestinal peptide in man. Am J Physiol 1989; 257:H1254–H1262.

27. Said SI, Bosher LP, Spath JA, Kontos HA. Positive inotropic action of a newly isolated vasoactive intestinal polypeptide (VIP). Clin Res 1972; 20:29.

28. Colston JT, Freeman GL. Beneficial influence of vasoactive intestinal peptide on ventriculovascular coupling in closed-chest dogs. Am J Physiol 1992; 262(Heart Circ Physiol 32):H1300–H1305.

29. Gozes I, Lilling G, Glazer R, Ticher A, Ashkenazi E, Davidson A, Rubinraut S, Fridkin M, Brenneman DE. Superactive lipophilic peptides discriminate multiple vasoactive intestinal peptide receptors. J Pharmacol Exp Ther 1995; 273:161–167.

30. Brenneman DE, Gozes I. A femtomolar-acting neuroprotective peptide. J Clin Invest 1996; 97:2299–2307.

31. Gozes I, Bardea A, Reshef A, Zamostiano R, Zhukovsky S, Rubinraut S, Fridkin M, Brenneman DE. Neuroprotective strategy for Alzheimer disease: intranasal administration of fatty neuropeptide. Proc Natl Acad Sci (USA) 1996; 93:427–432.

32. Said SI, V Mutt, eds. 1988. Vasoactive Intestinal Peptide and Related Peptides. Ann NY Acad Sci 527. New York: NY Acad Sci.

33. Price DL. New perspectives on Alzheimer's disease. Annu Rev Neurosci 1986; 9: 489–512.

34. Pincus DW, Di Cicco-Bloom EM, Black IB. Vasoactive intestinal peptide regulates mitosis, differentiation and survival of cultured sympathetic neuroblasts. Nature 1990; 343:564–567.

35. Pence JC, Shorter NA. In vitro differentiation of human neuroblastoma caused by vasoactive intestinal peptide. Cancer Res 1990; 50:5177–5183.

36. Hill JM, Mervis RF, Politi J, McCune SK, Gozes I, Fridkin M, Brenneman DE. Blockade of VIP during neonatal development induces neuronal damage and increases VIP and VIP receptors in brain. Ann NY Acad Sci 1994; 739:211–225.

37. Dey RD, Shannon WA, Said SI. Localization of VIP-immunoreactive nerves in airways and pulmonary vessels of dogs, cats, and human subjects. Cell Tissue Res 1981; 220: 231–238.

38. Aliakbari J, Sreedharan SP, Turck CW, Goetzl EJ. Selective localization of vasoactive

intestinal peptide and substance P in human eosinophils. Biochem Biophys Res Commun 1987; 148:1440–1445.

39. Lygren I, Revhaug A, Burhol PG, Giercksky K-E, Jenssen TG. Vasoactive intestinal polypeptide and somatostatin in leukocytes. Scand J Clin Lab Invest 1984; 44:347–351.

40. O'Dorisio MS, O'Dorisio TM, Cataland S, Balcerzak SP. Vasoactive intestinal polypeptide as a biochemical marker for polymorphonuclear leukocytes. J Lab Clin Med 1980; 96:666–672.

41. Goetzl EJ, Sreedharan SP, Turck CW. Structurally distinctive vasoactive intestinal peptide from rat basophilic leukemia cells. J Biol Chem 1988; 263:9083–9086.

42. Cutz E, Chan W, Track NS, Goth A, Said SI. Release of vasoactive intestinal polypeptide in mast cells by histamine liberators. Nature 1987; 275:661–662.

43. Danek A, O'Dorisio MS, O'Dorisio TM, George JM. Specific binding sites for vasoactive intestinal polypeptide on nonadherent peripheral blood lymphocytes. J Immunol 1983; 131:1173.

44. Ottaway CA, Greenberg GR. Interaction of vasoactive intestinal polypeptide with mouse lymphocytes: specific binding and the modulation of mitogen responses. J Immunol 1984; 132:417–423.

45. Beed EA, O'Dorisio S, O'Dorisio TM, Gaginella TS. Demonstration of functional receptor for vasoactive intestinal polypeptide on Molt 4b T lymphoblasts. Regulatory Peptides 1983; 6:1.

46. Finch RJ, Sreedharan SP, Goetzl EJ. High-affinity receptors for vasoactive intestinal peptide on human myeloma cells. J Immunol 1989; 142:1977–1981.

47. Guerrero JM, Prieto JC, Elorza L, Ramirez R, Goberna R. Interaction of vasoactive intestinal peptide with human blood mononuclear cells. Mol Cell Endocrinol 1981; 21:151.

48. Ottaway CA, Bernaerts C, Chan B, Greenberg GR. Specific binding of vasoactive intestinal peptide to human circulating mononuclear cells. Can J Physiol Pharmacol 1983; 61:664.

49. Wiik P, Opstad PK, Bøyum A. Binding of vasoactive intestinal polypeptide (VIP) by human blood monocytes: demonstration of specific binding sites. Regulatory Peptides 1985; 12:145–153.

50. Sakakibara H, Shima K, Said SI. Characterization of vasoactive intestinal peptide (VIP) receptors on rat alveolar macrophages. Am J Physiol 1994; 267:L256–L262.

51. Cox CP, Linden J, Said SI. VIP elevates platelet cyclic AMP (cAMP) levels and inhibits *in vitro* platelet activation induced by platelet-activating factor (PAF). Peptides 1984; 5:325–328.

51a. Ercal N, O'Dorisio MS, Vinik A, O'Dorisio TM, Kadrosfske M. Vasoactive intestinal peptide receptors in human platelet membrane. Ann NY Acad Sci 1988; 527:663–666.

52. Shima K, Sakakibara H, Said SI. Characterization of VIP- and helodermin-preferring receptors on rat platelets. Regulatory Peptides 1996; 63:99–103.

53. Ottaway CA. Selective effect of vasoactive intestinal peptide on the mitogenic response of murine T cells. Immunology 1987; 62:291–297.

54. Boundard F, Bastide M. Inhibition of mouse T-cell proliferation by CGRP and VIP: effects of these neuropeptides on IL-2 production and cAMP synthesis. J Neurosci Res 1991; 29:29–41.

55. Sun L, Ganea D. Vasoactive intestinal peptide inhibits interleukin (IL)-2 and IL-4 production through different molecular mechanisms in T cell activated via the T cell receptor/CD3 complex. J Neuroimmunol 1993; 48:59–70.

56. Stanisz AM, Befus D, Bienenstock J. Differential effects of vasoactive intestinal peptide, substance P, and somatostatin on immunoglobulin synthesis and proliferation by lymphocytes from Peyer's patches, mesenteric lymph nodes, and spleen. J Immunol 1986; 136:152–156.

57. Rola-Pieszczynski MD, Bolduc D, Pierre SS. The effects of vasoactive intestinal peptide on human natural killer cell function. J Immunol 1985; 135:2569–2573.

58. Moore TC, Spruck CH, Said SI. Depression of lymphocyte traffic in sheep by vasoactive intestinal peptide (VIP). Immunology 1988; 64:475–478.

59. Johnston JA, Taub DD, Lloyd AR, Conlon K, Oppenheim JJ, Kevlin DJ. Human T-lymphocyte chemotaxis and adhesion inducted by vasoactive intestinal peptide. J Immunol 1994; 13:1762–1769.

60. Ottaway CA. *In vitro* alteration of receptors for vasoactive intestinal peptide changes the *in vivo* localization of mouse T cells. J Exp Med 1984; 160:105–1069.

61. Matthew RC, Cook GA, Blum AM, Metwali A, Felman R, Weinstock JV. Vasoactive intestinal peptide stimulates T lymphocytes to release IL-5 is mansoni infection. J Immunol 1992; 148:3572–3581.

62. Wiik P. Vasoactive intestinal peptide inhibits the respiratory burst in human monocytes by a cyclic AMP-mediated mechanism. Regulatory Peptides 1989; 25:187–197.

63. Litwin DK, Wilson AK, Said SI. Vasoactive intestinal polypeptide (VIP) inhibits rat alveolar macrophage phagocytosis and chemotaxis in vitro. Regulatory Peptides 1992; 40:63–74.

64. Undem BJ, Dick EC, Buckner CK. Inhibition by vasoactive intestinal peptide of antigen-induced histamine release from guinea-pig minced lung. Eur J Pharmacol 1983; 88:247–249.

65. Tikiz H, Tunçel N, Gürer F, Bayçu C. Mast cell degranulation in hemorrhagic shock in rats and the effect of vasoactive intestinal peptide, aprotinin and H_1 and H_2 receptor blockers on degranulation. Pharmacology 1991; 43:47–52.

66. DiMarzo V, Tippins JR, Morris HR. Bradykinin- and chemotactic peptide fMLP-stimulated leukotriene biosynthesis in rat lungs and its inhibition by vasoactive intestinal peptide. Biochem Int 1988; 17:235–242.

67. Trotz ME, Said SI. Vasoactive intestinal peptide (VIP) and helodermin inhibit phospholipase A_2: a mechanism of anti-inflammatory activity. Regulatory Peptides 1993; 48:301–307.

68. Said SI, Dey RD. VIP in the airways. In: Kaliner MA, Barnes P, eds. The Airways: Neural Control in Health and Disease. Lung Biology in Health and Disease, Vol. 33. New York: Marcel Dekker; 1988:395–416.

69. Boomsma JD, Foda HD, Said SI. Vasoactive intestinal peptide (VIP) reverses endothelin-induced contractions of guinea pig trachea and pulmonary artery. Biomed Res 1991; 12:273–277.

70. Said SI. Vasoactive intestinal polypeptide. In: Kaliner MA, Barnes PJ, eds. Neuropeptides in the Respiratory Tract. New York: Marcel Dekker, 1994:143–159.

71. Misra BR, Misra HP. VIP, a singlet oxygen quencher. J Biol Chem 1990; 265:15371–15374.

72. Berisha H, Pakbaz H, Absood A, Foda HD, Said SI. Nitric oxide mediates oxidant tissue injury caused by paraquat and xanthine oxidase. Ann NY Acad Sci 1994; 723:422–425.

73. Berisha HI, Pakbaz H, Absood A, Said SI. Nitric oxide as a mediator of oxidant lung injury due to paraquat. Proc Natl Acad Sci (USA) 1994; 91:7445–7449.

74. Audigier S, Barberis C, Jard S. Vasoactive intestinal polypeptide increases inositol

phosphate breakdown in the rat superior cervical ganglion. Ann NY Acad Sci 1988; 527:579–581.

75. Malhotra RK, Wakade TD, Wakade AR. Vasoactive intestinal polypeptide and muscarine mobilize intracellular Ca^{2+} through breakdown of phosphoinositides to induce catecholamine secretion. J Biol Chem 1988; 263:2123–2126.

76. Fatatis A, Holtzclaw LA, Avidor R, Brenneman DE, Russell JT. VIP increases intracellular calcium in astroglia: synergism with alpha-adrenergic receptors. Proc Natl Acad Sci (USA) 1994; 91:2036–2040.

77. Tatsuno I, Yada T, Vigh S, Hidaka H, Arimura A. Pituitary adenylate cyclase activating polypeptide and vasoactive intestinal peptide increase cytosolic free calcium concentration in cultured rat hippocampal neurons. Endocrinology 1992; 131:73–81.

78. Sand O, Chen B, Li Q, Karlsen HE, Bjøro T, Haug E. Vasoactive intestinal peptide (VIP) may reduce the removal rate of cytosolic Ca^{2+} after transient elevations in clonal rat lactotrophs. Acta Physiol Scand 1989; 137:113–123.

79. Meisheri KD, Rüegg JC. Dependence of cyclic-AMP induced relaxation and Ca^{2+} and calmodulin in skinned smooth muscle of guinea pig *Taenia coli*. Pflügers Arch 1983; 399:315–320.

80. Pfitzer G, Rüegg JC, Zimmer M, Hofmann F. Relaxation of skinned coronary arteries depends on the relative concentrations of Ca^{2+}, calmodulin and active cAMP-dependent protein kinase. Pflügers Arch 1985; 405:70–76.

81. Szewczak SM, Behar J, Billett G. VIP-induced alterations in cAMP and inositol phosphates in the lower esophageal sphincter. Am J Physiol 1990; 259(Gastrointest Liver Physiol 22):G239–G244.

82. Saito K, Yamatani K, Manaka H, Takhashi K, Tominaga M, Sasaki H. Role of Ca^{2+} on vasoactive intestinal peptide-induced glucose and adenosine 3',5'-monophosphate production in the isolated perfused liver. Endocrinology 1992; 130:2267–2273.

83. Calderano V, Chiosi E, Greco R, Spina AM, Giovane A, Quagliuolo L, Servillo L, Balestrieri C, Illiano G. Role of calcium in chloride secretion mediated by cAMP pathway activation in rabbit distal colon mucosa. Am J Physiol 1993; 264(Gastrointest Liver Physiol 27):G252–G260.

84. Revhaug A, Lygren I, Jenssen TG, Giercksky K-E, Burhol PG. Vasoactive intestinal peptide in sepsis and shock. Ann NY Acad Sci 1988; 527:536–545.

85. Brandtzaeg P, Oktedalen O, Kierulf P, Opstad PK. Elevated VIP and endotoxin plasma levels in human gram-negative septic shock. Regulatory Peptides 1989; 24:37–44.

86. Mohney RP, Siegel RE, Zigmond RE. Galanin and vasoactive intestinal peptide messenger RNAs increase following axotomy of adult sympathetic neurons. J Neurobiol 1994; 25:108–118.

87. Hökfelt T, Zhang X, Wiesenfeld-Hallin. Messenger plasticity in primary sensory neurons following axotomy and its functional implications. Trends Neurosci 1994; 176:22–30.

88. Said SI. VIP and messenger plasticity. Trends Neurosci 1994; 17:339.

16

VIP Regulation of Neuronal Proliferation and Differentiation
Implications for Neuronal Repair

EMANUEL DiCICCO-BLOOM

University of Medicine and Dentistry of
 New Jersey/Robert Wood Johnson Medical School
Piscataway, New Jersey

I. Introduction and Perspective

Inflammation is a physiological process involved in disease pathogenesis as well as tissue repair. The inflammatory process, consisting of both humoral and cellular mediators, may act directly, or indirectly, on primary neurons and associated supporting cells. As fundamental regulators of body organ functions, neuronal processes, especially of the peripheral sensory and autonomic ganglia, are widely distributed within tissues and their related vasculature. VIP and the related peptide, pituitary adenylate cyclase-activating polypeptide (PACAP), are expressed in many peripheral neurons, serving a variety of roles in hemodynamic and physiological control (1). However, a number of studies in adult mammals indicate that VIP and PACAP are also newly expressed in neurons following physical injury to nerve processes and inflammation (2,3). The increases in neuronal expression of VIP and PACAP raise the possibility that the peptides play compensatory roles in damaged target tissues in which neurotransmitter balances are disturbed. Alternatively, peptide expression by injured neurons may represent a cellular signal important in organizing a neuronal repair program (2), a thesis we will explore.

   While studies of peptide functions in repair of neurons in the adult animal remain to be performed, our work on the role of peptides during neuronal development may provide some clues. In experimental models of nerve injury and regeneration, neurons frequently "re-express" molecules that originally appeared during early nervous system formation. These re-expressed molecules presumably serve similar functions during regeneration as they served during initial neuronal generation. For example, cytoskeletal and regulatory proteins mediating neuronal process outgrowth, such as neurofilaments, tubulins, microtubule-associated proteins (MAPs), growth-associated proteins (GAP43), and insulin-like growth factor (IGF) ligands and receptors (4–7), reappear or exhibit marked changes in expression. VIP and PACAP are also expressed during peripheral neuronal injury, though their functions remain unknown (2,3). Our observations indicate that VIP and PACAP peptides play important roles during development, stimulating neuronal precursor proliferation, nerve cell survival, and differentiation. The peptides appear to be widely expressed during the period when neurons are generated from dividing precursors (neurogenesis), and may be synthesized and released by the very neurons that respond to them, thus serving as autocrine/paracrine factors. In the review that follows, we define the role of peptides and underlying mechanisms in model neuronal culture systems. Subsequently, these functions are viewed from the perspective of adult neuronal repair.

II. Patterns of Peripheral and Central Neurogenesis

The nervous system regulates diverse body functions, from the relatively "simple" and automatic (vegetative), such as respiration, to the highly integrative and complex, such as cognition and memory. The neurons of the peripheral (PNS) and central nervous system (CNS) underlying these functional differences are produced from proliferating precursors or neuroblasts (8). Significantly, precursors of different neuronal populations exhibit distinct and highly characteristic patterns of neurogenesis: Precursors proliferate for varying periods of time, at different locations, and with lineage-specific relationships of mitosis to differentiation. For example, rat cerebral cortex precursors proliferate for 2–4 days in the densely aggregated neural tube ventricular zone. Following cessation of proliferation, neurons migrate to their final cortical position as they undergo differentiation (8). In contrast, peripheral sympathetic neuroblasts arise from the migratory neural crest cells that proliferate for 11 days at their final tissue destination, expressing multiple differentiated traits while actively dividing (8–11). Finally, cerebellar granule neurons, involved in motor coordination, are generated during the first three postnatal weeks, in a displaced ventricular zone overlying the cerebellum, and neurons migrate only after elaborating axonal processes (8).

The characteristic patterns of neurogenesis exhibited by different neuronal populations naturally lead to questions regarding underlying mechanisms. What is the basis of population specificity? Knowledge of governing mechanisms may provide new approaches to repair the damaged nervous system. Do specific precursors respond uniquely to common sets of signals, or alternatively, do local microenvironments provide different patterns of molecular regulators? In fact, both models appear to be involved. Recent studies suggest that there are two major classes of regulatory signals. First, there are population-specific molecules that act on restricted groups of cells, such as the action of ciliary neuronotrophic factor (CNTF) (12,13), neurotrophin-3 (NT-3) (14), and depolarization (15) on sympathetic neuroblasts, or tumor necrosis factor (16) and brain-derived neurotrophic factor (BDNF) (17) on cerebellar granule precursors. Second, there are broadly active regulators that signal via protein tyrosine kinase receptors, the well-known insulin-like growth factors (IGF), and epidermal (EGF) and fibroblast (bFGF) growth factors. These molecules stimulate proliferation of diverse neuronal precursors in the peripheral and central nervous system (10,11,18,19). Remarkably, emerging studies suggest that VIP and PACAP, signaling via G-protein-coupled receptors, represent new members of the broadly active neurogenetic regulators. The peptides influence growth and development of the early neural tube (see Brenneman, this volume), as well as precursors of cerebral cortex (20,21), cerebellum (22), and sensory (22) and sympathetic ganglia (23,24). While VIP and PACAP influence diverse neural progenitors, effects are region-specific, suggesting that neuroblasts exhibit distinct receptor/signaling pathways.

During PNS and CNS development, several interrelated processes are likely to contribute to the formation of stable populations. Specifically, the numerical growth of neuronal populations depends on three principal mechanisms: (a) stimulating neuroblasts to enter the cell division cycle, (b) promoting survival of dividing precursors, and (c) enhancing long-term neuronal survival by augmenting function of trophic factors (neurotrophins) and cognate *trk* receptors. Traditionally, proliferation, an increase in cell number, was considered the natural outcome of stimulating precursor cells to undergo cell division, or mitosis. Molecules stimulating cell cycle entry act by promoting an increase in cell mass and initiation of DNA synthesis (S-phase), allowing a doubling of chromosomes prior to cytokinesis (25). However, recent evidence indicates a second mechanism by which signals enhance proliferation: trophism. That is, dividing neuronal precursors may undergo cell death (14,26). By providing necessary trophic (survival) factors, dividing neuroblasts and their progeny are permitted to accumulate and continue proliferating (14,27). Trophism, a mechanism traditionally described as a postmitotic event during formation of the nervous system, thus plays a role far earlier during ontogeny than initially conceived, influencing neurogenesis itself. These distinct mechanisms governing proliferation can be distinguished experimentally (28). Our studies indicate that molecules influencing neuronal precursors may regulate mitosis or survival separately, or alternatively, activate both pathways simultaneously. For sympathetic neuroblasts, as will be discussed below, IGF and EGF stimulate mitosis (10,11), NT3 promotes precursor survival (14,28), and VIP and PACAP activate both mechanisms simultaneously (23, 24,27,28).

In addition to stimulating proliferation, population growth requires long-term survival of generated neurons. A major source of survival factors is the target population that neurons innervate, including other neurons or tissues and organs. Target-derived trophic factors, such as nerve growth factor (NGF), BDNF, and NT3, act on innervating nerve terminals via the *trk* family of receptors, including *trk*A, *trk*B, and *trk*C (see works cited in Ref. 14). Thus extracellular signals, such as the peptides, may enhance population stability by promoting expression and function of *trk* receptors. As one example, preliminary studies suggest that the peptides increase *trk*A expression during sympathetic development.

The roles of VIP and PACAP during nervous system development are multiple and diverse, regulating precursor mitosis, survival, and differentiation. Several characteristic features support the view that VIP and PACAP are developmental regulators, including patterns of ligand and receptor expression, sources of peptides, and the nature of responsive cells. For example, study of several tissues, such as sympathetic ganglia (23,27,29), cerebellar cortex (1,30,31), and the early embryo (see Brenneman, this volume), indicate that VIP and PACAP are present in the developing nervous system at critical periods of cell proliferation and differentiation. Indeed, peptide expression and cell responsiveness peak in parallel with periods of neuronal production (27,29) (see Brenneman, this volume). In light of the diversity of responsive neuronal systems influenced by the peptides, it is perhaps not surprising that the factors appear to originate from multiple distinct sources. Evidence suggests that peptides may be (a) released from presynaptic nerve terminals (32), (b) produced locally as autocrine or paracrine signals (20,21,23,27,29,33,34), or (c) transported across the placenta to the embryo from maternal sources (see Brenneman, this volume).

At the cellular level, the peptides appear to act both directly, as well as indirectly, on neurons and precursors. Peptides that act indirectly may stimulate production of various growth factors and cytokines by local glial cells (astrocytes; see Brenneman, this volume) and microglia (resident macrophages of the CNS; see Goetzl, this volume), which secondarily influence neuronal ontogeny. In this review, we will focus on direct peptide actions in a model sympathetic neuroblast system. The role of these peptides in regulating glial cell function are discussed by Brenneman (this volume). Since receptors of VIP and PACAP have been cloned only recently, little definitive information is currently available regarding mediating mechanisms. Nonetheless, there appears to be no strict association of specific receptors with direct or indirect regulation of mitosis, survival, differentiation, or cytokine release. Further, current data suggest that additional VIP and PACAP receptors will be forthcoming.

III. VIP and PACAP Regulation of Rat Sympathetic Neurogenesis: A Model of Direct Peptide Action

Neurons of the superior cervical ganglia (SCG) are generated by proliferation of precursors localized dorsolateral to the embryonic dorsal aorta (9,35). Sympathetic precursors originate from the neural crest, a transient, migratory population of cells from the embryonic neural folds that give rise to multiple neuronal and nonneuronal structures (8). Neural crest progenitors of sympathetic ganglia aggregate dorsolateral to the aorta on embryonic day 11.5 (E11.5) (9,35), and proliferate and differentiate simultaneously for up to 11 days (birth at E22) (27,29), under the influence of local microenvironmental signals. As early as E12, innervating nerve terminals, from the intermediolateral spinal motor column, form functional synapses in the embryonic ganglion (36), potentially providing presynaptic molecules that influence sympathetic neurogenesis. In addition, principal ganglion neurons establish nerve terminals in targets, including salivary glands, iris, and cranial blood vessels, by E15 (37), raising the unexpected possibility that signals from target tissues play a role in neurogenetic regulation. After birth, SCG neurons have an absolute requirement for target-derived trophic factor, NGF (38). From the perspective of defining molecular signals, the E15.5 SCG is an ideal model system, since 95% of ganglion cells at this age are neuroblasts in vivo, and in low-density, serum-free culture conditions, more than 99% of the cells are neuronal precursors, expressing multiple phenotypic markers (10,11,14). Using this model system, we have previously defined signals that stimulate and inhibit mitogenesis, as well as promote survival and differentiation. In this review, we will focus on the roles of VIP and PACAP during sympathetic ontogeny.

A. VIP and PACAP Are Multifunctional Regulators of Sympathetic Development

In initial studies, we examined the effects of VIP on E15.5 sympathetic neuroblast mitosis, survival, and neurite outgrowth (23). To examine mitogenesis, neuroblasts were incubated with [^{3}H]thymidine, a precursor incorporated into newly synthesized DNA during S-phase of the mitotic cycle. Extracellular signals that regulate the production of new cells from dividing precursors act by increasing the proportion of S-phase cells in the population (11,14,25). The proportion of mitotic cells is assessed by labeling index analysis. Following incubation with [^{3}H]thymidine, cultures are processed for immunocytochemistry, to identify neuroblasts, and autoradiography, to mark cells in S-phase (10). Tyrosine hydroxylase, the rate-limiting enzyme in catecholamine neurotransmitter biosynthesis, is used to identify sympathetic precursors (10,14). Traditional mitogens, such as IGF and EGF, increased the labeling index two- or threefold (10,11). VIP increased the labeling index from 6% to more than 15%, indicating that the peptide stimulated mitogenesis (23) (Fig. 1). Dose–response analysis indicated that maximal VIP effects were elicited by 1–10 μM peptide, while related family members, including secretin, peptide histidine-isoleucine, and growth hormone-releasing hormone, were without effects, suggesting that VIP action was specific. In further studies to clarify mitogenic mechanisms, effects of VIP were examined at 6 hr, a brief time period during which no mitotic cells were found to die. Nonetheless, VIP increased DNA synthesis. Thus, VIP served as a true mitogen, increasing the fraction of cells in S-phase of the mitotic cycle, and did not merely promote survival of the dividing population (28).

   However, in addition to stimulating mitosis, VIP also increased neuron survival (23,27). At 4 days in culture, VIP increased cell number three- to fourfold compared to controls. While it is likely that increased cell number was in part due to enhanced

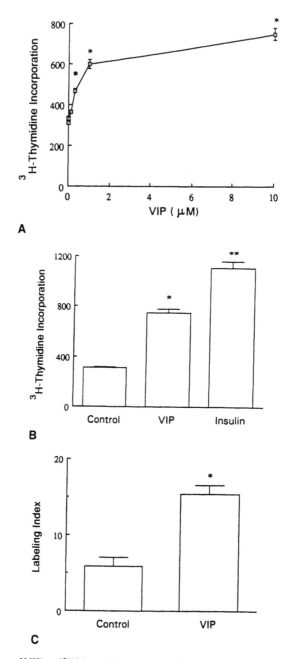

Figure 1 Effect of VIP on [³H]thymidine incorporation by cultured sympathetic neuroblasts. (A) Cultures were incubated for 48 hr in control medium or medium containing various concentrations of VIP. Each experimental value represents the mean incorporation of 4 culture wells and is expressed as mean cpm ± SEM. (B) Comparison of effects of VIP (10 μM) and insulin (10 μg/mL) on [³H]thymidine incorporation. (C) Effect of VIP on the labeling index of cultured neuroblasts. The labeling index is defined as the proportion of tyrosine hydroxylase-positive cells that incorporated [³H]thymidine into the nucleus. (From Ref. 23.)

mitogenesis, the peptide also increased neuron number in the presence of mitotic inhibition. Thus VIP appears to promote neuroblast survival independently of mitogenic stimulation. Finally, VIP increased neuritic process outgrowth fivefold (Fig. 2). Enhanced neuritogenesis was observed as early as 6 hr after plating, prior to cell death in vitro, indicating that effects were not due to selective survival of neurite-bearing cells. Similar differences in neurite outgrowth were present at 24 hr, indicating that effects were sustained.

The dramatic effects of VIP in culture indicated that the peptide promoted mitosis, survival, and differentiation simultaneously in sympathetic neuroblasts. In light of the absence of ganglion nonneuronal cells in this system, VIP effects were elicited by direct actions of the peptide on the precursors themselves. Indeed, preliminary studies suggest the presence of VIP-binding sites on cultured cells (33). Significantly, the multiple effects of the peptide occurred with peak concentrations in the micromolar range, suggesting that PACAP receptors may be involved. Indeed, recent reports indicate that the PACAP type I receptor mRNA is expressed in vivo in newborn ganglia, an observation we have confirmed in the embryonic SCG (39,40). Further, we have found that PACAP is 1000-fold more potent that VIP in culture (24) compatible with peptide action via PACAP type I receptors. To examine the ontogenetic relevance of peptide effects, studies were performed to characterize peptide expression and trophic activity during the neurogenetic period. Further, the role of VIP in supporting mitotic precursor survival was defined.

B. VIP Peptide Exhibits Peak Expression and Trophic Activity During the Period of Sympathetic Neurogenesis

To characterize VIP protein expression during development, radioimmunoassay was performed on SCG from E15.5 to birth (23,27). VIP expression exhibited a peak at E15.5, 0.5–1.0 pg/μg protein, and decreased threefold by birth. VIP concentrations of adult SCG were one-tenth of E15.5 levels (Fig. 3). A similar transient pattern of VIP expression has been observed using immunocytochemical analysis in sympathetic ganglia in vivo (29). The VIP expression pattern of SCG was highly specific, since peptide levels in dorsal root ganglia, also of neural crest origin, were undetectable in the embryo and increased to a peak in the adult. In sum, the embryonic peak of VIP expression in sympathetic ganglia suggested that the peptide plays a role in regulating neurogenesis in the animal.

To define relationships between peptide expression and trophic activity during ontogeny, the effects on survival of neurons from SCG of progressively older animals from E15.5 to birth were assessed (27). Significantly, ganglion neurons exhibited different trophic requirements as a function of developmental stage. In the absence of exogenous trophic factors, approximately 50% of E15.5 SCG neuroblasts survived for 48 hr in culture, whereas only 5% of neurons from postnatal day 1 (P1) SCG survived. Addition of VIP to E15.5 neuroblast cultures rescued a majority of the plated cells. Since 50% of neuroblasts survived in the absence of VIP, the peptide rescued a similar proportion (Fig. 4A). However, VIP was less effective in survival

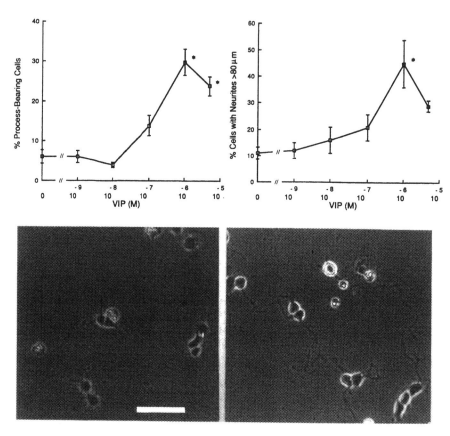

Figure 2   VIP elicits increased neurite outgrowth in cultured neuroblasts. Cells were incubated in control medium or in medium containing increasing concentrations of VIP. Ten hours after plating, neurite initiation was enhanced sixfold by VIP at 1 μM (top, left). After 24 hr, the percentage of cells bearing long processes was also increased by VIP (top, right). At 24 hr, neurons incubated in control medium had short neuritic processes (bottom, left), whereas processes were very long and branched in the presence of VIP (bottom, right). Bar = 50 μM. (From Ref. 23.)

promotion as a function of age: At E19.5, VIP rescued only 10% of neurons, and by birth, only 5%, suggesting that the peptide elicited trophic effects during a critical fetal period. Conversely, with advancing age, the percentage of cells supported by NGF increased (Fig. 4B), suggesting that as trophic support from local VIP diminishes, neuroblast survival depends on target derived NGF.

The stage-dependent ability of VIP to promote neuroblast survival correlated with the limited previous studies of mitosis in the SCG in vivo (27). To define the relationship of VIP trophic effects to neuroblast mitosis, we examined the labeling

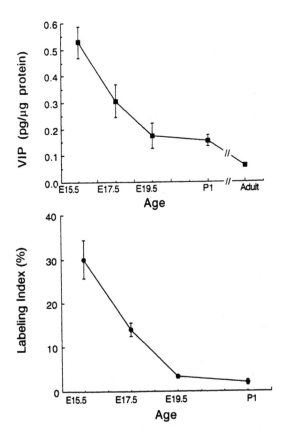

Figure 3 Developmental time-course of VIP expression and neurogenesis in the SCG. (Top) VIP content of pooled ganglia extracts were determined by RIA. Values were normalized to total protein in the sample as determined by the method of Lowry and are expressed as pg VIP/μg total protein $\pm$ SEM. (Bottom) Cells derived from SCG of different ages were cultured for 24 hr in the presence of the mitogen, insulin (10 μg/mL) and [^{3}H]-thymidine (1 μCi/mL), and assessed for labeling index. NGF (110 ng/mL) was included in medium to support maximal survival at all ages. (From Ref. 27.)

index of neuroblast cultures derived from SCG of different ages, in the presence of maximal mitogenic stimulation with insulin. NGF, which has no mitogenic activity in this system (10), was included to ensure maximal survival at older ages. The labeling index peaked at E15.5, at 30%, and decreased progressively to <2% at P1 (Fig. 3), indicating that the developmental restriction of neuroblast mitosis paralleled the critical period during which VIP peptide was expressed (Fig. 3) and promoted survival (Fig. 4). In turn, we considered the possibility that the peptide acted on mitotic cells or those that just completed cell division.

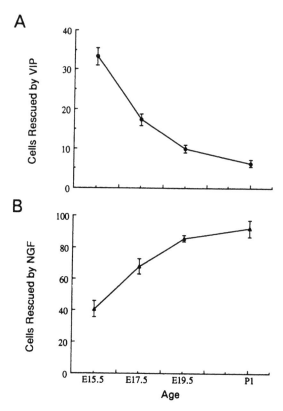

Figure 4 Developmental time course of neuronal cell number after 2 days in vitro in the presence of VIP (A) and NGF (B). Cells derived from SCG of different ages were cultured in medium containing VIP (3 μM) or NGF (110 ng/mL). Values derived from triplicate cultures are expressed as percent ± SEM cells initially plated less the percent surviving in control cultures. Thus, at E15.5, cell numbers after 2 days in the presence of VIP were 85% of cells initially plated. Since 51% of cells initially plated were present in control cultures, 34% were rescued by VIP at this age. (From Ref. 27.)

C. VIP Promotes Survival of Dividing Sympathetic Precursors

VIP peptide is expressed during the period of ongoing sympathetic neurogenesis (23,27,29). In studies described above, the peptide promoted survival of cells until 4 days in culture, apparently independent of mitosis per se. Does the peptide actually rescue cells that are dividing? To examine this question, E15.5 neuroblasts were incubated for 24 hr without and with trophic factors, and observed by time-lapse photomicroscopy (27). To characterize effects on mitotic cells, the fate of cells born in culture by cytokinesis (mitotic cells) as well as nonmitotic cells was recorded as a function of time (Table 1). During the first 24 hr, 80% of mitotic cells survived in the absence of exogenous factors. However, more than 75% of mitotic cells died in

Table 1 Effect of Combined VIP and NGF Treatment on Neuroblast Survival Analyzed by Serial Time-Lapse Microscopy

	0–24 hr		24–48 hr	
	Mitotic[a]	Nonmitotic	Mitotic	Nonmitotic
Control	80[b]	44.8	25	7.8
VIP	98.4	55	61.4	40.8
VIP + NGF	100	81.7	94.2	83.7

Cultures were incubated in control medium or in medium containing VIP (3 μM), or VIP and NGF (110 ng/mL) for 48 hr. All groups contained insulin (10 μg/mL), which maximally promotes mitogenesis. Cultures were analyzed by serial time-lapse microscopy and cell fate on the first and second day in vitro was determined.

[a]A mitotic cell was defined as a neuroblast which had undergone cytokinesis in vitro.

[b]For mitotic cells, results are expressed as percent of cells produced by cytokinesis during culture that survived to the indicated time. For the nonmitotic population, the data are expressed as the percent of cells initially plated (photographed at 3–4 hr) alive at the indicated time. (From Ref. 27.)

culture by 48 hr, indicating that mitotic cells developed trophic requirements in vitro. Addition of VIP rescued all mitotic cells at 24 hr, whereas 60% lived until 48 hr, indicating that the peptide promoted survival of mitotic precursors for a limited period. Moreover, addition of NGF to VIP allowed virtually all mitotic cells to survive for 48 hr, suggesting that newly born cells undergo a transition in trophic factor requirements with development (27) (Table 1). This analysis of single mitotic cells parallels results obtained above, characterizing trophic requirements of neurons from SCG of progressively older animals (Fig. 4). In sum, VIP appears to provide support to mitotic neuroblasts, which subsequently develop requirements for target-derived NGF.

For nondividing neuroblasts, trophic requirements were even greater: In control cultures, 50% of cells survived during the first 24 hr and <10% remained after 48 hr. Combined VIP and NGF rescued most of the cells, whereas VIP alone was less effective (Table 1). Greater trophic dependence of postmitotic neurons is well known. In aggregate, both mitotic and nonmitotic neuroblasts require trophic support, and VIP is a critical regulator prior to action of target-derived NGF.

D. Proliferation Requires Both Mitogenic and Trophic Stimulation

In the foregoing studies, maximal mitogenic stimulation was provided by insulin, which elicited a labeling index of 38%. However, cell number did not increase in vitro in the presence of insulin alone, due to concomitant cell death. By providing

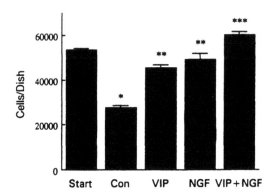

Figure 5   Effect of trophic factor treatment on neuronal cell number. Cells were cultured in control medium or in medium containing VIP, NGF, or VIP + NGF. All groups were grown in the presence of mitogenic insulin. Cells were counted after 48 hr. Start indicates the number of cells counted 4 hr after plating. Values from triplicate cultures are mean cells ± SEM. *Differs from start at $p < .05$. ** Differs from control at $p < .05$. *** Differs from VIP, NGF, and control at $p < .05$. (From Ref. 27.)

combined trophic factors, VIP and NGF, there was no change in labeling index, indicating that the factors did not stimulate mitosis directly. However, in the presence of both mitogenic and survival factors, an absolute increase in neuroblast number was observed at 48 hr: There were 16% more cells than the number initially plated (Fig. 5). In addition, combined factor treatment allowed multiple rounds of division by single precursors (Fig. 6). Time-lapse analysis further characterized the dynamics underlying this increase. Over 48 hr, 25% of neuroblasts divided, while approximately 10% underwent cell death, yielding a 15% net increase in cell number. Since cell number failed to increase in the presence of either insulin or trophic factors alone, our observations indicate that both mitogenic and trophic stimulation are required for proliferation of sympathetic neuroblasts (Fig. 5) (14,27). In aggregate, VIP enhances sympathetic proliferation by increasing mitogenesis and precursor survival during a critical embryonic period. In ongoing studies, the peptide also increases expression of the NGF *trk*A receptor protein and mRNA in SCG cultures, suggesting that VIP promotes the three principal mechanisms contributing to stable population growth.

E.   The Multiple Neurogenetic Effects of VIP Involve the cAMP Pathway

To begin defining second messengers mediating VIP ontogenetic effects, we examined cAMP production by radioimmunoassay (41). Exposure of neuroblasts to VIP (10 µM) increased cAMP content 18-fold, suggesting that this pathway may underlie peptide-induced mitogenesis and survival (Table 2). Further, activation of the cAMP pathway using multiple stimulators, including dibutyryl-cAMP, chlorophenylthio-cAMP, and forskolin, reproduced VIP induction of mitosis, neurite outgrowth, and

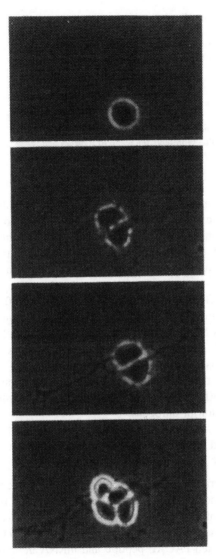

Figure 6 Serial time-lapse photographs of sympathetic neuroblast cultures. Cells were incubated in the presence of VIP and NGF and photographed at ~12 hr intervals for 2 days. A single neuroblast can be seen dividing twice, giving rise to 4 daughter cells. (From Ref. 27.)

Table 2 Effect of VIP and Forskolin on the cAMP Content of Sympathetic Neuroblast Cultures

	Control	VIP	Forskolin
cAMP/dish (pM)	2.3 ± 1.3	42.0 ± 1.1*	192.0 ± 9.4*

Twenty-four hours cultures were incubated with VIP (10 μM) or forskolin (30 μM) in the presence of IBMX (500 μM). cAMP content of culture extracts was determined by RIA after a 10-min incubation at 37°C. Each experimental value represents the mean of 4 dishes and is expressed as mean pM/dish ± SEM. (*) $p < .01$. (From Ref. 27.)

long-term survival, compatible with a role for cAMP signaling. However, other intracellular signals are likely involved, such as inositol phospholipids, since the combination of maximal cAMP activation, using forskolin, and VIP exposure produced an additive effect on mitogenic stimulation. In fact, in further studies, both VIP and PACAP activate turnover of inositol phospholipids in sympathetic neuroblasts (24).

Similar to VIP, IGF and cellular depolarization, using 30 mM KCl or the Na$^+$ channel agonist veratridine, also increased sympathetic neuroblast mitosis (15). Since VIP depolarizes SCG neurons, and depolarizing stimuli frequently activate voltage-sensitive Ca^{2+} channels, we examined the role of extracellular Ca^{2+} on peptide-induced mitosis (42). While Ca^{2+}-free medium and the voltage-sensitive Ca^{2+} channel antagonist nitrendipine (10^{-5}) blocked depolarization-induced mitosis, there was no effect on VIP mitogenesis, suggesting that extracellular Ca^{2+} plays little role in peptide regulation of ontogeny.

F. Sources and Relationships of VIP and PACAP in Sympathetic Neurogenesis

The multiple ontogenetic effects of VIP, and activation of second messenger pathways, required micromolar concentrations of the peptide. In recent studies, nanomolar concentrations of PACAP reproduced VIP regulation of mitosis, survival, differentiation, and intracellular signaling in the sympathetic neuroblast system (24). Further, we have detected PACAP peptide and mRNA in embryonic sympathetic ganglia (39), suggesting that both VIP and PACAP are involved in ganglion development. Potentially, the peptides affect different neuronal or glial subpopulations in the SCG, or act sequentially during a neurons ontogeny. While further study will be needed, the previous localization of VIP in innervating presynaptic nerve terminals from the intermediolateral spinal motor column (32) suggests that the peptide may be released, acting transsynaptically. However, VIP is localized in ganglion cells as well (29,33,34), raising the possibility of multiple sources. Combined developmental analysis of peptide expression, as well as in-vivo blockade of specific peptide activity, may be required to define contributions of these related peptide family members. The dose–response relationships of the peptides are compatible with activity mediated via the PACAP type I receptor. Recent studies have identified type I receptor mRNA in developing ganglia (39,40).

G. Peptide Activity and Signaling in Other Neurogenetic Model Systems

In addition to regulating peripheral neuroblasts, VIP and PACAP also influence precursors from the CNS. While sympathetic neurogenesis occurs prenatally in peripheral tissues, cerebellar granule neurons are generated postnatally in the external germinal layer overlying the cerebellum in the brain, as described earlier. The peptides can stimulate granule neuronal precursor mitosis in culture, apparently acting via the PACAP type I receptor (22). In fact, pharmacological studies indicate that PACAP type I receptors are present on the dividing granule neuronal precursors in vivo and in culture, and that peptide stimulation activates both the cAMP and inositiol phospholipid pathways independently (30,31). Both of these intracellular signals are common effectors for receptors coupled to G proteins, such as those of PACAP.

Recent studies in tumor cell lines related to sympathetic neuroblasts, however, raise the possibility that PACAP signaling in neurons may be far more complex. Emerging evidence suggests that there is no uniform relationship between a ligand–receptor system and transducing second messengers. Rather, receptors may activate a variety of intracellular signaling cascades as a function of cellular context. For example, studies of human neuroblastomas and rat pheochromocytoma (PC12) cell lines indicate that VIP and PACAP can stimulate neuronal differentiation, that is, neuritic process outgrowth (43,44). The peptides do elicit increases in both cAMP levels and phospholipid turnover in the cell lines, as we observed in primary sympathetic neurons (24). Unexpectedly, activation of cAMP signaling subsequently stimulated the mitogen-activated protein (MAP) kinase pathway (45). The MAP kinase pathway is interesting, since it was originally considered to be linked to receptors possessing protein tyrosine kinase activity, which mediate mitogenic stimulation. Later work, examining the sympathetic trophic factor, NGF and its cognate *trk* tyrosine kinase receptor, indicated that MAP kinase mediated neuronal differentiation, not proliferation, in these cells (46). Further, the MAP kinase cascade was essential for NGF stimulation of neuritic process outgrowth (46,47). The studies demonstrating PACAP stimulation of this same signaling cascade downstream of cAMP suggest that multiple molecules activate a diversity of intracellular messengers, acting in population-specific and developmental stage-dependent fashion. Further, complex cross-talk between intracellular signaling pathways is likely to occur (45).

IV. Potential Roles of VIP and PACAP in Neuronal Repair

The foregoing studies suggest that VIP and PACAP are expressed during the critical period of neurogenesis, and play regulatory roles in proliferation, survival, and neuronal process outgrowth. While peptide expression diminishes with age in peripheral ganglia, both are expressed in subpopulations of the adult SCG and dorsal root sensory ganglia (DRG), providing widespread innervation of blood vessels and organs, including the lungs (1). Significantly, following peripheral nerve injury or inflammation, the peptide levels increase in adult SCG and DRG, suggesting a role in the injury response (2,3).

The most extensive studies have been performed in the adult rat DRG, fibers of which are damaged during crush or section of the sciatic nerve. Few if any (0–3%) of DRG neuronal cell bodies express VIP protein or mRNA under normal conditions (3,48). Following injury, VIP immunoreactive sensory neurons increase to approximately 30% in the DRG, with a peak in protein occurring between 14 and 30 days (3,49). VIP expression occurs in neurons whose peripheral fibers have been transected (50). While a portion of the increase may reflect decreased axonal transport of the peptide from the soma to nerve terminales in targets, the majority reflects new synthesis, since comparable percentages of cells exhibited increased mRNA levels by in-situ hybridization histochemistry (51).

PACAP also appears to be a sensory neuronal peptide, involved in pain transmission (52,53). Under normal conditions, approximately 10–17% of DRG neurons express the peptide. After nerve section, PACAP peptide expression increases very rapidly: at 15 hr, 35% of neurons express the peptide, with increases to 50% at 3 days and 75% by 10 days (3,54). While expression levels fall thereafter, more than half the neurons maintain expression at one month. Thus, in comparison to VIP in the same study, PACAP exhibited a more rapid onset, with greater distribution, and more prolonged a time course (3), suggesting that the peptides serve different functions.

Finally, though time course studies remain to be performed, VIP expression in the adult SCG is also altered by axotomy. Forty-eight hours after nerve section, VIP immunoreactive protein increased 22-fold, accompanied by a fivefold increase in mRNA (55). Histologic analysis indicated that more neurons were expressing the peptide, and suggested that VIP appeared in a subset of those neurons whose processes were damaged (56).

The rapid onset of VIP and PACAP expression following peripheral nerve injury suggests that the peptides participate in the maintenance and repair program of the damaged neurons. Transection of the peripheral nerve processes severs connection of the neurons from target tissues which are known to provide necessary survival factors (8,14,27,38). Potentially, VIP or PACAP promote survival in these sympathetic neurons deprived of target NGF, as the peptides do during early development when the neurons lack the capacity (receptors) to respond to target trophic factors (23,27). Similarly, the peptides may stimulate neuronal process regrowth, potentially orchestrating the multiple gene products involved in initial nerve generation to reappear during mature regeneration (4,5,23).

Although we have focused on the direct effects of peptides on neuronal cell function, other cell types are likely targets for neuronal peptides, including supporting glial cells, immune cells, and local vasculature, to name just a few. Further, while

VIP and PACAP may serve to promote regeneration, other peptides, including galanin, cholecystokinin, and neuropeptide tyrosine, also increase in many neurons following injury, suggesting that different neuronal types require distinct sets of regulatory signals (2).

References

1. Rosselin G. Vasoactive Intestinal Peptide, Pituitary Adenylate Cyclase Activating Polypeptide, & Related Regulatory Peptides. Singapore: World Scientific, 1994.

2. Hokfelt T, Zhang X, Wiesenfeld-Hallin Z. Messenger plasticity in primary sensory neurons following axotomy and its functional implications. TINS 1994; 17:22–30.

3. Zhang Q, Shi T-J, Ji R-R, Zhang Y-T, Sundler F, Hannibal J, Fahrenkrug J, Hokfelt T. Expression of pituitary adenylate cyclase-activating polypeptide in dorsal root ganglia following axotomy: time course and coexistence. Br Res 1995; 705:149–158.

4. Oblinger MM, Szumlas RA, Wong J, Liuzzi FJ. Changes in neurofilament gene expression affects the composition of regenerating axonal sprouts elaborated by DRG neurons *in vivo*. J Neurosci 1989; 9:2645–2653.

5. Hoffman PN, Cleveland DW. Neurofilament and tubulin expression recapitulates the developmental program during axonal regeneration: induction of a specific tubulin isotype. Proc Natl Acad Sci (USA) 1988; 85:4530–4533.

6. Caroni P, Grandes P. Nerve sprouting in innervated adult skeletal muscle induced by exposure to elevated levels of insulin-like growth factors. J Cell Biol 1990; 110:1307–1317.

7. Ishii DN. Relationship of insulin-like growth factor II gene expression in muscle to synaptogenesis. Proc Natl Acad Sci (USA) 1989; 86:2898–2902.

8. Jacobson M. Developmental Neurobiology. New York: Plenum Press, 1991.

9. Rothman TP, Specht LA, Gershon MD, Joh TH, Teitelman G, Pickel VM, Reis DJ. Catecholamine biosynthetic enzymes are expressed in replicating cells of the peripheral but not the central nervous system. Proc Natl Acad Sci (USA) 1980; 77:6221–6225.

10. DiCicco-Bloom E, Black IB. Insulin growth factors regulate the mitotic cycle in cultured rat sympathetic neuroblasts. Proc Natl Acad Sci (USA) 1988; 85:4066–4070.

11. DiCicco-Bloom E, Townes-Anderson E, Black IB. Neuroblast mitosis in dissociated culture: regulation and relationship to differentiation. J Cell Biol 1990; 110:2073–2086.

12. Ernsberger A, Sendtener M, Rohrer H. Proliferation and differentiation of embryonic chick sympathetic neurons: effects of ciliary neurotrophic factor. Neuron 1989; 2:1275–1284.

13. Lee JM, DiCicco-Bloom E, Black IB. CNTF regulates neuroblast mitosis, survival and differentiation via multiple response mechanisms. Soc Neurosci 1991; 17:908.

14. DiCicco-Bloom E, Friedman WJ, Black IB. NT3 stimulates sympathetic neuroblast proliferation by promoting precursor survival. Neuron 1993; 11:1101–1111.

15. DiCicco-Bloom E, Black IB. Depolarization and insulin-like growth factor-I (IGF-I) differentially regulate the mitotic cycle in cultured rat sympathetic neuroblasts. Brain Res 1989; 491:403–406.

16. Chen LE, DiCicco-Bloom E, Black IB. Tumor necrosis factor (TNF) selectively inhibits mitosis of cultured brain neuroblasts. Soc Neurosci 1990; 16:803.

17. Wagner JP, Black IB, DiCicco-Bloom E. Stimulation of neurogenesis by peripheral

administration of basic fibroblast growth factor (bFGF) in the developing and adult rat brain. Soc Neurosci 1995; 21.

18. Cohen RN, DiCicco-Bloom E, Black IB. EGF and FGF regulate mitosis of cultured cerebellar granule cell precursors. Soc Neurosci 1990; 16:804.

19. Gao WQ, Heintz N, Hatten ME. Cerebellar granule cell neurogenesis is regulated by cell-cell interactions in vitro. Neuron 1991; 6:705–715.

20. Lu N, DiCicco-Bloom E. Lineage-specific regulation of neurogenesis: PACAP inhibits proliferation and promotes differentiation of cerebral cortical precursors. Soc Neurosci 1994; 20:1487.

21. Lu N, DiCicco-Bloom E. Pituitary adenylate cyclase activating polypeptide is an autocrine inhibitor of mitosis in cultured cortical precursor cells. Proc Natl Acad Sci (USA) 1997; 94:3357–3362.

22. DiCicco-Bloom E, Emsbo K, Black IB. VIP regulates development of multiple neuronal populations: sensory and cerebellar granule cells. Soc Neurosci 1992; 18:418.

23. Pincus DW, DiCicco-Bloom E, Black IB. Vasoactive intestinal peptide regulates mitosis, differentiation and survival of cultured sympathetic neuroblasts. Nature 1990; 343:564–567.

24. DiCicco-Bloom E, Deutsch P. Pituitary adenylate cyclase activating peptide (PACAP) potently stimulates mitosis, neuritogenesis and survival in cultured rat sympathetic neuroblasts. (Second Annual PACAP Symposium, New Orleans, 1991). Regulatory Peptides 1992; 37:319.

25. Baserga R. The Biology of Cell Reproduction. Cambridge, MA: Harvard University Press, 1985:251.

26. Yaginuma H, Tomita M, Takashita N, McKay SE, Cardwell C, Yin Q-W, Oppenheim RW. A novel type of programmed neuronal death in the cervical spinal cord of the chick embryo. J Neurosci 1996; 16:3685–3703.

27. Pincus DW, DiCicco-Bloom E, Black IB. Trophic mechanisms regulate mitotic neuronal precursors: role of vasoactive intestinal peptide (VIP). Brain Res 1994; 663:51–60.

28. Lu N, Black IB, DiCicco-Bloom E. A paradigm for distinguishing the roles of mitogenesis and trophism in neuronal precursor proliferation. Dev Brain Res 1996; 94:31–36.

29. Tyrrell S, Landis SC. The appearance of NPY and VIP in sympathetic neuroblasts and subsequent alterations in their expression. J Neurosci 1994; 14(7):4529–4547.

30. Basille M, Gonzales BJ, Leroux P, Jeandel L, Fournier A, Vandry H. Localization and characterization of PACAP receptors in the rat cerebellum during development: evidence for a stimulatory effect of PACAP on immature cerebellar granule cells. J Neurosci 1993; 57:329–338.

31. Basille M, Gonzalez BJ, Desrues L, Demas M, Fournier A, Vaudry H. Pituitary adenylate cyclase-activating polypeptide (PACAP) stimulates adenylyl cyclase and phospholipase C activity in rat cerebellar neuroblasts. J Neurochem 1995; 65:1318–1324.

32. Baldwin C, Sasek CA, Zigmond RE. Evidence that some preganglionic sympathetic neurons in the rat contain vasoactive intestinal peptide- or peptide histidine isoleucine amide-like immunoreactivities. Neuroscience 1991; 40:175–184.

33. Pincus DW, DiCicco-Bloom E, Black IB. Expression of vasoactive intestinal peptide (VIP) and receptor during early sympathetic ontogeny: potential autocrine role. Soc Neurosci 1991; 17:908.

34. Davidson A, DiCicco-Bloom E, Black IB, Draoui M, Zia F, Liling G, Fridkin M, Brenneman DE, Moody TW, Gozes I. The neuropeptide VIP: an autocrine regulator of cell growth. Soc Neurosci 1992; 18:418.

35. Cochard P, Goldstein M, Black IB. Initial development of the noradrenergic phenotype in autonomic neuroblasts of the rat embryo in vivo. Dev Biol 1979; 71:100–114.

36. Rubin E. Development of the rat superior cervical ganglion: ingrowth of preganglionic axons. J Neurosci 1985; 5:673–684.

37. Rubin E. Development of the rat superior cervical ganglion: initial stages of synapse formation. J Neurosci 1985; 5:697–704.

38. Coughlin MD, Collins MB. Nerve growth factor-independent development of embryonic mouse sympathetic neurons in dissociated cell culture. Dev Biol 1985; 110:392–401.

39. Lu N, DiCicco-Bloom E. PACAP ligand and receptor expression in embryonic cerebral cortex. Soc Neurosci 1996; 22:522.

40. May V, Brass KM. Pituitary adenylate cyclase activating polypeptide (PACAP) regulation of sympathetic neuron neuropeptide Y and catecholamine expression. J Neurochem 1995; 65:978–987.

41. Pincus DW, DiCicco-Bloom E, Black IB. Vasoactive intestinal peptide regulation of neuroblast mitosis and survival: role of cAMP. Brain Res 1990; 514:355–357.

42. Pincus DW, DiCicco-Bloom E, Black IB. Role of voltage-sensitive calcium channels in mitogenic stimulation of neuroblasts. Brain Res 1991; 553:211–214.

43. Deutsch PJ, Schadlow VC, Barzilai N. 38-Amino acid form of pituitary adenylate cyclase activating peptide induces process outgrowth in human neuroblastoma cells. J Neurosci Res 1993; 35:312–320.

44. Hernandez A, Kimball B, Romanchuk G, Mulholland W. Pituitary adenylate cyclase-activating peptide stimulates neurite growth in PC12 cells. Peptides 1995; 16:927–932.

45. Frodin MF, Peraldi P, Van Obberghen E. Cyclic AMP activates the mitogen-activated protein kinase cascade in PC12 cells. J Biol Chem 1994; 269:6207–6214.

46. Cowley S, Paterson H, Kemp P, Marshall CJ. Activation of MAP kinase kinase is necessary and sufficient for PC12 differentiation and for transformation of NIH 3T3 cells. Cell 1994; 77:841–852.

47. Pang L, Sawada T, Decker SJ, Saltiel AR. Inhibition of MAP kinase blocks the differentiation of PC-12 cells induced by nerve growth factor. J Biol Chem 1995; 270:13585–13588.

48. Kashiba H, Senba E, Ueda Y, Tohyama M. Colocalized but target-unrelated expression of vasoactive intestinal peptide and galanin in rat dorsal root ganglion neurons after peripheral nerve crush injury. Brain Res 1992; 582:47–57.

49. Shehab SAS, Atkinson ME. Vasoactive intestinal polypeptide (VIP) increases in the spinal cord after peripheral axotomy of the sciatic nerve originate from primary afferent neurons. Brain Res 1986; 372:37–44.

50. Shehab SAS, Atkinson ME, Payne JN. The origins of the sciatic nerve and the changes in neuropeptides after axotomy: a double labelling study using retrograde transport of true blue and vasoactive intestinal polypeptide immunohistochemistry. Brain Res 1986; 376:180–185.

51. Noguchi K, Senba E, Morita Y, Sato M, Tohyama M. Prepro-VIP and preprotachykinin mRNAs in the rat dorsal root ganglion cells following peripheral axotomy. Mol Brain Res 1989; 6:327–330.

52. Moller K, Zhang Y-Z, Hakanson R, Luts A, Sjolund B, Uddman R, Sundler F. Pituitary adenylate cyclase activating peptide is a sensory neuropeptide: immunocytochemical and immunochemical evidence. J Neurosci 1993; 57:725–732.

53. Zhang Y-Z, Sjolund B, Moller K, Hakanson R, Sundler F. Pituitary adenylate cyclase

activating peptide produces a marked and long-lasting depression of a c-fibre-evoked flexion reflex. J Neurosci 1993; 57:733–737.

54. Mulder H, Uddman R, Moller K, Zhang Y-Z, Ekblad E, Alulmets J, Sundler F. Pituitary adenylate cyclase activating polypeptide expression in sensory neurons. Neuroscience 1994; 63:307–312.

55. Hyatt-Sachs H, Schreiber RC, Bennett TA, Zigmond RE. Phenotypic plasticity in adult sympathetic ganglia *in vivo*: effects of deafferentiation and axotomy on the expression of vasoactive intestinal peptide. J Neurosci 1993; 13:1642–1653.

56. Mohney RP, Siegel RE, Zigmond RE. Galanin and vasoactive intestinal peptide messenger RNAs increase following axotomy of adult sympathetic neurons. J Neurobiol 1994; 25:108–118.

17

Neurotrophic Action of VIP
From CNS Ontogeny to Therapeutic Strategy

DOUGLAS E. BRENNEMAN and
JOANNA M. HILL

National Institute of Child Health and Human
  Development
National Institutes of Health
Bethesda, Maryland

PIERRE GRESSENS

Hôpital Robert-Debré
Paris, France

ILLANA GOZES

Sackler School of Medicine
Tel Aviv University
Tel Aviv, Israel

I. Introduction

Vasoactive intestinal peptide (VIP), a 28-amino acid neuropeptide widely distributed in the mammalian nervous system, has potent growth-related actions that influence cell division, neuronal survival, and neurodifferentiation. The recognition of a developmental action of VIP has its foundation in observations indicating that this neuropeptide has survival-promoting properties in primary neuronal cultures derived from the mouse central nervous system. During the past decade, the perceived role of VIP has expanded to include functions as a regulator of neurodevelopment and embryonic growth, a role that is not limited to single organ systems; rather, VIP has emerged as an integrative peptide that can coordinate organogenesis during critical periods.

The purpose of this review is to highlight evidence that indicates how VIP action elicits broad neurotrophic and growth-related effects, specifically in its role as a secretagogue for many molecules. In the CNS, VIP has been shown to cause the release of many substances from astroglia, including cytokines, growth factors, and a serine protease inhibitor. This diverse array of known VIP-regulated molecules provides a glimpse of the biochemical complexity of neuropeptide action during development. VIP's neurotrophic actions have been studied using a wide range of techniques and experimental preparations, encompassed by molecular and cellular studies, as well as effects on cultured whole embryos and VIP-related changes in mother and fetus during pregnancy. Combined, all data support the conclusion that VIP is a fundamental regulator of growth and development, while this peptide also provides cytoprotection from many neurotoxic substances.

Although most of this review will feature studies on the central nervous system, it is clear that the lung, the peripheral nervous system, and other organs utilize VIP in an analogous role as a growth regulator and neuroprotectant. As is the case for many developmental regulators, VIP is implicated directly in understanding and treatment of disease. In particular, VIP appears to be an autocrine growth factor for several tumors, including small-cell and non-small-cell lung carcinoma. The growth-inhibiting actions of VIP-related analogs on these tumors will be reviewed in the context of potential therapeutic agents. In addition, the emergence of VIP and VIP-related molecules as neuroprotectants in the treatment of neurodegenerative disease will be emphasized.

Important elements in establishing VIP as a growth and neurotrophic regulator include the localization of the peptide and its receptors in relevant neural structures at appropriate periods in development. With this perspective, we will describe the anatomical locations of the binding sites for VIP during prenatal development and discuss their possible functions and relationship to ontogenic events. Since the delineation of many of the neurotrophic actions of VIP have relied on the generation of specific VIP agonists and antagonists, pharmacological studies involving the use of these analogs will also be presented to provide insight into the nature of the receptors that mediate the neurotrophic actions of VIP. In addition, signal transduction studies on VIP's neurotrophic action will be summarized and related to current knowledge of cloned receptors for VIP and related peptides.

II. Neuronal Survival in the Central Nervous System

Often in the course of science, discoveries are made inadvertently while seeking other goals; such was the case in the discovery of the survival-promoting action of VIP. Earlier studies (1) indicated that treatment of developing spinal cord cultures with low concentrations of VIP could produce small increases in choline acetyltrans-ferase, an important enzyme in the synthesis of acetylcholine. While assessing control experiments testing whether these increases were due to an effect on cell survival, the discovery was made that VIP could prevent cell death in these cultures (2). The discovery was contingent on experimental conditions that probably account for the delayed recognition of the trophic effects: (a) The action of endogenous VIP was inhibited (2); (b) the peptide was tested in a system that was spontaneously degenerating (3); and (c) the cultures were utilized during a critical period of neurodevelopment (4,5). These experimental details are emphasized to underscore the possibility that similar strategies could unmask other neurotrophic substances. In the case of VIP, co-treating the cultures with tetrodotoxin (TTX) effectively blocked spontaneous synaptic activity and the exocytotic release of VIP from neurons (6).

The decrease in neuronal survival produced by this inhibition of VIP release allowed for the detection of survival-promoting activity of exogenous VIP. In addition, the spinal cord cultures utilized for these early experiments incurred a 50% cell loss during the second and third week in vitro (3), an event that has been used as a model for the naturally occurring neuronal cell death which is ubiquitous in the mammalian nervous system (7). The natural neuronal cell loss during the critical period in culture also provided an additional signal for the measurement of VIP-mediated increases in neuronal survival. Cell cultures from various regions of the CNS (spinal cord, cerebral cortex, and hippocampus) have been compared for their responsiveness to VIP neurotrophism. These studies have revealed that 35–50% of the total neurons in these cultures are apparently dependent on VIP for survival.

The survival-promoting action of VIP has interesting pharmacology. As shown in Fig. 1, the potency of VIP in promoting neuronal survival is very high, with EC_{50} of 30 pM. This effect is very unusual in that it was observed at a concentration far lower than that required for most VIP actions in other systems. In addition, a marked attenuation of the biological response was observed at concentrations that were greater than 1 nM, an amount at which most actions of VIP display a threshold of effectiveness. Several other actions of VIP in CNS cultures show similar dose

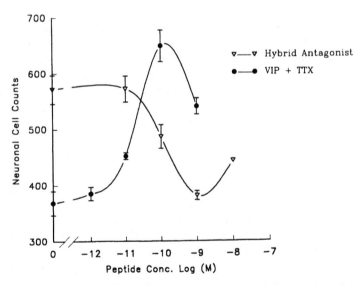

Figure 1 VIP increased the survival of dissociated spinal cord neurons obtained from fetal mice (filled circles). Cultures were treated for 5 days with VIP, in the presence of tetrodotoxin to block spontaneous electrical activity. Cell death produced by tetrodotoxin treatment alone is represented by the filled circle on the ordinate. The VIP hybrid antagonist (open triangles) decreased neuronal survival in electrically active spinal cord cultures. The EC_{50} is 0.1 nM. Cell counts from control cultures are represented by the open triangle on the ordinate.

responses. The proliferative response of astrocytes to VIP treatment has an EC_{50} in the 20–30 pM range (8). Most relevant to the survival-promoting action of VIP is its effect on cytokine release. Recent studies have shown that VIP-mediated release of interleukin-1α from astroglia has an EC_{50} of 50 pM (9), very similar to that observed for the effect of VIP on neuronal survival (2). During the critical period of cell death in spinal cord cultures, IL-1α has been shown to increase neuronal survival in electrically blocked spinal cord cultures (10).

The specificity of the survival-promoting effect of VIP has been demonstrated in that other members (secretin, growth hormone-releasing factor, PHI-27) of this family of peptides have not been effective (11). However, one other member of this family of peptides also has neurotrophic action: PACAP (pituitary adenylate cyclase activating peptide). PACAP (12), with about 70% amino acid identity to VIP, can increase neuronal survival in electrically blocked cultures. This peptide also has potent neuroprotectant effects (13). Currently, it is not known if VIP and PACAP work through a common receptor mechanism.

III. Mechanism of Neurotrophic Action

VIP neurotrophism in cultures entails both direct actions of the peptide (14) and indirect effects (15) elicited by the secretagogue activity of the peptide. In this section, the diversity of these secreted molecules in the CNS will be described, followed by a summary of current studies on the signal transduction mechanisms of VIP that relate to neurotrophic action.

Two observations indicated that VIP increases the survival of postmitotic CNS neurons. When developing neurons were cultured under conditions that resulted in very few surviving astroglia, there was no apparent effect of VIP on neuronal survival (15). In addition, when conditioned medium from VIP-stimulated glia was added to neuronal cultures, there was significant enhancement of survival (8), indicating that a secreted molecule(s) mediated the survival-promoting action of VIP. Much of our effect has focused on the identification of these secondary molecules and defining their role in development. A working model of the cellular interactions mediated by VIP is shown in Fig. 2. To date, the substances known to be released in response to VIP include cytokines (9); a novel growth factor (16); and a serine protease inhibitor, protease nexin I (17). VIP has been shown to release at least eight different cytokines from cerebral cortical astrocytes (18). These include interleukin-1α and -1β, interleukin-3, interleukin-6, tumor necrosis factor-alpha, interferon-gamma, macrophage colony-stimulating factor, and granulocyte colony-stimulating factor. Many of these substances have been shown to regulate cell division, gene expression, and neuronal survival (19). The variety and demonstrated biological activity of cytokines that are secreted after stimulation with VIP strongly suggest that their release is an important element in the control of central nervous system development by VIP. As previously mentioned, VIP has also been shown to increase the release of protease nexin-1 (PN-1), a serine protease inhibitor. This is particularly significant in that PN-1 has

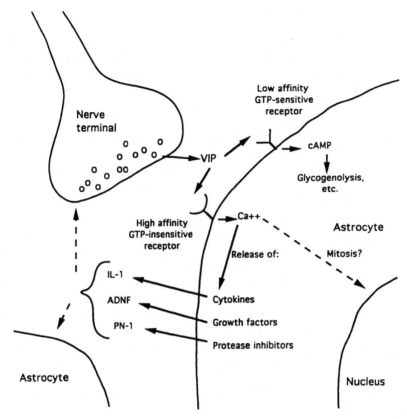

Figure 2 Schematic representation of VIP action in the central nervous system. VIP released from neurons acts on local receptors on both glia and other neurons. VIP can bind to at least two receptor subtypes on astrocytes: a low-affinity site that is hypothesized to be the GTP-sensitive receptor, and a high-affinity site thought to be the GTP-insensitive receptor. The low-affinity site is linked to adenylate cyclase and a number of functions including glycogenolysis. The high-affinity binding site is linked to calcium mobilization and has been associated with mitosis in astrocytes and secretagogue activity. Substances that are thought to be released by VIP include cytokines, growth factors, and protease inhibitors. Examples of these secreted substances are interleukin-1 (IL-1), activity-dependent neurotrophic factor (ADNF), and protease nexin I (PN-1). These factors may act on both adjacent glia and neurons.

marked effects on neurite extension (20) and neuronal survival (17). The presumed mechanism of PN-1 involves inhibition of thrombinlike protease activity in the CNS. Indeed, thrombin has been shown to decrease neuronal survival in developing spinal cord cultures (17). The role of thrombin in neuronal plasticity and neuronal survival is an active area of research in both developmental neurobiology and neuronal injury.

Growth factors comprise a third class of molecules released by VIP from glia.

A novel growth factor has recently been isolated from VIP-stimulated astroglia: activity-dependent neurotrophic factor or ADNF (16). Whereas all other molecules shown thus far to be released by VIP were previously identified substances, ADNF is novel, not only in structure but also as a representative of a new concept in neuro-protection: *extracellular stress proteins*. ADNF is a 14-kDa protein that has sequence similarity to heat shock protein 60, an intracellular molecule that has been recognized to be important in protein folding and cellular protection from thermal stress and toxic substances (21). Although the mode of action for ADNF is not yet established, the identification of this survival-promoting molecule opens a whole new realm of investigation into the regulation of neuroprotection and neuronal cell death. The pharmacology of ADNF is highly unusual in its extraordinary potency, exhibiting EC_{50}'s at or below femtomolar (10^{-15} M) concentrations. Furthermore, an active peptide fragment of ADNF has been discovered that has virtually identical potency to the parent molecule (see section on neuroprotection).

Previous studies indicated that the survival of postmitotic neurons in CNS cultures could be increased by a subnanomolar concentration of VIP. Therefore, one would predict from this pharmacological response that survival-promoting actions of VIP might involve a receptor and signal transduction mechanism that are unique, probably not involving cAMP-dependent phosphorylations, the conventional and accepted pathway for VIP's second messenger signal. Several lines of evidence indicate that VIP-mediated increases in neuronal survival and astroglial secretion are mediated through a calcium-dependent, protein kinase C-activated pathway. The first indication of a non-cAMP mechanism was from the great mismatch in EC_{50}'s for survival promotion (30 pM) (2) versus that of VIP-mediated increases in cAMP (3 μM) (22). This 100,000-fold difference in effective concentrations strongly suggested an alternative mechanism, an idea sustained by ensuing studies. In cultured astroglia cells, Fatatis and co-workers found that subnanomolar amounts of VIP elicited mobilization of intracellular calcium (23). This group found that one in five astrocytes responded by calcium increases after VIP treatment. The increase in calcium was apparently from an intracellular compartment, because the response to VIP persisted in the absence of extracellular calcium. Furthermore, thapsigargin pretreatment, which depletes intracellular calcium stores, abolished the VIP-induced calcium response. The VIP-elicited increases were accompanied by acute (1 min) and transient elevations of inositol triphosphate. Subsequent investigations indicated that subnanomolar VIP treatment resulted in the translocation of protein kinase C (PKC) from the cytoplasm to the nucleus (24). Examination of specific PKC isotypes by Western blot analysis showed that VIP treatment produced a marked increase in nuclear PKC-alpha and, to a lesser extent, PKC-delta and -zeta immunoreactivity. Although cAMP may not be involved in the secretion of survival-promoting substances from astroglia, this relationship may be restricted to specific cell types. In the case of insulin or amylase secretion, both cAMP and calcium-mobilizing effects may be involved (25,26). VIP action has been associated with other second messengers in the central nervous system. Treatment of pinealocytes with low concentrations of VIP produced increases in intracellular calcium, concomitant with elevations in cGMP and a cyclic nucleotide-gated cation channel (27). Thus, mechanisms of VIP-mediated secretion and other functions are probably cell-specific.

IV. VIP in Embryonic Growth and Ontogeny

Although the neurotrophic action of VIP was observed initially in CNS cultures, later work showed a dramatic stimulatory effect of VIP on embryonic growth. Previous in-vitro studies utilizing embryonic cells, as well as observations of widespread and abundant VIP-binding sites in the developing nervous system (28,29) (see below), provided impetus to investigate the roles of VIP during CNS ontogeny. Cultures of early postimplantation mouse embryos were utilized to test for potential effects of VIP on growth. This stage of embryonic development, which follows neural tube formation, is characterized by intense cell division in the primitive neuroepithelium. As shown in Fig. 3, a 4-hr incubation of embryonic day 9.5 (E9.5) mouse embryos in 10^{-10} to 10^{-7} M VIP stimulated growth (30). VIP caused a concentration-dependent increase in somite number, a marker of growth and maturation. An average of 5.2 new somites were produced with 10^{-7} M VIP, while control embryos acquired an average of 2.2 new somites during the culture period. Similar VIP growth-stimulating effects were seen on embryonic volume, DNA, and protein (11–36%, 103%, and 63% increase, respectively, compared with control). Labeling of cells in S-phase with

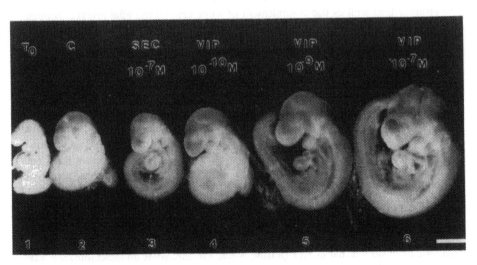

Figure 3 VIP enhanced growth of whole postimplantation embryos. T_0 shows the embryo size at the beginning of the 4-hr culture period. C is the size of controls after the 4-hr culture period. The VIP-related peptide, secretion, had no detectable effect on embryonic growth, whereas VIP produced a dose-dependent increase without apparent abnormalities. (Reprinted from Ref. 30, with permission.)

bromodeoxyuridine revealed increased labeling in both neural and nonneural tissues, indicating that cell division was the major mechanism for VIP-induced growth. Recent evidence has indicated that G_1 and S phases of the cell cycle are significantly shortened by VIP treatment of whole embryo cultures. This results in a dramatic reduction in the cell cycle duration (31). Although VIP-binding sites are localized to the developing nervous system (30), coordinated growth of both brain and body occurred, suggesting an indirect action of VIP on body growth through the VIP-stimulated release of additional regulatory factors. The growth-promoting effects were apparently specific for VIP and receptor-mediated, since addition of VIP to the culture medium resulted in a downregulation of binding sites. While the related peptides secretin and PACAP did not induce embryonic growth, a specific VIP hybrid antagonist partially blocked the VIP-stimulated increase in embryonic growth and the VIP-induced downregulation of VIP-binding sites.

Inhibition of VIP action during the early postimplantation period (E9-E11) with the VIP hybrid antagonist (see Section VII) produced growth retardation accompanied by microcephaly (32). At E17, the bodies of the treated fetuses had 89% of the DNA content and 72% of the protein content of control fetuses. However, the brains of the treated fetuses had only 52% of the DNA and 43% of the protein of control fetuses. Importantly, since blockade of VIP after E11 in the mouse did not retard growth, the regulatory effects of VIP on growth are apparently limited to a brief, early postimplantation period. Growth during this period, from E9 (closure of the neural tube) to E12 (onset of neocortical migration) in the mouse, is apparently under regulation by VIP. During this embryonic stage, precursor cells for the entire brain are produced, as well as the commitment of some cells to the glial lineage. Similar ontogenic events occur in the rat embryo from E10 to E12 and in the human embryo from 20 to 37 days of pregnancy (33).

In the rat, immunocytochemical and radioimmunoassay data have indicated that VIP is low or undetectable in the brain late in gestation, but increases rapidly after birth (34–36). The timetable for the appearance of VIP mRNA in the CNS, as indicated by in-situ hybridization histochemistry (ISHH) (28) and Northern blot (37), mirrored this pattern. VIP mRNA was not apparent in the brain until birth, after which it rapidly became widespread and abundant. Although ISHH (28) and Northern blot techniques (38) on rat prenatal tissues have identified abundant mRNA for VIP in the embryonic body as early as E14, even with the use of the more sensitive reverse transcription-PCR (rt-PCR) method, VIP mRNA was not detected in the E11 rat embryo (29), the developmental time comparable to E9 in the mouse when VIP is apparently an important regulator acting at CNS-binding sites. However, not only are VIP receptors present at that time in the rat embryo, radioimmunoassay (RIA) methods have shown that at E11 (a time when VIP mRNA is undetectable) the VIP content of the E11 rat embryo was four times higher than at E17 (a time when VIP mRNA is abundant in many body tissues) (29). This data strongly suggests that extraembryonic tissues contribute VIP to act on CNS receptors during early neurogenesis. For this reason we measured the VIP content of the maternal serum of rats throughout pregnancy. The RIA of maternal serum revealed a peak in VIP concentration at days E10 to E12 of pregnancy, with VIP rising to levels 6- to 10-fold higher than during the final third of pregnancy (29). This peak in maternal serum VIP levels coincided with the critical mid-gestational period, during which VIP has been shown to regulate embryonic growth, and suggests that it may act as a source of VIP during this stage. A mechanism for the transfer of VIP from the maternal blood to the embryo apparently exists, since radiolabeled VIP introduced into the maternal circulation was found in the embryo 15 min after injection. Furthermore, much of this material remained intact. This work indicates that VIP is among several maternal factors, including hormones and cytokines, that may influence the complex sequence of events resulting in coordinated prenatal development.

The VIP concentration of the E11 embryo (91 pg/mL) is higher than the concentration of VIP in the maternal serum (0.17 pg/mL), suggesting that there is a concentrating effect in the embryo and/or the uterine circulation. The measure of maternal serum VIP was obtained from blood collected from the entire body; however, higher concentrations of VIP may be reached within the uterine vasculature. The concentration of VIP in the human umbilical vasculature is reported to be about 2.5 times greater than in maternal venous blood, suggesting that VIP is selectively concentrated in the uterine circulation (39). Past work has indicated that VIP levels are held within narrow limits during mid-gestation (30,31). Whereas a dramatic increase in growth is demonstrated by treatment of embryos with 10^{-7} M VIP in whole embryo culture, the growth produced by 10^{-10} to 10^{-9} M VIP more clearly mimics in-utero growth (30). The retardation of growth experienced in control embryos in culture could be due to the removal of the embryo from its normal hormonal environment, including maternal VIP.

The fall of VIP in maternal serum in the rat, which begins at E13, is closely followed by changes in the embryo. VIP-binding sites appear in peripheral tissues and, in the CNS, VIP-binding sites triple (28,29). In the periphery, there is a rapid increase in the expression of VIP mRNA, including rat body at E16 (38), the stellate ganglia at E14.5 (40), and the superior cervical ganglion at E15 (14). Thus, the VIP synthesized by peripheral fetal rat tissues during the last third of pregnancy may provide a source of VIP to act on the abundant VIP-binding sites observed throughout the body during this time. However, using a highly sensitive ISHH method, Waschek et al. (41) have recently reported that, in the E11 mouse, VIP mRNA was detected in cells in the hindbrain. If VIP is being synthesized by these cells, centrally synthesized VIP may also act at CNS receptors in the developing mouse brain. In addition, although ISHH techniques failed to detect VIP mRNA in the rat placenta or preplacental tissues, rt-PCR of the E15 to E18 rat placenta revealed low levels of VIP mRNA, suggesting that this tissue may also synthesize VIP, which could reach the embryo-fetus (29). Whether rt-PCR will reveal low levels of VIP mRNA in earlier rat and mouse tissues is not yet known.

During postnatal brain development, VIP mRNA increases dramatically (28), with peak expression occurring during a period of synapse formation in the cerebral cortex (37). Similarly, VIP immunoreactivity increases rapidly during postnatal brain development, reaching a maximum 2 weeks after birth (34,35). As will be described in Section VII, blocking VIP in either pre- or postnatal development has deleterious effects on brain development.

V. VIP-Binding Sites in Brain and Development

Prerequisites for any growth-factor function are the bioavailability of the substance and the presence of relevant receptors. As described above, VIP is apparently available throughout development and may be derived from a combination of sources depending on the ontogenic stage. To date, most of our knowledge of developmental distribution of VIP receptors is based on the rat; however, data on the embryonic mouse will also be discussed. In this section a brief overview of molecular studies on VIP and PACAP receptors will be given, coupled with a survey of the both GTP-sensitive and GTP-insensitive binding sites for the brain and embryo.

Radioligand-binding assays, pharmacological experiments (22,28,42–45), and molecular cloning (46–48) have indicated multiple VIP receptors and binding sites. With regard to cloned receptors, the VIP1 (also called PACAP II) receptor consists of 459 amino acids and 7 transmembrane domains (46), while a second cloned VIP receptor (VIP2, also termed PACAP III) consists of 430 amino acids, with 50% homology with the VIP1 receptor (47). Stimulation of the VIP2-PACAP3 receptor probably mediates increases in cAMP and phospholipase C (48). VIP and the VIP-related peptide PACAP both bind to each of the two above-mentioned receptor proteins. A PACAP receptor exhibiting six different splice variants also has been cloned (49). This receptor is considered to be the PACAP type I receptor. Harmar and Lutz (50) have reviewed the current knowledge on the multiplicity of receptors and suggested two subtypes for PACAP type I receptor, IA and IB, corresponding to different states of the same receptor with differential coupling to G proteins that influence the pharmacology of the response. The IA responsiveness to PACAP27 and PACAP38 in elevating cAMP was much greater than it is to VIP; whereas in the same system, IB responded best to PACAP38, with decreasing responsiveness to PACAP27 and VIP in producing inositol phosphate.

Receptor-binding studies have indicated two types of PACAP receptor sites, type I preferring PACAP (IC_{50} = nM) versus VIP (IC_{50} = μM), and type II exhibiting a high and approximately equal affinity to both PACAP and VIP (51). Both types of receptor mediate an increase in cAMP. Type I was most abundant in the brain (52), while type II was most abundant in the lungs. We have recently demonstrated that the hybrid VIP antagonist was unable to displace. ^{125}I-PACAP from type I receptors (K_i > 10 μM) but was effective at type II receptors, suggesting a novel way to distinguish PACAP effects via type I and type II receptors. Furthermore, the hybrid antagonist recognized a VIP/helodermin-specific receptor, not recognized by PACAP (53).

In-vitro autoradiography with ^{125}I-VIP in the presence of a nonhydrolyzable guanosine 5'-triphosphate analogue, guanylyl-imido diphosphate (GMP-PNP) (10^{-5} M), has revealed both GTP-sensitive and GTP-insensitive VIP-binding sites throughout the brain. The GTP-insensitive sites have a higher affinity for VIP and likely consist of sites that are not linked to adenylate cyclase (44). The relationships among the cloned VIP/PACAP receptors and the GTP sensitivity in binding studies in the brain have yet to be established. The ratio of GTP-sensitive to GTP-insensitive binding sites in the brain differed from region to region. In most areas, including the neocortex, GMP-PNP inhibited between 40% and 60% of the VIP binding, indicating that the abundance of both binding sites was about equal. GTP had little effect on VIP binding in other brain regions, such as the medial geniculate, olfactory bulb, and ventral thalamic nuclei, indicating that these regions were especially enriched in the GTP-insensitive receptor. However, in the supraoptic nucleus, locus coeruleus, interpeduncular nucleus, olfactory tubercle, and periventricular hypothalamic nucleus, GTP-sensitive sites were more abundant, with 80% or more of VIP binding inhibited by GMP-PNP. Importantly, in-vitro autoradiography on brain sections co-incubated with ^{125}I-VIP and SNV, the superactive VIP analog acting at non-cAMP-linked binding sites (43), revealed that SNV displaced ^{125}I-VIP only at specific sites in the CNS. Those sites coincided with GTP-insensitive VIP-binding sites identified on adjacent sections (54). Co-treatment with both GTP and SNV resulted in the total displacement of radiolabeled VIP from brain sections. These data suggest that GTP-insensitive VIP-binding sites are not linked to adenylate cyclase and therefore are not either of the cloned VIP receptors.

Although little information is currently available on VIP receptors in the mouse throughout development, in-vitro autoradiography has localized binding sites in the E9 embryo (30). Transcripts for the VIP2 receptor mRNA (47), which is apparently linked to adenylate cyclase activation, have been detected as early as E14 in the brain (41). In the E9 mouse embryo, VIP-binding sites are largely, if not exclusively, localized to the CNS, with the highest binding occurring along the region of the floor plate of the developing neural tube (55). The floor plate of the neuroaxis governs tissue organization during this period of development, and the high density of VIP-binding sites along its extent suggests that VIP regulates the release of diffusible signals that are secreted from this region and coordinate early morphogenic events. The floor plate is composed exclusively of macroglia in the rat (56). This is consistent with the hypothesis that VIP actions on growth are similar to its neuroprotective actions and are mediated through factors relaxed from glial cells (18). In-vitro autoradiography of E9 embryos following whole embryo culture revealed that VIP treatment downregulated VIP-binding sites and further indicated that growth regulation occurs largely through GTP-insensitive binding sites (30).

During development of the rat, both GTP-sensitive and GTP-insensitive binding sites were shown to be abundant, exhibiting distribution patterns related to ontogenic events, and were primarily localized to the CNS early in development (28,29). VIP-binding sites have been observed as early as E11 in the rat. Although younger embryos were examined with in-vitro autoradiography, their small size made morphological localization of binding sites imprecise. In the E11 and E12 embryos, only the brain and neural tube exhibited moderately dense grains, with specific binding appearing to be primarily, if not exclusively, GTP-insensitive. In the E13 embryo, VIP-binding sites were also limited to the nervous system; however,

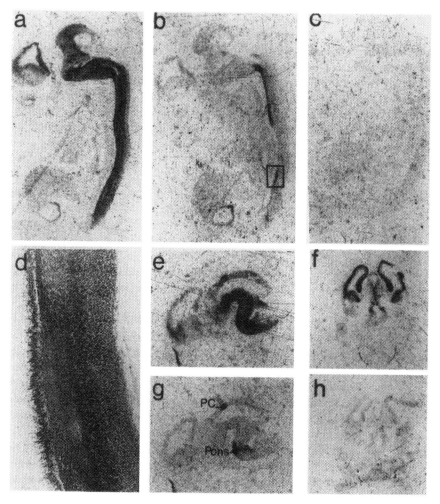

Figure 4   Autoradiographs of ^{125}I-VIP binding and photomicrographs of stained sections of E16 embryos illustrating the presence of GTP-sensitive and GTP-insensitive binding sites as well as their CNS localization. The total binding observed in the sagittal sections a, e, and f revealed the dense and widespread VIP-binding sites throughout the CNS. Plates b, g, and h were adjacent sections incubated in the presence of GTP and illustrate GTP-insensitive receptors which are localized to regions of gliogenesis and glial fasciculation in the floor plate of the spinal cord (b), brainstem (Pons), and in the posterior commissure (PC), in plate g. The region enclosed in the box in plate b is shown enlarged in plate d, and illustrates the fasciculation of glial fibers. (Reprinted from Ref. 28, with permission.)

binding sites were approximately tripled in density as compared to E11 and E12. Furthermore, about 50% of the binding in the E13 spinal cord was displaced with GTP. These changes in binding coincided with the decrease in VIP reported in maternal serum. The pattern of binding was maintained in the E14 spinal cord, where most VIP binding was GTP-sensitive and displaced with GTP and, as previously described in the E9 mouse, binding in the regions of the floor and roof plates was primarily GTP-insensitive.

From E12 to E16 in the rat (Fig. 4), binding sites were almost uniformly dense in the brainstem and spinal cord, while the brain began to exhibit regionally specific patterns of distribution. The highest binding occurred in neuroepithelial cell regions characterized by rapidly dividing cells: the floor and roof plates, the posterior commissure, and the intermediate medial thalamus. Beginning at E19, a new pattern emerged, and VIP binding became widespread and homogenous throughout the brain. This pattern remained until 2 weeks postnatal, when the adult pattern began to emerge. In the adult brain, there were regions (ventral thalamus, cortex, and suprachiasmatic nucleus) of overlap between VIP mRNA and VIP binding; however, other regions showed little temporal or spatial relationship between VIP gene expression and VIP binding.

The early postimplantation period of embryonic development, in which VIP regulates embryonic growth in the mouse, appears to overlap with the "initial stage of prenatal brain growth" described by Altman and Bayer (57). During the following "intermediate stage of prenatal brain growth," a change is seen in several measures related to VIP action. Some of the VIP-related parameters discussed here have been assessed only in the mouse and others only in the rat, and it is not known if these measures are typical of developing mammals. However, indications are that VIP, perhaps from extraembryonic sources including maternal tissues, regulates the coordinated growth of both brain and body of the embryo during the early postimplantation period of development. In addition, VIP regulation of growth occurs through VIP action on binding sites localized to the neural tube, including the floor-plate organizing center, to stimulate the release of additional growth-promoting factor(s), which in turn act on brain and body.

VI. VIP and Neuroprotection

A prevalent hope and aspiration is that neurotrophic molecules may be of therapeutic value against neurodegenerative disease or neoplasms. In this context, we believe that both VIP and substances related to the action of VIP have great potential as lead compounds for drug development, and as decisive probes for the investigation of growth and neurodevelopment. In this section, studies focusing on the neuroprotective properties of VIP in central nervous system will be reviewed.

As indicated previously, the discovery of VIP neurotrophism was made in a CNS culture system exposed to a recognized neurotoxin, tetrodotoxin (TTX), a substance that effectively blocks voltage-dependent sodium channels. Thus, these initial experiments not only focused on the original goal of determining the role of electrical activity in neurodevelopment, they also provided an ancillary goal in the investigation of neuroprotection. TTX-mediated cell death, which occurs in CNS cultures only during a restricted period of development in vitro, is believed to be an apoptotic process in which about half the neurons die. It is suggested that VIP is one of the extracellular signals that regulate CNS apoptosis.

Perhaps the most striking example of VIP neuroprotection in the CNS resides in studies of VIP and acquired immunodeficiency syndrome (AIDS). VIP was shown to effectively prevent neuronal cell death associated with the envelope protein (gp120) of the human immunodeficiency virus (59), the recognized etiological agent in AIDS. Exposure of hippocampal neurons to gp120 killed 30–40% of the neurons during a 5-day test period. Co-treatment with VIP prevented this cell loss. Although the mechanism of the cytoprotection is not entirely clear, subsequent studies strongly suggest that one of the substances released by VIP is probably involved. Treatment with femtomolar concentrations of ADNF, the glia-derived growth factor secreted upon VIP stimulation, protected neurons from gp120. Furthermore, a 14-amino acid peptide derived from ADNF mimicked the protection exhibited by the entire protein (16).

The ADNF-related peptide (ADNF-14) has also been tested for neuroprotective properties against other known neurotoxic agents. Treatment with femtomolar concentrations of ADNF-14 has been shown to prevent cell death with N-methyl-D-aspartate, an excitotoxic substance that stimulates a class of glutamate receptors. Overstimulation of these receptors has been hypothesized to be a factor mediating many types of neurotoxicity (60). The fact that ADNF-14 is effective in preventing this toxicity has many implications for its potential use in attenuating neuronal cell death associated with an excitotoxic mechanism. Gozes et al. have shown that a lipophilic analogue to VIP (SNV) (see section on VIP analog) can prevent neuronal cell death in cultures treated with beta-amyloid peptide, a putative neurotoxic substance implicated in the neurodegeneration of Alzheimer's disease (61). Studies with ADNF-14 now have shown that treatment of cerebral cortical cultures has prevented neuronal cell death produced by beta-amyloid peptide (16), suggesting again that VIP neuroprotection may be mediated, in part, by ADNF. Thus VIP, ADNF, and ADNF-related peptides are neuroprotective from a broad array of neurotoxic substances.

The molecular identity of the receptor that mediates the neuroprotective actions of VIP in the central nervous system has not yet been established. The two cloned VIP receptors are clearly linked to adenylate cyclase, and apparently do not interact with the VIP hybrid antagonist, as determined in cells transfected with either the VIP 1 or VIP 2 receptor (58). Therefore, it is unlikely that either of the currently recognized VIP receptors mediate neurotrophic VIP action. Alternatively, several PACAP receptors are linked to phosphotidyl inositol turnover and calcium mobilization, specifically those splice variants of the third intracellular loop that contain the "hop" cassette (49). At present, these receptors cannot be excluded as mediators of VIP neurotrophism. This is particularly relevant in that PACAP also exhibits neuroprotection. This response to PACAP exhibits a bimodal concentration effect, suggesting the interaction with two sets of receptors. Further studies are needed to resolve this aspect of VIP- and PACAP-associated neuroprotection.

VII. Pharmacology of VIP Analog

The elucidation of the biological role of any peptide is heavily contingent on the development of specific agonists and antagonists. Five VIP antagonists that have been utilized for this purpose include: (a) [4-Cl-D-Phe^6,Leu^{17}]-VIP, (b) [Ac-Tyrl,D-Phe2]-growth hormone releasing factor 1-29 amide, (c) VIP(10-28), (d) neurotensin$_{6-11}$-VIP$_{7-28}$ (hybrid antagonist), and (e) stearyl, norleucine17-hybrid antagonist (SNH). Since the actions of many of these compounds have been reviewed recently (61), our focus will be the newer compounds which appear to have specificity for VIP actions in the brain and in various lung cancer cell lines, and hence are most relevant to the scope of this discussion. The rationale for the hybrid VIP antagonist approach was that the chemical nature of the neurotensin N-terminal portion may augment membrane permeability of the VIP portion of the hybrid molecule (22). The hybrid antagonist was originally discovered as a potent inhibitor of VIP-stimulated sexual behavior (63). This compound also proved to be an important tool in confirming the neurotrophic action of VIP in CNS cultures. Incubation of the hybrid antagonist in spinal cord cultures produced a 35–50% reduction in the number of surviving neurons. The antagonist-induced cell death was competitively prevented by co-treating cultures with VIP (22). Similar findings were also observed with a C-terminal fragment, VIP$_{(10-28)}$ (2). The utility and specificity of the hybrid antagonist on central nervous system-associated receptor functions has been demonstrated: (a) The hybrid antagonist did not inhibit VIP-stimulated lymphoid cell proliferation (64); (b) the antagonist did not inhibit VIP-stimulated amylase secretion from pancreatic acini, exhibiting minimal agonistic activity (65); (c) prenatal administration of the hybrid early in development (E9-E11) produced severe microcephaly that was prevented by co-treatment with VIP, while producing smaller decreases in body growth (30); and (d) co-treatment with PACAP (66) did not block the hybrid-induced microcephaly (31). However, the hybrid antagonist was effective in inhibiting the growth of several lines of lung cancer cells (see next section). Taken together, the data suggest the existence of additional VIP receptor subtypes specific for peripheral tissues that are pharmacologically distinct from those in the central nervous system and in some lung cancer cell lines.

The prototype hybrid VIP antagonist (Neurotensin$_{6-11}$-VIP$_{7-28}$) was modified with two changes: the addition of the fatty acyl moiety stearic acid to the N-terminus (St-hybrid) and the replacement of the methionine in position 17 of the hybrid with norleucine (Nle-hybrid). Both of these analogs showed 10-fold greater potency than the original hybrid antagonist in neuronal killing. A molecule combining both the Nle substitution and the stearyl addition (SNH) showed remarkable biological activity: a 100-fold greater potency in killing spinal cord neurons than the original hybrid antagonist, and greater efficacy [significantly more neurons were killed (43)]. However, the increases in potency were only part of the benefit that was obtained with the second-generation hybrid antagonists. Quite unexpectedly, the new compounds exhibited a marked increase in receptor specificity which was not apparent in the parent hybrid antagonist. Although SNH had an exceptional ability to kill, it was a 1000-fold less potent than the original VIP hybrid antagonist in blocking VIP-stimulated cAMP accumulation. In contrast, the Nle-hybrid and the St-hybrid were each more potent than SNH in blocking VIP-stimulated cAMP accumulation, but less potent than the original hybrid antagonist. Thus, a single amino acid substitution and the addition of the fatty acid chain dramatically increased the potency and specificity of the parent peptide.

The importance of VIP to postnatal brain development has been revealed with the use of the hybrid VIP antagonist. The availability of various VIP antagonists has allowed for the assessment of VIP activities in vivo and delineation of the receptor sites and second messenger mechanisms involved. Daily administration of the VIP hybrid antagonist to newborn rat pups resulted in both cortical neurodystrophy (67) and delays in the onset of developmental milestones and behaviors (68). Golgi analysis of layer V cortical pyramidal neurons revealed a significant reduction in dendritic branch length and in the number of branches from the basilar tree after treatment with the VIP hybrid antagonist. In addition, many of the dendritic branches from the treated neurons were thickened, coarse, and vacuolated. Spines frequently appeared fused into clumps and often were malformed.

The morphological changes produced by treatment with the VIP hybrid antagonist strongly indicate that VIP has an important role in the formation and maintenance of dendritic aborizations during postnatal development. Additional support for this hypothesis was obtained by assessing the effect of the drug on the behavioral development of animals. Delays in motor development were observed in neonates treated with the hybrid VIP antagonist. Placing and righting behaviors were retarded (68). The development of circadian rhythmicity of motor behavior also was blocked (43). However, interference with circadian activity rhythms was limited to VIP antagonists that could inhibit VIP-mediated increases in cAMP. The lipophilic antagonist had no effect (43). Indeed, a major site of VIP synthesis is the suprachiasmatic nucleus (SCN) of the hypothalamus, the brain area controlling various biological rhythms (69). VIP mRNA oscillates during the day/night cycle, with peak levels observed at night (70–72). It has been demonstrated that intracerebral infusion of VIP antisense oligonucleotides into the SCN of rats inhibited multiple circadian outputs such as corticosterone and LH secretion (73,74). Furthermore, cAMP has been shown to reset some circadian clocks in the mammalian SCN in vitro (75). Utilizing a receptor-specific antagonist, the involvement of VIP-stimulated cAMP formation in the determination of motor daily rhythmicity in vivo was demonstrated in the developing rat (43). In contrast to their lack of effect on biological rhythmicity, the lipophilic analogs did promote neuronal survival and did promote learning and memory (76). Clearly, both receptor discrimination and assessment of VIP's role in postnatal development have advanced with the discovery and pharmacological characterization of these analogs.

An important addition to VIP pharmacology, as it relates to neurotrophism, has emerged with the discovery of a lipophilic VIP agonist: stearyl, norleucine[17]-VIP (SNV). The strategy for SNV design was substituting norleucine for methionine (amino acid 17 of VIP), with the goal of stabilizing the molecule against oxidation as well as increasing lipophilicity with the addition of a fatty acyl moiety (77,78). The resulting VIP analog, stearyl-Nle-VIP (SNV), exhibited a 100-fold greater potency than VIP, with maximal neuroprotective action observed at 1 pM. SNV was effective over a broader concentration range than VIP, which itself exhibited a peak of activity at 0.1 nM, decreasing at either 10-fold higher or 10-fold lower concentrations (43). Furthermore, SNV provided the most definitive evidence that VIP neurotrophism was operating through a mechanism that did not involve cAMP as a second messenger. SNV treatment of cerebral cortical cultures or spinal cord cultures increased neuronal survival via a cAMP-independent mechanism (43). As SNV comprised two structural modifications in the VIP molecule, each of the two compounds containing the individual changes (stearyl-VIP and Nle-VIP) was tested separately. Each of the analogs showed a 10-fold greater potency than VIP in protecting neurons from death. When tested for cAMP formation in cortical astrocytes (22,79), both SNV and stearyl-VIP were essentially inactive; whereas Nle-VIP increased cAMP levels at 10^{-7} M (43), a concentration at which VIP was inactive (22). SNV had no antagonistic effect on VIP-stimulated cAMP formation when tested over a wide concentration range. The net pharmacological effect of SNV was not only a dramatic increase in the potency to promote neuronal survival, but also an unexpected increase in both its specificity (excluding adenylate cyclase activation) and biological effectiveness (activity over a much broader range of concentration than VIP). The mechanism by which SNV exerts its neuroprotective effects is yet to be elucidated. However, it is known that SNV operates through the high-affinity VIP receptor, invoking the possibility of Ca^{2+} mobilization followed by secretion of growth factors. The addition of the lipophilic moiety confers stability to the molecule (80), which may lead to activity at lower apparent concentrations as compared to the parent molecule, VIP.

VIII. Therapeutic Potential of VIP Analogs

A central target in drug design is the formulation of compounds that efficaciously and specifically interact with a unique receptor site to elicit a desired response. Since neuropeptides in general, and VIP in particular, have been suggested as regulators of cell mitosis (81–84), the influence of VIP on cancer proliferation was investigated. Cells of the most prevalent lung cancer, non-small-cell lung cancer (NSCLC), have receptors for VIP. We have recently demonstrated that the NSCLC cells also express the VIP gene (85) and that VIP can serve as an autocrine growth factor in these malignancies (86). Moreover, the VIP hybrid antagonist was shown to inhibit lung cancer cell growth, in vitro utilizing an agarose cloning system, and in vivo in nude mice (86). The attenuation of tumor proliferation was dose-dependent and receptor-mediated, with cAMP as a possible second messenger. In small-cell lung carcinoma, VIP mRNA was much less abundant (85); however, the hybrid antagonist inhibited the growth of these cells in vitro (87), suggesting a paracrine mechanism of growth regulation by VIP.

Interestingly, in a recent publication, Sreedharan and his colleagues (88) describe the structure, expression, and chromosomal localization of the VIP1 receptor gene. Their results localized the human gene to chromosome 3 (3p22), a region associated with small-cell lung cancer. Allele loss in the area 3p23-p21 has been linked to many types of cancer, including small-cell lung carcinoma, possibly due to the presence of a functional tumor-suppressor gene in the region (89,90). Sreedharan et al. (48) have demonstrated a high level of expression of the human VIP1 receptor gene in lung tissue and suggested analysis of this gene-associated polymorphism for the acquisition of further information on the role of VIP and its receptors in human cancer. The results of Sreedharan et al. indicate that VIP may suppress lung cancer proliferation through the VIP1 receptor. However, results by Usdin et al. (58) indicate that the hybrid antagonist, under certain conditions, does not recognize the VIP1 receptor, suggesting the involvement of another VIP receptor site in the tumor inhibition mediated by the hybrid antagonist, perhaps a PACAP-preferring receptor. Studies on other VIP receptor genes, as well as on PACAP receptor genes, are necessary to decipher the role of VIP in cancer propagation.

While stimulating the division of some cell types, VIP also can inhibit normal cell proliferation, e.g., of lymphoid cells (64). Again, this dual action may be mediated by different receptors and may lead to differential activation of second messengers, exhibiting strong dependency on the concentration of the ligand.

In neuroblastoma, the most common solid malignancy of children less than 5 years of age, VIP has been suggested as an autocrine growth factor, inducing differentiation or stimulating cell division (91,92) in a temporal fashion, depending on the cell line tested and the environmental milieu. Our studies were conducted on human neuroblastoma (NMB) cells, originally obtained during a bone marrow biopsy of a 10-month-old girl (93). A 24-hr treatment with VIP stimulated neuroblastoma cell division, in a dose-dependent manner, as measured 48 hr after plating by cell counts and thymidine incorporation. VIP specifically bound to receptors on these neuroblastoma cells, and the receptor expression was temporally regulated, exhibiting higher affinity in younger cells (24 hr after plating versus 48 hr). Northern blot hybridizations indicated the existence of VIP mRNA in these cells (91), and VIP-like immunoreactivity was secreted by this neuroblastoma (92). Furthermore, recent findings have demonstrated specific inhibition of neuroblastoma cell multiplication by the VIP hybrid antagonist (92), possibly extending the use of this compound as an anticancer agent.

The most dramatic inhibition of cancer growth by the hybrid antagonist was observed on breast cancer (94). Moreover, similar effects of VIP were also found in other tumor lines, e.g., in a human colon cell line, where VIP exhibited a dose-related biphasic influence on cell proliferation (95) and the hybrid antagonist inhibited colon cell line proliferation (96). Thus, future experiments are aimed at both identifying other tumors that respond to the hybrid as well as at improving the molecule as a potential targeted anticancer drug.

VIP is a peptide that is vital for development and, as suggested below, its reduction may be associated with aging. VIP gene expression exhibits developmental changes with peaks of expression at day 16 of gestation in the developing rat embryonic intestinal system (37) and during postnatal brain development at the time of synapse formation. Anatomical studies combining immunohistochemistry, radio-immunoassays, receptor distribution, and other gene expression studies attest to the availability of VIP in brain areas important for cognitive functions, such as the cerebral cortex (97,98) and the hippocampus. However, VIP mRNA is significantly decreased in the cerebral cortex of aged animals (38), leading to decreased peptide concentration (99). Blockade of VIP actions by the VIP hybrid antagonist (100) or experimentally reduced VIP expression in transgenic animals (101) resulted in behavioral deficits associated with impairment of learning and memory. Furthermore, VIP may modulate brain activity through its ability to enhance cholinergic function (102–104), which is markedly deteriorated in Alzheimer's disease patients. Indeed, a major class of neurons that are known to be lost in Alzheimer's disease are cholinergic neurons (105). Although there is conflicting evidence concerning the role of VIP in dementia in general, and in Alzheimer's dementia in particular, a study in Alzheimer's and control brains showed a significant reduction of VIP immunoreactivity in the cerebral cortex, especially in the insular and angulate cortex, of Alzheimer's patients (106). However, it has never been determined whether this reduction was the cause or a result of the Alzheimer's deterioration of the cortex. Another potential player in the etiology of this disease is the beta-amyloid peptide that is excessively deposited in the brains of the patients, contributing to the neurodegenerative process which leads to senility (107).

The potent lipophilic analog to VIP [stearyl-norleucine[17] VIP (SNV)] (see Section V) has been investigated for neuroprotection in model systems related to Alzheimer's disease (61,76). In one model, using the beta-amyloid peptide (amino acids 25–35) as a neurotoxin, a 70% loss in the number of neurons in rat cerebral cortical cultures in comparison to controls was found. This cell death was completely prevented by co-treatment with 0.1 pM SNV. In a second in-vivo model of rats treated with a cholinergic blocker (ethylcholine aziridium) (108), SNV injected intracerebro-ventricularly or delivered intranasally prevented impairments in spatial learning and memory associated with cholinergic blockade. These studies suggested both a novel therapeutic strategy for the treatment of Alzheimer's deficiencies and a means for noninvasive peptide administration to the brain (109). With senile dementia of the Alzheimer's type afflicting 3–5 million people in the United States, and with the increasing size of the elderly population, a VIP-based novel drug design is of timely importance.

IX. Summary

VIP is an integrative peptide that exhibits multiple regulatory effects on the developing nervous system. Many of the neurotrophic functions of VIP are mediated through the coordinated release of other substances, including cytokines, growth factors, and protease inhibitors. The secretagogue action of VIP has been pharmacologically associated with the mobilization of intracellular calcium in astroglia. A growth-promoting action of VIP has been demonstrated in rodent embryos at mid-gestation, a period in pregnancy when significant increases VIP levels occur in the maternal plasma. VIP-binding sites in the embryo and brain are abundant, exhibiting patterns which change dramatically during ontogeny. The neurotrophic function of VIP in developing neural systems has engendered studies on the potential of VIP as a neuroprotective agent. Stable and potent agonists to VIP have been developed that protect neurons from clinically relevant toxic substances, including the HIV envelope protein (AIDS dementia complex) and beta-amyloid peptide (Alzheimer's disease). In addition, promising antagonists to VIP have been discovered that inhibit the autocrine growth function of VIP in some tumors. The further investigation of VIP actions will surely produce greater insight into the regulation of brain and embryonic development and provide exciting new therapeutic approaches to the treatment of human disease.

Acknowledgments

This manuscript was written when Professor Illana Gozes (The Lily and Avraham Gildor Chair for the Investigation of Growth Factors, Tel Aviv University) was a Fogarty-Scholar-in-Residence at the National Institutes of Health. The authors wish to thank Gretchen Gibney, Susan Lee, Raquel Castellon, and Janet Hauser for their helpful suggestions about this manuscript.

References

1. Brenneman DE. Role of electrical activity and trophic factors during cholinergic development in dissociated cultures. Can J Pharmacol Physiol 1986; 64:356–362.
2. Brenneman DE, Eiden LE. Vasoactive intestinal peptide and electrical activity influence neuronal survival. Proc Natl Acad Sci (USA) 1986; 83:1159–1162.
3. Brenneman DE, Fitzgerald S, Litzinger MJ. Neuronal survival during electrical blockade is increased by 8-bromo cyclic adenosine $3',5'$ monophosphate. J Pharmacol Exp Therap 1985; 233:402–408.
4. Brenneman DE, Neale E, Habig W, Nelson PG. Developmental and neurochemical specificity of neuronal death produced by electrical blockade in dissociated spinal cord cultures. Dev Brain Res 1983; 9:13–27.
5. Brenneman DE, Fitzgerald S, Nelson PG. Interaction between trophic action and electrical activity in spinal cord cultures. Dev Brain Res 1984; 15:211–217.
6. Brenneman DE, Eiden LE, Siegel RE. Neuropeptide regulation of neuronal development in spinal cord cultures. Peptides 1985; 6:35–39.
7. Clarke PGH, Oppenheim RW. Neuron death in vertebrate development: in vivo methods. In: Methods in Cell Biology. Vol. 46. New York: Academic Press, 1995: 277–321.

8. Brenneman DE, Nicol T, Warren D, Bowers LM. Vasoactive intestinal peptide: a neurotrophic releasing agent and an astroglial mitogen. J Neurosci Res 1990; 25:386–394.

9. Brenneman DE, Hill JM, Glazner GW, Gozes I, Phillips TM. Interleukin-1 alpha and vasoactive intestinal peptide: enigmatic regulation of neuronal survival. Int J Devel Neurosci 1995; 13:187–200.

10. Brenneman DE, Schultzberg M, Bartfai T, Gozes I. Cytokine regulation of neuronal survival. J Neurochem 1992; 58:454–460.

11. Brenneman DE, Foster GA. Structural specificity of peptides influencing neuronal survival during development. Peptides 1987; 8:687–694.

12. Miyata A, Arimura A, Dahl RR, Minamino N, Uehara A, Jiang L, Culler MD, Coy DH. Isolation of a neuropeptide corresponding to the N-terminal 27 residues of the pituitary adenylate cyclase activating polypeptide with 38 residues (PACAP 38). Biochem Biophys Res Commun 1990; 170:643–648.

13. Arimura A, Somogyyvari-Vigh A, Weill C, Fiore C, Tatsuno I, Bay V, Brenneman DE. PACAP functions as a neurotrophic factor. Ann NY Acad Sci 1994; 739:228–243.

14. Pincus DW, DiCicco-Bloom E, Black IB. Vasoactive intestinal peptide regulates mitosis, differentiation and survival of cultured sympathetic neuroblasts. Nature 1990; 43:564–567.

15. Brenneman DE, Neale EA, Foster GA, d'Autremont SW, Westbrook GL. Non-neuronal cells mediate neurotrophic action of VIP. J Cell Biol 1987; 104:1603–1610.

16. Brenneman DE, Gozes I. A femtomolar acting neuroprotective peptide. J Clin Invest 1966; 97:2299–2307.

17. Festoff BE, Nelson PG, Brenneman DE. Prevention of activity-dependent neuronal death: vasoactive intestinal polypeptide stimulates astrocytes to secret the thrombin-inhibiting, neurotrophic serpin, protease nexin I. J Neurobiol 1996; 30:255–266.

18. Brenneman DE, Hill JM, Gozes I, Phillips TM. Vasoactive intestinal peptide increases the secretion of eight cytokines from cerebral cortical astrocytes. Soc Neurosci Abstr 1994; 20:695.

19. Mehler MF, Kessler JA. Growth factor regulation of neuronal development. Dev Neurosci 1994; 16:180–195.

20. Monard D, Niday E, Limat A, Solomon F. Inhibition of protease activity can lead to neurite extension in neuroblastoma cells. Progr Brain Res 1983; 58:359–364.

21. Linquist S, Craig EA. The heat shock proteins. Annu Rev Genet 1988; 22:631–677.

22. Gozes I, McCune S, Jacobson L, Warren D, Moody TW, Fridkin M, Brenneman DE. An antagonist to vasoactive intestinal peptide affects cellular functions in the central nervous system. J Pharmacol Exp Ther 1991; 257:959–966.

23. Fatatis A, Holtzclaw LA, Avidor R, Brenneman DE, Russell JT. VIP increases intracellular calcium in astroglia: synergism with alpha-adrenergic receptors. Proc Natl Acad Sci (USA) 1994; 91:2036–2040.

24. Olah Z, Lehel C, Anderson WB, Brenneman DE, Agoston DV. Subnanomolar concentration of VIP induces the nuclear translocation of protein kinase C in neonatal rat cortical astrocytes. J Neurosci Res 1994; 39:355–363.

25. Straub SG, Sharp GW. A wortmannin-sensitive signal transduction pathway is involved in the stimulation of insulin release by vasoactive intestinal peptide and pituitary adenylate cyclase-activating polypeptide. J Biol Chem 1996; 271:1660–1668.

26. Schafer C, Steffen H, Printz H, Goke B. Effects of synthetic cyclic AMP analogs on amylase exocytosis from rat pancreatic acini. Can J Physiol Pharmacol 1994; 72:1138–1147.

27. Schaad NC, Vanecek J, Rodriguez IR, Klein DC, Holtzclaw L, Russell JT. Vasoactive intestinal peptide elevates pinealocyte intracellular calcium concentrations by enhancing influx: evidence for involvement of a cyclic GMP-dependent mechanism. J Pharmacol Exp Ther 1995; 47:923–933.

28. Hill JM, Agoston DV, Gressens P, McCune SK. Distribution of VIP mRNA and two distinct VIP binding sites in the developing rat brain: relation to ontogenic events. J Comp Neurol 1994; 342:186–205.

29. Hill JM, McCune SK, Alvero RJ, Glazner GW, Henins KA, Stanziale SF, Keimowitz J, Brenneman DE. Maternal vasoactive intestinal peptide and the regulation of embryonic growth. J Clin Invest 1996; 97:202–208.

30. Gressens P, Hill JM, Gozes I, Fridkin M, Brenneman DE. Growth factor function of vasoactive intestinal peptide in whole cultured mouse embryos. Nature 1993; 362:155–158.

31. Gressens P, Paindaveine B, Hill JM, Evrard P, Brenneman DE. Vasoactive intestinal peptide-induced shortening of S phase in whole cultured mouse embryos. Soc Neurosci Abstr 1995; 607:21.

32. Gressens P, Hill JM, Paindaveine B, Gozes I, Fridkin M, Brenneman DE. Severe microcephaly induced by blockade of vasoactive intestinal peptide function in the primitive neuroepithelium of the mouse. J Clin Invest 1994; 94:2020–2027.

33. Kaufman MH. 1992. The Atlas of Mouse Development. New York: Academic Press.

34. Emson PC, Gilbert RFT, Loren I, Fahrenkrug J, Sundler F, Schaffalitzky de Muckadel OB. Development of vasoactive intestinal polypeptide (VIP) containing neurones in the rat brain. Brain Res 1979; 177:437–444.

35. McGregor GP, Woodhams PL, O'Shaughnessy DJ, Ghatei MA, Polak JM, Bloom SR. Developmental changes in bombesin, substance P, somatostatin and vasoactive intestinal polypeptide in the rat brain. Neurosci Lett 1982; 28:21–27.

36. Maletti M, Besson J, Batialle D, Laburthe M, Rosselin PJ. Ontogeny and immunoreactive forms of vasoactive intestinal peptide (VIP) in rat brain. Acta Endocrinol (Copenh) 1980; 93:479–487.

37. Gozes I, Shani Y, Rostene WH. Developmental expression of the VIP gene in brain and intestine. Mol Brain Res 1987; 2:137–148.

38. Gozes I, Schachter P, Shani Y, Giladi E. Vasoactive intestinal peptide gene expression from embryos to aging rats. Neuroendocrinology 1988; 47:27–31.

39. Ottesen B, Ulrichsen H, Fahrenkrug J, Larsen JJ, Wagner G, Schierup L, Sondergaard F. Vasoactive intestinal polypeptide and the female genital tract: relationship to reproductive phase and delivery. Am J Obstet Gynecol 1982; 143:414–420.

40. Tyrrell S, Landis SC. The appearance of NPY and VIP in sympathetic neuroblasts and subsequent alterations in their expression. J Neurosci 1994; 14:4529–4547.

41. Waschek JA, Ellison J, Bravo DT, Handley V, Embryonic expression of vasoactive intestinal peptide (VIP) and VIP receptor genes. J Neurochem 1996; 66:1762–1765.

42. Sarrieau A, Najimi M, Chigr F, Kopp N, Jordan D, Rostene W. Localization and developmental pattern of vasoactive intestinal polypeptide binding sites in the human hypothalamus. Synapse 1994; 17:129–140.

43. Gozes I, Lilling G, Glazer R, Ticher A, Ashkenazi IE, Davidson A, Rubinraut S, Fridkin M, Brenneman DE. Superactive lipophilic peptides discriminate multiple VIP receptors. J Pharmacol Exp Ther 1995; 273:161–167.

44. Hill JM, Harris A, Hilton-Clarke DI. Regional distribution of guanine nucleotide-sensitive and guanine nucleotide-insensitive vasoactive intestinal peptide receptors in rat brain. Neuroscience 1992; 48:925–932.

45. Martin J-L, Feinstein DL, Yu N, Sorg O, Rossier C, Magistretti PJ. VIP receptor subtypes in mouse cerebral cortex: evidence for a differential localization in astrocytes, microvessels and synaptosomal membranes. Brain Res 1992; 587:1–12.

46. Ishihara T, Shigemoto R, Mori K, Takahashi K, Nagata S. Functional expression and tissue distribution of a novel receptor for vasoactive intestinal polypeptide. Neuron 1992; 8:811–819.

47. Lutz EM, Sheward WJ, West KM, Morrow JA, Fink G, Harmar AJ. The VIP2 receptor: molecular characterization of cDNA encoding a novel receptor for vasoactive intestinal peptide. FEBS Lett 1993; 334:3–8.

48. Sreedharan SP, Patel DR, Huang JX, Goetzl EJ. Cloning and functional expression of a human neuroendocrine vasoactive intestinal peptide. Biochem Biophys Res Commun 1993; 193:546–553.

49. Journot L, Waeber C, Pantaloni C, Holsboer F, Seeburg PH, Bockaert J, Spengler D. Differential signal transduction by six splice variants of the pituitary adenylate cyclase-activating peptide (PACAP) receptor. Biochem Soc Trans 1995, 23:133–137.

50. Harmar T, Lutz E. Multiple receptors for PACAP and VIP. TIPS 1994; 16:97–99.

51. Shivers BD, Gorcs TJ, Gottschall PE, Arimura A. Two high affinity binding sites for pituitary adenylate cyclase activating polypeptide have different tissue distributions. Endocrinology 1991; 128:3055–3065.

52. Lam HC, Takahashi K, Ghatei MA, Kanse SM, Polak JM, Bloom SR. Binding sites of a novel neuropeptide pituitary adenylate cyclase activating polypeptide in the rat brain and lung. Eur J Biochem 1990; 193:725–729.

53. Sone M, Smith DM, Ghatei MA, Gozes I, Brenneman DE, Fridkin M, Bloom SR. Pituitary adenylate cyclase activating polypeptide (PACAP)/vasoactive intestinal peptide (VIP) receptor subtypes in rat tissues: investigation of receptor binding and molecular identification by chemical cross linking. Biomed Res 1994; 15:145–153.

54. Hill JM, Dibbern DA Jr, Gozes I, Fridkin M, Brenneman DE. VIP analogue stimulates embryonic growth through GTP-insensitive binding sites. Soc Neurosci Abstr 1995; 21:1783.

55. Hill JM, Gressens P, Glazner GW, Brenneman DE. Floor plate neuroepithelium has binding sites for growth regulator, vasoactive intestinal peptide. Soc Neurosci Abstr. In press.

56. Altman J, Bayer SA. The development of the rat spinal cord. Adv Anat Embryol Cell Biol 1984; 85:1–166.

57. Altman J, Bayer SA, Atlas of prenatal rat brain development. Ann Arbor, MI: CRC Press, 1995:589.

58. Usdin TB, Bonner TI, Mezey E. Two receptors for vasoactive intestinal polypeptide with similar specificity and complementary distributions. Endocrinology 1994; 135:2662–2680.

59. Brenneman DE, Westbrook GL, Fitzgerald SP, Ennist DL, Elkins KL, Ruff R, Pert CB. Neuronal cell killing by the envelope protein of HIV and its prevention by vasoactive intestinal peptide. Nature 1988; 335:639–642.

60. Lipton SA, Rosenberg PA. Excitatory amino acids as a common pathway for neurologic disorders. N Engl J Med 1994; 330:613–622.

61. Gozes I, Bardea A, Reshef A, Zamostiatno R, Zhukovsky S, Rubinraut S, Fridkin M, Brenneman DE. Novel neuroprotective strategy for Alzheimer's disease: inhalation of a fatty neuropeptide. Proc Natl Acad Sci (USA) 1996; 93:427–432.

62. Gozes I, Fridkin M, Brenneman DE. A VIP hybrid antagonist: from neurotrophism to clinical applications. Cell Mol Neurobiol 1995; 15:675–687.

63. Gozes I, Meltzer E, Rubinrout S, Brenneman DE, Fridkin M. Vasoactive intestinal peptide potentiates sexual behavior: inhibition by novel antagonist. Endocrinology 1989; 125:2945–2949.

64. Gozes Y, Brenneman DE, Fridkin M, Asofsky R, Gozes I. A VIP antagonist distinguishes VIP receptors on spinal cord cells and lymphocytes. Brain Res 1991; 540:319–321.

65. Fishbein VA, Coy DH, Hocart SJ, Jiang N-Y, Mrozinski JE Jr, Mantey SA, Jensen RT. A chimeric VIP-PACAP analogue but not VIP pseudopeptides functions as VIP receptor agonists. Peptides 1994; 15:95–100.

66. Arimura A, Somogyvari-Vigh A, Weill C, Fiore RC, Tatsuno I, Bay V, Brenneman D. PACAP functions as a neurotrophic factor. Ann NY Acad Sci 1994; 739:228–243.

67. Hill JM, Mervis RF, Politi J, McCune SK, Gozes I, Fridkin M, Brenneman DE. Blockade of VIP during neonatal development induces neuronal damage and increases VIP and VIP receptors in brain. Ann NY Acad Sci 1994; 739:253–261.

68. Hill JM, Gozes I, Hill JL, Fridkin M, Brenneman DE. Vasoactive intestinal peptide antagonist retards the development of neonatal behaviors in the rat. Peptides 1991; 12: 187–192.

69. Card JP, Fitzpatrick-McElligott S, Gozes I, Baldino F. Localization of vasopressin, somatostatin and VIP messenger RNA in the rat suprachiasmatic nucleus. Cell Tissue Res 1988; 252:307–315.

70. Gozes I, Shani Y, Liu B, Burbach JP. Diurnal variation in vasoactive intestinal peptide messenger RNA in the suprachiasmatic nucleus of the rat. Neurosci Res Commun 1989; 5:83–86.

71. Alberts HE, Stopa EG, Zoeller RT, Kauer JS, King JC, Fink JS, Mobtaker H, Wolfe H. Day-night variation in prepro VIP/PHI mRNA within the suprachiasmatic nucleus. Mol Brain Res 1990; 7:85–89.

72. Glazer R, Gozes I. Diurnal oscillation in vasoactive intestinal peptide gene expression independent of environmental light entraining. Brain Res 1994; 644:164–168.

73. Scarbrough K, Wise PM. Intracerebral infusion of VIP antisense oligonucleotides into the suprachiasmatic nuclei affects the circadian corticosterone peak. Soc Neurosci Abstr 1992 18:2.

74. Harney JP, Scarborough K, Rosewell K, Wise PM. Effect of intracerebral infusion of VIP antisense oligonucleotides into the suprachiasmatic nuclei (SCN) on the LH surge in the rat. Soc Neurosci Abstr 1993; 19:573.

75. Prosser RA, Gillete MU. Cyclic changes in cAMP concentration and phosphodiesterase activity in a mammalian circadian clock studied *in vitro*. Brain Res 1991; 568:185–192.

76. Gozes I, Fridkin M, Brenneman DE. Stearyl-Nle-VIP: a non-invasive impotence drug and a potent agent of neuroprotection. Drugs of the Future 1995; 20:680–685.

77. Fauchere JL, Chateon M, Kier LB, Verloop A, Pliska V. Amino acid side chain parameters for correlation studies in biology and pharmacology. Int J Peptide Protein Res 1988; 32:269–278.

78. Gozes I, Reshef A, Salah D, Rubinrout S, Fridkin M. Stearyl-norleucine-VIP: a novel VIP analogue for noninvasive impotence treatment. Endocrinology 1994; 134:2121–2125.

79. Evans T, McCarthy KD, Harden TK. Regulation of cyclic AMP accumulation by peptide hormone receptors in immunocytochemically defined astroglial cells. J Neurochem 1984; 4:131–138.

80. Gozes I, Fridkin M. A fatty neuropeptide: potential drug for noninvasive impotence treatment in a rat model. J Clin Invest 1992; 90:810–814.

81. Gozes I, Brenneman DE. VIP molecular biology and neurobiological function. Mol Neurobiol 1989; 3:201–236.

82. Gozes I, Brenneman DE. Neuropeptides as growth and differentiation factors in general and VIP in particular. J Mol Neurosci 1993; 4:1–9.

83. Gozes I, Brenneman DE. Vasoactive intestinal peptide: from molecular genetics to neurotropism. In: Moody TW, ed. Proceedings of the 12th International Washington Spring Symposium on Growth Factors, Peptides and Receptors. New York: Plenum Press, 1993:15–20.

84. Gozes I, Brenneman DE, Lilling G, Davidson A, Moody TW. Neuropeptide regulation of mitosis. Ann NY Acad Sci 1994; 739:253–261.

85. Gozes I, Davidson A, Draui M, Moody TW. The VIP gene is expressed in non-small cell lung cancer cell lines. Biomed Res 1992; 13(suppl 2):37–39.

86. Moody TW, Zia F, Draoui M, Brenneman D, Fridkin M, Davidson A, Gozes I. A novel VIP antagonist inhibits non-small cell lung cancer growth. Proc Natl Acad Sci (USA) 1993; 90:4345–4349.

87. Moody TW, Zia F, Goldstein AL, Naylor PH, Sarin E, Brenneman DE, Koros AMC, Reubi JC, Korma LY, Fridkin M, Gozes I. VIP analogues inhibit small cell lung cancer growth. Biomed Res 1992; 3:131–135.

88. Sreedharan SP, Huang J-X, Cheung M-C, Goetzl EJ. Structure, expression and chromosomal localization of the type I human vasoactive intestinal peptide receptor gene. Proc Natl Acad Sci (USA) 1995; 92:2939–2943.

89. Naylor SL, Johnson BE, Minna JD, Sakaguchi AY. Loss of heterozygosity of chromosome 3p markers in small-cell lung cancer. Nature 1987; 329:451–454.

90. Daly MC, Xiang RH, Buchhagen D, Hensel CH, Garcia DK, Killary AM, Minna JD, Naylor SL. A homozygous deletion on chromosome 3 in a small cell lung cancer cell line correlates with a region of tumor suppressor activity. Oncogene 1993; 8:1721–1729.

91. Wollman Y, Lilling G, Goldstein MN, Fridkin M, Gozes I. Vasoactive intestinal peptide: a growth promoter in neuroblastoma cells. Brain Res 1993; 624:339–341.

92. Lilling G, Wollman Y, Goldstein MN, Rubinraut S, Fridkin M, Brenneman DE, Gozes I. Inhibition of human neuroblastoma growth by a specific VIP antagonist. J Mol Neurosci 1995; 5:231–239.

93. Brodeur GM, Sekhor GS, Goldstein MN. Chromosomal aberrations in human neuroblastomas. Cancer 1977; 40:2256–2263.

94. Zia H, Hida T, Jakowlew S, Birrer M, Gozes Y, Reubi JC, Fridkin M, Gozes I, Moody TW. Breast cancer growth is inhibited by vasoactive intestinal peptide (VIP) hybrid, a synthetic VIP receptor antagonist. Cancer Res 1996; 56:3486–3489.

95. Yu D, Seitz PK, Selvanayagam P, Rajaraman S, Townsend CM, Cooper CW. Effects of vasoactive intestinal peptide on adenosine 3′,5′-monophosphate, ornithine decarboxylase, and cell growth in a human colon cell line. Endocrinology 1992; 131:1188–1194.

96. Gozes I, Nesher Y, Lilling G, Fridkin M, Rubinraut S, Brenneman DE, Moody TW, Chaimof C. Inhibition of colon cancer by VIP antagonism through a cancer-associated VIP receptor. Proceedings of the 15th World Congress of Collegium Internationale Chirugiae Digestivae, Seoul, Korea, 1996; 819–821.

97. Baldino F, Fitzpatrick-McElligott S, Gozes I, Card PJ. Localization of VIP and PHI-27 messenger RNA in rat thalamic and cortical neurons. J Mol Neurosci 1989; 1:199–207.

98. Dussaillan TM, Sarrieau A, Gozes I, Berod A, Rostene W. Quantitative and qualitative distribution of cells expressing VIP/PHI precursor mRNA in the rat brain. Neuroscience 1992; 50:519–530.

99. Cha CI, Lee EY, Lee YI, Baik SH. Age-related change in the vasoactive intestinal polypeptide-immunoreactive neurons in the cerebral cortex of aged rats. Neurosci Lett 1995; 197:45–48.

100. Glowa JR, Panlilio LV, Brenneman DE, Gozes I, Fridkin M, Hill JM. Learning impairment following intracerebral administration of the HIV envelope protein gp120 or a VIP antagonist. Brain Res 1992; 570:49–53.

101. Gozes I, Glowa J, Brenneman DE, McCune SK, Lee E, Westphal H. Learning and sexual deficiencies in transgenic mice carrying a chimeric vasoactive intestinal peptide gene. J Mol Neurosci 1993; 4:185–193.

102. Lundberg JM, Hedlund B, Bartfai T. Vasoactive intestinal polypeptide enhances muscarinic ligand binding in cat submandibular salivary gland. Nature 1982; 295:147–149.

103. Hedlund B, Abens J, Bartfai T. Vasoactive intestinal polypeptide and muscarinic receptors: supersensitivity induced by long-term atropine treatment. Science 1983; 220: 519–521.

104. Eckenstein F, Baughman RW. Two types of cholinergic innervation in cortex, one co-localized with vasoactive intestinal polypeptide. Nature 1984; 309:153–155.

105. Price DL, Koliatsos VE, Clatterbuck RC. Cholinergic systems: human diseases, animal models, and prospects for therapy. Progr Brain Res 1993; 98:51–60.

106. Arai H, Moroji T, Kosaka K. Somatostatin and vasoactive intestinal polypeptide in postmortem brains from patients with Alzheimer-type dementia. Neurosci Lett 1984; 52:73–78.

107. Selkoe DJ. Physiological production of the beat-amyloid protein and the mechanism of Alzheimer's disease. Trends Neurosci 1993; 16:403–409.

108. Fisher A, Brandeis R, Karton I, Pittel Z, Gurwitz D, Haring R, Sapir M, Levy A, Heldman E, (+ −)-*cis*-2-methyl-spiro(1,3-oxathiolane-5,3′)quinuclidine, an M1 selective cholinergic agonist, attenuates cognitive dysfunctions in an animal model of Alzheimer's disease. J Pharmacol Exp Ther 1991; 257:392–403.

109. Shapira J. Research trends in Alzheimer's disease. J Gerontol Nurs 1994; 20:4–9.

Part Four

**MODULATION OF PEPTIDE ACTION:
THERAPEUTIC APPLICATIONS**

18

Long-Acting VIP Analogs

ERNST JAEGER

Max-Planck-Institut für Biochemie
Munich, Germany

I. Introduction

Vasoactive intestinal peptide (VIP) is a linear and carboxy terminally amidated peptide composed of 28 amino acid residues. It was discovered and characterized in 1970 by Said and Mutt (1,2), and the structure determined (3) was confirmed by conventional chemical synthesis (4). It is present in many animal species with considerable conservation of its amino acid composition (Table 1). VIP is structurally related to secretin, glucagon, PHI/PHM (peptide with N-terminal histidine and C-terminal isoleucine/methionine), GRF (growth hormone-releasing factor), helodermin, helospectin, PACAP (pituitary adenylate cyclase-activating peptide), and others and is therefore a member of a group of peptides called the secretin/glucagon family. The different degrees of amino acid homologies among the primary structures of these peptides are indicated in Table 1.

VIP is generally considered to function mainly as a neuropeptide and sometimes as a bloodborne hormone. It is widely distributed in the central and peripheral nervous system and has a wide variety of biological functions: relaxation of smooth muscle, systemic vasodilation, lowering of arterial blood pressure, increase of cardiac output, and others (for comprehensive reviews of the chemistry and biology of

Table 1 Amino Acid Sequences of Some Members of the Secretin/Glucagon Family of Peptides with Strongest Homologies to VIP (VIP Family)[a]

	5	10	15	20	25	30			
VIP (h,p,do,b,r)	**HSDAV**	**FTDNY**	**TRLRK**	**QMAVK**	**KYLNS**	**ILN***			
VIP (ch)	- - - - -	- - - - -	S - F - -	- - - - -	- - - - -	V - T*			
VIP (gp)	- - - L	- - T -	- - - - -	- - M -	- - - - -	V - T*			
PACAP-38	**HS**DGI	**FT**DSY	SRYRK	**QMAVK**	**KYL**AA	VLGKR	YKQRV	KNK*	(19)
Helodermin	**HS**DAI	**FT**QQY	SKLLA	KLALQ	**KYL**AS	ILGSR	TSPPP*		(15)
Helospectin	**HS**DAT	**FT**AEY	SKLLA	KLALQ	**KYL**ES	ILGSS	TSPRP	PS*	(15)
PHI (p)	HADGV	**FTS**DF	S**RL**LG	Q**L**SAK	**KYL**ES	LI*			(13)
PHM (h)	HADGV	**FTS**DF	SKLLG	Q**L**SAK	**KYL**ES	LM*			(12)
Secretin (p,b)	**HS**DGT	**FTS**EL	S**RLR**D	SARLQ	RLLQG	LV*			(8)
Glucagon (h,p,b)	**HS**QGT	**FTS**DY	SKYLD	SRRAQ	DFVQW	LMNT*			(6)

[a]Amino acid identities with VIP are in **boldface**; their numbers are in (*italics*); * = C-terminal carboxamide; b = bovine, ch = chicken, do = dog, gp = guinea pig, h = human, p = porcine, r = rat.

VIP, see refs. 5–10). It is probably involved in the pathogenesis of several diseases and is therefore an interesting candidate for the treatment of such disorders (11). In particular, VIP has important pharmacological actions on the respiratory tract, including relaxation of airway smooth muscle and attenuation of acute inflammation (12,13). Due to these properties VIP is, like helodermin, a promising therapeutic agent in bronchial asthma and lung injury (10,14). The preferable method of application for such purposes was found to be inhalation (15,16) rather than intravenous injection (17–20), thus avoiding undesired side effects.

Sufficient amounts of VIP for such pharmaceutical applications may be obtained by chemical synthesis. Since appropriate methods for the synthesis of peptides—even those with rather complex structures—have become available in the last two decades (21–23), optimized syntheses of VIP, either by conventional (4,24–26) or by solid-phase techniques (27–29), have been described. First attempts at the production of recombinant VIP and VIP derivatives have also been reported (30,31).

The molecule of VIP is, however, rather sensitive to degradation or modification at several sites during synthesis, purification, storage, or application, and a more or less rapid decrease or loss of biological activity may result. The therapeutic application of unmodified VIP is therefore strongly limited. This has been observed during several clinical trials for the treatment of asthma (15,16,20,32–34). It is very likely that the slight effectiveness of VIP during such trials is due to rapid degradation of the peptide by airway peptidases (35).

In order to overcome these limitations, many attempts at development of more stable and therefore longer-acting VIP analogs have been undertaken in several laboratories. The prerequisite and basis for the finding of adequate structures was a comprehensive knowledge of structure–activity relationships and of receptor-binding properties. Results of such studies and considerations about possible sites of instability of the VIP molecule are described in this chapter, as well as the conclusions which were drawn which will serve as a basis for the design and development of more stable VIP analogs with prolonged duration of action.

II. Structure–Activity Relationships

Native human VIP has a linear octacosapeptide-amide structure with the following amino acid sequence (3):

H-His^1-Ser^2-Asp^3-Ala^4-Val^5-Phe^6-Thr^7-Asp^8-Asn^9-Tyr^{10}-Thr^{11}-Arg^{12}-Leu^{13}-Arg^{14}-Lys^{15}-Gln^{16}-Met^{17}-Ala^{18}-Val^{19}-Lys^{20}-Lys^{21}-Tyr^{22}-Leu^{23}-Asn^{24}-Ser^{25}-Ile^{26}-Leu^{27}-Asn^{28}-NH$_2$

The relations between this chemical structure and the biological properties of VIP were studied either by the use of synthetic variants and fragments or by comparing bioactivity and receptor-binding behavior of VIP with that of closely related naturally occurring members of the secretin/VIP/PACAP group of peptides, i.e., the "VIP family" (11).

A. Biological Properties of VIP Analogs and Fragments

Numerous analogs of VIP have been synthesized in several laboratories, either by replacement or removal of one or several amino acid residues within the sequence or by shortening or extending the molecule at the amino or carboxylic terminus. The biological actions of these analogs or fragments were determined in different systems, for example, either in vitro by testing the relaxation of smooth muscle preparations (e.g., isolated strips of guinea pig or human trachea) or in vivo by evaluation of the vasodilatory activity in dogs or the bronchodilatory effects in guinea pigs. In addition, the degree of adenylate cyclase activation, e.g., in rat liver or lung membrane preparations, was determined. The ability of the analogs and fragments to replace radioiodinated VIP from such membranes or other organ preparations, e.g., isolated lungs, served as a method by which to evaluate the tightness of binding to VIP receptors.

As a result of such studies it became clear that the entire sequence of the VIP molecule is necessary for recognition by VIP receptors and that binding affinity and biological action of VIP are sensitive to removal of amino acid residues in the middle part of the sequence or at the N- or C-terminus as well (36,37). For example, the synthetic peptides VIP(1-6,19-28)-NH$_2$ and VIP(1-9,20-28)-NH$_2$, in which the VIP-sequences (1-6) or (1-9) were joined with the sequences (19-28) or (20-28), respectively, with omission of the middle portions (7-19) or (10-20) of the entire VIP, had no detectable biological activity on isolated *Taenia coli* from guinea pig. The shorter fragments alone, i.e., VIP(1-6), VIP(1-9), VIP(19-28), and VIP(20-28), were found to be inactive as well (36). The same was found for other synthetic partial sequences (37,38).

Further insight into the structural requirements for the interaction of VIP with its receptors (for type and heterogeneity of VIP receptors see Section II.B) was gained by another series of experiments utilizing synthetic VIP fragments (39) and analogs (40) as well. It could be confirmed that the entire sequence of VIP is necessary for binding to VIP receptors with high affinity. Shorter fragments either have very low affinity or do not interact at all with VIP receptors. If the VIP molecule is modified in the N-terminal region, the affinity for receptors is sharply decreased as well. In another approach, VIP analogs were synthesized (41) which are sequentially truncated from the amino or carboxylic end of the octacosapeptide-amide; their tracheal relaxant activity was determined as well as their receptor-binding affinity (Fig. 1). It again became evident that the primary structure of the VIP molecule cannot be minimized significantly without a marked decrease in biological potency. Even the sole elimination of His in position 1 resulted in a substantial loss of activity and potency of the resulting des-His^1-VIP. Any modification of His^1 (3-methyl-, 1-methyl-, desamino-His or D-His) was not tolerated as well, besides N$_\alpha$-acylation (see below). Other synthetic fragments with N-terminal (1-13, 1-19, 1-23), C-terminal (20-28), and midregion (18-23, 12-23, 7-23) sequences of VIP were also found to be inactive (37,41).

In general, all fragments of VIP prepared and tested so far have been found to

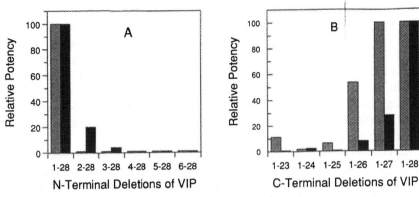

Figure 1 Comparison of the relative potency of VIP analogs with N-terminal (A) and C-terminal (B) amino acid deletions in a tracheal relaxation assay (striped bars) and in a guinea pig receptor binding assay (solid bars). Data represent the mean of determinations made on three tracheal smooth muscle tissues and on three lung membrane preparations. (Reproduced from Ref. 41, with permission of Munksgaard International Publishers Ltd., Copenhagen, Denmark: Copyright © 1991.)

possess none or only very little bioactivity in different systems. Other VIP fragments were even found to act as antagonists, like the fragments VIP(1-11)-OH (42), VIP(10-28)-NH$_2$ (43,44), or VIP(4-28)-NH$_2$ (45). As an inhibitor of phagocytosis in mouse peritoneal macrophages, VIP(22-28)-NH$_2$ was more active than the entire VIP, while VIP(1-12) was inactive and had no suppressive effect in this system (46).

On the other hand, however, modifications or elongations of the entire, unshortened sequence of VIP could be performed at the amino or carboxylic ends without negative effects on the biological properties. The replacement of the C-terminal carboxamide group by a carboxylic group (VIP free acid) or an extension of the peptide chain by -Gly-OH (47) or -Gly-Lys-OH in the VIP precursor analog [Leu^{17}]-VIP-Gly-Lys^{30}-OH (preVIP) (48–50) was tolerated and even accompanied by enhanced stability. When N$_\alpha$-acylation of His^1 by acetyl-, benzoyl-, or glycyl- was performed, equal or only slightly reduced potency was found (37). Even H$_\alpha$-acylation of VIP with the bulky and hydrophobic stearyl group (St) was tolerated: N$_\alpha$-stearyl-VIP (St-VIP) was as active as VIP in a variety of impotence models in rats during intradermal application (51). In the middle part of the peptide chain, methionine in position 17 can be oxidized to Met-S-oxide (52) or replaced by α-aminohexanoic acid (Ahx; former nomenclature: norleucine, Nle) (22), without affecting the biological properties; even an increase of the appropriate activities was reported later for [Ahx^{17}]-VIP (37) and St-[Ahx^{17}]-VIP (53). When position 17 and the adjacent positions 16 and 18 were replaced by Gly concomitantly, however, the activity was lost (36).

Some replacements of single amino acid residues within the N-terminal sequence region of VIP were accompanied by partial or full loss of activity. When aspartic acid in position 8 was substituted by glutamic acid, the resulting [Glu^8]-VIP was less potent than VIP in effecting pancreatic secretion (54). An almost complete loss of tracheal relaxant activity was found with the synthetic analogs [Asp^9]-VIP (incorporation of an additional negative charge in position 9 instead of the uncharged side chain of Asn), [β-Asp^3]-VIP (Asp in position 3 linked to Ala^4 via the β-carboxylic group), [Sar^4]-VIP (N-methyl group on the peptide bond between positions 3 and 4), and [Ser(SO$_3$H)]^3-VIP (replacement of Asp^3 by the nearly isosteric and also negatively charged serine-O-sulfonate) (55,56).

In contrast, it was shown, in an extensive program of analog synthesis and testing, that amino acid replacements in the middle and C-terminal parts of the sequence (besides incorporation of Ahx in position 17 instead of Met; see above) are tolerated and can even lead to VIP analogs with enhanced potency as airway smooth muscle relaxants (37,57–59). When the positions 11, 12, 13, 14, 26, and 28 of [Ahx^{17}]-VIP were systematically substituted by amino acids selected from the group Ala, Lys, Orn, Phe, Ser, Thr, and Val, several analogs were obtained with improved biological action of producing sustained bronchodilation. For example, replacement of Arg^{12} or Arg^{14} with Lys or Orn resulted in a 1.7- to 2.6-fold increase in potency. When Ile^{26} and Asn^{28} were substituted by Val and Tyr, respectively, potency was doubled as compared to [Ahx^{17}]-VIP. A combination of the most favorable changes and additional N$_\alpha$-acetylation led to a series of superagonists (Table 2). The most active analog, Ac-Lys^{12},Lys^{14},Ahx^{17},Val^{26},Thr^{28}]-VIP [6], was 10-fold more potent than native VIP (37). By variations of the N-terminal position 1 of analog 3 it was shown that replacement of Ac-His^1 by Ac-Ala^1, Ac-Gly^1 and other N-acyl-amino acids is also tolerated (60) and can even lead to compounds with further increased potency.

The design of the aforementioned analogs had been supported, besides other aspects, by consideration of the three-dimensional structure of the VIP molecule in solution, determined by nuclear magnetic resonance (NMR) and circular dichroism (CD) spectroscopic techniques. Amino acid replacements in the middle portions of the molecule were carried out with regard to the helix-promoting properties, e.g., of Lys, according to the rules of Chou and Fasman (61). However, the helicity of the most potent analog 6 (10-fold potency; see Table 2) in solution was found to be not greater than that of native VIP (58).

Additional attempts to further enhance the potency of VIP analogs as bronchodilators focused on the question of whether moderate C-terminal elongation of the peptide chain with concomitant incorporation of amino acid residues with helix-capping potential in this portion of the molecule might be effective (62). Indeed, several of the series of such compounds showed enhanced potency up to 10-fold of the parent compound Ac-[Lys^{12},Ahx^{17},Val^{26},Thr^{28}]-VIP [3] (Table 2) used for comparison. As an example, Ac-[Glu^8,Lys^{12},Ahx^{17},Ala^{19},Ala^{25},Leu^{26},Lys^{27,28},Gly^{29,30}, Thr^{31}]-VIP [7] was almost 10 times as potent in relaxing guinea pig tracheal smooth muscle in vitro. When several analogs of this series were examined by circular

Table 2 Biological Activities of Combination Replacement Analogs of VIP: In-Vitro Relaxation of Guinea Pig Tracheal Rings

No.	VIP analog[a]	EC_{50} (nM)	Potency (%)[b]
1	[Ahx^{17}]-VIP	6.4	156
2	Ac-[Ahx^{17},Val^{26},Thr^{28}]-VIP	3.5	286
3	Ac-[Lys^{12},Ahx^{17},Val^{26},Thr^{28}]-VIP	2.7	370
4	Ac-[Lys^{12},Ahx^{17},Thr^{25},Val^{26},Thr^{28}]-VIP	2.1	481
5	Ac-[Orn^{12},Ahx^{17},Val^{26},Thr^{28}]-VIP	1.74	575
6	Ac-[Lys^{12},Lys^{14},Ahx^{17},Val^{26},Thr^{28}]-VIP	0.98	1020

[a]For additional analogs of this series, see Ref. 62.
[b]Potency relative to native VIP (EC_{50} = 10 nM[c] = 100%).
[c]In a later publication (62), a value of 16 nM for native VIP was used for the calculation of the potencies listed here.
(Reproduced from Ref. 37, with permission of ESCOM Science Publishers B.V.: Copyright © 1988.)

dichroism (CD) spectroscopy, a general agreement was observed between enhanced potency and percentage of helical structure.

The preparation of cyclic analogs of the same series of peptides by site-specific cyclization with formation of lactam bridges during solid-phase synthesis did not yield compounds with useful potencies, despite the fact that cyclic analogs of other natural peptides have been shown to possess enhanced potency, receptor selectivity, and enzymatic stability. In the present case almost all cyclic VIP analogs prepared with cyclizations performed in the N-terminal region between His^1 and Asp^3, His^1 and Asp^8, His^1 and Glu^{16}, Asp^3 and Lys^{12} (37), and in the middle sequence between position 8 or 9 and 12 (analogs **8, 9, 10, 11** in Table 3) were found to have no or only little activity in the range up to about 10% the potency of [Ahx^{17}]-VIP. Only the analogs **12** and **13** with cyclizations in the C-terminal sequence region, namely, between positions 20 or 21 and 24 or 25, respectively, had relative potencies similar to the appropriate linear species (62). Compound **13** was also found to have enhanced helical characteristics when examined by CD spectroscopy. This was interpreted as an indication of rigidification of the molecule at this site, which supports a conformation at least similar to the receptor-active conformation. Therefore, this principle of a lactam bridge between positions 21 and 25 was further investigated and combined with the principle of C-terminal sequence elongation with concomitant incorporation of helix-capping residues, which was outlined above on analog **7**. The results, which have led to superactive and long-acting VIP analogs are described in Section IV.

Further insight (41) into the role of the single amino acid residues during receptor binding and transduction of the biological message was gained in a systematic study by performing a so-called alanine scan on the VIP analog Ac-[Lys^{12},Ahx^{17}, Val^{26},Thr^{28}]-VIP [**3**], which had a 3.7-fold enhanced potency over native VIP (63). Synthetic analogs of **3** with site-directed modifications through the use of systematic

Table 3 Biological Activity of Conformationally Restricted VIP Analogs: In-Vitro Relaxant Activity on Guinea Pig Tracheal Smooth Muscle

No.	Compound	IC_{50} (nM)	Relative Potency[a]
8	Ac-[Lys^{12},Ahx^{17},Val^{26},Thr^{28}]-VIP cyclo(Asp^8→Lys^{12})	14	0.19
9	Ac-[Orn^{12},Ahx^{17},Val^{26},Thr^{28}]-VIP cyclo(Asp^8→Orn^{12})	40	0.07
10	Ac-[Lys^8,Asp^{12},Ahx^{17},Val^{26},Thr^{28}]-VIP cyclo(Lys^8→Asp^{12})	38	0.07
11	Ac-[Asn^8,Asp^9,Lys^{12},Ahx^{17},Val^{26},Thr^{28}]-VIP cyclo(Asp^9→Lys^{12})	17	0.16
12	Ac-[Lys^{12},Ahx^{17},Asp^{24},Val^{26},Thr^{28}]-VIP cyclo(Lys^{20}→Asp^{24})	5.3	0.5
13	Ac-[Lys^{12},Ahx^{17},Asp^{25},Val^{26},Thr^{28}]-VIP cyclo(Lys^{21}→Asp^{25})	3.1	0.9

[a]Potencies relative to Ac-[Lys^{12},Ahx^{17},Val^{26},Thr^{28}]-VIP [**3**]; EC_{50} = 2.7 nM (1.0).
(Reproduced from Ref. 62, with permission of John Wiley & Sons, Inc.: Copyright © 1995.)

single-point alanine substitutions of all sequence positions were prepared. [This method was applied earlier in the case of other peptides (64), in order to determine their potential binding sites and biological pharmacophores]. The results obtained (Fig. 2A) by measuring the relative in vitro biological activities of all synthetic alanine-substituted analogs on guinea pig and on human tracheal smooth muscles indicated that analogs with alanine replacements in the positions Asp^3, Phe^6, Thr^7, Tyr^{10}, Tyr^{22}, and Leu^{23} exhibit the greatest loss in potency in this bioassay (40- to 400-fold; similar in the animal and the human model). It was concluded, that these positions are most important for receptor binding. Alternatively, the biopotency was widely retained when the sequence positions Thr^{11} through Lys^{21} were replaced by alanine, and this finding suggests that the side chains of these 11 residues are not necessary for binding to the receptor. It was noteworthy to find that in all 26 alanine-substituted compounds, none of the replacements produced a superagonist. Another series of compounds obtained by systematic single substitutions of all L-amino acids of the VIP molecule by their D-enantiomers (a so-called D-amino acid scan) did not produce compounds with useful activity (Fig. 2B). The evaluation of the in vivo properties of the series of compounds from the alanine scan for the detection of pharmaceutically useful long-acting VIP analogs is discussed in Section IV.

B. Biological Properties of VIP-Related Natural Peptides

Additional insight into structure–activity relations of VIP, besides using fragments and analogs as described above, can be gained by taking advantage of the existence of the large family of VIP-related naturally occurring peptides with high sequence homology to VIP (Table 1) and by taking their affinity to VIP receptors and their bioactivity into consideration.

It is well established that the actions of VIP are initiated by the binding of the neuropeptide to specific membrane receptors coupled to the stimulation of adenylate cyclase activity through a regulatory protein Gs, and this signal transduction pathway has carefully been studied in many tissues (65–67). VIP receptors could be identified

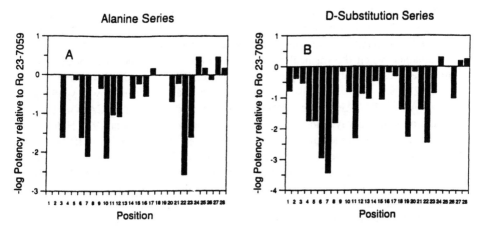

Figure 2 Smooth muscle relaxant activity of VIP analogs with systematic single alanine (A) or D-amino acid (B) substitutions, relative to the parent compound Ac-[Lys^{12},Ahx^{17},Val^{26}, Thr^{28}]-VIP [3]. Plots of the −log of the biological potency versus sequence position of substitution. (Reproduced from Ref. 63, with permission of ESCOM Science Publishers B.V.: Copyright © 1992.)

in membrane preparations and were solubilized from several sources, e.g., from the liver (68), from normal mammalian and human lungs (69,70), and from human lung tumor cells (71) including small cell carcinoma cell lines (72). Two classes of receptors with different affinity and specificity have been described (73,74), and the heterogeneity of the VIP receptors has been reviewed (75,76). A VIP receptor could be purified to apparent homogeneity from liver membranes (77), and the cloning and expression of a human VIP receptor has been described (78). Recently, two types of VIP receptors were cloned, designated VIP1 (79) and VIP2 (80) receptors. They belong to an emerging subfamily within the superfamily of G-protein-coupled receptors.

It was found that the VIP-related peptides secretin, PHI/PHM, GRF, helodermin, and PACAP compete with [^{125}I]-VIP for binding to receptors in various target cells (81–87), even if some of these peptides, e.g., PACAP (88,89), have their own specific receptors. As an example, it has been shown by studies of receptors in rat and human intestine (83–85) that the VIP-related peptides act as full agonists (besides human GRF). The human VIP receptor was found to be more discriminating than the rat receptor, since the cross-reactivity of the VIP-related peptides was lower in human than in rat tissues. In other systems it was found, for example, that PACAP usually binds with much higher affinity to VIP receptors than does PHI (89). However, the relative affinity of these peptides to the VIP receptors in different tissues and species is not yet fully established.

As a functional response to the binding to VIP receptors, the members of the VIP family also exhibit biological activities similar to VIP in some target cells, even with different potencies. As an example, the two different forms of PACAP, namely,

PACAP-27 and PACAP-38, are 100 times more potent than VIP and PHI as vasodilators in the rabbit, while this difference in potency is less in humans (90). On the other hand, PHI was found to be less biologically active than VIP (91). Interesting results were obtained when the tracheal relaxant activities of some members of the VIP family were tested. Rather low potencies of about 10% were found with PHI and PACAP-27 (S.I. Said, unpublished results). This was confirmed by a synthetic analog: [Ahx^{17}]-PACAP-27 was rather inactive as well (56). For PACAP-38, however, a prolonged airway smooth muscle relaxant activity was found (92). Helodermin proved to relax guinea pig trachea with about the same potency as VIP, but the duration of this action was longer (14,93) than that of VIP (Fig. 3). A long-lasting inhibitory effect of inhaled helodermin on guinea pig airway contraction has also been demonstrated (94). The reason for the prolonged duration of action of helodermin was at first attributed to the C-terminal sequence extension (29-35) with a -Pro-Pro-Pro-NH$_2$ sequence (93,95). This opinion was later revised by a synthetic approach, as outlined in Section IV. In conclusion, a comparison of the amino acid sequences of the VIP-related peptides with that of VIP may serve as a useful guide during the design and search for long-acting synthetic VIP analogs due to some similarities in receptor affinity and bioactivity within the VIP family. Together with concomitant considerations of possible sites of degradation, discussed in Section III,

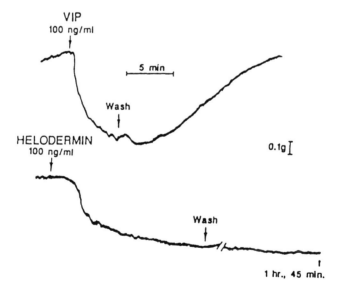

Figure 3 Tracheal relaxations induced by equal concentrations of VIP and helodermin (100 ng/mL). The tissue baths were washed where indicated. The relaxations were of approximately the same magnitude, but differed in duration: tracheal tone returned to baseline within 22 min in the case of VIP, while it remained fully relaxed 105 min after addition of helodermin. (Data from Refs. 14 and 93.)

some hybrid molecules between VIP and other members of the VIP family have been designed and tested. These results are also described in Section IV.

III. Sensitivity of the VIP Molecule

Despite its interesting and valuable biological properties, possible therapeutic application of VIP is strongly limited by the rapid decrease of its bioactivity observed during attempts at clinical use, as mentioned in Section I. Possible reasons for such limitations are peptide backbone fragmentations and/or modifications of the intact peptide. Such degradation reactions can be caused either by metabolic processes during in-vivo administrations or by chemical influences during preparation, storage, and application of the peptide. The particular sites of the VIP molecule which are sensitive to such cleavages or changes as a result of enzymatic or other biochemical and chemical processes are described in this section.

A. Sites of Enzymatic Degradation

The proteolytic inactivation of neurohormonal peptides in general, and of VIP in particular, involves several exopeptidases, endopeptidases, and desamidases. VIP is rapidly metabolized in the liver, the kidney, the brain, the respiratory system including the lung (11,96,97), and in the gastrointestinal tract (98). The enzymes mainly responsible for VIP degradation and inactivation include aminopeptidases, neutral endopeptidase (enkephalinase), trypsin, chymotrypsin, and mast cell tryptase and chymase.

The removal of N-terminal His with formation of des-His[1]-VIP by the action of an aminopeptidase, which is sensitive to the inhibitors amastatin and bestatin, was observed as a result of incubation of VIP with dispersed pig enterocytes at 37°C for only 30 sec (99). It was shown that the formation of des-His[1]-VIP can effectively terminate the action of VIP in the epithelial layer of the intestine, because this metabolite has only 1% of the bioactivity of the intact peptide.

The rapid decrease of VIP activity observed during attempts to treat asthmatic bronchoconstriction can be explained as a result of destruction of the peptide by tracheobronchial peptidases. From the following findings, it seemed very likely that one of these enzymes is neutral endopeptidase (NEP, EC 3.4.24.11; "enkephalinase"): The relaxant effect of VIP on isolated guinea pig tracheal strips was selectively potentiated and prolonged in the presence of phosphoramidon, an enkephalinase inhibitor (35,100), and VIP was enzymatically degraded by guinea pig tracheal extracts with high specificity. These results were confirmed by incubation of VIP with a mixture of peptidases from bovine tracheal mucosa and with purified human kidney NEP (101). With respect to a search for more stable VIP analogs, it is remarkable that the VIP-related peptide helodermin (Table 1) was found to be much more resistant to the enzymes of these systems (101) as well as under other conditions (102). The cleavage sites of VIP were determined by incubation with recombinant human NEP and subsequent isolation and characterization of the VIP fragments obtained (103). It was demonstrated that VIP was readily hydrolyzed at sites N-terminal to the hydrophobic amino acids at positions 4, 5, 13, 17, 22, and 26. The relative rates of hydrolysis were $Ala^4 = Val^5 > Tyr^{22} = Ile^{26} >> Leu^{13} = Met^{17}$. VIP was also a good substrate for purified endopeptidase 24.11 from porcine fundic muscle, and the pattern of metabolites obtained was similar to that produced by the human recombinant enzyme (R. Nau and J. M. Conlon, unpublished data).

In the respiratory tract, VIP is degraded not only by NEP, but also by the mast cell enzymes tryptase and chymase. A deficiency of VIP in the airways with the result of asthma is considered to be caused by these enzymes (14,104). As a result of enhanced degranulation of mast cells, increased amounts of tryptase and chymase are liberated, and VIP can be cleaved fast and effectively (105–107). The sites of VIP cleavage by both enzymes were determined in vitro as follows. In one study, human VIP was incubated with tryptase and chymase isolated from dog mast cell tumors (105). By separation and characterization of the resulting main fragments, the major cleavage sites of VIP by tryptase were found to be the peptide bonds Arg^{14}-Lys^{15} and Lys^{20}-Lys^{21}. The hydrolysis achieved with chymase occurred at the bond Tyr^{22}-Leu^{23} of VIP. It was shown later (107) that a similar pattern of VIP degradation is caused by a tryptase isolated from human lungs: Besides the main cleavages C-terminal to Arg^{14} and Lys^{20}, minor cleavages were found to occur at positions Arg^{12} and Lys^{21} at incubation times up to 10 min. When the VIP-related peptide PHM (Table 1) was incubated with human tryptase, the major site of hydrolysis was only C-terminal to Lys^{20} (10 min of incubation), while the other tryptic sites Lys^{12} and Lys^{21} of this peptide were not hydrolysed significantly despite prolonged incubation (60 min). This finding led to the design of a VIP/PHM hybrid molecule, which was also found to be more resistant to degradation by human tryptase (see Section IV, Ref. 132). By development and application of a HPLC assay based on the hydrolysis of N_α-dansyl-VIP (108), it was confirmed that the peptide chain of VIP is predominantly cleaved C-terminal to Arg^{14} and Lys^{20}. In this case, minor cleavages after Lys^{15} and Lys^{21} were also found.

In order to examine the potential influence of enzymatic metabolism on VIP in human airway smooth muscle when the peptide is administered by aerosol, the degradation was examined in bronchial alveolar lavage (BAL). Synthetic human VIP was incubated with freshly collected pig or human BAL fluid at 37°C and analyzed by HPLC over time (109). The pattern of metabolites, which were identified as peptide fragments, resulted from cleavage at two completely novel sites, the peptide bonds Ser^{25}-Ile^{26} and Thr^7-Asp^8. The enzyme responsible for this cleavage has not yet been identified. It appears, however, not to be an exopeptidase or to have tryptase- or chymase-like activity.

B. Sites of Nonenzymatic and Chemical Attack

Besides enzymatic cleavages, the hydrolysis of VIP can also be catalyzed by circulating high-affinity VIP-binding antibodies: Rather amazingly, it has been found (110) that VIP labeled with ^{125}I, namely, $[^{125}I$-$Tyr^{10}]$-VIP, was cleaved when treated with human immunoglobulin G (IgG) from an asthmatic patient containing antibodies to VIP. The VIP hydrolytic activity was retained in a Fab fragment prepared from IgG by papain treatment, and two major VIP fragments were produced by treatment of VIP with the antibody fraction. VIP(1-6) and VIP(17-28)-NH$_2$ were isolated and identified by fast atom bombardment (FAB) mass spectrometry and peptide sequencing, thus indicating that the scissile bond in VIP was Gln^{16}-Met^{17}. The catalytic cleavage of peptide bonds by autoantibodies as a newly detected effector mechanism was further investigated and confirmed by additional experiments (111–113). The antibodies from two other individuals were capable of cleaving more than one peptide bond, namely, in both cases, six bonds located between residues 14 and 22 in VIP, i.e., Arg^{14}-Lys^{15}, Gln^{16}-Met^{17}, Met^{17}-Ala^{18}, Lys^{20}-Lys^{21}, Lys^{21}-Tyr^{22}, and in one case the bond Thr^7-Asp^8 additionally (113). Compared to classical proteases, the anti-VIP antibodies hydrolyse peptide bonds slowly. Due to tight VIP binding, however, their kinetic efficiency approaches that of proteases (111). The unusual findings have recently been reviewed (114–116).

The spontaneous hydrolysis of VIP was also observed in neutral aqueous solution in the absence of antibodies or other catalytic additives (117). This phenomenon was detected by a sensitive radiometric assay using [^{125}I-Tyr^{10}]-VIP and further investigated on unlabeled VIP: When a solution of 300 μM VIP in water was kept for 48 hr at 38°C, sufficient quantities of six VIP fragments were generated for an unambiguous identification by N-terminal sequencing. Based on the identities of these fragments, five major cleavage sites located between the residues 17 and 25 of VIP were deduced, namely, Met^{16}-Ala^{17}, Lys^{20}-Lys^{21}, Lys^{21}-Tyr^{22}, Tyr^{22}-Leu^{23}, and Asn^{24}-Ser^{25}. Besides these fragments, two components with amino acid analyses of full-length VIP were isolated and identified as products of oxidation and isomerization, namely, [Met(O)^{17}]-VIP and [β-Asp^{24}]-VIP. It was suggested that the breakdown of VIP in dilute solution represents an autolytic process, possibly involving the amino acids His, Ser, and Asp at the N-terminus of VIP, which are known to participate in chemical catalysis at enzyme active sites (118) or an unusually reactive amino acid or group of amino acids.

Possible sites of chemical modification or degradation reactions in VIP are all Asp- and Asn-containing sequence sections, namely, Asp^3-Ala^4, Asp^8-Asn^9, Asn^{24}-Ser^{25}, and the C-terminal Asn^{28}-NH$_2$. It is well known that Asn-X and Asp-X sequences (X = Gly, Ala, Ser) in peptides and proteins can easily undergo spontaneous deamidation, α,β-isomerization, and razemisation via succinimide-linked reactions (119–121). These reactions can occur under physiological conditions (pH 7.4, 37°C) but also in weakly acidic or basic solutions and can involve cleavage of the peptide bond, particularly in the case of Asn-Ser. It has been found in the case of the VIP-related peptide secretin (Table 1) that α,β-isomerization via the cycloimide intermediate, which is formed slowly in dilute acetic acid, is one of the reactions responsible for the inactivation of this peptide (122). In the case of human growth hormone-releasing factor (GRF) with an N-terminal sequence similar to VIP (H-Tyr^1-Ala-Asp-Ala-Ile-Phe-Thr-Asn-Ser-Tyr^{10}- ...), the degradations at Asp^3-Ala^4 and Asn^8-Ser^9 via the succinimide pathway in acidic solution with formation of β-Asp³-, Asp⁸-, and β-Asp⁸-containing products were demonstrated and studied in detail (123). It may be assumed that VIP can also be degraded in the same way, and it was shown that possible products of such reactions, namely, [β-Asp³]-VIP and [Asp⁹]-VIP, are rather inactive in a guinea pig tracheal relaxant test (55).

The presence of such inactive conversion products of VIP in preparations which were kept under acidic conditions, e.g., during storage or in HPLC purification processes with trifluoro acetic acid-containing solvents, might therefore be one of the reasons for diminished activity of VIP preparations in clinical trials.

IV. Development of Stable and Long-Acting VIP Analogs

The results of structure–activity relationship and receptor-binding studies and the considerations about possible sites of degradation of the VIP molecule, discussed in the preceding sections, were utilized by several research groups as a basis for the design of potent and metabolically stable VIP receptor agonists with significantly prolonged duration of action, particularly when tested by in-vivo experiments. Conformational aspects were also considered. The ultimate goal of these efforts was to obtain long-acting and concomitantly superactive peptides for pharmaceutical applications.

For a rational design of such long-acting and stable VIP analogs, the following principles must be observed. (a) The unshortened peptide chain of 28 amino acid residues must be used as a basis; elongations at both ends are possible. (b) The homology of amino acid sequences among the members of the secretin/glucagon/VIP family of peptides and their respective variants in different species (Table 1) should be considered. The fact that His¹, Asp³, Phe⁶, Thr⁷, Tyr¹⁰, and Leu²³ are conserved within this group of peptides suggests that these amino acids are very likely involved in the binding process and should therefore not be replaced. (c) The amino acid residues or sequences which have been found by experiments to be essential for binding of the VIP molecule to its receptors (mainly in the middle and C-terminal regions) or for the transduction of the biological message (mainly located in the N-terminal sequence 1-9) must be retained. Several of these positions can be derived from the SAR studies described in Section II. (d) As many sites of VIP instability as possible should be eliminated according to the considerations discussed in Section III. In particular, the essential and therefore indispensable pharmacophores on the VIP molecule, as derived from SAR studies, were described as follows (60,63): (a) a lone-pair electron or *p*-structure at His¹; (b) a negative charge at Asp³; (c) aromatic rings from Phe⁶, Tyr¹⁰, and Tyr²²; (d) possible effects from the side chains of Thr⁷ and Leu²³. It is obvious that several of these positions correlate with those conserved in nature within the VIP family (see above).

In a first major approach to meet the goals mentioned, several VIP analogs with enhanced potency to relax guinea pig tracheal smooth muscle were found by the SAR studies described in Section II, but the duration of action of these compounds was not determined or reported. Another analog with 3.7-fold activity, namely, Ac-[Lys¹², Ahx¹⁷,Val²⁶,Thr²⁸]-VIP [**3** in Table 2] [Note: a 5.9-fold activity was reported later for this compound (62), when a value of 16 nM for native VIP was used as reference], and all the related analogs with single Ala-substitutions (see above), were tested in vivo in a guinea pig model for their bronchodilator potency by a trachea instillation route of administration (41). The analogs with alanine replacements in the positions Asp^3, Phe^6, Thr^7, Tyr^{10}, Tyr^{22}, and Leu^{23} showed approximately 16- to 200-fold reduced potency, and this result was also in direct correlation with the considerations about indispensable amino acids discussed above. The results obtained with all analogs administered at a dose of 300 μg each are compared in Table 4. In the same model, the duration of the bronchodilatory effect was determined in comparison to native VIP, which had an activity equal to analog 3 in this model, namely, 5 min. Peptides with alanine replacements in positions Val^5, Leu^{13}, Val^{19}, or Ile^{26} showed extended durations of action of 20, 30, 40, and 50 min, respectively. It was suggested that these findings be utilized for further design of metabolically stable and therefore long-acting VIP analogs.

Table 4 Relative in-Vivo Biological Activity of Alanine-Substituted Analogs of Ac-[Lys^{12},Ahx^{17},Val^{26},Thr^{28}]-VIP [3][a]

Analog	ID_{50} (μg[b])	Duration (min[c])	Analog	ID_{50} (μg[b])	Duration (min[d])
[3][d]	6	5			
Ala[1]	6	4	Ala[15]	7	5
Ala[2]	3	4	Ala[16]	1	5
Ala[3]	Inactive[e]		Ala[17]	6	5
Ala[5]	3	26	Ala[18]	1	39
Ala[6]	95	6	Ala[20]	35	5
Ala[7]	Inactive		Ala[21]	1	4
Ala[8]	3	3	Ala[22]	Inactive	
Ala[9]	6	5	Ala[23]	25	16
Ala[10]	Inactive		Ala[24]	1	4
Ala[11]	140	5	Ala[25]	1	4
Ala[12]	100	4	Ala[26]	76	50
Ala[13]	10	30	Ala[27]	6	5
Ala[14]	46	15	Ala[28]	2	4

[a]Bronchodilator potency and duration of action were determined in guinea pigs by a trachea instillation route (41). Data represent mean of determinations made on three guinea pigs.
[b]Dose of analog that inhibits histamine-induced (50 μg/kg, intravenously) bronchoconstriction by 50%. Pretreatment time for all compounds was 1 min.
[c]Time for maximal inhibitory activity to decrease to 40%. For comparative purposes, all analogs were instilled at a dose of 300 μg.
[d]Designated Ro 23-7059.
[e]Inactive: no significant inhibition at 1000 μg.
(Reproduced from Ref. 41, with permission of Munksgaard International Publishers Ltd., Copenhagen, Denmark: Copyright © 1991.)

In a recent report about further structure–activity studies of the VIP pharmacophore (124), it was shown by other variations of the same analog 3 that in certain cases even the side chains of the amino acids in the positions 1, 3, 6, 10, and 22, which are most important for interaction with the receptors, may be modified with a further positive effect on the bioactivity. When the side chain of Tyr[22] in 3 was systematically modified, two out of more than 40 synthetic compounds emerged with increased in-vitro tracheal relaxant activity. The analog with 3-fluoro,4-hydroxy-phenylalanine in position 22 [analog 14] was 1.5 times more potent than the parent compound 3. Another analog [15], with 3-methoxy,4-hydroxy-phenylalanine in the same position, was even 3.4-times more potent than 3, which is equivalent to a 20-fold increase in potency over native VIP. Rather interestingly, it could be shown in addition that these compounds with enhanced in-vitro potency also retained in-vivo activity. The result of the bronchodilator assay for analog 14 in comparison to the parent compound 3 is shown in Table 5. By an instillation route of administration to guinea pigs, 14 was slightly over two times as potent as compound 3, in accordance with the in-vitro results. Furthermore, analog 14 had a significantly longer duration of action, with bronchodilator activity observed for >20 min as compared to 5 min for 3. This increase of duration of action was interpreted to be the result of enhanced potency shifting the apparent duration rather than an effect caused by enhancement of stability.

In another approach (125), several types of cyclic analogs of VIP with internal cystine bridges were evaluated with the assumption that cyclization of the peptide chain might possibly create enhanced metabolic stability and resistance to enzymatic degradation. In particular, analogs of VIP with cysteine residues incorporated at selected sites within the sequence were synthesized by solid-phase methods and subsequently oxidized to the corresponding cyclic disulfides. The cyclic compounds were assayed as smooth muscle relaxants on isolated guinea pig trachea in vitro, as bronchodilators in guinea pigs in vivo, and for binding to VIP receptors in guinea pig lung membranes. While most of the compounds prepared with cystine-bridges in the

Table 5   Relative in-Vivo Biological Activity of a VIP Analog Containing a 3-Fluoro-tyrosine Residue in Sequence Position 22: Bronchodilator Potency and Duration of Action

No.	Compound	IC_{50} (μg^a)	Duration (min[b])
3	Ac-[Lys[12],Ahx[17],Val[26],Thr-[28]]-VIP	6.3	5
14	Ac-[Lys[12],Ahx[17],(3-F,4-OH)Phe[22],Val[26],Thr-[28]]-VIP	2.9	>20

[a]Compounds were given by intratracheal administration to anesthetized guinea pigs 1 min before challenging with histamine (50 mg/kg, i.v.).
[b]The duration of action was calculated as the time for inhibitory activity to decrease to 40%. For details see Ref. 41.
(Reproduced from Ref. 124, with permission of Munksgaard International Publishers Ltd., Copenhagen, Denmark: Copyright © 1995.)

N-terminal region were inactive (compare the results with lactam bridges, above), one cyclic VIP analog, namely, Ac-(D-Cys^6,D-Cys^{11},Lys^{12},Ahx^{17},Val^{26},Thr^{28})-VIP [16], with a disulfide bridge between D-Cys^6 and D-Cys^{11}, was found to be an agonist with slightly more than one-tenth the potency of native VIP in the in-vitro test. A second analog, Ac-[Lys^{12},Cys^{17},Val^{26},Cys^{28}]-VIP [17] (disulfide bridge from Cys^{17} to Cys^{28}), proved to be a full agonist with a potency about one-third that of VIP in the same model (Fig. 4, Table 6). In addition, this peptide was active as a bronchodilator in vivo with slightly diminished potency as compared to native VIP. Quite interestingly, however, this latter compound, 17, was found to have significantly longer duration of action, more than 40 min, in comparison to the appropriate acyclic linear reference compound 3, which has an activity lasting only 5 min (Table 6). Obviously, in this case the conformational restrictions created by the cyclic ring structures may have effected stabilization toward degradation, with the result of enhanced effective duration of action (125).

The development of other types of cyclic analogs with internal lactam bridges was described in Section II. It had been the goal of these efforts to enhance potency as well as stability against enzymatic degradation by creating conformational restrictions of these molecules in order to obtain highly active and long-acting analogs (62). It has been pointed out that the formation of a lactam bridge between Lys^{21} and Asp^{25} in one analog, namely, Ac-[Lys^{12},Ahx^{17},Asp^{25},Val^{26},Thr^{28}]-VIP cyclo(Lys^{21}→Asp^{25}) [13 in Table 3], was most successful. When this analog was tested as a bronchodilator

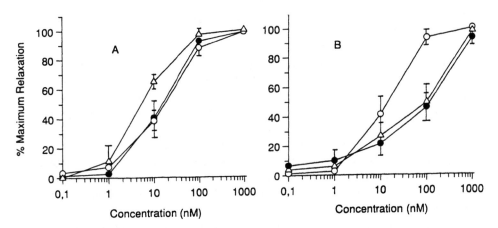

Figure 4 Cumulative relaxation concentration-response curves for native VIP and some VIP analogs in isolated guinea pig tracheal smooth muscle. Left: native VIP (○); [Ahx^{17}]-VIP (●); Ac-[Lys^{12},Ahx^{17},Val^{26},Thr^{28}]-VIP [3] (△). Right: [3] (○); Ac-[D-Cys^6,D-Cys^{11},Lys^{12},Ahx^{17}, Val^{26},Thr^{28}]-VIP cyclo(D-Cys^6→D-Cys^{11}) [16] (●); Ac-[Lys^{12},Cys^{17},Val^{26},Cys^{28}]-VIP cyclo-(Cys^{17}→Cys^{28}) [17] (△). Relaxant responses are expressed as a percentage of the maximum obtainable relaxation elicited by isoproterenol (0.1 mM). Data are mean ± SE ($n = 6$). (Reproduced from Ref. 125, with permission of Munksgaard International Publishers Ltd., Copenhagen, Denmark: Copyright © 1993.)

Table 6 In-Vitro and in-Vivo Biological Properties of Ac-[Lys[12], Cys[17],Val[26],Cys[28]]-VIP cyclo(Cys[17]→Cys[28]) [17] and of the linear parent compound Ac-Lys[12],Ahx[17],Val[26],Thr[28]]-VIP [3]

Compound no.	Bioassay guinea pig EC_{50} (nM)[a]	Binding assay guinea pig IC_{50} (nM)[b]	Inhibition of histamine-induced bronchoconstriction	
			Instillation ED_{50} (μg)[c]	Duration (min)[d]
3	2.7	15	6.3	5
17	60	2500	72	>40

[a]Concentration of compound which relaxed 1.5 g resting tension by 50%.
[b]Concentration of compound which inhibited binding of ^{125}I-VIP by 50%.
[c]Dose of compound which inhibited histamine (50 μg/kg, i.v.)-induced bronchoconstriction by 50%. Pretreatment time was 1 min.
[d]Time for inhibition to decrease to 40%. Dose of compound was 300 μg.
(Reproduced from Ref. 125, with permission of Munksgaard International Publishers Ltd., Copenhagen, Denmark: Copyright © 1993.)

in guinea pigs in vivo, it was also found to be quite potent, with an ED_{50} of 1.2 mg in comparison to the parent linear analog **3** (ED_{50} = 6.3 mg). It was important to find, in addition to increased potency, also a prolonged duration of action: a halftime of activity of 90 min was determined for the cyclic compound **13**, as compared to only 5 min observed for the parent linear peptide **3**.

The positive influence of the lactam bridge between Lys[21] and Asp[25] on activity and duration of action was further utilized by preparing a series of other VIP analogs (62), some of which are listed in Table 7. Their relative potencies to relax guinea pig trachea in vitro were found to be up to 53-fold in comparison to native VIP. Some other analogs with Lys[21]→Asp[25] bridges and additional substitutions were reported later (126). For example, Ac-[Ala[2],Glu[8],Lys[12],Ahx[17],Asp[25],Leu[26],Lys[27,28],Ala[29–31]]-VIP cyclo(Lys[21]→Asp[25]) [24] was found to be 263-fold more potent than the closely related linear analog. When tested in vivo for their bronchoconstriction-inhibitory activity, the cyclic analogs were equal to or even more potent than the linear species. Similarly, nearly all of the cyclic compounds within each pair showed an increased duration of action up to fivefold.

The most valuable compound of this series was obtained when the principle of a cyclic lactam bridge between positions 21 and 25 was applied to the C-terminally extended VIP variant **7**. The resulting analog (**23** in Table 7; see also Fig. 5) was designated Ro 25-1553 (62,127–129). This compound proved to be a highly potent relaxant of both guinea pig and human airway smooth muscle, being 53- and 385-fold, respectively, more potent than native VIP. When being evaluated in vitro in comparison to the β-adrenoceptor agonists salbutamol and salmeterol, the rank order of potency in relaxing isolated guinea pig tracheal smooth muscle was **23** >

Table 7 Biological Activity of Conformationally Restricted VIP Analogs with Lactam Bridges Between Sequence Positions 21 and 25: In-Vitro Relaxant Activity on Guinea Pig Smooth Muscle

No.	Compound	EC_{50} (nM)[a]	Relative potency[b]
18	Ac-[Lys^{12},Ahx^{17},Ala^{19},Asp^{25},Val^{26},Thr^{28}]-VIP cyclo(Lys^{21}→Asp^{25})	0.7	23
19	Ac-[Glu^8,Lys^{12},Ahx^{17},Ala^{19},Asp^{25},Leu^{26},Lys^{27,28}]-VIP cyclo(Lys^{21}→Asp^{25})	0.45	36
20	Ac-[Glu^8,Lys^{12},Ahx^{17},Ala^{19},Asp^{25},Leu^{26},Lys^{27,28},Ala$^{29-31}$]-VIP cyclo(Lys^{21}→Asp^{25})	0.61	26
21	Ac-[Lys^{12},Ahx^{17},Ala^{19},Asp^{25},Leu^{26},Lys^{27,28},Ala$^{29-31}$]-VIP cyclo(Lys^{21}→Asp^{25})	0.42	38
22	Ac-[Glu^8,Lys^{12},Ahx^{17},Asp^{25},Leu^{26},Lys^{27,28},Gly$^{29-30}$,Thr^{31}]-VIP cyclo(Lys^{21}→Asp^{25})	0.3	53
23[c]	Ac-[Glu^8,Lys^{12},Ahx^{17},Ala^{19},Asp^{25},Leu^{26},Lys^{27,28},Gly$^{29-30}$,Thr^{31}]-VIP cyclo(Lys^{21}→Asp^{25})	0.3	53

[a]EC_{50} values were determined by linear regression of log concentration–response curves on three tissues as described in Ref. 41.
[b]Potency relative to native VIP, EC_{50} = 16 nM (1.0).
[c]Designated Ro 25-1553.
(Reproduced from Ref. 62, with permission of John Wiley & Sons, Inc.: Copyright © 1995.)

```
      1    2    3    4    5    6    7    8    9   10   11  12   13   14   15   16
```
Ac-His-Ser-Asp-Ala-Val-Phe-Thr-Glu-Asn-Tyr-Thr-Lys-Leu-Arg-Lys-Gln-

```
     17   18   19   20   21   22   23   24   25   26   27   28   29   30   31
```
-Ahx-Ala-Ala-Lys-Lys-Tyr-Leu-Asn-Asp-Leu-Lys-Lys-Gly-Gly-Thr-NH$_2$

Figure 5 Amino acid sequence of VIP analog **23** (Ro 25-1553). The bond between Lys[21] and Asp[25] represents the side chain-to-side chain lactam ring. (Reproduced from Ref. 62, with permission of John Wiley & Sons, Inc.: Copyright © 1995.)

salbutamol > VIP > salmeterol (1:12:53:77), as evidenced by EC$_{50}$ values of 0.3, 3.5, 16, and 23 nM, respectively. When the ability to relax histamine-induced contractions was tested on isolated human bronchial tissue, all four compounds caused concentration-dependent relaxations (Fig. 6), but only compound **23** displayed full intrinsic activity. The rank order of potency in this test was **23** > salbutamol = salmeterol > VIP (1:29:34:385), as derived from EC$_{50}$ values of 20, 570, 680, and 7700 nM, respectively. When tested in vivo for the ability to prevent histamine-induced bronchoconstriction in guinea pigs, analog **23** was again superior to native VIP and the two β-agonists and was capable of producing nearly 100% inhibition. The time course of this bronchodilator activity of **23**, VIP, and salbutamol following a single intratracheally instilled dose of 300 µg is shown in Fig. 7. A long

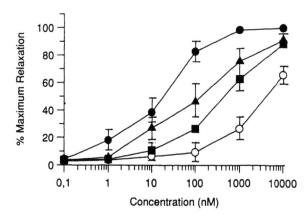

Figure 6 Cumulative relaxation concentration-response curves ($n = 6$) for the VIP analog **23** (Ro 25-1553) (●), salbutamol (▲), salmeterol (■), and native VIP (○) in isolated human bronchial smooth muscle preconcentrated with histamine (0.3 mM). Relaxant responses are expressed as a percentage of the maximum obtainable relaxation. (Reproduced from Ref. 127, with permission of ESCOM Science Publishers B.V.: Copyright © 1992.)

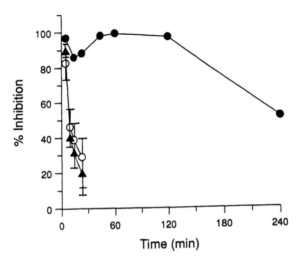

Figure 7 Comparison of the time course of inhibition of histamine-induced bronchoconstriction in anesthetized guinea pigs caused by the VIP analog **23** (●), native VIP (○), and salbutamol (▲). The time between administration of compound (300 µg) and subsequent challenge with histamine (50 µg/kg, i.v.) was varied. The duration of action was calculated as the time for inhibitory activity to decrease to 40%. Each point represents the mean ± SEM ($n = 6$). (Reproduced from Ref. 62, with permission of John Wiley & Sons, Inc.: Copyright © 1995.)

duration of action, more than 240 min, could be achieved, as compared to a rather short duration of approximately 5 min observed for native VIP and salbutamol in this model. Due to the long duration of bronchodilator potency of **23**, combined with antiinflammatory action evaluated later (Bolin DR, unpublished data), this unique VIP analog has the potential to be clinically applicable in the treatment of bronchospastic diseases if the useful properties described are also found similar in the in-vivo situation in humans. In a different strategy for the development of long-acting VIP analogs, attempts were made to find out the positions of proteolytic degradation (see Section III) and eliminate scissile peptide bonds by replacement of the appropriate amino acid residues. In one of these studies, aimed at finding VIP analogs with enhanced proteolytic stability, conformational aspects were also considered. Several compounds were designed on the basis of a putative π-helical structure (130) and were tested for VIP-like activity in two systems: The receptor-binding behavior was studied on lung membranes as receptor source due to the interest in the activity of VIP as a bronchodilator, and the stimulation of α-amylase release from guinea pig pancreatic acini was measured as well. Affinity and potency of one model compound out of five under investigation were found to be comparable to VIP in these systems. The proteolytic stability of the analogs was tested with respect to their ability to resist degradation by enzymes in a crude rat homogenate. However, none of the five analogs showed enhanced proteolytic stability in this test: The half-life times determined were either equal or lower than that of native VIP.

Another study, with the goal of eliminating enzymatic cleavage sites, was based on the elucidation of a novel degradative pathway of VIP within the lung (62). By ex-vivo experiments using bronchoalveolar lavage (BAL) fluid from both guinea pig and humans, one major cleavage site of VIP has been found to be the bond Ser^{25}-Ile^{26} (see Section III). In an attempt to stabilize this scissile bond, several compounds related to Ac-[Lys^{12},Ahx^{17},Val^{25},Thr^{28}]-VIP [3] (see Table 2) with amino acid replacements at the P_1 and P_1' positions 25 and 26, including inversion of chirality and methylation of the scissile peptide bond, were synthesized and their half-times in BAL fluid were determined as well as their biological activities in vitro. The results are shown in Table 8. All compounds 26 to 33 were between 1.7 and 3.0 times more stable to degradation than native VIP. Except for 29, 31, and 32, they were all equipotent in tracheal relaxant activity. When tested in vivo by intratracheal installation to guinea pigs, the analogs 26, 28, and 30 were found to be significantly more potent, with EC_{50} values of 0.67, 0.3, and 0.61 mg, respectively, as compared to 3. Furthermore, the in-vivo durations of action for these compounds were significantly prolonged, to 86, 59, and 55 min, respectively, as compared to 5 min for 3.

In an attempt to stabilize the scissile peptide bonds at the carboxylic sides of the VIP sequence positions Arg^{12}, Arg^{14}, Lys^{15}, Lys^{20}, and Lys^{21}, which are known to be sites of enzymatic cleavage by tryptase, an endopeptidase present in human airways

Table 8 Degradation Halftimes of VIP Analogs in Guinea Pig Bronchial Alveolar Lavage Fluid versus in-Vitro Relaxant Activity on Guinea Pig Tracheal Smooth Muscle

No.	Compound	Relative halftime[a]	EC_{50}[b]
3	Ac-[Lys^{12},Ahx^{17},Val^{26},Thr^{28}]-VIP	1.7	2.7
25	Ac-[Lys^{12},Ahx^{17},Thr^{25},Val^{26},Thr^{28}]-VIP	2.6	2.1
26	Ac-[Lys^{12},Ahx^{17},Ala^{25},Val^{26},Thr^{28}]-VIP	1.9	1.9
27	Ac-[Lys^{12},Ahx^{17},Ala^{26},Thr^{28}]-VIP	2.3	1.7
28	Ac-[Lys^{12},Ahx^{17},Lys^{26},Thr^{28}]-VIP	2.0	2.0
29	Ac-[Lys^{12},Ahx^{17},Asp^{26},Thr^{28}]-VIP	2.4	130
30	Ac-[Lys^{12},Ahx^{17},D-Ser^{25},Val^{26},Thr^{28}]-VIP	2.3	2.4
31	Ac-[Lys^{12},Ahx^{17},D-Val^{26},Thr^{28}]-VIP	2.3	30
32	Ac-[Lys^{12},Ahx^{17},N-Me-Val^{26},Thr^{28}]-VIP	3.0	81

[a]Degradation was followed by analytical HPLC. Peak areas were integrated and degradation halftimes were calculated by plotting the percentage of VIP remaining versus time and reported relative to native VIP. VIP = 1.0.
[b]EC_{50} values were determined by linear regression of log concentration–response curves on at least three tissues as described in Ref. 41.
(Reproduced from Ref. 62, with permission of John Wiley & Sons, Inc.: Copyright © 1995.)

(see Section II), analogs with replacements by D-enantiomers of Arg and Lys or by nonbasic amino acids at these positions were synthesized (131,132). The analogs [D-Arg[12],D-Arg[14],Ahx[17]]-VIP [35] and [Ahx[17],D-Lys[20],Ala[21]]-VIP [36] were about half as active as [Ahx[17]]-VIP in relaxing isolated strips of guinea pigs, but they were cleaved almost as fast as native VIP by isolated human tryptase, as shown by HPLC experiments: The desired stability was not achieved. Much more successful was the design of a hybrid molecule between [Ahx[17]]-VIP and PHM, in which the VIP sequence -Arg[12]-Leu[13]-Arg[14]-Lys[15]- was substituted by the PHM sequence -Lys[12]-Leu[13]-Leu[14]-Gly[15]-. The resulting VIP analog [Lys[12],Leu[14],Gly[15],Ahx[17]]-VIP [37] proved to be almost as active in vitro as [Ahx[17]]-VIP and native VIP in the tracheal relaxant test. Enhanced stability against degradation by human tryptase could be proved by HPLC studies (132): A substantial amount (approx. 20%) of uncleaved 37 was still present in the incubation mixture after 1 hr, while VIP was completely cleaved after 5 min under the same conditions. In an in-vivo test, 37 was shown to be nearly as potent as native VIP against histamine-induced bronchoconstriction in guinea pigs (133).

In order to eliminate the moieties -Asn[24]-Ser[25]- and -Asn[28]-NH$_2$ of VIP, which are sensitive to degradation by isomerisation and/or desamidation (see Section II), another hybrid molecule was designed (56) between [Ahx[17]]-VIP and PACAP-27 by substituting the C-terminal VIP-sequence -Asn[24]-Ser[25]-Ile[26]-Leu[27]-Asn[28]-NH$_2$ by the PACAP sequence -Ala[24]-Ala[25]-Val[26]-Leu[27]-NH$_2$. The resulting VIP/PACAP hybrid [Ahx[17],Ala[24],Ala[25],Val[26]]-VIP-(1-27)-amide [38] had about a twofold increased relaxant activity on guinea pig tracheal strips in vitro in comparison to natural VIP, while synthetic [Ahx[17]]-PACAP-27, despite a rather high (64%) sequence homology to VIP, was almost inactive. Furthermore, an increased duration of action of about 40 min (VIP: 10 min) was found for 38 in this test.

Encouraged by the result of longer duration of tracheal relaxant action obtained with the VIP/PHM and VIP/PACAP hybrid molecules mentioned, the investigations were extended to helodermin (HD), another member of the VIP family (134). The entire HD molecule with 35 amino acid residues was shown to have VIP-like but much longer-lasting tracheal relaxant activity (see Section II). This long duration of action was attributed to the carboxy-terminal extension (29-35) with a -Pro-Pro-Pro-NH$_2$ sequence (14,93). By a comparison of the amino acid sequences of HD and PHM, it was observed that the middle sequence, -Lys[12]-Leu[13]-Leu[14]-Ala[15]-, of HD is very similar to the corresponding PHM sequence, which caused the enhanced enzymatic stability of analog 37. Therefore it seemed worthwhile to find out if the prolonged action of HD is retained if the entire HD sequence is shortened at the C-terminus to obtain a HD segment with a chain length identical to that of VIP, i.e., with 28 amino acid residues. The synthetic HD fragment-(1-28)-amide, with the sequence H-His-Ser-Asp-Ala-Ile-Phe-Thr-Gln-Gln-Tyr-Ser-Lys-Leu-Leu-Ala-Lys-Leu-Ala-Leu-Gln-Lys-Tyr-Leu-Ala-Ser-Ile-Leu-Gly-NH$_2$ [39], a VIP analog representing only 54% sequence homology to VIP, is almost as active in relaxing isolated strips of guinea pig trachea in vitro as the entire HD and as VIP. The duration of this action proved to be nearly as long, namely, 125 min, as that of natural HD (136

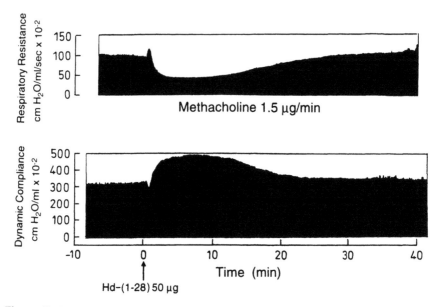

Figure 8   Reversal by helodermin-(1-28)-amide [VIP analog **38**] of methacholine-induced bronchoconstriction in guinea pig in vivo. Methacholine was infused i.v. (1.5 μg/min) throughout the experiment. Peptide was given (50 μg) at arrow. Bronchodilation is evident in decrease in respiratory resistance (upper panel) and increase in dynamic pulmonary compliance (lower panel). (From Ref. 134.)

min), while VIP acts relatively briefly (15 min) under the same conditions. The protective effect of **39** against histamine- or methacholine-induced bronchoconstriction was also demonstrated on guinea pigs in vivo (Fig. 8). As a result it was also concluded that the HD sequence (12-15), i.e., Lys-Leu-Leu-Ala, is essential for the prolonged action of HD rather than the C-terminal sequence (29-35), i.e., -Ser-Arg-Thr-Ser-Pro-Pro-Pro-NH$_2$, while the same sequence segment (12-15) in VIP, i.e., Arg-Leu-Arg-Lys, is one major feature responsible for the rapid loss of tracheal relaxant activity observed with VIP.

V. Conclusions

The recognition of a possible potential therapeutic value of VIP for the treatment of diseases such as bronchial asthma has led to a search for stable analogs with enhanced stability against metabolic processes and with full agonist or even superagonist properties. The basis for the development of such long-acting VIP analogs was (a) structure–activity relationship studies; (b) recognition of sequence similarities within members of the VIP family of peptides with respect to their receptor-binding properties and biological actions; (c) conformational considerations; and (d) knowledge about the sites of the VIP molecule which are sensitive to degradation and modification.

A series of VIP analogs with equal or enhanced biopotency and significantly prolonged duration of action, even in in-vivo experiments, was found. The most valuable compound of this series, with a very high therapeutic potential for the treatment of asthma, is analog **23** (Fig. 5), a 31-residue peptide amide with a structure-stabilizing cycling moiety, designated Ro-25-1553. However, the chemical synthesis of this compound, containing an arginine residue and an internal lactam bridge, is rather difficult and expensive. Another compound, namely, analog **39**, with much simpler structure (no arginine, no asparagine, no cyclization) but also long-acting VIP-like properties, is easier and less costly to synthesize. It may serve as an asthma therapeutic as well.

References

1. Said SI, Mutt V. Polypeptide with broad biological activity. Isolation from small intestine. Science 1970; 169:1217–1218.
2. Said SI, Mutt V. Isolation from porcine-intestinal wall of a vasoactive octacosapeptide related to secretin and glucagon. Eur J Biochem 1972; 28:199–204.
3. Mutt V, Said SI. Structure of the porcine vasoactive intestinal octacosapeptide. Eur J Biochem 1974; 42:581–589.
4. Bodanszky M, Klausner YS, Lin CY, Mutt V, Said SI. Synthesis of vasoactive intestinal peptide (VIP). J Am Chem Soc 1974; 96:4973–4978.
5. Said SI, ed. Vasoactive Intestinal Peptide. New York: Raven Press, 1982.
6. Said SI. Vasoactive intestinal polypeptide (VIP): current status. Peptides 1984; 5:143–150.
7. Said SI. Vasoactive intestinal peptide. J Endocrinol Invest 1986; 9:191–200.
8. Said SI, Mutt V, eds. Vasoactive intestinal peptide and related peptides. Ann NY Acad Sci 1988; 527:1–688.
9. Dockray G. Vasoactive intestinal polypeptide and related peptides. In: Walsh JH, Dockray GJ, eds. Gut Peptides: Biochemistry and Physiology. New York: Raven Press, 1994:447–472.
10. Maggi CA, Giachetti A, Dey RD, Said SI. Neuropeptides as regulators of airway function: vasoactive intestinal peptide and the tachykinins. Physiol Rev 1995; 75:277–322.
11. Said SI. Vasoactive intestinal peptide. Biological role in health and disease. Trends Endocrinol Metab 1991; 2:107–112.
12. Said SI. Vasoactive intestinal peptide (VIP) and related peptides as anti-asthma and anti-inflammatory agents. Biomed Res 1992; 13(suppl 2):257–262.
13. Said SI. VIP as a modulator of lung inflammation and airway constriction. Am Rev Respir Dis 1991; 143:S22–S24.
14. Said SI. VIP in asthma. Ann NY Acad Sci 1991; 629:305–318.
15. Barnes PJ, Dixon CMS. The effect of inhaled vasoactive intestinal peptide on bronchial reactivity to histamine in humans. Am Rev Respir Dis 1984; 130:162–166.
16. Mojarad M, Grode TL, Cox C, Kimmel G, Said SI. Differential responses of human asthmatics to inhaled vasoactive intestinal peptide (VIP). Am Rev Respir Dis 1985; 131:A281.
17. Morice A, Unwin RJ, Sever PS. Vasoactive intestinal peptide causes bronchodilation

and protects against histamine-induced bronchoconstriction in asthmatic subjects. Lancet 1983:1225–1227.

18. Morice AH, Unwin RJ, Sever PS. Vasoactive intestinal peptide as bronchodilator in asthmatic subjects. Peptides 1984; 5:439–440.

19. Morice AH, Sever PS. Vasoactive intestinal peptide as a bronchodilator in severe asthma. Peptides 1986; 7(suppl 1):279–280.

20. Palmer JBD, Cuss FMC, Warren JB, Blank M, Bloom SR, Barnes PJ. Effect of infused vasoactive intestinal peptide on airway function in normal subjects. Thorax 1986; 41: 663–666.

21. Jones J, ed. The chemical synthesis of peptides. In: Halpern J, Green MLH, Mukayama T, eds. International Series of Monographs on Chemistry. Oxford: Clarendon Press 1991; 23:1–228.

22. Bodanszky M, ed. The Principles of Peptide Synthesis. 2d ed. Berlin-Heidelberg: Springer Verlag, 1993.

23. Pennington MW, Dunn BM. Peptide synthesis protocols. In Walker JM, ed. Methods in Molecular Biology 1994; 35:1–316.

24. Wendlberger G, Thamm P, Gemeiner M, Bataille D, Wünsch E. Total synthesis of the 17-norleucine analogue of porcine VIP. In: Brunfeldt K, ed. Peptides 1980. Proc. 16th European Peptide Symposium. Copenhagen: Scriptor, 1981:290–295.

25. Wünsch E, Wendlberger G. Vasoactive intestinal polypeptide (VIP). Synthese von biologisch aktivem 17-Norleucin-Analogon. Wiener tierärztl. Monatsschrift 1986; 73: 164–168.

26. Wendlberger G, Schötz A, Knaup G, Hübener G, Stocker H, Wünsch E. Problems in large scale syntheses of VIP. Proceedings 4th Akabori-Conference, Japan. Protein Research Foundation, 1992:60–65.

27. Coy DH, Gardner J. Solid-phase synthesis of porcine vasoactive intestinal peptide. Int J Peptide Protein Res 1980; 15:73–78.

28. Colombo R. A new solid-phase synthesis of porcine vasoactive intestinal peptide using N_α-9-fluorenylmethyloxycarbonyl amino acids. Experientia 1982; 38:773–775.

29. Remmer HA, Jaeger E, Rücknagel P, Jung G. Comparison of Fmoc-solid-phase methods for the synthesis of pure [Ahx[17]]-VIP. In: Epton R, ed. Innovations and Perspectives in Solid Phase Synthesis. Proceedings, 3rd Internat Symposium on SPS, Oxford, UK. Birmingham: Mayflower Worldwide, 1994:657–660.

30. Wulff BS, Georg B, Fahrenkrug J. Expression and characterization of VIP and two VIP mutants in NIH 3T3 cells. FEBS Lett 1994; 341:43–48.

31. Raingeaud J, Lavergne F, Lelievre V, Muller JM, Julien R, Cenatiempo Y. Production, analysis and bioactivity of recombinant vasoactive intestinal peptide analogs. Biochimie 1996; 78:14–25.

32. Altiere RJ, Kung M, Diamond L. Comparative effects of aerosolized vasoactive intestinal peptide (VIP) on histamine (HIST) induced bronchoconstriction in human subjects. Pharmacologist 1983; 25:123.

33. Bundgaard A, Enehjelm SD, Aggesterup S. Pretreatment of exercise-induced asthma with inhaled vasoactive intestinal peptide (VIP). Eur J Respir Dis 1983; 64(suppl 128): 427–429.

34. Barnes PJ. New therapeutic approaches. In: Barnes PJ, ed. Asthma. Edinburgh, London: Churchill Livingstone. Br Med Bull, 1992; 48:231–247.

35. Liu L-W, Sata T, Kubota E, Paul S, Iwanaga T, Foda H, Said SI. VIP is enzymatically degraded in the trachea, probably by an enkephalinase. Clin Res 1987; 35:647A.

36. Pipkorn R, Hakanson R. The effect of VIP on guinea pig taenia coli requires the whole sequence. Peptides 1984; 5:267–269.
37. Bolin DR, Sytwu I-I, Cottrell M, Garippa RJ, Brooks CC, O'Donnell M. Synthesis and airway smooth muscle relaxant activity of linear and cyclic vasoactive intestinal peptide analogs. In: Marshall GR, ed. Peptides: Chemistry and Biology. Proceedings of the 10th American Peptide Symposium. Leiden: ESCOM, 1988:441–443.
38. Bodanszky M, Bodanszky A. VIP and related peptides: structure-activity relationships. In: Said SI, ed. Vasoactive Intestinal Peptide. New York: Raven Press, 1982:11–22.
39. Couvineau A, Rouyer-Fessard C, Fournier A, St Pierre S, Pipkorn R, Laburthe M. Structural requirements for VIP interaction with specific receptors in human and rat intestinal membranes: effect of nine partial sequences. Biochem Biophys Res Commun 1984; 121:493–498.
40. Robberecht P, Coy DH, De Neff P, Camus J-C, Cauvin A, Waelbroeck M, Christophe J. Interaction of vasoactive intestinal peptide (VIP) and N-terminally modified analogs with rat pancreatic, hepatic and pituitary membranes. Eur J Biochem 1986; 159:45–49.
41. O'Donnel M, Garippa RJ, O'Neill NC, Bolin DR, Cottrell MO. Structure-activity studies of vasoactive intestinal polypeptide. J Biol Chem 1991; 266:6389–6392.
42. Goosens J-F, Pommery N, Lohez M, Pommery J, Helbecque N, Cotelle P, Lhermitte M, Henichart J-P. Antagonistic effect of vasoactive intestinal peptide fragment, vasoactive intestinal peptide(1-11), on guinea pig trachea smooth muscle relaxation. Mol Pharmacol 1991; 41:104–109.
43. Turner JT, Jones SB, Bylund DB. A fragment of vasoactive intestinal peptide, VIP(10-28), is an antagonist of VIP in the colon carcinoma cell line, HT29. Peptides 1986; 7:849–854.
44. Sutliff VE, Raufman JP, Jensen RT, Gardner JD. Actions of vasoactive intestinal peptide and secretin on chief cells prepared from guinea pig stomach. Am J Physiol 1989; 251: G96–G102.
45. Xia M, Spreedharan P, Goetzl EJ. Predominant expression of type II vasoactive intestinal peptide receptors by human T lymphoblastoma cells: Transduction of both Ca2* and cyclic AMP signals. J Clin Immunol 1996; 16:21–30.
46. Ichinose M, Sawada M, Maeno T. Inhibitory effect of vasoactive intestinal peptide (VIP) on phagocytosis in mouse peritoneal macrophages. Regulatory Peptides 1994; 54: 457–466.
47. Fahrenkrug J, Ottesen B, Palle C. Non-amidated forms of VIP (glycine extended VIP and VIP-free acid) have full bioactivity on smooth muscle. Regulatory Peptides, 1989; 26:235–239.
48. Hamada M, Uchida Y, Inoue M, Homma T, Saitoh T, Nomura A, Ohtsuka M, Kameyama M, Hacegawa S, Itoh O, Tachibana S. Possibilities of clinical uses of a novel precursor analogue of vasoactive intestinal polypeptide as an inhaled bronchodilator. Am Rev Respir Dis 1988; 137(suppl N4,pt2):35.
49. Sakamoto T, Hamada M, Uchida Y, Nomura A, Kameyama M, Itoh O, Tachibana S, Hasegawa S. A novel precursor analogue of VIP is stable in guinea pig airway. Am Rev Respir Dis 1989; 139(suppl N5,pt2):A614.
50. Uchida Y, Nomura A, Ohtsuka M, Hamada M, Hasegawa S, Goto K, Tachibana S, Itoh O. A novel precursor analog of vasoactive intestinal polypeptide (VIP) as a potent bronchodilator: a pharmacological study. Am Rev Respir Dis 1988; 137:375A.
51. Gozes I, Fridkin M. A fatty neuropeptide: potential drug for noninvasive impotence treatment in a rat model. J Clin Invest 1992; 90:810–814.

52. Mutt V. Vasoactive intestinal polypeptide and related peptides. Ann NY Acad Sci 1988; 527:1–19.

53. Gozes I, Reshef A, Salah D, Rubinraut S, Fridkin M. Stearyl-norleucine-vasoactive intestinal peptide (VIP): a novel VIP analog for noninvasive impotence treatment. Endocrinology 1994; 134:2121–2125.

54. Takeyama M, Koyama K, Yajima H, Moriga M, Aono M, Murakami M. Studies on Peptides. XCVII. Synthesis of porcine Glu^8-vasoactive intestinal polypeptide (VIP). Chem Pharm Bull 1980; 28:2265–2269.

55. Jaeger E, Remmer HA, Thamm P, Said SI, Sharaf H. Structure-activity studies on vasoactive intestinal peptide: Synthesis of analogues modified at residues Asp^3-Ala^4 and Asn^9. Regulatory Peptides 1992; 40:175.

56. Remmer HA, Jaeger E, Jung G, Rücknagel P, Abdel-Racek T, Said SI. Structure-activity studies on VIP, III. Synthesis and properties of [Ser(SO$_3$)^3,Ahx^{17}]-VIP, [Ahx^{17}]-PACAP-27 and the VIP/PACAP-hybrid [Ahx^{17},Ala^{24},Ala^{25},Val^{26}]-VIP-(1-27)-amide. In: Maia HLS, ed. Peptides 1994. Proceedings 23rd European Peptide Symposium (Braga/Portugal, 1994). Leiden: ESCOM, 1995:367–368.

57. Bolin DR, Meienhofer JA, Montclair U, Sytwu I-I. Synthetic vasoactive intestinal peptide analogs. U.S. Patent 4,605,641, 1986.

58. Fry DC, Madison VS, Bolin DR, Greeley DN, Toome V, Wegrzynski BB. Solution structure of an analogue of vasointestinal peptide as determined by two-dimensional NMR and circular dichroism spectroscopies and constrained molecular dynamics. Biochemistry 1989; 28:2399–2409.

59. Bolin DR, Cottrell JM, Fry DC, Madison VS, O'Neill N, Garippa R, O'Donnell M. Synthesis, biological activity, and conformation of analogs of vasoactive intestinal peptide. Regulatory Peptides 1989; 26:145.

60. Bolin DR, Cottrell JM, O'Neill N, Garippa RJ, O'Donnell M. N-terminal analogs of vasoactive intestinal peptide: identification of a binding pharmacophore. In: Rivier JE, Marshall GR, eds. Peptides: Chemistry and Biology. Proceedings of the 11th American Peptide Symposium. Leiden: ESCOM, 1990:208–210.

61. Chou PY, Fasmann GD. Empirical predictions of protein conformation. Annu Rev Biochem 1978; 47:251–276.

62. Bolin DR, Michalewsky J, Wasserman MA, O'Donnell M. Design and development of a vasoactive intestinal peptide analog as a novel therapeutic for bronchial asthma. Biopolymers (Peptide Science) 1995; 37:57–66.

63. Bolin DR, Cottrell JM, Senda R, Merritt D, Garippa R, O'Neill N, O'Donnell M. Identification of the binding pharmacophores of vasoactive intestinal peptide (VIP). In: Smith JA, Rivier JE, eds. Peptides: Chemistry and Biology. Proceedings of the 12th American Peptide Symposium (Cambridge, MA, 1991). Leiden: ESCOM, 1992: 150–151.

64. Nutt RF, Ciccarone TM, Brady SF, Colton CD, Paleveda WJ, Lyle TA, Williams TM, Veber DR, Wallace A, Winquist RJ. Structure-activity studies of atrial natriuretic factor. In: Marshall GR, ed. Peptides: Chemistry and Biology. Proceedings of the 10th American Peptide Symposium (St. Louis, MO, 1986). Leiden: ESCOM, 1988:444–446.

65. Christophe J, Svoboda M, Lambert M, Waelbroeck M, Winand J, Dehaye JP, Vandermeers MC, Robberecht P. Effector mechanisms of peptides of the VIP family. Peptides 1986; 7(suppl 1):101–107.

66. D'Orisio MS. Biochemical characteristics of receptors for vasoactive intestinal peptide in nervous, endocrine and immune systems. Fed Proc 1987; 46:192–195.

67. Laburthe M, Kitabki P, Couvineau A, Amiranoff B. Peptide receptors and signal transduction in the digestive tract. In: Brown DR, ed. Gastrointestinal regulatory peptides (Handbook of Experimental Pharmacology). Berlin, Heidelberg, New York: Springer Verlag, 1993; 106:148–153.

68. Couvineau A, Amiranoff B, Laburthe M. Solubilization of the liver vasoactive intestinal peptide receptor. J Biol Chem 1986; 261:14482–14489.

69. Paul S, Said SI. Characterization of receptors for vasoactive intestinal peptide solubilized from the lung. J Biol Chem 1987; 262:158–162.

70. Provow S, Velicelebi G. Characterization and solubilization of vasoactive intestinal receptors from rat lung membranes. Endocrinology 1987; 120:2442–2452.

71. Shaffer MM, Carney DM, Korman LY, Lebovic GS, Moody TW. High affinity binding of VIP to human lung cancer cell lines. Peptides 1987; 8:1101–1106.

72. Luis J, Said SI. Characterization of VIP- and helodermin-preferring receptors on human small cell lung carcinoma cell lines. Peptides 1990; 11:1239–1244.

73. Luis J, Martin J-M, El Battari A, Margaldi J, Pichon J. The vasoactive intestinal polypeptide (VIP) receptor: recent data and hypothesis. Biochimie 1988; 70:1311–1322.

74. Rosselin G. The receptors of the VIP family peptides (VIP, secretin, GRP, PHI, PHM, GIP, glucagon and oxyntomodulin). Peptides 1986; 7(suppl 1):89–100.

75. Robberecht P, Cauvin A, Gourlet P, Christophe J. Heterogeneity of VIP receptors. Arch Int Pharmacodyn 1990; 303:51–66.

76. Harmar T, Lutz E. Multiple Receptors for PACAP and VIP. TIPS 1994; 15:97–99.

77. Couvineau A, Voisin T, Guijarro L, Laburthe M. Purification of vasoactive intestinal peptide receptor from porcine liver by a newly designed one-step affinity chromatography. J Biol Chem 1990; 265:13386–13390.

78. Sreedharan SP, Robichon A, Peterson KE, Goetzl EJ. Cloning and expression of the human vasoactive intestinal peptide receptor. Proc Natl Acad Sci (USA) 1991; 88:4986–4990.

79. Ishihara T, Shigemoto R, Mori K, Takahashi K, Nagata S. Functional expression and tissue distribution of a novel receptor for vasoactive intestinal polypeptide. Neuron 1992; 8:811–819.

80. Lutz EM, Sheward WJ, West KM, Morrow JA, Fink G, Hamar AJ. The VIP2 receptor: molecular characterisation of a cDNA encoding a novel receptor for vasoactive intestinal peptide. FEBS Lett 1993; 334:3–8.

81. Bataille D, Gespach M, Amiranoff B, Tatemoto K, Vauclin N, Mutt V, Rosselin G. Porcine peptide having N-terminal histidine and C-terminal isoleucine amide (PHI). Vasoactive intestinal peptide (VIP) and secretin-like effects in different tissues from the rat. FEBS Lett 1980; 114:240–242.

82. Raufmann J-P, Jensen RT, Sutliff E, Pisano J, Gardner JD. Action of Gila monster venom on dispersed acini from guinea pig pancreas. Am J Physiol 1982; 242:G470–G476.

83. Laburthe M, Amiranoff B, Boige N, Rouyer-Fessard C, Tatemoto K, Moroder L. Interaction of GRF with VIP receptors and stimulation of adenylate cyclase in rat and human intestinal epithelial membranes. Comparison with PHI and secretin. FEBS Lett 1983; 159:89–92.

84. Laburthe M, Couvineau A, Rouyer-Fessard, Moroder L. Interaction of PHM, PHI and 24-glutamine PHI with human VIP receptors from colonic epithelium: comparison with rat intestinal receptors. Life Sci 1985; 36:991–995.

85. Laburthe M, Couvineau A, Rouyer-Fessard C. Study of species specificity in growth hormone-releasing factor (GRF) interaction with vasoactive intestinal peptide (VIP) receptors using GRF and intestinal VIP receptors from rat and human: evidence that Ac-Tyr¹hGRF is a competitive VIP antagonist in the rat. Mol Pharmacol 1986; 29:23–27.

86. Christophe J, Svoboda M, Waelbroack M, Winand J, Robberecht P. Vasoactive intestinal peptide receptors in pancreas and liver. Structure-function relationship. Ann NY Acad Sci 1988; 527:238–256.

87. Laburthe M, Couvineau A. Molecular analysis of vasoactive intestinal peptide receptors: a comparison with receptors for VIP-related peptides. Ann NY Acad Sci 1988; 527:296–313.

88. Robberecht P, Gourlet P, Cauvin A, Buscail L, de Neef P, Arimura A, Christophe J. PACAP and VIP receptors in rat liver membranes. Am J Physiol 1991; 260:G97–G102.

89. Arimura A. Receptors for pituitary adenylate cyclase-activating peptide: comparison with vasoactive intestinal peptide receptors. Trends Endocrinol Metab 1992; 3:288–310.

90. Nielsson SFE. PACAP-27 and PACAP-38: vascular effects in the eye and some other tissues in the rabbit. Eur J Pharmacol 1994; 253:17–25.

91. Lundberg JM, Fahrenkrug J, Hökfelt T, Martling C-R, Larsson O, Tatemoto K, Anggard A. Co-existence of peptide HI (PHI) and VIP in nerves regulating blood flow and bronchial smooth muscle tone in various mammals including man. Peptides 1984; 5:593–606.

92. Foda HD, Sharaf HH, Jacobson G, Said SI. Pituitary adenylate cyclase activating peptide (PACAP), a novel VIP-like peptide, has prolonged airway smooth muscle relaxant activity. Am Rev Respir Dis 1991; 143:A618.

93. Foda HD, Said SI. Helodermin, a C-terminally extended VIP-like peptide, evokes long-lasting tracheal relaxation. Biomed Res 1989; 10:107–110.

94. Yoshihara S, Ichimura T, Yanaihara N. Lasting inhibitory effect of helodermin inhalation on guinea pig airway contraction. Biomed Res 1992; 13(suppl 2):367–371.

95. Li M, Hoshino M, Zheng L-Q, Naruse S, Yanaihara C, Ohshima K, Iguchi K, Mochizuki T, Yanaihara N. Helodermin analogues: structure-function studies of helodermin. Biomed Res 1993; 14(suppl 3):61–69.

96. Keltz TN, Straus E, Yalow RS. Degradation of vasoactive intestinal polypeptide by tissue homogenates. Biochem Biophys Res Commun 1980; 92:669–674.

97. Straus E, Keltz TN, Yalow RS. Enzymatic degradation of VIP. In: Said SI, ed. Vasoactive Intestinal Peptide. New York: Raven Press 1982:333–339.

98. Conlon JM. Proteolytic inactivation of neurohormonal peptides in the gastrointestinal tract. In: Brown DR, ed. Gastrointestinal Regulatory Peptides. Handbook of Experimental Pharmacology. Vol. 106. Berlin, Heidelberg, New York: Springer-Verlag, 1993:177–190.

99. Nau R, Ballmann M, Conlon JM. Binding of vasoactive intestinal polypeptide to dispersed enterocytes results in rapid removal of the NH_2-terminal histidyl residue. Mol Cell Endocrinol 1987; 52:97–103.

100. Liu LW, Sata T, Kubota E, Paul S, Said SI. Airway relaxant effect of vasoactive intestinal peptide (VIP): selective potentiation by phosphoramidon, an enkephalinase inhibitor. Am Rev Respir Dis 1987; 135:A86.

101. Liu L-W, Trotz M, Erdös EG, Said SI. Vasoactive intestinal peptide (VIP) and helodermin: degradation by airway enzymes. Am Rev Respir Dis 1991; 143:A618.

102. Hachisu M, Hiranuma T, Tani S, Lizuka T.Enzymatic degradation of helodermin and vasoactive intestinal polypeptide. J Pharmacobio-Dynam 1991; 14:126–131.

103. Goetzl EJ, Streedharan SP, Turck CW, Bridenbaugh R, Malfroy B. Preferential cleavage of amino- and carboxyl-terminal oligopeptides from vasoactive intestinal peptide by human recombinant enkephalinase (neutral endopeptidase, EC 3.4.24.11). Biochem Biophys Res Commun 1989; 158:850–854.

104. Barnes PJ. Neuronal mechanisms in asthma. In: Barnes EJ, ed. Asthma. Edinburgh, London: Churchill Livingstone. Br Med Bull 1992; 48:149–168.

105. Caughey GH, Leidig F, Viro NF, Nadel JA. Substance P and vasoactive intestinal peptide degradation by mast cell tryptase and chymase. J Pharmacol Exp Ther 1988; 244:133–137.

106. Franconi GM, Graf PD, Lazarus SC, Nadel JA, Caughey GH. Mast cell tryptase and chymase reverse airway smooth muscle relaxation induced by vasoactive intectinal peptide in the ferret. J Pharmacol Exp Ther 1989; 248:947–951.

107. Tam ET, Caughey GH. Degradation of airway neuropeptides by human lung tryptase. Am J Respir Cell Mol Biol 1990; 3:27–32.

108. Delaria K, Muller D. High-performance liquid chromatographic assay for tryptase based on the hydrolysis of dansyl-vasoactive intestinal peptide. Anal Biochem 1996; 236:74–81.

109. Bolin DR, Cottrell J, Michalewsky J, Garippa R, O'Neill N, Simko B, O'Donnell M. Degradation of vasoactive intestinal peptide in bronchial alveolar lavage fluid. Biomed Res 1992; 13(suppl 2):25–30.

110. Paul S, Volle DJ, Beach CM, Johnson DR, Powell MJ, Massey RJ. Catalytic hydrolysis of vasoactive intestinal peptide by human autoantibody. Science 1989; 244:1158–1162.

111. Paul S. A new effector mechanism for antibodies: catalytic cleavage of peptide bonds. Cold Spring Harbor Symp Quant Biol 1989; 54:283–286.

112. Paul S, Volle DJ, Powell MJ, Massey RJ. Site specificity of a catalytic vasoactive intestinal peptide antibody. J Biol Chem 1990; 265:11910–11913.

113. Paul S, Mei S, Mody B, Eklund SH, Beach CM, Massey RJ, Hamel F. Cleavage of vasoactive intestinal peptide at multiple sites by autoantibodies. J Biol Chem 1991; 266:16128–16134.

114. Paul S, Ebadi M. Vasoactive intestinal peptide: its interactions with calmodulin and catalytic antibodies. Neurochem Int 1993; 23:197–214.

115. Said SI. Vasoactive intestinal peptide: involvement of calmodulin and catalytic antibodies. Neurochem Int 1993; 32:215–219.

116. Bolin DR. Vasoactive intestinal peptide: role of calmodulin and catalytic antibodies. Neurochem Int 1993; 32:221–227.

117. Mody R, Tramontano A, Paul S. Spontaneous hydrolysis of vasoactive intestinal peptide in neutral aqueous solution. Int J Peptide Protein Res 44:441–447.

118. Nishi N, Tsutsumi A, Morishige M, Kiyama S, Fujii N, Takeyama M, Yajima H. Apparent autolysis of the N-terminal tetrapeptide of vasoactive intestinal polypeptide (VIP). Chem Pharm Bull 1983; 31:1067–1072.

119. Geiger T, Clarke S. Deamidation, isomerization and racemization at asparaginyl and aspartyl residues in peptides. J Biol Chem 1987; 262:785–794.

120. Stephenson RC, Clarke S. Succinimide formation from aspartyl and asparaginyl peptides as a model for the spontaneous degradation of proteins. J Biol Chem 1989; 264: 6164–6170.

121. Tyler-Cross R, Schirch V. Effects of amino acid sequence, buffers, and ionic strength on the rate and mechanism of deamidation of asparagine residues in small peptides. J Biol Chem 1991; 266:22549–22556.

122. Jaeger E, Knof S, Scharf R, Lehnert P, Schulz I, Wünsch E. Chemical evidence for the mechanism of inactivation of secretin. Scand J Gastroenterol 1978; 13(suppl 49):93.

123. Bongers J, Heimer EP, Lambros T, Pan YE, Campbell RM, Felix AM. Degradation of aspartic acid and asparagine residues in human growth hormone-releasing factor. Int J Peptide Protein Res 1992; 39:364–374.

124. Bolin DR, Cottrell J, Garippa R, Michalewsky J, Rinaldi N, Simko B, O'Donnell M. Structure-activity studies on the vasoactive intestinal peptide pharmacophore. I. Analogs of tyrosine[22]. Int J Peptide Protein Res 1995; 46:279–289.

125. Bolin DR, Cottrell J, Garippa R, O'Neill N, Simko B, O'Donnell M. Structure-activity studies of vasoactive intestinal peptide (VIP): cyclic disulfide analogs. Int J Peptide Protein Res 1993; 41:124–132.

126. Bolin DR, Cottrell JM, Garippa R, Rinaldi N, Senda R, Simko B, O'Connell M. Comparison of cyclic and linear analogs of vasoactive intestinal peptide. In: Kaumaya PTP, Hodges RS, eds. Peptides: Chemistry, Structure and Biology. Proceedings of the 14th American Peptide Symposium (Columbus, OH, 1995). England: Mayflower Scientific, 1996:174–175.

127. Bolin DR, Cottrell JM, Michalewsky J, Garippa R, Rinaldi N, O'Donnell M, Selig W. Ro 25-1553: A potent, metabolically stable vasoactive intestinal peptide agonist. In: Hodges RS, Smith JA, eds. Peptides: Chemistry, Structure and Biology. Proceedings 13th American Peptide Symposium (Edmonton, 1993). Leiden: ESCOM, 1994:843–845.

128. O'Donnell M, Garippa RJ, Rinaldi N, Selig WM, SImko B, Renzetti L, Tannu SA. Ro 25-1553: a novel, long acting vasoactive intestinal peptide agonist. Part I: in vitro and in vivo bronchodilator studies. J Pharmacol Exp Ther 1994; 270:1282–1288.

129. O'Donnell M, Garippa RJ, Rinaldi N, Selig WM, Tocker JE, Tannu SA, Wasserman MA, Welton A, Bolin DR. Ro 25-1553: a novel, long-acting vasoactive intestinal peptide agonist. Part II: effect on in vitro and in vivo models of pulmonary anaphylaxis. J Pharmacol Exp Ther 1994; 270:1289–1294.

130. Musso GF, Patthi S, Ryskamp TC, Provow S, Kaiser ET, Velicelebi G. Development of helix-based vasoactive intestinal peptide analogues: identification of residues required for receptor interaction. Biochemistry 1988; 27:8174–8181.

131. Jaeger E, Remmer HA, Abdel-Razek TT, Said SI. Structure-activity studies on VIP, II. Synthesis of analogues modified at positions Arg[12], Arg[14]-Lys[15], Met[17] and Lys[20]-Lys[21], including a potent VIP/PHM-hybrid. In: Rosselin G, ed. 1st International Symposium on VIP, PACAP & Related Regulatory Peptides. Singapore, London: World Scientific, 1994:89–92.

132. Remmer HA. Synthese und Analytik von hochreinem Vasoaktivem Intestinal-Peptid (VIP) und medizinisch relevanten Analoga. Dr.rer.nat. dissertation, Eberhard-Karls-Universität, Tübingen, Germany, 1994.

133. Abdel-Razek T, Foda HD, Jaeger E, Said SI. A vasoactive intestinal peptide (VIP) analog with potent bronchodilator activity in guinea pigs in vivo. Am J Respir Crit Care Med 1995; 151:A348.

134. Jaeger E, Bauer S, Joyce MW, Foda HD, Berisha HI, Said SI. Structure-activity studies on VIP, IV. The synthetic agonist helodermin-fragment-(1-28)-amide is a potent VIP-agonist with prolonged duration of tracheal relaxant activity. In: Arimura A, Said SI, eds. VIP, PACAP, and Related Peptides. Ann NY Acad Sci 1996; 805:499–504.

19

Relevance of Catalytic Anti-VIP Antibodies to the Airway

SUDHIR PAUL

University of Nebraska Medical Center
Omaha, Nebraska

I. Introduction

Antibodies are molecules that are capable of binding the target antigen with high affinity. The ability of the immune system to make antibodies with combining sites specialized to recognize different antigens arises from unique mechanisms that permit amino acid sequence diversification of the variable regions of antibody light-chain (L-chain) and heavy chain (H-chain) subunits. Conservative estimates suggest that more than 10^{10} different antibody combining sites can be generated. Some of the diversity is already built into the genome, in the form of multiple variable (V), diversity (D), and joining (J) genes encoding antibody V regions. Initial sequence diversification occurs during formation of exons encoding the V_L and V_H domains, involving imprecise joining of the V and J genes, and the V, D, and J genes, respectively. The V regions mutate rapidly during the final stages of antibody maturation (mutation rate 10^{-3}–10^{-4}/base pair/cell division), and B lymphocytes producing mutants that bind antigens tightly are selected, because antigen binding to surface-expressed antibody stimulates cell proliferation (1).

As in other high-affinity ligand–receptor interactions, specific antigen–antibody binding occurs by precise surface complementarity between the molecules, mediated by electrostatic forces, the hydrophobic effect, van der Waals interactions, and weak long-range forces. It is now recognized that antibodies are capable of chemical catalysis; i.e., antibodies can combine the abilities to recognize the stable form of the antigen, the ground state, and the high-energy intermediate formed en route to product formation, the transition state. Efficient transition-state recognition by antibodies, as by enzymes, derives from the chemical reactivity of activated amino acids in the combining site, as suggested by the catalytic characteristics of human autoantibodies (2–6), and of antibodies synthesized in response to experimental immunization with the substrate (7,8), transition-state mimics (9,10), and anti-idiotypic antibodies to antienzyme antibodies (11).

II. Evidence for Catalysis by Natural Antibodies

Information on catalytic antibodies has been obtained via two approaches. The first approach holds that immunization with transition state analogs is *necessary* to elicit catalytic antibody synthesis (reviewed in Ref. 12). The second approach has developed from empirical evidence that catalytic activities can be elaborated by entirely *natural* means in antibodies, as summarized below.

1. The first complementarity-determining region (CDR) of certain antibody L chains contains a detectable sequence homology with the region surrounding the active-site Ser residue of serine proteases (13).
2. Antibodies produced by immunization with phosphopyridoxyltyrosine significantly accelerate the reaction of pyridoxal phosphate and tyrosine via the proximity effect (14).
3. An esterase-like activity is found in antibodies raised by immunization with the stable ground-state structures of dinitrophenol and testosterone conjugated to carrier proteins (15,16).
4. Autoantibodies in asthma cleave vasoactive intestinal peptide (VIP) at various peptide bonds (2,3).
5. Autoantibodies in patients with asthma and gastrointestinal disease cleave VIP at the Ala^{18}–Val^{19} bond (4).
6. Autoantibodies found in lupus catalyze the cleavage of DNA (5,6).
7. Autoantibodies found in Hashimoto's thyroiditis catalyze the cleavage of thyroglobulin (17).
8. Antibodies in healthy humans and unimmunized mice cleave synthetic peptide substrates with low efficiency (18).
9. The catalytic activity of autoantibodies to VIP is located in the L-chain subunit (19,20).
10. Monoclonal L chains isolated from multiple myeloma patients express amidase activity (21).
11. L chains isolated from multiple myeloma patients cleave Arg-vaso-pressin (22).

12. Most L chains from multiple myeloma patients express amidase activity, and some express VIP cleaving activity (23).
13. An L chain isolated from a multiple myeloma patient expresses the ability to cleave the HIV-1 protein gp120 (24).
14. Monoclonal IgG and its L-chain subunit synthesized in response to immunization with VIP catalyze the cleavage of this polypeptide (7,8,25).
15. Site-directed mutagenesis of the anti-VIP L chain suggests that a Ser-His dyad serves as the catalytic site (26).
16. Linkage of the anti-VIP V_L domain with its natural V_H partner but not an irrelevant V_H domain improves the specificity of the resultant F_V construct for VIP (27).
17. Monoclonal antibodies (28) and their L-chain subunits (29) raised by immunization with heme-porphyrin complexes express thermostable peroxidase activity.
18. Antiidiotypic antibodies to enzymes, which are well known to be part of the natural immune repertoire expressed in physiological and disease-producing immune responses (30), display catalytic activity (11,31), presumably because the combining site of the antiidiotypic antibody is a molecular image of the enzyme active site.
19. An antibody raised to acetylcholinesterase expresses acetylcholinesterase activity, apparently due to an "accidental" reactivity with acetylcholine (32).
20. Several antibodies raised to transition-state analogs (TSAs) express unexpected catalytic activities that are not programmed by the structure of the TSA (12,33), reflecting the so-called accidental synthesis of catalytic sites in the antibodies.
21. Mouse strains that are genetically predisposed to autoimmune disease produce far more esterolytic antibody clones following immunization with a transition-state analog compared to control mouse strains (34), supporting the bias toward catalytic antibody synthesis in autoimmune disease.

III. Mechanism of Catalysis by Anti-VIP Antibodies

Proteolysis by specific antibodies is characterized by high-affinity antibody–substrate binding mediated by contacts at residues from the V_L and V_H domains. The initial ground-state complexation is followed by cleavage of one or more peptide bonds at which contacts with the catalytic residues in the antibody are established in the transition state. The purified H and L subunits of antibodies are independently capable of binding antigens, albeit with lower affinity than the parent antibody (35,36). X-ray crystallography of antibody–antigen complexes have shown that the V_L and V_H domains are both involved in making stabilizing contacts with the antigen (37). The precise contribution of the two V domains varies in individual antibody–

antigen complexes, but the V_H domain contributes somewhat more to the initial antigen binding, because of the greater sequence variability and greater length of CDRH3 compared to CDRL3 (38). The ability to hydrolyze peptide bonds, on the other hand, appears to reside solely in the V_L domain. This conclusion is supported by the ability of polyclonal autoantibody L chains (19), monoclonal L chains (25), recombinant L chains (8), and recombinant V_L domains (23) to hydrolyze vasoactive intestinal polypeptide (VIP), thyroglobulin, gp120, and peptide-methylcoumarinamide substrates. The H chains of polyclonal and monoclonal antibodies to VIP are capable of VIP binding but are devoid of the catalytic activity (19,25).

Study of an L chain raised by immunization with VIP has suggested the development of an enzyme-like arrangement of amino acids responsible for catalysis. The proteolytic activity of this L chain is inhibited by the serine protease inhibitor diisopropylfluorophosphate (DFP), and molecular modeling suggested that the geometry of Ser[27a], His[93], and Asp[1] approximates that of the catalytic triad found in serine proteases (Fig. 1) (8,26). Site-directed mutagenesis studies have indicated that the turnover number of the L chain is reduced by 100-fold by substitution of Ser[27a] or His[93] by Ala (Fig. 2), and about 10-fold by substitution of Asp[1] by Ala (26). Hydrogen bonding with the His[93] imidazole is likely to confer enhanced nucleophilicity to the Ser[27a] hydroxyl group, permitting its participation in peptide bond cleavage by a serine protease-like mechanism. The contribution of Asp[1] in the catalysis is weak. The N-terminal location of Asp[1] can be anticipated to confer translational and rotational mobility to this residue, which probably interferes with precise positioning necessary for hydrogen bonding with His[93]. This factor may also

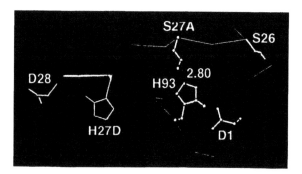

Figure 1 Computer-generated model of anti-VIP light chain (GenBank accession no. L34775) showing the relative positions of Asp[1], Ser[27a], and His[93] residues posited to participate in peptide bond hydrolysis. Ser[26], His[27d], and Asp[28] are potential substrate-contacting residues with relative accessibilities of 40%, 30%, and 44%, respectively. Distances from Ser[27a] OG are: Ser[26] OG, 8.5 Å; His[27d] NE2 and ND1, 9.9 Å and 10.4 Å, respectively; Asp[28] OD1 and OD2, 16.2 Å and 15.2 Å, respectively. Construction of the model is described in Gao et al. (26). Light-chain amino acids are numbered according to Kabat et al. (38). Side chain atoms are designated according to IUPAC-IUB recommendations.

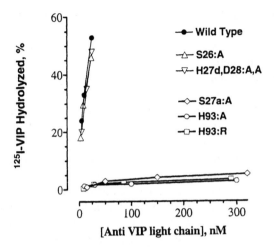

Figure 2 VIP hydrolysis by an antibody light chain (clone c23.5) and its mutants. Reaction conditions, 100 pM ^{125}I-VIP; 3 hr incubation in pH 7.7 buffer. Peptide hydrolysis was estimated as radioactivity soluble in 10% trichloroacetic acid.

contribute to the low turnover of the L chain compared to conventional serine proteases. Other examples of molecules that partially mimic the mechanism utilized by serine proteases are available. The esterase activity of an antibody may be entirely due to a Ser–His dyad (10). A fungal esterase has been described (39) in which a Ser–His dyad is responsible for catalysis in conjunction with a backbone carboxyl group.

The development of a serine protease-like mechanism in a particular L chain does not imply that a single conserved site is responsible for the catalytic activities found in other L chains. This statement is based on the observation of proteolytic activity both in κ and λ L chains, in various V_L subgroups families, in L chains originating from different V_L germline genes, and in L chains that do not contain Ser and His residues at the position found in the anti-VIP L chain (20,23). Therefore, the possibility that L chains are capable of developing diverse proteolytic mechanisms must be considered. The diversity could in principle arise from the existence of different germline genes containing rudimentary versions of different types of catalytic sites, or from the de novo development of the catalytic sites due to sequence diversification known to occur in L-chain CDRs over the course of the immune response.

The H chain, although itself devoid of proteolytic activity, can influence the activity of the L chains, as shown by study of F_V constructs composed of the catalytic anti-VIP V_L domain linked to its natural V_H domain partner (anti-VIP F_V) or an irrelevant V_H domain (hybrid F_V; the V_H domain is from an antilysozyme antibody). The anti-VIP V_H domain exerted beneficial effects and the anti-lysozyme V_H domain exerted detrimental effects on the catalytic activity, as evaluated by the values of VIP-binding affinity and catalytic efficiency (measured as the k_{cat}/K_m parameter).

IV. Neutralizing Potency of Proteolytic Antibodies

There are two reasons to expect proteolytic antibodies to be biologically more potent than reversibly binding antibodies. First, cleavage of the target polypeptide can be anticipated to produce fragments with biological activity distinct from the parent polypeptide. Second, a single catalyst molecules can cleave multiple substrate molecules, whereas noncatalytic antibodies can act only stoichiometrically. The catalytic efficiency of proteolytic autoantibodies and recombinant antibody fragments is sufficiently great to suggest that they are functionally important. The case of VIP is discussed here. This polypeptide elicits its biological effects in the airways and other tissues at picomolar to nanomolar concentrations, which are lower than the K_m values observed for antibodies (nanomolar range) and proteolytic enzymes (micromolar–millimolar range). At VIP concentrations lower than the K_m of the catalyst, the rate of proteolysis is given by the ratio k_{cat}/K_m (catalytic efficiency), where $1/K_m$ approximates the VIP-binding affinity, and k_{cat} denotes turnover number. Recombinant antibody fragments with catalytic efficiency comparable to highly evolved conventional proteases have been isolated (20), and autoantibodies found in blood display even greater catalytic efficiency. Antibodies are found in physiological fluids at far greater concentrations than conventional enzymes. The IgG concentration in serum is about 70 μM. In multiple myeloma patients, the antibody and antibody light-chain products of the tumor cells accumulate in serum and urine to millimolar levels (40). At these concentrations, it appears most unlikely that the proteolytic activity is biologically inconsequential.

Different proteolytic antibodies and antibody L chains show considerable variation in their catalytic properties. The range of turnover number observed using available recombinant antibody fragments varies from 0.003/min to 2.1/min. The half-life of IgG in blood is about 20 days, which corresponds to cleavage of 86 molecules and 60,480 molecules per molecule of the IgG with turnovers of 0.003 and 2.1/min, respectively. It can be readily seen, therefore, that even at excess substrate concentration, at which only the turnover number determines the rate of catalysis, the proteolytic antibodies are likely to express considerably more potent VIP-neutralizing activity than stoichiometric, reversibly binding molecules. It is worth noting that the antibodies cited here were identified based on their antigen-binding affinity, a procedure that favors tight binding to the antigen ground state, but is unlikely to select catalysts with the best turnover. A strong case can be made that higher-turnover antibodies will be found by applying direct screening assays for catalysis.

The entire sequence of VIP is believed to be responsible for its biological activity (41). Thus, proteolytic cleavage of VIP will inactivate the peptide completely. In other instances, the site of cleavage will determine the biological effect of the antibodies, since the different product peptides might show different levels and type of activities, as has been described for synthetic peptides corresponding to subsequences of the tachykinins (42).

A rigid lock-and-key type of fit may not be conducive for catalysis. An imperfect surface alignment between a polypeptide and an antibody raised against a structurally related but different antigen could afford sufficient conformational freedom to permit facile catalysis. The strength of substrate interactions at the antibody catalytic residues may not be related directly to the macroscopic strength of the antigen–antibody binding reaction. For example, interactions of substrate at antibody catalytic residues could be weak, and strong overall binding may simply concentrate the substrate at the catalytic site. We observed that antibodies to VIP also bind the calmodulin-dependent enzyme myosin light-chain kinase (MLCK) (43). Calmodulin apparently recognizes basic amphiphilic helices expressed by both VIP (44) and MLCK, and VIP inhibits calmodulin-dependent MLCK activity (45), suggesting that the two polypeptides bind a common site in calmodulin. A previous report describes binding of several calmodulin-dependent enzymes by antibodies to melittin, a calmodulin-binding peptide (46). It may be profitable, therefore, to explore the MLCK-binding and -cleaving activities of the human autoantibodies.

V. Biological Role of VIP in the Airways

A. Smooth Muscle Relaxation

In the human airway, an inhibitory noncholinergic, nonadrenergic system (I-NANC) is the principal neurally mediated relaxation system (47). A transmitter role for VIP in I-NANC transmission is supported by these observations: potent airway relaxation induced by VIP, presence of VIP in nerves supplying the airways, tetrodotoxin-sensitive release of VIP during electrically stimulated airway relaxation, and inhibition of airway relaxation by anti-VIP antibodies in vitro (48,49). Other neuropeptides of the VIP family, peptide histidine methionine (PHM) and pituitary adenylate cyclase-activating peptide (PACAP) (Fig. 3), could also serve as I-NANC mediators (50,51). PHM is co-synthesized and co-released with VIP, PACAP is a newly discovered peptide found in airway nerves (51). Both peptides relax airway smooth muscle. Nitric oxide (NO) is another candidate I-NANC mediator. In the GI tract, VIP stimulates NO synthesis (52). NO synthase is a calmodulin-dependent enzyme. VIP binds calmodulin and has been shown to regulate the activity of two calmodulin-dependent enzymes, phosphodiesterase (53) and MLCK (45). It is possible, therefore, that VIP directly influences NO synthesis in the airways.

```
VIP     H S D A V F T D N Y T R L R K Q M A V K K Y L N S I L N

PHM     H A D G V F T S S Y R R I L G Q L S A K K Y L E S L M

PACAP   H S D G I F T D S Y S R Y R K Q M A V K K Y L A A V L G K R Y K Q R V K N K
```

Figure 3 Sequence of VIP and related peptides. Underlined residues are identities. K-R substitutions are treated as identities.

B. Inflammation

VIP is an antiinflammatory agent (54–62). It inhibits IgE and IgG4 production (55), release of histamine and eicosanoid mediators from the lung (56,57), and interleukin production by T-helper cells (58). It inhibits antigen-induced and LTD4-induced bronchoconstriction in guinea pig airways (57,59,60), protects against lung injury caused by reactive oxygen radicals (61), and inhibits phagocytosis and chemotaxis by alveolar macrophages (62).

VI. Autoantibodies to VIP and Airway Dysfunction

There is a statistical association of the occurrence of autoantibodies to VIP and asthma. Two types of antibodies are found in asthma patients. The first type is characterized by high V_{max} values. The second variety is noncatalytic and is characterized by high affinity for VIP (63). Hydrolysis of VIP can be expected to inactivate the molecule, since VIP fragments relax airways minimally or not at all. In the case of catalytically nonproductive binding, the antibodies may compete with VIP receptors on target cells. Airway dysfunction could arise, therefore, from simple high-affinity peptide binding. Chronic exposure to the antibodies may lead to other kinds of damage, e.g., by internalization of antibodies via Fc receptors on inflammatory cells and neurons or by binding of antibody to cell-surface VIP. At least three cell types in the airway express VIP receptors: smooth muscle cells, inflammatory cells, and epithelial cells. VIP binding to these cells mediates airway relaxation, decreased release of constrictor mediators, and increased or decreased secretion of glycoconjugates, respectively (64). VIP is an amphiphilic peptide, capable of spontaneous insertion into lipid bilayers (65). Thus, in addition to the receptor-bound peptide, VIP inserted directly into membrane bilayers may serve as a target for the antibodies. Preliminary study of bronchoalveolar lavage fluid and airway sections from animals immunized passively with a catalytic anti-VIP antibody or a control nonimmune antibody has suggested provocation of inflammatory cell infiltration and epithelial cell metaplasia and stripping by the former type of antibody (S. Paul and R. Kalaga, unpublished).

The tendencies toward increased airway smooth muscle constriction and inflammation in asthma are consistent with a deficit of VIP brought about by autoantibodies. Thus, a VIP deficit may underlie decreased relaxation and enhanced contraction of smooth muscle. Similarly, increased inflammation could be explained by decreased levels of VIP. Active immunization of cats with VIP decreases the relaxation of airway smooth muscle in response to electrical stimulation (66). Our preliminary observations suggest a complete inhibition of electrically stimulation tracheal relaxation in vitro under conditions of nonadrenergic and noncholinergic receptor blockade by pretreatment of the tissue with catalytic antibody to VIP (S.I. Said, S. Paul, and co-workers, unpublished). One report shows that nerves in the airways of patients with *severe* asthma are deficient in VIP compared to healthy airways (67). Another study describes indistinguishable levels of VIP in nonasthmatic and asthmatic airways (68). Plasma levels of VIP in response to exercise are reportedly increased to a greater extent in patients with exercise-induced asthma compared to nonasthmatics (69). We have shown an increased incidence of autoantibodies to VIP in individuals who exercise habitually (70) and in asthma patients (63). A hypothesis that takes these apparently contradictory observations into account is: *An increased antibody-mediated clearance produces a deficit in the VIP system, which predisposes the airway toward exaggerated bronchoconstriction. Binding and catalytic hydrolysis of VIP by antibodies could cause compensatory release from neurons, e.g., by disturbing feedback regulation of VIP release mediated by peptide binding to presynaptic receptors. This may eventually deplete the peptide stores in neurons.* Since VIP is stored mainly in neuronal vesicles, any changes in airway VIP levels induced by antibodies are likely to represent alterations in neuronal stores of the peptide. Prejunctional autoreceptors for VIP regulating VIP (71–73), and acetylcholine (74) release have been deduced in studies using guinea pig intestinal tissue and neuroblastoma cell cultures. VIP binding to these receptors is thought to decrease the release of the peptide from the cells. Decreased immunoreactivity for VIP in diabetic skin is well documented (75). An early increase precedes VIP depletion in insulin-dependent diabetics (76), suggesting that compensatory synthetic responses may precede exhaustion of the peptide stores.

Airway immunostaining for nitric oxide synthase and the concentration of nitric oxide in exhaled air is increased in asthmatics (77,78). Whether the increased NO synthesis is a factor in airway inflammation and regulation of smooth muscle tone is not yet clear. Similarly, it is not known whether changes in NO synthesis are secondary to another mediator such as VIP or autoantibodies to VIP. Neuropeptides related to VIP (PHM, PACAP) could compensate for decreased VIP levels brought about by antibody binding and catalysis. The incidence of PHM and PACAP autoantibodies in asthma has not been studied.

Chymase, tryptase, and neutral endopeptidase have been implicated in degradation of VIP in the airway, especially in the presence of an inflammatory response (79–81). The affinity of cellular receptors for VIP is in the low nanomoles range. Effective physiological clearance of VIP may, therefore, require catalysts displaying low K_m values. The K_m's of conventional enzymes for VIP (82,83) are approximately three orders of magnitude greater than the K_m's of catalytic autoantibodies, and their kinetic efficiency (k_{cat}/K_m) is lower than that of the antibodies by 10- to 100-fold. Autoantibodies to VIP may efficiently clear this peptide. An increased level of the reducing agent glutathione in airway lavage fluid from asthmatics has recently been noted (84). Under reducing conditions, formation of L chains will occur by reduction of S–S bonds in antibodies. Since the L chains hydrolyze VIP more efficiently than intact IgG (19), their potential role in clearance of VIP in the airways also warrants consideration.

VII. Airway Selectivity of VIP Autoantibodies

A role for the antibodies in asthma is plausible if antibody synthesis occurs selectively in the airways or there is selective antibody uptake from blood by the airways.

Upon local challenge with antigens, lymphocytes in airway-associated lymph nodes produce antigen-specific antibodies (85–88). The predominant response is of the IgG isotype, although other isotypes are also formed. The local response is far stronger than the systemic response if antigen is directly instilled or aerosolized into the airways (85,86), it can occur even without adjuvant (87), and it is strong enough to lead to detectable circulating antibody levels (88). Antibody synthesis occurs in cells in tracheobronchial lymph nodes and in mucosal lymphocytes in the lung parenchyma. These considerations lead to the following hypothesis: *Natural intrapulmonary immunization with VIP or a homologous polypeptide leads to local synthesis of antibodies that bind and hydrolyze VIP. The concentration of the antibodies in the airway is greater than in other VIP-containing tissues, resulting in airway-selective interference in the physiological functions of VIP.*

Studies of radiolabeled antibodies administered into the blood have suggested that they reach tissue interstitial fluid by passive diffusion across the endothelial barrier (and the basement membrane) and by active transport by pinocytosis (89,90). IgG antibodies are thought to transude freely from blood into the airways (91). Intraperitoneal administration of an anti-VIP IgG preparation in experimental animals resulted in its accumulation in BAL fluids. Near equivalent anti-VIP activity levels were evident in the serum and BAL fluid when normalized for the albumin concentration, suggesting unimpeded transudation from blood to the airway lining fluid (Fig. 4). Tissue biodistribution and imaging studies (Fig. 5) using radiolabeled anti-VIP antibody showed that its uptake occurred at the highest levels in the trachea and lung, even though tissues like the intestine contain greater amounts of VIP. The airway selectivity was not observed using the recombinant L chain of the anti-VIP antibody, suggesting that molecular mass may be an important determinant for the airway selective localization. It may be that the blood–trachea/lung barrier is more porous for IgG antibodies than other blood–tissue barriers. Slower egress of anti-VIP from pulmonary interstitial fluid via drainage into lymphatics is an unlikely explanation, because the pulmonary capillaries are thought to be relatively leaky to proteins compared to the capillaries in other tissues, reflected by the higher protein concentration of lymph leaving the lungs (92). Formation of soluble complexes of VIP and anti-VIP is unlikely to contribute to preferential localization of the antibody in the airways, since the mass of the immune complex is only slightly greater than of uncomplexed antibody (153 and 150 kDa, respectively). Therefore, we cannot exclude the possibility that a proportion of the antibody binds receptor-bound VIP on various cell surfaces, or is internalized by the cells via Fc receptor-mediated endocytosis. Regardless of the mechanism of the observed selective uptake of anti-VIP antibody by the airways, it is reasonable to hypothesize that this phenomenon may induce an airway-selective effect of systemic autoimmunity to VIP.

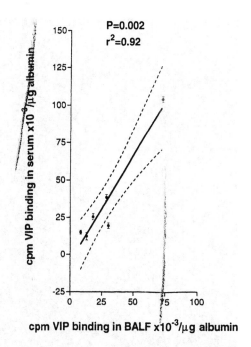

Figure 4 Transudation of monoclonal anti-VIP antibody into airway lining fluid. Monoclonal antibody (clone c23.5, doses ranging from 10 to 21 mg IgG) was injected intraperitoneally into 6 guinea pigs. Serum and bronchoalveolar lavage fluid collected on day 7 was analyzed for albumin content by ELISA and for anti-VIP content by radioimmunoassay for [125]-I-VIP-binding activity.

VIII. Therapeutic Use of VIP in Airway Disorders

In experimental animals, VIP produces strong bronchodilation. The results with VIP administered to humans have been disappointing. It has been suggested that proteolytic breakdown and the presence of thicker cellular barriers may explain the low bronchodilator potency of VIP in the human. Success in preparation of protease-resistant analogs of VIP with improved bronchodilator potency has been reported (93). Conjugation of VIP with fatty acids (94) also improves the permeability and biological efficacy of the peptide. We observed that expression of VIP in liposomes enhances and prolongs the vascular smooth muscle relaxation action of this peptide in hamsters (65). Preliminary studies suggest that liposomal VIP formulations also offer superior airway smooth muscle relaxation in vitro compared to aqueous VIP. Perhaps the liposomal VIP formulations will prove effective in alleviating airway dysfunction, especially in patients who are positive for autoantibodies to VIP.

Figure 5   Planar whole-body image of a mouse 70 min after intravenous injection of ^{125}I-IgG (50 μCi, clone c23.5). Posterior image acquired with a pinhole collimator. Uptake is to lung, and to a lesser extent, to the heart and liver.

Acknowledgment

This work was supported by U.S. Public Health Science grants HL44126 and AI31268. The author is grateful to my laboratory colleagues for their contributions. Whole-body imaging was done with the help of Dr. Janina B.-Kortlywicz.

References

1.   Honjo T, Alt FW. Immunoglobulin Genes. 2d ed. London: Academic Press, 1995.
2.   Paul S, Volle DJ, Beach CM, Johnson DR, Powell MJ, Massey RJ. Catalytic hydrolysis of vasoactive intestinal peptide by human autoantibody. Science 1989; 244:1158–1162.
3.   Paul S, Sun M, Mody R, et al. Cleavage of vasoactive intestinal peptide at multiple sites by autoantibodies. J Biol Chem 1991; 266:16128–16134.
4.   Suzuki H, Imanishi H, Nakai T, Konishi YK. Human autoantibodies that catalyze the hydrolysis of vasoactive intestinal polypeptide. Biochem (Life Sci Adv) 1992; 11:173–177.
5.   Shuster AM, Gololobov GV, Kvashuk OA, Bogomolova AE, Smirnov IV, Gabibov AG. DNA hydrolyzing autoantibodies. Science 1992; 256:665–667.

6. Gololobov GV, Chernova EA, Schourov DV, Smirnov IV, Kudelina IA, Gabibov AG. Cleavage of supercoiled plasmid DNA by autoantibody Fab fragment: application of the flow linear dichroism technique. Proc Natl Acad Sci (USA) 1995; 92:254–257.

7. Paul S, Sun M, Mody R, Tewary HK, Mehrotra S, Gianferrara T, Meldal M, Tramontano A. Peptidolytic monoclonal antibody elicited by a neuropeptide. J Biol Chem 1992; 267:13142–13145.

8. Gao QS, Sun M, Tyutyulkova S, et al. Molecular cloning of a proteolytic antibody light chain. J Biol Chem 1994; 269:32389–32393.

9. Wirsching P, Ashley JA, Benkovic SJ, Janda KD, Lerner RA. An unexpectedly efficient catalytic antibody operating by ping-pong and induced fit mechanisms. Science 1991; 252:680–685.

10. Zhou GW, Guo J, Huang W, Fletterick RJ, Scanlan TS. Crystal structure of a catalytic antibody with a serine protease active site. Science 1994; 265:1059–1064.

11. Izadyar L, Friboulet A, Remy MH, Roseto A, Thomas D. Monoclonal anti-idiotypic antibodies as functional internal images of enzymes active sites: production of a catalytic antibody with a cholinesterase activity. Proc Natl Acad Sci (USA) 1993; 90:8876–8880.

12. Tramontano A. Immune recognition, antigen design, and catalytic antibody production. Appl Biochem Biotechnol 1994; 47:257–275.

13. Erhan S, Greller LD. Do immunoglobulins have proteolytic activity? Nature 1974; 251: 353–355.

14. Raso V, Stollar BD. The antibody-enzyme analogy. Comparison of enzymes and antibodies specific for phosphopyridoxyltyrosine. Biochemistry 1975; 14:591–599.

15. Kohen F, Kim JB, Barnard G, Linder HR. Asteroid immunoassay based on antibody-enhanced hydrolysis of steroid umbelliferone conjugate. FEBS Lett 1979; 100:137–140.

16. Kohen F, Kim JB, Linder HR, Eshhar Z, Green B. Monoclonal immunoglobulin G augments hydrolysis of an ester of the homologous hapten: An esterase-like activity of the antibody-containing site. FEBS Lett 1980; 111:427–431.

17. Li L, Kaveri S, Tyutyulkova S, Kazatchkine M, Paul S. Catalytic activity of anti-thyroglobulin antibodies. J Immunol 1995; 154:3328–3332.

18. Kalaga R, Li L, O'Dell J, Paul S. Unexpected presence of polyreactive catalytic antibodies in IgG from unimmunized donors and decreased levels in rheumatoid arthritis. J Immunol 1995; 155:2695–2702.

19. Sun M, Mody B, Ecklund SH, Paul S. Vasoactive intestinal peptide hydrolysis by antibody light chains. J Biol Chem 1991; 266:15571–15574.

20. Tyutyulkova S, Gao Q, Thompson A, Rennard A, Paul S. Efficient vasoactive intestinal polypeptide hydrolyzing antibody light chains selected from an asthma patient by phage display. Biochim Biophys Acta 1996; 1316:217–223.

21. Matsuura K, Yamamoto K, Sinohara H. Amidase activity of human Bence Jones proteins. Biochem Biophys Res Commun 1994; 204:57–62.

22. Matsuura K, Sinohara H. Catalytic cleavage of vasopressin by human Bence Jones proteins at the arginylglycinamide bond. FEBS Lett. In press.

23. Paul S, Li L, Kalaga R, Wilkins-Stevens P, Stevens FJ, Solomon A. Natural catalytic antibodies: peptide hydrolyzing activities of Bence Jones proteins and V_L fragment. J Biol Chem 1995; 270:15257–15261.

24. Kalaga R, Huang H, Stevens FJ, Solomon A, Paul S. gp120 hydrolysis by catalytic antibody light chain (abstr). Presented at the 9th International Congress of Immunology, San Francisco, CA, July 23–29, 1995.

25. Sun M, Gao QS, Li L, Paul S. Proteolytic activity of an antibody light chain. J Immunol 1994; 153:5121–5126.

26. Gao QS, Sun M, Rees A, Paul S. Site-directed mutagenesis of proteolytic antibody light chain. J Mol Biol 1995; 253:658–664.

27. Gao QS, Paul S. Molecular cloning of anti-ground state proteolytic antibody fragments. In: Paul S, ed. Antibody Engineering Protocols, Methods of Molecular Biology Series 51. Totowa, NJ: Humana, 1995:281–296.

28. Savitsky AP, Nelen MI, Yatsmirsky AK, Demcheva MV, Ponomarev GV, Sinikov IV. Kinetics of oxidation of o-dianisidine by hydrogen peroxide in the presence of antibody complexes of iron(III) coproporphyrin. Appl Biochem Biotechnol 1994; 47:317–327.

29. Takagi M, Kohda K, Hamuro T, Harada A, Yamaguchi H, Kamachi M, Imanaka T. Thermostable peroxidase activity with a recombinant antibody L chain-porphyrin Fe(III) complex. FEBS Lett 1995; 375:273–276.

30. Dolman KM, Van de Wiel BA, Kam C-M, et al. Proteinase 3: substrate specificity and possible pathogenetic effect of Wegener's granulomatosis autoantibodies (C-ANCA) by dysregulation of the enzyme. Adv Exp Med Biol 1993; 336:55–60.

31. Crespeau H, Laouar A, Rochu D. Un abzyme DNase polyclonal produit par la méthode de image interne anti-idiotypique. CR Acad Sci Paris de la Vie/Life Sci 1994; 317:819–823.

32. Johnson G, Moore SW. Anti-acetylcholinesterase antibodies display cholinesterase-like activity. Eur J Immunol 1995; 25:25–29.

33. Thomas NR. Hapten design for the generation of catalytic antibodies. Appl Biochem Biotechnol 1994; 47:345–373.

34. Tawfik D, Chap R, Green B, Sela M, Eshhar Z. Unexpectedly high occurrence of catalytic antibodies in MRL/*lpr* and SJL mice immunized with a transition state analog. Is there a linkage to autoimmunity? Proc Natl Acad Sci (USA) 1995; 92:2145–2149.

35. Ward ES, Gussow D, Griffiths AD, Jones PT, Winter G. Binding activities of a repertoire of single immunoglobulin variable domains secreted for *Escherichia coli*. Nature 1989; 341:544–546.

36. Sun M, Li L, Gao QS, Paul S. Antigen recognition by an antibody light chain. J Biol Chem 1994; 269:734–738.

37. Davies DR, Chacko S. Antibody structure. Acc Chem Res 1993; 26:421–427.

38. Kabat EA, Wu TT, Perry HM, Gottesman KS, Foeller C. Sequences of Proteins of Immunological Interest. 5th ed. Washington, DC: U.S. Department of Health and Human Services, 1991.

39. Wei Y, Schottel JL, Derewenda U, Swenson L. A novel variant of the catalytic triad in the *Streptomyces scabies* esterase. Nature Struct Biol 1995; 2:218–223.

40. Nelson M, Brown RD, Gibson J, Joshua DE. Measurement of free kappa and lambda chains in serum and the significance of their ratio in patients with multiple myeloma. Br J Haematol 1992; 81:223–230.

41. Maggi CA, Giachette A, Dey RD, Said SI. Neuropeptides as regulators of airway function: vasoactive intestinal peptide and the tachykinins. Physiol Rev 1995; 75:227–322.

42. Martins MA, Shore SA, Gerard NP, Gerard C, Drazen JM. Peptidase modulation of the pulmonary effects of tachykinins in tracheal superfused guinea pig lungs. J Clin Invest 1990; 85:170–176.

43. Tyutyulkova S, Paul S. Cross-reactivity of anti-VIP monoclonal antibody with myosin light chain kinase (abstr). FASEB J 1993; 7:317.

44. Stallwood D, Brugger CH, Baggenstoss BA, Stemmer PM, Shiraga H, Landers DF,

Paul S. Identity of a membrane-bound vasoactive intestinal peptide-binding protein with calmodulin. J Biol Chem 1992; 267:19617–19621.

45. Shiraga H, Stallwood D, Ebadi M, Pfeiffer R, Landers D, Paul S. Inhibition of calmodulin-dependent myosin light chain kinase by growth hormone releasing factor and vasoactive intestinal peptide. Biochem J 1994; 300:901–905.

46. Kaetzel MA, Dedman JR. Affinity-purified melittin antibody recognizes the calmodulin-binding domain on calmodulin target proteins. J Biol Chem 1987; 262:3726–3729.

47. Richardson J, Beland J. Nonadrenergic inhibitory nervous system in human airways. J Appl Physiol 1976; 41:764–771.

48. Said SI. Influence of neuropeptides on airway smooth muscle. Am Rev Respir Dis 1987; 136:S52–S58.

49. Matsuzaki Y, Hamasaki Y, Said SI. Vasoactive intestinal peptide: a possible transmitter of nonadrenergic relaxation of guinea pig airways. Science 1980; 210:1252–1253.

50. Palmer JBD, Cuss FMC, Barnes PJ. VIP and PHM and their role in nonadrenergic inhibitory responses in isolated human airways. J Appl Physiol 1986; 61:1322–1328.

51. Uddman R, Luts A, Arimura A, Sundler F. Pituitary adenylate cyclase-activating peptide (PACAP), a new vasoactive intestinal peptide (VIP)-like peptide in the respiratory tract. Cell Tissue Res 1991; 265:197–201.

52. Grider JR. Interplay of VIP and nitric oxide in regulation of the descending relaxation phase of peristalsis. Gastrointest Liver Physiol 1993; 27:G334–G340.

53. Barnette MS, Weiss B. Inhibition of calmodulin-stimulated phosphodiesterase by vasoactive intestinal peptide. J Neurochem 1985; 45:640–643.

54. Said SI. VIP as a modulator of lung inflammation and airway constriction. Am Rev Respir Dis 1991; 143:S22–S24.

55. Kimata H, Yoshida A, Fujimoto M, Mikawa H. Effect of vasoactive intestinal peptide, somatostatin, and substance P on spontaneous IgE and IgG4 production in atopic patients. J Immunol 1993; 150:4630–4640.

56. Undem BJ, Dick EC, Buckner CK. Inhibition of vasoactive intestinal peptide of antigen-induced histamine release from guinea pig minced lung. J Pharmacol 1983; 88:247–250.

57. Ciabattoni G, Montuschi P, Curro D, Togna G, Preziosi P. Effects of vasoactive intestinal peptide on antigen-induced bronchoconstriction and thromboxane release in guinea-pig lung. Br J Pharmacol 1993; 109:243–250.

58. Sun L, Ganea D. Vasoactive intestinal peptide inhibits interleukin (IL)-2 and IL-4 production through different molecular mechanisms in T cells activated via the T cell receptor/CD3 complex. J Neuroimmunol 1993; 48:59–70.

59. Conroy DM, Samhoun MN, Piper PJ. Effects of vasoactive intestinal peptide, helodermin and galanin on responses of guinea-pig lung parenchyma to histamine, acetylcholine and leukotriene D_4. Br J Pharmacol 1991; 104:1012–1018.

60. Conroy DM, Samhoun MM, Piper PJ. Vasoactive intestinal peptide and helodermin inhibit the release of cyclo-oxygenase products induced by leukotriene D_4 and bradykinin from guinea-pig perfused lung. Eur J Pharmacol 1992; 218:43–50.

61. Berisha H, Foda H, Sakakibara H, Trotz M, Pakbaz H, Said SI. Vasoactive intestinal peptide prevents lung injury due to xanthine/xanthine oxidase. Am J Physiol 1990; 259: L151–L155.

62. Litwin DK, Wilson AK, Said SI. Vasoactive intestinal polypeptide (VIP) inhibits rat alveolar macrophage phagocytosis and chemotaxis *in vitro*. Regulatory Peptides 1992; 40:63–74.

63. Paul S, Said SI, Thompson AB, Volle DJ, Agrawal DK, Foda H, de la Rocha S. Characterization of autoantibodies to vasoactive intestinal peptide in asthma. J Neuroimmunol 1989; 23:133–142.

64. Barnes PJ. Neural control of airways in health and disease. Am Rev Respir Dis 1986; 134:1289–1314.

65. Noda Y, Rodriguez-Sierra J, Liu J, Landers D, Mori A, Paul S. Partitioning of vasoactive intestinal polypeptide into lipid bilayers. Biochim Biophys Acta 1994; 1191:324–330.

66. Hakoda H, Zhouquiu X, Aizawa H, Inoue H, Hirata M, Ito Y. Effects of immunization against VIP on neurotransmission in cat trachea. Am J Physiol 1991; 261:L341–L348.

67. Ollerenshaw S, Jarvis D, Woolcock AS, Sullivan C, Scheibner T. Absence of immunoreactive vasoactive intestinal polypeptide in tissue from the lungs of patients with asthma. N Engl J Med 1989; 320:1244–1248.

68. Howarth PH, Djukanovic R, Wilson JW, Holgate ST, Springall DR, Polak JM. Mucosal nerves in endobronchial biopsies in asthma and non-asthma. Int Arch Allergy Appl Immunol 1991; 94:330–333.

69. Hvidsten D, Jenssen TG, Bolle R, Burhol PG. Plasma gastrointestinal regulatory peptides in exercise induced asthma. Eur J Respir Dis 1986; 68:326–331.

70. Paul S, Said SI. Human autoantibody to vasoactive intestinal peptide: increased incidence in muscular exercise. Life Sci 1988; 43:1079–1084.

71. Grider JR, Makhlouf GM. Prejunctional inhibition of vasoactive intestinal peptide release. Am J Physiol 1987; 253:G7–G12.

72. Yau WM, Youther ML, Verdun PR. A presynaptic site of action of substance P and vasoactive intestinal polypeptide on myenteric neurons. Brain Res 1985; 330:382–385.

73. O'Dorisio MS, Fleshman DJ, Qualman SJ, O'Dorisio TM. VIP: Autocrine growth factor in neuroblastoma. Regulatory Peptides 1992; 37:213–226.

74. Wollman Y, Lilling G, Goldstein MN, Fridkin M, Gozes I. Vasoactive intestinal peptide: a growth promoter in neuroblastoma cells. Brain Res 1993; 624:339–341.

75. Editorial. VIP and the skin. Lancet 1991; 337:886–887.

76. Properzi G, Francavilla S, Poccia G, et al. Early increase precedes a depletion of VIP and PGP-9.5 in the skin of insulin-dependent diabetics—correlation between quantitative immunohistochemistry and clinical assessment of peripheral neuropathy. J Pathol 1993; 169:269–277.

77. Hamid Q, Springall DR, Riveros-Moreno V, Chanez P, Howart P, Redington A, Bousquet J, Godard P, Holgate S, Polak JM. Induction of nitric oxide synthase in asthma. Lancet 1993; 342:1510–1513.

78. Persson MG, Zetterstrom O, Agrenius V, Ihre E, Gustafsson LE. Single-breath nitric oxide measurements in asthmatic patients and smokers. Lancet 1994; 343:146–147.

79. Thompson DC, Diamond L, Altiere RJ. Enzymatic modulation of vasoactive intestinal peptide and nonadrenergic noncholinergic inhibitory responses in guinea pig tracheae. Am Rev Respir Dis 1990; 142:1119–1123.

80. Tam EK, Franconi GM, Nadel JA, Caughey GH. Protease inhibitors potentiate smooth muscle relaxation induced by vasoactive intestinal peptide in isolated human bronchi. Am J Respir Cell Mol Biol 1990; 2:449–452.

81. Craig ML, Martins MA, Drazen JM. Peptidase modulation of vasoactive intestinal peptide pulmonary relaxation in tracheal superfused guinea pig lungs. J Clin Invest 1993; 91:235–243.

82. Caughey GH, Leidig F, Viro NF, Nadel JA. Substance P and vasoactive intestinal peptide degradation by mast cell tryptase and chymase. J Pharmacol Exp Ther 1988; 244:133–137.

83. Hachisu M, Hiranuma T, Tani S, Iizuka T. Enzymatic degradation of helodermin and vasoactive intestinal polypeptide. J Pharmacobio-Dyn 1991; 14:126–131.

84. Smith LJ, Houston T, Anderson J. Increased levels of glutathione in bronchoalveolar lavage fluid from patients with asthma. Am Rev Respir Dis 1993; 147:1461–1464.

85. van der Brugge-Gamelkoorn GJ, Claassen E, Smimia T. Anti-TNP-forming cells in bronchus-associated lymphoid tissue (BALT) and paratracheal lymph node (PTLN) of the rat after intratracheal priming and boosting with TNP-KLH. Immunology 1986; 57: 405–409.

86. Butler JE, Swanson PA, Richerson HB, Ratajczak HV, Richards DW, Suelzer MT. The local and systemic IgA and IgG antibody responses of rabbits to a soluble inhaled antigen. Am Rev Respir Dis 1982; 126:80–85.

87. Shopp GM, Bice DE. The pulmonary antibody-forming cell response in the guinea pig after intratracheal immunization. Exp Lung Res 1987; 13:193–203.

88. Jones SE, Davila DR, Haley PJ, Bice DE. The effects of age on immune responses in the antigen-instilled dog lung. Antibody responses in the lung and lymphoid tissues following primary and secondary antigen instillation. Mech Ageing Dev 1993; 68:191–207.

89. Renkin EM. Multiple pathways of capillary permeability. Circulation Res 1977; 41: 735–743.

90. Waldmann TA, Strober W. Metabolism of immunoglobulins. Prog Allergy 1969; 13: 1–110.

91. Reynolds HY. Identification and role of immunoglobulins in respiratory secretions. Eur J Respir Dis 1987; 71:103–116.

92. Guyton AC. Textbook of Medical Physiology. Philadelphia: Saunders, 1986:371–372.

93. O'Donnell M, Garippa RJ, Rinaldi N, et al. A novel, long-acting vasoactive intestinal peptide agonist. Part I: *in vitro and in vivo* bronchodilator studies. J Pharmacol Exp Ther 1994; 270:1282–1288.

94. Gozes I, Reshef A, Salah D, Rubinraut S, Fridkin M. Stearyl-norleucine-vasoactive intestinal peptide (VIP): a novel VIP analog for noninvasive impotence treatment. Endocrinology 1994; 134:2121–2125.

20

Enzymatic Degradation of Bradykinin

RANDAL A. SKIDGEL and ERVIN G. ERDÖS

University of Illinois College of Medicine at Chicago
Chicago, Illinois

I. Introduction

The early work of Frey and Werle in the 1920s and 1930s (Frey et al., 1950, 1968; Erdös, 1989) revealed the presence of a hypotensive factor, which they called kallikrein, in urine, blood, pancreas, and salivary glands. In the late 1930s, Werle and colleagues characterized this factor as an enzyme that released an active component from a blood-borne precursor which then gradually disappeared. They called the active decapeptide "kallidin" (Frey et al., 1950; Erdös, 1989), while Rocha e Silva (1963) named the nonapeptide released from plasma kininogen "bradykinin" (Bk). The disappearance of its activity indicated that there was an inactivating factor in blood. However, the importance and nature of this inactivator was not explored until the early 1960s, when an enzyme in plasma was discovered that inactivated Bk (Erdös and Sloane, 1962). The enzyme is a carboxypeptidase, named "carboxypeptidase N" and also called kininase I (Erdös, 1979). A second Bk-degrading enzyme in blood and kidney was discovered soon after and named kininase II, which was later shown to be identical with the angiotensin-converting enzyme (ACE) (Yang and Erdös, 1967; Yang et al., 1970b, 1971; Erdös, 1979). For convenience sake, the kininase I and kininase II designations have been used to categorize enzymes that cleave either 1 (Arg^9) or 2 (Phe^8-Arg^9) amino acids from the C terminus of Bk (Fig. 1,

459

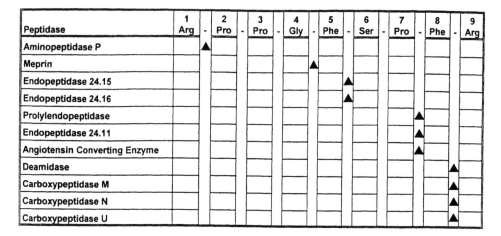

Figure 1 Hydrolysis of bradykinin by peptidases. Triangles show the primary bond cleaved, as established using purified enzymes and bradykinin in vitro.

Table 1 Kinetics of Bradykinin Hydrolysis by Peptidases[a]

Enzyme class	Peptidase	Source	K_m (μM)	k_{cat} (min^{-1})	k_{cat}/K_m (μM^{-1}min^{-1})	Ref.[b]
Kininase I type	CPM	Human	16	147	9.2	1
	CPN	Human	19	58	3.1	2
	CPU	Human	10,000	7,260	0.7	3
Kininase II type	Somatic ACE	Human	0.18	660	3,667	4
	ACE N domain	Human	0.54	300	555	4
	ACE C domain	Human	0.24	480	2,000	4
Endopeptidases	EP 24.11	Human	120	4,770	39.8	5
	EP 24.15	Rat	67	2,028	30.3	6
	Meprin	Mouse	520	1,320	2.5	7
	Prolylendopeptidase	Pig	7.5	95	12.7	8
Aminopeptidase	Aminopeptidase P	Rat	21	720	34.3	9

[a]Kinetic constants for bradykinin with deamidase (cathepsin A/protective protein) and endopeptidase 24.16 have not been reported. Abbreviations: CP, carboxypeptidase; EP, endopeptidase; ACE, angiotensin I-converting enzyme; ACE N domain, angiotensin I-converting enzyme in which the two zinc-binding histidine residues in the C-domain active site (His^{959} and His^{963}) have been mutated to Lys, leaving only the N-domain active site functional; ACE C domain, angiotensin I-converting enzyme in which the two zinc-binding histidine residues in the N-domain active site (His^{361} and His^{365}) have been mutated to Lys, leaving only the C-domain active site functional.

[b]References: 1. Skidgel et al., 1989; 2. Skidgel et al., 1984b; 3. Tan and Eaton, 1995; 4. Jaspard et al., 1993; 5. Gafford et al., 1983; 6. Orlowski et al., 1989; 7. Wolz et al., 1991; 8. Ward et al., 1987; 9. Orawski and Simmons, 1995.

Table 1). For example, neutral endopeptidase 24.11 (NEP) has been called a kininase II-type enzyme because it removes the C-terminal Phe^8-Arg^9 as does ACE, but it is more correctly classified as an endopeptidase and has essentially no sequence homology to ACE except for the HEXXH zinc-binding motif. In the case of kininase I-type enzymes, B-type carboxypeptidases would belong here. For example, the active subunit of carboxypeptidase N has 41% sequence identity with carboxypeptidase M, which efficiently cleaves Bk as well (see below). Deamidase/cathepsin A also readily removes the C terminal Arg of Bk and could be classified as a kininase I-type enzyme, but it is a serine carboxypeptidase with no homology to carboxypeptidase M or N. It also is not specific for C-terminal Arg, but hydrolyzes other peptides with penultimate or C-terminal hydrophobic residues in preference. Several peptidases cleave Bk at sites different from those mentioned above, and although they can be considered to be kininases, they do not fall into either the "I" or "II" category. Thus, in this chapter, instead of kininase I or II, the more specific names of enzymes will be used.

II. Kininase II, Angiotensin I-Converting Enzyme (ACE)

ACE, or kininase II, is a major kininase. It is widely distributed in the body, it is on the vascular endothelium, and Bk has the lowest K_m with this enzyme ($0.1-1\mu M$; Table 1). ACE has been found in many species, from mammalia to housefly. Based on physical properties, enzymic characteristics, and immunological cross-reactivity, most ACE enzymes in various tissues are quite similar and likely represent the two-domain somatic ACE (Skidgel and Erdös, 1993). There is also a testicular enzyme, germinal ACE, which contains only the C-terminal half of somatic ACE (Soubrier et al., 1993a,b). From human material, an active, truncated ACE was purified which contains only the N-terminal half (the N domain) of the somatic ACE (Deddish et al., 1994).

A. Localization

ACE is present in vascular beds bound to the plasma membrane of endothelial cells as an ectoenzyme, where it cleaves circulating peptides (Vane, 1969). It is in lower concentration also in subendothelial structures. Epithelial and neuroepithelial cells contain ACE as well. Polyclonal antibodies against either the human kidney or lung enzyme could not distinguish ACE prepared from endothelial, epithelial, or neuro-epithelial cells, or plasma (Skidgel and Erdös, 1993). ACE activity is higher in cultured arterial cells than in venous endothelial cells (Johnson et al., 1982). For example, the lung, because of heavy vascularization, and the capillary beds in the retina and brain (Igić, 1985), are rich in ACE. Monoclonal antibodies also located it in the vascular endothelium of the human heart and in subendothelial structures in blood vessels (Falkenhahn et al., 1995). Epithelial cells in general have more ACE than endothelial cells. The human kidney, for example, contains five to six times more ACE per unit wet weight than does the lung. In humans and in most other animals, the proximal tubular brush border (Caldwell et al., 1976; Hall et al., 1976; Ward et al., 1975, 1976, 1977; Schulz et al., 1988) is a rich source of renal ACE (Erdös and Yang, 1967), the exception being the rat kidney. ACE activity is also very high in other microvillar structures of brush-border epithelial linings, for example, in the small intestine, choroid plexus, and placenta (Skidgel et al., 1987a,b; Skidgel and Erdös, 1993).

In the rat heart, ACE distribution is very uneven, as established by auto-radiography with labeled inhibitor. Its concentration is high in cardiac valves and in coronaries, but low in endocardium (Yamada et al., 1991). In general, the ACE content of the heart varies from species to species, being high in guinea pig but low in human (Dragovic et al., 1996). Skeletal muscle also contains ACE.

In the rat brain, high ACE activity was found in the subfornical organ, area postrema, substantia nigra, locus ceruleus, cerebellum, striatum, and posterior pitu-itary (Yang and Neff, 1972; Yang et al., 1973; Saavedra and Chevillard, 1982; Defendini et al., 1983; Skidgel et al., 1987a). In the human brain, ACE activity is most concentrated in the caudate nucleus (Skidgel et al., 1987a). Possibly the highest concentration of ACE in any tissue was found in the choroid plexus (Igić et al., 1977) on the ventricular surface of the epithelial cells (Rix et al., 1981; Defendini et al., 1983). ACE in the cerebrospinal fluid may originate from the choroid plexus. By using a radiolabeled ACE inhibitor, a very detailed study of ACE distribution in monkey brain was carried out with computer-aided autoradiography (Chai et al., 1991). A similar technique was employed by Correa et al. (1985) to study ACE in rat brain.

ACE is also present in many other human and animal tissues, including fish gills, the electric organ of *Torpedo marmorata* (Lipke and Olson, 1988), and in flies (Cornell et al., 1995; Lamango et al., 1996). Human male reproductive tract (Erdös et al., 1985), chorionic membranes, and cultured chorion cells (Alhenc-Gelas et al., 1984) are also rich in ACE. ACE seems to be a well-conserved enzyme; for example, the deduced amino acid sequences of human and mouse ACE are 83% identical (Bernstein et al., 1989; Soubrier et al., 1988; Soubrier 1933a,b); even the *Drosophila* ACE has over 60% sequence similarity (Tatei et al., 1995).

B. Physical and Structural Properties

ACE purified from various human tissues is a single-chain glycoprotein, with a molecular weight ranging from 140,000 to 170,000 in SDS-PAGE (Skidgel and Erdös, 1993). This discrepancy of estimates is probably due to the heterogeneity in carbohydrate content and variability in electrophoretic conditions in different labora-tories. ACE is heavily glycosylated; the human renal enzyme contains approximately 25% carbohydrate by weight, with a 1.63:3.0:4.5:4.4 ratio of fucose:mannose:N-acetylglucosamine:galactose (Weare et al., 1982). There are differences, however, in the carbohydrate moieties; for example, human renal ACE has only traces of sialic acid, while the human lung enzyme contains up to 20 sialic acid residues per molecule, in agreement with the lung being the source of plasma ACE (Stewart et al.,

1981; Weare et al., 1982). The high sialic acid content protects it from uptake by liver lectins. The distribution of potential glycosylation sites between the N and C domains is uneven; the N domain contains 10, whereas the C domain has 7 (Soubrier et al., 1988). This is in agreement with the extensive glycosylation (37% by weight) on the naturally occurring active N domain in human ileal fluid (Deddish et al., 1994). The molecular weight of ACE deduced from the cDNA sequence (without carbohydrate) is 146,600 (Skidgel and Erdös, 1993).

ACE is a zinc metalloenzyme, and as such, it is inhibited by chelating agents and sulfhydryl compounds (Erdös and Yang, 1970; Erdös, 1979). ACE has a pH optimum above neutral, but it drops steeply at acid pH, due primarily to the dissociation of zinc from the active center, caused by the protonation of the zinc-binding histidine residues (Ehlers and Riordan, 1990).

The idea that Bk and Ang I could be hydrolyzed by the same enzyme protein (Yang et al., 1970b, 1971) was not accepted at first because Ang I conversion proceeds almost entirely in the presence of chloride ions, while Bk was still cleaved in the absence of chloride at approximately 30–35% of the maximal rate (Igić et al., 1973; Erdös, 1979). Studies with rabbit lung ACE showed the anion activation depends on both the structure of the substrate and the pH of the medium (Ehlers and Riordan, 1990). Accordingly, the substrates were divided into three classes, and Bk is a class II substrate with a lower activation constant for chloride than Ang I, a class I substrate. Weare (1982) described that ACE has two anion-binding sites, and one of them is the primary activation site.

The concentration of chloride in vivo appears to be high enough for ACE to be fully active in most tissues, although at some sites, chloride concentrations may fluctuate enough to regulate ACE activity with some substrates.

By using mutated ACE with a single active site on either the N or C domain (Wei et al., 1991a,b, 1992; Soubrier et al., 1993a,b), differences in the activation by Cl^- were detected. In general, the N domain is activated by much lower Cl^- concentration (e.g., 20 mM) than the C domain (e.g., 800 mM). This applies to the hydrolysis of several substrates, including Bk, but definitely not to all of them. The heptapeptide enkephalin congener Met^5-enkephalin-Arg^6-Phe^7 is converted to enkephalin even in the absence of chloride. Cl^- paradoxically activates the hydrolysis of this substrate by the N domain, while it inhibits its cleavage by the C domain (Deddish et al., 1997), leading to up to a fivefold difference in the hydrolysis rate.

ACE probably has many functions outside the vascular endothelium, where it cleaves circulating or locally released kinins (Vane, 1969), for example, on the renal proximal tubular brush border (Skidgel et al., 1987b). ACE can inactivate kinins that enter the nephron after glomerular filtration, which otherwise would interfere with the autoregulation in the kidney. Kinins are believed to be released into the nephron at the level of the distal tubules (Scicli et al., 1978), where they subsequently stimulate prostaglandin synthesis and salt and water excretion (Carretero and Scicli, 1995).

In addition to tissue, where it is membrane-bound, soluble ACE is present in blood, urine, lung edema fluid, amniotic fluid, seminal plasma, cerebrospinal fluid, and in homogenates of prostate and epididymis (Skidgel and Erdös, 1993) and in inflammatory exudates. The level of ACE in human plasma is very low; the estimate is in the order of 10^{-9} M (Alhenc-Gelas et al., 1983). Among the laboratory animals, guinea pig plasma has the highest activity (Yang et al., 1971) and in humans, plasma ACE activity varies significantly from person to person (Soubrier et al., 1993a).

C. Release of ACE from Plasma Membranes

The enzyme is membrane-bound, with a majority of the protein containing the active sites of both domains projecting into the extracellular space. The deduced amino acid sequence of human ACE contains a potential 17 amino acid membrane-spanning region near the C terminus. That it functions as a transmembrane anchor was proved in transfection experiments, where a mutant ACE lacking the putative C-terminal transmembrane domain was secreted primarily into the medium, but cells transfected with the full-length cDNA synthesized mainly membrane-bound ACE (Wei et al., 1991a,b). As a transmembrane enzyme, it has to be solubilized prior to purification with a detergent (Erdős and Yang, 1967) or by cleaving it from the membrane with an enzyme such as trypsin (Nishimura et al., 1976). The trypsin-liberated ACE has a lower molecular weight than the one mobilized from the membrane with detergent (Erdős and Gafford, 1983), because of the removal of a small anchor peptide by trypsin. Since germinal ACE contains a single C domain anchored to cell membrane, it was used to study ACE release. Germinal ACE was expressed in cells, and its proteolytic release was linked to activation of a protein kinase C (Ehlers et al., 1995). Cleavage sites in the peptide chain for mobilizing the enzyme were established (Beldent et al., 1993; Ramchandran et al., 1994). The so-called α-secretase could be one of the enzymes which solubilizes ACE (Oppong and Hooper, 1993). In some lung tissues, a metalloprotease is responsible for the solubilization of membrane-bound ACE (Oppong and Hooper, 1993).

D. Cloning and Sequencing

The molecular cloning and sequencing of the cDNA for the human (Soubrier et al., 1988) and mouse (Bernstein et al., 1989) enzymes revealed that ACE has two domains, each with a zinc-binding site and an active center (Soubrier et al., 1993a). This two-domain ACE was named somatic ACE, while the one extracted first from rabbit testicles, which contains only the C-domain active site, was called germinal ACE (Lattion et al., 1989; Ehlers et al., 1989; Kumar et al., 1989). Each active site contains two histidine residues and a glutamic acid to coordinate the zinc atom (Soubrier et al., 1993a), and a catalytic glutamic acid which is presumed to be a base donor. The overall sequence identity of the two domains is 67%, while it is higher (89%) around the active centers (Soubrier et al., 1993a). Although both domains are catalytically active (Wei et al., 1991a), inhibitors react differently with the active sites at the N and C domains (Perich et al., 1991, 1994; Deddish et al., 1996). For example, captopril has a lower K_i for the N-domain active site, while at optimal Cl^- concentration, the K_i of some other inhibitors is lower for the C-domain active site (Wei et al., 1992). Investigations with recombinant ACE or purified separate N and C domains showed by using radiolabeled inhibitor (Wei et al., 1992; Deddish et al., 1996) that inhibitors dissociate from the N-domain active site faster than from the C domain.

Germinal ACE present in epididymis has only the C domain, the C-terminal half of endothelial ACE plus 67 unique amino acids at the N terminus (Soubrier et al., 1988, 1993a; Ehlers et al., 1989). This is probably the ancient form of ACE, and the two-domain form developed by gene duplication (Soubrier et al., 1993a,b). The ACE present in *Drosophila*, or housefly, is a single-domain enzyme, similar in one respect to testicular ACE and in another respect to both domains, that structurally represents a more ancient, soluble form of ACE (Cornell et al., 1995; Lamango et al., 1996). Even Brazilian snake plasma (*Bothrops jararaca*) contains ACE activity shown with Bk substrate (Lavras et al., 1980).

E. Variations in ACE Activity

A polymorphism in the ACE gene sequence is due to either the presence or the absence of a 287-base-pair fragment in intron 16 of the gene. The importance of this insertion/deletion polymorphism (ACE II/DD) is currently being investigated by many laboratories. How this polymorphism could affect Bk metabolism is not known, but in ACE II homozygotes, ACE activity is lower in T lymphocytes and in plasma than in ACE DD subjects (Costerousse et al., 1993; Soubrier et al., 1993a,b; Cambien and Soubrier, 1995).

The level of circulating ACE and its concentration in cells are affected by a variety of conditions. Under normal conditions, plasma ACE very likely originates from vascular endothelial cells (Skidgel and Erdös, 1993). In sarcoidosis, the lymph nodes contain a high concentration of ACE, which, after release, raises the enzyme level in the circulation (Lieberman, 1974, 1985; Bunting et al., 1987; Silverstein et al., 1976, 1979). Other granulomatous diseases can also elevate circulating ACE, for example, Gaucher's disease or leprosy (Silverstein et al., 1979; Lieberman, 1985; Dhople et al., 1985).

ACE is very low in monocytes and macrophages, but it can be induced by corticosteroids (Friedland et al., 1978), although paradoxically, its level decreases in plasma of sarcoid patients treated with the hormone. Glucocorticoids also elevate ACE concentration in endothelial cells (Mendelsohn et al., 1982). Thyroid hormones strongly affect ACE level in plasma (Reiners et al., 1988); it is enhanced by hyperthyroidism and is lower in hypothyroidism (Grönhagen-Riska 1979; Brent et al., 1984). In Addison's disease, and in silicosis, asbestosis, and berylosis, the patients' ACE level is elevated (Falezza et al., 1985; Bunting et al., 1987). Chronic administration of an ACE inhibitor leads to higher ACE level in plasma (Fyhrquist et al., 1983). In the brain, where it is localized on cell membranes, ACE concentration is lower in Huntington's chorea (Butterworth 1986; Arregui et al., 1978). ACE appears in inflammatory exudates, such as synovial fluid (Elisseeva et al., 1981; Chercuite et al., 1987).

Because much of circulating ACE is presumably released from vascular endothelial cells in the lung, injury to the lungs affects the ACE level in plasma. In acute lung injury there is an early increase in blood ACE (Heck and Niederle, 1983) and ACE appears in pleural effusion and lavage fluid in perfused lungs (Igić et al., 1972, 1973; Dragovic et al., 1993). In neurogenic inflammation of rat nasal mucosa, extravasation was enhanced by both ACE and NEP inhibitors (Peterson et al., 1993). In malignant lung tumors and in leukemia, ACE activity in plasma is lower than normal (Heck and Niederle, 1983; Schweisfurth et al., 1985a,b). Blood levels of ACE also tend to be lower in Hodgkin's disease and multiple myeloma (Romer and Emmertsen, 1980). In adult respiratory distress syndrome, ACE activity decreases in serum (Bedrossian et al., 1978; Johnson et al., 1985b).

Immunohistochemical techniques were also employed in studies on changes in ACE concentration. By employing monoclonal antibodies elicited to intact somatic ACE, one group showed that the N domain of human ACE is the immunodominant one because antibodies reacted with only N-domain epitopes (Danilov et al., 1994). Inhibitory monoclonal antibodies were also raised which inhibited only the N domain of ACE, also with Bk as substrate (Danilov et al., 1994). These monoclonal antibodies to ACE were used in the hypoxic rat lung model of pulmonary hypertension to localize ACE. ACE staining was reduced in capillary endothelium of hypoxic animals, but increased in the small muscularized pulmonary arteries (Morrell et al., 1995).

F. Hydrolysis of Bradykinin

The degree of participation of the two active sites of ACE in Bk hydrolysis was established by using recombinant wild-type and mutant forms of ACE, purified single N-domain ACE, and germinal ACE as a source of C domain (Jaspard et al., 1993; Deddish et al., 1996). In the mutants, one active site was eliminated by deletion or by mutation of the two zinc-binding histidine residues. In agreement with previous findings, Bk was cleaved by the sequential removal of the C-terminal dipeptides to Bk_{1-7} (Fig. 1) and Bk_{1-5}. Bk is a preferred substrate of ACE over Ang I, because its specificity constant (k_{cat}/K_m) is about 20 times higher with soluble purified ACE (Table 1). The N domain alone contributes 26% of the Bk-degrading activity, while the C domain is responsible for 76% of the activity in producing Bk_{1-7} (Jaspard et al., 1993). The C-domain active site is more dependent on Cl^- and is activated more by this anion in general, but certainly not with all substrates (Deddish et al., 1996). The hydrolysis of Ang I by the Cl^--sensitive C domain was activated at optimal Cl^- concentration about 100-fold, while under similar conditions, that of Bk was activated only 5-fold. The k_{cat}/K_m for Bk is 3.6 times higher with the C domain than with the N domain (Jaspard et al., 1993; Table 1). When the properties of mutated ACE, bound to the plasma membrane of cultured cells, were studied, the ratios were reversed with Ang I substrate, which was cleaved under these conditions faster by the N domain (Jaspard and Alhenc-Gelas, 1995).

The inactivation of Bk was investigated further using a naturally occurring shorter version of ACE containing only the N domain. Bk was hydrolyzed about twice as fast by the C domain (germinal ACE) than by the purified N domain (Deddish et al., 1996).

ACE, and possibly NEP, can break down bradykinin even after internalization of the plasma membrane. This was induced by Bk binding to the B_2 receptor. One of the peptide fragments recovered was the product Bk_{1-7} (Munoz and Leeb-Lundberg, 1992).

Although ACE is frequently called a peptidyl dipeptidase, it can release a C-terminal tripeptide from des-Arg[9]-Bk (Inokuchi and Nagamatsu, 1981; Oshima et al., 1985) (Fig. 2), the latter being a product of carboxypeptidase N or M. The K_m of des-Arg[9]-Bk is much higher than that of Bk, resulting in a much lower k_{cat}/K_m for des-Arg[9]-Bk (Oshima et al., 1985; Inokuchi and Nagamatsu, 1981). Bk_{1-8} is cleaved orders of magnitude slower by both the N and C domains of ACE than Bk_{1-9} or Bk_{1-7} (Deddish et al., 1996). ACE also liberates protected C-terminal di- and tripeptides (e.g., Gly-Leu-Met-NH$_2$) of substance P and even the protected N-terminal tripeptide ($<$ Glu-His-Trp) of LH-RH (Skidgel and Erdös, 1985, 1993). This N-terminal tripeptide is cleaved mainly by the N domain (Deddish et al., 1994; Ehlers and Riordan, 1991).

G. Inhibition of ACE and Bradykinin Potentiation

The potentiation of Bk effects in vitro and in vivo by inhibitors of kininases initially appeared to be due to prolongation of the half-life of this peptide (Frey et al., 1968; Erdös, 1966; Erdös and Wohler, 1963; Ferreira, 1965). However, recent studies suggest that the potentiation of Bk actions by ACE inhibitors may involve additional mechanisms (see below). The widespread use of ACE inhibitors as therapeutic agents (Gavras et al., 1978; Gavras and Gavras, 1987, 1993), and the development of B_1 and B_2 Bk receptor blockers (Regoli and Barabe, 1980; Linz et al., 1995; Bhoola et al., 1992) led to investigations of the importance of kinins as mediators of the effects of ACE inhibition. This is unlikely to result from raising the level of circulating kinins

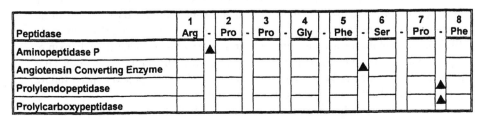

Peptidase	1 Arg	-	2 Pro	-	3 Pro	-	4 Gly	-	5 Phe	-	6 Ser	-	7 Pro	-	8 Phe
Aminopeptidase P		▲														
Angiotensin Converting Enzyme									▲							
Prolylendopeptidase													▲			
Prolylcarboxypeptidase															▲	

Figure 2 Hydrolysis of des-Arg[9]-bradykinin by peptidases. Triangles show the primary bond cleaved. Neutral endopeptidase 24.11 would cleave the Gly[4]-Phe[5] bond in des-Arg[9]-bradykinin, as this bond is hydrolyzed in bradykinin under prolonged incubation conditions (Gafford et al., 1983). Possibly, meprin, endopeptidase 24.15, neutral endopeptidases 24.11, 24.15, and 24.16 could cleave des-Arg[9]-bradykinin at the same sites as in bradykinin, but this has not been proven yet.

because kinins are broken down by other enzymes as well (Erdös and Yang, 1970; Erdös, 1979; Bönner, 1995). ACE inhibitors most likely inhibit kinin breakdown locally, if kinins are indeed paracrine agents (Carretero and Scicli, 1989, 1995) and as such are released, for example, from endothelial cells (Wiemer et al., 1991; Hecker et al., 1993).

ACE is inhibited by a variety of compounds. Reagents which react with the Zn cofactor of the enzyme (e.g., EDTA, *o*-phenanthroline, SH compounds), and snake venom peptides, are among them (Erdös and Yang, 1970; Erdös, 1979). Circulating blood plasma can also inhibit ACE; the inhibition appears to be abolished by dilution of plasma (Erdös, 1979). ACE substrates can also be competitive inhibitors, and even the N-terminal tripeptide (Arg-Pro-Pro) of Bk inhibits in vitro (Oshima and Erdös, 1974). From studies of the sequence of snake venom inhibitors (Ferreira et al., 1970), synthetic ACE inhibitors were designed to retain the C-terminal proline (Erdös, 1979; Cushman and Ondetti, 1980; Kato and Suzuki, 1970; Patchett and Cordes, 1985; Wyvratt and Patchett, 1985; Ondetti, 1994).

Because of their dual action, namely, blocking the Ang II release and the inactivation of Bk, some of the beneficial effects of ACE inhibitors can be interpreted in different ways. This issue is especially relevant when considering the effects of ACE inhibitors in heart and vascular diseases (Ambrosioni et al., 1995; CONSENSUS, 1987; Gavras and Gavras, 1987, 1993; Pfeffer, 1993, 1995; Parratt, 1994) and also in interpreting animal experiments. For example, enalapril lowers the blood pressure of hypertensive rats; this effect is partially blocked by the administration of a Bk receptor antagonist (Carbonell et al., 1988). Captopril enhances skin microvascular blood flow; this was attributed to NO and prostaglandins released by Bk (Warren and Loi, 1995).

Ang II constricts the coronary arteries and can cause cellular proliferation of subendothelial tissues. Bk is a vasodilator and antiproliferative, at least in the rat (Farhy et al., 1992; Scicli, 1994; Margolius, 1995; Hartman, 1995; Swartz et al., 1980; Sunman and Sever, 1993), partly because it releases NO and prostaglandins (Auch-Schwelk et al., 1993; De Meyer et al., 1995). Thus, ACE inhibitors can have a dual activity on the coronary circulation. B_2-type high-affinity Bk receptors in rat cardiomyocytes are coupled to IP_3 production through a G-protein (Minshall et al., 1995a). ACE inhibitors also ameliorated cardiac arrhythmias induced by digoxin or reperfusion, in part via Bk (Linz et al., 1995). ACE inhibitors given soon after myocardial infarction reduced mortality and the development of severe heart failure (Pfeffer, 1995; Hall et al., 1994; Gavras and Gavras, 1993).

In a spontaneously hypertensive rat strain, ACE inhibitors prevented the development of hypertension and left ventricular hypertrophy (Gohlke et al., 1994a,b). ACE inhibitors also lowered the blood pressure of DOCA-salt hypertensive rats (Chen et al., 1996). These actions can also be blocked at the B_2 receptor by the antagonist HOE 140 (Gohlke et al., 1994a,b). ACE inhibitors indeed could prevent wall thickening in rat heart after restenosis, but not the neointimal formation in other species (Shaw et al., 1995). In rats made diabetic, there is a hypertrophy of the vessel wall. ACE inhibitors have an antitrophic effect mediated in part by Bk and NO and by blocking Ang II release (Rumble et al., 1996). Inhibition of ACE in skeletal muscle enhances insulin-sensitive glucose uptake; this is considered to be a Bk effect (Dietze et al., 1996), and may improve muscle metabolism (Dietze, 1982). Bk potentiates glucose uptake in part via the release of prostaglandins, and ACE inhibitors reduce lactate release from the heart (Linz et al., 1995) by enhancing the effects of Bk (Linz et al., 1996). The reduction of hyperfiltration in experimental diabetes of rats after ACE inhibition was also attributed to kinins as mediators (Komers and Cooper, 1995).

Coughing is a frequently encountered side effect of the use of ACE inhibitors (Israili and Hall, 1992; Semple, 1995), which may be due to the prolongation of the half-life of Bk or may be caused by potentiation of Bk's effect at or near the receptor site. Angioneurotic edema, a rare but serious side effect, may also be caused by Bk (Slater et al., 1988).

The potentiation of Bk effects by ACE inhibitors appears to go beyond protecting Bk against breakdown by ACE, because it is not necessarily due to inhibition of enzymatic inactivation alone (Erdös and Skidgel, 1996). ACE inhibitors potentiated Bk effects even when a Bk analog was used that is resistant to hydrolysis by ACE (Minshall et al., 1995b, 1997a; Auch-Schwelk et al., 1993). It remains to be established how ACE inhibitors affect Bk receptors (although indirectly, Minshall et al., 1997b) and how they enhance the action of BK apart from prolonging its half-life (Erdös and Skidgel, 1996).

III. Neutral Endopeptidase 24.11

NEP is also a Zn-metallopeptidase, and it contains a single active site with the canonical HEXXH sequence (Howell et al., 1994; Skidgel, 1993; Erdös and Skidgel, 1989). The primary sequence of NEP is well conserved among species (Devault et al., 1987); there are only six nonconservative differences between the human and rat NEP (Malfroy et al., 1987, 1988). The enzyme is a single-chain protein of 742 amino acids. It is a transmembrane protein bound to plasma membrane by an uncleaved N-terminal signal peptide (Roy et al., 1993).

A. Localization

NEP is widely distributed in the body, but in contrast to ACE, its concentration in vascular endothelial cells is low (Johnson et al., 1985a; Llorens-Cortes et al., 1992; Howell et al., 1994), and varies within various vascular beds (Graf et al., 1995). Epithelial cells (Johnson et al., 1985a), especially in microvillar structures, are rich in NEP (Turner, 1987; Ronco et al., 1988). Its concentration in the proximal tubules (Kerr and Kenny, 1974) is high, about the same as that of ACE (Schulz et al., 1988). Brush borders in the placenta (Johnson et al., 1984), intestine, and the choroid plexus (Turner, 1987; Ronco et al., 1988) also have a lot of NEP. Its presence in the areas of the brain has been studied extensively because neuropeptides such as enkephalins (Schwartz et al., 1985) and substance P (Matsas et al., 1983) are among its substrates (Turner, 1987).

Because of its wide distribution, NEP's role in vivo can differ in each organ with the different substrates. The skeletal muscle of the dog contains high concentrations of NEP, in contrast to other species where it is lower (Dragovic et al., 1996). Similar to ACE, the importance of the highly active NEP in the male genital tract, especially in prostate glands (Erdös et al., 1985), is not yet known. NEP has also been the subject of many studies, but under the names CALLA (common acute lymphoblastic leukemia antigen) or CD 10 (Letarte et al., 1988; LeBien and McCormack, 1989). CALLA is present in lymphoblasts but is absent from mature lymphocytes. In contrast, it is in neutrophils but absent from progenitor cells (Connelly et al., 1985; Painter et al., 1988). Because of its presence in neutrophils and synovial fibroblasts (Werb and Clark, 1989) and in respiratory tract (Nadel, 1994), it may be an important factor in breaking down inflammatory peptides.

B. Enzymatic Properties and Inhibition

NEP cleaves peptides of less than about 3 kDa at the N terminus of hydrophobic amino acids. NEP is the second kininase II-type enzyme that releases the C-terminal Phe^8-Arg^9 of Bk (Gafford et al., 1983; Fig. 1). Although it was discovered as an endopeptidase that cleaves the B chain of insulin (Kerr and Kenny, 1974), many active peptide substrates were found later. Among them are enkephalins (Schwartz et al., 1985), endothelin (Vijayaraghaven et al., 1990), atrial natriuretic peptide, substance P, Ang I and Ang II, and a chemotactic peptide (Erdös and Skidgel, 1989; Connelly et al., 1985; Gafford et al., 1983; Matsas et al., 1983).

NEP cleaves Bk at the Pro^{-7}-Phe^8 bond (Fig. 1), which was first shown qualitatively by HPLC (Almenoff et al., 1981), and prolonged incubation results in the further hydrolysis of the Gly^4-Phe^5 bond as well (Gafford et al., 1983). The kinetics of hydrolysis were established by using purified human renal NEP (Gafford et al., 1983; Table 1). The k_{cat} of Bk is indeed higher with NEP than with ACE, but because of the higher K_m (120 μM versus <1 μM), the specificity constant k_{cat}/K_m of NEP is lower (500 versus 4,770 3.98 μm^{-1} min^{-1}) (Table 1).

NEP may be an important kininase in the epithelial cells of the respiratory tract (Johnson et al., 1985a; Dusser et al., 1988; Nadel, 1992, 1994), skeletal muscles (Dragovic et al., 1996), neutrophils (Connelly et al., 1985; Painter et al., 1988; Skidgel et al., 1991b), renal proximal tubules (Kerr and Kenny, 1974; Ura et al., 1987; Skidgel et al., 1987b), and possibly in human coronary vessels (Graf et al., 1995). At low concentrations, inhibitors of NEP (e.g., thiorphan, phosphoramidon, or the clinically tested candoxatril) are fairly specific and do not inhibit ACE, and consequently they are employed to study the importance of this peptidase in vivo and in situ (Elsner et al., 1992; Schwartz et al., 1985; Bralet et al., 1991; Schilero et al., 1994; Nadel, 1992, 1994; Bertrand et al., 1993).

As Bk has an antiproliferative effect (Farhy et al., 1992), the high concentration of NEP in some cells of solid malignant tumors indicates that Bk may be a substrate there. Transplanted malignant tumors of rat liver, SK HEP1 malignant human liver cells (Dragovic et al., 1994), and primary liver tumors (Dragovic et al., 1997) are very rich in NEP. Because its concentration in normal liver cells is very low, the difference in NEP activity between normal and cancerous tissue can be a thousandfold (Dragovic et al., 1994).

NEP levels in circulating blood plasma are very low but increase 60- to 80-fold in adult respiratory distress syndrome patients afflicted with septic pneumonia (Johnson et al., 1985b). Chronic cholestasis also elevates plasma NEP activity (Swan et al., 1993). In contrast to plasma NEP, the level in amniotic fluid is high (Spillantini et al., 1990). Ura et al. (1987) attributed most of the kininase activity in rat urine to NEP. An NEP inhibitor enhanced the diuretic and natriuretic effect of kinins in the rat kidney, mediated in part by an interaction of atrial natriuretic peptide and kinins (Bralet et al., 1991; Smits et al., 1990). NEP was present in high concentration in the developing edema of the perfused rat lung, while its level in the perfusate was low (Dragovic et al., 1993). NEP activity increased in serum and urine after proximal tubular injury, and in end-stage renal disease (Nortier et al., 1993; Deschodt-Lanckman et al., 1989).

C. Inhibition of Neutral Endopeptidase 24.11 and Its Role as a Kininase

Although inhibition of NEP in bronchial epithelium prominently enhances the bronchoconstriction induced by substance P, it also potentiates this effect of Bk (Barnes, 1987; Bertrand et al., 1993). NEP inhibitors could also enhance the bronchoconstriction by Bk in asthmatic patients (Barnes, 1987; Crimi et al., 1995). Bk's indirect effects can also be influenced by NEP inhibition; for example, Bk releases substance P from nerve endings, which is then cleaved by NEP (von Euler, 1963; Bhoola et al., 1992; Barnes, 1994; Skidgel, 1994; Nadel, 1994). Bk stimulates ciliary activity via prostaglandin E_2 release in rabbit, and this process is also strongly affected by the NEP in the airways (Tamaoki et al., 1989).

Permeability of the microvessels in the airways in guinea pigs was potentiated by both ACE and NEP inhibitors (Lötvall, 1990; Lötvall et al., 1990). In the perfused rat lung, administration of an NEP and an ACE inhibitor together potentiated the effect of a subthreshold dose of Bk to cause edema (Dragovic et al., 1993). Bk injected into the left ventricle of rats increased the blood flow in the microcirculation of airways (Yamawaki et al., 1994), which was enhanced by inhibition of NEP by phosphoramidon. In the rat heart, sensory nerve stimulation enhanced myocardial blood flow that was abolished by a Bk receptor blocker but potentiated by phosphoramidon, again owing to the inactivation of Bk by NEP (Piedimonte et al., 1994). Consequently, besides ACE, NEP also participates in the local inactivation of Bk.

Renal effects of Bk were enhanced in dogs by the combined administration of ACE and NEP inhibitors (Seymour et al., 1993, 1994). This type of dosing lowered arterial blood pressure and vascular resistance in nonanesthetized dogs (Seymour et al., 1994). Co-administration of ACE and aminopeptidase P inhibitors reduced the blood pressure in hypertensive rats more than the inhibition of a single enzyme alone (Scicli et al., personal communication). Because the combined administration of inhibitors of two peptidases increases their efficiency, several compounds were synthesized that inhibit both NEP and ACE or ACE and an aminopeptidase (Gros et al., 1991; Gonzalez-Vera 1995; Flynn et al., 1995; French et al., 1994; Bralet et al., 1994; Marie et al., 1995; Trippodo et al., 1995). For example, a dual inhibitor of ACE and NEP potentiated the hypotensive effect of injected Bk in rats more than a single agent inhibiting only one enzyme would (French et al., 1995).

IV. Carboxypeptidases

Carboxypeptidases cleave the C-terminal amino acid of peptides and proteins. Carboxypeptidases involved in Bk metabolism can be divided into two groups, the serine carboxypeptidases and the metallocarboxypeptidases. Most investigations have focused on the role of metallocarboxypeptidases, as, historically, carboxypeptidase N was the first human kininase discovered (Erdös and Sloane, 1962).

A. Metallocarboxypeptidases

Metallocarboxypeptidases hydrolyze peptides with the participation of a tightly bound zinc cofactor (Skidgel, 1996); carboxypeptidase A-type or carboxypeptidase B-type enzymes belong here. Carboxypeptidase B-type enzymes cleave only basic C-terminal Arg or Lys amino acids (Skidgel, 1988), and because Bk and congeners contain a C-terminal Arg, they are substrates of carboxypeptidase B-type enzymes.

General Enzymatic Properties

Owing to the essential function of zinc in the active centers, these carboxypeptidases are inhibited by chelating agents such as 1,10-phenanthroline which bind the metal ion (Skidgel, 1996). All of the B-type carboxypeptidases can be inhibited relatively specifically in micromolar or nanomolar concentration by the small synthetic compounds derived from a split product of substrate hydrolysis, arginine, DL-2-mercaptomethyl-3-guanidinoethylthiopropanoic acid (MGTA), and guanidinoethyl-mercaptosuccinic acid (GEMSA) (Plummer and Ryan, 1981; Skidgel, 1991, 1996). The pH of the incubation and the form of the enzyme (soluble versus membrane-bound) can also affect this inhibition. For example, the IC_{50} of GEMSA for membrane-bound carboxypeptidase M at pH 5.5 is 100-fold lower than for the solubilized, purified enzyme at pH 7.5 (Deddish et al., 1989; Skidgel, 1991).

 Replacement of zinc with other divalent metal ions, such as by cobalt or cadmium, alters the activity of these enzymes. Co^{2+} enhances the carboxypeptidase activity by 1.3- to 10-fold; the extent of this activation depends on the source of enzyme and substrate (Folk and Gladner, 1960, 1961; Marinkovic et al., 1977; Skidgel, 1988; Deddish et al., 1989). The activation by Co^{2+} depends on the pH of the incubation mixture: The cleavage of the dansyl-Ala-Arg substrate by carboxypeptidase M is enhanced 5.5-fold at pH 5.0, but only 1.4-fold at pH 7.5 (Deddish et al., 1989; Skidgel, 1991). The enhancement of carboxypeptidase N activity by Co^{2+} was due to an increase in the V_{max}, as the K_m also increased (Skidgel et al., 1984b). In contrast to cobalt, cadmium inhibits the peptidase activity of all of the metallocarboxypeptidases except for carboxypeptidase U (plasma carboxypeptidase B), which it stimulates (Skidgel, 1996).

Sequence Comparisons

Metallocarboxypeptidases can be divided into two groups based on a high sequence identity with the pancreatic carboxypeptidases or with the group of regulatory carboxypeptidases. Pancreatic carboxypeptidases A and B, mast cell carboxypeptidase A, and carboxypeptidase U (Skidgel, 1996) are in the first group, and carboxypeptidases N, M, D, and E are in the second one. Within each group there is a significant sequence similarity (40–58%), but between groups the identity is much less (14–20%). These data, and the conservation of active-site residues, indicate that the metallocarboxypeptidases arose from the same ancestral gene which duplicated and diverged to give rise to the pancreatic carboxypeptidase-like enzymes and the regulatory carboxypeptidases M, N, D, and E (Tan et al., 1989a; Avilés et al., 1993; Skidgel, 1996).

Carboxypeptidase N

Erdös and Sloane (1962) showed for the first time that a carboxypeptidase in plasma inactivated Bk and that an endogenous carboxypeptidase regulates the activity of a peptide hormone. Although the enzyme has been named and renamed through the years (e.g., anaphylatoxin inactivator, creatine kinase conversion factor, plasma carboxypeptidase B, arginine carboxypeptidase, lysine carboxypeptidase, protaminase), the term *carboxypeptidase N* will be used here.

Physical and Structural Properties

Purified native human carboxypeptidase N, a tetrameric enzyme, has an estimated molecular mass of 280 kDa (Oshima et al., 1975; Levin et al., 1982; Plummer and Hurwitz, 1978). Under denaturing conditions, it dissociates into three major bands of 83, 55, and 48 kDa (Plummer and Hurwitz, 1978; Levin et al., 1982). The 83-kDa protein is a noncatalytic subunit, whereas the 55-kDa and 48-kDa proteins represent two forms of the same active subunit (Levin et al., 1982; Skidgel, 1995). The 83-kDa subunit is heavily glycosylated (about 28% by weight), containing Asn-linked complex carbohydrate chains (Plummer and Hurwitz, 1978; Levin et al., 1982), but the active subunit lacks carbohydrate. The subunits are held together by noncovalent forces; 3 M guanidine has been in use to dissociate the subunits, which then can be separated by gel filtration (Levin et al., 1982; Skidgel, 1995).

The active 50-kDa subunit* has sequence similarities to other metallocarboxypeptidases ranging from 14% to 49% as revealed by molecular cloning (Gebhard et

---

*Although reported molecular mass for the active subunit ranges between 48 and 55 kDa as determined in SDS-PAGE, the calculated molecular mass of the active subunit based on its sequence is 50 kDa. For the sake of clarity, the 50 kDa value will be used here.

al., 1989), but the 83-kDa subunit is entirely different (Levin et al., 1982; Skidgel, 1995). Cloning and sequencing of the 83-kDa subunit showed that its 59-kDa core protein has no similarity to the 50-kDa active subunit or to any other carboxypeptidases (Tan et al., 1990). Consistent with its high carbohydrate content, the sequence contains seven potential Asn-linked glycosylation sites and, in addition, a serine/threonine-rich region that may be a site for attachment of O-linked carbohydrate. The most interesting feature is a domain comprising 12 leucine-rich tandem repeats of 24 amino acids each (Tan et al., 1990; Skidgel and Tan, 1992). This type of repeat was detected in the sequence of the leucine-rich α_2-glycoprotein, but many other mammalian or nonmammalian proteins (e.g., platelet GPIb, GP V, and GP IX, proteoglycans, RNAse inhibitor, luteinizing hormone receptor, oligodendrocyte/myelin glycoprotein, and U2 snRNP-A') and *Drosophila*, yeast, bacterial, and viral proteins have it (Tan et al., 1990; Skidgel and Tan, 1992; Kobe and Deisenhofer, 1993). The leucine-rich repeat region is probably an important structural element of these proteins (Kobe and Deisenhofer, 1993), as all of them participate in some sort of binding interaction (e.g., protein–protein or protein–membrane). Possibly the leucine-rich repeat region in the 83-kDa subunit mediates its interaction with the 50-kDa active subunit to form a heterodimer, and the N- and C-terminal domains of the 83-kDa subunit link the two heterodimers into a tetramer (Tan et al., 1990; Skidgel and Tan, 1992).

Because of its lack of carbohydrate, small size, and relative instability at 37°C (Levin et al., 1982; Skidgel, 1988), the active subunit alone would not survive in the circulation for long. In vitro, the 83-kDa subunit stabilizes the active subunit at 37°C and at low pH (Levin et al., 1982), indicating that the 83-kDa subunit, although it has no enzymatic activity, is a regulatory subunit, a carrier and stabilizer of the active subunit in the blood plasma.

Enzymatic Properties

Carboxypeptidase N hydrolyzes a variety of synthetic and naturally occurring substrates containing C-terminal Arg or Lys. Carboxypeptidase N generally cleaves Lys faster than Arg (due to a higher k_{cat}), but the penultimate residue also plays an important role, with alanine being preferred in many cases (Oshima et al., 1975; Skidgel, 1995). The specificity constants (k_{cat}/K_m) of the synthetic substrates are as follows ($\mu M^{-1} min^{-1}$): benzoyl-Gly-Lys, 0.7; benzoyl-Gly-Arg, 0.4; benzoyl-Ala-Lys, 60.3; benzoyl-Ala-Arg, 29.8; 3-(2-furyl)acryloyl-Ala-Lys, 17.1; 3-(2-furyl)acryloyl-Ala-Arg, 7.1 (Skidgel, 1995). Other endogenous substrates include anaphylatoxins, which contain C-terminal arginine, and creatine kinase, which has C-terminal lysine and is released from damaged heart muscles.*

---

*Because of its ability to cleave C-terminal lysine residues from proteins, carboxypeptidase N can also decrease the cellular binding of plasmin and plasminogen, which bind to C-terminal lysine residues of cell surface proteins (Redlitz et al., 1995). In this regard, a major cellular plasminogen receptor was identified as α-enolase (Miles et al., 1991), a known substrate of plasma carboxypeptidase N (Weavers et al., 1984).

The pH optimum of carboxypeptidase N is in the neutral range (7.5), and it retains little activity at pH 5.5—about 7% of that measured at pH 7.5 with dansyl-Ala-Arg (Erdös et al., 1964; Deddish et al., 1989). However, added Co^{2+} still activates the enzyme at pH 5.5 to 156% of the value found at pH 7.5 without Co^{2+} (Deddish et al., 1989).

The enzymatic properties of the isolated catalytic 50-kDa subunit (Levin et al., 1982) are in good agreement with those of the intact 280-kDa tetramer, especially with low-molecular-weight substrates. The 83-kDa subunit can allosterically modify the interaction of the enzyme with some substrates (e.g., anaphylatoxin C3a) and inhibitors (protamine) (Skidgel et al., 1986; Tan et al., 1989b).

Localization

Carboxypeptidase N is synthesized in the liver (Oshima et al., 1975) and released into the circulation, where it is present at a relatively high concentration of approximately 30 μg/mL (10^{-7} M) (Erdös, 1979). Apparently, no other cells or organs synthesize the enzyme, because Northern blot analysis of various organs gave negative results (Tan and Skidgel, unpublished). Although it was reported that carboxypeptidase N antigen is present on the cell membrane of cultured bovine pulmonary arterial endothelial cells (Ryan and Ryan, 1983), very likely these cells took it up from the serum in the medium. We found that cell membranes of cultured human pulmonary arterial endothelial cells have high levels of a carboxypeptidase, but this is the M-type carboxypeptidase; no carboxypeptidase N was detected by immunoprecipitation with specific antisera (Nagae et al., 1993). The carboxypeptidase—probably N—secreted by nasal mucosa originates from plasma, coming from plasma extravasation and interstitial fluid exudation across the epithelium (Ohkubo et al., 1995).

Hydrolysis of Bradykinin

Carboxypeptidase N was discovered as a kininase that cleaved the C-terminal Arg[9] from Bk (Erdös and Sloane, 1962; Fig. 1). Although the turnover number (k_{cat}) of Bk is lower than those of other naturally occurring substrates, Bk has the lowest K_m (19μM) of all the peptide substrates tested (Skidgel, 1995; Table 1).

Bk is rapidly inactivated during a single passage through the pulmonary circulation (Vane, 1969) by ACE on vascular endothelial cells (Erdös, 1979). The rate of Bk degradation in the blood by carboxypeptidase N is slower; however, this pathway becomes more significant when ACE is nonfunctional, for example, in patients treated with an ACE inhibitor.

Carboxypeptidase N is an important enzyme; it is present in blood in a relatively high concentration, and no person has ever been found who completely lacks the enzyme—even patients with low enzyme levels are rare (Erdös et al., 1965; Mathews et al., 1980). Genetically determined low blood levels (about 20% of normal), caused by decreased hepatic synthesis, were associated with repeated attacks of angioedema in one patient, possibly due to the increased half-life of kinins and/or anaphylatoxins (Mathews et al., 1980, 1986).

Conditions that affect hepatic plasma protein synthesis also alter plasma carboxypeptidase N levels: Cirrhosis of the liver decreases it, and pregnancy increases it (Erdös et al., 1965). A wide variety of diseases (e.g., cardiovascular disease, diabetes, allergic conditions) do not change carboxypeptidase N levels (Erdös et al., 1965; Mathews et al., 1980). Higher carboxypeptidase N activity has been found in certain types of cancer and in the blood and synovial fluid of arthritic patients (Erdös et al., 1965; Mathews et al., 1980; Schweisfurth et al., 1985b; Chercuitte et al., 1987). The relationship of carboxypeptidase N levels to these disease states, if any, is not known.

The evidence for a protective function of carboxypeptidase N in humans is indirect. The so-called protamine-reversal syndrome is a condition in which low carboxypeptidase N activity could be involved. Protamine is given frequently to neutralize the effects of heparin after extracorporeal circulation. In some patients, this triggers a catastrophic reaction consisting of pulmonary vasoconstriction, broncho-constriction, and systemic hypotension (Lowenstein et al., 1983; Morel et al., 1987). This reaction has been attributed to the release of thromboxane and the generation of anaphylatoxins and kinins subsequent to the activation of the complement cascade and factor XII, which in turn activates plasma kallikrein (Morel et al., 1987; Colman, 1987). Protamine is a potent inhibitor of carboxypeptidase N (Tan et al., 1989b); consequently, a decreased degradation of anaphylatoxins and kinins may be part of the protamine reversal syndrome. In addition, the carboxypeptidase N level decreases to half of the normal after initiation of cardiopulmonary bypass, because of dilution of the blood (Rabito et al., 1992). However, the fact that this syndrome is relatively rare (incidence ≈ 1%) indicates that other factors must be involved. Because heparin binds protamine and reverses the inhibition of carboxypeptidase N (Tan et al., 1989b), only protamine given in excess would be harmful. Administration of heparin indeed reversed protamine reactions in two patients, and it was proposed that this was due to reactivation of carboxypeptidase N (Lock and Hessel, 1990). In addition, Mathews et al. (1980) indicated that carboxypeptidase N levels of 20% of normal or greater might be sufficient for its protective role; thus, a 50% reduction may not cause significant symptoms. However, in patients with abnormally low carboxypeptidase N levels and in addition in those who produce excessive amounts of anaphylatoxins or kinins, a 50% reduction could have serious consequences, especially if protamine is given in excess.

Another function of carboxypeptidase N is its alteration of the receptor speci-ficity of kinins. Most of the well-known actions of Bk are mediated by the B_2 receptor, and that depends on the presence of the C-terminal arginine (Bhoola et al., 1992). The conversion of Bk to des-Arg^9-Bk by carboxypeptidase N was originally considered to be an inactivation step. However, des-Arg^9-Bk can stimulate a variety of responses: thymidine incorporation, protein accumulation, and prostaglandin pro-duction in human fetal lung fibroblasts (Goldstein et al., 1984) or rabbit skin fibro-blasts (Marceau and Tremblay, 1986); DNA synthesis, phospholipase C and protein kinase C activity in cultured mesangial cells (Issandou and Darbon, 1991); PGE_2 synthesis in W138 human lung fibroblasts (Crecelius et al., 1986); PGE_2 synthesis and bone resorption by osteoblasts (Ljunggren and Lerner, 1990); cGMP production in bovine aortic endothelial cells (Wiemer and Wirth, 1992); interleukin-1 and tumor necrosis factor release from macrophages (Tiffany and Burch, 1989); and increased lymphocyte chemokinetic activity (MacFadden and Vickers, 1988). Des-Arg^9-Bk is a ligand for a different receptor called B$_1$ (Regoli and Barabe, 1980), which was recently cloned and sequenced (Menke et al., 1994). The B$_1$ receptor system is upregulated in response to injury or inflammation and may be part of the acute-phase reaction (Bhoola et al., 1992). Even many isolated tissues respond to des-Arg^9-Bk, but only after incubation for several hours in a tissue bath (Bhoola et al., 1992; DeBlois et al., 1991); this upregulation of B$_1$ receptors is blocked by cycloheximide and actinomycin D (Regoli and Barabe, 1980). Noxious stimuli, such as Triton X-100 or endotoxin and interleukin-1 and -2 as well, induce the B$_1$ receptor (Crecelius et al., 1986; Bhoola et al., 1992; DeBlois et al., 1991, 1992), but dexamethasone or cortisol inhibit the upregulation (DeBlois et al., 1988). The B$_2$ receptor system does not respond to the same stimuli. Thus, the conversion of Bk to des-Arg^9-Bk by an enzyme such as carboxypeptidase N produces an agonist for the B$_1$ receptor system which may play an important role in inflammatory or pathological responses. That this can occur in vivo is indicated by the fact that blood levels of des-Arg^9-Bk are over threefold higher than that of native Bk in normotensive individuals and in patients with low-renin essential hypertension (Odya et al., 1983).

Carboxypeptidase M

Significant carboxypeptidase B-like activity was discovered in membrane fractions obtained from human and animal tissues and cells (Skidgel et al., 1984b; Johnson et al., 1984) and was named carboxypeptidase M to denote the fact that it is membrane-bound (Skidgel et al., 1989).

Physical and Structural Properties

Human carboxypeptidase M has a molecular mass of 62 kDa, in SDS-PAGE with or without reduction, showing that it is a single-chain protein (Skidgel et al., 1989). Carboxypeptidase M is a glycoprotein, and its mass is reduced to 47.6 kDa by chemical deglycosylation (Skidgel et al., 1989). Thus it has 23% carbohydrate content by weight, in agreement with the presence of six potential Asn-linked glycosylation sites in the deduced protein sequence (Tan et al., 1989a).

In subcellular fractions of cells or tissues, the majority of carboxypeptidase M is firmly membrane-bound (Skidgel, 1988; Skidgel et al., 1984b, 1989). The primary sequence does not contain a true hydrophobic transmembrane spanning region (Tan et al., 1989a), but it does have a weakly hydrophobic region of 15 amino acids at the C terminus, similar to other proteins that are membrane-bound via a glycosylphosphatidylinositol (GPI) anchor (Low, 1987). Indeed, carboxypeptidase M can be released from membrane preparations by bacterial phosphatidylinositol-specific phospholipase C (PI-PLC), a characteristic feature of GPI-anchored proteins, and direct evidence for the presence of a GPI anchor on carboxypeptidase M was obtained by labeling with [^{3}H]-ethanolamine (Deddish et al., 1990; Tan et al., 1995).

Carboxypeptidase M is also found in soluble form in various body fluids such as urine, seminal plasma, amniotic fluid, and bronchoalveolar lavage fluid (Skidgel et al., 1984a,b, 1988; Dragovic et al., 1995; McGwire and Skidgel, 1995). The mechanism of release of the enzyme is unknown, but the fact that the hydrophobic portion of the anchor is removed indicates that either a protease or phospholipase is involved (Deddish et al., 1990).

Enzymatic Properties

Carboxypeptidase M has a neutral pH optimum and cleaves only C-terminal Arg or Lys from a variety of synthetic peptide substrates as well as naturally occurring peptides such as Bk (Fig. 1), Arg^6- and Lys^6-enkephalins, and dynorphin A$_{1-13}$ (Skidgel et al., 1989). Carboxypeptidase M cleaves C-terminal arginine preferentially over lysine, and the penultimate residue can prominently affect the rate of hydrolysis. Thus, Met^5-Arg^6-enkephalin has the highest specificity constant (k_{cat}/K_m = 20.3 μM^{-1}min^{-1}), and changing the C-terminal amino acid to lysine (Met^5-Lys^6-enkephalin) decreases it over 10-fold due to a large increase in the K_m (Skidgel et al., 1989). Although the K_m of Leu^5-Arg^6-enkephalin (63 μM) is similar to that of Met^5-Arg^6-enkephalin (46 μM), the k_{cat} is almost ninefold lower, indicating the effect of the penultimate residue on hydrolysis. Of the substrates tested, Bk (with C-terminal Phe^8-Arg^9) has the lowest K_m (16 μM) and the second highest k_{cat}/K_m (9.2 μM^{-1} min^{-1}) (Table 1).

Carboxypeptidase M is activated by cobalt chloride and inhibited by *o*-phenanthroline, MGTA, GEMSA, and cadmium acetate, as are most B-type carboxypeptidases (Tan et al., 1995).

Localization

Carboxypeptidase M is widely distributed, with the lung and placenta having the highest concentrations. However, significant amounts are also present in kidney, blood vessels, intestine, brain, and peripheral nerves (Skidgel, et al., 1984b, 1991a; Skidgel, 1988; Nagae et al., 1992, 1993). Carboxypeptidase M mRNA is high in human placenta, lung, and kidney, intermediate in brain and liver, and low in heart, skeletal muscle, and pancreas (Nagae et al., 1993). In immunohistochemical studies, carboxypeptidase M was localized on the surface of type I pneumocytes and in pulmonary macrophages in lung (Nagae et al., 1993). In Madin Darby canine kidney (MDCK) cells derived from distal tubules, carboxypeptidase M was detected by using immunogold on the luminal side (Deddish et al., 1990). In the brain, glia which appear to be oligodendrocytes or astrocytes stain positively, and in peripheral nerves it is concentrated on the outer aspects of myelin sheaths and Schwann cell membranes (Nagae et al., 1992). Carboxypeptidase M is also found in soluble form in various body fluids, as indicated above (Skidgel et al., 1984a, 1988; McGwire and Skidgel, 1995). Recently, it was discovered that monoclonal antibodies against a differentiation-dependent cell surface antigen on white blood cells were specifically reacting with carboxypeptidase M (de Saint-Vis et al., 1995; Rehli et al., 1995). In one case, it was shown that carboxypeptidase M, almost undetectable on peripheral blood monocytes, is highly expressed in differentiated macrophages (Rehli et al.,

1995). In another study, carboxypeptidase M was present on pre-B lymphocytes, downregulated on circulating B lymphocytes, but reexpressed on activated germinal center B cells (de Saint-Vis et al., 1995).

Hydrolysis of Bradykinin

Because membrane-bound carboxypeptidase M is a widely distributed ectoenzyme, it is well situated to metabolize Bk at the cell surface. After its release from cell membranes it also can function as a soluble enzyme. As with carboxypeptidase N, des-Arg^9-Bk, the agonist for the B$_1$ receptor, is produced by carboxypeptidase M (Fig. 1). However, because of its location on plasma membranes, carboxypeptidase M can carry out this role in a local environment outside the circulation, for example, at sites of inflammation where the B$_1$ receptor may be upregulated (Skidgel, 1992, 1996; Bhoola et al., 1992).

Epidermal growth factor (EGF; urogastrone) is a substrate of carboxypeptidase M with relevance to the kinin system. EGF is a 53-amino acid (6-kDa) peptide with a C-terminal Arg residue (Carpenter and Wahl, 1990). Recent studies showed that purified carboxypeptidase M readily converts EGF to des-Arg^{53}-EGF (McGwire and Skidgel, 1995). Incubation of EGF with cells which have high carboxypeptidase M activity resulted in rapid conversion (61% in 2 hr) to des-Arg^{53}-EGF as the only metabolite; the hydrolysis was completely blocked by the carboxypeptidase inhibitor, MGTA. Similar results were obtained with urine or amniotic fluid (McGwire and Skidgel, 1995). Because carboxypeptidase M cleaves both EGF and Bk, it is of interest that EGF stimulates the growth of breast stromal cells, whereas Bk decreases growth and causes a dose-dependent inhibition of EGF-stimulated DNA synthesis (Patel and Schrey, 1992). This effect of Bk is mediated through the B$_1$ receptor, implying that it is first converted to des-Arg^9-Bk by a carboxypeptidase in this system. EGF also potentiates the contractile response to des-Arg^9-Bk in rabbit aortic rings (DeBlois et al., 1992).

The potential physiological and pathophysiological functions of carboxypeptidase M are many. For example, carboxypeptidase M could be involved in inflammatory and pathological processes by regulating the activity of kinins and anaphylatoxins that mediate many responses to inflammation. In the kidney, carboxypeptidase M may control the activity of kinins which are released by kallikrein liberated from the distal tubules (Scicli et al., 1978). In one study, hypertensive patients excreted significantly more kininases than normal individuals, and the major kininase was a kininase I-type enzyme (Iimura et al., 1987). (As a basic carboxypeptidase, it was probably carboxypeptidase M.) Thus, increased release of carboxypeptidase M into the urine could be an early sign of renal damage due to hypertension or other diseases.

Carboxypeptidase M may also have important functions in the lung, as indicated by the high level of activity present in membrane fractions from the lungs of all species tested (i.e., bovine, guinea pig, baboon, dog, rat, and human) (Chodimella et al., 1991; Nagae et al., 1993). The presence of carboxypeptidase M on the large surface area of the type I epithelial cells indicates that it may be a protective agent. The pulmonary synthesis of carboxypeptidase M or its release from the membrane may be upregulated in disease states, as enzyme levels in bronchoalveolar lavage fluid were elevated almost fivefold in patients with pneumocystic or bacterial pneumonia or lung cancer (Dragovic et al., 1995). Carboxypeptidase M was also released into the edema fluid of rat lung in an experimental model of lung injury (Dragovic et al., 1993).

Bk can cause pulmonary edema and bronchoconstriction when administered to animals. The carboxypeptidase inhibitor MGTA does not enhance either of these responses by itself (Ichinose and Barnes, 1990; Chodimella et al., 1991; Dragovic et al., 1993), but it further potentiates the response after inhibition of NEP and ACE (Chodimella et al., 1991; Dragovic et al., 1993), suggesting the involvement of all three enzymes in the in-vivo metabolism of Bk. A persistent dry cough is one of the major side effects encountered after the administration of ACE inhibitors to hypertensive patients (Israili and Hall, 1992; Semple, 1995). The speculation that this can result from increased concentration of peptides such as Bk in the respiratory tract (Morice et al., 1987) highlights the potential importance of carboxypeptidase M in the lungs of patients taking ACE inhibitors.

Another indication of the importance of pulmonary carboxypeptidase M comes from studies showing that the carboxypeptidase inhibitor MGTA enhances the non-cholinergic bronchoconstrictor response to capsaicin and vagus nerve stimulation in guinea pigs (Desmazes et al., 1992). While the response to MGTA was attributed to inhibition of carboxypeptidase M activity in the airways, it could have also been due to inhibition of the enzyme in the vagus nerve, where it is present in high concentration (Nagae et al., 1992).

Finally, the functions of carboxypeptidase M in many other locations remain to be explored. For example, in the placenta it may protect the fetus from maternally derived peptides. Its location in CNS myelin and Schwann cells in peripheral nerves is intriguing and may indicate a role for the growth or protection of neurons.

Carboxypeptidase U

Carboxypeptidase U is an unstable blood-borne carboxypeptidase that is activated during coagulation. In 1989, several groups reported that human serum has a much higher (about two- to threefold) arginine carboxypeptidase activity than plasma, and that the difference could not be explained by changes in carboxypeptidase N activity (Sheikh and Kaplan, 1989; Hendriks et al., 1989; Campbell and Okada, 1989). Further investigations showed that the enzyme is a unique carboxypeptidase (Campbell and Okada, 1989; Hendriks et al., 1990, 1992). Interestingly, Erdös and colleagues in 1964 (Erdös et al., 1964) noticed that serum had a higher carboxypeptidase activity than plasma, but the reason for the difference was not investigated further. In unrelated studies, the cDNA sequence of a plasminogen-binding protein had significant homology with pancreatic carboxypeptidase B (Eaton et al., 1991). Although many of the properties are quite similar to those of carboxypeptidase U, Eaton and co-workers named the one they isolated "plasma carboxypeptidase B." Based on similarities in properties and partial sequence information, most investigators now believe they are identical (Wang et al., 1994, Shinohara et al., 1994). For the purposes of this review, "carboxypeptidase U" will be used to avoid the possible confusion that could arise from the name "plasma carboxypeptidase B," employed for many years as a misnomer for carboxypeptidase N. The same protein was recently isolated and named TAFI, for "thrombin-activatable fibrinolysis inhibitor" (Bajzar et al., 1995).

Physical and Structural Properties

In the initial report, the size of the partially purified enzyme was estimated to be 435 kDa (Hendriks et al., 1990). Complete purification on a plasminogen affinity column yielded a 60-kDa protein (Eaton et al., 1991). This turned out to be the proenzyme, which, when activated by trypsin, yielded an active protein with a molecular mass of 35 kDa (Eaton et al., 1991). The size of the enzyme initially reported by Hendriks et al. (1990) could thus represent carboxypeptidase U bound to plasminogen and possibly other proteins in a multimeric complex (Wang et al., 1994).

Procarboxypeptidase U is a glycoprotein, as shown by a reduction from 60 kDa to 45 kDa after enzymatic deglycosylation (Eaton et al., 1991); however, the four potential Asn-linked glycosylation sites in the sequence are all in the propeptide segment, so the active enzyme is not a glycoprotein (Eaton et al., 1991).

The reason why carboxypeptidase U is unstable in serum or after partial purification is not fully understood, although it likely involves proteolysis by enzymes activated during coagulation. The purified proenzyme can be activated by trypsin, as stated above, but further incubation leads to additional cleavage, reducing the size of the activated carboxypeptidase from 35 to 25 kDa, and inactivating it (Eaton et al., 1991).

Enzymatic Properties

Carboxypeptidase U does not show a preference for C-terminal Arg or Lys in short synthetic substrates (Hendriks et al., 1989, 1990, 1992; Wang et al., 1994; Eaton et al., 1991; Tan and Eaton, 1995). However, the enzyme cleaves C-terminal Arg faster than Lys with larger naturally occurring peptides. For example, with enkephalin-Met[5]-Lys[6], k_{cat}/K_m = 0.08 $\mu M^{-1} min^{-1}$, but with enkephalin-Met[5]-Arg[6], k_{cat}/K_m = 0.28 $\mu M^{-1} min^{-1}$, and with enkephalin-Leu[5]-Lys[6], k_{cat}/K_m = 0.15 $\mu M^{-1} min^{-1}$, whereas enkephalin-Leu[5]-Arg[6] is cleaved faster with a k_{cat}/K_m = 0.71 $\mu M^{-1} min^{-1}$ (Tan and Eaton, 1995). The very low k_{cat}/K_m values with carboxypeptidase U are due to the very high K_m values (63–220 mM; Tan and Eaton, 1995) which are about 1000-fold higher than with carboxypeptidases M or N (Skidgel et al., 1984b, 1989). Carboxypeptidase U has a neutral pH optimum but, in contrast to other metallocarboxypeptidases, Co^{2+} inhibits and Cd^{2+} stimulates its peptidase activity (Hendriks et al., 1989; Tan and Eaton, 1995).

Localization

Carboxypeptidase U has been found only in blood, where it exists as an inactive proenzyme (Campbell and Okada, 1989; Hendriks et al., 1989; Eaton et al., 1991). The enzyme is activated during coagulation, presumably by a serine protease activated during clotting (Campbell and Okada, 1989; Hendriks et al., 1989; Eaton et al.,

1991). Thrombin (Bajzar et al., 1995) and plasmin (Wang et al., 1994) activate procarboxypeptidase U, but it remains to be determined whether these are the major activation pathways in vivo. As with other plasma proteins, it is probably synthesized in the liver, consistent with the cloning of its cDNA from a human liver library (Eaton et al., 1991).

Hydrolysis of Bradykinin

Because carboxypeptidase U is a relatively new member of the metallocarboxypeptidase family, its functions are still being explored. However, it is likely important in the fibrinolytic pathway by regulating lysine-mediated plasminogen binding to proteins and cells, for example, by cleaving C-terminal lysine residues from α_2-antiplasmin, histidine-rich glycoprotein, fibrin, annexin II, or α-enolase (Redlitz et al., 1995). Recently, it was shown to regulate clot lysis in dogs in vivo (Redlitz et al., 1996).

Concerning its functions as a kininase, Sheikh and Kaplan (1989) used Bk as substrate, and they noted that the conversion of Bk to des-Arg^9-Bk was fivefold higher in serum that could be accounted for by the activity of carboxypeptidase N (Sheikh and Kaplan, 1989). More recently, it was shown that after activation of procarboxypeptidase U in plasma with trypsin, the enzyme removed the C-terminal Arg from Bk (Shinohara et al., 1994; Fig. 1). The kinetics of Bk hydrolysis were also determined for the purified enzyme after trypsin activation (Tan and Eaton, 1995). Of all synthetic and naturally occurring substrates tested, Bk had the lowest K_m and highest k_{cat}/K_m. However, the reported K_m is extraordinarily high (10 mM; Table 1), even though it is lower than that of other substrates tested, with K_m values ranging from 63 to 290 mM. In comparison, the K_m values of Bk with other peptidases are much lower, 0.18 to 520 μM (Table 1). This raises the question of the physiological relevance of carboxypeptidase U in the degradation of Bk; however, the fact that serum degrades Bk fivefold faster than can be accounted for by the carboxypeptidase N level should indicate that carboxypeptidase U would be a significant kininase under certain circumstances, such as in clotted blood. In any case, carboxypeptidase U would not cleave Bk in circulating blood because it is a proenzyme in plasma, but after injury or during pathological processes which trigger the activation of the coagulation cascade, carboxypeptidase U may affect kinin activity.

Carboxypeptidase D

Mice that are genetically deficient in carboxypeptidase E (Song and Fricker, 1995) and cell lines that completely lack carboxypeptidase E mRNA or protein (McGwire et al., 1997) have a significant amount of B-type carboxypeptidase activity of a low pH optimum. Purification and characterization proved that this activity was due to a unique enzyme, named carboxypeptidase D (Song and Fricker, 1995; McGwire et al., 1997).

Physical and Structural Properties

Carboxypeptidase D is a single-chain glycoprotein with a molecular mass of about 180 kDa in SDS-PAGE, which is about three times larger than other regulatory carboxypeptidase B-type enzymes (Song and Fricker, 1995; McGwire et al., 1996, submitted). A short N-terminal sequence analysis (10 residues) revealed that 7 residues are identical with duck gp180 (Song and Fricker, 1995), a 180-kDa hepatitis B virus-binding glycoprotein that was recently cloned and sequenced (Kuroki et al., 1995). The gp180 sequence contains three tandem carboxypeptidase-like domains, with each domain exhibiting 26–44% sequence identity with carboxypeptidases E, M, and N, but it was not characterized as an enzyme (Kuroki et al., 1995). Cloning and sequencing of the cDNA for human (Tan et al., 1997) or rat (L. D. Fricker, personal communication) carboxypeptidase D revealed about 80% sequence identity with the duck protein. This sequence contains a potential transmembrane spanning sequence of 21 hydrophobic residues near the C terminus and a potential cytoplasmic domain of 60 residues that is highly conserved; there are only two differences between the duck and human or rat sequences, which are identical (Tan et al., 1997; L. D. Fricker, personal communication). When the regions containing putative active-site residues (e.g., zinc-binding residues, catalytic glutamic acid, and substrate-binding residues) are compared, the sequence similarity is much higher. Here, out of a total of 76 amino acids, carboxypeptidase D domain 2 has 70% identity with M, 74% with N, and 79% with E. All of the putative active residues are strictly conserved among these enzymes. In contrast, the identity in these regions of carboxypeptidase D domain 2 with domain 3 is only 40%, and of the eight putative active-site residues identified in the other carboxypeptidases, only one is conserved in D domain 3, making it unlikely that carboxypeptidase D-3 is active. Thus, it is possible that the enzyme contains two functional active-site domains, similar to ACE, which contains two homologous but nonidentical active sites (see above).

Carboxypeptidase D is a true membrane-bound enzyme, as detergent is required to solubilize it from membranes (Song and Fricker, 1995; McGwire et al., 1997). However, soluble forms of the enzyme have been detected in tissue extracts which are smaller in size (Song and Fricker, 1995).

Enzymatic Properties

Carboxypeptidase D cleaves synthetic peptide substrates containing a C-terminal Arg (and presumably Lys, although it has not been directly tested) with a pH optimum of 6.0–6.5 (Song and Fricker, 1995; McGwire et al., 1997). As with other members of this family, it shows a strong preference for a penultimate Ala over other residues such as Phe, Leu, Gly, Ile, or Pro (Song and Fricker, 1995). Carboxypeptidase D can also convert dynorphin B-14 to dynorphin B-13 (Song and Fricker, 1995). The enzyme is activated by cobalt and inhibited by MGTA, GEMSA and *o*-phenanthroline (Song and Fricker, 1995; McGwire et al., 1997). *p*-Chloromercuriphenylsulfonate, which is a good inhibitor of the other low-pH carboxypeptidase B-type enzyme, carboxypeptidase E, only partially inhibits carboxypeptidase D (Song and Fricker, 1995; McGwire et al., 1997).

Localization

Carboxypeptidase D is relatively widely distributed, being found in almost all duck tissues (Kuroki et al., 1994) and in many mammalian tissues and cells as well, including brain, pituitary, placenta, pancreas, kidney, fibroblasts, monocyte/macrophages (J774A.1 cells), and pituitary-derived cells (AtT20) (Song and Fricker, 1995; McGwire et al., 1997). The subcellular distribution of the enzyme has not been studied in detail, but evidence from duck gp180 indicates that it is both on the cell surface and on intracellular membranes (Kuroki et al., 1994).

Hydrolysis of Bradykinin

In preliminary studies, purified human recombinant carboxypeptidase D rapidly converted bradykinin to des-Arg^9-bradykinin at a rate similar to that of angiotensin converting enzyme (McGwire, Tan, and Skidgel, unpublished). Because of its localization, the enzyme has the potential to be involved in both extra- and intracellular kinin metabolism, but its acidic pH optimum would preclude it from having high activity at the normal neutral pH of extracellular fluid. Therefore, it might be involved in hydrolysis of bradykinin after receptor binding, endocytosis, and subsequent acidification of the endosome.

At present, other potential functions of carboxypeptidase D are still unknown, but hypothetically it could be involved in peptide and protein processing in the constitutive secretory pathway, analogous to the participation of carboxypeptidase E in prohormone processing in the regulated secretory pathway. Whether mammalian carboxypeptidase D can also act as a hepatitis B virus-binding protein is unknown, but this might be a fruitful area for further investigation.

B. Serine Carboxypeptidases

Serine carboxypeptidases can also cleave Bk, but their involvement in kinin metabolism in vivo has yet to be explored. Two lysosomal serine carboxypeptidases can potentially affect kinin activity in pathological conditions—prolylcarboxypeptidase and deamidase (cathepsin A/lysosomal protective protein).

The active sites of serine carboxypeptidases contain a catalytic triad of Ser, Asp, His, characteristic of other serine proteases, but in a unique order within the primary sequence. This is the same as in the prolylendopeptidase (or prolyloligopeptidase) family of serine proteases (Tan et al., 1993). While deamidase clearly belongs to the serine carboxypeptidase family owing to its sequence identity, prolylcarboxypeptidase has the active-site motif characteristics of both the serine carboxypeptidases and the prolylendopeptidase family. This indicates that prolylcarboxypeptidase links these two families of peptidases (Tan et al., 1993).

Physical and Structural Properties

Prolylcarboxypeptidase (Yang et al., 1968, 1970a) is a soluble single-chain protein of 58 kDa, which, as indicated by gel filtration, forms a dimer in its native state (Odya et al., 1978a; Tan et al., 1993). It is a glycoprotein containing about 12% carbohydrate by weight and six potential Asn-linked glycosylation sites (Tan et al., 1993). The cDNA sequence indicates that the protein contains a 30-residue signal peptide and a 15-amino acid propeptide (Tan et al., 1993).

The other serine carboxypeptidase, deamidase, was purified to homogeneity from platelets (Jackman et al., 1990). In gel filtration the enzyme has a molecular weight of 94,000, but in non-reducing PAGE the MW is 52,000, indicating that it exists as a homodimer. After reduction on SDS-PAGE, the 52-kDa protein dissociates into two chains of 33 and 21 kDa. The 33-kDa chain was labeled with [^{3}H]-diisopropylfluorophosphate, at the active-site serine residue (Jackman et al., 1990). The first 25 residues of each chain are identical with the sequences of the two chains of lysosomal protective protein, which binds and maintains the activity and stability of β-galactosidase and neuraminidase in lysosomes (Galjart et al., 1988). A defect in this protein is the cause of a severe genetically determined disease, galactosialidosis (d'Azzo et al., 1982). Deamidase can also be isolated from lysosomes in a high-molecular-weight complex with the other proteins (500–600 kDa) in active form (van der Horst et al., 1989; Potier et al., 1990).

Enzymatic Properties

Prolylcarboxypeptidase and deamidase are inhibited by compounds that react with the active-site serine residue, such as [^{3}H]-diisopropylfluorophosphate (Jackman et al., 1990; Tan et al., 1993). However, they are not inhibited by serine protease inhibitors such as aprotinin, etc., nor by chelating agents or specific cysteine protease inhibitors (Odya et al., 1978a; Jackman et al., 1990; Tan et al., 1993). Deamidase is effectively inhibited by nonspecific SH-reactive compounds such as *p*-chloromercuriphenylsulfonate or $HgCl_2$ (Jackman et al., 1990), by inhibitors of chymotrypsin-type enzymes such as Cbz-Gly-Leu-Phe-CHCl$_2$, and by chymostatin (Jackman et al., 1990). Prolylcarboxypeptidase is inhibited by the prolylendopeptidase inhibitor, Cbz-Pro-prolinal, because it specifically cleaves peptides with penultimate proline residues (Tan et al., 1993).

Both prolylcarboxypeptidase and deamidase have acidic pH optima (~5.0) when hydrolyzing short synthetic peptide substrates (Jackman et al., 1990; Tan et al., 1993). Interestingly, with longer naturally occurring peptides, both enzymes retain significant activity in the neutral range. For example, at pH 7.0, prolylcarboxypeptidase cleaves angiotensin II at 63% of the rate observed at pH 5 and deamidase cleaves Bk at 72% of the rate at pH 5.5 (Odya et al., 1978a; Jackman et al., 1990; Tan et al., 1993). In addition, the conversion of C-terminally amidated peptides (such as -Met^{11}-NH$_2$ in substance P) by deamidase to free acid (e.g., Met^{11}-OH) is optimal at neutral pH (Jackman et al., 1990).

Prolylcarboxypeptidase cleaves peptides only if the penultimate residue is proline. Of the short synthetic substrates (Odya et al., 1978a), Cbz-Pro-Ala was cleaved fastest, and Cbz-Pro-Pro and Cbz-Pro-OH-Pro were not cleaved at all (Odya et al., 1978a; Tan et al., 1993).

Deamidase/cathepsin A deamidates peptides such as substance P, neurokinin A, and enkephalinamides but also cleaves peptides with free COOH termini by carboxypeptidase action. It prefers peptides that contain hydrophobic residues in the P$_1'$ and/or P$_1$ position. For example, it readily cleaves substance P free acid, converts angiotensin I to II (Jackman et al., 1990), cleaves the chemotactic peptide fMet-Leu-Phe (Jackman et al., 1995), and is most active in degrading endothelin (Jackman et al., 1992, 1993).

Localization

Both deamidase and prolylcarboxypeptidase are in lysosomes in the cells, but can be released in response to stimulation (e.g., platelets, white blood cells) and appear in the extracellular medium or biological fluids. For example, deamidase and prolylcarboxypeptidase were found in urine (Yang et al., 1970a; Miller et al., 1991) and prolylcarboxypeptidase was released into synovial fluid (Kumamoto et al., 1981). In addition, lysosomal enzymes can be on the plasma membrane after exocytosis, where they may be bound to other transmembrane or membrane-associated proteins (Skidgel et al., 1991b).

Both enzymes are widely distributed, and mRNA expression is highest in human placenta, lung, and liver for prolylcarboxypeptidase (Tan et al., 1993), and mouse kidney and placenta for deamidase/protective protein (Galjart et al., 1990), as revealed by Northern analysis. Isolated or cultured human cells also contain high concentrations of the enzymes. For example, deamidase is highly active in macrophages as well as in platelets, endothelial cells, and fibroblasts (Jackman et al., 1990, 1992, 1993, 1995). Prolylcarboxypeptidase is also present in white blood cells and fibroblasts, and is expressed at high levels in endothelial cells (Kumamoto et al., 1991; Skidgel et al., 1981).

Hydrolysis of Bradykinin

The function of serine carboxypeptidases in regulating kinin actions has not been explored in detail. In the lysosomes, they would not normally have access to extracellular kinins, but would gain access after release. The kinin system is upregulated in inflammation, and leukocytes and other blood cells can be the source of these enzymes. Because they retain significant activity at neutral pH with naturally occurring substrates such as Bk or angiotensin (Jackman et al., 1990; Odya et al., 1978a), serine carboxypeptidases function as peptidases even in an extracellular fluid with a pH in the neutral range.

In addition, these lysosomal serine carboxypeptidases could block signal transduction stimulated by kinins by cleaving them after ligand-mediated receptor endocytosis (Erdös et al., 1989; Munoz and Leeb-Lundberg, 1992). This usually involves fusion of the endosomes with lysosomes and results in peptide degradation and receptor recycling (Yamashiro and Maxfield, 1988).

Besides the substrates of deamidase which contain C-terminal hydrophobic residues (e.g., endothelin 1, fMet-Leu-Phe, furylacryloyl-Phe-Phe, angiotensin I) (Jackman et al., 1990, 1992, 1995), Bk, with a C-terminal Arg^9, is also a good substrate (Jackman et al., 1990; Fig. 1). A hydrophobic amino acid in the penultimate position allows deamidase to remove nonhydrophobic amino acids by carboxypeptidase action. Thus, Bk ($-Phe^8-Arg^9$) and angiotensin$_{1-9}$($-Phe-His$) (Jackman et al.,

1990) and the synthetic substrate, dansyl-Phe-Leu-Arg, used in routine assays, are also cleaved well by the enzyme (Jackman et al., 1995).

Prolylcarboxypeptidase does not cleave Bk because it lacks a penultimate Pro residue, but the C-terminal -Pro[7]-Phe[8] of the ligand for the B_1 Bk receptor, des-Arg[9]-Bk, is readily split by the enzyme (Yang et al., 1968, 1970a; Odya et al., 1978a; Fig. 2). Metallocarboxypeptidases such as carboxypeptidase M in tissues, N in exudates, or deamidase from mononuclear and polymorphonuclear cells could generate the B_1 agonist (Fig. 1), which then binds to the B_1 receptor that is upregulated in inflammation (Bhoola et al., 1992). Prolylcarboxypeptidase also released from cells would then cleave the C-terminal Phe from des-Arg[9]-Bk and inactivate it (Fig. 2). Prolylcarboxypeptidase is released into blood during endotoxin shock (Sorrells and Erdös, 1972), and the B_1 receptor can be upregulated by endotoxin. Prolylcarboxypeptidase then could control the activity of des-Arg[9]-Bk, the ligand of this receptor in these conditions.

V. Aminopeptidases That Hydrolyze Kinins

Aminopeptidases remove amino acids sequentially from the N terminus of peptides and proteins. Bk contains a Pro[2] that makes it resistant to most aminopeptidases, but the aminopeptidase P cleaves peptides with a proline in the second position; consequently, it can inactivate Bk at the Arg[1]-Pro[2] bond.

Aminopeptidases can release Bk from kallidin (or Lys[1]-Bk) that is a product of tissue kallikrein, whereas Bk is liberated by plasma kallikrein (Bhoola et al., 1992). Early studies showed that Lys[1]-Bk is converted to Bk by an aminopeptidase in the blood, and subsequently, similar activity was detected in a variety of tissues (Erdös, 1979; Erdös et al., 1963; Webster and Pierce, 1963).

Dipeptidyl aminopeptidase IV removes dipeptides from the N termini of peptides with a Pro in the second position (McDonald and Barrett, 1986), but it cannot cleave Bk (Arg[1]-Pro[2]-Pro[3]-...) because it does not hydrolyze Pro-Pro bonds. Nevertheless, it could release Pro-Pro from Bk if it lost its first amino acid (Arg[1]) as a potential product of Bk metabolism after passing through the lung (Ryan et al., 1968).

A. Aminopeptidase P

That an aminopeptidase contributes to Bk metabolism was found first in vitro (Erdös et al., 1963; Erdös and Yang, 1966), then in the isolated perfused rat lungs in situ (Ryan et al., 1968). This enzyme (first called "prolidase") was extracted from pig kidney (Erdös and Yang, 1966; Dehm and Nordwig, 1970), but its purification and its properties were described in detail only recently (Hooper et al., 1990; Simmons and Orawski, 1992; Orawski and Simmons, 1995; Vergas Romero et al., 1995).

Physical and Structural Properties

Aminopeptidase P is membrane-bound via a glycosylphosphatidylinositol (GPI) anchor and can be solubilized with phosphatidylinositol-specific phospholipase C

(PI-PLC) (Hooper and Turner, 1988; Hooper et al., 1990; Simmons and Orawski, 1992; Orawski and Simmons, 1995). Under denaturing conditions in SDS-PAGE, the enzyme, as a single-chain protein, has Mr = 90,000–95,000 and contains about 17–25% carbohydrate by weight (Hooper et al., 1990; Simmons and Orawski, 1992; Orawski and Simmons, 1995). There are six consensus N-linked glycosylation sequences, all of which are coupled to carbohydrates in the N-terminal half of the protein (Vergas Romero et al., 1995). After gel filtration chromatography, the purified enzyme has a multimeric structure with a molecular mass of 220–360 kDa (Hooper et al., 1990; Simmons and Orawski, 1992; Orawski and Simmons, 1995), which varies depending on the salt concentration (Orawski and Simmons, 1995). Partial protein sequencing of aminopeptidase P, purified from guinea pig lung and kidney (Denslow et al., 1994), and the full sequence of the pig kidney enzyme (Vergas Romero et al., 1995) show that aminopeptidase P has some sequence similarity to human and *E. coli* prolidase as well as to *E. coli* aminopeptidase P, belonging to a newly recognized family of proline peptidases.

There is a soluble (cytosolic) form of aminopeptidase P in kidney extracts (Dehm and Nordwig, 1970) which has been purified from human lung, human erythrocytes, rat brain, and human platelets (Sidorowicz et al., 1984a,b; Harbeck and Mentlein, 1991; Vanhoof et al., 1992). The soluble enzyme differs from the membrane-bound form; for example, the soluble aminopeptidase P of rat brain has an Mr = 71,000 in SDS-PAGE and a native Mr = 143,000 in gel filtration (Harbeck and Mentlein, 1991).

Enzymatic Properties

Membrane-bound aminopeptidase P cleaves peptides containing proline in the second position such as Bk and neuropeptide Y with a neutral pH optimum (Simmons and Orawski; 1992, Orawski and Simmons, 1995). The tripeptide substrate Gly-Pro-Hyp has been used to measure its activity, but the N-terminal tripeptide of Bk (Arg-Pro-Pro) is cleaved faster (Simmons and Orawski, 1992; Orawski and Simmons, 1995). Longer peptides are also substrates, but not shorter ones; the dipeptide Arg-Pro is not cleaved (Simmons and Orawski, 1992; Orawski and Simmons, 1995). Aminopeptidase P contains one zinc per mole of enzyme (Hooper et al., 1990) and can be activated by Mn^{2+} with some substrates; it is inhibited by chelating agents. Other inhibitors include SH compounds and SH-reactive reagents, e.g., *p*-chloromercuriphenylsulfonate (Hooper et al., 1990; Simmons and Orawski, 1992; Orawski and Simmons, 1995). Many ACE inhibitors also inhibit aminopeptidase P, although with a K_i in the micromolar range (Hooper et al., 1992); Mn^{2+} can enhance their inhibitory effect (Orawski and Simmons, 1995). This inhibition is likely due to proline or proline-like structures and effective zinc-binding moieties in ACE inhibitors. Inhibitors with large substituents on the proline ring inhibit less or not at all (Hooper et al., 1992). The specific inhibitor of aminopeptidase P, apstatin, has a K_i for aminopeptidase P of 2.6 μM, but it does not inhibit other kininases such as ACE, neutral endopeptidases 24.11 and 24.15, or prolylendopeptidase (Scicli et al., personal communication).

The properties of the soluble form of aminopeptidase P are similar to those of the membrane-bound enzyme, including pH optimum, inhibition, and activation by Mn^{2+} (Sidorowicz et al., 1984a,b; Harbeck and Mentlein, 1991; Vanhoof et al., 1992). The exception is that the soluble form readily cleaves NH_2-X-Pro dipeptides, whereas the membrane-bound form does not (Sidorowicz et al., 1984a,b; Harbeck and Mentlein, 1991; Vanhoof et al., 1992; Simmons and Orawski, 1992; Orawski and Simmons, 1995). This may indicate that the soluble form is a different gene product.

Localization

Aminopeptidase P-type activity is in many tissues, but differences in the distribution of the soluble and membrane-bound forms have not been studied. Lung and kidney contain high concentrations of membrane-bound aminopeptidase P, and the soluble or cytosolic enzyme is also present in lung and kidney as well as in brain, erythrocytes, and platelets (Dehm and Nordwig, 1970; Sidorowicz et al., 1984a,b; Harbeck and Mentlein, 1991; Vanhoof et al., 1992; Hooper et al., 1990; Simmons and Orawski, 1992; Orawski and Simmons, 1995). In the lung, the membrane-bound form is probably localized on the surface of pulmonary vascular endothelial cells, where it has access there to circulating peptides such as Bk (Ryan, 1989).

Hydrolysis of Bradykinin

In the first detailed studies on Bk metabolism, it was discovered that removal of the N-terminal Arg^1 residue is one of the modes of inactivation by human erythrocytes (Erdös et al., 1963) and hog kidney (Erdös and Yang, 1966). This activity was initially attributed to a prolidase. It is known by now that prolidase only cleaves X-Pro dipeptides and that aminopeptidase P is the major enzyme that removes the N-terminal Arg^1 from Bk (Fig. 1).

The role of aminopeptidase P in the degradation of Bk in vivo has been investigated in some detail only recently. Perfusing Bk through rat lungs removed the Arg^1 residue, among other reactions (Ryan et al., 1968). Two recent studies have assessed the relative contributions of ACE and aminopeptidase P to the metabolism of Bk in the perfused rat lungs. It was concluded that 70% of kininase activity in rat lung is due to ACE, and aminopeptidase P is responsible for the rest of the activity (Ryan et al., 1994; Prechel et al., 1995). Nevertheless, even when ACE activity was inhibited, 75% of the Bk was inactivated during a single passage by aminopeptidase P alone (Ryan et al., 1994; Prechel et al., 1995). In studies in vivo, the specific aminopeptidase P inhibitor, apstatin, doubled the hypotensive action of Bk in rats (Scicli et al., personal communication). But apstatin was much less effective than lisinopril, an ACE inhibitor, confirming that ACE is the major kininase in rat lungs. The relevance of these findings to other species is questionable because rat lungs contain extremely high aminopeptidase P activity (Ryan, 1989). For example, aminopeptidase P concentration in rat lungs is 200-fold higher than in rabbit lungs, 30-fold higher than in pig lungs, and 100-fold higher than in cat lungs (Ryan, 1989). The relative level of aminopeptidase P in human lungs is unknown. The in-vivo contribution of either membrane-bound or soluble aminopeptidase P to the degradation of Bk in other locations (e.g., kidney, brain, intestine, platelets, red cells) is also not known.

Aminopeptidase P can degrade des-Arg[9]-Bk (Fig. 2), as well as Bk (Simmons and Orawski, 1992; Orawski and Simmons, 1995). The specificity constant of des-Arg[9]-Bk with ACE is worse than that of Bk due to the much higher K_m (120–240 μM versus <1 μM for Bk; Inokuchi and Nagamatsu, 1981; Oshima et al., 1985). Thus, aminopeptidase P could be more important to cleave des-Arg[9]-Bk than ACE. However, des-Arg[9]-Bk's catabolic pathway in vivo has not yet been well characterized, although this peptide is a substrate for prolylcarboxypeptidase and prolylendopeptidase.

Other aminopeptidases cannot cleave Bk, owing to Pro[2], but Bk may interact with one or more of these enzymes. Bk can inhibit aminopeptidase N (CD13), with a K_i of 9.4 μM, even though it is not cleaved by it (Xu et al., 1995). Thus, hypothetically, Bk might enhance the half-life of other biologically active peptides (e.g., enkephalins) that are natural substrates of aminopeptidase N.

VI. Endopeptidases

Endopeptidases cleave in the interior of the peptide chain; the specificity is determined by the amino acids at the site of cleavage. Endopeptidase 24.11 is such an enzyme, although it also acts as a peptidyl dipeptidase with Bk. Several other endopeptidases can cleave Bk, at least in vitro (Fig. 1), but their roles as kininases in vivo are not yet well understood. They will be discussed only briefly here.

A. Meprin

Meprin was purified from a mouse kidney membrane fraction; it is the mouse homolog of human PABA-peptide hydrolase and rat endopeptidase 2 (Beynon et al., 1981; Butler et al., 1987; Dumermuth et al., 1991; Wolz and Bond, 1995). Meprin contains two unique but related subunits, named α and β, and it is a member of the astacin family of metalloproteases (Gorbea et al., 1991; Dumermuth et al., 1991; Jiang et al., 1992; Johnson and Hersh, 1992; Wolz and Bond, 1995; Bond and Beynon, 1995). The enzyme is an oligomeric, cell-surface protein, bound via the transmembrane β subunit to which the α subunits are either disulfide-linked or noncovalently bound (Gorbea et al., 1991; Johnson and Hersh, 1994; Marchand et al., 1994). Meprin so far has been detected only in kidney and intestine, and its expression varies from species to species (Gorbea et al., 1991; Jiang et al., 1992, 1993). In contrast to endopeptidases 24.15 and 24.11 and ACE, which hydrolyze only short peptides, meprin cleaves large protein substrates such as azocasein (Butler et al., 1987). The α subunit of meprin hydrolyzes peptide and protein substrates longer than seven amino acids and cleaves the Gly[4]-Phe[5] bond of Bk (Butler et al., 1987; Fig. 1). Of the peptides tested, Bk was hydrolyzed fastest among those cleaved at a single site (Wolz et al., 1991; Table 1, Fig. 1). This finding led to the synthesis of Phe[5](4-nitro)-

Bk used in a spectrophotometric assay for meprin (Wolz and Bond, 1990, 1995). The β subunit does not cleave Bk even after activation (Kounnas et al., 1991; Wolz and Bond, 1995). Meprin-α has a very broad substrate specificity and does not have strict requirements for residues adjacent to the cleavage site but seems to prefer Pro in the P_2' or P_3' position (Wolz et al., 1991). Meprin-α cleaves biologically active peptides such as α-MSH, neurotensin, and LH-RH, and also Ang I and II, but rather slowly (Wolz et al., 1991).

Meprin is also highly concentrated on the proximal tubular brush border membranes. Here meprin may act as a kininase by cleaving kinins filtered through the glomerulus. However, the high concentration of other effective kininases, such as NEP and ACE, in this location makes it unlikely that meprin plays a dominant role in degrading kinins. Whether it might function as a kininase in intestinal brush borders is not known, but this location is also rich in other kininases (Deddish et al., 1994).

B. Prolylendopeptidase

Prolylendopeptidase (also called post-proline cleaving enzyme or prolyloligopeptidase) is a cytoplasmic enzyme present in many tissues and highly concentrated in the brain and kidney (Wilk, 1983). Its molecular mass is about 70–77 kDa, and it acts optimally at a pH of 7.5 (Wilk, 1983). The enzyme is a serine protease (Wilk, 1983), but cloning and sequencing of the porcine brain enzyme showed that it differs from classical serine proteases around the active site serine (Rennex et al., 1991). It belongs to a new family of serine proteases (Rawlings et al., 1991) with a catalytic triad similar to others but in a unique order of Ser ... Asp ... His (Polgar, 1992).

Prolylendopeptidase cleaves at the C-terminal side of prolyl residues in peptides of about 30 or less amino acids (Wilk, 1983). It hydrolyzes Bk at the Pro^7-Phe^8 bond (Fig. 1), the same site as ACE and NEP (Wilk, 1983). With the purified porcine kidney enzyme, Bk has a K_m of 7.5 μM and a V_{max} of 1.37 μmol/min/mg (Ward et al., 1987). Prolylendopeptidase can also cleave des-Arg^9-Bk by releasing Phe^8 (Fig. 2; Ward et al., 1987); thus, it inactivates ligands of both the B_1 and B_2 Bk receptors. Prolylendopeptidase cleaves a variety of other peptides containing proline, including Ang II (Wilk, 1983), and converts angiotensin I to angiotensin$_{1-7}$, which has activity of its own (Welches et al., 1991).

The importance of prolylendopeptidase as a kininase is not yet established. Because it is a cytosolic enzyme, whether it would have access to Bk under normal conditions is unknown, although activity was reported in serum (Wilk, 1983). Its role in Bk metabolism could be a minor one. The potent prolylendopeptidase inhibitor, Cbz-Pro-prolinal, may help to decipher the role of prolylendopeptidase in peptide metabolism (Wilk and Orlowski, 1983), but it is also a relatively potent inhibitor of prolylcarboxypeptidase (Tan et al., 1993), which cleaves some of the same substrates (e.g., Ang II, des-Arg^9-Bk). A new prolylendopeptidase inhibitor (JTP-4819; IC_{50} = 0.83 nM in rat brain supernatant) enhanced cognitive function in rats in vivo (Toide et al., 1995), but its specificity and mechanism of action are not yet established.

C. Endopeptidase 24.15

After endopeptidase 24.15 was purified from rat brain homogenates (Orlowski et al., 1983), it was realized that the enzyme was identical with two other enzymes described and purified earlier: Pz-peptidase, which cleaves a synthetic collagenase substrate; and endooligopeptidase A, which was first described as a kininase (for reviews, see Tisljar, 1993; Barrett et al., 1995; Erdös, 1979).

Endopeptidase 24.15 has a neutral pH optimum and, as a Zn metalloenzyme, contains the consensus HEXXH motif (Orlowski et al., 1983, 1989; Pierotti et al., 1990; Tisljar, 1993). The enzyme is inhibited by sulfhydryl reactive agents and is stabilized by low concentrations of thiols (Orlowski et al., 1983, 1989; Tisljar and Barrett, 1990). One group classified it as a cysteine peptidase, but others proposed renaming it "thimet oligopeptidase" (Tisljar, 1993; Barrett et al., 1995). The enzyme is clearly a metallopeptidase, and it is likely sensitive to thiols because of a free cysteine, five residues removed from the catalytic center, which is not involved in the catalysis (Pierotti et al., 1990). This situation is analogous to that of carboxypeptidase E, a zinc carboxypeptidase that is inhibited by *p*-chloromercuriphenylsulfonate (Skidgel, 1988).

Endopeptidase 24.15 is a single-chain enzyme of 645 amino acids with a molecular mass of 73 kDa (Pierotti et al., 1990; Tisljar, 1993). Although the enzyme contains a single potential glycosylation site, there is no evidence that it is glycosylated (Pierotti et al., 1990). Endopeptidase 24.15 is primarily a cytosolic enzyme, and as such it lacks a signal peptide (Pierotti et al., 1990), but up to 20% of the enzyme may be membrane-associated (Acker et al., 1987). The enzyme is distributed in most tissues, and is highly concentrated in brain and testes (Tisljar, 1993; Barrett et al., 1995).

Endopeptidase 24.15 has a broad substrate specificity, but cleaves substrates in preference at the carboxyl side of hydrophobic, aromatic amino acids, especially those with an aromatic residue at the P_3' position (Orlowski et al., 1983). For example, Bk is hydrolyzed at the Phe^5-Ser^6 bond with the aromatic Phe^8 residue in the P_3' position (Orlowski et al., 1983; Fig. 1; Table 1). Endopeptidase 24.15 hydrolyzes other biologically active peptides as well, provided they contain less than 20 residues, such as neurotensin, substance P, and LH-RH (Tisljar, 1993). It converts large opioid peptides to enkephalins (Orlowski et al., 1983; Chu and Orlowski, 1985).

Because it is a cytosolic enzyme, the role of endopeptidase 24.15 as a kininase is questionable. In subcellular fractions of rat brain (Acker et al., 1987), about 20% of it is membrane-bound, but it is not known whether it is on the outside of the plasma membrane or on an intracellular membrane. Some studies with the endopeptidase 24.15 inhibitor N-[1(RS)-carboxy-3-phenylpropyl]-Ala-Ala-Phe-pAB (cFP-AAF-pAB) (Orlowski et al., 1988) indicated that the enzyme inactivates Bk in vivo or in situ. For example, the inhibitor enhanced Bk's effect on rat uterus (Schriefer and Molineaux, 1993) or blocked Bk degradation by rat hypothalamic slices (McDermott et al., 1987). Infusion of cFP-AAF-pAB into normotensive rats dropped blood pressure up to 50 mm Hg which was blocked by a B_2 receptor antagonist (Genden and

Molineaux, 1991). However, these striking results were due primarily to inhibition of ACE (Yang et al., 1994; Telford et al., 1995), because endopeptidase 24.11 cleaved the inhibitor in vivo to produce cFP-AA, which inhibits ACE (Cardozo and Orlowski, 1993). In addition, cFP-AAF-pAB inhibits endopeptidase 24.11 although at higher concentrations (K_i = 17 µM) (Orlowski et al., 1988). Thus, the function of endopeptidase 24.15 in the metabolism of Bk in vivo without the contributions of ACE and endopeptidase 24.11 has yet to be established.

D. Endopeptidase 24.16

Endopeptidase 24.16 was first described as an enzyme in rat brain membranes that degrades neurotensin (Checler et al., 1983) and has been called neurotensin-degrading enzyme and neurolysin (Checler et al., 1995). Although many properties are similar to those of endopeptidase 24.15, purification and characterization of the enzyme proved that it is different (Checler et al., 1986; Millican et al., 1991; Checler et al., 1995). Recently, it was shown that the enzyme is likely identical with the soluble Ang II-binding protein and rabbit microsomal endopeptidase (Rosenberg et al., 1988; Sugiura et al., 1992) and is at least partly localized in the mitochondrial intermembrane space, where it interacts noncovalently with the inner membrane (Serizawa et al., 1995; Barrett et al., 1995). Its primary sequence reveals that it belongs to the same family of enzymes as endopeptidase 24.15 (Serizawa et al., 1995; Barrett et al., 1995). A variant of this enzyme was called endopeptidase 24.16B (Rodd and Hersh, 1995). Purified endopeptidase 24.16 cleaves Bk at the same site as endopeptidase 24.15 (Fig. 1), at the Phe[5]-Ser[6] bond (Millican et al., 1991). The cytosolic and mitochondrial localization of endopeptidase 24.16 appears to exclude its importance as a major kininase.

VII. Resistant Bradykinin Agonists

Studies on Bk metabolism led to the search for Bk agonists that act on Bk receptors but would be resistant to breakdown, mainly by carboxypeptidases (kininase I) and kininase II-type enzymes. Bk analogs [e.g., Hyp[3]-Tyr-(Met)[8]-Bk and D-Arg-(Hyp[3])-Bk] have been tested as agonists on B_2 receptors in tissues and cells, because they are inactivated much more slowly than Bk (Auch-Schwelk et al., 1993; Minshall et al., 1995a,b, 1997a). Bk coupled to high-molecular-weight dextran is still active, but more resistant to breakdown by ACE than the nonapeptide itself (Odya et al., 1978b; Minshall et al., 1995b, 1997a,b). Another derivative called RMP-7 was tested in animal experiments to enhance drug uptake through blood–brain and blood–ocular barriers. This compound, Arg-Pro-Hyp-Thi-Gly-Ser-Pro-4-MeTyr-ψ(CH$_2$NH)-Arg (Straub et al., 1994), was also designed to minimize breakdown by ACE and carboxypeptidases. It had been well established a long time ago that Bk enhances capillary permeability (Erdös and Yang, 1970). The impetus for the current application of the peptide was given in experiments where the uptake of labeled dextran was enhanced in rat brain tumors by infusing high doses of Bk (10 µg/mg/min) (Inamura and Black, 1994). RMP-7 was equally effective but in a much lower dose (0.1 µg/mg/ min); it increased the uptake by rat glioma of a variety of agents, including dextran, carboplatin, and methothraxate (Inamura et al., 1994; Nomura et al., 1994; Matsukado et al., 1996). Permeability was increased about 10-fold by RMP-7. This pseudopeptide was also effective in promoting the transport of sucrose and an antiviral agent through the blood–ocular barrier into the guinea pig retina and lens (Elliott et al., 1995). In mice, using lanthanum uptake as marker, it was deduced that the blood–brain barrier permeability increased because RMP-7 modulated the tight junctions of vascular endothelial cells (Sanovich et al., 1995).

VIII. Epilogue

This rather brief and certainly not all-inclusive summary of the catabolism of Bk should point out the complexities of the system and indicate further directions for research. The fact that kinins are rapidly inactivated has been known since their discovery (Erdös, 1989), but the characterization of the enzymes involved and the development of inhibitors are still leading us to learn more and more about the functions of kinins. The inactivation at the C-terminal end of the peptides appears to be the most important, but certainly multiple pathways are available in the body to hydrolyze Bk.

The significance of the potentiation of the cardiac actions of Bk by ACE inhibitors and the beneficial therapeutic effects and the mechanism of this potentiation, which probably goes beyond a single enzyme inhibition, are currently being intensively explored. The studies on the metabolism of Bk led also to the synthesis of B_2 receptor agonist derivatives that are more resistant to inactivator enzymes than the parent peptide. These agents are useful when studying the effects of the peptide, for example, signal transduction or investigating how to enhance the uptake of some drugs by tumors. The introduction of multivalent inhibitors which can block several enzymes, thus inhibiting two pathways of simultaneous inactivation of peptides, may lead to unraveling additional functions of kinins very likely closely integrated with the actions of other peptide agonists.

Acknowledgment

Some of the studies on carboxypeptidases M and N described here were supported in part by National Institutes of Health grants DK41431, HL36082, and HL36473. We thank Sara Thorburn for her very skilled help in preparing the manuscript.

References

Acker GR, Molineaux C, Orlowski M. (1987). Synaptosomal membrane-bound form of endopeptidase-24.15 generates Leu-enkephalin from dynorphin$_{1-8}$, α- and β-neoendor phin, and Met-enkephalin from Met-enkephalin-Arg[6]-Gly[7]-Leu[8]. J Neurochem 48: 284–292.

Alhenc-Gelas F, Weare JA, Johnson RL Jr, Erdös EG. (1983) Measurement of human converting enzyme level by direct radioimmunoassay. J Lab Clin Med 101:83–96.

Alhenc-Gelas F, Yasui T, Allegrini J, Pinet F, Acker G, Corvol P, Ménard J. (1984). Angiotensin I-converting enzyme in foetal membranes and chorionic cells in culture. J Hypertension 2(suppl 3):247–249.

Almenoff J, Wilk S, Orlowski M. (1981). Membrane bound pituitary metalloendopeptidase: apparent identity to enkephalinase. Biochem Biophys Res Commun 102:206–214.

Ambrosioni E, Borghi C, Magnani B. (1995). The effect of the angiotensin-converting-enzyme inhibitor zofenopril on mortality and morbidity after anterior myocardial infarction. N Engl J Med 332:80–85.

Arregui A, Emson PC, Spokes EG. (1978). Angiotensin-converting enzyme in substantia nigra: reduction of activity in Huntington's disease and after intrastriatal kainic acid in rats. Eur J Pharmacol 52:121–124.

Auch-Schwelk W, Bossaller C, Claus M, Graf K, Gräfe M, Fleck E. (1993). ACE inhibitors are endothelium dependent vasodilators of coronary arteries during submaximal stimulation with bradykinin. Cardiovasc Res 27:312–317.

Avilés FX, Vendrell J, Guasch A, Coll M, Huber R. (1993). Advances in metallo-procarboxypeptidases. Emerging details on the inhibition mechanism and on the activation process. Eur J Biochem 211:381–389.

Bajzar L, Manuel R, Nesheim ME. (1995). Purification and characterization of TAFI, a thrombin-activatable fibrinolysis inhibitor. J Biol Chem 270:14477–14484.

Barnes PJ. (1987). Airway neuropeptides and asthma. TiPS 8:24–27.

Barnes PJ. (1994). Neuropeptides and asthma. In: Kaliner MA, Barnes PJ, Kunkel GHH, Baraniuk JN, eds. Neuropeptides in Respiratory Medicine. New York: Marcel Dekker: 501–542.

Barrett AJ, Brown MA, Dando PM, Knight CG, McKie N, Rawlings ND, Serizawa A. (1995). Thimet oligopeptidase and oligopeptidase M or neurolysin. Meth Enzymol 248: 529–556.

Bedrossian CWM, Woo J, Miller WC, Cannon DC. (1978). Decreased angiotensin-converting enzyme in the adult respiratory distress syndrome. Am J Clin Pathol 70:244–247.

Beldent V, Michaud A, Wei L, Chauvet MT, Corvol P. (1993). Proteolytic release of human angiotensin-converting enzyme. Localization of the cleavage site. J Biol Chem 268: 26428–26434.

Bernstein KE, Martin BM, Edwards AS, Bernstein EA. (1989). Mouse angiotensin-converting enzyme is a protein composed of two homologous domains. J Biol Chem 264:11945–11951.

Bertrand C, Geppetti P, Baker J, Petersson G, Piedimonte G, Nadel JA. (1993). Role of peptidases and NK$_1$ receptors in vascular extravasation induced by bradykinin in rat nasal mucosa. J Appl Physiol 74:2456–2461.

Beynon RJ, Shannon JD, Bond JS. (1981). Purification and characterization of a metalloendoproteinase from mouse kidney. Biochem J 199:591–598.

Bhoola KD, Figueroa CD, Worthy K. (1992). Bioregulation of kinins: kallikreins, kininogens, and kininases. Pharmocol Rev 44:1–80.

Bond JS, Beynon RJ. (1995). The astacin family of metalloendopeptidases. Protein Sci 4: 1247–1261.

Bönner, G. (1995). Do kinins play a significant role in the antihypertensive and cardioprotective effects of angiotensin I-converting enzyme inhibitors? In: Laragh JH, Brenner BJ, eds. Hypertension: Pathophysiology, Diagnosis, and Management. 2d ed. New York: Raven Press: 2877–2893.

Bralet J, Mossiat C, Gros C, Schwartz JC. (1991). Thiorphan-induced natriuresis in volume-expanded rats: roles of endogenous atrial natriuretic factor and kinins. J Pharm Exp Ther 258:807–811.

Bralet J, Marie C, Mossiat C, Lecomte, J-M, Gros C, Schwartz JC. (1994). Effects of alatriopril, a mixed inhibitor of atriopeptidase and angiotensin I-converting enzyme, on cardiac hypertrophy and hormonal responses in rats with myocardial infarction. Comparison with captopril. J Pharm Exp Ther 270:8–14.

Brent GA, Hershman JM, Reed AW, Sastre A, Lieberman J. (1984). Serum angiotensin-converting enzyme in severe nonthyroidal illnesses associated with low serum thyroxine concentration. Ann Intern Med 100:680–683.

Bunting PS, Szalai JP, Katic M. (1987). Diagnostic aspects of angiotensin converting enzyme in pulmonary sarcoidosis. Clin Biochem 20:213–219.

Butler PE, McKay MJ, Bond JS. (1987). Characterization of meprin, a membrane-bound metalloendopeptidase from mouse kidney. Biochem J 241:29–235.

Butterworth J. (1986). Changes in nine enzyme markers for neurons, glia, and endothelial cells in agonal state and Huntington's disease caudate nucleus. J Neurochem 47:583–587.

Caldwell PRB, Seegal BC, Hsu KC, Das M, Soffer RL. (1976). Angiotensin-converting enzyme: vascular endothelial localization. Science 191:1050–1051.

Cambien F, Soubrier F. (1995). The angiotensin-converting enzyme: molecular biology and implication of the gene polymorphism in cardiovascular diseases. In: Laragh JH, Brenner BM, eds. Hypertension: Pathophysiology, Diagnosis, and Management. 2d ed. eds. New York: Raven Press: 1667–1682.

Campbell W, Okada H. (1989). An arginine specific carboxypeptidase generated in blood during coagulation or inflammation which is unrelated to carboxypeptidase N or its subunits. Biochem Biophys Res Commun 162:933–939.

Carbonell LF, Carretero OA, Stewart JM, Scicli AG. (1988). Effect of a kinin antagonist on the acute antihypertensive activity of enalaprilat in severe hypertension. Hypertension 11: 239–243.

Cardozo C, Orlowski M. (1993). Evidence that enzymatic conversion of N-[1(R,S)-carboxy-3-phenylpropyl]-Ala-Ala-Phe-*p*-aminobenzoate, a specific inhibitor of endopeptidase 24.15, to N-[1(R,S)-carboxy-3-phenylpropyl]-Ala-Ala is necessary for inhibition of angiotensin converting enzyme. Peptides 14:11259–11262.

Carpenter G, Wahl MI. (1990). The epidermal growth factor family. In Sporn MB, Roberts AB, eds. Peptide Growth Factors and Their Receptors I. Handbook of Experimental Pharmacology. Vol. 95. New York: Springer-Verlag: 69–171.

Carretero OA, Scicli AG. (1989). Kinins paracrine hormone. In: Fritz H, Schmidt I, Dietze G, eds. The Kallikrein-Kinin System in Health and Disease. Braunschweig, Munich: Limbach-Verlag: 63–78.

Carretero OA, Scicli AG. (1995). The kallikrein-kinin system as a regulator of cardiovascular and renal function. In: Laragh JH, Brenner BM, eds. Hypertension: Pathophysiology, Diagnosis, and Management. 2d ed. New York: Raven Press: 983–999.

Chai SY, McKinley MJ, Paxinos G, Mendelsohn FAO. (1991). Angiotensin-converting enzyme in the monkey (*Macaca fascicularis*) brain visualized by *in vitro* autoradiography. Neuroscience 42:483–495.

Checler F, Vincent JP, Kitabgi P. (1983). Degradation of neurotensin by rat brain synaptic membranes: involvement of a thermolysin-like metalloendopeptidase (enkephalinase), angiotensin-converting enzyme, and other unidentified peptidases. J Neurochem 41: 375–384.

Checler F, Vincent JP, Kitabgi P. (1986). Purification and characterization of a novel neurotensin-degrading peptidase from rat brain synaptic membranes. J Biol Chem 261:11274–11281.

Checler F, Barelli H, Dauch P, Dive V, Vincent B, Vincent JP. (1995). Neurolysin: purification and assays. Meth Enzymol 248:593–614.

Chen K, Zhang X, Dunham EW, Zimmerman BG. (1996). Kinin-mediated antihypertensive effect of captopril in deoxycorticosterone acetate-salt hypertension. Hypertension 27: 85–89.

Chercuitte F, Beaulieu AD, Poubelle P, Marceau F. (1987). Carboxypeptidase N (kininase I) activity in blood and synovial fluid from patients with arthritis. Lif Sci 41:1225–1232.

Chodimella V, Skidgel RA, Krowiak EJ, Murlas CG. (1991). Lung peptidases, including carboxypeptidase, modulate airway reactivity to intravenous bradykinin. Am Rev Respir Dis 144:869–874.

Chu TG, Orlowski M. (1985). Soluble metalloendopeptidase from rat brain: action on enkephalin-containing peptides and other bioactive peptides. Endocrinology 116:1418–1425.

Colman RW. (1987). Humoral mediators of catastrophic reactions associated with protamine neutralization. Anesthesiology 66:595–596.

Connelly JC, Skidgel RA, Schulz WW, Johnson AR, Erdös EG. (1985). Neutral endopeptidase 24.11 inhuman neutrophils: cleavage of chemotactic peptide. Proc Natl Acad Sci (USA) 82:8737–8741.

CONSENSUS Trial Study Group (1987). Effects of enalapril on mortality in severe congestive heart failure. N Engl J Med 316:1429–1435.

Cornell MJ, Williams TA, Lamango NS, Coates D, Corvol P, Soubrier F, Hoheisel J, Lehrach H, Isaac RE. (1995). Cloning and expression of an evolutionary conserved single-domain angiotensin-converting enzyme from *Drosophila melanogaster*. J Biol Chem 270:13613–13619.

Correa FMA, Plunkett LM, Saavedra JM, Hichens M. (1985). Quantitative autoradiographic determination of angiotensin-converting enzyme (kininase II) binding in individual rat brain nuclei with ^{125}I-351A, a specific enzyme inhibitor. Brain Res 347:192–195.

Costerousse O, Allegrini J, Lopez M, Alhenc-Gelas F. (1993). Angiotensin I-converting enzyme in human circulating mononuclear cells: genetic polymorphism of expression in T-lymphocytes. Biochem J 290:33–40.

Crecelius DM, Stewart JM, Vavrek RJ, Balasubramaniam TM, Baenziger NL. (1986). Interaction of bradykinin (BK) with receptors on human lung fibroblasts. Fed Proc 45:454.

Crimi N, Polosa R, Pulvirenti G, Magri S, Santonocito G, Prosperini G, Mastruzzo C, Mistretta A. (1995). Effect of an inhaled neutral endopeptidase inhibitor, phosphoramidon, on baseline airway calibre and bronchial responsiveness to bradykinin in asthma. Thorax 50:505–510.

Cushman DW, Ondetti MA. (1980). Inhibitors of angiotensin converting enzyme. Prog Med Chem 17:42–104.

Danilov S, Jaspard E, Churakova T, Towbin H, Savoie F, Wei L, Alhenc-Gelas F. (1994). Structure-function analysis of angiotensin I-converting enzyme using monoclonal antibodies. Selective inhibition of the amino-terminal active site. J Biol Chem 269:26806–26814.

D'Azzo A, Hoogeveen A, Reuser AJJ, Robinson D, Galjaard H. (1982). Molecular defect in combined β-galactosidase and neuraminidase deficiency in man. Proc Natl Acad Sci (USA) 79:4535–4539.

De Meyer GRY, Bult H, Kockx MM, Herman AG. (1995). Effect of angiotensin-converting enzyme inhibition on intimal thickening in rabbit collared carotid artery. J Cardiovasc Pharmacol 26:614–620.

de Saint-Vis B, Cupillard L, Pandrau-Garcia D, Ho S, Renard N, Grouard G, Duvert V, Thomas X, Galizzi JP, Banchereau J, Saeland S. (1995). Distribution of carboxypeptidase-M on lymphoid and myeloid cells parallels the other zinc-dependent proteases CD10 and CD13. Blood 86:1098–1105.

DeBlois D, Bouthillier J, Marceau F. (1988). Effect of glucocorticoids, monokines and growth factors on the responses of the isolated rabbit aorta to des-Arg⁹-bradykinin (BK). FASEB J 2:A1145.

DeBlois D, Bouthillier J, Marceau F. (1991). Pulse exposure to protein synthesis inhibitors enhances vascular responses to des-Arg⁹-bradykinin: possible role of interleukin-1. Br J Pharmacol 103:1057–1066.

DeBlois D, Drapeau G, Petitclerc E, Marceau F. (1992). Synergism between the contractile effect of epidermal growth factor and that of des-Arg⁹-bradykinin or of α-thrombin in rabbit aortic rings. Br J Pharmacol 105:959–967.

Deddish PA, Skidgel RA, Erdös EG. (1989). Enhanced Co²⁺ activation and inhibitor binding of carboxypeptidase M at low pH. Similarity to carboxypeptidase H (enkephalin convertase). Biochem J 261:289–291.

Deddish PA, Skidgel RA, Kriho VB, Li X-Y, Becker RP, Erdös EG. (1990). Carboxypeptidase M in cultured Madin-Darby canine kidney (MDCK) cells. Evidence that carboxypeptidase M has a phosphatidylinositol glycan anchor. J Biol Chem 265:15083–15089.

Deddish PA, Wang J, Michel B, Morris PW, Davidson NO, Skidgel RA, Erdös EG (1994). Naturally occurring active N-domain of human angiotensin I converting enzyme. Proc Natl Acad Sci (USA) 91:7807–7811.

Deddish PA, Wang L-X, Jackman HL, Michel B, Wang J, Skidgel RA, Erdös EG. Single-domain angiotensin I converting enzyme (kininase II; ACE): characterization and properties. J Pharmacol Exp Ther 1996; 279:1582–1589.

Deddish PA, Jackman HL, Skidgel RA, Erdös EG. (1997). Differences in the hydrolysis of enkephalin congeners by the two domains of angiotensin converting enzyme. Biochem Pharmacol 53:1459–1463.

Defendini R, Zimmerman EA, Weare JA, Alhenc-Gelas F, Erdös EG. (1983). Angiotensin-converting enzyme in epithelial and neuroepithelial cells. Neuroendocrinology 37:32–40.

Dehm P, Nordwig A. (1970). The cleavage of prolyl peptides by kidney peptidases. Partial purification of a "X-prolyl-aminopeptidase" from swine kidney microsomes. Eur J Biochem 17:364–371.

Denslow ND, Ryan JW, Nguyen HP. (1994). Guinea pig membrane-bound aminopeptidase P is a member of the proline peptidase family. Biochem Biophys Res Commun 205: 1790–1795.

Deschodt-Lanckman M, Michaux F, De Prez E, Abramowicz D, Vanherweghem J-L, Goldman M. (1989). Increased serum levels of endopeptidase 24.11 ("enkephalinase") in patients with end-stage renal failure. Life Sci 45:133–141.

Desmazes NA, Lockhart A, Lacroix H, Dusser DJ. (1992). Carboxypeptidase M-like enzyme modulates the noncholinergic bronchoconstrictor response in guinea pigs. Am J Respir Cell Mol Biol 7:477–484.

Devault A, Lazure C, Nault C, Le Moual H, Seidah NG, Chretien M, Kahn P, Powell J, Mallett J, Beaumont A, Roques BP, Crine P, Boileau G. (1987). Amino acid sequence of rabbit kidney neutral endopeptidase 24.11 (enkephalinase) deduced from a complementary DNA. EMBO J 6:1317–1322.

Dhople AM, Howell PC, Williams SL, Zeigler JA, Storrs EE. (1985). Serum angiotensin-converting enzyme in leprosy. Indian J Leprosy 57:282–287.

Dietze GJ, Wicklmayr M, Rett K, Jacob S, Henriksen EJ. (1996). Potential role of bradykinin in forearm muscle metabolism in humans. Diabetes 45:S110–S114.

Dragovic T, Igić R, Erdös EG, Rabito SF. (1993). Metabolism of bradykinin by peptidases in the lung. Am Rev Respir Dis 147:1491–1496.

Dragovic T, Deddish PA, Tan F, Weber G, Erdös EG. (1994). Increased expression of neprilysin (neutral endopeptidase 24.11) in rat and human hepatocellular carcinomas. Lab Invest 70:107–113.

Dragovic T, Schraufnagel DE, Becker RP, Sekosan M, Votta-Velis EG, Erdös EG. (1995). Carboxypeptidase M activity is increased in bronchoalveolar lavage in human lung disease. Am J Respir Crit Care Med 152:760–764.

Dragovic T, Minshall R, Jackman HL, Wang LX, Erdös EG. (1996). Kininase II-type enzymes, their putative role in muscle energy metabolism. Diabetes 45:S34–S37.

Dragovic T, Sekosan M, Becker RP, Erdös EG. (1997). Detection of neutral endopeptidase 24.11 (neprilysin) in human hepatocellular carcinomas by immunocytochemistry. Anti-cancer Res. In press.

Dumermuth E, Sterchi EE, Jiang W, Wolz RL, Bond JS, Flannery AV, Beynon RJ. (1991). The astacin family of metallopeptidases. J Biol Chem 266:21381–21385.

Dusser DJ, Nadel JA, Sekizawa K, Graf PD, Borson DB. (1988). Neutral endopeptidase and angiotensin converting enzyme inhibitors potentiate kinin-induced contraction of ferret trachea. J Pharmacol Exp Ther 244:531–536.

Eaton DL, Malloy BE, Tsai SP, Henzel W, Drayna D. (1991). Isolation, molecular cloning, and partial characterization of a novel carboxypeptidase B from human plasma. J Biol Chem 266:21833–21838.

Ehlers MRW, Riordan JF. (1990). Angiotensin-converting enzyme. Biochemistry and molecular biology. In: Laragh JH, Brenner BM, eds. Hypertension: Pathophysiology, Diagnosis, and Management. New York: Raven Press: 1217–1231.

Ehlers MRW, Riordan JF. (1991). Angiotensin converting enzyme: zinc- and inhibitor-binding stoichiometries of the somatic and testis isozymes. Biochemistry 30:7118–7126.

Ehlers MRW, Fox EA, Strydom DJ, Riordan JF. (1989). Molecular cloning of human testicular angiotensin-converting enzyme: the testis isozyme is identical to the C-terminal half of endothelial angiotensin-converting enzyme. Proc Natl Acad Sci (USA) 86:7741–7745.

Ehlers MRW, Scholle RR, Riordan JF. (1995). Proteolytic release of human angiotensin-converting enzyme expressed in Chinese hamster ovary cells is enhanced by phorbol ester. Biochem Biophys Res Commun 206:541–547.

Elisseeva YE, Pavlikhina LV, Orekhovich VN, Giacomello A, Salerno C, Fasella P. (1981). Evidence for the presence of dipeptidyl carboxypeptidase and its inhibitors in inflammatory synovial fluids. Biochim Biophys Acta 658:165–168.

Elliott PJ, Mackic JB, Graney WF, Bartus RT, Zlokovic BV. (1995). RMP-7, a bradykinin agonist, increases permeability of blood-ocular barriers in the guinea pig. Invest Ophthalm Visual Sci 36:2542–2547.

Elsner D, Müntze A, Kromer EP, Riegger GAJ. (1992). Effectiveness of endopeptidase inhibition (candoxatril) in congestive heart failure. Am J Cardiol 70:494–498.

Erdös EG. (1966). Hypotensive peptides: bradykinin, kallidin, and eledoisin. Adv Pharmacol 4:1–90.

Erdös EG. (1979). Kininases. In: Erdös EG, ed. Bradykinin, Kallidin and Kallikrein. Handbook of Experimental Pharmacology. Supplement to Vol. XXV. Heidelberg: Springer-Verlag: 427–487.

Erdös EG. (1989). From measuring the blood pressure to mapping the gene: the development of the ideas of Frey and Werle. In Fritz H, Schmidt I, Dietze G, eds. The Kallikrein and Kinin System in Health and Disease. E.K. Frey and E. Werle Memorial Volume. Braunschweig: Limbach-Verlag: 261–276.

Erdös EG, Gafford JT. (1983). Human converting enzyme. Clin Exp Hypertension A5:1251–1262.

Erdös EG, Skidgel RA. (1989). Neutral endopeptidase 24.11-enkephalinase and related regulators of peptide hormones. FASEB J 3:145–151.

Erdös EG, Skidgel RA. (1996). Metabolism of bradykinin by peptidases in health and disease. In Farmer SG, ed. The Kinin System. Series: Handbook of Immunopharmacology. London: Academic Press.

Erdös EG, Sloane EM. (1962). An enzyme in human blood plasma that inactivates bradykinin and kallidins. Biochem Pharmacol 11:585–592.

Erdös EG, Wohler JR. (1963). Inhibition *in vivo* of the enzymatic inactivation of bradykinin and kallidin. Biochem Pharmacol 12:1193–1199.

Erdös EG, Yang HYT. (1966). Inactivation and potentiation of the effects of bradykinin. In: Erdös EG, Back N, Sicuteri F, eds. Hypotensive Peptides. New York: Springer-Verlag: 235–251.

Erdös EG, Yang HYT. (1967). An enzyme in microsomal fraction of kidney that inactivates bradykinin. Life Sci 6:569–574.

Erdös EG, Yang HYT. (1970). Kininases. In: Erdös EG, ed. Bradykinin, Kallidin and Kallikrein. Handbook of Experimental Pharmacology. Vol. XXV. Heidelberg: Springer-Verlag: 289–323.

Erdös EG, Renfrew AG, Sloane EM, Wohler JR. (1963). Enzymatic studies on bradykinin and similar peptides. Ann NY Acad Sci 104:222–235.

Erdös EG, Sloane EM, Wohler IM. (1964). Carboxypeptidase in blood and other fluids—I. Properties, distribution, and partial purification of the enzyme. Biochem Pharmacol 13: 893–905.

Erdös EG, Wohler IM, Levine MI, Westerman P. (1965). Carboxypeptidase in blood and other fluids. Values in human blood in normal and pathological conditions. Clin Chim Acta 11:39–43.

Erdös EG, Schulz WW, Gafford JT, Defendini R. (1985). Neutral metalloendopeptidase in human male genital tract. Comparison to angiotensin I converting enzyme. Lab Invest 52:437–447.

Erdös EG, Wagner BA, Harbury CB, Painter RG, Skidgel RA, Fa X-G. (1989). Down-regulation and inactivation of neutral endopeptidase 24.11 (enkephalinase) in human neutrophils. J Biol Chem 264:14519–14523.

Falezza G, Santonastaso CL, Parisi T, Muggeo M. (1985). High serum levels of angiotensin-converting enzyme in untreated Addison's disease. J Clin Endocrinol Metab 61:496–498.

Falkenhahn M, Franke F, Bohle RM, Zhu Y-C, Stauss HM, Bachmann S, Danilov S, Unger T. (1995). Cellular distribution of angiotensin-converting enzyme after myocardial infarction. Hypertension 25:219–226.

Farhy RD, Hoo K-L, Carretero OA, Scicli AG. (1992). Kinins mediate the antiproliferative effect of ramipril in rat carotid artery. Biochem Biophys Res Commun 182:283–288.

Ferreira SH. (1965). A bradykinin-potentiating factor (BPF) present in the venom of *Bothrops jararaca*. Br J Pharmacol Chemother 24:163–169.

Ferreira SH, Bartelt DC, Greene LJ. (1970). Isolation of bradykinin-potentiating peptides from *Bothrops jararaca* venom. Biochemistry 9:2583–2593.

Flynn GA, French JF, Dage RC. (1995). Dual inhibitors of angiotensin-converting enzyme and neutral endopeptidase: design and therapeutic rationale. In: Laragh JH, Brenner BM, eds. Hypertension: Pathophysiology, Diagnosis, and Management. 2d ed. New York: Raven Press: 3099–3114.

Folk JE, Gladner JA. (1960). Cobalt activation of carboxypeptidase A. J Biol Chem 235: 60–63.

Folk JE, Gladner JA. (1961). Influence of cobalt and cadmium on the peptidase and esterase activities of carboxypeptidase B. Biochim Biophys Acta 48:139–147.

French JF, Flynn GA, Giroux EL, Mehdi S, Anderson B, Beach DC, Koehl JR, Dage RC. (1994). Characterization of a dual inhibitor of angiotensin I-converting enzyme and neutral endopeptidase. J Pharm Exp Ther 268:180–186.

French JF, Anderson BA, Downs TR, Dage RC. (1995). Dual inhibition of angiotensin-converting enzyme and neutral endopeptidase in rats with hypertension. J Cardiovasc Pharmacol 26:107–113.

Frey EK, Kraut H, Werle E. (1950). Kallikrein Padutin. Stuttgart: Ferdinand Enke Verlag: 1–209.

Frey EK, Kraut H, Werle E, Vogel R, Zickgraf-Rüdel G, Trautschold I. (1968). Das Kallikrein-Kinin-System und seine Inhibitoren. Stuttgart: Ferdinand Enke Verlag: 1–290.

Friedland J, Setton C, Silverstein E. (1978). Induction of angiotensin-converting enzyme in human monocytes in culture. Biochem Biophys Res Commun 83:843–849.

Fyhrquist F, Grönhagen-Riska C, Hortling L, Forslund T, Tikkanen I. (1983). Regulation of angiotensin-converting enzyme. J Hypertension 1(suppl 1):25–30.

Gafford JT, Skidgel RA, Erdös EG, Hersh LB. (1983). Human kidney "enkephalinase," a neutral metalloendo-peptidase that cleaves active peptides. Biochemistry 22:3265–3271.

Galjart NJ, Gillemans N, Harris A, van der Horst GTJ, Verheijen FW, Galjaard H, d'Azzo A. (1988). Expression of cDNA encoding the human "protective protein" associated with lysosomal β-galactosidase and neuraminidase: homology to yeast proteases. Cell 54: 755–764.

Galjart NJ, Gillemans N, Meijer D, d'Azzo A. (1990). Mouse "protective protein." cDNA cloning, sequence comparison, and expression. J Biol Chem 265:4678–4684.

Gavras I, Gavras H. (1987). The use of ACE inhibitors in hypertension. In: Kostis JB, DeFelice EA, eds. Angiotensin Converting Enzyme Inhibitors. New York: Alan R. Liss: 93–122.

Gavras I, Gavras H. (1993). ACE inhibitors: a decade of clinical experience. Hosp Pract July 15:61–71.

Gavras H, Faxon DP, Berkoben J, Brunner HR, Ryan TJ. (1978). Angiotensin converting enzyme inhibition in patients with congestive heart failure. Circulation 58:770–776.

Gebhard W, Schube M, Eulitz M. (1989). cDNA cloning and complete primary structure of the small, active subunit of human carboxypeptidase N (kininase I). Eur J Biochem 178: 603–607.

Genden EM, Molineaux CJ. (1991). Inhibition of endopeptidase-24.15 decreases blood pressure in normotensive rats. Hypertension 18:360–365.

Gohlke P, Linz W, Schölkens BA, Kuwer I, Bartenbach S, Schnell A, Unger T. (1994a). Angiotensin-converting enzyme inhibition improves cardiac function. Role of bradykinin. Hypertension 23:411–418.

Gohlke P, Kuwer I, Bartenbach S, Schnell A, Unger T. (1994b). Effect of low-dose treatment with perindopril on cardiac function in stroke-prone spontaneously hypertensive rats: role of bradykinin. J Cardiovasc Pharmacol 24:462–469.

Goldstein RH, Wall M. (1984). Activation of protein formation and cell division by bradykinin and des-Arg[9]-bradykinin. J Biol Chem 259:9263–9268.

Gonzalez-Vera W, Fournie-Zaluski M-C, Pham I, Laboulandine I, Roques B-P, Michel J-B. (1995). Hypotensive and natriuretic effects of RB 105, a new dual inhibitor of angiotensin converting enzyme and neutral endopeptidase in hypertensive rats. J Pharmacol Exp Ther 272:343–351.

Gorbea CM, Flannery AV, Bond JS. (1991). Homo- and heterotetrameric forms of the membrane-bound metalloendopeptidases meprin A and B. Arch Biochem Biophys 290: 549–553.

Graf K, Koehne P, Gräfe M, Zhang M, Auch-Schwelk W, Fleck E. (1995). Regulation and differential expression of neutral endopeptidase 24.11 in human endothelial cells. Hypertension 26:230–235.

Grönhagen-Riska C. (1979). Angiotensin-converting enzyme. I. Activity and correlation with serum lysozyme in sarcoidosis, other chest or lymph node diseases and healthy persons. Scand J Respir Dis 60:83–93.

Gros C, Noël N, Souque A, Schwartz J-C, Danvy D, Plaquevent JC, Duhamel L, Duhamel P, Lecomte J-M, Bralet J. (1991). Mixed inhibitors of angiotensin-converting enzyme (EC 3.4.15.1) and enkephalinase (EC 3.4.24.11): rational design, properties, and potential cardiovascular applications of glycopril and alatriopril. Proc Natl Acad Sci (USA). 88: 4210–4214.

Hall AS, Tan L-B, Ball SG. (1994). Inhibition of ACE/kininase II, acute myocardial infarction, and survival. Cardiovasc Res 28:190–198.

Hall ER, Kato J, Erdös EG, Robinson CJG, Oshima G. (1976). Angiotensin I-converting enzyme in the nephron. Life Sci 18:1299–1303.

Harbeck HT, Mentlein R. (1991). Aminopeptidase P from rat brain. Purification and action on bioactive peptides. Eur J Biochem 198:451–458.

Hartman JC. (1995). The role of bradykinin and nitric oxide in the cardioprotective action of ACE inhibitors. Ann Thoracic Surg 60:789–782.

Heck I, Niederle N. (1983). Angiotensin-converting-enzym-aktivität während zytostatischer Therapie bei Patienten mit primär inoperablem Bronchialkarzinom. Klin Wochenschr 61:923–927.

Hecker M, Bara AT, Busse R. (1993). Relaxation of isolated coronary arteries by angiotensin-converting enzyme inhibitors: role of endothelium-derived kinins. J Vasc Res 30: 257–262.

Hendriks D, Scharpé S, van Sande M, Lommaert MP. (1989). Characterisation of a carboxy-peptidase in human serum distinct from carboxypeptidase N. J Clin Chem Clin Biochem 27:277–285.

Hendriks D, Wang W, Scharpé S, Lommaert MP, van Sande M. (1990). Purification and characterization of a new arginine carboxypeptidase in human serum. Biochim Biophys Acta 1034:86–92.

Hendriks D, Wang W, van Sande M, Scharpé S. (1992). Human serum carboxypeptidase U: a new kininase? Agents Actions Suppl 38/I:407–413.

Hooper NM, Turner AJ. (1988). Ectoenzymes of the kidney microvillar membrane. Aminopeptidase P is anchored by a glycosyl-phosphatidylinositol moiety. FEBS Lett 229: 340–344.

Hooper NM, Hryszko J, Turner AJ. (1990). Purification and characterization of a pig kidney aminopeptidase P. A glycosyl-phosphatidylinositol-anchored ectoenzyme. Biochem J 267:509–515.

Hooper NM, Hryszko J, Oppong SY, Turner AJ. (1992). Inhibition by converting enzyme inhibitors of pig kidney aminopeptidase P. Hypertension 19:281–285.

Howell S, Boileau G, Crine P. (1994). Neutral endopeptidase (EC 3.4.24.11): constructed molecular forms show new angles of an old enzyme. Biochem Cell Biol 72:67–69.

Ichinose M, Barnes PJ. (1990). The effect of peptidase inhibitors on bradykinin-induced bronchoconstriction in guinea-pigs *in vivo*. Br J Pharmacol 101:77–80.

Igić R. (1985). Kallikrein and kininases in ocular tissues. Exp Eye Res 41:117–120.

Igić R, Erdös EG, Yeh HSJ, Sorrells K, Nakajima T. (1972). Angiotensin I converting enzyme of the lung. Circ Res 31(suppl II):51–61.

Igić R, Nakajima T, Yeh HSJ, Sorrells K, Erdös EG. (1973). Kininases. In: Acheson G, ed. Pharmacology and the Future of Man—Proceedings of the 5th International Congress of Pharmacology. Basel: Karger: 307–319.

Igić R, Robinson CJG, Erdös EG. (1977). Angiotensin I converting enzyme activity in the choroid plexus and in the retina. In: Buckley JP, Ferrario CM, eds. Central Actions of Angiotensin and Related Hormones. New York: Pergamon Press: 23–27.

Iimura O. (1987). Pathophysiological significance of kallikrein-kinin system in essential hypertension. In Iimura O, Margolius HS, eds. Renal Function, Hypertension and Kallikrein-Kinin System. Tokyo: Hokusen-Sha: 3–18.

Inamura T, Black KL. (1994). Bradykinin selectively opens blood-tumor barrier in experimental brain tumors. J Cerebral Blood Flow Metab 14:862–870.

Inamura T, Nomura T, Bartus RT, Black KL. (1994). Intracarotid infusion of RMP-7, a bradykinin analog: a method for selective drug delivery to brain tumors. J Neurosurg 81: 752–758.

Inokuchi J-I, Nagamatsu A. (1981). Tripeptidyl carboxypeptidase activity of kininase II (angiotensin-converting enzyme). Biochim Biophys Acta 662:300–307.

Israili ZH, Hall WD. (1992). Cough and angioneurotic edema associated with angiotensin-converting enzyme inhibitor therapy. A review of the literature and pathophysiology. Ann Intern Med 117:234–242.

Issandou M, Darbon JM. (1991). Des-Arg^9 bradykinin modulates DNA synthesis, phospholipase C, and protein kinase C in cultured mesangial cells. J Biol Chem 266:21037–21043.

Jackman HL, Tan F, Tamei H, Beurling-Harbury C, Li X-Y, Skidgel RA, Erdös EG. (1990). A peptidase in human platelets that deamidates tachykinins: probable identity with the lysosomal "protective protein." J Biol Chem 265:11265–11272.

Jackman HL, Morris PW, Deddish PA, Skidgel RA, Erdös EG. (1992). Inactivation of endothelin I by deamidase (lysosomal protective protein). J Biol Chem 267:2872–2875.

Jackman HL, Morris PW, Rabito SF, Johansson GB, Erdös EG. (1993). Inactivation of endothelin 1 by an enzyme of the vascular endothelial cells. Hypertension 21:925–928.

Jackman HL, Tan F, Schraufnagel D, Dragovic T, Dezsö B, Becker RP, Erdös EG. (1995). Plasma membrane-bound and lysosomal peptidases in human alveolar macrophages. Am J Respir Cell Mol Biol 13:196–204.

Jaspard E, Alhenc-Gelas F. (1995). Catalytic properties of the two active sites of angiotensin I-converting enzyme on the cell surface. Biochem Biophys Res Commun 211:528–534.

Jaspard E, Wei L, Alhenc-Gelas F. (1993). Differences in the properties and enzymatic specificities of the two active sites of angiotensin I-converting enzyme (kininase II). Studies with bradykinin and other natural peptides. J Biol Chem 268:9496–9503.

Jiang W, Gorbea CM, Flannery AV, Beynon RJ, Grant GA, Bond JS. (1992). The α subunit of meprin A. Molecular cloning and sequencing, differential expression in inbred mouse strains, and evidence for divergent evolution of the α and β subunits. J Biol Chem 267:9185–9193.

Jiang W, Sadler PM, Jenkins NA, Gilbert DJ, Copeland NG, Bond JS. (1993). Tissue-specific expression and chromosomal localization of the α subunit of mouse meprin A. J Biol Chem 268:10380–10385.

Johnson AR, John M, Erdös EG. (1982). Metabolism of vasoactive peptides by membrane-enriched fractions from human lung tissue, pulmonary arteries, and endothelial cells. NY Acad Sci 384:72–89.

Johnson AR, Skidgel RA, Gafford JT, Erdös EG. (1984). Enzymes in placental microvilli: angiotensin I converting enzyme, angiotensinase A, carboxypeptidase, and neutral endopeptidase ("enkephalinase"). Peptides 5:789–796.

Johnson AR, Ashton J, Schulz W, Erdös EG. (1985a). Neutral metalloendopeptidase in human lung tissue and cultured cells. Am Rev Respir Dis 132:564–568.

Johnson AR, Coalson JJ, Ashton J, Larumbide M, Erdös EG. (1985b). Neutral metalloendopeptidase in serum samples from patients with adult respiratory distress syndrome: comparison with angiotensin-converting enzyme. Am Rev Respir Dis 132:1262–1267.

Johnson GD, Hersh LB. (1992). Cloning a rat meprin cDNA reveals the enzyme is a heterodimer. J Biol Chem 267:13505–13512.

Johnson GD, Hersh LB. (1994). Expression of meprin subunit precursors. Membrane anchoring through the β subunit and mechanism of zymogen activation. J Biol Chem 269:7682–7688.

Kato H, Suzuki T. (1970). Amino acid sequence of bradykinin-potentiating peptide isolated from the venom of *Agkistrodon halys blomhoffii*. Proc Jpn Acad 46:176–181.

Kerr MA, Kenny AJ. (1974). The purification and specificity of a neutral endopeptidase from rabbit kidney brush border. Biochem J 137:477–488.

Kobe B, Deisenhofer J. (1993). Crystal structure of porcine ribonuclease inhibitor, a protein with leucine-rich repeats. Nature 366:751–756.

Komers R, Cooper ME. (1995). Acute renal hemodynamic effects of ACE inhibition in diabetic hyperfiltration: role of kinins. Am J Physiol 268:F588–F594.

Kounnas MZ, Wolz RL, Gorbea CM, Bond JS. (1991). Meprin-A and -B. Cell surface endopeptidases of the mouse kidney. J Biol Chem 266:17350–17357.

Kumamoto K, Stewart TA, Johnson AR, Erdös EG. (1981). Prolylcarboxypeptidase (angiotensinase C) in human cultured cells. J Clin Invest 67:210–215.

Kumar RS, Kusari J, Roy SN, Soffer RL, Sen GC. (1989). Structure of testicular angiotensin-converting enzyme. A segmental mosaic enzyme. J Biol Chem 264:16754–16758.

Kuroki K, Eng F, Ishikawa T, Turck C, Harada F, Ganem D. (1995). gp180, A host cell glycoprotein that binds duck hepatitis B virus particles, is encoded by a member of the carboxypeptidase gene family. J Biol Chem 270:15022–15028.

Lamango NS, Sajid M, Isaac RE. (1996). The endopeptidase activity and the activation by Cl⁻ of angiotensin converting enzyme is evolutionarily conserved: purification and properties of an angiotensin converting enzyme from the housefly, *Musca domestica*. Biochem J 314:639–646.

Lattion AL, Soubrier F, Allegrini J, Hubert C, Corvol P, Alhenc-Gelas F. (1989). The testicular transcript of the angiotensin I-converting enzyme encodes for the ancestral, non-duplicated form of the enzyme. FEBS Lett 252:99–104.

Lavras AAC, Fichman M, Hiraichi E, Tobo T, Boucault MA. (1980). The kininases of *Bothrops jararaca* plasma. Acta Physiol Latinoam 30:269–274.

LeBien TW, McCormack RT. (1989). The common acute lymphoblastic leukemia antigen (CD10)—emancipation from a functional enigma. Blood 73:625–635.

Letarte M, Vera S, Tran R, Addis JBL, Onizuka RJ, Quackenbush EJ, Jongeneel CV, McInnes RR. (1988). Common acute lymphocytic leukemia antigen is identical to neutral endopeptidase. J Exp Med 168:1247–1253.

Levin Y, Skidgel RA, Erdös EG. (1982). Isolation and characterization of the subunits of human plasma carboxypeptidase N (kininase I). Proc Natl Acad Sci (USA) 79:4618–4622.

Lieberman J. (1974). Elevation of serum angiotensin converting enzyme (ACE) level in sarcoidosis. Am J Med 59:365–372.

Lieberman J. (1985). Angiotensin-converting enzyme (ACE) and serum lysozyme in sarcoidosis. In: Lieberman J, ed. Sarcoidosis. Orlando, FL: Grune & Stratton: 145–159.

Linz W, Wiemer G, Gohlke P, Unger T, Schölkens BA. (1995). Contribution of kinins to the cardiovascular actions of angiotensin-converting enzyme inhibitors. Pharmacol Rev 47:25–49.

Linz W, Wiemer G, Schölkens BA. (1996). Role of kinins in the pathophysiology of myocardial ischemia. In vitro and in vivo studies. Diabetes 45:S51–S58.

Lipke DW, Olson KR. (1988). Distribution of angiotensin-converting enzyme-like activity in vertebrate tissues. Physiol Zool 61:420–428.

Ljunggren Ö, Lerner UH. (1990). Evidence for BK_1 bradykinin-receptor-mediated prostaglandin formation in osteoblasts and subsequent enhancement of bone resorption. Br J Pharmacol 101:382–386.

Llorens-Cortes C, Huang H, Vicart P, Gasc J-M, Paulin D, Corvol P. (1992). Identification and characterization of neutral endopeptidase in endothelial cells from venous or arterial origins. J Biol Chem 267:14012–14018.

Lock R, Hessel EA. (1990). Probable reversal of protamine reactions by heparin administration. J Cardiothor Anesth 4:605–608.

Lötvall J. (1990). Tachykinin- and bradykinin-induced airflow obstruction and airway microvascular leakage. Modulation by neutral endopeptidase and angiotensin converting enzyme. Göteborg Sweden: University of Göteborg.

Lötvall JO, Tokuyama K, Barnes PJ, Chung KF. (1990). Bradykinin-induced airway microvascular leakage is potentiated by captopril and phosphoramidon. Eur J Pharmacol 200:211–217.

Low MG. (1987). Biochemistry of the glycosyl-phosphatidylinositol membrane protein anchors. Biochem J 244:1–13.

Lowenstein E, Johnston EW, Lappas DG, D'Ambra MN, Schneider RC, Daggett WM, Akins CW, Philbin DM. (1983). Catastrophic pulmonary vasoconstriction associated with protamine reversal of heparin. Anesthesiology 59:470–473.

MacFadden RG, Vickers K. (1988). Bradykinin stimulates lymphocyte movement in vitro. FASEB J 2:A1179.

Malfroy B, Schofield PR, Kuang WJ, Seeburg PH, Mason AJ, Henzel WJ. (1987). Molecular cloning and amino acid sequence of rat enkephalinase. Biochem Biophys Res Commun 144:59–66.

Malfroy B, Kuang WJ, Seeburg PH, Mason AJ, Schofield PR. (1988). Molecular cloning and amino acid sequence of human enkephalinase (neutral endopeptidase). FEBS Lett 229:206–210.

Marceau F, Tremblay B. (1986). Mitogenic effect of bradykinin and of des-Arg[9]-bradykinin on cultured fibroblasts. Life Sci 39:2351–2358.

Marchand P, Tang J, Bond JS. (1994). Membrane association and oligomeric organization of the alpha and beta subunits of mouse meprin A. J Biol Chem 269:15388–15393.

Margolius HS. (1995). Kallikreins and kinins. Some unanswered questions about system characteristics and roles in human disease. Hypertension 26:221–229.

Marie C, Mossiat C, Lecomte JM, Schwartz JC, Bralet J. (1995). Hemodynamic effects of acute and chronic treatment with aladotril, a mixed inhibitor of neutral endopeptidase and angiotensin I converting enzyme, in conscious rats with myocardial infarction. J Pharmacol Exp Ther 275:1324–1331.

Marinkovic DV, Marinkovic JN, Erdös EG, Robinson CJG. (1977). Purification of carboxy-peptidase B from human pancreas. Biochem J 163:253–260.

Mathews KP, Pan PM, Gardner NJ, Hugli TE. (1980). Familial carboxypeptidase N deficiency. Ann Intern Med 93:443–445.

Mathews KP, Curd JG, Hugli TE. (1986). Decreased synthesis of serum carboxypeptidase N (SCPN) in familial SCPN deficiency. J Clin Immunol 6:87–91.

Matsas R, Fulcher IS, Kenny AJ, Turner AJ. (1983). Substance P and (Leu) enkephalin are hydrolyzed by an enzyme in pig caudate synaptic membranes that is identical with the endopeptidase of kidney microvilli. Proc Natl Acad Sci (USA) 80:3111–3115.

Matsukado K, Inamura T, Nakano S, Fukui M, Bartus RT, Black KL. (1996). Enhanced tumor uptake of carboplatin and survival in glioma-bearing rats by intracarotid infusion of bradykinin analog, RMP-7. Neurosurgery 39:125–133.

McDermott JR, Gibson AM, Turner JD. (1987). Involvement of endopeptidase 24.15 in the inactivation of bradykinin by rat brain slices. Biochem Biophys Res Commun 146:154–158.

McDonald JK, Barrett AJ. (1986). Mammalian proteases: a glossary and bibliography. Exo-peptidases. Vol 2. London: Academic Press: 1–357.

McGwire GB, Skidgel RA. (1995). Extracellular conversion of epidermal growth factor (EGF) to des-Arg[53]-EGF by carboxypeptidase M. J Biol Chem 270:17154–17158.

McGwire GB, Tan F, Michel B, Rehli M, Skidgel RA. (1997). Identification of a membrane-bound carboxypeptidase as the mammalian homolog of duck gp180, a hepatitis B-virus binding protein. Life Sci 60:715–724.

Mendelsohn FAO, Lloyd CJ, Kachel C, Funder JW. (1982). Induction by glucocorticoids of angiotensin-converting enzyme production from bovine endothelial cells in culture and rat lung in vivo. J Clin Invest 70:684–692.

Menke JG, Borkowski JA, Bierilo KK, MacNeil T, Derrick AW, Schneck KA, Ransom RW, Strader CD, Linemeyer DL, Hess JF. (1994). Expression cloning of a human B₁ bradykinin receptor. J Biol Chem 269:21583–21586.

Miles LA, Dahlberg CM, Plescia J, Felez J, Kato K, Plow EF. (1991). Role of cell-surface lysines in plasminogen binding to cells: Identification of α-enolase as a candidate plasminogen receptor. Biochemistry 30:1682–1691.

Miller JJ, Changaris DG, Levy RS. (1991). Angiotensin carboxypeptidase activity in urine from normal subjects and patients with kidney damage. Life Sci 48:1529–1535.

Millican PE, Kenny AJ, Turner AJ. (1991). Purification and properties of a neurotensin-degrading endopeptidase from pig brain. Biochem J 276:583–591.

Minshall RD, Nakamura F, Becker RP, Rabito SF. (1995a). Characterization of bradykinin B2 receptors in adult myocardium and neonatal rat cardiomyocytes. Circ Res 76:773–780.

Minshall RD, Vogel SM, Miletich DJ, Erdös EG. (1995b). Potentiation of the inotropic actions of bradykinin on the isolated guinea pig left atria by the angiotensin-converting enzyme/ kininase II inhibitor, enalaprilat. Circulation 92(suppl):I-221.

Minshall RD, Erdös EG, Vogel SM. (1997a). Angiotensin I converting enzyme inhibitors potentiate bradykinin's inotropic effects independent of blocking its inactivation. Am J Cardiol 80(3A):132A–136A.

Minshall RD, Tan F, Nakamura F, Rabito SF, Becker RP, Marcic B, Erdös EG. (1997b). Potentiation of the actions of bradykinin by angiotensin I converting enzyme (ACE) inhibitors. The role of expressed human bradykinin B_2 receptors and ACE in CHO cells. Circul Res. In press.

Morel DR, Zapol WM, Thomas SJ, Kitain EM, Robinson DR, Moss J, Chenoweth DE, Lowenstein E. (1987). C5a and thromboxane generation associated with pulmonary vaso- and broncho-constriction during protamine reversal of heparin. Anesthesiology 66:597–604.

Morice AH, Brown MJ, Lowry R, Higenbottam T. (1987). Angiotensin-converting enzyme and the cough reflex. Lancet 2:1116–1118.

Morrell NW, Atochina EN, Morris KG, Danilov SM, Stenmark KR. (1995). Angiotensin-converting enzyme expression is increased in small pulmonary arteries of rats with hypoxia-induced pulmonary hypertension. J Clin Invest 96:1823–1833.

Munoz CM, Leeb-Lundberg LMF. (1992). Receptor-mediated internalization of bradykinin. DDT_1, MF-2 smooth muscle cells process internalized bradykinin via multiple degradative pathways. J Biol Chem 267:303–309.

Nadel JA. (1992). Membrane-bound peptidases: endocrine, paracrine, and autocrine effects. Am J Respir Cell Mol Biol 7:469–470.

Nadel JA. (1994). Modulation of neurogenic inflammation by peptidases. In: Kaliner MA, Barnes PJ, Kunkel GHH, Baraniuk JN, eds. Neuropeptides in Respiratory Medicine. New York: Marcel Dekker: 351–371.

Nagae A, Deddish PA, Becker RP, Anderson CH, Abe M, Skidgel RA, Erdös EG. (1992). Carboxypeptidase M in brain and peripheral nerves. J Neurochem 59:2201–2212.

Nagae A, Abe M, Becker RP, Deddish PA, Skidgel RA, Erdös EG. (1993). High concentration of carboxypeptidase M in lungs: presence of the enzyme in alveolar type I cells. Am J Respir Cell Mol Biol 9:221–229.

Nishimura K, Hiwada K, Ueda E, Kokubu T. (1976). Solubilization of angiotensin-converting enzyme from rabbit lung using trypsin treatment. Biochim Biophys Acta 452:144,150.

Nomura T, Inamura T, Black KL. (1994). Intracarotid infusion of bradykinin selectively increases blood-tumor permeability in 9L and C6 brain tumors. Brain Res 659:62–66.

Nortier J, Abramowicz D, Najdovski T, Kinnaert P, Vanherweghem JL, Goldman M, Deschodt-Lanckman M. (1993). Urinary endopeptidase 24.11 as a new marker of proximal tubular injury. In: Bianchi C, Bocci V, Carone FA, Rabkin R, eds. Kidney, Proteins and Drugs: An Update. Contrib. Nephrol. Vol. 101. Basel: Karger: 169–176.

Odya CE, Marinkovic D, Hammon KJ, Stewart TA, Erdös EG. (1978a). Purification and properties of prolylcarboxypeptidase (angiotensinase C) from human kidney. J Biol Chem 253:5927–5931.

Odya CE, Levin Y, Erdös EG, Robinson CJG. (1978b). Soluble dextran complexes of kallikrein, bradykinin and enzyme inhibitors. Biochem Pharmacol 27:173–179.

Odya CE, Wilgis FP, Walker JF, Oparil S. (1983). Immunoreactive bradykinin and [des-Arg[9]]-bradykinin in low-renin essential hypertension—before and after treatment with enalapril (MK421). J Lab Clin Med 102:714–721.

Ohkubo K, Baraniuk JN, Merida M, Hausfeld JM, Okada H, Kaliner MA. (1995). Human nasal mucosal carboxypeptidase—activity, location, and release. J Allergy Clin Immunol 96: 924–931.

Ondetti MA. (1994). From peptides to peptidases: a chronicle of drug discovery. Annu Rev Pharmacol 34:1–16.

Oppong SY, Hooper NM. (1993). Characterization of a secretase activity which releases angiotensin-converting enzyme from the membrane. Biochem J 292:597–603.

Orawski AT, Simmons WH. (1995). Purification and properties of membrane-bound aminopeptidase P from rat lung. Biochemistry 34:11227–11236.

Orlowski M, Michaud C, Chu TG. (1983). A soluble metalloendopeptidase from rat brain. Purification of the enzyme and determination of specificity with synthetic and natural peptides. Eur J Biochem 135:81–88.

Orlowski M, Michaud C, Molineaux CJ. (1988). Substrate-related potent inhibitors of brain metalloendopeptidase. Biochemistry 27:596–602.

Orlowski M, Reznik S, Ayala J, Pierotti AR. (1989). Endopeptidase 24.15 from rat testes. Isolation of the enzyme and its specificity toward synthetic and natural peptides, including enkephalin-containing peptides. Biochem J 261:951–958.

Oshima E, Erdös EG. (1974). Inhibition of the angiotensin I converting enzyme of the lung by a peptide fragment of bradykinin. Experientia 30:733.

Oshima G, Kato J, Erdös EG. (1975). Plasma carboxypeptidase N, subunits and characteristics. Arch Biochem Biophys 170:132–138.

Oshima G, Hiraga Y, Shirono K, Oh-ishi S, Sakakibara S, Kinoshita T. (1985). Cleavage of des-Arg⁹-bradykinin by angiotensin I-converting enzyme from pig kidney cortex. Experientia 41:325–328.

Painter RG, Dukes R, Sullivan J, Carter R, Erdös EG, Johnson AR. (1988). Function of neutral endopeptidase on the cell membrane of human neutrophils. J Biol Chem 263:9456–9461.

Parratt JR. (1994). Cardioprotection by angiotensin-converting enzyme inhibitors—the experimental evidence. Cardiovasc Res 28:183–189.

Patchett AA, Cordes EH. (1985). The design and properties of N-carboxyalkyldipeptide inhibitors of angiotensin converting enzyme. In: Meister A, ed. Advances in Enzymology and Related Areas of Molecular Biology. Vol. 57:1–84.

Patel KV, Schrey MP. (1992). Inhibition of DNA synthesis and growth in human breast stromal cells by bradykinin: evidence for independent roles of B_1 and B_2 receptors in the respective control of cell growth and phospholipid hydrolysis. Cancer Res 52:334–340.

Perich RB, Jackson B, Attwood MR, Prior K, Johnston CI. (1991). Angiotensin-converting enzyme inhibitors act at two different binding sites on angiotensin-converting enzyme. Pharm Pharmacol Lett 1:41–43.

Perich RB, Jackson B, Johnston CI. (1994). Structural constraints of inhibitors for binding at two active sites on somatic angiotensin converting enzyme. Eur J Pharmacol 266: 201–211.

Petersson G, Bacci E, McDonald DM, Nadel JA. (1993). Neurogenic plasma extravasation in the rat mucosa is potentiated by peptidase inhibitors. J Pharmacol Exp Ther 264: 509–514.

Pfeffer MA. (1993). Angiotensin-converting enzyme inhibition in congestive heart failure: benefit and perspective. Am Heart J 126:789–793.

Pfeffer MA. (1995). ACE inhibition in acute myocardial infarction. N Engl J Med 332: 118–120.

Piedimonte G, Nadel JA, Long CS, Hoffman JIE. (1994). Neutral endopeptidase in the heart. Neutral endopeptidase inhibition prevents isoproterenol-induced myocardial hypoperfusion in rats by reducing bradykinin degradation. Circ Res 75:770–779.

Pierotti A, Dong K-W, Glucksman MJ, Orlowski M, Roberts JL. (1990). Molecular cloning and primary structure of rat testes metalloendopeptidase EC 3.4.24.15. Biochemistry 29:10323–10329.

Plummer TH Jr, Hurwitz MY. (1978). Human plasma carboxypeptidase N. Isolation and characterization. J Biol Chem 253:3907–3912.

Plummer TH Jr, Ryan TJ. (1981). A potent mercapto bi-product analogue inhibitor for human carboxypeptidase N. Biochem Biophys Res Commun 98:448–454.

Polgar L. (1992). Structural relationship between lipases and peptidases of the prolyl oligopeptidase family. FEBS Lett 311:281–284.

Potier M, Michaud L, Tranchemontagne J, Thauvette L. (1990). Structure of the lysosomal neuraminidase–β-galactosidase–carboxypeptidase multienzymic complex. Biochem J 267:197–202.

Prechel MM, Orawski AT, Maggiora LL, Simmons WH. (1995). Effect of a new aminopeptidase P inhibitor, apstatin, on bradykinin degradation in the rat lung. J Pharmacol Exp Ther 275:1136–1142.

Rabito SF, Anders R, Soden W, Skidgel RA. (1992). Carboxypeptidase N concentration during cardiopulmonary bypass in humans. Can J Anaesth 39:54–59.

Ramchandran R, Sen GC, Misono K, Sen I. (1994). Regulated cleavage-secretion of the membrane-bound angiotensin converting enzyme. J Biol Chem 269:2125–2130.

Rawlings ND, Polgar L, Barrett AJ. (1991). A new family of serine-type peptidases related to prolyl oligopeptidase. Biochem J 279:907–908.

Redlitz A, Tan AK, Eaton DL, Plow EF. (1995). Plasma carboxypeptidases as regulators of the plasminogen system. J Clin Invest 96:2534–2538.

Redlitz A, Nicolini FA, Malycky JL, Topol EJ, Plow EF. (1996). Inducible carboxypeptidase activity. A role in clot lysis in vivo. Circulation 93:1328–1330.

Regoli D, Barabe J. (1980). Pharmacology of bradykinin and related kinins. Pharmacol Rev 32:1–46.

Rehli M, Krause SW, Kreutz M, Andreesen R. (1995). Carboxypeptidase M is identical to the MAX.1 antigen and its expression is associated with monocyte to macrophage differentiation. J Biol Chem 270:15644–15649.

Reiners C, Gramer-Kurz E, Pickert E, Schweisfurth H. (1988). Changes of serum angiotensin I-converting enzyme in patients with thyroid disorders. Clin Physiol Biochem 6:44–49.

Rennex D, Hemmings BA, Hofsteenge J, Stone SR. (1991). cDNA cloning of porcine brain prolyl endopeptidase and identification of the active-site seryl residue. Biochemistry 30:2195–2203.

Rix E, Ganten D, Schüll B, Unger T, Taugner R. (1981). Converting-enzyme in the choroid plexus, brain, and kidney: immunocytochemical and biochemical studies in rats. Neurosci Lett 22:125–130.

Rocha e Silva M. (1963). The physiological significance of bradykinin. Ann NY Acad Sci 104:190–211.

Rodd D, Hersh LB. (1995). Endopeptidase 24.16B. A new variant of endopeptidase 24.16. J Biol Chem 270:10056–10061.

Romer FK, Emmertsen K. (1980). Serum angiotensin converting enzyme in malignant lymphomas, leukaemia and multiple myeloma. Br J Cancer 42:314–318.

Ronco P, Pollard H, Galceran M, Delauche M, Schwartz JC, Verroust P. (1988). Distribution of enkephalinase (membrane metalloendopeptidase, E.C. 3.4.24.11) in rat organs. Detection using a monoclonal antibody. Lab Invest 58:210–217.

Rosenberg E, Kiron MAR, Soffer RL. (1988). A soluble angiotensin II-binding protein from rabbit liver. Biochem Biophys Res Commun 151:466–472.

Roy P, Chatellard C, Lemay G, Crine P, Boileau G. (1993). Transformation of the signal peptide membrane anchor domain of a type II transmembrane protein into a cleavable signal peptide. J Biol Chem 268:2699–2704.

Rumble JR, Komers R, Cooper ME. (1996). Kinins or nitric oxide, or both, are involved in the antitrophic effects of angiotensin converting enzyme inhibitors on diabetes-associated mesenteric vascular hypertrophy in the rat. Hypertens 14:601–607.

Ryan JW. (1989). Peptidase enzymes of the pulmonary vascular surface. Am J Physiol 257: L53–L60.

Ryan JW, Roblero J, Stewart JM. (1968). Inactivation of bradykinin in the pulmonary circulation. Biochem J 110:795–797.

Ryan JW, Berryer P, Chung AY, Sheffy DH. (1994). Characterization of rat pulmonary vascular aminopeptidase P in vivo: role in the inactivation of bradykinin. J Pharmacol Exp Ther 269:941–947.

Ryan US, Ryan JW. (1983). Endothelial cells and inflammation. In: Ward PA, ed. Clinics in Laboratory Medicine. Vol. 3. Philadelphia: Saunders: 577–599.

Saavedra JM, Chevillard C. (1982). Angiotensin-converting enzyme is present in the subfornical organ and other circumventricular organs of the rat. Neurosci Lett 29: 123–127.

Sanovich E, Bartus RT, Friden PM, Dean RL, Le HQ, Brightman MW. (1995). Pathway across blood-brain barrier opened by the bradykinin agonist, RMP-7. Brain Res 705:125–135.

Schilero GJ, Almenoff P, Cardozo C, Lesser M. (1994). Effects of peptidase inhibitors on bradykinin-induced bronchoconstriction in the rat. Peptides 15:1445–1449.

Schriefer JA, Molineaux CJ. (1993). Modulatory effect of endopeptidase inhibitors on bradykinin-induced contraction of rat uterus. J Pharmacol Exp Ther 266:700–706.

Schulz WW, Hagler HK, Buja LM, Erdös EG. (1988). Ultrastructural localization of angiotensin I converting enzyme (EC 3.4.15.1) and neutral metallo-endopeptidase (EC 3.4.24.11) in the proximal tubule of the human kidney. Lab Invest 59:789–797.

Schwartz JC, Costentin J, Lecomte JM. (1985). Pharmacology of enkephalinase inhibitors. TiPS 6:472–485.

Schweisfurth H, Heinrich J, Brugger E, Steinl C, Maiwald L. (1985a). The value of angiotensin-converting enzyme determinations in malignant and other diseases. Clin Physiol Biochem 3:184–192.

Schweisfurth H, Schmidt M, Brugger E, Maiwald L, Thiel H. (1985b). Alterations of serum carboxypeptidases N and angiotensin I-converting enzyme in malignant diseases. Clin Biochem 18:242–246.

Scicli AG. (1994). Increases in cardiac kinins as a new mechanism to protect the heart. Hypertension 23:419–421.

Scicli AG, Gandolfi R, Carretero OA. (1978). Site of formation of kinins in the dog nephron. Am J Physiol 234:F36–F40.

Semple PF. (1995). Putative mechanisms of cough after treatment with angiotensin converting enzyme inhibitors. J Hypertens 13(suppl 3):S17–S21.

Serizawa A, Dando PM, Barrett AJ. (1995). Characterization of a mitochondrial metallopeptidase reveals neurolysin as a homologue of thimet oligopeptidase. J Biol Chem 270: 2092–2098.

Seymour AA, Asaad MM, Lanoce VM, Langenbacher KM, Fennell SA, Rogers WL. (1993). Systematic hemodynamics, renal function and hormonal levels during inhibition of neutral endopeptidase 3.4.24.11 and angiotensin-converting enzyme in conscious dogs with pacing-induced heart failure. J Pharmacol Exp Ther 266:872–883.

Seymour AA, Sheldon JH, Smith PL, Asaad M, Rogers WL. (1994). Potentiation of the renal responses to bradykinin by inhibition of neutral endopeptidase 3.4.24.11 and angiotensin-converting enzyme in anesthetized dogs. J Pharmacol Exp Ther 269:263–270.

Shaw LA, Rudin M, Cook NS. (1995). Pharmacological inhibition of restenosis: learning from experience. TiPS 16:401–404.

Sheikh IA, Kaplan AP. (1989). Mechanism of digestion of bradykinin and lysylbradykinin (kallidin) in human serum. Role of carboxypeptidase, angiotensin-converting enzyme and determination of final degradation products. Biochem Pharmacol 38:993–1000.

Shinohara T, Sakurada C, Suzuki T, Takeuchi O, Campbell W, Ikeda S, Okada N, Okada H. (1994). Pro-carboxypeptidase R cleaves bradykinin following activation. Int Arch Allergy Immunol 103:400–404.

Sidorowicz W, Canizaro PC, Behal FJ. (1984a). Kinin cleavage by human erythrocytes. Am J Hematol 17:383–391.

Sidorowicz W, Szechinski J, Canizaro PC, Behal FJ. (1984b). Cleavage of the Arg^1-Pro^2 bond of bradykinin by a human lung peptidase: isolation, characterization and inhibition by several β-lactam antibiotics. Proc Soc Exp Biol Med 175:503–509.

Silverstein E, Friedland J, Lyons HA, Gourin A. (1976). Markedly elevated angiotensin converting enzyme in lymph nodes containing non-necrotizing granulomas in sarcoidosis. Proc Natl Acad Sci (USA) 73:2137–2141.

Silverstein E, Friedland J, Vuletin JC. (1978). Marked elevation of serum angiotensin-converting enzyme and hepatic fibrosis containing long-spacing collagen fibrils in type 2 acute neuronopathic Gaucher's disease. Am J Clin Pathol 69:467–470.

Silverstein E, Pertschuk LP, Friedland J. (1979). Immunofluorescent localization of angiotensin converting enzyme in epithelioid and giant cells of sarcoidosis granulomas. Proc Natl Acad Sci (USA) 76:6646–6648.

Simmons WH, Orawski AT. (1992). Membrane-bound aminopeptidase P from bovine lung. Its purification, properties, and degradation of bradykinin. J Biol Chem 267:4897–4903.

Skidgel RA. (1988). Basic carboxypeptidases: regulators of peptide hormone activity. TiPS 9:299–304.

Skidgel RA. (1991). Assays for arginine/lysine carboxypeptidases: carboxypeptidase H (E; enkephalin convertase), M and N. In: Conn PM, ed. Methods in Neurosciences: Peptide Technology. Vol. 6. Orlando, FL: Academic Press: 373–385.

Skidgel RA. (1992). Bradykinin-degrading enzymes: structure, function, distribution, and potential roles in cardiovascular pharmacology. J Cardiovasc Pharmacol 20:S4–S9.

Skidgel RA. (1993). Basic science aspects of angiotensin converting enzyme and its inhibitors. In: Gwathmey JK, Briggs GM, Allen PD, eds. Heart Failure: Basic Science and Clinical Aspects. New York: Marcel Dekker: 399–427.

Skidgel RA. (1994). Pulmonary peptidases: general principles of peptide metabolism and molecular biology of angiotensin converting enzyme, neutral endopeptidase 24.11, and carboxypeptidase M. In: Kaliner MA, Barnes PJ, Kunkel GHH, Baraniuk JN, eds. Neuropeptides in Respiratory Medicine. New York: Marcel Dekker: 301–312.

Skidgel RA. (1995). Human carboxypeptidase N (lysine carboxypeptidase). In: Barrett AJ, ed. Methods in Enzymology Vol. 248: Proteolytic Enzymes, part E; Aspartic, Metallo and Other Peptidases. Orlando, FL: Academic Press: 653–663.

Skidgel, RA. (1996). Structure and function of mammalian zinc carboxypeptidases. In: Hooper NM, ed. Zinc Metalloproteases in Health and Disease. London: Taylor & Francis:241–283.

Skidgel RA, Erdös EG. (1985). Novel activity of human angiotensin I converting enzyme: release of the NH_2- and COOH-terminal tripeptides from the luteinizing hormone-releasing hormone. Proc Natl Acad Sci (USA) 82:1025–1029.

Skidgel RA, Erdös EG. (1993). Biochemistry of angiotensin converting enzyme. In: Robertson JIS, Nicholls MG, eds. The Renin Angiotensin System. Vol. 1. London: Gower Medical: 10.1–10.10.

Skidgel RA, Tan F. (1992). Structural features of two kininase I-type enzymes revealed by molecular cloning. Agents and Actions (Suppl) 38/I:359–367.

Skidgel RA, Wickstrom E, Kumamoto K, Erdös EG. (1981). Rapid radioassay for prolylcarboxypeptidase (angiotensinase C). Anal Biochem 118:113–119.

Skidgel RA, Davis RM, Erdös EG. (1984a). Purification of a human urinary carboxypeptidase (kininase) distinct from carboxypeptidase A, B, or N. Anal Biochem 140:520–531.

Skidgel RA, Johnson AR, Erdös EG. (1984b). Hydrolysis of opioid hexapeptides by carboxypeptidase N. Presence of carboxypeptidase in cell membranes. Biochem Pharmacol 33: 3471–3478.

Skidgel RA, Kawahara MS, Hugli TE. (1986). Functional significance of the subunits of carboxypeptidase N (kininase I). Adv Exp Med Biol 198A:375–380.

Skidgel RA, Defendini R, Erdös EG. (1987a). Angiotensin I converting enzyme and its role in neuropeptide metabolism. In: Turner AJ, ed. Neuropeptides and Their Peptidases. Chichester, UK: Ellis-Horwood: 165–182.

Skidgel RA, Schulz WW, Tam LT, Erdös (1987b). Human renal angiotensin I converting enzyme and neutral endopeptidase. Kidney Int 31:S45–S48.

Skidgel RA, Deddish PA, Davis RM. (1988). Isolation and characterization of a basic carboxypeptidase from human seminal plasma. Arch Biochem Biophys 267:660–667.

Skidgel RA, Davis RM, Tan F. (1989). Human carboxypeptidase M: purification and characterization of a membrane-bound carboxypeptidase that cleaves peptide hormones. J Biol Chem 264:2236–2241.

Skidgel RA, Anders RA, Deddish PA, Erdös EG. (1991a). Carboxypeptidase (CP) M and H in small intestine. FASEB J 5:A1578.

Skidgel RA, Jackman HL, Erdös EG. (1991b). Metabolism of substance P and bradykinin by human neutrophils. Biochem Pharmacol 41:1335–1344.

Slater EE, Merrill DD, Guess HA, Roylance PJ, Cooper WD, Inman WHW, Ewan PW. (1988). Clinical profile of angioedema associated with angiotensin-converting enzyme inhibition. JAMA 260:967–970.

Smits GJ, McGraw DE, Trapani AJ. (1990). Interaction of ANP and bradykinin during endopeptidase 24.11 inhibition: renal effects. Am J Physiol 258:F1417–F1424.

Song L, Fricker LD. (1995). Purification and characterization of carboxypeptidase D, a novel carboxypeptidase E-like enzyme, from bovine pituitary. J Biol Chem 270:25007–25013.

Sorrells K, Erdös EG. (1972). Prolylcarboxypeptidase in biological fluids. In: Hinshaw LB, Cox BG, eds. The Fundamental Mechanisms of Shock. New York: Plenum: 393–397.

Soubrier F, Alhenc-Gelas F, Hubert C, Allegrini J, John M, Gregear G, Corvol P. (1988). Two putative active centers in human angiotensin I-converting enzyme revealed by molecular cloning. Proc Natl Acad Sci (USA) 85:9386–9390.

Soubrier F, Hubert C, Testut P, Nadaud S, Alhenc-Gelas F, Corvol P. (1993a). Molecular biology of the angiotensin I converting enzyme: I. Biochemistry and structure of the gene. J Hypertens 11:471–476.

Soubrier F, Wei L, Hubert C, Clauser E, Alhenc-Gelas F, Corvol P. (1993b). Molecular biology of the angiotensin I converting enzyme: II. Structure-function. Gene polymorphism and clinical implications. J Hypertens 11:599–604.

Spillantini MG, Sicuteri F, Salmon S, Malfroy B. (1990). Characterization of endopeptidase 3.4.24.11 (enkephalinase) activity in human plasma and cerebrospinal fluid. Biochem Pharmacol 39:1353–1356.

Stewart TA, Weare JA, Erdös EG. (1981). Purification and characterization of human converting enzyme (kininase II). Peptides 2:145–152.

Straub JA, Akiyama A, Parmar P. (1994). In vitro plasma metabolism of RMP-7. Pharmaceut Res 11:1673–1676.

Sugiura N, Hagiwara H, Hirose S. (1992). Molecular cloning of porcine soluble angiotensin-binding protein. J Biol Chem 267:18067–18072.

Sunman W, Sever PS. (1993). Non-angiotensin effects of angiotensin converting enzyme inhibitors. Clin Sci 85:661–670.

Swan MG, Vergalla J, Jones EA. (1993). Plasma endopeptidase 24.11 (enkephalinase) activity is markedly increased in cholestatic liver disease. Hepatology 18:556–558.

Swartz SL, Williams GH, Hollenberg NK, Levine L, Dluhy RG, Moore TJ. (1980). Captopril-induced changes in prostaglandin production. Relationship to vascular responses in normal man. J Clin Invest 65:1257–1264.

Tamaoki J, Kobayashi K, Sakai N, Chiyotani A, Kanemura T, Takizawa T. (1989). Effect of bradykinin on airway ciliary motility and its modulation by neutral endopeptidase. Am Rev Respir Dis 140:430–435.

Tan AK, Eaton DL. (1995). Activation and characterization of procarboxypeptidase B from human plasma. Biochemistry 34:5811–5816.

Tan F, Chan SJ, Steiner DF, Schilling JW, Skidgel RA. (1989a). Molecular cloning and sequencing of the cDNA for human membrane-bound carboxypeptidase M: comparison with carboxypeptidases A, B, H and N. J Biol Chem 264:13165–13170.

Tan F, Jackman H, Skidgel RA, Zsigmond EK, Erdös EG. (1989b). Protamine inhibits plasma carboxypeptidase N, the inactivator of anaphylatoxins and kinins. Anesthesiology 70: 267–275.

Tan F, Weerasinghe DK, Skidgel RA, Tamei H, Kaul RK, Roninson IB, Schilling JW, Erdös EG. (1990). The deduced protein sequence of the human carboxypeptidase N high molecular weight subunit reveals the presence of leucine-rich tandem repeats. J Biol Chem 265:13–19.

Tan F, Morris PW, Skidgel RA, Erdös EG. (1993). Sequencing and cloning of human prolylcarboxypeptidase (angiotensinase C): similarity to both serine carboxypeptidase and pro-lylendopeptidase families. J Biol Chem 268:16631–16638.

Tan F, Deddish PA, Skidgel RA. (1995). Human carboxypeptidase M. In: Barrett AJ, ed. Methods in Enzymology, Vol. 248: Proteolytic Enzymes, part E; Aspartic, Metallo and Other Peptidases. Orlando, FL: Academic Press: 663–675.

Tan F, Rehli M, Krause SW, Skidgel RA. (1997). The sequence of human carboxypeptidase D reveals it to be a member of the regulatory carboxypeptidase family with three tandem active site domains. Biochem J. In press.

Tatei K, Cai H, Ip YT, Levine M. (1995). Race: a *Drosophila* homologue of the angiotensin-converting enzyme. Mech Dev 51:157–168.

Telford SE, Smith AI, Lew RA, Perich RB, Madden AC, Evans RG. (1995). Role of angiotensin converting enzyme in the vascular effects of an endopeptidase 24.15 inhibitor. Br J Pharmacol 114:1185–1192.

Tiffany CW, Burch RM. (1989). Bradykinin stimulates tumor necrosis factor and interleukin-1 release from macrophages. FEBS Lett 247:189–192.

Tisljar U. (1993). Thimet oligopeptidase—a review of a thiol dependent metallo-endopeptidase also known as Pz-peptidase endopeptidase 24.15 and endo-oligopeptidase. Biol Chem Hoppe-Seyler 374:91–100.

Tisljar U, Barrett AJ. (1990). Thiol-dependent metalloendopeptidase characteristics of Pz-peptidase in rat and rabbit. Biochem J 267:531–533.

Toide K, Iwamoto Y, Fujiwara T, Abe H. (1995). JTP-4819: a novel prolyl endopeptidase inhibitor with potential as a cognitive enhancer. J Pharmacol Exp Ther 274:1370–1378.

Trippodo NC, Robl JA, Asaad MM, Bird JE, Panchal BC, Schaeffer TR, Fox M, Giancarli MR, Cheung HS. (1995). Cardiovascular effects of the novel dual inhibitor of neutral endopeptidase and angiotensin-converting enzyme BMS-182657 in experimental hypertension and heart failure. J Pharmacol Exp Ther 275:745–752.

Turner AJ. (1987). Endopeptidase-24.11 and neuropeptide metabolism. In: Turner AJ, ed. Neuropeptides and Their Peptidases. Chichester, UK: Ellis-Horwood: 183–201.

Ura N, Carretero OA, Erdös EG. (1987). Role of renal endopeptidase 24.11 in kinin metabolism in vitro and in vivo. Kidney Int 32:507–513.

van der Horst GTJ, Galjart NJ, d'Azzo A, Galjaard H, Verheijen FW. (1989). Identification and *in vitro* reconstitution of lysosomal neuraminidase from human placenta. J Biol Chem 264:1317–1322.

Vane JR. (1969). The release and fate of vaso-active hormones in the circulation. Br J Pharmacol 35:209–242.

Vanhoof G, De Meester I, Goossens F, Hendriks D, Scharpe S, Yaron A. (1992). Kininase activity in human platelets: cleavage of the Arg^1-Pro^2 bond of bradykinin by aminopeptidase P. Biochem Pharmacol 44:479–487.

Vergas Romero C, Neudorfer I, Mann K, Schäfer W. (1995). Purification and amino acid sequence of aminopeptidase P from pig kidney. Eur J Biochem 229:262–269.

Vijayaraghavan J, Scicli AG, Carretero OA, Slaughter C, Moomaw C, Hersh LB. (1990). The hydrolysis of endothelins by neutral endopeptidase 24.11 (enkephalinase). J Biol Chem 265:14150–14155.

von Euler US. (1963). Substance P in subcellular particles in peripheral nerves. Ann NY Acad Sci 104:449–463.

Wang W, Hendriks DF, Scharpé SS. (1994). Carboxypeptidase U, a plasma carboxypeptidase with high affinity for plasminogen. J Biol Chem 269:15937–15944.

Ward PE, Gedney CD, Dowben RM, Erdös EG. (1975). Isolation of membrane-bound renal kallikrein and kininase. Biochem J 151:755–758.

Ward PE, Erdös EG, Gedney CD, Dowben RM, Reynolds RC. (1976). Isolation of membrane-bound renal enzymes that metabolize kinins and angiotensins. Biochem J 157:643–650.

Ward PE, Schulz W, Reynolds RC, Erdös EG. (1977). Metabolism of kinins and angiotensins in the isolated glomerulus and brush border of rat kidney. Lab Invest 36:599–606.

Ward PE, Bausback HH, Odya CE. (1987). Kinin and angiotensin metabolism by purified renal post-proline cleaving enzyme. Biochem Pharmacol 36:3187–3193.

Warren JB, Loi RK. (1995). Captopril increases skin microvascular blood flow secondary to bradykinin, nitric oxide, and prostaglandins. FASEB J 9:411–418.

Weare JA. (1982). Activation/inactivation of human angiotensin I converting enzyme following chemical modifications of amino groups near the active site. Biochem Biophys Res Commun 104:1319–1326.

Weare JA, Gafford JT, Lu HS, Erdös EG. (1982). Purification of human kidney angiotensin I converting enzyme using reverse immunoadsorption chromatography. Anal Biochem 123:310–319.

Webster ME, Pierce JV. (1963). The nature of the kallidins released from human plasma by kallikreins and other enzymes. Ann NY Acad Sci 104:91–107.

Wei L, Alhenc-Gelas F, Corvol P, Clauser E. (1991a). The two homologous domains of human angiotensin I-converting enzyme are both catalytically active. J Biol Chem 266:9002–9008.

Wei L, Alhenc-Gelas F, Soubrier F, Michaud A, Corvol P, Clauser E. (1991b). Expression and characterization of recombinant human angiotensin I-converting enzyme. Evidence for a C-terminal transmembrane anchor and for a proteolytic processing of the secreted recombinant and plasma enzymes. J Biol Chem 266:5540–5546.

Wei L, Clauser E, Alhenc-Gelas F, Corvol P. (1992). The two monological domains of human angiotensin I-converting enzyme interact differently with competitive inhibitors. J Biol Chem 267:13398–13405.

Welches WR, Santos RAS, Chappell MC, Brosnihan KB, Greene LJ, Ferrario CM. (1991). Evidence that prolyl endopeptidase participates in the processing of brain angiotensin. J Hypertens 9:631–638.

Werb Z, Clark EJ. (1989). Phorbol diesters regulate expression of the membrane neutral metalloendopeptidase (EC 3.4.24.11) in rabbit synovial fibroblasts and mammary epithelial cells. J Biol Chem 264:9111–9113.

Wevers RA, Boegheim JPJ, Hommes OR, van Landeghem AAJ, Mul-Steinbusch MWFJ, van der Stappen JWJ, Soons JBJ. (1984). A study on post-synthetic modifications in alfa-alfa enolase (EC 4.2.1.11) brought about by a human serum protein. Clin Chim Acta 139:127–135.

Wiemer G, Wirth K. (1992). Production of cyclic GMP via activation of B_1 and B_2 kinin receptors in cultured bovine aortic endothelial cells. J Pharmacol Exp Ther 262:729–733.

Wiemer G, Schölkens BA, Becker RHA, Busse R. (1991). Ramiprilat enhances endothelial autocoid formation by inhibiting breakdown of endothelium-derived bradykinin. Hypertension 18:558–563.

Wilk S. (1983). Minireview: prolyl endopeptidase. Life Sci 33:2149–2157.

Wilk S, Orlowski M. (1983). Inhibition of rabbit brain prolyl endopeptidase by N-benzyloxy-carbonyl-prolyl-prolinal, a transition state aldehyde inhibitor. J Neurochem 41:69–75.

Wolz RL, Bond JS. (1990). Phe[5](4-nitro)-bradykinin: a chromogenic substrate for assay and kinetics of the metalloendopeptidase meprin. Anal Biochem 191:314–320.

Wolz RL, Bond JS. (1995). Meprins A and B. Meth Enzymol 248:325–345.

Wolz RL, Harris RB, Bond JS. (1991). Mapping the active site of meprin-A with peptide substrates and inhibitors. Biochemistry 30:8488–8493.

Wyvratt MJ, Patchett AA. (1985). Recent developments in the design of angiotensin-converting enzyme inhibitors. Med Res Rev 5:483–531.

Xu Y, Wellner D, Scheinberg DA. (1995). Substance P and bradykinin are natural inhibitors of CD13/aminopeptidase N. Biochem Biophys Res Commun 208:664–674.

Yamada H, Fabris B, Allen AW, Jackson B, Johnston CI, Mendelsohn FAO. (1991). Localization of angiotensin converting enzyme in rat heart. Circ Res 68:141–149.

Yamashiro DJ, Maxfield FR. (1988). Regulation of endocytic process by pH. TiPS 9:190–194.

Yamawaki I, Geppetti P, Bertrand C, Chan B, Nadel JA. (1994). Airway vasodilation by bradykinin is mediated via B_2 receptors and modulated by peptidase inhibitors. Am J Physiol 266:L156–L162.

Yang HYT, Erdös EG. (1967). Second kininase in human blood plasma. Nature 215:1402–1403.

Yang HYT, Neff NH. (1972). Distribution and properties of angiotensin converting enzyme of rat brain. J Neurochem 19:2433–2450.

Yang HYT, Neff NH. (1973). Differential distribution of angiotensin converting enzyme in the anterior and posterior lobe of the rat pituitary. J Neurochem 21:1035–1036.

Yang HYT, Erdös EG, Chiang TX. (1968). New enzymatic route for the inactivation of angiotensin. Nature 218:1224–1226.

Yang HYT, Erdös EG, Chiang TS, Jenssen TA, Rodgers JG. (1970a). Characteristics of an enzyme that inactivates angiotensin II (angiotensinase C). Biochem Pharmacol 19:1201–1211.

Yang HYT, Erdös EG, Levin Y. (1970b). A dipeptidyl carboxypeptidase that converts angiotensin I and inactivates bradykinin. Biochim Biophys Acta 214:374–376.

Yang HYT, Erdös EG, Levin Y. (1971). Characterization of a dipeptide hydrolase (kininase II; angiotensin I converting enzyme). J Pharmacol Exp Ther 177:291–300.

Yang X-P, Saitoh S, Scicli AG, Mascha E, Orlowski M, Carretero OA. (1994). Effects of a metalloendopeptidase-24.15 inhibitor on renal hemodynamics and function in rats. Hypertension 23:I-235–I-239.

21

Enzymatic Modulation of Tachykinins and VIP in the Lung

CRAIG M. LILLY

Harvard Medical School
Boston, Massachusetts

STEPHANIE A. SHORE

Harvard School of Public Health
Boston, Massachusetts

I. Introduction

The physiological effects of neuropeptides in the lung are receptor-mediated events that depend on the apposition of ligand and receptor (1). The availability of the ligand, including the amount of release, the relative position of the site of release to the receptor, and the presence of mechanisms that can sequester or inactivate the ligand, is controlled at several levels. It is now known that the physiological activity of vasoactive intestinal peptide (VIP) and the tachykinins depends in part on the activity of enzymes that limit their microenvironmental availability by hydrolytic inactivation (2,3). Enzymes present in the extracellular space can hydrolyze VIP or tachykinins into inactive products. These enzymes have prereceptor access to the ligand and can limit the effective tissue distribution of VIP and the tachykinins. In this way enzymes can function as regulatory elements that modulate peptide tissue effects (4). Advances in our understanding of these mechanisms were initiated by studies that identified enzymes with high affinity for VIP and the tachykinins and identified the inactivating hydrolysis or cleavage sites in the peptide sequence. The pulmonary relevance of these enzymes was supported by their identification in the lung and definition of their anatomic loci in and near the airways. The availability of effective inhibitors of these enzyme systems then allowed investigation of the physiological role of enzymatic inactivation of VIP and the tachykinins. Alterations in the activity of these enzyme systems that occur with airway inflammation are correlated with changes in the sensitivity of the lung to the contractile and relaxant effects of these agents. Extending the results of animal and human tissue experiments to human inflammatory airway disease presents an exciting and challenging opportunity.

II. Tachykinins

We will review the evidence that identifies the types and anatomic locations of tachykinins in the airways, the factors that release these substances, and their effects. The role of enzymatic inactivation will be explored by identifying the enzyme systems that can efficiently hydrolyze the tachykinins and by reviewing the studies that define the physiological role of the relevant systems. Alterations in enzyme activity that occur with airway inflammation will be correlated with the tissue responses to the relevant ligand and followed by studies in the human system.

A. Synthesis, Location, and Release of Tachykinins in the Airways

The tachykinin family of small peptides is distinguished by the common amidated carboxyl-terminal sequence Phe-X-Gly-Leu-Met-NH$_2$. Of the three recognized mammalian tachykinins, only substance P (SP) and neurokinin A (NKA) have been identified in the lungs and airways. Tachykinins are neurotransmitters and neuronally secreted paracrine effectors that are synthesized in nerve cell bodies of the nodose, jugular, and thoracic dorsal root ganglia (5,6), appropriately processed and transported by axoplasmic flow to the terminal ramifications of axon dendrites, where they are released at distinct sites in the pulmonary microenvironment. SP and NKA are peptide products of the preprotachykinin I (PPT-I) gene (7,8). Alternative post-transcriptional processing of the PPT-I gene product results in the production of three distinct mRNA species designated PPT-I α, β, and γ (9). Translation of the α form generates only SP, while translation of the β and γ forms also generates NKA. An amino-terminal truncated form of NKA, designated NKA$_{3-10}$, and the carboxyl-terminal extended forms of NKA, neuropeptide γ, and neuropeptide K can be generated by alternative posttranslational processing of the β and γ forms. NKA$_{3-10}$ and neuropeptide K have been identified in lung tissue of guinea pigs and humans (10,11). In lung tissue SP and NKA immunolocalize to C-fiber afferent neurons that are found in the greatest density beneath and between airway epithelial cells (6). This immunoreactivity also identifies neurons that impinge upon blood vessels, within layers of airway smooth muscle, and at parasympathetic ganglia (11–15). This neuronal distribution implies an effector role in addition to the well-recognized sensory function of this class of neurons. In most species studied, SP is identified in peripheral neuronal processes; however, in the cat it has also been identified in the neurons of airway ganglia (16). SP immunoreactive neurons can be identified in airway neurons from the nose to peripheral bronchioles and have been identified at the alveolar level in some mammalian species (12,17). Immunohistochemical studies demonstrating greater SP immunoreactive neuronal density in the central than in the peripheral airways have been confirmed by measurements of airway SP content (12, 14,18). It is clear from these studies that measurable amounts of SP are present in samples of peripheral human lung. In addition to alterations in SP presence as a function of airway size, age-related differences have also been observed. In an autopsy study of subjects 3 years old or younger, SP airway immunoreactivity was observed at all airway levels, including the alveolar ducts, where immunoreactivity in the neurons subtending the small airways of older subjects was infrequent (15). PPT-1 mRNA has also been demonstrated in rat alveolar macrophages, where its expression is increased following activation of the cells with endotoxin (19).

The C-fiber neurons that contain SP and NKA can be stimulated to release these peptides by exogenous substances such as irritants, by endogenous substances, or by mechanical factors such as lung inflation or tissue edema (20,21). A listing of established endogenous and exogenous substances that can release tachykinins from C fibers is presented in Table 1. In response to such a stimulus the C fiber generates a centrally directed action potential that can result in central nervous system reflex activity. As these action potentials proceed up the axon, they pass the terminal ramifications of axon dendrites, and antidromic conduction can allow activation of additional nerve terminals subtended by the stimulated neuron. The arrival of the action potential at the nerve terminal allows the release of neuropeptides and other neurotransmitters into the microenvironment surrounding the nerve terminal. Since each neuron has multiple ramifications, a stimulus can result in the release of effector molecules at diverse sites by this "axonal reflex" mechanism. Tachykinins can be released from airway neurons by a variety of irritant mediators of airway constriction, which suggests that neuronal reflexes may modulate or influence airway responses.

B. Airway Responses to Tachykinins

Tachykinins act on airway structures to induce bronchoconstriction, increase transepithelial ion and water flux, promote the cough reflex, alter ciliary beat frequency, augment mucus secretion, and increase airway blood flow, vascular permeability, and

Table 1 C-Fiber Stimulants

Endogenous	Exogenous
Histamine	Cigarette smoke
Cholinergic agonists	Capsaicin
Bradykinin	Formaldehyde
Leukotrienes	TDI
Prostaglandins	Acrolein
Platelet-activating factor	SO_2
Arachidonic acid	Ether

the resultant airway edema (22–28). These airway-active, neuronally released substances transduce constriction directly through activation of neurokinin receptors (29) and indirectly through mechanisms that involve effects on cholinergic nerves (30) and mast cells (31,32). Indirect effects of the tachykinins are easily demonstrated in many species, but their importance varies according to the species and preparation. In rabbit tracheal smooth muscle, SP-induced constriction can be partially inhibited by the cholinergic antagonist atropine, thus implying an interaction between tachykinins and cholinergic neurons (30). Such an interaction is supported by the observation that subthreshold concentrations of SP and NKA increase the contractile response of ferret tracheal rings to electrical field stimulation but not to exogenous acetylcholine (23). In contrast, SP-induced constriction in the guinea pig is not modified by the presence of effective anticholinergics, cyclooxygenase inhibitors, or antihistamines (33). Even though the contractile effects of tachykinins are direct in this species, the release of tachykinins from neuronal stores activates airway mast cells and induces the release of histamine (31). The relative importance of direct and indirect mechanisms is different not only between species but among strains of rats. In relatively responsive Fisher 344 rats, NKA- and SP-induced constriction was associated with neurokinin (NK)1- and NK2-dependent serotonin release from mast cells, while in the less responsive BDE strain, activation of the NK2 receptor was more important (34). SP has been shown to elicit bronchodilation in preconstricted Sprague-Dawley rat tracheal rings (35) by an NK1 receptor-mediated induction of dilator prostaglandin production by airway epithelial cells. These observations have recently been extended to F344 rats (36), and a similar mechanism of SP inhibitory effect has been demonstrated in murine airways (37).

Three types of tachykinin receptors have been cloned and described, NK1, NK2, and NK3. In addition to the indirect effects of the tachykinins described above, SP and NKA can elicit airway responses directly, by activating NK1 and NK2 receptors, and the relative importance of these receptors is different in different species (38). The availability of specific and selective neurokinin receptor agonists and antagonists has facilitated our understanding of the roles of distinct classes of tachykinin receptors in the airway. The NK1 receptor is important in the genesis of neurogenic airway edema, vasodilatation, and mucus secretion. The NK2 receptor appears to be more potent than the NK1 receptor for transducing direct tachykinin-induced constriction in both guinea pig and human airway smooth muscle (39,40). In guinea pig lung strips, NK1, NK2, and NK3 agonists each cause tension to develop. This effect of the NK3 agonist is not produced by NK1, NK2 antagonists, or their combination, which suggests that NK3 receptors can transduce constriction in the guinea pig lung periphery (41). Studies in an intact guinea pig perfused lung preparation demonstrated that the NK2 receptor is predominantly responsible for the contractile response to endogenous tachykinins released by capsaicin (29). Tachykinins are airway active mediators with potent physiological action that exert their effects by direct receptor activation and by complex interactions with airway neurons, epithelial cells, smooth muscle cells, and inflammatory cells.

C. Tissue Location of Relevant Enzymes (NEP and ACE)

The tissue location of regulatory enzymes determines their access to and ability to inactivate SP and NKA. The tissue location of NEP has been determined by immuno-histochemistry, by the activity of differential tissue fractions to metabolize preferred florogenic substrates, and by the effects of specific inhibitors. Organ comparative studies in the rat demonstrated significant NEP activity in the lung (42). NEP is also present in the lungs of pigs (43) and humans (44). In the ferret, NEP immunolocalizes to the ciliated airway epithelium, with activity in columnar and basal epithelial cells, the interstitium, and airway smooth muscle cells (23). In human lungs, NEP immuno-localizes to the airway epithelium and is distinct at the epithelial surface of alveolar septae (44). An airway locus for NEP is supported by the finding that airway homogenates contain greater hydrolytic activity for preferred NEP substrates than do pulmonary vascular homogenates, and in-vitro studies document greater activity in fibroblast than in endothelial cell lines (44). The relative importance of epithelial versus interstitial and smooth muscle-associated NEP has been explored in experi-ments comparing the effects of NEP inhibitors on SP contractile effects in the presence and absence of the airway epithelial cell layer (23,45–47). Epithelial removal enhances SP and NKA contractile activity and diminishes the effects of NEP inhibition more significantly in guinea pig (48) than in ferret airway preparations (23). These findings demonstrate that epithelial and interstitial/smooth muscle distri-bution of NEP activity can vary as a function of species. The presence of an NEP inhibitor effect in the absence of the airway epithelium in some guinea pig cohorts (48) but not in others (49) suggests that genetic or environmental factors can influence the physiological activity of NEP by altering its tissue distribution. The weight of evidence places NEP in the lung interstitium and at the epithelial surface of the airway.

The lung is well recognized as a prominent site of ACE activity. Although NEP and ACE co-localize in the kidney, their tissue distribution in other organs is divergent (50). In the lung, ACE immunoactivity is concentrated along the luminal surface of the vascular endothelium in a distribution similar to that of the endothelial marker factor VIII (51). Further evidence supporting a vascular locus of pulmonary ACE is the observation that captopril-inhibitable hydrolysis of ACE-preferred sub-strate is greater in lung vascular than in airway homogenates. These data correlate with the in-vitro finding that ACE activity is greater in endothelial than in fibroblast cell lines. These tissue locations correlate with the relative physiological role of these enzymes for regulating SP released at different sites within the lung.

D. Limitation of Tachykinin Effect by Enzymatic Hydrolysis

The contractile effects of SP and NKA are limited in part by enzymatic inactivation by neutral endopeptidase (25,52,53) (E.C. 3.4.24.11). This regulatory enzyme is constitutively expressed by many airway resident and inflammatory cells and is present at the epithelial surface of the lung (44,54). NEP is an integral membrane enzyme that extends into the extracellular space from a hydrophobic domain near the amino terminus (55). The glycosylated hydrophilic extracellular domain contains a zinc-requiring hydrolytic site that has hydrolytic activity at the amino-terminal side of the hydrophobic amino acid residues Phe, Leu, Ile, Val, Tyr, and Trp (56,57). This location is optimal for altering the local extracellular concentrations of susceptible small peptides that are released from adjacent cells.

NKA

Purified human NEP, isolated from the brush border of the kidney, hydrolyzes NKA primarily at Gly^8-Leu^9 and Ser^5-Phe^6 (58) (Fig. 1). The relevance of these hydrolysis sites to NKA inactivation in the lung is supported by hydrolysis profile studies in which radiolabeled NKA was applied to the lung. Radiolabeled products consistent with hydrolysis at these sites were isolated from lung perfusate (59). A hydrolysis site with lower NEP affinity has been described at Phe^6-Val^7. Purified human NEP has a high affinity for NKA, with a K_m of 113 μM. When the hydrolytic efficiency of NEP was compared for 22 putative substrates, the tachykinins SP and NKA were found to be inactivated most efficiently (58,60). The concept that NEP hydrolysis of NKA results in loss of NKA's biological activity is supported by the observation that NKA hydrolysis products are markedly less efficient contractile agonists than intact NKA (61). NKA is known to be a substrate for the corpus striatum isozyme of angiotensin converting enzyme (ACE, E.C.3.4.15.1). It is interesting that the lung and kidney isozymes of ACE lack activity for NKA (55,62).

The physiological importance of NKA hydrolysis is suggested by the inactivating effect of hydrolysis on NKA and by the presence of NEP near sites of NKA

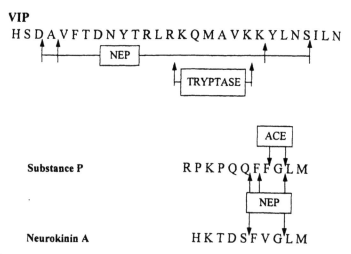

Figure 1 Established enzymatic hydrolysis sites with kinetics of potential physiological significance.

release and is confirmed by studies demonstrating augmented NKA physiological effects in the presence of compounds that inhibit NEP. When exogenous intravenous NKA is administered to anaesthetized, mechanically ventilated guinea pigs, its contractile effects are increased in the presence of the NEP inhibitors thiorphan and SCH 32,615 (25,61,63). The NEP inhibitor phosphoramidon not only augments NKA-induced bronchoconstriction but also enhances proximal airway microvascular leakage (64). These same inhibitors enhance NKA-induced constriction in an ex-vivo tracheally perfused lung model (52). In this model, the addition of a combination of noncytotoxic enzyme inhibitors failed to enhance NKA-induced bronchoconstriction beyond that observed with NEP inhibitors alone. The lack of effect of alternative inhibitors including ACE inhibitors supports a primary role for NEP in modulating the contractile effects of NKA. It is unlikely that carboxypeptidases inactivate NKA in vivo, because they have limited activity on peptides such as NKA that are amidated at the carboxyl terminus. Regulation by aminopeptidases is also unlikely, as the products of aminopeptidase action on NKA are equipotent to NKA as contractile agonists (61).

Substance P

Pulmonary hydrolytic inactivation of SP is more complex than inactivation of NKA because SP released into the lung can be inactivated by at least two distinct enzyme systems. The NEP hydrolysis sites on SP are the Gln^6-Phe^7, Phe^7-Phe^8, and Gly^9-Leu^{10} bonds (65,66) (Fig. 1). The K_m for NEP cleavage of SP is 31.9 μM, and purified human kidney NEP can hydrolyze 1 mM SP at a rate of 7.7 μmol/min/mg. This activity is completely inhibited by phosphoramidon, thiorphan, and SCH 32,615 but is not affected by the ACE inhibitor captopril. Unlike NKA, SP is also a substrate for the pulmonary isozyme of ACE. Purified rat lung ACE cleaves SP at the Phe^8-Gly^9 bond with a k_{cat}/K_m 60-fold less than that reported for the preferred ACE substrate angiotensin I (67). An additional hydrolysis site with less favorable kinetics has been identified at Gly^9-Leu^{10} (68). This differential hydrolytic activity has been observed in human lung preparations, where SP cleavage at the Phe^8-Gly^9 site is approximately three times more efficient than hydrolysis at the Gly^9-Leu^{10} bond (62). ACE purified from human kidney prefers the Phe^8-Gly^9 site by a similar 4:1 ratio and can hydrolyze 0.1 mM SP at a rate of 1.2 μmol/min/mg (69). This activity is completely inhibited by captopril but is unaffected by the specific NEP inhibitor SCH 32,615. The SP hydrolysis products of NEP and ACE are markedly less active than intact SP as contractile agonists in intact, mechanically ventilated guinea pigs (70). SP is also known to be a substrate for dipeptidyl(amino)peptidase IV (DAP IV, E.C.3.4.14.5), an enzyme present in rat and human serum that is sensitive to the inhibitor diprotinin (71). SP hydrolysis by DAP IV at the Pro^2-Lys^3 and Pro^4-Gln^5 sites results in SP fragments that are more potent contractile agonists than intact SP (70). However, the fact that these SP fragments are sensitive to inactivation by plasma aminopeptidase M suggests that any contractile activity would be short-lived (72). A regulatory role for DAP IV is suggested by the finding that rats that are genetically deficient in

DAP IV are more sensitive than controls to the salivary effects of SP. The role of DAP IV for limiting SP effects in the lower airways remains to be defined. In addition to NEP, ACE, and DAP IV, SP can be inactivated by hydrolysis at its Phe[7]-Phe[8] bond by the mast cell protease chymase (73). The k_{cat}/K_m of 3.9×10^4 for SP hydrolysis by chymase derived from canine mastocytoma cells suggests that SP hydrolysis by chymase may be significant only when the tissue presence of chymase is high.

The importance of NEP and ACE for modulating SP pulmonary activity has been clarified by studies determining the effects of enzyme inhibitors on the cleavage and physiological actions of SP. Interpretation of these studies depends on the specificity of the inhibitors employed. Enzyme systems identified by their hydrolytic site, substrate preference, and inhibitor sensitivity which are considered here to be distinct may in fact represent families of closely related enzymes that may be distinguished in the future by other characteristics. The importance of NEP for modulating the effects of SP is implied by studies that demonstrate that NEP inhibitors can augment the contractile effects of exogenously applied SP in tracheal rings (48,49) and in intact animals (74,75). The relevance of NEP for limiting the contractile activity of endogenously released tachykinins was demonstrated by documention of the ability of NEP inhibitors to augment the effects of tachykinins released by electrical field stimulation or by capsaicin in large airway preparations (23,48) on hilar bronchi (76) and in tracheally perfused guinea pig lungs (52).

The relative importance of ACE and NEP for regulating SP effect depends on the site of SP release or entry into the lung. When SP is presented to the lung by the vasculature, SP-contractile effects are augmented by the presence of an ACE inhibitor (77,78). Vascular SP-induced constriction is further enhanced when an NEP inhibitor is added, thus demonstrating the ability of NEP to modulate this process (14). In mechanically ventilated guinea pigs, vascularly delivered, SP-induced constriction is not affected by the presence of the specific NEP inhibitor SCH 32,615 alone, and SP levels in peripheral blood are not altered by this inhibitor. In contrast, the ACE inhibitor captopril given as a single agent augments both serum levels of SP and intravenous SP-induced constriction (78). These findings support the concept that ACE activity determines the serum levels of intravenously delivered SP and its access to the lung when released or infused into the vasculature. This physiological role of ACE correlates with its anatomic location in the pulmonary vasculature (51). When tissue access is allowed by inhibition of ACE, modulation of intravascularly delivered SP effect by NEP can then be demonstrated.

This correlation of modulatory effect with anatomic location of the relevant enzyme system is further supported by inhibitor studies in lungs, where SP is delivered to the tissues via the airway epithelium. When SP is delivered to the airways of tracheally perfused guinea pig lungs, inhibitors of NEP alone reduce hydrolysis at the known NEP hydrolysis sites, increase the presence of intact active SP in the perfusate, and augment SP-induced contractile effects (52). In this system, ACE inhibition with captopril had no effect. When SP is delivered by aerosol, inhibitors of NEP increase its contractile effects (74). In addition, when SP is applied to isolated guinea pig, ferret, and human airways, its contractile effects are enhanced by inhibitors of NEP but not by inhibitors of ACE (23,79,80). This modulatory role of NEP also correlates with its anatomical location at the epithelial surface and in the airway interstitium (44).

Inhibitor studies also support a regulatory role for NEP in other actions of SP that occur at or near the airway epithelium. Inhibitors of NEP augment the secretion of macromolecules from tracheal glands induced by the topical application of SP (22). In the guinea pig, intraperitoneal injection of an NEP inhibitor is associated with spontaneous cough, and the ablation of this effect by capsaicin-induced tachykinin depletion implies that the effect is dependent on the endogenous release of tachykinins (81). The tussive effects of SP and capsaicin are augmented by NEP inhibitors, and the application of human recombinant NEP to guinea pig airways significantly reduced SP-induced cough (24,82). Inhibitors of NEP but not ACE, serine proteases, or aminopeptidases increased ciliary beat frequency in cultured rabbit airway epithelial cells (83). SP-induced chloride secretion from cultured canine epithelial cells is augmented by inhibitors of NEP, which also implies that SP has important, regulated effects on airway epithelial cells (26). Clinical studies have associated ACE inhibitors with the induction of cough (56,84). Since the release of SP by capsaicin is associated with cough production, it has been speculated that ACE inhibitors may induce cough, at least partly by preventing the hydrolytic inactivation of SP by ACE or by ACE effects of bradykinin. These observations imply that NEP and ACE are located in a position to be competitive with SP receptors and can degrade ligand released into the extracellular tissue environment.

E. Inflammatory Modulation of Tachykinin Enzymatic Hydrolysis

There is reason to believe that the process of airway inflammation leads to alterations in lung tissue hydrolytic activity and altered responses to exogenous and endogenous tachykinins. In this section we will consider the evidence that different forms of inflammation can change NEP-like activity in the airways, the mechanisms by which these changes might occur, and the importance of other peptidases that may become relevant when the activity of NEP is diminished.

There is increasing evidence that viral airway infection can decrease the activity of NEP. Tracheal rings from ferrets inoculated with influenza A virus are more sensitive than control rings to the contractile effects of SP (85,86). Since infected rings were less sensitive than control rings to the enhancing effects of NEP inhibitors, at least part of the difference in SP effect could be ascribed to a loss of NEP activity. Similar findings in intact guinea pigs infected with Sendi virus suggest that decreased NEP activity occurs after viral infection in vivo (87). These findings are supported by permeability studies in rat trachea, where infection with *Mycoplasma pneumoniae*, coronavirus, or Sendai virus was associated with augmented SP-induced permeability and loss of NEP inhibitor effect (88). To further support this indirect evidence, direct measurements in lung homogenates demonstrate loss of NEP activity after airway viral infection (85). To further document the site of loss,

NEP activity was determined in tissue fractions composed predominantly of epithelial, submucosal, and smooth muscle components. Selective loss in the epithelial fraction correlated with the observation that contractile responses to SP that was applied to the epithelial surface were augmented. Mucus secretory responses were unaltered, as was NEP activity in the submucosal fraction that surrounds these glands (89). The selective loss of NEP activity in the epithelial fraction is evidence that the tissue location of NEP activity is biologically relevant and that this activity can be differentially altered by disease.

Toluene 2,4-diisocyanate (TDI) is a widely used plasticizer that has been associated with the development of asthma symptoms and airflow obstruction in sensitive exposed individuals. TDI-exposed guinea pigs demonstrate airway hyperresponsiveness and are more sensitive to the contractile effects of SP. Unlike unexposed guinea pigs, TDI-exposed animals do not develop augmented contractile responses when NEP inhibitors are added, which suggests that NEP activity has been lost. Direct measurements of NEP in lung homogenates confirm that NEP hydrolytic activity is decreased 30% in TDI-exposed lungs (90). The complete inability of NEP inhibitors to effect SP-induced constriction after losing only 30% of total NEP activity implies that the physiologically relevant fraction of NEP activity was differentially lost. Cigarette smoke is another known airway toxin that can enhance the sensitivity of guinea pig airways to the contractile effects of SP. The pulmonary resistance response to SP in animals exposed to the smoke from one and two cigarettes is not effected by the presence of phosphoramidon, thus implying that, unlike the effect of NEP in unexposed animals, NEP is not degrading SP (89). Exposure of tissue homogenates to cigarette smoke was associated with loss of directly measured NEP activity. In addition to the effects of acute exposure, chronic exposure to cigarette smoke increases the tussive effects of acute capsaicin exposure, a finding that is compatible with decreased NEP activity (91). Ozone is an environmental toxin that is known to cause airway hyperresponsiveness and the recruitment of neutrophils to the airway epithelium (92). SP-induced contractile responses are augmented by ozone exposure, and the potentiating effects of the NEP inhibitor phosphoramidon are diminished in ozone-exposed guinea pigs. Diminished NEP activity has been directly demonstrated in homogenates of ozone-exposed trachea (93). In each instance the effects of toxin exposure on NEP activity are concordant with the changes in SP effect.

The hypothesis that tachykinins are involved in the acute phase response to antigen exposure is suggested by the ability of phosphoramidon to augment antigen-induced constriction in mechanically ventilated, sensitized guinea pigs (94). Similar findings have been demonstrated in tracheal rings taken from sensitized animals (95). These studies show that NEP activity is not diminished acutely by the process of sensitization and that NEP is active during the acute contractile response. To determine whether the airway hyperresponsiveness and allergic inflammation that occur in the hours following antigen exposure are associated with altered NEP activity, we determined the NEP inhibitor sensitivity and SP hydrolysis product profile of hyperresponsive, inflamed, tracheally perfused guinea pig lungs (96). We found that lungs with antigen-induced airway inflammation were less sensitive to NEP inhibitors and generated smaller quantities of NEP hydrolysis fragments than control lungs. When levels of the guinea pig homolog of NEP mRNA was compared in groups of lungs, lungs from SP-responsive animals with lower perfusate levels of SP NEP hydrolysis fragments had lower levels of NEP mRNA. These findings suggest that severe allergen-induced airway inflammation is associated with decreased NEP activity and increased sensitivity to SP.

Mechanisms for Altered NEP Activity

A number of mechanisms may contribute to the changes in NEP activity described above. First, NEP can be lost when the cells expressing it undergo apoptosis or are shed into the airway lumen. The possibility of NEP loss due to desquamation of the airway epithelium is suggested by histological studies in severe asthma, where epithelial sloughing occurs (97). Removal or loss of the epithelial layer has been associated with loss of NEP activity in several studies (23,45–47). Since airway infections can result in the loss of NEP activity in the absence of epithelial shedding, it is likely that decreased transcription or loss from the cell surface is responsible.

NEP activity can be lost following oxidant stress, reversibly inhibited, or lost from the cell surface by endocytosis or proteolytic cleavage. The protective effects of superoxide dismutase on cigarette smoke-related loss of NEP activity in the guinea pig model implies that oxidants may be responsible. In addition, the NEP activity of tracheal homogenates is also markedly reduced by the presence of the oxidant N-chlorosuccinimide (89). Neutrophil NEP activity can be diminished by the addition of phorbol esters that activate protein kinase C (PKC) (54,98). In this system the fact that the lost NEP activity can be restored by lysing the cells suggests that NEP is lost by endocytosis. In airway epithelial cells, exposure to the oxidant hypochlorous acid causes a dose-related loss of NEP activity that can be recovered after cell sonification (99). These findings have been extended to tracheal rings, where hypochlorous acid exposure causes increased contractile responses and loss of NEP activity (100,101).

NEP production can be limited by decreased expression or alternative splicing of its mRNA or less avidly transcribed following PKC activation. Inflammatory stimuli can decrease cellular capacity of NEP production at the level of transcription. The phorbol ester 12-O-tetra-decaoylphorbol 13-acetate decreases the expression of NEP mRNA in rabbit synovial fibroblast and mammary epithelial cell lines by a mechanism that is thought to depend on PKC activation (102). NEP availability can be influenced by alternative splicing, as COS cells transfected with an NEP RNA isoform-missing sequence from exon 16 exhibit markedly reduced levels of NEP activity (103).

In addition to mechanisms that diminish NEP availability, there is reason to believe that the recruitment of inflammatory cells including neutrophils and lymphocytes (104) that bear NEP on their surfaces can increase NEP availability. Under conditions of chronic inflammation, hypertrophy of fibroblasts or smooth muscle cells results in increased airway NEP, as these cell types are known to express NEP (44). Extracellular mediators of inflammation including TNF-α, GM-CSF, and complement have the ability to increase NEP activity by facilitating its translocation to the cell surface of neutrophils (98,105).

There is increasing evidence that NEP activity can be increased by corticosteroid treatment. Oxidative loss of NEP can be ameliorated by treatment with dexamethasone by a mechanism that depends in part on augmented NEP mRNA expression. Dexamethasone and budesonide have been shown to induce fivefold increases in NEP activity and increased mRNA expression in transformed human airway epithelial cells (106), and dexamethasone treatment of rats is associated with diminished capsaicin-induced vascular permeability (107), an effect that was abolished by inhibitors of NEP and ACE (108). ACE activity is also stimulated by dexamethasone and reduced by adrenalectomy, which indicates that even physiological levels of corticosteroids may be important for regulating SP responses (109). In addition to internalization on the cell membrane, NEP activity may be lost after proteolytic cleavage of the enzyme and its liberation from the cell surface. The higher levels of NEP in the serum of patients with the adult respiratory distress syndrome may result from the hydrolytic mobilization of NEP by elastase released from the neutrophils in acute lung injury (110).

While NEP and ACE are the predominant enzymes regulating tachykinin metabolism in healthy animals, in animals exposed to oxidants and after allergic inflammation, it is possible that tachykinin effects may be limited by alternative enzyme systems. Plasma is a rich source of aminopeptidases and DAP IV, which can hydrolyze SP, and these enzymes may be important at sites of plasma leakage (55,72). Although neutrophil NEP can hydrolyze SP, the predominant enzyme inactivating SP in these cells is cathepsin G (111). In the microenvironment surrounding activated mast cells, chymase may limit the local effects of SP (112). The potential importance of alternative enzyme systems that act at limited tissue loci awaits systems that reliably detect local mediator effects.

F. Human Experience

Inhibitor studies of isolated human airways have implied that tachykinins transduce contractile effects predominantly through NK2 receptors of the A subtype (38). Recent studies demonstrate that neurokinin receptors can also transduce tachykinin-induced constriction (113,114). Topical application of FK 224, a cyclic peptide with in-vitro activity against NK1 and NK2 receptors, was shown to limit the contractile and tussive effects of the tachykinin-releasing peptide bradykinin (115). It is of interest that airway delivery of this agent did not alter the contractile effects of nebulized NKA, thereby implying that its mechanism of action may have been idiosyncratic or may have involved the antagonism of bradykinin (116). Nebulized FK 224 had no beneficial effects on asthma symptoms or lung function in a group of subjects with mild to moderate asthma (117). Intravenous CP 99,994, a selective NK1 antagonist, was similarly ineffective at limiting the contractile and tussive effects of nebulized hypertonic saline (118). Insight into this issue has been provided by a recent study demonstrating that allergen-induced airway hyperresponsiveness and eosinophilia in primates could be blocked with the combination of potent NK1 and NK2 blockers, while the single-agent antagonists had no effect (119). This finding implies that effective strategies may depend on the availability of potent antagonists that have activity at both receptors.

Studies of NEP activity in human subjects have also been difficult. The NEP inhibitor thiorphan had no effect on baseline specific airway resistance in normal or stable asthmatic subjects. Subjects with mild asthma are more sensitive to the contractile effects of inhaled NKA than are normal subjects, but both groups demonstrate equivalent potentiation of the contractile effects of NKA in the presence of the NEP inhibitor thiorphan (113,114). Although differences in NEP have not been detected between asymptomatic mild asthmatics and normals, this does not exclude a potentially significant role for tachykinins in severe asthma or during active disease.

III. VIP

A. Synthesis, Tissue Location, and Release of VIP

The gene that encodes VIP and peptide histadine methionine is located on chromosome 6 and occupies 9 kb between 6q26 and 6q27 (120,121). VIP is a member of the glucagon superfamily that includes glucagon, secretin, PHM-27, gastric inhibitory peptide, and growth hormone-releasing hormone. VIP is encoded on one of seven exons, and only minor variations in the coding sequence occur across species. The 5′ sequence is highly conserved and contains a cyclic AMP-responsive consensus sequence (122).

Like the tachykinins, VIP is a neuropeptide, synthesized in nerve cell bodies and released from axon dendrites. Neurons with VIP immunoreactivity are prominent in the wall of the trachea and major bronchi (123,124). VIP immunoreactivity is demonstrable in airway ganglia as well as airway neurons (125), and functional studies suggest that neurons of esophageal ganglia may also project onto airway structures (126). VIP-containing axon dendrites are located at or near bronchial smooth muscle, seromucous glands, and airway smooth muscle (12,14,124,127,128). VIP presence in the lung has been confirmed by direct measurement in extracts of lung tissue (12,18,129). In human lungs, VIP is largely but not exclusively co-localized with acetylcholine (130), while it is co-localized with nitric oxide synthase-reactive neurons in the ferret (131). In adult lungs VIP is immunohistochemically identified in networks of neurons that envelop airways; its presence decreases as a function of airway size but can be identified in bronchioles as small as 200 μM in diameter (15). These immunohistochemical findings have been corroborated in a study demonstrating that direct measurement of VIP in purified extracts of human trachea were greater than VIP levels in peripheral lung (18).

The content and distribution of VIP in the airway differs significantly among species. VIP-containing neurons are approximately 10 times more abundant in cat than in guinea pig or rat airways (12). In rodents and guinea pigs, VIP neuronal density is more concentrated on mucus glands than on airway blood vessels or smooth muscle. In humans, smooth muscle and vascular structures are more densely innervated by VIP-containing neurons (14). There are important age-related differences in the pulmonary distribution of VIP. VIP-containing neurons do not extend beyond the level of the bronchioles in the airways of subjects older than 10 years (132). These neurons localize to ganglia and extend beyond the respiratory bronchioles to alveolar ducts in lungs of subjects less than 3 years of age (15).

Changes in the distribution of VIP-containing neurons may occur in disease. A controversial report asserted that the airways of asthmatic subjects who died contained few if any VIP-immunoreactive neurons, which were easily observed in nonasthmatic airways (133). A follow-up study in mild asthmatics that measured VIP-reactive neurons in bronchoscopically obtained airway tissue demonstrated density of VIP-containing neurons equivalent to normals (134). A study that measured directly the VIP content of tracheal and lung parenchymal samples from normals and patients who died of asthma found no difference in the total tissue content of VIP between these groups (18). In addition, airway VIP content in children who succumbed to the respiratory distress syndrome did not differ from that of age-matched controls whose cause of death was not related to respiratory illness (129). One explanation for the discrepancies among these studies is that the immunohistochemical techniques employed may have detected non-VIP material that is antigenically similar to authentic VIP.

VIP is released from its neuronal stores in direct relation to the degree of neuronal activation (135). Recovery of VIP in a guinea pig tracheal smooth muscle system was directly proportional to the observed nonadrenergic noncholinergic smooth muscle relaxation. In this system both VIP recovery and relaxant effect were reduced by application of tetrodotoxin or by the presence of neutralizing VIP antiserum (135,136). It is interesting that anti-VIP serum does not block NANC-inhibitory responses in human tracheal smooth muscle preparations (137). Indeed, in humans it is likley that nitric oxide rather than VIP is the predominant mediator of nonadrenergic, noncholinergic airway relaxation. VIP can be released by the C-fiber stimulant capsaicin (138) and by electrical field stimulation, which implies that VIP is present in sensory neurons.

B. Enzymatic Modulation of VIP Physiological Effects

VIP is a preferred substrate for NEP, mast cell tryptase, and mast cell chymase. These three enzymes are present in the pulmonary extracellular space and are active at neutral pH. VIP is a 28-amino acid peptide that is amidated at the carboxyl terminus (139,140). Its longer primary sequence allows more complex cleavage patterns than those of the tachykinins. NEP hydrolyzes VIP at several sites (Fig. 1). Human recombinant NEP is most active at Asp^3-Ala^4, Ala^4-Val^5, Lys^{21}-Trp^{22}, and Ser^{25}-Ile^{26}. NEP can also cleave VIP at Arg^{12}-Lys^1, and Gln^{16}-Met^{17}, but hydrolysis at these additional sites has less favorable kinetics (141). When the relaxant effects of the expected products of VIP hydrolysis were studied in a tracheally perfused lung model, the hydrolysis products had markedly less activity than intact VIP (142). This finding supports the concept that enzymatic hydrolysis can alter the physiological activity of VIP.

Mast cell tryptase gains access to the extracellular microenvironment after it is released from activated mast cells. Canine mastocytoma tryptase hydrolyses VIP with a k_{cat}/K_m of 2.2×10^5 sec^{-1} M^{-1} and a K_m of 3.3 mM (Fig. 1). Hydrolysis product analysis indicates that VIP is cleaved by tryptase at two sites, Arg[14]-Lys[15] and Lys[20]-Lys[21] (73). The kinetics of this reaction compare favorably with those of other potential substrates screened for susceptibility to human skin and lung mast cell proteases (143,144). A second mast cell protease that can be purified from canine mastocytoma cells can efficiently inactivate VIP. Mast cell chymase cleaves VIP at Tyr[22]-Leu[23] with a K_m of 3.3 mM and a k_{cat}/K_m of 5.4×10^4 sec^{-1} M^{-1} (73). The VIP hydrolytic kinetics for chymase are less favorable than those obtained for tryptase but are compatible with a physiologically significant role for this enzyme, which can be present in appreciable local concentration after its liberation from airway mast cells. Like the VIP hydrolysis products of NEP, the hydrolysis products of tryptase and chymase are inactive. They fail to relax vascular, gastrointestinal, or tracheal smooth muscle (73), or tracheally perfused lungs (142).

The physiological relevance of enzymatic inactivation of VIP is suggested by studies of purified enzyme on exogenous VIP effects. The addition of purified tryptase or chymase attenuates VIP-induced relaxation of ferret tracheal smooth muscle (145). Similarly, the addition of papain or chymotrypsin diminished VIP-induced smooth muscle relaxation in the guinea pig (146). These observations were extended by determining which inhibitors of endogenous enzymes could augment VIP-induced relaxant effects. Inhibitors of NEP were shown to augment VIP-induced relaxation of guinea pig tracheal rings (147), and soybean trypsin inhibitor was independently demonstrated to be effective. In this preparation, inhibitors of ACE and aminopeptidases were found to be ineffective (146). The importance of airway epithelial NEP is also suggested by studies showing that NEP inhibition and epithelial removal are equally effective for augmenting the relaxant effects of VIP (148). In the cat, all effective combinations of inhibitors contained an NEP inhibitor (149). Combinations of inhibitors are also effective in human bronchial rings; VIP-induced relaxation is potentiated by the combination of an NEP inhibitor with a serine protease inhibitor (150). In addition, epithelial removal augmented the relaxant effects of VIP in human airway smooth muscle (151). In isolated, tracheally perfused guinea pig lungs, the combination of a serine protease and an NEP inhibitor effectively augmented VIP-induced relaxation, while single-agent inhibitors did not. VIP hydrolysis fragment analysis in this system identified NEP and mast cell tryptase as the enzymes responsible for inactivating VIP (96). Recent work in our laboratory implicates these same enzymes in the hydrolytic inactivation of VIP in human bronchial smooth muscle.

C. Inflammatory Modulation of VIP Hydrolysis

The airway effects of VIP are regulated by enzymatic hydrolysis, and inflammation induced by chronic exposure to antigens alters the activity and distribution of VIP regulatory enzymes. The contractile effects of SP are augmented by airway inflammation including allergic inflammation, but the relaxant effects of VIP are diminished by allergic airway inflammation. This inflammation-related loss of VIP relaxant activity is specific to VIP, as these lungs retain their sensitivity to the relaxant effects of isoproterenol (96). The concept that airway inflammation is related to changes in enzymatic inactivation of VIP is demonstrated by the lower perfusate levels of VIP in inflamed lungs. This finding suggests that inflamed lungs more readily degrade VIP. Comparison of VIP hydrolysis fragments from normal lungs (142) indicates that NEP and mast cell tryptase inactivate VIP, while only tryptase fragments are present in the perfusate of inflamed lungs (96). To demonstrate that tryptase activity is the dominant mechanism of VIP inactivation in inflamed lungs, addition of a tryptase inhibitor restored VIP relaxant activity and the level of intact active VIP in the perfusate.

Alteration in the activity of regulatory enzymes in the setting of allergic inflammation provides a possible mechanism for the phenomenon of nonspecific airway responsiveness in inflamed lungs (152). It is known that a variety of allergic mediators release tachykinins (10,48,153,154); inflammatory loss of NEP activity should enhance constriction from tachykinins released by these other bronchoactive allergic mediators. This mechanism is in concordance with the observation that augmented methacholine and antigen-induced contractile responses in guinea pigs with allergen-induced airway inflammation are diminished by tachykinin depletion by prior capsaicin exposure (155,156). However, these studies demonstrating the loss of the homeostatic relaxant effects of VIP in inflamed lungs extend this mechanism by suggesting an inflammation-induced imbalance between endogenous relaxant and contractile substances (96).

In summary, VIP is an endogenous airway relaxant that localizes to neurons that subtend vessels, glands, and smooth muscle. It is released by neuronal activation that can be accomplished by electrical field stimulation or capsaicin exposure. The pulmonary relaxant effects of VIP are regulated in a complex way by two physiologically competitive enzyme systems. Enzymatic modulation of the vascular and glandular effects of VIP remain to be fully explored. There is now convincing evidence that alterations in the enzymatic inactivation and regulation of VIP can occur in the setting of allergic inflammation, but the effects of alternative forms of inflammation on VIP hydrolysis are not fully defined.

D. Human Studies

In human bronchial smooth muscle, VIP causes relaxation with a slow time course relative to that reported for beta-agonists. The magnitude of this effect (in the absence of enzyme inhibitors) is also modest compared with that of beta-agonists. The maximal relaxant effect of VIP is greatly enhanced when enzyme inhibitors are present, but still takes many minutes to develop (150). Studies of VIP effects on lung function tests in humans were encouraged by studies demonstrating that inhaled VIP diminished the bronchoconstrictor effects of histamine in intact dogs and guinea pigs (157,158). The effects of VIP on histamine-induced changes in airway conductance were determined in a group of six atopic subjects with mild asthma. These subjects were significantly less responsive to the contractile effects of histamine after inhaling 100 µg of VIP, but this effect was modest compared with that of 200 µg of the beta-agonist salbutamol (159). Although the effect of VIP may have been related to the release of endogenous amines in response to VIP-induced vascular effects, the authors did not note alterations in heart rate or blood pressure when VIP was inhaled. There may be important differences among individual subjects, as another group of investigators did not observe a protective effect of 375 µg of inhaled VIP in two subjects who demonstrated less histamine-induced airflow limitation after nebulized isoproterenol (160).

Attempts to demonstrate relaxant effects of intravenous VIP on baseline airway tone have been limited by the vasodepressor effects of intravenous VIP (161,162). The difference in VIP effect that occurs as a function of the route of administration could be due either to a greater presence or sensitivity of VIP receptors in the vascular space than in the airways or to an inability of VIP to reach airway smooth muscle due to enzymatic inactivation. An alternative explanation for not detecting VIP-induced airway relaxation in these studies is patient selection bias. It may be that the beta-agonist-responsive patients selected in these studies (162) do not respond to VIP as readily as subjects who are less beta-agonist-responsive. In addition to the lack of potency of single-agent VIP, the slower kinetics of VIP-induced relaxation of bronchial smooth muscle (150) make it likely that any protective effects of VIP will have slower onset of action than those of the first-generation beta-agonists.

In summary, it has been difficult to test the relaxant effects of VIP in the experimental settings where preclinical testing has demonstrated the greatest efficacy. These studies suggest that VIP is most effective when delivered with protease inhibitors to subjects with significantly elevated airway tone. Although the reported clinical studies to date have not been promising, they have not excluded the possibility that VIP or a more potent analog or cyclic form that is resistant to enzymatic inactivation has therapeutic efficacy.

References

1. Casale TB. Neuropeptides and the lung. J Allergy Clin Immunol 1991; 88:1–16.
2. Nadel JA. Membrane-bound peptidases—endocrine, paracrine, and autocrine effects. Am J Respir Cell Mol Biol 1992; 7:469–470.
3. Lilly CM, Drazen JM, Shore SA. Peptidase modulation of airway effects of neuropeptides. Proc Soc Exp Biol Med 1993; 203:388–404.
4. Barnes PJ. Modulation of neurotransmission in airways. Physiol Rev 1992; 72: 699–729.
5. Kummer W, Fischer A, Kurkowski R, Heym C. The sensory and sympathetic innerva-

tion of guinea-pig lung and trachea as studied by retrograde neuronal tracing and double-labelling immunohistochemistry. Neuroscience 1992; 49:715–737.

6. Lundberg JM, Hokfelt T, Martling CR, Saria A, Cuello C. Substance P-immunoreactive sensory neves in the lower respratory tract of various mammals including man. Cell Tissue Res 1984; 235:251–261.

7. Nawa H, Kotani H, Nakanishi S. Tissue-specific generation of two preprotachykinin mRNAs from one gene by alternative RNA splicing. Nature 1984; 312:729–734.

8. Krause JE, Chirgwin JM, Carter MS, Xu ZS, Hershey AD. Three rat preprotachykinin mRNAs encode the neuropeptides substance P and neurokinin A. Proc Natl Acad Sci USA 1987; 84:881–885.

9. MacDonald MR, Takeda J, Rice CM, Krause JE. Multiple tachykinins are produced and secreted upon post-translational processing of the three substance P precursor proteins, alpha-, beta-, and gamma-preprotachykinin: expression of the preprotachykinins in AtT-20 cells infected with vaccinia virus recombinants. J Biol Chem 1989; 264:15578–15592.

10. Saria A, Martling CR, Yan Z, Theodorsson-Norheim E, Gamse R, Lundberg JM. Release of multiple tachykinins from capsaicin-sensitive sensory nerves in the lung by bradykinin, histamine, dimethylphenyl piperazinium, and vagal nerve stimulation. Am Rev Respir Dis 1988; 137:1330–1335.

11. Hua XY, Theodorsson-Norheim E, Brodin E, Lundberg JM, Hokfelt T. Multiple tachykinins (neurokinin A, neuropeptide K and substance P)X in capsaicin-sensitive sensory neurons in the guinea-pig. Regulatory Peptides 1985; 13:1–19.

12. Ghatei MA, Sheppard MN, O'Shaughnessy DJ, Adrian TE, Mcgregor GP, Polak JM, Bloom SR. Regulatory peptides in the mammalian respiratory tract. Endocrinology 1982; 111:1248–1254.

13. Sundler F, Brodin E, Ekblad E, Hakanson R, Uddman R. Sensory nerve fibers: distribution of substance P, neurokinin A and calcintonin gene-related peptide. In: Hakanson R, Sundler F, eds. Tachykinin Antagonists. New York: Elsevier, 1985:3–14.

14. Springall DR, Bloom SR, Polak JM. Distribution, nature, and origin of peptide containing nerves in mammalian airways. In: Kaliner MA, Barnes PJ, eds. The Airways: Neural Control in Health and Disease. New York: Marcel Dekker, 1988:299–341.

15. Hislop AA, Wharton J, Allen KM, Polak JM, Haworth SG. Immunohistochemical localization of peptide-containing nerves in human airways: age-related changes. Am J Respir Cell Mol Biol 1990; 3:191–198.

16. Dey RD, Hoffpauir J, Said SI. Co-localization of vasoactive intestinal peptide- and substance P-containing nerves in cat bronchi. Neuroscience 1988; 24:275–281.

17. Baraniuk JN, Lundgren JD, Okayama M, Goff J, Mullol J, Merida M, Shelhamer JH, Kaliner MA. Substance P and neurokinin A in human nasal mucosa. Am J Respir Cell Mol Biol 1991; 4:228–236.

18. Lilly CM, Bai TR, Shore SA, Hall AE, Drazen JM. Neuropeptide content of lungs from asthmatic and nonasthmatic patients. Am J Respir Crit Care Med 1995; 151:548–553.

19. Killingsworth CR, Shore SA, Paulauskis JD. Rat alveolar macrophages express preprotachykinin gene-1 mRNA. Am J Respir Crit Care Med 1995; 151:A134.

20. Coleridge JC, Coleridge HM. Afferent vagal C fibre innervation of the lungs and airways and its functional significance. Rev Physiol Biochem Pharmacol 1984; 99: 1–110.

21. Lee LY, Morton RF. Histamine enhances vagal pulmonary C-fiber responses to capsaicin and lung inflation. Respir Physiol 1993; 93:83–96.

22. Borson DB, Corrales R, Varsano S, Gold M, Viro N, Caughey G, Ramachandran J, Nadel JA. Enkephalinase inhibitors potentiate substance P-induced secretion of 35SO4-macromolecules from ferret trachea. Exp Lung Res 1987; 12:21–36.

23. Sekizawa K, Tamaoki J, Nadel JA, Borson DB. Enkephalinase inhibitor potentiates substance P- and electrically induced contraction in ferret trachea. J Appl Physiol 1987; 63:1401–1405.

24. Kohrogi H, Graf PD, Sekizawa K, Borson DB, Nadel JA. Neutral endopeptidase inhibitors potentiate substance P- and capsaicin-induced cough in awake guinea pigs. J Clin Invest 1988; 82:2063–2068.

25. Shore SA, Drazen JM. Degradative enzymes modulate airway responses to intravenous neurokinins A and B. J Appl Physiol 1989; 67:2504–2511.

26. Tamaoki J, Sakai N, Isono K, Kanemura T, Chiyotani A, Yamauchi F, Takizawa T, Konno K. Effect of neutral endopeptidase inhibition on substance-P-induced increase in short-circuit current of canine cultured tracheal epithelium. Int Arch Allergy Appl Immunol 1991; 95:169–173.

27. Wong LB, Miller IF, Yeates DB. Pathways of substance P stimulation of canine tracheal ciliary beat frequency. J Appl Physiol 1991; 70:267–273.

28. Piedimonte G, Hoffman JIE, Husseini WK, Hiser WL, Nadel JA. Effect of neuropeptides released from sensory nerves on blood flow in the rat airway microcirculation. J Appl Physiol 1992; 72:1563–1570.

29. Lilly CM, Besson G, Israel E, Rodger IW, Drazen JM. Capsaicin-induced airway obstruction in tracheally perfused guinea pig lungs. Am J Respir Crit Care Med 1994; 149:1175–1179.

30. Tanaka DT, Grunstein MM. Maturation of neuromodulatory effect of substance P in rabbit airways. J Clin Invest 1990; 85:345–350.

31. Lilly CM, Hall AE, Rodger IW, Kobzik L, Haley KJ, Drazen JM. Substance P-induced histamine release in tracheally perfused guinea pig lungs. J Appl Physiol 1995; 78:1234–1241.

32. Joos GF, Pauwels RA. The in vivo effect of tachykinins on airway mast cells of the rat. Am Rev Respir Dis 1993; 148:922–926.

33. Shore SA, Drazen JM. Enhanced airway responses to substance P after repeated challenge in guinea pigs. J Appl Physiol 1989; 66:955–961.

34. Joos GF, Kips JC, Pauwels RA. In vivo characterization of the tachykinin receptors involved in the direct and indirect bronchoconstrictor effect of tachykinins in two inbred rat strains. Am J Respir Crit Care Med 1994; 149:1160–1166.

35. Szarek JL, Stewart NL, Spurlock B, Schneider C. Sensory nerve- and neuropeptide-mediated relaxation responses in airways of Sprague-Dawley rats. J Appl Physiol 1995; 78:1679–1687.

36. Szarek JL, Spurlock B. Contraction and relaxation responses to substance P in airways of F344 rats. Pulmon Pharmacol. In press.

37. Manzini S. Bronchodilatation by tachykinins and capsaicin in the mouse main bronchus. Br J Pharmacol 1992; 105:968–972.

38. Maggi CA. Tachykinin receptors and airway pathophysiology. Eur Respir J 1993; 6: 735–742.

39. Naline E, Devillier P, Drapeau G, Toty L, Bakdach H, Regoli D, Adviner C. Characterization of neurokinin effects and receptor selectivity in human isolated bronchi. Am Rev Respir Dis 1989; 140:679–686.

40. Ellis JL, Undem BJ, Kays JS, Ghanekar SV, Barthlow HG, Buckner CK. Pharmacologi-

cal examination of receptors mediating contractile responses to tachykinins in airways isolated from human, guinea pig and hamster. J Pharmacol Exp Ther 1993; 267:95–101.

41. Killingsworth CR, Shore SA. Tachykinin receptors mediating contraction of guinea pig lung strips. Regulatory Peptides 1995; 57:149–161.

42. Llorens C, Schwartz JC. Enkephalinase activity in rat peripheral organs. Eur J Pharmacol 1981; 69:113–116.

43. Kenny AJ, Bowes MA, Gee NS, Matsas R. Endopeptidase-24.11: a cell-surface enzyme for metabolizing regulatory peptides. Biochem Soc Trans 1985; 13:293–295.

44. Johnson AR, Ashton J, Schulz WW, Erdos EG. Neutral metalloendopeptidase in human lung tissue and cultured cells. Am Rev Respir Dis 1985; 132:564–568.

45. Djokic TD, Nadel JA, Dusser DJ, Sekizawa K, Graf PD, Borson DB. Inhibitors of neutral endopeptidase potentiate electrically and capsaicin-induced noncholinergic contraction in guinea pig bronchi. J Pharmacol Exp Ther 1989; 248:7–11.

46. Fine JM, Gordon T, Sheppard D. Epithelium removal alters responsiveness of guinea pig trachea to substance P. J Appl Physiol 1989; 66:232–237.

47. Frossard N, Rhoden KJ, Barnes PJ. Influence of epithelium on guinea pig airway responses to tachykinins: role of endopeptidase and cyclooxygenase. J Pharmacol Exp Ther 1989; 248:292–298.

48. Maggi CA, Patacchini R, Perretti F, Meini S, Manzini S, Santicioli P, Del Bianco E, Meli A. The effect of thiorphan and epithelium removal on contractions and tachykinin release produced by activation of capsaicin-sensitive afferents in the guinea-pig isolated bronchus. Naunyn Schmiedebergs Arch Pharmacol 1990; 341:74–79.

49. Devillier P, Advenier C, Drapeau G, Marsac J, Regoli D. Comparison of the effects of epithelium removal and of an enkephalinase inhibitor on the neurokinin-induced contractions of guinea-pig isolated trachea. Br J Pharmacol 1988; 94:675–684.

50. Erdos EG, Schulz WW, Gafford JT, Defendini R. Neutral metalloendopeptidase in human male genital tract: comparison to angiotensin I-converting enzyme. Lab Invest 1985; 52:437–447.

51. Takada Y, Hiwada K, Unno M, Kokubu T. Immunocytochemical localization of angiotensin converting enzyme at the ultrastructural level in the human lung and kidney. Biomed Res 1982; 3:169–174.

52. Martins MA, Shore SA, Drazen JM. Peptidase modulation of the pulmonary effects of tachykinins. Int Arch Allergy Appl Immunol 1991; 94:325–329.

53. Shore SA, Martins MA, Drazen JM. Effect of the NEP inhibitor SCH32615 on airway responses to intravenous substance-P in guinea pigs. J Appl Physiol 1992; 73:1847–1853.

54. Erdos EG, Wagner B, Harbury CB, Painter RG, Skidgel RA, Fa XG. Down-regulation and inactivation of neutral endopeptidase 24.11 (enkephalinase) in human neutrophils. J Biol Chem 1989; 264:14519–14523.

55. Turner AJ, Matsas R, Kenny AJ. Are there neuropeptide-specific peptidases? Biochem Pharmacol 1985; 34:1347–1356.

56. Erdos EG, Skidgel RA. Neutral endopeptidase 24.11 (enkephalinase) and related regulators of peptide hormones. FASEB J 1989; 3:145–151.

57. Kenny J. Endopeptidase-24.11—putative substrates and possible roles. Biochem Soc Trans 1993; 21:663–668.

58. Hooper NM, Kenny AJ, Turner AJ. The metabolism of neuropeptides. Neurokinin A (substance K) is a substrate for endopeptidase-24.11 but not for peptidyl dipeptidase A (angiotensin-converting enzyme). Biochem J 1985; 231:357–361.

59. Martins MA, Shore SA, Gerard NP, Gerard C, Drazen JM. Peptidase modulation of the pulmonary effects of tachykinins in tracheal superfused guinea pig lungs. J Clin Invest 1990; 85:170–176.
60. Martling CR, Theodorsson-Norheim E, Norheim I, Lundberg JM. Bronchoconstrictor and hypotensive effects in relation to pharmacokinetics of tachykinins in the guinea-pig—evidence for extraneuronal cleavage of neuropeptide K to neurokinin A. Naunyn Schmiedebergs Arch Pharmacol 1987; 336:183–189.
61. Shore SA, Drazen JM. Relative bronchoconstrictor activity of neurokinin-A and neurokinin-A fragments in guinea pigs. J Appl Physiol 1991; 71:452–457.
62. Thiele EA, Strittmatter SM, Snyder SH. Substance K and substance P as possible endogenous substrates of angiotensin converting enzyme in the brain. Biochem Biophys Res Commun 1985; 128:317–324.
63. Joos GF, Pauwels RA. Mechanisms involved in neurokinin-induced bronchoconstriction. Arch Int Pharmacodyn Ther 1990; 303:132–146.
64. Lotvall JO, Elwood W, Tokuyama K, Barnes PJ, Chung KF. Differential effects of phosphoramidon on neurokinin-A-induced and substance-P-induced airflow obstruction and airway microvascular leakage in guinea-pig. Br J Pharmacol 1991; 104:945–949.
65. Matsas R, Fulcher IS, Kenny AJ, Turner AJ. Substance P and [Leu]enkephalin are hydrolyzed by an enzyme in pig caudate synaptic membranes that is identical with the endopeptidase of kidney microvilli. Proc Natl Acad Sci USA 1983; 80:3111–3115.
66. Matsas R, Kenny AJ, Turner AJ. The metabolism of neuropeptides. The hydrolysis of peptides, including enkephalins, tachykinins and their analogues, by endopeptidase-24.11. Biochem J 1984; 223:433–440.
67. Cascieri MA, Bull HG, Mumford RA, Patchett AA, Thornberry NA, Liang T. Carboxyl-terminal tripeptidyl hydrolysis of substance P by purified rabbit lung angiotensin-converting enzyme and the potentiation of substance P activity in vivo by captopril and MK-422. Mol Pharmacol 1984; 25:287–293.
68. Skidgel RA, Erdos EG. Cleavage of peptide bonds by angiotensin I converting enzyme. Agents Actions Suppl 1987; 22:289–296.
69. Skidgel RA, Engelbrecht S, Johnson AR, Erdos EG. Hydrolysis of substance P and neurotensin by converting enzyme and neutral endopeptidase. Peptides 1984; 5:769–776.
70. Shore SA, Drazen JM. Airway responses to substance P and substance P fragments in the guinea pig. Pulmon Pharmacol 1988; 1:113–118.
71. Wang LH, Ahmad S, Benter IF, Chow A, Mizutani S, Ward PE. Differential processing of substance-P and neurokinin-A by plasma dipeptidyl(amino)peptidase-IV, aminopeptidase-M and angiotensin converting enzyme. Peptides 1991; 12:1357–1364.
72. Ahmad S, Wang LH, Ward PE. Dipeptidyl(amino)peptidase-IV and aminopeptidase-M metabolize circulating substance-P in vivo. J Pharmacol Exp Ther 1992; 260:1257–1261.
73. Caughey GH, Leidig F, Viro NF, Nadel JA. Substance P and vasoactive intestinal peptide degradation by mast cell tryptase and chymase. J Pharmacol Exp Ther 1988; 244:133–137.
74. Dusser DJ, Umeno E, Graf PD, Djokic T, Borson DB, Nadel JA. Airway neutral endopeptidase-like enzyme modulates tachykinin-induced bronchoconstriction in vivo. J Appl Physiol 1988; 65:2585–2591.
75. Thompson JE, Sheppard D. Phosphoramidon potentiates the increase in lung resistance mediated by tachykinins in guinea pigs. Am Rev Respir Dis 1988; 137:337–340.

76. Warner EA, Krell RD, Buckner CK. Pharmacologic studies on the differential influence of inhibitors of neutral endopeptidase on nonadrenergic, noncholinergic contractile responses of the guinea pig isolated hilar bronchus to transmural electrical stimulation and exogenously applied tachykinins. J Pharmacol Exp Ther 1990; 254:824–830.

77. Shore SA, Stimler-Gerard NP, Coats SR, Drazen JM. Substance P-induced bronchoconstriction in the guinea pig. Enhancement by inhibitors of neutral metalloendopeptidase and angiotensin-converting enzyme. Am Rev Respir Dis 1988; 137:331–336.

78. Drazen JM, Shore SA, Gerard NP. Substance P-induced effects in guinea pig lungs: effects of thiorphan and captopril. J Appl Physiol 1989; 66:1364–1372.

79. Stimler-Gerard NP. Neutral endopeptidase-like enzyme controls the contractile activity of substance P in guinea pig lung. J Clin Invest 1987; 79:1819–1825.

80. Honda I, Kohrogi H, Yamaguchi T, Ando M, Araki S. Enkephalinase inhibitor potentiates substance P-induced and capsaicin-induced bronchial smooth muscle contractions in humans. Am Rev Respir Dis 1991; 143:1416–1418.

81. Ujiie Y, Sekizawa K, Aikawa T, Sasaki H. Evidence for substance P as an endogenous substance causing cough in guinea pigs. Am Rev Respir Dis 1993; 148:1628–1632.

82. Kohrogi H, Nadel JA, Malfroy B, Gorman C, Bridenbaugh R, Patton JS, Borson DB. Recombinant human enkephalinase (neutral endopeptidase) prevents cough induced by tachykinins in awake guinea pigs. J Clin Invest 1989; 84:781–786.

83. Kondo M, Tamaoki J, Takizawa T. Neutral endopeptidase inhibitor potentiates the tachykinin-induced increase in ciliary beat frequency in rabbit trachea. Am Rev Respir Dis 1990; 142:403–406.

84. Lunde H, Hedner T, Samuelsson O, Lotvall J, Andren L, Lindholm L, Wilholm BE. Dyspnoea, asthma, and bronchospasm in relation to treatment with angiotensin converting enzyme inhibitors. Br Med J 1994; 308:18–21.

85. Jacoby DB, Tamaoki J, Borson DB, Nadel JA. Influenza infection causes airway hyperresponsiveness by decreasing enkephalinase. J Appl Physiol 1988; 64:2653–2658.

86. Murray TC, Jacoby DB. Viral infection increases contractile but not secretory responses to substance-P in ferret trachea. J Appl Physiol 1992; 72:608–611.

87. Dusser DJ, Jacoby DB, Djokic TD, Rubinstein I, Borson DB, Nadel JA. Virus induces airway hyperresponsiveness to tachykinins: role of neutral endopeptidase. J Appl Physiol 1989; 67:1504–1511.

88. Borson DB, Brokaw JJ, Sekizawa K, Mcdonald DM, Nadel JA. Neutral endopeptidase and neurogenic inflammation in rats with respiratory infections. J Appl Physiol 1989; 66:2653–2658.

89. Dusser DJ, Djokic TD, Borson DB, Nadel JA. Cigarette smoke induces bronchoconstrictor hyperresponsiveness to substance P and inactivates airway neutral endopeptidase in the guinea pig. Possible role of free radicals. J Clin Invest 1989; 84:900–906.

90. Sheppard D, Thompson JE, Scypinski L, Dusser D, Nadel JA, Borson DB. Toluene diisocyanate increases airway responsiveness to substance P and decreases airway neutral endopeptidase. J Clin Invest 1988; 81:1111–1115.

91. Karlsson JA, Zackrisson C, Lundberg JM. Hyperresponsiveness to tussive stimuli in cigarette smoke-exposed guinea-pigs: a role for capsaicin-sensitive, calcitonin gene-related peptide-containing nerves. Acta Physiol Scand 1991; 141:445–454.

92. Gordon T, Venugopalan CS, Amdur MO, Drazen JM. Ozone-induced airway hyperreactivity in the guinea pig. J Appl Physiol 1984; 57:1034–1038.

93. Murlas CG, Lang Z, Williams GJ, Chodimella V. Aerosolized neutral endopeptidase

reverses ozone-induced airway hyperreactivity to substance P. J Appl Physiol 1992; 72:1133–1141.

94. Dusser DJ, Umeno E, Graf PD, Djokic T, Borson DB, Nadel JA. Airway neutral endopeptidase-like enzyme modulates tachykinin-induced bronchoconstriction in vivo. J Appl Physiol 1988; 65:2585–2591.

95. Tudoric N, Coon RL, Flynn NM, Bosnjak ZJ. Inhibition of neutral endopeptidase augments anaphylactic constriction of guinea pig tracheal smooth muscle. Int Arch Allergy Immunol 1993; 100:170–177.

96. Lilly CM, Kobzik L, Hall AE, Drazen JM. Effects of chronic airway inflammation on the activity and enzymatic inactivation of neuropeptides in guinea pig lungs. J Clin Invest 1994; 93:2667–2674.

97. Dunnill MS. The pathology of asthma, with special reference to changes in the bronchial mucosa. J Clin Pathol 1960; 13:27–33.

98. Shipp MA, Stefano GB, Switzer SN, Griffin JD, Reinherz EL. CD10 (CALLA)/neutral endopeptidase 24.11 modulates inflammatory peptide-induced changes in neutrophil morphology, migration, and adhesion proteins and is itself regulated by neutrophil activation. Blood 1991; 78:1834–1841.

99. Lang ZH, Murlas CG. HOCl exposure of a human airway epithelial cell line decreases its plasma membrane neutral endopeptidase. Lung 1991; 169:311–323.

100. Lang Z, Murlas CG. Neutral endopeptidase of a human airway epithelial cell line recovers after hypochlorous acid exposure: dexamethasone accelerates this by stimulating neutral endopeptidase mRNA synthesis. Am J Respir Cell Mol Biol 1992; 7: 300–306.

101. Murlas CG, Lang Z, Chodimella V. Dexamethasone reduces tachykinin but not ACh airway hyperreactivity after O3. Lung 1993; 171:109–121.

102. Werb Z, Clark EJ. Phorbol diesters regulate expression of the membrane neutral metalloendopeptidase (EC 3.4.24.11) in rabbit synovial fibroblasts and mammary epithelial cells. J Biol Chem 1989; 264:9111–9113.

103. Iijima H, Gerard NP, Squassoni C, Ewig J, Face D, Drazen JM, Kim YA, Shriver B, Hersh LB, Gerard C. Exon 16 del: a novel form of human neutral endopeptidase (CALLA). Am J Physiol 1992; 262:L725–L729.

104. Shipp MA, Richardson NE, Sayre PH, Brown NR, Masteller EL, Clayton LK, Ritz J, Reinherz EL. Molecular cloning of the common acute lymphoblastic leukemia antigen (CALLA) identifies a type II integral membrane protein. Proc Natl Acad Sci USA 1988; 85:4819–4823.

105. Werfel T, Sonntag G, Weber MH, Gotze O. Rapid increases in the membrane expression of neutral endopeptidase (CD10), aminopeptidase N (CD13), tyrosine phosphatase (CD45), and Fc gamma-RIII (CD16) upon stimulation of human peripheral leukocytes with human C5a. J Immunol 1991; 147:3909–3914.

106. Borson DB, Gruenert DC. Glucocorticoids induce neutral endopeptidase in transformed human tracheal epithelial cells. Am J Physiol 1991; 260:L83–L89.

107. Piedimonte G, Mcdonald DM, Nadel JA. Glucocorticoids inhibit neurogenic plasma extravasation and prevent virus-potentiated extravasation in the rat trachea. J Clin Invest 1990; 86:1409–1415.

108. Piedimonte G, Mcdonald DM, Nadel JA. Neutral endopeptidase and kininase-II mediate glucocorticoid inhibition of neurogenic inflammation in the rat trachea. J Clin Invest 1991; 88:40–44.

109. Mendelsohn FA, Lloyd CJ, Kachel C, Funder JW. Induction by glucocorticoids of angiotensin converting enzyme production from bovine endothelial cells in culture and rat lung in vivo. J Clin Invest 1982; 70:684–692.

110. Johnson AR, Coalson JJ, Ashton J, Larumbide M, Erdos EG. Neutral endopeptidase in serum samples from patients with adult respiratory distress syndrome: comparison with angiotensin-respiratory distress syndrome: comparison with angiotensin-converting enzyme. Am Rev Respir Dis 1985; 132:1262–1267.

111. Skidgel RA, Jackman HL, Erdos EG. Metabolism of substance P and bradykinin by human neutrophils. Biochem Pharmacol 1991; 41:1335–1344.

112. Caughey GH. Roles of mast cell tryptase and chymase in airway function. Am J Physiol 1989; 257:L39–L46.

113. Cheung D, Bel EH, Den Hartigh J, Dijkman JH, Sterk PJ. The effect of an inhaled neutral endopeptidase inhibitor, thiorphan, on airway responses to neurokinin A in normal humans in vivo. Am Rev Respir Dis 1992; 145:1275–1280.

114. Cheung D, Timmers MC, Zwinderman AH, Den Hartigh J, Dijkman JH, Sterk PJ. Neutral endopeptidase activity and airway hyperresponsiveness to neurokinin A in asthmatic subjects in vivo. Am Rev Respir Dis 1993; 148:1467–1473.

115. Ichinose M, Nakajima N, Takahashi T, Yamauchi H, Inoue H, Takishima T. Protection against bradykinin-induced bronchoconstriction in asthmatic patients by neurokinin receptor antagonist. Lancet 1992; 340:1248–1251.

116. Joos GF, Kips JC, Peleman RA, Pauwels RA. Tachykinin antagonists and the airways. Arch Int Pharmacodyn Ther 1995; 329:205–219.

117. Lunde H, Hedner J, Svedmyr N. Lack of 4 weeks treatment with the neurokinin receptor antagonist FK224 in mild to moderate asthma. Eur Respir J 1994; 7:151s.

118. Fahy JV, Wong HH, Geppetti P, Nadel JA, Boshey HA. Effect of an NK1 receptor antagonist (CP-99,994) on hypertononic saline-induced bronchoconstriction and cough in asthmatic subjects. Am J Respir Crit Care Med 1994; 149:A1057.

119. Turner CJ, Andersen DK, Patterson RF, Keir SO, Lee P, Watson JW. Dual antagonism of NK1 and NK2 receptors by CP 99,994, and SR 48,968 prevents airway hyperresponsiveness in primates. Am J Respir Crit Care Med 1996; 153:A160.

120. Gotoh E, Yamagami T, Yamamoto H, Okamoto H. Chromosomal assignment of human VIP/PHM-27 gene to 6q26–q27 region by spot blot hybridization and in situ hybridization. Biochem Int 1988; 17:555–562.

121. Gozes I, Avidor R, Yahav Y, Katznelson D, Croce CM, Huebner K. The gene encoding vasoactive intestinal peptide is located on human chromosome 6p21–6qter. Hum Genet 1987; 75:41–44.

122. Tsukada T, Fink JS, Mandel G, Goodman RH. Identification of a region in the human vasoactive intestinal polypeptide gene responsible for regulation by cyclic AMP. J Biol Chem 1987; 262:8743–8747.

123. Uddman R, Alumets J, Densert O, Hakanson R, Sundler F. Occurrence and distribution of VIP nerves in the nasal mucosa and tracheobronchial wall. Acta Otolaryngol (Stockh) 1978; 86:443–448.

124. Dey RD, Shannon WA Jr, Said SI. Localization of VIP-immunoreactive nerves in airways and pulmonary vessels of dogs, cat, and human subjects. Cell Tissue Res 1981; 220:231–238.

125. Dey RD, Altemus JB, Michalkiewicz M. Distribution of vasoactive intestinal peptide- and substance P-containing nerves originating from neurons of airway ganglia in cat bronchi. J Comp Neurol 1991; 304:330–340.

126. Canning BJ, Undem BJ. Evidence that antidromically stimulated vagal afferents activate inhibitory neurones innervating guinea-pig trachealis. J Physiol (Lond) 1994; 480:613–625.

127. Polak JM, Bloom SR. Regulatory peptides and neuron-specific enolase in the respiratory tract of man and other mammals. Exp Lung Res 1982; 3:313–328.

128. Laitinen A, Partanen M, Hervonen A, Pelto-Huikko M, Laitinen LA. VIP like immunoreactive nerves in the human respiratory tract. Histochemistry 1985; 82:313–319.

129. Ghatei MA, Sheppard DJ, Henzen-Logman S, Blank MA, Polak JM, Bloom SR. Bombesin and vasoactive intestinal polypeptide in the developing lung: marked changes in the acute respiratory distress syndrome. J Clin Endocrinol Metab 1983; 57:1226–1232.

130. Barnes PJ, Baraniuk JN, Belvisi MG. Neuropeptides in the respiratory tract, Part I. Am Rev Respir Dis 1991; 144:1187–1198.

131. Dey RD, Mayer B, Said SI. Colocalization of vasoactive intestinal peptide and nitric oxide synthase in neurons of the ferret trachea. Neuroscience 1993; 54:839–843.

132. Geppetti P, De Rossi M, Renzi D, Amenta F. Age-related changes in vasoactive intestinal polypeptide levels and distribution in the rat lung. J Neural Transm 1988; 74:1–10.

133. Ollerenshaw S, Jarvis D, Woolcock A, Sullivan C, Scheibner T. Absence of immunoreactive vasoactive intestinal polypeptide in tissue from the lungs of patients with asthma. N Engl J Med 1989; 320:1244–1248.

134. Howarth PH, Djukanovic R, Wilson JW, Holgate ST, Springall DR, Polak JM. Mucosal nerves in endobronchial biopsies in asthma and non-asthma. Int Arch Allergy Appl Immunol 1991; 94:330–333.

135. Matsuzaki Y, Hamasaki Y, Said SI. Vasoactive intestinal peptide: a possible transmitter of nonadrenergic relaxation of guinea pig airways. Science 1980; 210:1252–1253.

136. Lei YH, Barnes PJ, Rogers DF. Regulation of NANC neural bronchoconstriction in vivo in the guinea-pig-involvement of nitric oxide, vasoactive intestinal peptide and soluble guanylyl cyclase. Br J Pharmacol 1993; 108:228–235.

137. Belvisi MG, Stretton CD, Miura M, Verleden GM, Tadjkarimi S, Yacoub MH, Barnes PJ. Inhibitory NANC nerves in human tracheal smooth muscle: a quest for the neurotransmitter. J Appl Physiol 1992; 73:2505–2510.

138. Ingenito EP, Mark L, Lilly C, Davidson B. Autonomic regulation of tissue resistance in the guinea pig. J Appl Physiol 1995; 78:1382–1387.

139. Said SI, Mutt V. Relationship of spasmogenic and smooth muscle relaxant peptides from normal lung to other vasoactive compounds. Nature 1977; 265:84–86.

140. Said SI. Critique—vasoactive intestinal peptide—involvement of calmodulin and catalytic antibodies. Neurochem Int 1993; 23:215–219.

141. Goetzl EJ, Sreedharan SP, Turck CW, Bridenbaugh R, Malfroy B. Preferential cleavage of amino- and carboxyl-terminal oligopeptides from vasoactive intestinal polypeptide by human recombinant enkephalinase (neutral endopeptidase, EC 3.4.24.11). Biochem Biophys Res Commun 1989; 158:850–854.

142. Lilly CM, Martins MA, Drazen JM. Peptidase modulation of vasoactive intestinal peptide pulmonary relaxation in tracheal superfused guinea pig lungs. J Clin Invest 1993; 91:235–243.

143. Tanaka T, McRae BJ, Cho K, Cook R, Fraki JE, Johnson DA, Powers JC. Mammalian tissue trypsin-like enzymes. Comparative reactivities of human skin tryptase, human lung tryptase, and bovine trypsin with peptide 4-nitroanilide and thioester substrates. J Biol Chem 1983; 258:13552–13557.

144. Tam EK, Caughey GH. Degradation of airway neuropeptides by human lung tryptase. Am J Respir Cell Mol Biol 1990; 3:27–32.

145. Franconi GM, Graf PD, Lazarus SC, Nadel JA, Caughey GH. Mast cell tryptase and chymase reverse airway smooth muscle relaxation induced by vasoactive intestinal peptide in the ferret. J Pharmacol Exp Ther 1989; 248:947–951.

146. Thompson DC, Diamond L, Altiere RJ. Enzymatic modulation of vasoactive intestinal peptide and nonadrenergic noncholinergic inhibitory responses in guinea pig tracheae. Am Rev Respir Dis 1990; 142:1119–1123.

147. Hachisu M, Hiranuma T, Tani S, Iizuka T. Enzymatic degradation of helodermin and vasoactive intestinal polypeptide. J Pharmacobiodyn 1991; 14:126–131.

148. Farmer SG, Togo J. Effects of epithelium removal on relaxation of airway smooth muscle induced by vasoactive intestinal peptide and electrical field stimulation. Br J Pharmacol 1990; 100:73–78.

149. Thompson DC, Altiere RJ, Diamond L. The effects of antagonists of vasoactive intestinal peptide on nonadrenergic noncholinergic inhibitory responses in feline airways. Peptides 1988; 9:443–447.

150. Tam EK, Franconi GM, Nadel JA, Caughey GH. Protease inhibitors potentiate smooth muscle relaxation induced by vasoactive intestinal peptide in isolated human bronchi. Am J Respir Cell Mol Biol 1990; 2:449–452.

151. Hulsmann AR, Jongejan RC, Raatgeep HR, Stijnen T, Bonta IL, Kerrebijn KF, Dejongste JC. Epithelium removal and peptidase inhibition enhance relaxation of human airways to vasoactive intestinal peptide. Am Rev Respir Dis 1993; 147:1483–1486.

152. Drazen JM, Shore SA, Gerard NP. Enzymatic degradation of neuropeptides: a possible mechanism of airway hyperresponsiveness. Prog Clin Biol Res 1989; 297:45–55.

153. Martins MA, Shore SA, Drazen JM. Release of tachykinins by histamine, methacholine, PAF, LTD4, and substance-P from guinea pig lungs. Am J Physiol 1991; 261:L449–L455.

154. Manzini S, Meini S. Involvement of capsaicin-sensitive nerves in the bronchomotor effects of arachidonic acid and melittin: a possible role for lipoxin-A4. Br J Pharmacol 1991; 103:1027–1032.

155. Matsuse T, Thomson RJ, Chen XR, Salari H, Schellenberg RR. Capsaicin inhibits airway hyperresponsiveness but not lipoxygenase activity or eosinophilia after repeated aerosolized antigen in guinea pigs. Am Rev Respir Dis 1991; 144:368–372.

156. Manzini S, Maggi CA, Geppetti P, Bacciarelli C. Capsaicin desensitization protects from antigen-induced bronchospasm in conscious guinea-pigs. Eur J Pharmacol 1987; 138:307–308.

157. Cox CP, Lerner MR, Wells JH, Said SI. Inhaled vasoactive intestinal peptide (VIP) prevents bronchoconstriction induced by inhaled histamine. Am Rev Respir Dis 1983; 127:249.

158. Said SI, Geumei A, Hara N. Bronchodilator effect of VIP in vivo: protection against bronchoconstriction by histamine or prostaglandin$_{F2\alpha}$. In: Said SI, ed. Vasoactive Intestinal Peptide. New York: Raven Press, 1982:185–191.

159. Barnes PJ, Dixon CM. The effect of inhaled vasoactive intestinal peptide on bronchial reactivity to histamine in humans. Am Rev Respir Dis 1984; 130:162–166.

160. Altiere RJ, Kung M, Diamond L. Comparative effects of inhaled isoproterenol and vasoactive intestinal peptide on histamine-induced bronchoconstriction in human subjects. Chest 1984; 1:153–154.

161. Morice A, Unwin RJ, Sever PS. Vasoactive intestinal peptide causes bronchodilatation and protects against histamine-induced bronchoconstriction in asthmatic subjects. Lancet 1983; 2:1225–1227.

162. Palmer JB, Cuss FM, Warren JB, Blank M, Bloom SR, Barnes PJ. Effect of infused vasoactive intestinal peptide on airway function in normal subjects. Thorax 1986; 41: 663–666.

22

Tachykinin Antagonists in Asthma and Inflammation

CARLO ALBERTO MAGGI

Menarini Ricerche
Florence, Italy

I. Introduction

Tachykinins (TKs) are the members of a family of peptides which share the common C-terminal sequence Phe-Xaa-Gly-Leu-MetNH$_2$. The first peptide of this family, substance P (SP), was discovered by Von Euler and Gaddum (1) as a hypotensive and spasmogenic extract from the equine intestine, but it was other 40 years before the definition of the chemical structure of this peptide (2). Meantime, Erspamer and co-workers had discovered and chemically characterized a number of nonmammalian peptides endowed with powerful biological activity in mammalian tissues and grouped them on the basis of the common C-terminal sequence Phe-Xaa-Gly-Leu-MetNH$_2$. When the primary sequence of SP was established, it became evident that SP is the mammalian counterpart of the amphibian peptides discovered by Erspamer and co-workers, and the "tachykinin peptide family" was broadened to include its mammalian and nonmammalian members (3). The term "tachykinin" was coined to describe the relatively fast development of the contractile action produced by these peptides in smooth muscles (3).

Since the early work of Lembeck (4), the hypothesis has been put forward that SP could be a neurotransmitter, especially a primary afferent transmitter involved in signaling painful/noxious events to the spinal cord and, in parallel, responsible for the phenomenon of antidromic vasodilatation.

In the 1970s, research on the distribution and physiological significance of TKs expanded rapidly, with the demonstration that SP is a neuropeptide, widely distributed in the central and peripheral nervous system (5,6). In the middle 1980s the discovery of neurokinin A (NKA) and neurokinin B (NKB) gave further impulse to this line of research (7), culminating with the isolation of the genes which encode the mammalian TKs (8) and the TK receptor proteins (9), and with the development of a number of potent and selective peptide and nonpeptide receptor antagonists (10). The current status of TKs as neurotransmitters in the mammalian central and peripheral nervous system has been reviewed recently (11). Owing to the putative importance of TKs in various pathophysiological events, there is a general expectation that TK receptor antagonists may prove to be useful drugs for treatment of a variety of human diseases (10). In particular, the prominent inflammatory and proinflammatory actions exerted by TKs in the airways and in other regions of the body provide the basis for speculating a role of TK receptor antagonists as a new class of drugs to be used for treatment of asthma and/or other inflammatory diseases.

II. Tachykinin Receptors and Their Roles in Asthma and Inflammation

A. Three Receptors Mediate the Actions of Tachykinins

Since the middle of the 1980s, pharmacological evidence has been accumulated to indicate the existence of three distinct receptors which mediate the biological actions encoded by the common C-terminal sequence of TKs. The three receptors have been termed NK_1, NK_2, and NK_3, for which SP, NKA, and NKB possess preferential affinity, respectively (10,12,13). The three receptors have now been cloned from several different species. They belong to the superfamily of rhodopsin-like G-protein-coupled receptors with seven putative transmembrane-spanning hydrophobic segments (9,14). The stimulation of inositol phospholipid hydrolysis is the main second messenger system activated by the three receptors (15), although clear examples exist indicating that the signal generated through a selective stimulation of different TK receptors in the same preparation can diverge into a remarkable specialization (16).

The rank order of potency of natural TKs in producing certain biological effects (or affinity in radioligand-binding assays) was the original criterion for distinguishing the three mammalian TK receptors (12,17). A number of physiological paradigms exist in which different TKs are released from nerve terminals (e.g., SP and NKA released from peripheral endings of sensory nerves in guinea pig airways) (18,19), and different TK receptors are expressed by the same target cells (e.g., NK_1 and NK_2 receptors by smooth muscle cells in guinea pig trachea and bronchi) (20,21). When such an arrangement occurs, it appears quite logical to speculate that the different endogenous ligands will act preferentially on their "own" or "preferred" receptor to deliver distinct components of the TKergic message. On the other hand, a certain degree of promiscuity exists between different ligands and receptors, illustrating the complexities of in TKergic transmission. At one extreme, examples are available in which no preferred "endogenous ligand" is available to stimulate a given TK receptor. Thus, a subclass of myenteric neurons in guinea pig intestine expresses NK_3 receptors, while NKB is not found at this level; it is believed that SP/NKA activates NK_3 receptors at the location (22,23).

For the purpose of this chapter, related to the role of TKs in asthma and inflammation, two peptides, SP and NKA, are mostly important, because NKB is apparently missing (or nonexpressed) in the peripheral nervous system. With regard to TK receptors, it appears that activation of NK_1 (SP-preferring) or NK_2 (NKA-preferring) receptors accounts for the major part, if not all, of the inflammatory actions of TKs in the peripheral nervous system.

B. Vascular and Proinflammatory Actions of Tachykinins

Many of the actions of TKs which are potentially relevant to asthma/inflammation are produced by activating NK_1 receptors. The NK_1 receptors mediate TK-induced vasodilatation, which is endothelium-dependent and involves the release of both nitric oxide (NO) and a hyperpolarizing/vasodilator factor different from NO (24–26). Following deprivation of the endothelium, a vasoconstrictor response to TKs can be evidenced in certain blood vessels such as in the guinea pig or rabbit pulmonary artery, involving either NK_1 or NK_2 receptors (27,28). TKs also produce changes in tone of isolated veins; again, an endothelium-dependent vasodilation via NK_1 receptors (29) and an endothelium-independent vasoconstriction via NK_1 or NK_3 receptors (29–31) can be involved. NK_1 receptors likewise predominate in mediating the increase in postcapillary venular permeability leading to plasma protein extravasation (PPE), recognized as a distinct feature of neurogenic inflammation since the work of Jancso et al. (32,33). PPE, in response to sensory nerve stimulation, occurs in the skin but also in joints, eye and conjunctiva, upper and lower airways, genitourinary tract, nasal mucosa, and many other visceral tissues from various species; PPE can be evoked either by acute application of capsaicin or by electrical stimulation of nerve trunks (saphenous, vagus, pelvic, etc.), but also following electrical stimulation of dorsal roots (see Refs. 34 and 35 for reviews). The importance of TKs in producing PPE was appreciated early, because exogenously administered SP increases vascular permeability in the same regions of the body where PPE is elicited by capsaicin (36,37). Experiments with receptor selective agonists (38,39) and antagonists (40,41) have indicated that the NK_1 receptor is the major if not the only mediator of PPE produced by both exogenous and endogenous TKs released by capsaicin or nerve stimulation. NO generation has been implicated in PPE produced by SP in rat blisters (42), while PPE induced in rat airways and urinary bladder in response to i.v. injection of SP or of a selective NK_1 receptor agonist was unchanged by NO synthase inhibitors (43). The mechanism linking TK release to NK_1 receptor stimulation and increase of vascular permeability has not been exactly determined. In particular, PPE occurs at the level of postcapillary venules through the formation of gaps between endothelial cells, which express NK_1 receptors; thus, the simplest explanation is to assume that released TKs act directly on the endothelium of postcapillary venules to increase vascular permeability. On the other hand, there is little direct evidence for the presence of SP-containing nerves at the postcapillary venules level, while a dense innervation is observed at the arteriolar level (44); thus, SP released at the arteriolar level may reach the endothelium of postcapillary venules through the bloodstream. Alternatively, the possibility remains that TKs act on other target cells to induce the release of mediators which in turn produce the PPE (see Ref. 35 for review). Although the vast majority of studies has consistently indicated the NK_1 receptor as the main if not exclusive mediator of PPE, the indication has been provided that NK_2 receptors could also be involved in producing PPE in lower guinea pig airways (45).

NK_1 receptors are also predominant mediators of the stimulant action of TKs on exocrine gland secretion: In the airways, including human airways, mucus secretion from both seromucous glands and goblet cells is NK_1 receptor-mediated (46–50).

NK_1 receptors mediate variable changes of bronchomotor tone in different species; thus, contractile NK_1 receptors on tracheobronchial smooth muscle are evident in, e.g., guinea pig isolated airways (20), while epithelial NK_1 receptors mediate the release of bronchodilator substance(s) in other species (51,52). NK_2 receptors have a more consistent bronchoconstrictor role in different species: They are apparently expressed directly on tracheobronchial smooth muscle cells; in some species, activation of NK_2 receptors by endogenous TKs provides the dominant component of the NANC bronchoconstriction, both in vitro and in vivo (21,53,54). Even more importantly, NK_2 receptors mediate constriction of human isolated airway smooth muscle and are the obvious target for a possible antibronchospastic approach using TK receptor antagonists (55,56).

C. Immune and Trophic Actions of Tachykinins

There is an extensive literature documenting the ability of TKs to stimulate/recruit leukocytes and inflammatory cells. At the present time, the relevance of these effects for the role of TKs in asthma and inflammtion can neither be overlooked nor emphasized. In most instances, what is simply missing in this field is a convincing piece of evidence to link the effects produced by TKs, as demonstrated by studies on isolated cells in highly artificial conditions, with the overall changes in intensity/time course of inflammatory responses observed following administration of TK receptor antagonists in vivo. In other words, it is conceivable that a part of the antiinflammatory effect produced by TK receptor antagonists in experimental models of inflammation depends on a blockade of TK receptors expressed on leukocyte/inflammatory cells, but the exact importance of these mechanisms in vivo can hardly be determined with the currently available techniques. Moreover, many of the in vitro studies on this topic have been performed using natural TKs and nonselective antagonists, so a firm characterization of the receptor involved is not always possible.

Neutrophils and Eosinophils

In addition to vasodilatation and leakage of plasma proteins, the adhesion of granulocytes (neutrophils and eosinophils) to endothelial cells and their subsequent migration into inflamed tissues is an integral and important component of neurogenic inflammation (57). Exposure to the TK-releasing agent, capsaicin, or inhalation of exogenous TKs induces granulocytes accumulation in guinea pig or rat airways (58–61); experiments with selective agonists and antagonists indicate that NK_1 but not NK_2 receptors mediate this effect in rat airways (61). Likewise, exogenous TKs induce granulocyte accumulation in mouse or guinea pig skin (62–65). The mechanism(s) through which TKs induce granulocyte migration into the tissues is controversial. In particular, two elements are in discussion, as follows: (a) the molecular mechanisms through which TKs induce granulocyte migration and (b) whether this effect is a direct one on granulocytes or may be produced indirectly.

With regard to the first issue, some studies have provided convincing evidence for involvement of NK_1 receptors (61,65). In these studies, tissue accumulation of granulocytes induced by inflammatory stimuli was blocked by using potent and selective nonpeptide NK_1 receptor antagonists and appropriate controls were performed to validate the inhibitory action of these agents. Other studies have provided evidence that SP-induced granulocyte infiltration in mouse skin is critically dependent on mast cell degranulation (63,64), a receptor-independent effect of SP (see Section II.C). Of particular interest are the results presented by Walsh et al. (62). They found that intradermally injected SP produces edema and granulocyte (neutrophils and eosinophils) accumulation in guinea pig skin with different potency; edema formation was induced by lower concentrations of SP and was blocked by NK_1 receptor antagonists, while granulocyte accumulation required higher doses of the peptide and was not affected by the antagonists. Walsh et al. (62) also found that a specific inhibitor of 5-lipoxygenase blocked the eosinophil accumulation induced by SP and speculated that mast cell-derived products of 5-lipoxygenase may play an important role in granulocyte infiltration.

Clearly, the possibility remains open that in different models and/or for different stimuli, both mechanisms (NK_1 receptor activation and mast cell degranulation) are involved in mediating granulocyte migration into tissues. It is also worth noting that, in different rat strains, airway neurogenic inflammation receives a variable contribution from mast cell degranulation, in addition to NK_1 receptor activation (66–68), indicating that genetic factors also influence the magnitude of and mechanisms involved in the inflammatory responses to TKs (68).

A complex question is whether stimulation of granulocyte infiltration into tissues by TKs is a direct effect on granulocytes or is produced indirectly; clearly, the results of studies implicating mast cells degranulation in granulocyte migration by SP (62–64) favor the second hypothesis. The obvious question of whether granulocytes express TK receptors has been only partially solved by studies addressing the direct effects of SP and other TKs on granulocyte function in vitro: TKs are chemotactic (69–72), stimulate granulocyte metabolism (73), exert a "priming" effect toward responses evoked by other stimulants (74–77), stimulate lyso-PAF:acetyl CoA trans-ferase activity (78), antibody-dependent cell-mediated cytotoxicity (76), inter-leukin-8 synthesis, and release (75) on granulocytes. The true significance of the effects observed in these studies is uncertain: With few exceptions (71,74,77), the observed effects are produced by micromolar concentrations of SP, which are un-likely to be relevant for the in vivo action of this neuropeptide during neurogenic inflammation. On the other hand, these studies illustrate at least a potential of TKs to influence granulocyte function directly. Other investigators have examined the possi-bility that granulocyte recruitment by TKs during neurogenic inflammation is an indirect effect. In particular, since the adhesion to endothelial cells is the first step for granulocyte migration into inflamed tissues, some investigators have addressed this aspect of the problem specifically. Matis et al. (79) showed that micromolar concen-trations of SP, via mast cell degranulation, induce expression of the adhesion mole-cule ELAM-1 on foreskin postcapillary venules. Working with confluent mono-layers, Zimmerman et al. (80) showed that picomolar concentrations of SP are capable of stimulating neutrophil adherence to human umbilical vein endothelial cells (HUVEC); they found that preincubating SP with neutrophils but not with HUVEC yielded the proadhesive response, and that antibodies directed toward several endothelial cell adhesion molecules, including ICAM-1, were ineffective in preventing the response to SP. Zimmerman et al. concluded that the effect of SP is likely to be mediated by an interaction with granulocytes themselves, rather than through the expression of adhesion molecules by endothelial cells (80). This inter-pretation may be supported by the observation that SP induces expression of the leukocyte adhesion glycoprotein CD11 (81), although antibodies directed toward this or other adhesion molecules failed to prevent the proadhesive action of SP in the study of Zimmerman et al. (80). In contrast, Walsh et al. (62) reported that a mono-clonal antibody directed toward the granulocyte adhesion molecule C-18 blocked SP-induced granulocyte infiltration in guinea pig skin. Finally, Nakagawa et al. (82) found that SP increases expression of the adhesion molecule ICAM-1 by HUVEC and also that an anti-ICAM-1 antibody blocked the transendothelial granulocyte migration induced by SP. Therefore, there are data supporting the idea that SP can induce the expression of adhesion molecules from either granulocytes and endo-thelial cells, although the true importance of either mechanism for SP-induced granulocyte infiltration remains to be assessed.

Von Essen et al. (83) addressed another aspect of the problem by investigating the possibility that an action of SP on bronchial epithelial cell may induce the release of factor(s) which are in turn responsible for granulocyte migration during neuro-genic inflammation. They reported that bovine bronchial epithelial cells challenged with picomolar concentrations of SP produce neutrophil chemotactic activity (83). More recently, the same group produced evidence that the adherence of neutrophils to bronchial epithelial cells, a process thought to be of importance in retaining granulocytes at the site of inflammation, is increased by SP via NK_1 receptor stimulation and expression of adhesion molecules (84).

Summarizing this section, granulocyte recruitment at sites of inflammation is

an essential component of neurogenic inflammation, and this effect can be reproduced by TKs in vivo. There is evidence that NK_1 receptors mediate this effect, at
least in certain models, but the possible involvement of mast cell degranulation
should also be considered. Although TKs produce a number of direct effects on
granulocytes, it is uncertain whether these direct effects are of any physiological
relevance. Contrasting data are available on the possibility that endothelial cells are
the primary target of action of SP in inducing migration of granulocytes in vivo.

Lymphocytes, Macrophages, and Monocytes

A number of studies have documented the presence of TK-containing nerves in
various lymphoid organs, such as the thymus, spleen, lymph nodes, tonsils, lymphoid
aggregates in the gut, lung, and nasal mucosa (85–91), and in the walls of lymphatic
vessels (92,93); SP and NKA have been detected in rat thymus, spleen, and lymph
nodes by radioimmunoassay (89,94,95). At these sites TK-containing nerve fibers
show a perivascular distribution but some fibers also distribute within the follicles of
lymph nodes and aggregates of lymphoid cells (89–91,96). TK-containing nerve
fibers are of lower density in visceral (abdominal) than in cervical (submaxillary
submandibular) or somatic (axillar popliteal) lymph nodes (87,88). TK-containing
nerve fibers in lymphoid organs represent the peripheral endings of capsaicin-
sensitive primary afferent neurons: Early studies showed that only part of the TK
content of lymphoid organs measurable by radioimmunoassay is depleted by pretreatment with capsaicin (94,95). The observation that cells expressing basal levels of
preprotachykinin I mRNA are present in lymphoid organ (97) suggests that extraneuronal sources of TK peptides are also present at this level. Evidence has also been
presented of an increased expression of preprotachykinin I mRNA in lymphoid
tissues after immunization (98).

Receptors for TKs have been described in a variety of lymphoid organs (96,
99,100). SP (NK_1) receptors were observed in germinal centers of canine mesenteric
lymph nodes and in lymph nodes from canine gastric antral mucosa, ileum, and colon
(96,101–103). A dramatic (1000- to 2000-fold) and selective increase of SP (NK_1)-
binding sites has been reported in the colon of patients suffering from ulcerative
colitis or Crohn's disease. Furthermore, an "ectopic" expression of the receptors was
observed as compared to the distribution observed in the normal human colon,
including germinal centers of lymph nodules which border the muscularis mucosa
(101–103). These findings led to speculation that the local release of TKs might be a
factor contributing to the immune disorder underlying chronic inflammatory bowel
diseases.

The presence and characterization of TK receptors on lymphocytes have been
the object of several studies: The human B-lymphoblastic cell line IM-9 expresses a
well-characterized NK_1 receptor and is commonly used to study the affinity of novel
ligands for the human type NK_1 receptor (104,105). Two studies reported the presence of specific SP receptor on human peripheral blood T lymphocytes (K_d for SP =
185 nM), but not on B lymphocytes or monocytes (106), and on rat T cells from

spleen and Peyer's patches (K_d for SP = 0.5 nM) (107). Other groups have failed to demonstrate specific SP receptors on human lymphocytes (108,109). Recently, molecular evidence for the expression of NK_1 receptor by granuloma T lymphocytes in a murine model of schistosomiasis has been presented (110).

Therefore, a simple answer to the question of whether lymphocytes in general express TK receptors is not possible. When considering the remarkable heterogeneity and functional plasticity of lymphoid cells, not to mention possible species and regional differences, the contrast between the published data could turn out to be more apparent than substantial. At the present time it may be concluded that lymphoid cells, under certain circumstances, do indeed express a SP (NK_1) receptor (104,110); this expression is probably regulated by a number of factors which are presently poorly understood, and generalizations on this matter are very difficult. Kavelaars et al. (109) have recently reviewed a number of studies on this issue and concluded that, in addition to a "classical" NK_1 receptor, other pathways for SP signaling exist in leukocytes which may involve both a "nonneurokinin" receptor for B cells and monocytes (see below) and possibly also a receptor-independent pathway similar to the mechanism proposed for SP-induced mast cell degranulation (see Section II.C) (109). The latter two pathways would both require relatively high (micromolar) concentrations of SP to be activated; this obviously poses a question about their physiological significance, although a case could be made for an "emergency system" activated only when a massive release of the neuropeptide occurs.

With regard to the functional effect of TKs on lymphocytes, several studies have reported that SP, or other TKs, stimulate lymphocyte proliferation (109,111–113). SP has also been reported to elevate intracellular calcium in human T lymphocytes (114) and to stimulate production of interleukin-2 (115,116), while the migration of T and B lymphocytes between blood and lymph is not altered by SP in vivo (117).

The effects of SP on human lymphocytes from diseased patients have also been examined in a few studies. Agro and Stanisz (118) failed to detect any stimulant action of SP on proliferation of lymphocytes from blood or synovial fluid of patients with rheumatoid arthritis. On the other hand, Covas et al. (119) reported a marked tendency (albeit not statistically significant) toward an increased proliferative response in peripheral blood mononuclear cells (lymphocytes and monocytes) from patients with rheumatoid arthritis. SP was shown to alter the expression of membrane markers in peripheral mononuclear cells from these patients (120). Finally, a paradoxical inhibition of lymphocyte proliferation by SP has been reported recently in AIDS patients, as opposed to the stimulant effect observed in healthy subjects or in patients with asymptomatic HIV infection (121).

Various studies have documented the ability of SP to enhance production of immunoglobulins (see Ref. 122 for review). Since T lymphocytes and macrophages can express TK receptors, and these cells are important for B-lymphocyte differentiation and immunoglobulin production, the question arises whether the effect of SP on B lymphocytes is a direct one. Studies involving pure populations of B lymphocytes and B-lymphocyte cloned cells cell lines have shown that B cells are a direct target for SP action; however, SP alone is not a sufficient stimulus for immunoglobulin production but rather enhances the response to a first stimulant such as lipopolysaccharide (122,123). Thus SP has been defined as a late-acting B-cell differentiation cofactor. The ability of SP to induce the expression of mitogen-activated kinase (MAP kinase) in human peripheral blood B lymphocytes has also been demonstrated recently (109).

Several studies have also addressed the question of effects of SP and other TKs on monocytes and macrophages. SP stimulates the release of inflammatory cytokines (124,125), activates MAP kinase (126), and stimulates chemotaxis of human blood monocytes (71). Human monocytes do not express a conventional SP (NK$_1$) receptor: The effects of SP appear to be mediated by an atypical SP receptor which is also partially recognized by bombesin (the K_i values for displacement of ^{125}I-SP binding were 224 and 729 nM for SP and bombesin, respectively) (126,127).

With regard to macrophages, Hartung et al. (128) first described the existence of a SP-binding site on guinea pig macrophages with a dissociation constant of 19 nM; various natural TKs stimulate interleukin-1 secretion from macrophages (129). Moreover, macrophages themselves are a source of SP (130,131), and expression of SP mRNA is upregulated by lipopolysaccharide in macrophages (132). TKs, via NK$_2$ and, to a minor extent, via NK$_1$ receptors, stimulate superoxide anion production and prostanoids release from guinea pig alveolar macrophages (133–136); in macrophages obtained from ovalbumin-sensitized guinea pigs, a remarkable (one to three orders of magnitude for different TKs) leftward shift in the concentration–response curve to NK$_2$ receptor agonists has been observed (135).

These findings are of special interest because of the proposed key role of alveolar macrophages in the local inflammatory processes associated with bronchial asthma (137), and the reported association between TK-containing nerve fibers and macrophages in the lung (87,88,138). Activation of alveolar macrophages has been implicated in the genesis of bronchial hyperresponsiveness induced by inhalation of SP (139), and recent findings have indicated that TKs via NK$_2$ receptor induce the production of gelatinase from alveolar macrophages, a response which could be important for inducing the epithelial lesions observed in bronchial hyperresponsiveness and asthma (140).

As compared to a relatively extensive in vitro literature, few studies have addressed the importance of TKs in modulating the immune response in vivo. Based on the ability of capsaicin pretreatment to deplete SP from sensory nerves in vivo, Helme's group postulated that a change in the immune response observed in these animals may indirectly indicate a role of TKs in this function. Indeed, they found that capsaicin pretreatment induces a marked ($> 80\%$) reduction in the number of cells secreting antigen-specific antibodies in the regional lymphnode after s.c. antigenic stimulus in rats, and that the reduced immune response was reversed by exogenous SP application (141). Further studies from the same group indicated that the restorative effect of SP was reproduced by other TKs and suggested the involvement of NK$_2$ receptors in this response (142,143). Other groups have likewise shown that exogenous administration of SP in vivo stimulates the development of immunoglobulin-producing cells (144,145).

Summarizing this section, it appears established that TKs are present in primary and secondary lymphoid organs, that sensory nerves represent their major source at this level although some nonneuronal cell types also contribute, and that TK receptors are present in lymphoid organs where they can undergo dramatic changes in expression during pathology. Lymphoid cells can express the NK_1 receptor, at least under certain circumstances, yet certain effects of TKs may involve atypical interactions with lymphocytes. Obviously, indirect effects of TKs on lymphocyte function, especially in vivo, must be considered for, e.g., effects involving monocytes/macrophages. The true relevance of TK actions in regulating immune function remains substantially speculative at present.

Mast Cells

The observation that SP degranulates mast cells, producing histamine release (146,147), has prompted a number of studies aiming to establish the possible relevance of this mechanism in the genesis of axon reflex and neurogenic inflammation (see Ref. 148). Anatomical evidence supports a communication between sensory nerves and mast cells: A close spatial relationship between sensory nerves containing SP/CGRP-LI and mast cells has been described in the rat diaphragm and mesentery (149), rat intestine (150), rat thymus, lymph nodes, and airways (87,88), pig skin and airway mucosa (151), and rat knee joint (152), and a similar arrangement was proposed to exist in the human gastrointestinal mucosa (153). Upon electron microscopy, close contacts were observed between nerve profiles and mast cells in rat jejunum, with a space of less than 20 nm between the two membranes; dense core vesicles were frequently seen in the nerve approximating to the site of contact (150,154).

Studies on isolated mast cells have led to various general conclusions about mast cell degranulation by SP: (a) It requires the N-terminal region of the SP molecule and does not involve NK_1, NK_2, or NK_3 receptors; (b) it is produced by relatively high concentrations of SP, in the micromolar range; (c) it is not shared by other TKs which lack the basic N-terminal residues of SP; (d) it is not uniformly observed in mast cells from different species, nor from mast cells of the same species but from different organs (155–161). Regarding the mechanisms involved, a direct activation of G proteins (receptor-independent mechanism) by the cluster of positively charged amino acid residues at the N-terminal end of the SP molecule is the currently available explanation for this effect; this mechanism would be common to the mast cell-degranulating activity of compound 48/80, bradykinin, mastoparan, and other cationic amphiphilic peptides (see Ref. 162 for review). Certain mast cell lines may express specific TK receptors which recognize the C-terminal sequence of TKs (163,164), but the general significance of these findings remains to be assessed. Recently, Ansel et al. (165) showed that SP, at concentrations ranging between 1 and 100 nM, activates the expression of tumor necrosis factor-α mRNA in murine peritoneal mast cells; owing to the relatively low concentration of peptide used, it is possible that a typical TK receptor is involved, but no attempt at characterization was presented.

The ability of SP to induce mast cell degranulation has also been repeatedly evidenced in vivo. SP injection in the human skin produces wheal and flare along with histamine release, and the flare is blocked by antihistamine drugs (147,166,167); and intradermal injection of other TKs, which do not possess the basic N-terminal residues of SP and do not degranulate mast cells in vitro, produce wheal (increased vascular permeability) but not flare (147,159). Part of the inflammatory reaction (e.g., PPE) produced by exogenous SP in vivo involves mast cell degranulation and release of other mediators (histamine, serotonin, prostanoids, etc.) (38,168,169).

Therefore, a solid body of literature documents the point that SP, released from sensory nerves in the peripheral nervous system, could degranulate mast cells via a receptor-independent mechanism. The question then remains as to whether this process does occur in vivo, and whether it retains a pathophysiological significance. Since the mechanism of SP-induced mast cell degranulation is relatively undefined and no specific blocker of this process exists, a straightforward reply to the above question has not yet been obtained.

On the other hand, a number of studies have tried to address the issue indirectly by exploring whether stimulation of sensory nerves does indeed produce mast cell degranulation in vivo. Antidromic electrical stimulation of sensory nerves produces mast cell degranulation in, e.g., rat skin and dura mater, although this response does not seem a necessary prerequisite for initiating/maintaining the concomitant increase in plasma protein extravasation (170–175). From this approach, the possible contribution of CGRP and/or other neuropeptides or transmitters released antidromically together with TKs from sensory nerves cannot be excluded. In a recent study, Tausk and Undem (176) showed that application of exogenous SP effectively evokes histamine release from human skin explants, while application of capsaicin at a concentration (10 μM) assumed to produce a maximal release of SP from sensory nerves did not produce this effect; their conclusion was that SP released from sensory nerves does not achieve sufficient concentrations to produce mast cell degranulation in human skin (176).

Wallace et al. (177) showed that the hyperaemic response produced by topical application of capsaicin to rat gastric mucosa is significantly greater in rats in which mastocytosis was induced by prior infection with *Nippostrongylus brasiliensis* than in normal rats; the augmented response was sensitive to pretreatment with a H_1 histamine-receptor blocker or a mast cell stabilizer as well as to depletion of mucosal mast cells by dexamethasone, and was completely abolished by pretreatment with the functional capsaicin antagonist ruthenium red or by capsaicin pretreatment. More recently, Mathison and Davison (178) reported an increased vasodilator response to topically applied capsaicin in the jejunum of *N. brasiliensis*-sensitized rats which is largely prevented by neonatal capsaicin administration, ruthenium red, pretreatment with H_1 histamine-receptor antagonists, or depletion of mucosal mast cells by dexamethasone. These findings underline the concept that interactions between mast cells and peripheral endings of sensory nerves unraveled by studies in "normal" animals may not be necessarily representative of mechanisms operating during disease or pathophysiological conditions.

Trophic Actions

TKs, like other neuropeptides, produce a number of effects on various cell types involved in tissue growth/repair which can be collectively labeled "trophic." The common denominator of these "trophic" actions is either the nature of the effect produced (e.g., stimulation of cell proliferation) or the specialized function of the cell type involved (e.g., fibroblasts or keratinocytes), indicating a possible role in the processes of tissue growth/repair. It is quite clear that all the inflammatory or pro-inflammatory actions of TKs described in the previous sections (vasodilatition, plasma protein extravasation, mast cell degranulation, etc.) could be considered "trophic actions" of TKs as well.

As described above, NK_1 receptors on endothelial cells mediate the vasodilator response to TKs and also the increase in postcapillary venular permeability which determines plasma protein extravasation. Ziche et al. (179,180) showed that TKs, via NK_1 receptors, also stimulate the proliferation and migration of endothelial cells in culture, two effects which are central to the process of angiogenesis. In fact, topical application of SP or NK_1 receptor agonists into the avascular rabbit cornea powerfully stimulate its neovascularization in vivo (179). The same group also showed that SP induces cGMP accumulation in endothelial cells (181) and that exposure to SP activates a calcium-dependent nitric oxide (NO) synthase in endothelial cells (182). Thus, NO synthase inhibitors selectively prevent the neovascularization of rabbit cornea induced by a NK_1 receptor selective agonist, as well as the proliferation and migration of endothelial cells in culture induced by SP (182). Thus, NO produced by endothelial cells following NK_1 receptor occupancy functions as an autocrine regulator of the microvascular events necessary for neovascularization, a process which is central not only for tissue development and repair but also for tumor growth and progress of certain chronic inflammatory diseases (e.g., rheumatoid synovial hypertrophy). Fan et al. (183) showed that co-administration of SP and interleukin-1α at doses which are ineffective on their own promote an intense neovascularization in a rat sponge model of angiogenesis, and that this effect is prevented by NK_1 receptor antagonists.

The contribution of the nervous system and especially of sensory nerves to the maintenance of normal tissue trophism and repair to injury is a well-established clinical concept. Nilsson et al. (184) first reported that natural TKs, SP and NKA, stimulate the proliferation of human skin fibroblasts and smooth muscle cells in culture, thus providing a possible pathway through which these sensory neuropeptides could influence wound healing. The stimulant action of TKs on smooth muscle cell proliferation is coupled with formation of inositol phosphates (185) and, in cultured rabbit airway smooth muscle cells, is mediated by NK_1 receptors (186); this effect could be important for airway smooth muscle hyperplasia and remodeling observed in chronic asthma.

The ability of TKs to stimulate fibroblast proliferation has been the object of several investigations. By using a number of receptor selective agonists and antagonists, Ziche et al. (187–189) showed that this effect is produced by NK_1 receptors in human skin fibroblasts; moreover, migration of fibroblasts occurs in parallel to their proliferation (189,190). Kahler et al. (191) showed that TKs, via NK_1 receptor activation, also stimulate the production of arachidonic acid metabolites having inhibitory influence on the proliferation of human skin fibroblasts and proposed that a direct mitogenic effect of SP is counterbalanced by SP-induced prostanoid release.

TKs also stimulate the proliferation of keratinocytes by inducing phospho-inositide accumulation and elevating intracellular Ca, as well as by inducing protein kinase C translocation (192–194), but others have failed to replicate these results (195). Endogenous SP would favor wound healing in rat skin without inducing keratinocyte proliferation (196).

III. Tachykinin Receptor Antagonists

The first generation of TK receptor antagonists was developed in the first half of the 1980s, based mainly on the insertion of multiple D-Trp residues in the backbone of the full-length SP sequence or in shorter C-terminal fragments of this undecapeptide. The corresponding ligands, the best known of which is Spantide, have had some utility for in vitro studies, but present a number of drawbacks which strongly limit their usefulness. In particular, Spantide and congeners have low potency and selec-tivity, possess local anesthetic and mast cell degranulating activities, and, in general, do not possess enough in vivo potency/duration of action to enable a firm understand-ing of the pathophysiological roles of TKs (see Ref. 10 for review).

The so-called second generation of TK receptor antagonists (10) still comprises ligands of peptide nature (linear or cyclic) but show improved potency (nanomolar affinity) and resistance to peptidases as compared to first-generation antagonists; they are also characterized by improved selectivity for only one of the three TK receptors which had been meanwhile recognized (10). Several of these ligands, such as GR 82,334 (197) or FK 888 (198) for NK_1 receptors, L 659877 (199) or MEN 10376 (200) for NK_2 receptors, have been used extensively for studying the role and distribution of the corresponding TK receptors. On the other hand, second-generation TK receptor antagonists lack, in general, sufficient potency and duration of action to be considered as drug candidates.

The discovery of CP 96345 (201) is considered the starting point of the third generation of TK receptor antagonists. The major part of these ligands is of nonpep-tide nature: They are characterized by a remarkably high potency for TK receptors (nanomolar or subnanomolar affinity) which enables them to produce an efficient and long-lasting occlusion of TK receptors in vivo. Owing to their non peptide nature, many of these ligands are also effective when administered by nonparenteral routes and are considered solid candidates for testing in humans.

A number of these third-generation antagonists are now available which pos-sess marked selectivity for NK_1 (e.g., SR 140333) (202), NK_2 (e.g., SR 48968, GR159897) (203,204), or NK_3 receptor (e.g., SR 142801) (205). Owing to its high potency, long duration of action, and bioavailability after systemic administration, the polycyclic peptide-derived NK_2 receptor antagonist MEN 10627 (206) can be included in this group.

It has been argued that, at least in certain instances, blocking more than one tachykinin may result in a compound with a better therapeutic profile than receptor-selective ligands (207); thus a compound capable of blocking both NK_1 and NK_2 receptors could be able to inhibit different components of TKergic contribution to neurogenic inflammation in the airways and may perhaps turn out to be superior to receptor-selective antagonists in treatment of, e.g., asthma/bronchial hyperreactivity. Examples of compounds with "balanced" antagonist activity at NK_1 and NK_2 receptors have been reported, such as FK 224 (208), MEN 10581 and MEN 10619 (209), and S16474 (210). However, the latter compounds are in general less potent than receptor-selective compounds; in other words, while the best receptor-selective antagonists available possess nanomolar or subnanomolar potency in blocking NK_1, NK_2, or NK_3 receptors, the reported compounds with a "balanced" antagonist affinity for NK_1/NK_2 receptors do so with micromolar potency.

In the next sections, the preclinical evidence suggesting a possible usefulness of TK receptor antagonists in asthma and inflammatory diseases will be considered as it has emerged from studies investigating the effects of TK receptor antagonists in animal models of disease.

IV. Tachykinin Receptor Antagonists in Asthma: Preclinical Evidence

In the early 1980s, work by Szolcsanyi and Bartho' (211,212) and by Lundberg and Saria (213,214) identified the existence of an "efferent" component in the airway innervation exerted by a particular subset of primary afferent neurons which are characterized by their sensitivity to the actions of capsaicin. These early investigations (211–214) showed that the capsaicin-sensitive primary afferent neurons release mediators in the airways to produce a constellation of biological effects which is commonly referred to as neurogenic inflammation. The possible relevance of neurogenic inflammation in airway pathophysiology has been the object of several recent review articles (15,35,215,216).

The following major statements can be made to summarize a number of key elements which characterize neurogenic inflammation in the airways.

1. Capsaicin-sensitive nerves are present in the airways of several mammalian species which are of both vagal and spinal afferent origin; these nerves store and release tachykinins (SP and NKA) as well as calcitonin gene-related peptide (CGRP) and, possibly, other mediators as well (215,216).
2. The adequate stimuli for eliciting neuropeptide release from these sensory nerves are of different natures, yet they can almost invariably be classified as inflammatory or potentially threatening to tissue integrity (34,35).
3. Chemosensitivity is a striking characteristic of capsaicin-sensitive affer-

ents in the airways (217). Several mediators of inflammation excite these sensory nerves to produce neurogenic inflammation; moreover, various environmental chemicals which have a recognized pathogenic relevance for airway diseases (e.g., cigarette smoke, toluene diisocyanate) activate this mechanism as well.

4. Activation of NK_1 and NK_2 receptor by endogenous TKs accounts for many features of neurogenic inflammation as it can be demonstrated in the airways of experimental animal (216).

The availability of potent and selective antagonists for blocking NK_1 or NK_2 receptors has enabled us to prove the involvement of TKs in various experimental models of asthma/bronchial hyperreactivity. Inhalation of cigarette smoke induces plasma protein extravasation (PPE) in the guinea pig trachea and bronchi which is blocked by NK_1 receptor antagonism (218,219); similarly, NK_1 receptor antagonists block airway PPE produced by challenge with hypertonic saline (220), bradykinin (221), delayed-type hypersensitivity reaction (222), and cold air inhalation (223). PPE induced by acute allergen challenge in sensitized animals was reported to be unaffected (221) or inhibited (224) by NK_1 receptor antagonists. The latter study introduced the concept that the early phase of PPE in response to acute antigen challenge does not involve TK release from sensory nerves, while the involvement of neurogenic inflammation becomes evident at a later stage; moreover, the "late" contribution of TKs to allergen-induced PPE is sharply evidenced when inhibiting neutral endopeptidase (224). In parallel to PPE, recruitment of leucocytes either induced by challenge with hypertonic saline (61) or present during delayed-type hypersensitivity reaction (222) were shown to be blocked by NK_1 receptor antagonists in rodent airways.

In-vivo studies have demonstrated that the powerful and long-lasting non-adrenergic, noncholinergic (NANC) bronchoconstriction produced by electrical stimulation of the vagus nerve or administration of capsaicin in guinea pigs chiefly involves the activation of NK_2 receptors, while NK_1 receptors provide a relatively minor contribution to the overall bronchoconstrictor response (53,54,225–227). In keeping with these results, the a predominant role of NK_2 over NK_1 receptors, as indicated by the inhibitory effect of receptor-selective antagonists, was demonstrated for the sensory nerves-dependent component of antigen- (228), hyperpnea- (229), and cold air-induced bronchoconstriction in guinea pigs (230). In addition to direct participation in inducing bronchoconstriction, NK_2 receptors appear to be involved in bronchial hyperresponsiveness to acetylcholine or histamine induced by PAF (231), antigen challenge (232,233), and inhalation of a subthreshold dose of capsaicin (233).

Several studies have reported an antitussive activity of TK receptor antagonists. Advenier et al. (234) showed that the nonpeptide NK_2 receptor antagonist, SR 48968 inhibits, in a naloxone-resistant manner, the cough-induced inhalation of citric acid in guinea pigs. A subsequent study from the same group (235) showed that the NK_1 receptor antagonist SR 140333 was devoid of antitussive activity in the same model, although it potentiated the maximal inhibitory effect of the NK_2 receptor antagonist. On the other hand, Ujiie et al. (236) found that administration of phosphoramidon, an inhibitor of neutral endopeptidase, produced cough which was inhibited by aerosol administration of FK 888, a selective NK_1 receptor antagonist. The same group also reported that FK 888 inhibits cough induced by inhalation of histamine, acetylcholine, and ovalbumin (in sensitized animals) at concentrations which do not affect the bronchoconstriction induced by the same agents (237). Thus, it appears that TK receptor antagonists could represent a novel class of nonopioid antitussive agents or that an antitussive action could be at least part of the overall effect of these agents in the respiratory tract, although there are discrepancies in the literature on the type of TK receptor involved.

Altogether, the preclinical evidence collected thus far indicates a role for these neuropeptides in airway pathophysiology and suggests that TK receptor antagonists could have a place in respiratory medicine. On the other hand, as we are dealing with compounds which do not themselves stimulate receptors, it is quite evident that their true relevance in therapeutics will be, by and large, directly proportional to the relevance that TKs have in the general economy of a disease such as asthma.

Human airways possess TKs/CGRP-containing nerve profiles (238–240) which are of extrinsic origin, as they disappear after lung transplantation (241). The density of TK innervation in human lungs is lower than in rodents (238) and was reported to be increased or unchanged in asthmatics (242,243). Various TKs, including SP, NKA, and neuropeptide K, have been detected in extracts from human lung (244) and in bronchoalveolar lavage fluid. SP-LI has been found to be elevated in the bronchial and nasal lavage fluid of allergic patients after allergen provocation (245,246), in the pulmonary edema fluid of patients with adult respiratory distress syndrome (247), and in the sputum of patients with asthma or chronic bronchitis (248).

TK receptors are present in the muscle, submucosal glands, microvascular endothelium, and respiratory epithelium of the human trachea and lung (249,250). Adcock et al. (251) reported about 50% increase in expression of NK_1 receptor mRNA in lungs from asthmatic subjects, an effect which was downregulated by in-vitro incubation with dexamethasone, while the expression of NK_2 receptor mRNA was found to be unaltered in asthmatics. Recently, Bai et al. (252) reported a fourfold increase in expression of NK_2 receptor mRNA in lung samples of asthmatics as compared to nonsmoking controls, whereas they found the expression of NK_1 receptor mRNA to be similar in the two groups; the expression of both NK_1 and NK_2 receptor mRNA was increased twofold in smokers without airflow obstruction as compared to nonsmokers (252).

TKs are powerful contractile agents of human isolated airways (55,56,253), promote mucus secretions (254), and potentiate cholinergic contraction to electrical nerve stimulation (255). Consistent with the exclusive involvement of NK_2 receptors in producing contraction of human isolated airway smooth muscle (55,56,253,256), NKA but not SP produces bronchoconstriction in humans when administered by either the i.v. or aerosol route (257,258).

To date, only a few studies have appeared about the effects of TK receptor antagonists on airway function in humans. Ichinose et al. (259) reported that inhaled FK 224, a mixed NK_1/NK_2 receptor antagonist, prevents bradykinin-induced bronchoconstriction in asthmatics. However, Joos et al. (260) failed to detect any effect of FK 224 on baseline lung function and bronchoconstriction to NKA in mild asthmatic patients, and Lund et al. (261) did not observe any beneficial effect of FK 224 on lung function and symptoms during 4 weeks of administration in asthmatics. Fahy et al. (262) failed to observe any inhibitory effect of the NK_1 receptor antagonist CP 99994 toward bronchoconstriction or cough induced by hypertonic saline in subjects with mild asthma.

Clearly, clinical studies aiming at defining a possible usefulness of TK receptor antagonists in respiratory medicine are still in their infancy. It appears that the efficacy of test antagonists in occluding TK receptors in the airways should be established first by using pharmacological provocation tests, to enable a meaningful interpretation of data obtained with these ligands.

V. Tachykinin Receptor Antagonists in Animal Models of Inflammation

A. Inflammation in the Skin

At skin level, TKs (SP and NKA) are present in nerve fibers which are in close spatial arrangement with the cutaneous microvasculature. Single free nerve endings have been observed in the epidermis and dermis, and they distribute to hair follicles and sweat glands and among keratinocytes in the epidermis (263–265).

Intradermal injection of SP produces wheal and flare, while other TKs produce wheal but not flare (147,159); the flare is produced by SP through its N-terminal basic sequence, leading to mast cell degranulation and histamine release, that is an action not shared by the other mammalian TKs (see Section II.C).

The wheal produced by intradermal administration of TKs is considered as expression of increased vascular permeability to plasma proteins (266–270). Early studies suggested a role for NK_3 receptors in producing this effect, because neurokinin B (NKB) was the most potent natural TK-producing PPE in rat skin after intradermal administration (266,267), but subsequent investigations using receptor-selective agonists revealed the importance of NK_1 receptors (268–270). Devillier et al. (159) reported the rank order of potency SP > NKA > NKB for producing wheal after intradermal administration in human skin, indicative of an NK_1 receptor-mediated response. Similar conclusions were reported by Wallengren and Hakanson (271) and by Fuller et al. (272). NK_1 receptors have been localized by autoradiography at discrete anatomical sites (capillary endothelial cells, dermal papillae, sweat glands, hair follicles) of rat and human skin (273).

As mentioned in Section II.C, the mitogenic effects of TKs on skin fibroblasts and keratinocytes may also be relevant for inflammatory skin diseases, particularly those having a proliferative component such as inappropriate wound healing (e.g., hypertrophic scarring) and psoriasis.

Injection of capsaicin in the skin produces pain, reddening at the site of injection, flare reaction, and vasodilation, and these responses are absent when capsaicin is injected into chronically denervated skin (32,33). This was a clear-cut demonstration of the existence of neurogenic inflammation in human skin and the specific site of action of capsaicin as a sensory neuron stimulant agent. Intradermal capsaicin administration into human skin produces flare but not edema (274–276). The failure of capsaicin to produce edema was explained by Raud et al. (277) as being due to released endogenous CGRP exerting an inhibitory effect on microvascular permeability. They showed that both CGRP and capsaicin inhibit histamine-induced wheal at doses which did not affect flare. Local desensitization of the human skin by repeated applications of capsaicin produces blockade of the flare reaction to intradermal injection of SP and other neuropeptides such as somatostatin or VIP, histamine, or platelet-activating factor, while the wheal produced by these agents was unchanged (278–280).

The recent availability of receptor antagonists has enabled direct testing of the hypothesis that TKs are involved in cutaneous vasodilatation and increased vascular permeability produced during neurogenic inflammation. The available evidence indicates that CGRP but not TKs is mainly responsible for antidromic vasodilatation (281): CGRP(8-37), at a dose which blocks vasodilation produced by exogenous CGRP, markedly reduces antidromic vasodilatation in rat skin, while the NK_1 receptor antagonist RP 67,580, which blocks the SP-induced vasodilatation, was without effect on antidromic vasodilatation. On the other hand, TKs via NK_1 receptors mediate the increase in microvascular permeability at skin level, as demonstrated by the potent inhibitory effect exerted by various NK_1 receptor antagonists on PPE induced by antidromic electrical stimulation of the saphenous nerve (282–284) or cutaneous application of irritants (mustard oil, xylene, capsaicin) (283,285–287). In other studies the same approach has been used to demonstrate an involvement of NK_1 receptors in skin PPE induced by other inflammatory/tissue-damaging stimuli: thus, NK_1 receptor antagonists have been shown to inhibit cutaneous PPE induced by passive cutaneous anaphylaxis in guinea pigs (288), ultraviolet irradiation (289), and thermal injury in rats (290).

To summarize this section, the available preclinical evidence suggests a possible role of TKs as mediators of acute inflammatory and allergic reactions in the skin. A role in inappropriate wound healing and proliferative skin disorders such as psoriasis could also be hypothesized.

B. Arthritis

Nerve fibers containing TKs/CGRP are present in in the synovial membrane and around blood vessels of normal joints (92,93,291,292). In the proliferating synovium of rats with adjuvant-induced arthritis, the amount of nervous tissue is reduced and SP/CGRP-containing nerves are found only at the junction of the bone and synovial membrane (293); similar findings were reported in the synovium of patients with rheumatoid arthritis (291,294). Although the meaning of this finding is not unequivocal, it may indicate an increased release of sensory neuropeptides in these patients, thus reducing the stores in the nerves to levels below those detectable by immuno-histochemistry.

There are several lines of evidence implicating TKs in the genesis of inflammatory joint diseases. Experimental arthritis in rats is accompanied by increased levels of SP in peripheral nerves, dorsal roots, and dorsal spinal cord (295–297), reflecting an overexpression of preprotachykinin mRNA and an augmented synthesis of TKs in dorsal root ganglia (298,299). In parallel, the expression of NK_1 mRNA in the dorsal horn of the spinal cord is increased during experimental arthritis in rats (300,301); this, and the evidence for increased release of SP from central endings of primary afferents in the spinal cord (302,303), indicate an overall upregulation of TKergic sensory transmission to the spinal cord in experimental arthritis. In keeping with this concept is the demonstration of a significant analgesic effect of NK_1 receptor antagonists in rats with experimentally induced arthritis (304,305).

The study of Levine et al. (306) provided a strong basis for further studies on the possible role of TKs as mediators of arthritis; they speculated that, if the sensory nervous system plays a role in the genesis and severity of arthritis, there should be a correlation between intensity of the disease and density of sensory innervation of different joints. After having established that the ankle joint has a lower nociceptive threshold than the knee joint in the rat, they demonstrated that the ankle has a significantly greater afferent projection to the spinal cord and higher SP content compared to the knee. Furthermore, the infusion of SP in the knee joint produced a significant increase in severity of induced arthritis and severe destruction of the joint and periarticular bone (306).

These observations have focused the attention of researchers on the possible importance of locally released TKs as mediators of inflammation in the joints. Antidromic articular nerve stimulation increases the release of SP in the intraarticular space (307), resulting in PPE and antidromic vasodilation (308–311); CGRP could also be involved in vasodilation, while PPE in the joints appears to be mediated by TKs and is blocked by NK_1 receptor antagonists (312–316). In the rat knee joint, the PPE produced by exogenous SP requires quite high concentrations of the peptide and involves, in part, mast cell degranulation (312–314). Previous induction of inflammation by carrageenan enhanced and prolonged the PPE produced by exogenous SP in rat knee joint (317); in particular, a peak PPE response was produced by 100 nM SP in the inflamed joint as compared to the 10 μM concentration which is required to produce a maximal response in the normal joint. The mechanisms responsible for the enhancement and prolongation of the response to SP in inflamed joints are unsettled, yet the concentrations of SP which are effective in producing PPE in the inflamed joint are presumably closer to those released from sensory nerves. Conversely, the low intensity and transient nature of the response to SP in normal joints could be viewed, whatever mechanisms is involved, as a protective mechanisms toward development of a chronic inflammation. Other groups (318,319) failed to observe significant PPE in response to administration of SP *or* CGRP alone in rat knee joint; rather, a remarkable response was observed following co-perfusion of the two neuropeptides or in response to capsaicin (319). In addition to vasodilatation and increase in microvascular permeability, another local effect of TKs is potentially relevant for the pathogensis of arthritis: Lotz et al. (320) showed that SP induces collagenase release from rheumatoid synoviocytes and stimulates their proliferation through a specific receptor-mediated mechanism.

Furher supporting a local role of TKs in joint inflammation, NK_1 receptors have been detected at autoradiography on the microvascular endothelium of either rat and human synovium (321,322); no significant differences were observed between synovium from naive and monoarthritic rats nor between that from patients with rheumatoid arthritis or osteoarthritis (321,322).

Elevated levels of sensory neuropeptides have been detected in the synovial fluid of patients suffering from rheumatoid arthritis or arthritis (323–325), and a recent study demonstrated a significant reduction of synovial fluid SP levels in patients treated with nonsteroidal antiinflammatory agents (326). Neutral endopeptidase, known to degrade TKs, has also been localized in the human synovium of arthritic joints, where it is probably localized in synovial fibroblasts (327).

Summarizing this section, there is ample preclinical evidence supporting the idea that TKs, and especially SP, acting via NK_1 receptors, could be involved in initiating and maintaining joint inflammation. From this a beneficial effect of NK_1 receptor antagonists in rheumatoid arthritis could be speculated, a hypothesis awaiting clinical evaluation.

C. Inflammatory Bowel Diseases and Cystitis

Two sources of TKs exist at the intestinal level. In addition to the peripheral endings of the capsaicin-sensitive primary afferent neurons, several classes of enteric neurons express the preprotachykinin I gene, yielding substance P (SP) and neurokinin A (NKA) as the mature neurotransmitters stored in secretory vesicles (328,329). Contrary to other peripheral organs, the bulk of the TK-immunoreactive material present in the gut does not originate from primary afferent nerves but is contributed by intrinsic neurons (330). TKs have an established role as enteric neurotransmitters involved in neuromuscular excitation and neuro-neuronal communication (see Refs. 328 and 329 for reviews); the recognition that TKs may play a role in the regulation of ion and water transport across the intestine and the well-established inflammatory and proinflammatory action of TKs (vasodilation, increase in vascular permeability, recruitment/stimulation of inflammatory cells, mast cell degranulation, etc.), have prompted a number of studies aimed at defining a pathophysiological role of these peptides in intestinal inflammatory diseases. By using receptor-selective antagonists, evidence has been presented implicating the action of endogenous TKs in colonic hypersecretion/inflammation induced by, e.g., interleukin-1β (331), antigen challenge (332,333), enteritis produced by *Clostridium difficile* toxin A (334), or irritant-induced ileitis (335). These preclinical data, along with evidence for changes in tissue levels of TKs and upregulation of TK receptors (336,337), suggest a possible role of TK receptor antagonists in therapy for human inflammatory bowel diseases.

Capsaicin administration, electrical stimulation of pelvic nerves, or antidromic stimulation of dorsal roots produces PPE in the ureter, urinary bladder, and urethra (37,338–342). Exogenously administered TKs likewise produce a marked inflammatory reaction in the ureter, urinary bladder, and urethra (37,340,343). The reaction produced by antidromic electrical nerve stimulation or capsaicin administration shows a distinct regional gradient, its intensity being greater in the bladder base than in the dome and much larger in the ureters and urethra than in the bladder (340,341). A similar regional difference in the intensity of PPE is observed in response to exogenously administered TKs, although its boundaries are not as sharp as those observed in response to capsaicin, especially in the urinary bladder (343,344).

As happens in other regions of the body, NK_1 receptor activation is responsible for PPE produced by TKs in the ureter, bladder, and urethra; this has been supported by experiments with selective agonists (343) and antagonists (40,43,342). SP-induced PPE in the rat bladder is unchanged by antihistamine, methysergide, bradykinin receptor antagonist, or NO synthase inhibitors (40,43,345). Various stimuli of pathophysiological relevance produce PPE in the urinary bladder and urethra via capsaicin-sensitive primary afferents: These include irritant-induced cystitis (xylene, cyclophosphamide) (346,347), mechanical irritation produced by catheteterization (345,348), and intraluminal application of hypertonic media (349). The early phase of PPE induced by chemical irritation (xylene, cyclophosphamide) and PPE induced by bladder catheterization are prevented by NK_1 receptor antagonists, indicating the involvement of endogenous TKs via NK_1 receptor (345,350). Altogether these data suggest a possible involvement of TKs in the genesis of inflammation and irritative symptoms of cystitis.

D. Neurogenic Inflammation in the Dura Mater

Various studies have addressed the possible relevance of neurogenic inflammation and of local release of TKs in the dura mater in the genesis of migraine. While being often considered a relatively inert tissue, the dura mater receives a dense innervation which includes peptidergic fibers containing TK/CGRP-LI (351–354). Application of capsaicin or electrical stimulation produces release of sensory neuropeptides from dural venous sinuses (355,356).

Markowitz et al. (173) showed that chemical, electrical, or immunological stimulation induces PPE in the rat or guinea pig dura mater, measured either by the Evans blue leakage technique or by the extravasation of ^{125}I bovine serum albumin. The response to electrical stimulation of the trigeminal ganglion was abolished by capsaicin pretreatment; intravenous administration of SP or NKA but not of CGRP produces PPE in the dura mater (173). Several studies have showed that NK_1 receptor antagonists block neurogenic plasma extravasation in the dura mater, providing direct demonstration that endogenous TKs mediate this response (357–359).

Morphological data also show that electrical stimulation of trigeminal afferents induces endothelial changes in the dura mater suggestive of increased vascular permeability, mast cell degranulation, and formation of platelet aggregates (172,

360). Moreover, antimigraine drugs such as ergot alkaloids or sumatriptan prevent the PPE induced by trigeminal stimulation in the dura mater and the elevation of plasma CGRP induced by trigeminal ganglion stimulation (361–363). A role for SP as an endogenous vasodilator in cerebral circulation and the possible impact of this effect in producing pain has been discussed by Beattie et al. (364). Overall, these data provide a background for speculation that neurogenic inflammation in the dura mater/ cerebral circulation mediated through the local release of sensory neuropeptides may play an important role in the pathogenesis of migraine and vascular headache (365), and that NK_1 receptor antagonists may have antimigraine properties (364).

VI. Conclusions

The evidence reviewed in this chapter indicates that at several locations of the body a TKergic innervation exists from which peptides of the TK family are released, especially during inflammation and in response to potentially harmful/tissue-threatening stimuli. The released TKs produce a number of powerful biological effects which are proinflammatory or frankly inflammatory; in various viscera, including the airways, a powerful contraction of smooth muscles also occurs. These events, which are receptor-mediated and collectively termed "neurogenic inflammation," provide a mechanism through which the sensory nervous system contributes to initiation and maintenance of the inflammatory process. The preclinical evidence collected thus far, and reviewed in this chapter, indicates that blockade of neurogenic inflammation can be effectively achieved, in a number of animal models of disease, by occluding postjunctional TK receptors with TK receptor antagonists. From the above, a possible therapeutic usefulness of TK receptor antagonists can be envisaged in diseases as diverse as asthma/bronchial hyperreactivity, psoriasis, rheumatoid arthritis, inflammatory bowel diseases, cystitis, and migraine. This hypothesis is currently under evaluation.

References

1. Von Euler US, Gaddum JH. An unidentified depressor substance in certain tissue extracts. J Physiol (Lond) 1931; 72:74–86.
2. Chang MM, Leeman SE, Niall HD. Aminoacid sequence of substance P. Nature 1971; 232:86–87.
3. Erspamer V, Melchiorri P. Active polypeptides of the amphibian skin and their synthetic analogs. Pure Appl Chem 1973; 35:463–494.
4. Lembeck F. Zur Frage der zentralen Ubertragung afferenter Impulse III. Mitteilung. Das Vorkommen und die Bedeutung der Substanz P in den dorsalen Wurzeln des Ruckenmarks. Arch Exp Pathol Pharmakol 1953; 219:197–213.
5. Hokfelt T, Kellerth JO, Nilsson G, Pernow B. Substance P: localization in the central nervous system and in some primary sensory neurons. Science 1975; 190:889–890.
6. Hokfelt T, Kellerth JO, Nilsson G, Pernow B. Experimental immunohistochemical

studies on the localization and distribution of substance P in cat primary sensory neurons. Brain Res 1975; 100:235–252.

7. Maggio JE. Tachykinins. Annu Rev Neurosci 1988; 11:13–21.

8. Nakanishi S. Substance P precursor and kininogen: their structures, gene organizations and regulation. Physiol Rev 1987; 67:1117–1142.

9. Nakanishi S. Mammalian tachykinin receptors. Annu Rev Neurosci 1991; 14:123–136.

10. Maggi CA, Patacchini R, Rovero P, Giachetti A. Tachykinin receptors and tachykinin receptor antagonists. J Autonom Pharmacol 1993; 13:23–93.

11. Otsuka M, Yoshioka K. Neurotransmitter function of mammalian TKs. Physiol Rev 1993; 73:229–307.

12. Regoli D, Boudon A, Fauchere JL. Receptors and antagonists for substance P and related peptides. Pharmacol Rev 1994; 46:551–599.

13. Maggi CA. The mammalian tachykinin receptors. Gen Pharmacol 1995; 26:911–944.

14. Gerard NP, Bao L, Ping HX, Gerard C. Molecular aspects of the tachykinin receptors. Regulatory Peptides 1993; 43:21–35.

15. Maggi CA. Tachykinins and CGRP: receptor subtypes, signal transduction and airways effects. In: Raeburn D, Gyembicz MA, eds. Airway Smooth Muscle: A Reference Source. Volume III: Peptide Receptors, Ion Channels and Signal Transduction. Basel: Birkhauser, 1995:67–86.

16. Maggi CA, Zagorodnyuk V, Giuliani S. Specialization of tachykinin NK_1 and NK_2 receptors in producing fast and slow atropine-resistant neurotransmission to the circular muscle of the guinea-pig colon. Neuroscience 1994; 63:1137–1152.

17. Lee CM, Campbell NJ, Williams BJ, Iversen LL. Multiple tachykinin binding sites in peripheral tissues and in brain. Eur J Pharmacol 1986; 130:209–217.

18. Saria A, Martling CR, Yan Z, Theodorsson-Norheim E, Gamse R, Lundberg JM. Release of multiple tachykinins from capsaicin-sensitive sensory nerves in the lung by bradykinin, histamine, DMPP and vagal nerve stimulation. Am Rev Respir Dis 1988; 137:1330–1335.

19. Maggi CA, Patacchini R, Perretti F, Meini S, Manzini S, Santicioli P, Del Bianco E, Meli A. The effect of thiorphan and epithelium removal on contractions and tachykinin release produced by activation of the capsaicin-sensitive afferents in the guinea-pig isolated bronchus. Naunyn Schmiedeberg's Arch Pharmacol 1990; 341:74–79.

20. Maggi CA, Patacchini R, Quartara L, Rovero P, Santicioli P. Tachykinin receptors in the guinea-pig isolated bronchi. Eur J Pharmacol 1991; 197:167–174.

21. Maggi CA, Patacchini R, Rovero P, Santicioli P. Tachykinin receptors and noncholinergic bronchoconstriction in the guinea-pig isolated bronchi. Am Rev Respir Dis 1991; 144:363–367.

22. Laufer R, Wormser U, Friedman ZY, Gilon C, Chorev M, Selinger Z. Neurokinin B is a preferred agonist for a neuronal substance P receptor and its action is antagonized by enkephalin. Proc Natl Acad Sci (USA) 1985; 82:7444–7448.

23. Too HP, Cordova JL, Maggio JE. A novel radioimmunoassay for neuromedin K. I. Absence of neuromedin K-like immunoreactivity in guinea-pig ileum and urinary bladder. II. Heterogeneity of tachykinins in guinea-pig tissues. Regulatory Peptides 1989; 26:93–105.

24. Pacicca C, Von der Weid PY, Beny JL. Effects of nitro-L-arginine on endothelium-dependent hyperpolarizations and relaxations of pig coronary arteries. J Physiol (Lond) 1992; 457:247–256.

25. Zhang G, Yamamoto Y, Miwa K, Suzuki H. Vasodilatation induced by substance P in guinea-pig carotid arteries. Am J Physiol 1994; 266:H1132–H1137.
26. Kuroiwa M, Aoki H, Kobayashi S, Nishimura J, Kanaide H. Mechanism of endothelium-dependent relaxation induced by substance P in the coronary artery of the pig. Br J Pharmacol 1995; 116:2040–2047.
27. D'Orleans-Juste P, Dion S, Drapeau G, Regoli D. Different receptors are involved in the endothelium-mediated relaxation and the smooth muscle contraction of the rabbit pulmonary artery in response to SP and related neurokinins. Eur J Pharmacol 1986; 125:37–44.
28. Maggi CA, Patacchini R, Perretti F, Tramontana M, Manzini S, Geppetti P, Santicioli P. Sensory nerves, vascular endothelium and neurogenic relaxation of the guinea-pig isolated pulmonary artery. Naunyn Schmiedeberg's Arch Pharmacol 1990; 342:78–84.
29. Patacchini R, Maggi CA, Tachykinin NK_1 receptors mediate both vasoconstrictor and vasodilator responses in the rabbit isolated jugular vein. Eur J Pharmacol 1995; 283:233–240.
30. Mastrangelo P, Mathison R, Huggel HJ, Dion S, D'Orleans-Juste P, Rhaleb NE, Drapeau G, Rovero P, Regoli D. The rat isolated portal vein: a preparation sensitive to neurokinins, particularly neurokinin B. Eur J Pharmacol 1987; 134:321–326.
31. Nantel F, Rouissi N, Rhaleb NE, Dion S, Drapeau G, Regoli D. The rabbit jugular vein is a contractile NK_1 receptor system. Eur J Pharmacol 1990; 179:457–462.
32. Jancso N, Jancso-Gabor A, Szolcsanyi J. Direct evidence for neurogenic inflammation and its prevention by denervation and by pretreatment with capsaicin. Br J Pharmacol 1967; 31:138–151.
33. Jancso N, Jancso-Gabor A, Szolcsanyi J. The role of sensory nerve endings in neurogenic inflammation induced in human skin and in the eye and paw of the rat. Br J Pharmacol 1968; 32:32–41.
34. Maggi CA, Meli A. The sensory-efferent function of capsaicin-sensitive sensory neurons. Gen Pharmacol 1988; 19:1–43.
35. Maggi CA. Tachykinins and calcitonin gene-related peptide (CGRP) as co-transmitters released from peripheral endings of sensory nerves. Prog Neurobiol 1995; 45:1–98.
36. Lembeck F, Holzer P. Substance P as neurogenic mediator of antidromic vasodilatation and neurogenic plasma extravasation. Naunyn Schmiedeberg's Arch Pharmacol 1979; 310:175–183.
37. Saria A, Lundberg JM, Skofitsch G, Lembeck F. Vascular protein leakage in various tissues induced by substance P, capsaicin, bradykinin, serotonin, histamine and by antigen challenge. Naunyn Schmiedeberg's Arch Pharmacol 1983; 324:212–218.
38. Jacques L, Couture R, Drapeau G, Regoli D. Capillary permeability induced by intravenous neurokinins, receptor characterization and mechanism of action. Naunyn Schmiedeberg's Arch Pharmacol 1989; 340:170–179.
39. Abelli L, Maggi CA, Rovero P, Del Bianco E, Regoli D, Drapeau G, Giachetti A. Effect of synthetic tachykinin analogues on airway microvascular leakage in rats and guinea-pigs: evidence for the involvement on NK_1 receptors. J Autonom Pharmacol 1991; 11: 267–275.
40. Eglezos A, Giuliani S, Viti G, Maggi CA. Direct evidence that capsaicin-induced plasma protein extravasation is mediated through tachykinin NK_1 receptors. Eur J Pharmacol 1991; 209:277–279.
41. Xu XJ, Dalsgaard CJ, Maggi CA, Wiesenfeld-Hallin Z. NK_1, but not NK_2, tachykinin receptors mediate plasma extravasation induced by antidromic C-fiber stimulation in rat

hindpaw: demonstrated with the NK_1 antagonist CP 96,345 and the NK_2 antagonist MEN 10207. Neurosci Lett 1992; 139:249–252.

42. Ralevic V, Khalil Z, Helme RD, Dusting GJ. Role of NO in the actions of SP and other mediators of inflammation in rat skin microvasculature. Eur J Pharmacol 1995; 284: 231–239.

43. Santicioli P, Giuliani S, Maggi CA. Failure of L-nitro arginine, a nitric oxide synthase inhibitor, to affect hypotension and plasma protein extravasation produced by tachykinin NK_1 receptor activation in rats. J Autonom Pharmacol 1993; 13:193–199.

44. Baluk P, Nadel JA, Mc Donald D. Substance P-immunoreactive sensory axons in the rat respiratory tract: a quantitative study of their distribution and role in neurogenic inflammation. J Comp Neurol 1992; 319:586–598.

45. Tousignant C, Chan CC, Guevremont D, Brideau C, Hale JJ, Mac Coss M, Rodger IW. NK_2 receptor mediate plasma extravasation in guinea-pig lower airways. Br J Pharmacol 1993; 108:383–386.

46. Webber SE. Receptors mediating the effects of substance P and neurokinin A on mucus secretion and smooth muscle tone of the ferret trachea: potentiation by an enkephalinase inhibitor. Br J Pharmacol 1989; 98:1197–1206.

47. Rogers DF, Aursudkij B. Barnes PJ. Effect of tachykinins on mucus secretion in human bronchi in vitro. Eur J Pharmacol 1989; 174:283–286.

48. Kuo HP, Rohde JA, Tokuyama K, Barnes PJ, Rogers DF. Capsaicin and sensory neuropeptide stimulation of goblet cell secretion in guinea-pig trachea. J Physiol (Lond) 1990; 431:629–641.

49. Meini S, Mak JCW, Rohde JAL, Rogers DF. Tachykinin control of ferret airways: mucus secretion, bronchoconstriction and receptor mapping. Neuropeptides 1993; 24: 81–89.

50. Ranmarine SI, Hirayama Y, Barnes PJ, Rogers DF. "Sensory-efferent" neural control of mucus secretion: characterization using tachykinins receptor antagonists in ferret trachea in vitro. Br J Pharmacol 1994; 113:1183–1190.

51. Manzini S. Bronchodilation by tachykinins and capsaicin in the mouse main bronchus. Br J Pharmacol 1992; 105:968–972.

52. Devillier P, Acker M, Advenier C, Marsac J, Regoli D, Frossard N. Activation of an epithelial NK_1 receptor induces relaxation of rat trachea through release of prostaglandin E_2. J Pharmacol Exp Ther 1992; 263:767–772.

53. Maggi CA, Giuliani S, Ballati L, Lecci A, Manzini S, Patacchini R, Renzetti AR, Rovero P, Quartara L, Giachetti A. In vivo evidence for tachykininergic trasmission using a new NK_2 receptor selective antagonist, MEN 10376. J Pharmacol Exp Ther 1991; 257:1172–1178.

54. Ballati L, Evangelista S. Maggi CA, Manzini S. Effect of selective tachykinin receptor antagonists on capsaicin- and tachykinin-induced bronchospasm in anaesthetized guinea-pigs. Eur J Pharmacol 1992; 214:215–221.

55. Dion S, Rouissi N, Nantel F, Drapeau G, Regoli D, Naline E, Advenier C. Receptors for neurokinins in human bronchus and urinary bladder are of the NK_2 type. Eur J Pharmacol 1990; 178:215–219.

56. Astolfi M, Treggiari S, Giachetti A, Meini S, Maggi CA, Manzini S. Characterization of the tachykinin NK_2 receptor in the human bronchus: influence of amastatin-sensitive pathway. Br J Pharmacol 1994; 111:570–574.

57. McDonald D. Neurogenic inflammation in the rat trachea. I Changes in venules, leucocytes and epithelial cells. J Neurocytol 1988; 17:583–603.

58. Sagara H, Yukawa T, Arima M, Terashi Y, Yoshihara S, Abe T, Ichimura T, Makino S. Capsaicin and substance P induce eosinophil infiltration in bronchi in guinea-pigs. Am Rev Respir Dis 1991; 143:A616.

59. Hsiue TR, Garland A, Ray DW, Hershenson MB, Leff AR, Solway J. Endogenous sensory neuropeptides release enhances nonspecific airway responsiveness in guinea-pigs. Am Rev Respir Dis 1992; 146:148–153.

60. Kudlacz EM, Knippenberg RW. In vitro and in vivo effects of tachykinins on immune cell function in guinea-pig airways. J Neuroimmunol 1994; 50:119–125.

61. Baluk P, Bertrand C, Geppetti P, Mc Donald D, Nadel JA, NK$_1$ receptors mediate leukocyte adhesionin neurogenic inflammation in the rat trachea. Am J Physiol 1995; 268:L263–L269.

62. Walsh DT, Weg VB, Williams TJ, Nourshargh S. Substance P induced inflammatory responses in guinea-pig skin: the effect of specific NK$_1$ receptor antagonists and the role of endogenous mediators. Br J Pharmacol 1995; 114:1343–1350.

63. Matsuda H, Kawakita K, Kiso Y, Nakano T, Kitamura Y. Substance P induces granulocyte infiltration through degranulation of mast cells. J Immunol 1989; 142:927–931.

64. Yano H, Wershil BK, Arizono N, Galli SJ. Substance P-induced augmentation of cutaneous vascular permeability and granulocyte infiltration in mice is mast cell dependent. J Clin Invest 1989; 84:1276–1286.

65. Perretti M, Ahluwalia A, Flower RJ, Manzini S. Endogenous tachykinins play a role in IL-1 -induced neutrophil accumulation: involvement of NK$_1$ receptors. J Immunol 1993; 80:73–77.

66. Norris AA, Leeson ME, Jackson DM, Holroyde MC. Modulation of neurogenic inflammation in rat trachea. Pulmon Pharmacol 1991; 3:180–184.

67. Woie K, Koller ME, Heyeraas KJ, Reed RK. Neurogenic inflammation in rat trachea is accompanied by increased negativity of interstitial fluid pressure. Circ Res 1993; 73: 840–845.

68. Germonpré PR, Joos GF, Everaert E, Kips JC, Pauwels RA. Characterization of neurogenic inflammation in the airways of two highly inbred rat strains. Am J Respir Crit Care Med 1995; 152:1796–1804.

69. Marasco WA, Showell HJ, Becker EL. Substance P binds to the formylpeptide chemotaxis receptor on the rabbit neutrophil. Biochem Biophys Res Commun 1981; 99:1065–1072.

70. Roch-Arveiller M, Regoli D, Chanaud B, Lenoir M, Muntaner O, Stralzko S, Giroud JP. Tachykinins: effects on motility and metabolism of rat polymorphonuclear leucocytes. Pharmacology 1986; 33:266–273.

71. Wiedermann CJ, Wiedermann FJ, Apperl A, Kieselbach G, Konwalinka G, Braunsteiner H. In vitro polymorphonuclear leukocyte chemokinesis and human monocyte chemotaxis are different activities of N-terminal and C-terminal SP. Naunyn Schmiedeberg's Arch Pharmacol 1989; 340:185–190.

72. Carolan EJ, Casale TB. Effects of neuropeptides on neutrophil migration through noncellular and endothelial barriers. J Allergy Clin Immunol 1993; 92:589–598.

73. Serra MC, Bazzoni F, Della Bianca V, Greskowiak M, Rossi F. Activation of human neutrophils by substance P. Effect on oxidative metabolism exocytosis, cytosolic calcium concentration and inositol phosphate formation. J Immunol 1988; 141:2118–2124.

74. Perianin A, Snyderman R, Malfroy B. Substance P primes human neutrophil activation: a mechanism for neurological regulation of inflammation. Biochem Biophys Res Commun 1989; 161:520–524.

75. Serra MC, Calzetti F, Ceska M, Cassatella MA. Effect of substance P on superoxide anion and IL-8 production by human PMNL. Immunology 1994; 82:63–69.

76. Wozniak A, Betts WH, Mc Lennan G, Scicchitano R. Activation of human neutrophils by tachykinins, effects on FMLP and PAF-stimulated superoxidse anion production and antibody-dependent cell-mediated cytotoxicity. Immunology 1993; 78:629–634.

77. Wiedermann CJ, Niedermuhlbichler M, Zilian U, Geissler D, Lindley I, Braunsteiner H. Priming of normal human neutrophils by tachykinins: tuftsin-like inhibition of in vitro chemotaxis stimulated by formylpeptide or interleukin-8. Regulatory Peptides 1991; 36:359–368.

78. Brunelleschi S, Ceni E, Giotti A, Fantozzi R. Tachykinins stimulate lyso-PAF: acetyl CoA acetyltransferase activity in neutrophils. Eur J Pharmacol 1990; 186:367–368.

79. Matis WL, Lavker RM, Murphy GF. Substance P induces the expression of endothelial-leukocyte adhesion molecule by microvascular endothelium. J Invest Dermatol 1990; 94:492–495.

80. Zimmerman BJ, Anderson DC, Granger DN. Neuropeptides promote neutrophil adherence to endothelial cell monolayers. Am J Physiol 1992; 263:G678–G682.

81. Shipp MA, Stefano GB, Switzer SN, Griffin JD, Reinherz EL. CD10 (CALLA)/neutral endopeptidase 24.11 modulates inflammatory peptide-induced changes in neutrophil morphology, migration and adhesion proteins and is itself regulated by neutrophil activation. Blood 1991; 78:1834–1841.

82. Nakagawa N, Sano H, Iwamoto I. Substance P induces the expression of intercellular adhesion molecule-1 on vascular endothelial cell and enhances neutrophil transendothelial migration. Peptides 1995, 16:721–725.

83. Von Essen SG, Rennard SI, O'Neill D, Ertl RF, Robbins RA, Koyama S, Rubinstein I. Bronchial epithelial cells release neutrophil chemotactic activity in response to tachykinins. Am J Physiol 1992; 263:L226–L231.

84. De Rose V, Robbins RA, Snider RM, Spurzem JR, Thiele GM, Rennard SI, Rubinstein I. Substance P increases neutrophil adhesion to bronchial epithelial cells. J Immunol 1994; 152:1339–1346.

85. Fink T, Weihe E. Multiple neuropeptides in nerves supplying mammalian lymph nodes: messenger candidates for sensory and autonomic neuroimmunomodulation. Neurosci Lett 1988; 90:39–44.

86. Weihe E, Muller S, Fink T, Zentel HJ. Tachykinins, CGRP and neuropeptide Y in nerves of the mammalian thymus: interactions with mast cells in autonomic and sensory neuroimmunomodulation? Neurosci Lett 1989; 100:77–82.

87. Weihe E, Nohr D, Michel S, Muller S, Zentel HJ, Fink T, Krekel J. Molecular anatomy of the neuro-immune connection. Int J Neurosci 1991; 59:1–23.

88. Weihe E, Nohr D, Muller S, Buchler M, Friess H, Zentel HJ. The tachykinins neuro-immune connection in inflammatory pain. Ann NY Acad Sci 1988; 632:283–295.

89. Nilsson G, Alving K, Ahlstedt S, Hokfelt T, Lundberg JM. Peptidergic innervation of rat lymphoid tissue and lung: relation to mast cells and sensitivity to capsaicin and immunization. Cell Tissue Res 1990; 262:125–133.

90. Lorton D, Bellinger DL, Felten SY, Felten DL. Substance P innervation of the rat thymus. Peptides 1990; 11:1269–1275.

91. Lorton D, Bellinger DL, Felten SY, Felten DL. Substance P innervation of spleen in rats: nerve fibres associated with lymphocytes and macrophages in specific compartments in the spleen. Brain Behav Immun 1991; 5:29–40.

92. Hukkanen M, Konttinen YT, Rees RG, Gibson SJ, Santavirta S, Polak JM. Innervation

of bone from healthy and arthritis rats by SP and CGRP containing sensory fibers. J Rheumatol 1992; 19:1252–1259.

93. Hukkanen M, Konttinen YT, Rees RG, Santavirta S, Terenghi G, Polak JM. Distribution of nerve endings and sensory neuropeptides in rat synovium, meniscus and bone. Int J Tissue React 1992; 14:1–10.

94. Geppetti P, Maggi CA, Zecchi-Orlandini S, Santicioli P, Meli A, Frilli S, Spillantini MG, Amenta F. Substance P-like immunoreactivity in capsaicin-sensitive structures of the rat thymus. Regulatory Peptides 1987; 18:321–329.

95. Geppetti P, Theodorsson-Norheim E, Ballerini G, Alessandri M, Maggi CA, Santicioli P, Amenta F, Fanciullacci M. Capsaicin-sensitive, tachykinin-like immunoreactivity in the thymus of rats and guinea-pigs. J Neuroimmunol 1988; 19:3–10.

96. Popper P, Mantyh CR, Vigna SR, Maggio JE, Mantyh PW. The localization of sensory nerve fibers and receptor binding sites for sensory neuropeptides in canine mesenteric lymph nodes. Peptides 1988; 9:257–267.

97. Ericsson A, Geenen V, Robert F, Legros J, Vrindts-Gevaert Y, Franchimont P, Brene S, Persson H. Expression of preprotachykinin A and NPY-mRNA in the thymus. Mol Endocrinol 1990; 4:1211–1218.

98. Bost KL. Inducible preprotachykinin mRNA expression in mucosal lymphoid organs following oral immunization with *Salmonella*. J Neuroimmunol 1995; 62:59–67.

99. Shigematsu K, Saavedra MJ, Kurihara M. Specific SP binding sites in rat thymus and spleen: in vitro autoradiographic study. Regulatory Peptides 1986; 16:147–156.

100. Tang SC, Fend F, Muller L, Braunsteiner H, Wiedermann CJ. High affinity substance P binding sites of NK_1 receptor type autoradiographically associated with vascular sinuses and high endothelial venules of human lymphoid tissues. Lab Invest 1993; 69:86–93.

101. Mantyh CR, Gates TS, Zimmermann RP, Welton ML, Passaro EP, Vigna SR, Maggio JE, Kruger L, Mantyh PW. Receptor binding sites for SP but not substance K or neuromedin K are expressed in high concentrations by arterioles, venules and lymph nodules in surgical specimens obtained from patients with ulcerative colitis and Crohn disease. Proc Natl Acad Sci (USA) 1988; 85:3235–3239.

102. Mantyh PW, Mantyh CR, Gates T, Vigna SR, Maggio JE. Receptor binding sites for SP and substance K in the canine gastrointestinal tract and their possible role in inflammatory bowel disease. Neuroscience 1988; 25:817–837.

103. Mantyh PW, Catton MD, Boehmer CG, Welton ML, Passaro EP, Maggio JE, Vigna SR. Receptors for sensory neuropeptides in human inflammatory diseases: implications for the effector role of sensory neurons. Peptides 1989; 10:627–645.

104. Payan DG, Brewster DR, Goetzl EJ. Stereospecific receptors for SP on cultured human IM9 lymphoblasts. J Immunol 1984; 133:3260–3265.

105. Payan DG, McGillis JP, Organist ML. Binding characteristics and affinity labeling of protein constituents of the human IM-9 lymphoblast receptor for SP. J Biol Chem 1986; 261:14321–14329.

106. Payan DG, Brewster DR, Missirian-Bastian A, Goetzl EJ. Substance P recognition of a subset of human T lymphocytes. J Clin Invest 1984; 74:1532–1539.

107. Stanisz A, Schicchitano P, Dazin J, Bienenstock J, Payan DG. Distribution of substance P receptors on murine spleen and Peyer's patch T and B cells. J Immunol 1987; 139: 749–754.

108. Roberts AI, Taunk J, Ebert EC. Human lymphocyets lack substance P receptors. Cell Immunol 1992; 141:457–465.

109. Kavelaars A, Jeurissen F, Heijnen CJ. Substance P receptors and signal transduction in leucocytes. Immunomethods 1994; 5:41–48.

110. Cook GA, Elliott D, Metwali A, Blum AM, Sandor M, Lynch R, Weinstock JV. Molecular evidence that granuloma T lymphocytes in murine schistosomiasis mansoni express an authentic SP (NK₁) receptor. J Immunol 1994; 152:1830–1835.

111. Payan DG, Brewster DR, Goetzl EJ. Specific stimulation of human T lymphocytes by SP. J Immunol 1983; 131:1613–1615.

112. Stanisz AM, Befus D, Bienenstock J. Differential effects of VIP, SP and somatostatin on immunoglobulin synthesis and proliferations by lymphocytes from Peyer's patches mesenteric lymph nodes and spleen. J Immunol 1986; 136:152–156.

113. Casini A, Geppetti P, Maggi CA, Surrenti C. Effects of CGRP, NKA and NKA(4-10) on the mitogenic response of human peripheral blood mononuclear cells. Naunyn Schmiedeberg's Arch Pharmacol 1989; 339:354–358.

114. Kavelaars A, Jeurissen F, von Frijtag Drabbe Kunzel J, van Roijen JH, Rijkers GT, Heijnen CJ. Substance P induces a rise in intracellular calcium concentration in human T lymphocytes in vitro: evidence of a receptor independent mechanism. J Neuroimmunol 1993; 42:61–70.

115. Calvo CF, Chavanel G, Senik A. Substance P enhances IL-2 expression in activated human T cells. J Immunol 1992; 148:3498–3504.

116. Rameshwar P, Gascon P, Ganea D. Stimulation of IL-2 production in murine lymphocytes by substance P and related tachykinins. J Immunol 1993; 151:2484–2496.

117. Heerwagen C, Pabst R, Westermann J. The neuropeptide substance P does not influence the migration of B, T, CD8+ and CD4+ (naive and memory) lymphocytes from blood to lymph in the normal rat. Scand J Immunol 1995; 42:480–486.

118. Agro A, Stanisz AM. Are lymphocytes a target for substance P modulation arthritis? Semin Arthr Rheum 1992; 21:252–258.

119. Covas MJ, Pinto LA, Pereira da Silva JA, Victorino RMM. Effects of the neuropeptide substance P on lymphocyte proliferation in rheumatoid arthritis. J Intern Med Res 1995; 23:431–438.

120. Yokoyama MM, Fujimoto K. Role of lymphocyte activation by substance P in rheumatoid arthritis. Int J Tissue React 1990; 12:1–9.

121. Covas MJ, Pinto LA, Victorino RMM. Disturbed immunoregulatory properties of the neuropeptide substance P on lymphocyte proliferation in HIV patients. Clin Exp Immunol 1994; 96:384–388.

122. Bost KL, Pascual DW. Substance P: a late-acting B lymphocyte differentiation cofactor. Am J Physiol 1992; 262:C537–C545.

123. Pascual DW, Xu-Amano J, Kiyono H, Mc Ghee JR, Bost KL. Substance P acts directly upon cloned B lymphoma cells to enhance IgA and IgM production. J Immunol 1991; 146:2130–2136.

124. Lotz M, Vaughan JH, Carson DA. Effect of neuropeptides on production of inflammatory cytokines by human monocytes. Science 1988; 241:1218–1221.

125. Laurenzi MA, Persson MAA, Dalsgaard CJ, Haegerstrand A. The neuropeptide substance P stimulates production of interleukin-1 in human blood monocytes: activated cells are preferentially influenced by the neuropeptide. Sc and J Immunol 1990; 31:529–534.

126. Jeurissen F, Kavelaars A, Korstjens M, Broeke D, Franklin RA, Gelfand EW, Heijnen CJ. Monocytes express a non-neurokinin substance P receptor that is functionally coupled with MAP kinase. J Immunol 1994; 152:2987–2994.

127. Kavelaars A, Broeke K, Jeurissen F, Kardux J, Meijer A, Franklin R, Gelfand EW, Heijnen CJ. Activation of human monocytes via a non-neurokinin substance P receptor that is coupled to Gi protein, calcium, phospholipase D, MAP kinase and IL-6 production. J Immunol 1994; 153:3691–3699.

128. Hartung HP, Wolters K, Toyker KV. Substance P: binding properties and studies on cellular responses in guinea-pig macrophages. J Immunol 1986; 136:3856–3863.

129. Kimball SE, Persico FJ, Vaught JL. Substance P, neurokinin A and neurokinin B induce generation of interleukin-1-like activity by P388D1 cells. J Immunol 1988; 141:3564–3569.

130. Pascual DW, Bost KL. A monoclonal anti-substance P antibody recognizes macrophage generated immunoreactive substance P and modulates IL-1 secretion by the same cells. FASEB J 1990; 4:A305.

131. Pascual DW, Bost KL. Substance P production by P388D1 macrophages: a possible autocrine function for this neuropeptide. Immunology 1990; 71:52–56.

132. Bost KL, Breeding SAL, Pascual DW. Modulation of the mRNAs encoding SP and its receptor in rat macrophages by LPS. Reg Immunol 1992; 4:105–112.

133. Brunelleschi S, Vanni L, Ledda F, Giotti A, Maggi CA, Fantozzi R. Tachykinins activate guinea-pig alveolar macrophages: involvement of NK_2 and NK_1 receptors. Br J Pharmacol 1990; 100:417–420.

134. Brunelleschi S, Ceni E, Fantozzi R, Maggi CA. Evidence for tachykinin NK_{2B}-like receptors in guinea-pig alveolar macrophages. Life Sci Pharmacol Lett 1992; 51:PL177–PL181.

135. Brunelleschi S, Parenti A, Ceni E, Giotti A, Fantozzi R. Enhanced responsiveness of ovalbumin-sensitized guinea-pig alveolar macrophages to tachykinins. Br J Pharmacol 1992; 107:964–969.

136. Murris-Espin M, Pinelli E, Pipy B, Leophonte P, Didier A. Substance P and alveolar macrophages: effects on oxidative metabolism and eicosanoid production. Allergy 1995; 50:334–339.

137. Rankin JA. The contribution of alveolar macrophages to hyperreactive airway disease. J Allergy Clin Immunol 1989; 83:722–729.

138. Nohr D, Weihe E. Tachykinin- CGRP- and protein gene product 9.5-immunoreactive nerve fibers in alveolar wall of mammals. Neurosci Lett 1991; 134:17–20.

139. Boichot E, Lagente V, Paubert-Braquet M, Frossard N. Inhaled SP induces activation of alveolar macrophages and increases airway responses in the guinea-pig. Neuropeptides 1993; 25:307–313.

140. D'Ortho MP, Jarreau PH, Delacourt C, Pezet S, Lafuma C, Harf A, Macquin-Mavier I. Tachykinins induce gelatinase production by guinea-pig alveolar macrophages: involvement of NK_2 receptors. Am J Physiol 1995; 269:L631–L636.

141. Helme RD, Eglezos A, Dandie GW, Andrews PV, Boyd RL. The effect of SP on the regional lymphonode antibody response to antigenic stimulation in capsaicin-pretreated rats. J Immunol 1987; 139:3470–3473.

142. Eglezos A, Andrews PV, Boyd RL, Helme RD. Effects of capsaicin treatment on immunoglobulin secretion in the rat: further evidence for involvement of tachykinin-containing afferents. J Neuroimmunol 1990; 26:131–138.

143. Eglezos A, Andrews PV, Boyd RL, Helme RD. Tachykinin-mediated modulation of the primary antibody response in rats: evidence for mediation by an NK_2 receptor. J Neuroimmunol 1991; 32:11–18.

144. Scicchitano R, Bienenstock J, Stanisz AM. In vivo immunomodulation by the neuropeptide Substance P. Immunology 1988; 63:733–735.

145. Ijaz MK, Dent D, Babiuk LA. Neuroimmunomodulation of in vivo anti-rotavirus humoral immune response. J Neuroimmunol 1990; 26:159–171.

146. Johnson AR, Erdos EG. Release of histamine from mast cells by vasoactive agents. Proc Soc Exp Biol Med 1973; 142:1252–1256.

147. Foreman JC, Jordan CC, Oehme P, Renner H. Structure-activity relationships for some SP-related peptides that cause wheal and flare reactions in human skin. J Physiol (Lond) 1983; 335:449–465.

148. Holzer P. Local effector functions of capsaicin-sensitive sensory nerve endings: involvement of tachykinins, CGRP and other neuropeptides. Neuroscience 1988; 24:739–768.

149. Skofitsch G, Savitt JM, Jacobowitz DM. Suggestive evidence for a functional unit between mast cells and substance P fibers in the rat diaphragm and mesentery. Histochemistry 1985; 82:5–8.

150. Stead RH, Tomioka M, Quinonez G, Simon GT, Felten SY, Bienenstock J. Intestinal mucosa mast cells in normal and nematode-infected rat intestines are in intimate contact with peptidergic nerves. Proc Natl Acad Sci (USA) 1987; 84:2975–2979.

151. Alving K, Sundstrom C, Matran R, Panula P, Hokfelt T, Lundberg JM. Association between histamine-containing mast cells and sensory nerves in the skin and control and capsaicin-treated pigs. Cell Tissue Res 1991; 264:529–538.

152. Renda T, Vaccaro R, Casu C. CGRP-LI nerve endings in rat knee joint. Ann NY Acad Sci 1992; 657:484–485.

153. Stead RH, Dixon MF, Bramwell NH, Riddell RH, Bienenstock J. Mast cells are closely apposed to nerves in the human gastrointestinal mucosa. Gastroenterology 1989; 97:575–585.

154. Bienenstock J, McQueen G, Sestini P, Marshall JS, Stead RH, Perdue MH. Mast cell/nerve interactions in vitro and in vivo. Am Rev Respir Dis 1991; 143:S55–S58.

155. Shibata H, Mio M, Tasaka K. Analysis of the mechanism of histamine release by SP. Biochim Biophys Acta 1985; 846:1–7.

156. Repke H, Bienert M. Structural requirements for mast cells triggering by SP-like peptides. Agents Actions 1988; 23:207–210.

157. Arock M, Devillier P, Luffau G, Guillosson JJ, Renoux M. Histamine-releasing activity of endogenous peptides on mast cells derived from different sites and species. Int Arch Allergy Appl Immunol 1989; 89:229–235.

158. Assem ESK, Ghanem NS, Abdullah NA, Repke H, Foreman JC, Hayes NA. Substance P and Arg-Pro-Lys-Pro-NH-C_{12}-H_{25}-induced mediator release from different mast cells subtypes of rat and guinea-pig. Immunopharmacology 1989; 17:119–128.

159. Devillier P, Regoli D, Asseraf A, Descours B, Marsac J, Renoux M. Histamine release and local responses of rat and human skin to SP and other mammalian tachykinins. Pharmacology 1986; 32:340–347.

160. Devillier P, Drapeau G, Renoux M, Regoli D. Role of the N-terminal arginine in the histamine-releasing activity of SP, bradykinin and related peptides. Eur J Pharmacol 1989; 168:53–60.

161. Pearce FL, Kassessinoff TA, Liu WL. Characteristics of histamine secretion induced by neuropeptides: implications for the relevance of peptide-mast cell interactions in allergy and inflammation. Int Arch Allergy Appl Immunol 1989; 88:129–131.

162. Mousli M, Bueb JL, Bronner C, Rouot B, Landry Y. G protein activation: a receptor independent mode of action for cationic amphiphilic neuropeptides and venom peptides. Trends Pharmacol Sci 1990; 11:358–362.

163. Krumins SA, Bloomfield CA. Evidence of NK_1 and NK_2 tachykinin receptors and their involvement in histamine release in a murine mast cell line. Neuropeptides 1992; 21:65–72.

164. Krumins SA, Bloomfield CA. C-terminal substance P fragments elicit histamine release from a murine mast cell line. Neuropeptides 1993; 24:5–10.

165. Ansel JC, Brown JR, Payan DG, Brown MA. Substance P selectively increases TNFα gene expression in murine mast cells. J Immunol 1993; 150:4478–4485.

166. Hagermark O, Hokfelt T, Pernow B. Flare and itch induced by Substance P in human skin. J Invest Dermatol 1978; 71:233–235.

167. Coutts AA, Jorizzo JJ, Greaves MW, Burnstock G. Mechanism of vasodilatation due to Substance P in humans. Br J Dermatol 1981; 105:354–355.

168. Abelli L, Nappi F, Perretti F, Maggi CA, Manzini S, Giachetti A. Microvascular leakage induced by Substance P in rat urinary bladder: involvement of cyclooxygenase metabolites of arachidonic acid. J Autonom Pharmacol 1992; 12:269–276.

169. Lam FY, Ferrell WR. Mediators of Substance P induced inflammation in the rat knee joint. Agents Actions 1990; 31:298–307.

170. Kiernan JA. The involvement of mast cells in vasodilatation to axon reflex in injured skin. Quart J Exp Physiol 1972; 57:311–318.

171. Kiernan JA. Study of chemically induced acute inflammation in the skin of the rat. Quart J Exp Physiol 1977; 62:151–156.

172. Dimitriadou V, Buzzi MG, Moskowitz MA, Theoharides TC. Trigeminal sensory fibers stimulation induces morphological changes reflecting secretion in rat dura mater mast cells. Neuroscience 1991; 44:97–112.

173. Markowitz S, Saito K, Moskowitz MA. Neurogenically mediated leakage of plasma protein occurs from blood vessels in dura mater but not brain. J Neurosci 1987; 7:4129–4136.

174. Kowalski ML, Kaliner M. Neurogenic inflammation, vascular permeability and mast cells. J Immunol 1988; 140:3905–3912.

175. Kowalski ML, Sliwinska-Kowalska M, Kaliner M. Neurogenic inflammation, vascular permeability and mast cells II. Additional evidence indicating that mast cells are not involved in neurogenic inflammation. J Immunol 1988; 145:1214–1221.

176. Tausk F, Undem B. Exogenous but not endogenous substance P releases histamine from isolated human skin fragments. Neuropeptides 1995; 29:351–355.

177. Wallace JL, McKnight GW, Befus AD. Capsaicin-induced hyperaemia in the stomach: possible contribution of mast cells. Am J Physiol 1992; 263:G209–G214.

178. Mathison R, Davison JS. Vasodilation in jejunal arterioles induced by primary afferent nerves in sensitized rats is mast cell dependent. J Autonom Nervous System 1993; 43(suppl):93–94.

179. Ziche M, Morbidelli L, Pacini M, Geppetti P, Alessandri G, Maggi CA. Substance P stimulates neovascularization in vivo and proliferation of cultured endothelial cells. Microvasc Res 1990; 40:264–278.

180. Ziche M, Morbidelli L, Geppetti P, Maggi CA, Dolara P. Substance P induces migration of capillary endothelial cells: a novel NK_1 receptor mediated activity. Life Sci Pharmacol Lett 1991; 48:PL-7–PL-11.

181. Ziche M, Morbidelli L, Parenti A, Amerini S, Granger HJ, Maggi CA. Substance P

increases cyclic GMP levels on coronary postcapillary venular endothelial cells. Life Sci Pharmacol Lett 1993; 53:PL229–PL234.

182. Ziche M, Morbidelli L, Masini E, Amerini S, Granger HJ, Maggi CA, Geppetti P, Ledda F. Nitric oxide mediates angiogenesis in vivo and endothelial cell growth and migration in vitro promoted by substance P. J Clin Invest 1994; 94:2036–2044.

183. Fan TPD, Hu DE, Guard S, Gresham GA, Watling KJ. Stimulation of angiogenesis by substance P and interleukin 1 in the rat and its inhibition by NK_1 or interleukin 1 receptor antagonists. Br J Pharmacol 1993; 110:43–49.

184. Nilsson J, Von Euler AM, Dalsgaard CJ. Stimulation of connective tissue cell growth by SP and substance K. Nature 1985; 315:61–63.

185. Hultgardh-Nilsson A, Nilsson J, Jonzon B, Dalsgaard CJ. Coupling between inositol phosphate formation and DNA synthesis in smooth muscle cells stimulated with neurokinin A. J Cell Physiol 1988; 137:141–145.

186. Noveral JP, Grunstein MM. Tachykinin regulation of airway smooth muscle cell proliferation. Am J Physiol 1995; 269:L339–L343.

187. Ziche M, Morbidelli L, Pacini M, Dolara P, Maggi CA. NK_1 receptors mediate the proliferative response of human fibroblasts to tachykinins. Br J Pharmacol 1990; 100: 11–14.

188. Morbidelli L, Maggi CA, Ziche M. Effect of tachykinin selective receptor antagonists on the growth of cultured human skin fibroblasts. Neuropeptides 1992; 22:45.

189. Parenti A, Amerini S, Ledda F, Maggi CA, Ziche M. Tachykinin NK_1 receptor mediates the migration of adherent human skin fibrobalsts in culture. Naunyn Schmiedeberg's Arch Pharmacol. In press.

190. Kahler CM, Sitte BA, Reinisch N, Wiedermann CJ. Stimulation of the chemotactic migration of human fibroblasts by substance P. Eur J Pharmacol 1993; 249:281–286.

191. Kahler CM, Herold M, Wiedermann CJ. Substance P: a competence factor for human fibroblast proliferation that induces the release of growth-regulatory arachidonic acid metabolites. J Cell Physiol 1993; 156:579–587.

192. Tanaka T, Danno K, Ikai K, Imamura S. Effects of SP and substance K on the growth of cultured keratinocytes. J Invest Dermatol 1988; 90:399–401.

193. Wilkinson DI. Mitogenic effect of SP and CGRP on keratinocytes. J Cell Biol 1989; 107:509.

194. Koizumi H, Tanaka H, Fukaya T, Ohkawara A. SP induces intracellular Ca increase and translocation of protein kinase C in epidermis. Br J Dermatol 1992; 127:595–599.

195. Pincelli C, Fantini F, Romualdi P, Sevignani C, Lesa G, Benassi L, Giannetti A. SP is diminished and VIP is augmented in psoriatic lesions and these effects exert disparate effects on the proliferaion of cultured human keratinocytes. J Invest Dermatol 1992; 98: 421–427.

196. Benrath J, Zimmermann M, Gillardon F. SP and nitric oxide mediate wound healing of ultraviolet photodamaged rat skin: evidence for an effect of nitric oxide on keratinocyte proliferation. Neurosci Lett 1995; 200:17–20.

197. Hagan RM, Ireland SJ, Bailey F, McBride C, Jordan CA, Ward P. A spirolactam conformationally-constrained analogue of physalaemin which is a peptidase resistant, selective NK-1 receptor agonist. Br J Pharmacol, 1991; 102:168P.

198. Fujii T, Murai M, Morimoto H, Maeda Y, Yamaoka M, Hagiwara D, Miyake H, Ikari N, Matsuo M. Pharmacological profile of a high affinity dipeptide NK_1 receptor antagonist FK 888. Br J Pharmacol 1992; 107:785–789.

199. McKnight AT, Maguire JJ, Elliott NJ, Fletcher AE, Foster AC, Tridgett R, Williams BJ,

Longmore J, Iversen LL. Pharmacological specificity of novel, synthetic, cyclic peptides as antagonists at tachykinin receptors. Br J Pharmacol 1991; 104:335–360.

200. Maggi CA, Giuliani S, Ballati L, Lecci A, Manzini S, Patacchini R, Renzetti AR, Rovero P, Quartara L, Giachetti A. In vivo evidence for tachykininergic trasmission using a new NK_2 receptor selective antagonist, MEN 10376. J Pharmacol Exp Ther 1991; 257:1172–1178.

201. Snider RM, Constantine JW, Lowe JA III, Longo KP, Lebel WS, Woody HA, Drozda SE, Desai MC, Vinick FJ, Spencer RW, Hess HJ. A potent nonpeptide antagonist of the substance P (NK_1) receptor. Science 1991; 251:435–437.

202. Emonds-Alt X, Doutremepuich JD, Heaulme M, Neliat G, Santucci V, Steinberg R, Vilain P, Bichon D, Ducoux JP, Proietto V, Van Broeck D, Soubrier P, Le Fur G, Breliere JC. In vitro and in vivo biological activities of SR 140,333, a novel potent nonpeptide tachykinin NK_1 receptor antagonist. Eur J Pharmacol 1993; 250:403–413.

203. Emonds-Alt X, Vilain P, Goulaouic P Proietto V, Van Broeck D, Advenier C, Naline E, Neliat G, Le Fur G, Breliere JC. A potent and selective nonpeptide antagonist of the neurokinin A (NK_2) receptor. Life Sci Pharmacol Lett 1992; 50:PL101–PL106.

204. Beresford IJM, Sheldrick RLG, Ball DI, Turpin MP, Walsh DM, Hawcock AB, Coleman RA, Hagan RM, Tyers MB. GR159897, a potent nonpeptide antagonist at tachykinin NK_2 receptors. Eur J Pharmacol 1995; 272:241–248.

205. Emonds-Alt X, Bichon D, Ducoux JP, Heaulme M, Miloux B, Poncelet M, Proietto V, Van Broeck D, Vilain P, Neliat G, Soubrier P, Le Fur G, Breliere JC. SR 142,801 the first potent nonpeptide antagonist of the tachykinin NK_3 receptor. Life Sci Pharmacol Lett 1995; 56:PL27–PL32.

206. Maggi CA, Astolfi M, Giuliani S, Goso C, Manzini S, Meini S, Patacchini R, Pavone V, Pedone C, Quartara L, Renzetti AR, Giachetti A. MEN 10627, a novel polycyclic peptide antagonist of tachykinin NK_2 receptors. J Pharmacol Exp Ther 1994; 271:1489–1500.

207. Maggi CA. Tachykinin receptors in the airways and lung: what should we block? Pharmacol Res 1990; 22:527–539.

208. Morimoto H, Murai M, Maeda Y, Yamaoka M, Nishikawa M, Kiyotoh S, Fujii T. FK 224 a novel cyclopeptide substance P antagonist with NK_1 and NK_2 receptor selectivity. J Pharmacol Exp Ther 1992; 262:398–402.

209. Patacchini R, Quartara L, Astolfi M, Goso C, Giachetti A, Maggi CA. Activity of cyclic pseudopeptide antagonists at peripheral tachykinin receptors. J Pharmacol Exp Ther 1995; 272:1082–1087.

210. Robineau P, Lonchampt M, Kucharczyk N, Krause JE, Regoli D, Fauchere JL, Prost JF, Canet E. In vitro and in vivo pharmacology of S 16474, a novel dual tachykinin NK_1 and NK_2 receptor antagonist. Eur J Pharmacol 1995; 294:677–684.

211. Szolcsanyi J, Bartho L. Capsaicin-sensitive non-cholinergic excitatory innervation of the guinea-pig tracheobronchial smooth muscle. Neurosci Lett 1982; 34:247–250.

212. Szolcsanyi J. Tetrodotoxin-resistant noncholinergic neurogenic contraction evoked by capsaicinoids and piperine on the guinea-pig trachea. Neurosci Lett 1983; 42:83–88.

213. Lundberg JM, Saria A. Bronchial smooth muscle contraction induced by stimulation of capsaicin-sensitive sensory neurons. Acta Physiol Scand 1982; 116:473–476.

214. Lundberg JM, Saria A. Capsaicin-induced desensitization of airway mucosa to cigarette smoke, mechanical and chemical irritants. Nature 1983; 302:251–253.

215. Lundberg JM, Saria A. Polypeptide-containing neurons in airway smooth muscle. Annu Rev Physiol 1987; 49:557–572.

216. Maggi CA, Giachetti A, Dey RD, Said SI. Neuropeptides as regulators of airway function: with special reference to VIP and the tachykinins. Physiol Rev 1995; 75:277–322.

217. Maggi CA. The pharmacology of the efferent function of sensory nerves. J Autonom Pharmacol 1991; 11:173–208.

218. Delay-Goyet P, Lundberg JM. Cigarette smoke-induced airway oedema is blocked by the NK_1 antagonist CP 96345. Eur J Pharmacol 1991; 203:157–158.

219. Delay-Goyet P, Franco-Cereceda A, Gonsalves SF, Clingan CA, Lowe JA III, Lundberg JM. CP 96345 antagonism of NK_1 receptors and smoke-induced protein extravasation in relation to its cardiovascular effects. Eur J Pharmacol 1992; 222:213–218.

220. Piedimonte G, Bertrand C, Geppetti P, Snider RM, Desai MC, Nadel JA. A new NK_1 receptor antagonist (CP-99994) prevents the increase in tracheal vascular permeability produced by hypertonic saline. J Pharmacol Exp Ther 1993; 266:270–273.

221. Sakamoto T, Barnes PJ, Chung KF. Effect of CP96345, a nonpeptide NK_1 receptor antagonist, against substance P-bradykinin, and allergen-induced airway microvascular leakage and bronchoconstriction in the guinea-pig. Eur J Pharmacol 1993; 231:31–38.

222. Buckley TL, Nijkamp FP. Mucosal exudation associated with a pulmonary delayed-type hypersensitivity reaction in the mouse. Role for the tachykinins. J Immunol 1994; 153:4169–4178.

223. Yoshihara S, Chan B, Yamawaki I, Geppetti P, Ricciardolo FLM, Massion PP, Nadel JA. Plasma extravasation in the rat trachea induced by cold air is mediated by tachykinin release from sensory nerves. Am J Respir Crit Care 1995; 151:1011–1017.

224. Bertrand C, Geppetti P, Baker , Yamawaki I, Nadel JA. Role of neurogenic inflammation in antigen-induced vascular extravasation in guinea-pig trachea. J Immunol 1993; 150:1479–1485.

225. Hirayama Y, Lei YH, Barnes PJ, Rogers DF. Effects of two novel tachykinin antagonists FK 224 and FK888 on neurogenic airway plasma exudation, bronchoconstriction and systemic hypotensin in guinea-pigs in vivo. Br J Pharmacol 1993; 18:844–851.

226. Lou YP, Lee LY, Satoh H, Lundberg JM. Postjunctional inhibitory effect of the NK_2 receptor antagonist, SR48968, on sensory NANC bronchoconstriction in the guinea-pig. Br J Pharmacol 1993; 109:765–773.

227. Bertrand C, Nadel JA, Graf PD, Geppetti P. Capsaicin increases airflow resistance in guinea-pigs in vivo by activating both NK_2 and NK_1 tachykinin receptors. Am Rev Respir Dis 1993; 148:909–914.

228. Bertrand C, Geppetti P, Graf PD, Foresi A, Nadel JA. Involvement of neurogenic inflammation in antigen-induced bronchoconstriction in guinea-pigs. Am J Physiol 1993; 265:L507–L511.

229. Solway J, Kao BM, Jordan JE, Gitter B, Rodger IW, Howbert JJ, Alger LE, Necheles J, Leff AR, Garland A. Tachykinin receptor antagonists inhibit hyperpnea-induced bronchoconstriction in guinea-pigs. J Clin Invest 1993; 92:315–323.

230. Yoshihara S, Geppetti P, Hara M, Linden A, Ricciardolo FLM, Chan B, Nadel JA. Cold air-induced bronchoconstriction is mediayed by tachykinin and kinin release in guinea-pigs. Eur J Pharmacol 1996; 296:291–296.

231. Perretti F, Ballati L, Manzini S, Maggi CA, Evangelista S. Antibronchospastic activity of MEN 10627, a novel tachykinin NK_2 receptor antagonist in guinea-pig airways. Eur J Pharmacol 1995; 273:129–135.

232. Boichot E, Germain N, Lagente V, Advenier C. Prevention by the tachykinin NK_2 receptor antagonist, SR 48968, of antigen-induced airway hyperresponsiveness in sensitized guinea-pigs. Br J Pharmacol 1995; 114:259–261.

233. Mizuguchi M, Fujimura M, Amemiya T, Nishi K, Ohka T, Matsuda T. Involvement of NK$_2$ receptors rather than NK$_1$ receptors in bronchial hyperresponsiveness induced by allergic reaction in guinea-pigs. Br J Pharmacol 1996; 117:443–448.

234. Advenier C, Girard V, Naline E, Vilain P, Emonds-Alt X. Antitussive effect of SR 48968, a nonpeptide tachykinin NK$_2$ receptor antagonist. Eur J Pharmacol 1993; 250: 169–171.

235. Girard V, Naline E, Vilain P, Emonds-Alt X, Advenier C. Effect of the two tachykinin antagonists, SR 48968 and SR 140333, on cough induced by citric acid in the unanaesthetized guinea-pig. Eur Respir J 1995; 8:1110–1114.

236. Ujiie Y, Sekizawa K, Aikawa T, Sasaki H. Evidence for substance P as an endogenous substance causing cough in guinea-pigs. Am Rev Respir Dis 1993; 148:1628–1632.

237. Sekizawa K, Ebihara T, Sasaki H. Role of substance P in cough during bronchoconstriction in awake guinea-pigs. Am J Respir Crit Care Med 1995; 151:815–821.

238. Lundberg JM, Hokfelt T, Martling CR, Saria A, Cuello C. Substance P immunoreactive sensory nerves in the lower respiratory tract of various mammals including man. Cell Tissue Res 1984; 235:251–261.

239. Martling CR, Saria A, Fischer JA, Hokfelt T, Lundberg JM. CGRP and the lung: neuronal coexistence with SP release by capsaicin and vasodilatory effects. Regulatory Peptides 1988; 20:125–139.

240. Komatsu T, Yamamoto , Shimokata K, Nagura H. Distribution of SP-immunoreactive and CGRP-immunoreactive nerves in normal human lungs. Int Arch Allergy Appl Immunol 1991; 95:23–28.

241. Springall DR, Polak JM, Howard L, Power RF, Krausz T, Manickam S, Banner NR, Khagani A, Rose M, Yacoub MH. Persistence of intrinsic neurones and possible phenotypic changes after extrinsic denervation of human respiratory tract by heart lung transplantation. Am Rev Respir Dis 1990; 141:1538–1546.

242. Ollerenshaw SL, Jarvis D, Sullivan CE, Woolcock AJ. Substance P immunoreactive nerves in airways from asthmatics and nonasthmatics. Eur Respir J 1991; 4:673–682.

243. Howarth PH, Djukanovic R, Wilson JW, Holgate ST, Springall DR, Polak JM. Mucosal nerves in endobronchial biopsies in asthma and non-asthma. Int Arch Allergy Clin Immunol 1991; 94:330–333.

244. Martling CR, Theodorsson-Norheim E, Lundberg JM. Occurrence and effects of multiple tachykinins, SP, NKA and neuropeptide K in human airways. Life Sci 1987; 40: 1633–1643.

245. Nieber K, Baumgarten C, Witzel A, Rathsack R, Oehme P, Brunnee T, Kleine-Tebbe J, Kunkel G. The possible role of SP in the allergic reaction based on two different provocation models. Int Arch Allergy Appl Immunol 1991; 94:334–338.

246. Nieber K, Baumgarten CR, Rathsack R, Furkert J, Oehme P, Kunkel G. Substance P and β-endorphin-like immunoreactivity in lavage fluids of subjects with and without allergic asthma. J Allergy Clin Immunol 1992; 90:646–652.

247. Espiritu RF, Pittet JF, Matthay MA, Goetzl EJ. Neuropeptides in pulmonary edema of adult respiratory distress syndrome. Inflammation 1992; 16:509–517.

248. Tomaki M, Ichinose M, Miura M, Hirayama Y, Yamauchi H, Nakajima N, Shirato K. Elevated substance P content in induced sputum from patients with asthma and patients with chronic bronchitis. Am J Respir Crit Care Med 1995; 151:613–617.

249. Fischer A, Kummer W, Couraud JY, Adler D, Branscheid D, Heym C. Immunohistochemical localization of receptors for VIP and SP in human trachea. Lab Invest 1992; 67:387–393.

250. Walsh DA, Salmon M, Featherstone R, Wharton J, Church MK, Polak JM. Differences in the distribution and characteristics of tachykinin NK_1 binding sites between human and guinea-pig lung. Br J Pharmacol 1994; 113:1407–1415.

251. Adcock IM, Peters M, Gelder C, Shirasaki H, Brown CR, Barnes PJ. Increased tachykinin receptor gene expression in asthmatic lung and its modulation by steroids. J Mol Endocrinol 1993; 11:1–7.

252. Bai TR, Zhou D, Weir T, Walker B, Hegele R, Hayashi S, Mc Kay K, Bondy GP, Fong T. Substance P (NK_1) and neurokinin A (NK_2) receptor gene expression in inflammatory airway disease. Am J Physiol 1995; 269:L309–L317.

253. Naline E, Devillier P, Drapeau G, Toty L, Bakdach H, Regoli D, Advenier C. Characterization of neurokinin effects and receptor selectivity in human isolated bronchi. Am Rev Respir Dis 1989; 140:679–686.

254. Barnes PJ, Dewar A, Rogers DF. Human bronchial secretion: effect of SP, muscarinic and adrenergic stimulation in vitro. Br J Pharmacol 1986; 89:767P.

255. Black JL, Johnson PRA, Alouan L, Armour CL. Neurokinin A with potassium channel blockade potentiates contraction to electrical stimulation in human bronchus. Eur J Pharmacol 1990; 180:311–317.

256. Sheldrick RLG, Rabe KF, Fischer A, Magnussen H, Coleman RA. Further evidence that tachykinin-induced contraction of human isolated bronchus is mediated only by NK_2 receptors. Neuropeptides 1995; 29:281–292.

257. Joos G, Pauwels R, Van der Straeten M. Effect of inhaled substance P and neurokinin A on the airways of normal and asthmatic subjects. Thorax 1987; 42:779–783.

258. Evans TW, Dixon CM, Clarke B, Conradson TB, Barnes PJ. Comparison of neurokinin A and substance P on cardiovascular and airway function in man. Br J Clin Pharmacol 1988; 25:273–275.

259. Ichinose M, Nakajima N, Takahashi T, Yamauchi H, Inoue H, Takishima T. protection against bradykinin-induced bronchoconstriction in asthmatic patients by neurokinin receptor antagonist. Lancet 1992; 340:1248–1251.

260. Joos GF, Kips JC, Peleman RA, Pauwels RA. Tachykinin antagonists and the airways. Arch Int Pharmacodyn 1995; 329:205–219.

261. Lunde H, Hedner J, Svedmyr N. Lack of 4 weeks treatment with the neurokinin receptor antagonist FK 224 in mild to moderate asthma. Eur Respir J 1994; suppl 18. 7:151s.

262. Fahy JV, Wong HH, Geppetti P, Reis JM, Harris SC, MAc Lean DB, Nadel JA, Boushey HA. Effect of an NK_1 receptor antagonist (CP 99994) on hypertonic saline-induced bronchoconstriction and cough in male asthmatic subjects. Am J Respir Crit Care Med 1995; 152:879–884.

263. Bjorklund H, Dalsgaard CJ, Jonsson CE, Hermansson A. Sensory and autonomic innervation of nonhairy and hairy human skin. Cell Tissue Res 1986; 243:51–57.

264. Wallengren J, Ekman R, Sundler F. Occurrence and distribution of neuropeptides in the human skin. Acta Dermatol Venereol 1987; 67:185–192.

265. Alvarez FJ, Cervantes C, Blasco L, Villalba R, Martinez-Murillo R, Polak JM, Rodrigo J. Presence of CGRP and SP immunoreactivity in intraepidermal free nerve endings of cat skin. Brain Res 1988; 442:391–395.

266. Gamse R, Saria A. Potentiation of tachykinin-induced plasma protein extravasation by CGRP. Eur J Pharmacol 1985; 114:61–66.

267. Couture R, Kerouac R. Plasma protein extravasation induced by mammalian tachykinins in rat skin: influence of anesthetic agents and an acetylcholine antagonist. Br J Pharmacol 1987; 91:265–273.

268. Andrews PV, Helme RD, Thomas KL. NK-1 receptor mediation of neurogenic plasma extravasation in rat skin. Br J Pharmacol 1989; 97:1232–1238.

269. Jacques L, Couture R, Drapeau G, Regoli D. Capillary permeability induced by intravenous neurokinins, receptor characterization and mechanism of action. Naunyn Schmiedeberg's Arch Pharmacol 1989; 340:170–179.

270. Ahluwalia A, Giuliani S, Maggi CA. Demonstration of septide-sensitive inflammatory response in rat skin. Br J Pharmacol 1995; 116:2170–2174.

271. Wallengren J, Hakanson R. Effects of SP, NKA and CGRP in human skin and their involvement in sensory nerve mediated responses. Eur J Pharmacol 1987; 143:267–273.

272. Fuller RW, Conradson TB, Dixon CMS, Crossman DC, Barnes PJ. Sensory neuropeptide effects in human skin. Br J Pharmacol 1987; 92:781–788.

273. Deguchi M, Niwa M, Shigematsu K, Fujii T, Namba K, Ozaki M. Specific [^{125}I]Bolton-Hunter SP binding sites in human and rat skin. Neurosci Lett 1989; 99:287–292.

274. Lundblad L, Lundberg JM, Anggard A, Zetterstrom O. Capsaicin-pretreatment inhibits the flare component of the cutaneous allergic reaction in man. Eur J Pharmacol 1985; 113:461–462.

275. Lundblad L, Lundberg JM, Anggard A, Zetterstrom O. Capsaicin-sensitive nerves and the cutaneous allergy reaction in man. Allergy 1987; 42:20–25.

276. Helme RD, McKernan S. Neurogenic flare responses following topical application of capsaicin in humans. Ann Neurol 1985; 18:505–509.

277. Raud J, Lundeberg T, Brodda-Jensen G, Thedorsson E, Hedqvist P. Potent antiinflammatory action of CGRP. Biochem Biophys Res Commun 1991; 180:1429–1435.

278. Carpenter SE, Lynn B. Vascular and sensory responses of human skin to mild injury after topical treatment with capsaicin. Br J Pharmacol 1981; 73:755–758.

279. Anand P, Bloom SR, McGregor GP. Topical capsaicin pretreatment inhibits axon reflex vasodilatation caused by somatostatin and VIP in human skin. Br J Pharmacol 1983; 78: 665–669.

280. McCusker MT, Chung KF, Roberts NM, Barnes PJ. Effect of topical capsaicin on the cutaneous responses to inflammatory mediators and to antigen in man. J Allergy Clin Immunol 1989; 83:1118–1124.

281. Delay-Goyet P, Satoh H, Lundberg JM. Relative involvement of SP and CGRP mechanisms in antidromic vasodilatation in the rat skin. Acta Physiol Scand 1992; 146: 537–538.

282. Garret C, Carruette A, Fardin V, Moussaoui S, Peyronel JF, Blanchard JC, Laduron PM. Pharmacological properties of a potent and selective nonpeptide substance P antagonist. Proc Natl Acad Sci (USA) 1991; 88:10208–10212.

283. Lembeck F, Donnerer J, Tsuchiya M, Nagahisa A. The non peptide tachykinin antagonist, CP-96,345 is a potent inhibitor of neurogenic inflammation. Br J Pharmacol 1992; 105:527–530.

284. Xu XJ, Dalsgaard CJ, Maggi CA, Wiesenfeld-Hallin Z. NK$_1$, but not NK$_2$, tachykinin receptors mediate plasma extravasation induced by antidromic C-fiber stimulation in rat hindpaw: demonstrated with the NK$_1$ antagonist CP 96,345 and the NK$_2$ antagonist MEN 10207. Neurosci Lett 1992; 139:249–252.

285. Moussaoui SM, Montier F, Carruette A, Blanchard JC, Laduron PM, Garret C. A nonpeptide NK$_1$ receptor antagonist RP 67580, inhibits neurogenic inflammation post-synaptically. Br J Pharmacol 1993; 109:259–264.

286. Amann R, Schuligoi R, Holzer P, Donnerer J. The nonpeptide NK$_1$ receptor antagonist SR 140333 produces long lasting inhibition of neurogenic inflammation but does not

influence acute chemo- or thermonociception in rats. Naunyn Schmiedeberg's Arch Pharmacol 1995; 352:201–205.

287. Inoue H, Nagata N, Koshihara Y. Involvement of SP as a mediator of capsaicin-induced mouse ear oedema. Inflamm Res 1995; 44:470–474.

288. Wilsoncroft P, Euzger H, Brain SD. Effect of a NK_1 receptor antagonist on oedema formation induced by tachykinins carrageenin and an allergic response in guinea-pig skin. Neuropeptides 1994; 26:405–411.

289. Eschenfelder CC, Benrath J, Zimmermann M, Gillardon F. Involvement of substance P in ultraviolet irradiation-induced inflammation in rat skin. Eur J Neurosci 1995; 7: 1520–1526.

290. Siney L, Brain SD. Involvement of sensory neuropeptides in the development of plasma extravasation in rat dorsal skin following thermal injury. Br J Pharmacol. In press.

291. Mapp PI, Kidd BL, Gibson SJ, Terry JM, Revell PA, Ibrahim NBN, Blake DR, Polak JM. Substance P-, CGRP- and C-flanking peptide of neuropeptide Y-immunoreactive fibres are present in normal synovium but depleted in patients with rheumatoid arthritis. Neuroscience 1990; 37:143–153.

292. Buma P, Verschuren C, Versleyen D, Van der Kraan P, Oestricher AB. CGRP, SP and gap-43/B-50 immunoreactivity in the normal and arthrotic knee joint of the mouse. Histochemistry 1992; 98:327–339.

293. Konttinen YT, Rees R, Hukkanen M, Gronblad M, Tolvanen E, Gibson SJ, Polak JM, Brewerton DA. Nerves in inflammatory synovium: immunohistochemical observations on the adjuvant arthritic rat model. J Rheumatol 1990; 17:1586–1591.

294. Pereira da Silva JA, Carmo-Fonseca M. Peptide containing nerves in human synovium: immunohistochemical evidence for decreased innervation in rheumatoid arthritis. J Rheumatol 1990; 17:1592–1599.

295. Lembeck F, Donnerer J, Colpaert FC. Increase of Substance P in primary afferent nerves during chronic pain. Neuropeptides 1981; 1:175–180.

296. Colpaert FC, Donnerer J, Lembeck F. Effects of capsaicin on inflammation and on the Substance P content of nervous tissues in rats with adjuvant arthritis. Life Sci 1983; 32:1827–1834.

297. Schoenen J, Van Hees J, Gybels J, De Castro-Costa M, Vanderhaegen JJ. Histochemical changes of Substance P, FRAP, serotonin and succinic dehydrogenase in the spinal cord of rats with adjuvant arthritis. Life Sci 1985; 36:1247–1254.

298. Minami M, Kuraishi Y, Kawamura M, Yamaguchi T, Masu Y, Nakanishi S, Satoh M. Enhancement of preprotachykinin A gene expression by adjuvant-induced inflammation in the rat spinal cord: possible involvement of SP-containing spinal neurons in nociception. Neurosci Lett 1989; 98:105–110.

299. Hanesch U, Blecher F, Stiller RU, Emson PC, Schaible HG, Heppelmann B. The effect of unilateral inflammation at the rat's ankle joint on the expression of preprotachykinin A mRNA and preprosomatostatin mRNA in dorsal root ganglion cells—a study using non radioactive in situ hybridization. Brain Res 1995; 700:279–284.

300. Schafer MKH, Nohr D, Krause JE, Weihe E. Inflammation-induced upregulation of NK_1 receptor mRNA in dorsal horn neurones. Neuroreport 1993; 4:1007–1010.

301. Krause JE, Di Maggio DA, McCarson KE. Alterations in NK_1 receptor gene expression in models of pain and inflammation. Can J Physiol Pharmacol 1995; 73:854–859.

302. Oku R, Satoh M, Takagi H. Release of Substance P from the spinal dorsal horn is enhanced in polyarthritic rats. Neurosci Lett 1987; 74:315–319.

303. Schaible HG, Jarrott B, Hope PJ, Duggan AW. Release of immunoreactive Substance P

in spinal cord during development of acute arthritis in the knee joint of the cat: a study with antibody microprobes. Brain Res 1990; 529:214–223.

304. Neugebauer V, Weiretter F, Schaible HG. Involvement of Substance P and NK_1 receptors in the hyperexcitability of dorsal horn neurons during development of acute arthritis in rat's knee joint. J Neurophysiol 1995; 73:1574–1583.

305. Ren K, Iadarola MJ, Dubner R. An isobolographic analysis of the effect of NMDA and NK_1 tachykinin receptor antagonists on inflammatory hyperalgesia in the rat. Br J Pharmacol 1996; 117:196–202.

306. Levine JD, Clark R, Devor M, Helms C, Moskowitz MA, Basbaum AI. Intraneuronal Substance P contributes to the severity of experimental arthritis. Science 1984; 226: 547–549.

307. Yaksh TL. Substance P release from knee joint afferent terminals: modulation by opioids. Brain Res 1988; 458:319–324.

308. Ferrell WR, Russell NJW. Plasma extravasation in the cat knee joint induced by antidromic articular nerve stimulation. Pflugers Archiv 1985; 404:91–93.

309. Ferrell WR, Russell NJW. Extravasation in the knee induced by antidromic stimulation of articular C fibre afferents of the anesthetized cat. J Physiol (Lond) 1986; 379: 407–416.

310. Coderre TJ, Basbaum AI, Levine JD. Neural control of vascular permeability: interactions between primary afferents, mast cells and sympathetic efferents. J Neurophysiol 1989; 62:48–58.

311. Khoshbaten A, Ferrell WR. Alterations in cat knee joint blood flow induced by electrical stimulation of articular afferents and efferents. J Physiol (Lond) 1990; 430:77–86.

312. Lam FY, Ferrell WR. Inhibition of carrageenen induced inflammation in the rat knee joint by SP antagonist. Ann Rheum Dis 1989; 48:928–932.

313. Lam FY, Ferrell WR. Mediators of Substance P induced inflammation in the rat knee joint. Agents Actions 1990; 31:298–307.

314. Lam FY, Ferrell WR. Specific neurokinin receptors mediate plasma extravasation in the rat knee joint. Br J Pharmacol 1991; 103:1263–1267.

315. Lam FY, Ferrell WR. Effects of interactions of naturally-occurring neuropeptides on blood flow in the rat knee joint. Br J Pharmacol 1993; 108:694–699.

316. Hirayama H, Yasumitsu R, Kawamura A, Fujii T. NK_1 receptors mediate tachykinin-induced plasma extravasation in the rat knee joint. Agents Actions 1993; 40:171–175.

317. Scott DT, Lam FY, Ferrell WR. Acute inflammation enhances SP-induced plasma protein extravasation in the rat knee joint. Regulatory Peptides 1992; 39:227–235.

318. Green PG, Basbaum AI, Levine JD. Sensory neuropeptide interactions in the production of plasma extravasation in the rat. Neuroscience 1992; 50:745–749.

319. Cambridge H, Brain SD. CGRP increases blood flow and potentiates plasma protein extravasation in the rat knee joint. Br J Pharmacol 1992; 106:746–750.

320. Lotz M, Carson DA, Vaughan JH. Substance P activation of rheumatoid synoviocytes: neural pathway in pathogenesis of arthritis. Science 1987; 235:893–895.

321. Walsh DA, Mapp PI, Wharton J, Rutherford RAD, Kidd BL, Revell PA, Blake DR, Polak JM. Localisation and characterisation of SP binding to human synovial tissue in rheumatoid arthritis. Ann Rheum Dis 1992; 51:313–317.

322. Walsh DA, Salmon M, Mapp PI, Wharton J, Garrett N, Blake DR, Polak JM. Microvascular substance P binding to normal and inflamed rat and human synovium. J Pharmacol Exp Ther 1993; 267:951–960.

323. Devillier P, Weill B, Renoux C. Elevated levels of tachykinin-like immunoreactivity in joint fluids from patients with rheumatic inflammatory diseases. N Engl J Med 1986; 314:1323.

324. Marshall KW, Chin B, Inman RD. Substance P and arthritis: analysis of plasma and synovial fluid levels. Arthr Rheum 1990; 33:87–90.

325. Marabini S, Matucci-Cerinic M, Geppetti P, Del Bianco E, Marchesoni A, Tosi S, Cagnoni M, Partsch G. Substance P and somatostatin levels in rheumatoid arthritis, osteoarthritis and psoriatic arthritis synovial fluid. Ann NY Acad Sci 1991; 632:435–436.

326. Sacerdote P, Carrabba M. Galante A, Pisati R, Manfredi B, Panerai AE. Plasma and synovial fluid interleukin-1, interleukin-6 and substance P concentrations in rheumatoid arthritis patients: effect orf the nonsteroidal antinflammatory drugs indomethacin diclofenac and naproxen. Inflamm Res 1995; 44:486–490.

327. Mapp PI, Walsh DA, Kidd BL, Cruwys SC, Polak JM, Blake DR. Localization of the enzyme neutral endopeptidase to the human synovium. J Rheumatol 1992; 19:1838–1844.

328. Costa M, Furness JB, Llewellyn-Smith IJ. Histochemistry of the enteric nervous system. In Johnson LR, ed. Physiology of the Gastrointestinal Tract. 2d ed. New York: Raven Press, 1987:1–40.

329. Bartho L, Holzer P. Search for a physiological role of Substance P in gastrointestinal motility. Neuroscience 1985; 16:1–32.

330. Holzer P, Gamse R, Lembeck F. Distribution of Substance P in the rat gastrointestinal tract—lack of effect of capsaicin pretreatment. Eur J Pharmacol 1980; 61:303–307.

331. Eutamene H, Theodorou V, Fioramonti J, Bueno L. Implication of NK_1 and NK_2 receptors in rat colonic hypersecretion induced by interleukin 1β: role of nitric oxide. Gastroenterology 1995; 109:483–489.

332. Wang YZ, Palmer JM, Cooke HJ. Neuroimmune regulation of colonic secretion in guinea-pigs. Am J Physiol 1991; 23:G307–G314.

333. Kraneveld AD, Buckley TL, Van Heuven-Nolsen D, Van Schaik Y, Koster ASj, Nijkamp FP. Delayed type hypersensitivity-induced increase in vascular permeability in the mouse small intestine: inhibition by depletion of sensory neuropeptides and NK_1 receptor blockade. Br J Pharmacol 1993; 114:1483–1489.

334. Pothoulakis C, Castagliuolo I, LaMont JT, Jaffer A, O'Keane JC, Snider RM, Leeman SE. CP96345, a substance P antagonist, inhibits rat intestinal responses to *Clostridium difficile* toxin A but not to cholera toxin, Proc Natl Acad Sci (USA) 1994; 91:947–951.

335. Miller MJS, Sadowka-Krowicka H, Jeng AJ, Chotinaruemol S, Wong M, Clark DA, Ho W, Sharkey KA. Substance P levels in experimental ileitis in guinea-pigs: effects of misoprostol. Am J Physiol 1993; 265:G321–G330.

336. Goldin E, Karmeli F, Selinger Z, Rachmilevitz D. Colonic Substance P levels are increased in ulcerative colitis and decreased in chronic severe constipation. Dig Dis Sci 1989; 34:754–757.

337. Mantyh CR, Vigna SR, Bollinger RR, Mantyh PW, Maggio JE, Pappas TN. Differential expression of Substance P receptors in patients with Crohn's disease and ulcerative colitis. Gastroenterology 1995; 109:850–860.

338. Saria A, Lundberg JM, Hua XY, Lembeck F. Capsaicin induced SP release and sensory control of vascular permeability in the guinea-pig ureter. Neurosci Lett 1983; 41: 167–172.

339. Koltzenburg M, McMahon SB. Plasma extravasation in the rat urinary bladder follow-

ing mechanical, electrical and chemical stimuli: evidence for a new population of chemosensitive primary sensory neurons. Neurosci Lett 1986; 72:352–356.

340. Maggi CA, Santicioli P, Abelli L, Parlani M, Capasso M, Conte B, Giuliani S, Meli A. Regional differences in the effects of capsaicin and tachykinins on motor activity and vascular permeability of the rat lower urinary tract. Naunyn Schmiedeberg's Arch Pharmacol 1987; 335:636–645.

341. Pinter E, Szolcsanyi J. Plasma extravasation in the skin and pelvis organs evoked by antidromic stimulation of the lumbosacral dorsal roots of the rat. Neuroscience 1995; 68:603–614.

342. Nagahisa A, Kanai Y, Suga O, Taniguchi K, Tsuchiya M, Lowe JA III, Hess HJ. Antiinflammatory and analgesic activity of a nonpeptide substance P receptor antagonist. Eur J Pharmacol 1992; 217:191–195.

343. Abelli L, Somma V, Maggi CA, Regoli D, Astolfi M, Parlani M, Rovero P, Conte B, Meli A. Effects of tachykinins and selective tachykinin receptor agonists on vascular permeability in the rat lower urinary tract: evidence for the involvement of NK_1 receptors. J Autonom Pharmacol 1989; 9:253–263.

344. Maggi CA, Parlani M, Astolfi M, Santicioli P, Rovero P, Abelli V, Somma V, Giuliani S, Regoli D, Patacchini R, Meli A. Neurokinin receptors in the rat lower urinary tract. J Pharmacol Exp Ther 1988; 246:308–315.

345. Giuliani S, Santicioli P, Lippe ITh, Lecci A, Maggi CA. Effect of bradykinin and tachykinin receptor antagonist on xylene-induced cystitis in rats. J Urol 1993; 150: 1014–1017.

346. Maggi CA, Abelli L, Giuliani S, Santicioli P, Geppetti P, Somma V, Frilli S, Meli A. The contribution of sensory nerves to xylene-induced cystitis in rats. Neuroscience 1988; 26:709–723.

347. Maggi CA, Lecci A, Santicioli P, Del Bianco E, Giuliani S. Cyclophosphamide cystitis in rats: involvement of capsaicin-sensitive primary afferents. J Autonom Nerv System 1992; 38:201–208.

348. Abelli L, Conte B, Somma V, Parlani M, Geppetti P, Maggi CA. Mechanical irritation induces neurogenic inflammation in the rat urethra. J Urol 1991; 146:1624–1626.

349. Maggi CA, Abelli L, Giuliani S, Somma V, Furio M, Patacchini R, Meli A. Motor and inflammatory effect of hyperosmolar solutions on the rat urinary bladder in relation to capsaicin-sensitive sensory nerves. Gen Pharmacol 1990; 21:97–103.

350. Ahluwalia A, Maggi CA, Santicioli P, Lecci A, Giuliani S. Characterisation of the capsaicin-sensitive component of cyclophosphamide-induced inflammation in the rat urinary bladder. Br J Pharmacol 1994; 111:1017–1022.

351. Edvinsson L, Uddman R. Adrenergic, cholinergic and peptidergic nerve fibres in dura mater—involvement in headache? Cephalalgia 1981; 1:175–179.

352. Edvinsson L, Rosendal-Helgesen S, Uddman R. Substance P: localization, concentration and release in cerebral arteries, choroid plexus and dura mater. Cell Tissue Res 1983; 234:1–7.

353. Suzuki N, Hardebo JE, Owman C. Origins and pathways of cerebrovascular nerves storing SP and CGRP in rat. Neuroscience 1989; 31:427–438.

354. Keller JT, Marfurt CF. Peptidergic and serotoninergic innervation of the rat dura mater. J Comp Neurol 1991; 309:515–534.

355. Geppetti P, Del Bianco E, Santicioli P, Lippe ITh, Maggi CA, Sicuteri F. Release of sensory neuropeptides from dural venous sinuses of guinea-pig. Brain Res 1990; 510: 58–62.

356. Zagami AS, Goadsby PJ, Edvinsson L. Stimulation of the superior sagittal sinus in the cat causes release of vasoactive peptides. Neuropeptides 1990; 16:69–75.

357. Shepheard SL, Williamson DJ, Hill RG, Hargreaves RJ. The nonpeptide NK1 receptor antagonist, RP 67,580, blocks neurogenic plasma extravasation in the dura mater of rats. Br J Pharmacol 1993; 108:11–12.

358. Lee WS, Moussaoui SM, Moskowitz MA. Blockade by oral or parenteral RPR 100893 (a nonpeptide NK_1 receptor antagonist) of neurogenic plasma protein extravasation within guinea-pig dura mater and conjunctiva. Br J Pharmacol 1994; 112:920–924.

359. Shepheard SL, Williamson DJ, Williams J, Hill RG, Hargreaves RJ. Comparison of the effects of sumatriptan and the NK_1 antagonist CP 99994 on plasma extravasation in dura mater and c-fos mRNA expression in trigeminal nucleus caudalis of rats. Neuropharmacology 1995; 34:255–261.

360. Dimitriadou V, Buzzi MG, Theoharides TC, Moskowitz MA. Ultrastructural evidence for neurogenically mediated changes in blood vessels of the rat dura mater and tongue following antidromic trigeminal stimulation. Neuroscience 1992; 48:187–203.

361. Saito K, Markowitz S, Moskowitz MA. Ergot alkaloids block neurogenic extravasation in dura mater: proposed action in vascular headaches. Ann Neurol 1988; 24:732–737.

362. Buzzi MG, Moskowitz MA. The antimigraine drug sumatriptan (GR 43175) selectively blocks neurogenic plasma extravasation from blood vessels in dura mater. Br J Pharmacol 1990; 99:202–206.

363. Buzzi MG, Carter WB, Shimizu T, Heath H III, Moskowitz MA. Dihydroergotamine and sumatriptan attenuate levels of CGRP in plasma in rat superior sagittal sinus during electrical stimulation of the trigeminal ganglion. Neuropharmacology 1991; 30:1193–1200.

364. Beattie DT, Connor HE, Hagan RM. Recent develement in tachykinin NK_1 receptor antagonists: prospects for the treatment of migraine headache. Can J Physiol Pharmacol 1995; 73:871–877.

365. Moskowitz MA, Buzzi MG. Neuroeffector functions of sensory fibres: implications for headache mechanisms and drug actions. J Neurol 1991; 238:S18–S22.

23

Therapeutic Effects and Immunomodulatory Activities of Natriuretic Peptides

THOMAS FLÜGE, MARKUS MEYER, and WOLF-GEORG FORSSMANN

Lower Saxony Institute for Peptide Research
Hannover, Germany

I. Introduction

In 1956, osmiophilic bodies were described in the atrial myocytes by Bruno Kisch (1). Later it was shown that changes in dietary sodium and water alter the granularity of these cells, suggesting that the atria may be involved in endocrine control of extracellular fluid volume (2,3). Evidence of an atrial natriuretic factor in the granules revealed the inducement of a profound natriuresis and diuresis by intravenous administration of rat atrial but not ventricular extracts into intact animals (4). The hypothesis of an endocrine function of the atria led to the identification of peptides with various actions as well as precursors in atrial tissue (5,6). Because of its natriuretic and vasorelaxant effects, the peptide first isolated and sequenced from rat and porcine atria was initially called *cardionatrin* (5) or *cardiodilatin* (CDD) (6). The name *atrial natriuretic peptide* (ANP) was introduced after the human peptide was characterized (7). It was then obvious that all these peptides are the precursor or processed forms of a homologous molecule type (8). Additional members of the natriuretic peptide (NP) family, such as *brain natriuretic peptide* (BNP) and *C-type natriuretic peptide* (CNP), were isolated in porcine brain in 1988 and 1990, respectively (9,10). The active A- (CDD/ANP), B- (BNP), and C-type (CNP) natriuretic peptides share a common central ring structure. The circulating form of CDD/ANP (also known as CDD/ANP-99-126) consists of the amino acids 99 to 126 of its prohormone CDD/ANP-1-126 (11). The loop is formed by a disulfide bridge between cysteine residues at positions 105 and 121 of this prohormone (Fig. 1). A hydrolytic cleavage of the ring causes a complete loss of biological activity (12–14). In 1988, the search for NPs synthesized outside the heart led to the isolation and structural analysis of another peptide extracted from human urine, which was called *urodilatin* (URO, INN: ularitide) (15). This A-type NP consists of the entire molecule of human CDD/ANP-99-126 enlarged by a four-residue NH_2-terminal extension: threonine-alanine-proline-arginine (CDD/ANP-95-126). URO is produced in the kidney and stems from the same gene code as CDD/ANP in the heart, but a different posttranslational processing yields the 32-residue peptide (Fig. 1) (16). In contrast to CDD/ANP, no immunoreactivity of this peptide could be found in human plasma (17–20).

In the past decade, the three types of NPs and/or their receptors have been demonstrated in various organs such as the heart, lung, kidney, gastrointestinal tract, adrenal glands, central and peripheral nervous system, and lymphoid tissues, suggesting additional functions of these peptides (16,21–23). Two of the specific NP receptors (NPR), NPR-A and NPR-B, induce an increase of intracellular cyclic guanosine 3′,5′-monophosphate (cGMP) via a particulate, membrane-bound guanylyl cyclase (GC) (21,24–28). Due to their catalytic activities, these receptors are also called GC-A and GC-B (29). The NPR-C-receptor, however, is not associated with an intracellular GC catalytic domain and is supposed to remove NPs from the extracellular space as a "clearance receptor" (30) or to store these peptides and release them slowly (21,28,31). However, there may be other functions of this or another closely related receptor as well (28,32,33). Antimitogenic effects of some NPs, for instance, were described as mediated via the NPR-C (32,33). The affinity of the NPs to the different NPRs are shown in Table 1 (34). In 1993, Valentin et al. demonstrated that URO binds to and activates renal receptors similarly to CDD/ANP (35). Molecular targets for cGMP include protein kinases, phosphodiesterases, and ion channels (36). Activation of cGMP-dependent protein kinase induces an inhibition of sodium reabsorption via an amiloride-sensitive channel and a relaxation of vascular smooth muscle via a decrease in intracellular Ca^{2+} concentrations (28,37, 38). However, dissociation of CDD/ANP-mediated vasorelaxation from increases in cGMP levels appears to be possible. Some atrial peptide analogs block the CDD/ANP-induced elevations of cGMP concentrations, but do not diminish the CDD/ANP-induced vasorelaxation, and are devoid of intrinsic vasodilatory activity (28). The activation of cGMP-sensitive phosphodiesterases results in the degradation of cyclic adenosine 3′,5′-monophosphate (cAMP), whereas the direct effects on the amiloride-sensitive sodium channel are independent of phosphorylation (28).

Inactivation of NPs occurs not only by binding to the NPR-C clearance receptor but also by enzymatic degradation with cleavage of the loop structure. The main enzyme responsible for this process is the neutral endopeptidase (NEP, E.C.3.4.24.11) (39–42), which has been demonstrated in many organs, especially in a high concentration in the brush border of the proximal renal tubule and in lung

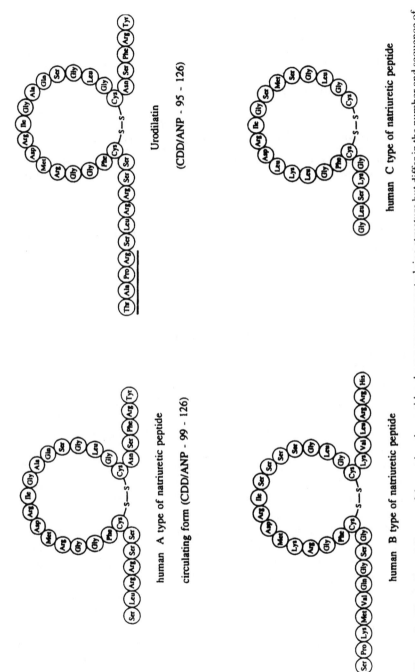

Urodilatin

(CDD/ANP - 95 - 126)

human A type of natriuretic peptide

circulating form (CDD/ANP - 99 - 126)

human C type of natriuretic peptide

human B type of natriuretic peptide

Figure 1 The A, B, and C types of the natriuretic peptides share a common central ring structure but differ in the number and sequence of their amino acids.

Table 1 The Different NPs and Their Receptors[a]

Subtype	Affinity	Localization
NPR-A	CDD/ANP, URO > BNP >> CNP	Lung, heart, kidney, adrenal, intestine, brain
NPR-B	CNP >> CDD/ANP, URO ≥ BNP	Lung/epithelium, heart, kidney, vascular endothelium/media, intestine, brain, pituitary, adrenal medulla
NPR-C	CDD/ANP, URO > CNP > BNP	Lung, heart, kidney, adrenal, brain

[a]Natriuretic peptide receptors A, B, and C (NPR-A, NPR-B, NPR-C), cardiodilatin/atrial natriuretic peptide (CDD/ANP), urodilatin (URO), brain natriuretic peptide (BNP), C-type natriuretic peptide (CNP).

epithelial cells (42–44). In comparison to CDD/ANP, URO was shown to be more resistant to the proteolytic inactivation by the NEP (40,45,46).

II. Therapeutic Applications

A. Acute Renal Failure

In the physiological regulation of renal function, particularly in the control of sodium and water excretion, both URO and CDD/ANP have been described as important mediators (47–49), but there are some results which favor URO instead of CDD/ANP as a paracrine intrarenal mechanism. Drummer et al. showed a parallelism between the circadian rhythm of urinary sodium and URO excretion (50), while the natriuresis induced by an acute saline infusion correlated more closely to the excretion of URO compared to CDD/ANP plasma levels (51). In a study by our group on the effects of a long-term sodium load in healthy volunteers, a stepwise increase in sodium intake was accompanied by a concomitant elevation in URO excretion (52). After adaptation to the different sodium states, intravenous URO induced a sodium load-associated diuresis and natriuresis (52). The higher NEP resistance of URO may thereby enhance the amount of this peptide reaching the distal tubule and the collecting duct without being degraded proximally to exert its renal effects. In accordance with this concept, clinical studies revealed strong diuretic and natriuretic effects of URO compared to CDD/ANP, while hemodynamic problems caused by vasodilation were observed less frequently (53,54).

Acute renal failure (ARF) is a life-threatening clinical situation which often complicates the postoperative period, particularly following cardiac surgery and transplantation of heart (HTx) and liver (LTx) (55–59). Pathophysiological mechanisms of ARF include cardiac low-output syndrome and acute cyclosporine nephrotoxicity, both accompanied by renal vasoconstriction (57,60). Early studies on CDD/ANP revealed its inhibitory effects on angiotensin- and norepinephrine-induced vascular contractility (61), and a preglomerular vasodilation combined with a postglomerular vasoconstriction resulting in an increase in glomerular filtration rate

(GFR) (62). Therefore, intravenous CDD/ANP was able to increase creatinine clearance and reduce the need for hemodialysis in patients with established ARF (63). The administration of URO, on the other hand, may be even more beneficial in this situation because stronger renal effects are combined with fewer cardiovascular problems. In rats, URO was shown to induce a GFR elevation in the early phase of toxic ARF (64). The first clinical study was performed using URO for prophylaxis of ARF following HTx (65). The peptide was administered in a long-term, low-dose infusion in addition to the routine medication. In comparison with controls, patients treated with URO demonstrated significantly better renal function: a reduction in the peak plasma creatinine, lower peak blood urea nitrogen (BUN), and a lower incidence of hemodialysis. Adequate diuresis was maintained in spite of the reduction of furosemide by more than 60% on each day of URO infusion (65). These beneficial effects of URO could even be confirmed in the treatment of incipient ARF (plasma creatinine $\geq$ 200% of preoperative value, oliguria/anuria demonstrated by a diuresis of $<$ 0.5 mL/kg bw/hr regardless of maximum diuretic treatment) following LTx and cardiac surgery (66,67). In some studies, however, URO neither prevented nor reversed ARF (68). Therefore, the use of URO in ARF remains controversial.

B. Congestive Heart Failure

The hemodynamic changes of intravenous infusion of URO or CDD/ANP were studied extensively in dogs and rats both before and after induction of heart failure (69–72). In healthy animals, both NPs caused a decrease in arterial blood pressure, right atrial pressure, and cardiac output (69–72). After induction of cardiac failure due to rapid right ventricular pacing (low-output failure), no or only a small effect on arterial blood pressure and cardiac output could be demonstrated (71,72). In dogs with high-output failure due to an arteriovenous fistula, however, the influence of the two NPs on arterial and right atrial pressure remained unchanged compared to control animals (73). A randomized, double-blind, placebo-controlled study on 12 patients with congestive heart failure (CHF, NYHA functional class II–III) revealed a significant URO-induced decrease of systolic blood pressure and central venous pressure, while diastolic blood pressure and heart rate remained unchanged (74). The effects on blood pressure must be interpreted mainly as a result of a reduction in cardiac output, whereas possible mechanisms for the decrease of right ventricular preload include volume contraction due to diuresis and venous pooling from direct venodilation (74). The influence of NPs on myocardiac function may follow a biphasic characteristic. A prolonged infusion of CDD/ANP in patients with chronic heart failure induced an initial increase in cardiac output of 20%, followed by a gradual fall in this parameter (75).

As expected, the renal effects of CDD/ANP and URO showed an increase in GFR, diuresis, and natriuresis in healthy animals (71,72), but striking differences between the two NPs could be documented in the case of heart failure. Whereas the efficacy of URO on renal excretory functions was well preserved in animal models and in patients (69,70,72,74), a complete absence of attenuation of the CDD/ANP-induced effects on the kidneys was described (70,71,76–78). Furthermore, the renal action of CDD/ANP infusions was only transient (< 4 hr) in heart failure (75), while the effects of URO persisted throughout an infusion period of 10 hr (74). The reduced proteolytic degradation of URO in the kidney is only one possible explanation for these results.

C. Bronchial Asthma

Besides their effects on the cardiovascular system and kidney, binding sites for NPs and stimulation of cGMP generation were also described in airway smooth muscle cells (79,80). In-vitro CDD/ANP relaxes bronchial segments of various animals (79,81) and inhibits the action of different bronchoconstrictors on the tracheo-bronchial tree (81,82). In guinea pigs and sheep, the potency of CDD/ANP to induce bronchodilation is more pronounced in the central than in the peripheral airways (81,83). While conflicting results were obtained from in-vitro studies with human tissue (84,85), a dose-dependent bronchodilating effect of CDD/ANP was demonstrated in asthmatic subjects by several investigators (86,87).

In order to compare the effects of various NPs, we analyzed the relaxation induced by CDD/ANP, URO, BNP, and CNP on guinea pig tracheal smooth muscle in vitro. In addition, we included phosphorylated derivatives, such as P-CDD/ANP and P-URO, in this investigation, because phosphorylation was shown to increase the NEP resistance of CDD/ANP (41). Carbachol was used to induce 80% of the maximum bronchoconstriction. After the tracheal tension reached a plateau, dose–response curves of the various peptides in the range of 5×10^{-10} to 10^{-6} M were recorded. Isoproterenol was added at the end of each experiment to demonstrate the maximum relaxation. Results were calculated as a percentage of this maximum response. The statistical evaluation revealed highly significant differences in the relaxant effects of the different peptides ($p < .0001$, ANOVA, repeated measures). At 10^{-6} M, the rank order of potency was URO > P-URO > P-CCD = BNP $\geqslant$ CDD > CNP, which equaled a relaxant effect of between $65.8 \pm 2.8\%$ (SEM) and $13.2 \pm 1.6\%$ of the maximum possible relaxation for URO and CNP, respectively (Fig. 2) (88). These results reflect functional evidence for NPR-A receptors, but no importance of NPR-B-binding sites for the interaction of NPs with bronchial smooth muscle cells. In accordance with data on the vasorelaxant potency of P-URO compared to URO, a significant decrease in the URO effects on the guinea pig tracheal smooth muscle preparations was documented following phosphorylation (89). The diminution in relaxant activity of CDD/ANP after phosphorylation may be counterbalanced by the higher NEP resistance of P-CDD/ANP, because there was no significant difference between the two peptides.

To correlate NP-induced relaxation and cGMP generation, guinea pig tracheal rings were incubated with various NPs for 5 min, followed by freezing of the tissue in liquid nitrogen and homogenization. Immunoreactive cGMP was related to the protein content of each ring. Again the ANOVA (repeated measures) revealed highly significant differences between the NPs (Fig. 3). The rank order of potency did not completely match the relaxant effects: CDD/ANP = URO = P-URO $\geqslant$ P-CDD/ANP

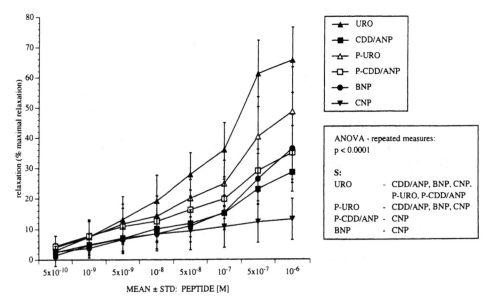

Figure 2 Cumulative dose–response curves of the relaxation of guinea pig tracheal smooth muscle preparations induced by the various natriuretic peptides in percent of the maximum response following isoproterenol 10^{-6} M.

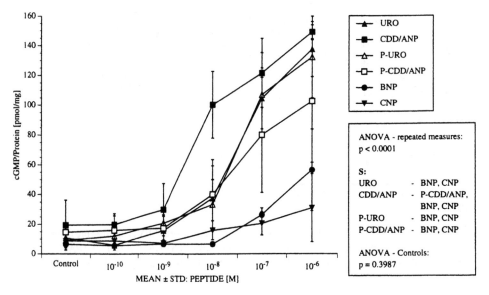

Figure 3 Cumulative dose–response curves of the intracellular cGMP generation in guinea pig tracheal smooth muscle preparations induced by the various natriuretic peptides in relation to protein content.

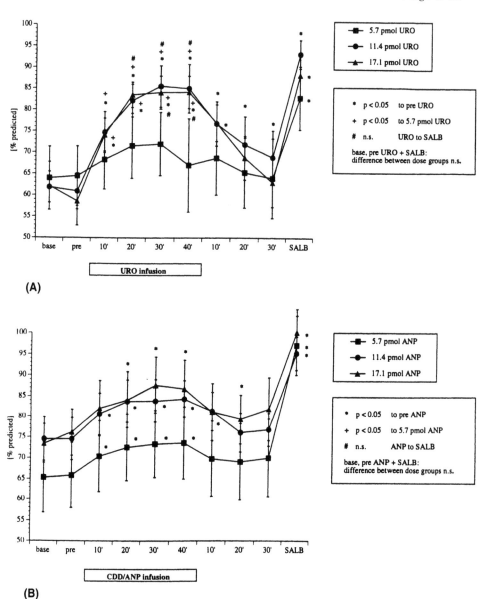

(A)

(B)

Figure 4 Time course of the forced expiratory volume in 1 sec (FEV$_1$) before (pre), during, and after an intravenous infusion of three different doses of (A) URO or (B) CDD/ANP and following an inhalation of 1.25 mg of salbutamol (SALB) in clinically stable asthmatics.

> BNP ≥ CNP. Since the cGMP-generation was not followed over time in this study, no comparison of the area under the curve (AUC) was possible for the different NPs. Discrepancies between cGMP increase and relaxing activities may be explained by additional intracellular mechanisms independent of this nucleotide (28).

The stronger relaxing effects of URO compared to CDD/ANP prompted us to evaluate their action on the bronchial system in vivo. At first, the protective effect of different doses of intravenous CDD/ANP, URO, or vehicle against an acetylcholine-induced bronchoconstriction was compared in spontaneously breathing, halothane-anesthetized Wistar rats. The inhalation of acetylcholine induced significant alterations of the spontaneous breathing parameters evaluated by whole-body plethysmography without significant differences between the treatment groups. Forced parameters detect airflow changes with a greater sensitivity and were measured in hyperventilation-induced temporary apnea after the challenge. The forced expiratory volume in 0.1 sec ($FEV_{0.1\%}$) and the parameters of the forced expiratory flow-volume curve (PEF, MMEF, FEF_{75}, FEF_{50}, FEF_{25}) revealed a significant protective effect of URO compared to controls ($p < .05$). CDD/ANP was without significant influence in this animal model (90).

These results encouraged us to compare the effects of URO and CDD/ANP in patients with bronchial asthma. Thirty-six clinically stable asthmatics showing a β_2-agonist-induced increase of the forced expiratory volume in 1 sec (FEV_1) by ≥ 15% participated in this clinical trial. Aerosol medication had to be discontinued for at least 8 hr prior to each study day. After baseline measurements of lung function parameters (FEV_1, VC, PEF, MEF_{75}, MEF_{50}, MEF_{25}), an intravenous infusion of 5.7, 11.4, or 17.1 pmol/kg/min URO or CDD/ANP was administered for 40 min in the morning. All measurements were repeated every 10 min during the infusion, for 30 min thereafter, and after the inhalation of 1.25 mg of salbutamol (SALB). Both peptides showed significant effects. While 11.4 pmol/kg/min URO dilated the central airways (FEV_1, PEF, MEF_{75}) slightly more potently than the peripheral bronchioles (MEF_{50}, MEF_{25}), 17.1 pmol/kg/min URO was as effective as SALB at all levels of the tracheobronchial tree (Figs. 4 and 5). CDD/ANP reached only 50% of the SALB effect without a predominant localization of its action. The cardiovascular parameters revealed a significantly stronger vasorelaxant activity of CDD/ANP (91–93).

These studies revealed in-vitro and in-vivo bronchodilating and broncho-protective properties of URO significantly superior to those of CDD/ANP. In the future, the stimulation of intracellular cGMP generation by NPs may become an established alternative pathway to induce bronchodilation in addition to cAMP-mediated therapy. A combination of both mechanisms may cause additive or hyper-additive effects.

III. Immunomodulatory Activities

Information on the immunomodulatory effects of NPs is sparse. In 1991, Moss et al. showed that the CDD/ANP fragment 4-28 enhances natural killer cell activity (94),

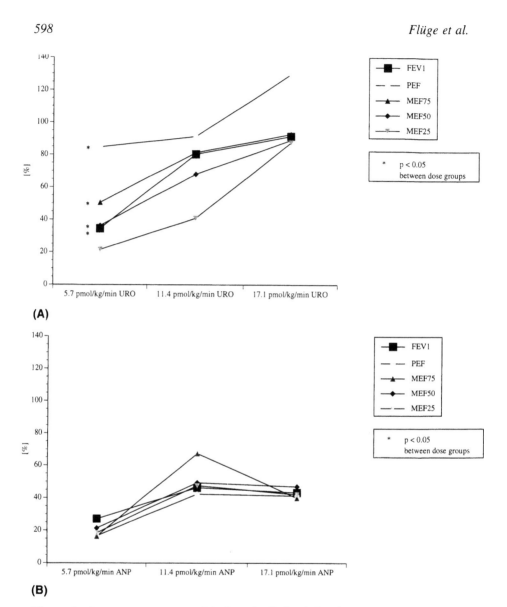

Figure 5 Dose–response curve of the bronchodilation induced by (A) URO or (B) CDD/ANP in relation to the individual effect of 1.25 mg of salbutamol (SALB, 100%): forced expiratory volume in 1 sec (FEV₁), peak expiratory flow (PEF), maximal expiratory flow at 75%, 50%, or 25% of forced vital capacity (MEF₇₅, MEF₅₀, MEF₂₅).

and Pella et al. revealed that the protective effect of CDD/ANP on cells damaged by oxygen radicals is mediated through elevated cGMP levels, reduction of Ca^{2+} inflow, and probably G proteins (95). In 1992, Wiedermann et al. demonstrated that CDD/ANP is a potent signal priming the polymorphonuclear neutrophil respiration burst to secrete superoxide anion (O_2^-) and speculated that the inhibition of stimulated chemotaxis of these cells may be due to their enhanced adhesiveness (96). More recently, Vollmar et al. showed that in mouse macrophages an exposure to bacterial lipopolysaccharides (LPS) resulted in a significant elevation of CDD/ANP and CNP mRNA and intra- and extracellular immunoreactivity to these peptides, while no changes were observed for BNP (97,98). In the same cells, a simultaneous exposure to LPS and CDD/ANP significantly decreases the nitric oxide (NO) synthesis, whereas CNP fails to affect LPS-induced NO generation (99,100). Further results indicate that CDD/ANP may affect transcriptional mechanisms of inducible nitric oxide synthase (NOS) synthesis, and that an inhibition of the NP receptor-mediated cGMP formation blocks the CDD/ANP-induced decrease of NO synthesis. As the levels of NPs, especially CDD/ANP, change in pulmonary diseases through altered release and/or metabolism in the lung and upstream, not only vascular or bronchial tone but also inflammation may be influenced differentially by the "cytokine-like activity" of the various NPs (98). The pathophysiological significance of these mechanisms and the therapeutic consequences in the context of bronchial asthma in which airway inflammation correlates to bronchial hyperreactivity are unknown.

The immunomodulation by NPs may also be important with regard to parenchymal inflammation in the lung. In 1988, Imamura et al. demonstrated protective effects of CDD/ANP against pulmonary edema caused by Triton-X, CHAPS (both surfactants), paraquat, arachidonic acid, and epinephrine in guinea pig isolated perfused lungs (101,102). Interestingly, the antiedematous actions of this peptide against Triton-X- and paraquat-induced lung edema seemed to be independent of the cGMP generation (101). Looking at these mechanisms, the impact of NPs on pulmonary shifts of electrolytes, protein, and fluid must also be considered. In rat alveolar type II cells, CDD/ANP does not directly modulate the activity of the Na^+-K^+-ATPase, but influences the effects of a β-adrenergic agonist on this pump (103). Investigations by Zimmermann et al. suggest that in rats CDD/ANP at pathophysiological levels (plasma level 1232 ± 199 pg/mL) induced by an infusion rate of 0.1 μg/kg/min shifts protein out of the circulation in peripheral vascular beds and the lungs and may thus contribute to pulmonary edema. On the other hand, pharmacological levels of this peptide (plasma level 7734 ± 675 pg/mL) caused by an infusion rate of 0.5 μ/kg/min may protect the lungs by preventing increased pulmonary albumin escape (104). CNP stimulates Cl^- secretion (105), and its level in tracheal mucosal secretions has been reported as 10,000 times higher than in the brain, suggesting nonneural functions (28,106). When CNP is added to either the apical (mucosal) or the basolateral (serosal) surface of cultured and polarized human airway epithelial cells and bovine isolated tracheal epithelium, cGMP levels are markedly enhanced (28). This peptide may be involved in the regulation of fluid and ion secretions within the lumen found in various regions of the body (28).

In conclusion, the data presently available already demonstrate potent effects of the NPs on the cardiovascular system, the kidney, and the tracheobronchial tree. In particular, URO may be a new tool to improve the clinical situation of patients with severe bronchial asthma by interfering with pathophysiological mechanisms involved in this disease, such as bronchoconstriction and bronchial inflammation. However, information on the immunomodulatory activities of NPs is still incomplete. Further studies may focus not only on this topic, but also on the interactions of the different NPs with other peptide systems in various parts of the body. Peptidergic or nonpeptidergic agonists of the NPRs not inactivated by the NEP and other proteases should improve the administration and extend the range of application for the clinician.

References

1. Kisch B. Electron microscopy of the atrium of the heart. I. Guinea pig. Exp Med Surg 1956; 14:99–112.
2. Marie JP, Guillemot H, Hatt PY. [Degree of granularity of the atrial cardiocytes. Morphometric study in rats subjected to different types of water and sodium load (author's transl)]. Pathol Biol Paris 1976; 24:549–554.
3. de Bold AJ. Heart atria granularity effects of changes in water-electrolyte balance. Proc Soc Exp Biol Med 1979; 161:508–511.
4. de Bold AJ, Borenstein HB, Veress AT, Sonnenberg H. A rapid and potent natriuretic response to intravenous injection of atrial myocardial extract in rats. Life Sci 1981; 28:89–94.
5. Flynn TG, de Bold ML, de Bold AJ. The amino acid sequence of an atrial peptide with potent diuretic and natriuretic properties. Biochem Biophys Res Commun 1983; 117: 859–865.
6. Forssmann WG, Hock D. Lottspeich F, Henschen A, Kreye V, Christmann M, Reinecke M, Metz J, Carlquist M, Mutt V. The right auricle of the heart is an endocrine organ: cardiodilatin as a peptide hormone candidate. Anat Embryol 1983; 168:307–313.
7. Kangawa K, Matsuo H. Purification and complete amino acid sequence of alpha-human atrial natriuretic polypeptide (alpha-hANP). Biochem Biophys Res Commun 1984; 118:131–139.
8. Forssmann WG, Birr C, Carlquist M, Christmann M, Finke R, Henschen A, Hock D, Kirchheim H, Kreye V, Lottspeich F, et al. The auricular myocardiocytes of the heart constitute an endocrine organ. Characterization of a porcine cardiac peptide hormone, cardiodilatin-126. Cell Tissue Res 1984; 238:425–430.
9. Sudoh T, Minamino N, Kangawa K, Matsuo H. C-type natriuretic peptide (CNP): a new member of natriuretic peptide family identified in porcine brain. Biochem Biophys Res Commun 1990; 168:863–870.
10. Sudoh T, Kangawa K, Minamino N, Matsuo H. A new natriuretic peptide in porcine brain. Nature 1988; 332:78–81.
11. Forssmann K, Hock D, Herbst F, Schulz-Knappe P, Talartschik J, Scheler F, Forssmann WG. Isolation and structural analysis of the circulating human cardiodilatin (alpha ANP). Klin Wochenschr 1986; 64:1276–1280.
12. Atlas SA, Kleinert HD, Camargo MJ, Januszewicz A, Sealey JE, Laragh JH, Schilling

JW, Lewicki JA, Johnson LK, Maack T. Purification, sequencing and synthesis of natriuretic and vasoactive rat atrial peptide. Nature 1984; 309:717–719.

13. Chartier L, Schiffrin E, Thibault G. Effect of atrial natriuretic factor (ANF)-related peptides on aldosterone secretion by adrenal glomerulosa cells: cricial role of the intramolecular disulphide bond. Biochem Biophys Res Commun 1984; 122:171–174.

14. Misono KS, Fukumi H, Grammer RT, Inagami T. Rat atrial natiuretic factor: complete amino acid sequence and disulfide linkage essential for biological activity. Biochem Biophys Res Commun 1984; 119:524–529.

15. Schulz-Knappe P, Forssmann K, Herbst F, Hock D, Pipkorn R, Forssmann WG. Isolation and structural analysis of "urodilatin," a new peptide of the cardiodilatin-(ANP)-family, extracted from human urine. Klin Wochenschr 1988; 66:752–759.

16. Forssmann WG, Nokihara K, Gagelmann M, Hock D, Feller S, Schulz-Knappe P, Herbst F. The heart is the center of a new endocrine, paracrine, and neuroendocrine system. Arch Histol Cytol 1989; 52(suppl):293–315.

17. Goetz KL. Renal natriuretic peptide (urodilatin?) and atriopeptin: evolving concepts. Am J Physiol 1991; 261:F921–F932.

18. Kentsch M, Ludwig D, Drummer C, Gerzer R, Muller-Esch G. Haemodynamic and renal effects of urodilatin in healthy volunteers. Eur J Clin Invest 1992; 22:319–325.

19. Drummer C, Fiedler F, Bub A, Kleefeld D, Dimitriades E, Gerzer R, Forssmann WG. Development and application of a urodilatin (CDD/ANP-95-126)-specific radio-immunoassay. Pflugers Arch 1993; 423:372–377.

20. Forssmann WG, Meyer M, Schulz-Knappe P. Urodilatin: from cardiac hormones to clinical trials. Exp Nephrol 1994; 2:318–323.

21. Koller KJ, Goeddel DV. Molecular biology of the natriuretic peptides and their receptors. Circulation 1992; 86:1081–1088.

22. Ruskoaho H. Atrial natriuretic peptide: synthesis, release, and metabolism. Pharmacol Rev 1992; 44:479–602.

23. Vollmar AM, Schulz R. Atrial natriuretic peptide in lymphoid organs of various species. Comp Biochem Physiol 1990; 96:459–463.

24. Heim JM, Kiefersauer S, Fülle H-J, Gerzer R. Urodilatin and beta-ANF. Binding properties and activiation of particulate guanylate cyclase. Biochem Biophys Res Commun 1989; 163:37–41.

25. Koike J, Nonoguchi H, Terada Y, Tomita K, Marumo F. Effect of urodilatin on cGMP accumulation in the kidney. J Am Soc Nephrol 1993; 3:1705–1709.

26. Chinkers M, Garbers DL, Chang MS, Lowe DG, Chin HM, Goeddel DV, Schulz S. A membrane form of guanylate cyclase is an atrial natriuretic peptide receptor. Nature 1989; 338:78–83.

27. Lowe DG, Chang MS, Hellmiss R, Chen E, Singh S, Garbers DL, Goeddel DV. Human atrial natriuretic peptide receptor defines a new paradigm for second messenger signal transduction. EMBO J 1989; 8:1377–1384.

28. Drewett JG, Garbers DL. The family of guanylyl cyclase receptors and their ligands. Endocr Rev 1994; 15:135–162.

29. Garbers DL. Guanylyl cyclase receptors and their endocrine, paracrine, and autocrine ligands. Cell 1996; 71:1–4.

30. Maack T, Suzuki M, Almeida FA, Nussenzveig D, Scarborough RM, McEnroe GA, Lewicki JA. Physiological role of silent receptors of atrial natriuretic factor. Science 1987; 238:675–678.

31. Maack T. Receptors of atrial natriuretic factor. Annu Rev Physiol 1992; 54:11–27.

32. Cahill PA, Hassid A. Clearance receptor-binding atrial natriuretic peptides inhibit mitogenesis and proliferation of rat aortic smooth muscle cells. Biochem Biophys Res Commun 1991; 179:1606–1613.

33. Levin ER. Natriuretic peptide C-receptor: more than a clearance receptor. Am J Physiol 1993; 264:E483–E489.

34. Suga S, Nakao K, Hosoda K, Mukoyama M, Ogawa Y, Shirakami G, Arai H, Saito Y, Kambayashi Y, Inouye K, et al. Receptor selectivity of natriuretic peptide family, atrial natriuretic peptide, brain natriuretic peptide, and C-type natriuretic peptide. Endocrinology 1992; 130:229–239.

35. Valentin JP, Sechi LA, Qui C, Schambelan M, Humphreys MH. Urodilatin binds to and activates renal receptors for atrial natriuretic peptide. Hypertension 1993; 21:432–438.

36. Lincoln TM, Cornwell TL. Intracellular cyclic GMP receptor proteins. FASEB J 1993; 7:328–338.

37. Zeidel ML, Seifter JL, Lear S, Brenner BM, Silva P. Atrial peptides inhibit oxygen consumption in kidney medullary collecting duct cells. Am J Physiol 1986; 251:F379–F383.

38. Cornwell TL, Lincoln TM. Regulation of intracellular Ca^{2+} levels in cultured vascular smooth muscle cells. Reduction of Ca^{2+} by atriopeptin and 8-bromo-cyclic GMP is mediated by cyclic GMP-dependent protein kinase. J Biol Chem 1989; 264:1146–1155.

39. Vogt-Schaden M, Gagelmann M, Hock D, Herbst F, Forssmann WG. Degradation of porcine brain natriuretic peptide (pBNP-26) by endoprotease-24.11 from kidney cortical membranes. Biochem Biophys Res Commun 1989; 161:1177–1183.

40. Kenny AJ, Bourne A, Ingram J. Hydrolysis of human and pig brain natriuretic peptides, urodilatin, C-type natriuretic peptide and some C-receptor ligands by endopeptidase-24.11. Biochem J 1993; 291:83–88.

41. Gagelmann M, Feller S, Hock D, Schulz-Knappe P, Forssmann WG. Biochemistry of the differential release, processing and degradation of cardiac and related peptide hormones. In: Kaufmann W, Wambach G, eds. Endocrinology of the Heart. Berlin, Heidelberg: Springer-Verlag, 1989:27–40.

42. Roques BP, Noble F, Dauge V, Fournie Zaluski MC, Beaumont A. Neutral endopeptidase 24.11: structure, inhibition, and experimental and clinical pharmacology. Pharmacol Rev 1993; 45:87–146.

43. Lindberg BF, Bengtsson HI, Lundin S, Andersson KE. Degradation and inactivation of rat atrial natriuretic peptide 1-28 by neutral endopeptidase-24.11 in rat pulmonary membranes. Regulatory Peptides 1992; 42:85–96.

44. Perrella MA, Margulies KB, Wei CM, Aarhus LL, Heublein DM, Burnett JCJ. Pulmonary and urinary clearance of atrial natriuretic factor in acute congestive heart failure in dogs. J Clin Invest 1991; 87:1649–1655.

45. Gagelmann M, Hock D, Forssmann WG. Urodilatin (CDD/ANP-95-126) is not biologically inactivated by a peptidase from dog kidney cortex membranes in contrast to atrial natriuretic peptide/cardiodilatin (alpha-hANP/CDD-99-126). FEBS Lett 1988; 233:249–254.

46. Abassi ZA, Golomb E, Agbaria R, Roller PP, Tate J, Keiser HR. Hydrolysis of iodine labelled urodilatin and ANP by recombinant neutral endopeptidase EC. 3.4.24.11. Br J Pharmacol 1994; 113:204–208.

47. Goetz KL, Drummer C, Zhu JL, Leadley R, Fiedler F, Gerzer R. Evidence that urodilatin, rather than ANP, regulates renal sodium excretion. J Am Soc Nephrol 1990; 1:867–874.

48. Emmeluth C, Drummer C, Gerzer R, Bie P. Roles of cephalic Na+ concentration and urodilatin in control of renal Na+ excretion. Am J Physiol 1992; 262:F513–F516.

49. Martin DR, Pevahouse JB, Trigg DJ, Vesely DL, Buerkert JE. Three peptides from the ANF prohormone NH(2)-terminus are natriuretic and/or kaliuretic [published erratum appears in Am J Physiol 1990 Dec;259(6 Pt 3):following table of contents]. Am J Physiol 1990; 258:F1401–F1408.

50. Drummer C, Fiedler F, Konig A, Gerzer R. Urodilatin, a kidney-derived natriuretic factor, is excreted with a circadian rhythm and is stimulated by saline infusion in man. J Am Soc Nephrol 1991; 1:1109–1113.

51. Drummer C, Gerzer R, Heer M, Molz B, Bie P, Schlossberger M, Stadaeger C, Rocker L, Strollo F, Heyduck B, et al. Effects of an acute saline infusion on fluid and electrolyte metabolism in humans. Am J Physiol 1992; 262:F744–F754.

52. Meyer M, Richter R, Brunkhorst R, Wrenger E, Schulz-Knappe P, Kist A, Brabant G, Koch KM, Rechkemmer G, Forssmann WG. Urodilatin is involved in sodium homeostasis and exerts sodium-state dependent natriuretic and diuretic effects. Am J Physiol 1996; 271:F489–F497.

53. Saxenhofer H, Raselli A, Weidmann P, Forssmann WG, Bub A, Ferrari P, Shaw SG. Urodilatin, a natriuretic factor from kidneys, can modify renal and cardiovascular function in men. Am J Physiol 1990; 259:F832–F838.

54. Meyer M, Schulz-Knappe P, Kuse ER, Hummel M, Hetzer R, Pichlmayr R, Forssmann WG. [Urodilatin. Use of a new peptide in intensive care]. Anaesthesist 1995; 44:81–91.

55. Greenberg A. Renal failure in cardiac transplantation. Cardiovasc Clin 1990; 20:189–198.

56. Myers BD, Carrie BJ, Yee RR, Hilberman M, Michaels AS. Pathophysiology of hemodynamically mediated acute renal failure in man. Kidney Int 1980; 18:495–504.

57. Hilberman M, Myers BD, Carrie BJ, Derby G, Jamison RL, Stinson EB. Acute renal failure following cardiac surgery. J Thorac Cardiovasc Surg 1979; 77:880–888.

58. Platz KP, Mueller AR, Blumhardt G, Bachmann S, Bechstein WO, Kahl A, Neuhaus P. Nephrotoxicity following orthotopic liver transplantation. A comparison between cyclosporine and FK506. Transplantation 1994; 58:170–178.

59. McCauley J, Van Thiel DH, Starzl TE, Puschett JB. Acute and chronic renal failure in liver transplantation. Nephron 1990; 55:121–128.

60. Myers BD. Cyclosporine nephrotoxicity. Kidney Int 1986; 30:964–974.

61. Kleinert HD, Maack T, Atlas SA, Januszewicz A, Sealey JE, Laragh JH. Atrial natriuretic factor inhibits angiotensin-, norepinephrine-, and potassium-induced vascular contractility. Hypertension 1984; 6:I143–I147.

62. Marin Grez M, Fleming JT, Steinhausen M. Atrial natriuretic peptide causes preglomerular vasodilatation and post-glomerular vasoconstriction in rat kidney. Nature 1986; 324:473–476.

63. Rahman SN, Kim GE, Mathew AS, Goldberg CA, Allgren R, Schrier RW, Conger JD. Effects of atrial natriuretic peptide in clinical acute renal failure. Kidney Int 1994; 45:1731–1738.

64. Schramm L, Heidbreder E, Schaar J, Lopau K, Zimmermann J, Gotz R, Ling H, Heidland A. Toxic acute renal failure in the rat: effects of diltiazem and urodilatin on renal function. Nephron 1994; 68:454–461.

65. Hummel M, Kuhn M, Bub A, Bittner H, Kleefeld D, Marxen P, Schneider B, Hetzer R, Forssmann WG. Urodilatin: a new peptide with beneficial effects in the postoperative therapy of cardiac transplant recipients. Clin Invest 1992; 70:674–682.

66. Wiebe K, Meyer M, Wahlers T, Zenker D, Schulze F-P, Michels P, Dalichau H, Mohr FW, Borst H-G, Forssmann WG. Acute renal failure following cardiac surgery is reverted by administration of urodilatin (INN: ularitide). Eur J Med Res 1996; 1: 259–265.

67. Cedidi C, Meyer M, Kuse ER, Schulz-Knappe P, Ringe B, Frei U, Pichlmayr R, Forssmann WG. Urodilatin: a new approach for the treatment of therapy-resistant acute renal failure after liver transplantation. Eur J Clin Invest 1994; 24:632–639.

68. Langrehr JM, Kahl A, Meyer M, Neumann U, Knoop M, Jonas S, Bechstein WO, Frei U, Forssmann WG. Prophylactic use of low-dose urodilatin following liver transplantation: a randomized placebo-controlled study. Clin Transplantation 1997 (in press).

69. Abassi ZA, Powell JR, Golomb E, Keiser HR. Renal and systemic effects of urodilatin in rats with high-output heart failure. Am J Physiol 1992; 262:F615–F621.

70. Villarreal D, Freeman RH, Johnson RA. Renal effects of ANF (95-126), a new atrial peptide analogue, in dogs with experimental heart failure. Am J Hypertens 1991; 4: 508–515.

71. Riegger GA, Elsner D, Kromer EP, Daffner C, Forssmann WG, Muders F, Pascher EW, Kochsiek K. Atrial natriuretic peptide in congestive heart failure in the dog: plasma levels, cyclic guanosine monophosphate, ultrastructure of atrial myoendocrine cells, and hemodynamic hormonal, and renal effects. Circulation 1988; 77:398–406.

72. Riegger GA, Elsner D, Forssmann WG, Kromer EP. Effects of ANP-(95-126) in dogs before and after induction of heart failure. Am J Physiol 1990; 259:H1643–H1648.

73. Wambach G. [Cardiac peptides and their importance in heart failure]. Z Kardiol 1991; 80(suppl 8):41–46.

74. Elsner D, Muders F, Muntze A, Kromer EP, Forssmann WG, Riegger GA. Efficacy of prolonged infusion of urodilatin [ANP-(95-126)] in patients with congestive heart failure. Am Heart J 1995; 129:766–773.

75. Münzel T, Drexler H, Holtz J, Kurtz S, Just H. Mechanisms involved in the response to prolonged infusion of atrial natriuretic factor in patients with chronic heart failure. Circulation 1991; 83:191–201.

76. Cody RJ, Atlas SA, Laragh JH, Kubo SH, Covit AB, Ryman KS, Shaknovich A, Pondolfino K, Clark M, Camargo MJ, et al. Atrial natriuretic factor in normal subjects and heart failure patients. Plasma levels and renal, hormonal, and hemodynamic responses to peptide infusion. J Clin Invest 1986; 78:1362–1374.

77. Molina CR, Fowler MB, McCrory S, Peterson C, Myers BD, Schroeder JS, Murad F. Hemodynamic, renal and endocrine effects of atrial natriuretic peptide infusion in severe heart failure. J Am Coll Cardiol 1988; 12:175–186.

78. Saito Y, Nakao K, Nishimura K, Sugawara A, Okumura K, Obata K, Sonoda R, Ban T, Yasue H, Imura H. Clinical application of atrial natriuretic polypeptide in patients with congestive heart failure: beneficial effects on left ventricular function. Circulation 1987; 76:115–124.

79. Ishii K, Murad F. ANP relaxes bovine tracheal smooth muscle and increases c-GMP. Am J Physiol 1989; 256(Cell Physiol 25):C495–C500.

80. Ishii Y, Watanabe T, Watanabe M, Hasegawa S, Uchiyama Y. Effects of atrial natriuretic peptide on type II alveolar epithelial cells of the rat lung. J Anat 1989; 166:86–95.

81. Hamel R, Ford-Hutchinson AW. Relaxant profile of synthetic atrial natriuretic factor in guinea pig pulmonary tissues. Eur J Pharmacol 1986; 121:151–155.

82. Potvin W, Varma DR. Bronchodilator activity of atrial natriuretic peptide in guinea pigs. Can J Physiol Pharmacol 1989; 67(10):1213–1218.

83. Banerjee MR, Newman JH. Acute effects of atrial natriuretic peptide on lung mechanics and hemodynamics in awake sheep. J Appl Physiol 1990; 69:728–733.
84. Candenas ML, Naline E, Puybasset L, Devillier P, Advenier C. Effect of atrial natriuretic peptide and of atriopeptins on the human isolated bronchus. Comparison with the reactivity of the guinea pig isolated trachea. Pulmon Pharmacol 1991; 4:120–125.
85. Hulks G, Crabb KG, McGrath JC, Thomson NC. In vitro effects of atrial natriuretic factor and sodium nitroprusside on bronchomotor tone in human bronchial smooth muscle (abstr). Am Rev Respir Dis 1991; 143:A344.
86. Hulks G, Jardine A, Connell JMC, Thomson NC. Bronchodilator effect of atrial natriuretic peptide in asthma. Br Med J 1989; 299(6707):1081–1082.
87. Chanez P, Mann C, Bousquet J, Chabrier PE, Godard P, Braquet P, Michel FB. Atrial natriuretic factor (ANF) is a potent bronchodilator in asthma. J Allergy Clin Immunol 1990; 86 (3 pt 1):321–324.
88. Flüge T, Forssmann WG. Relaxant effects of natriuretic peptides and their phosphorylated derivatives on the guinea pig tracheal smooth muscle (abstr). Am J Respir Crit Care Med 1995; 151 (4):A404.
89. Dorner T, Gagelmann M, Feller S, Herbst F, Forssmann WG. Phosphorylation and dephosphorylation of the natriuretic peptide urodilatin (CDD-/ANP-95-126) and the effect on biological activity. Biochem Biophys Res Commun 1989; 163:830–835.
90. Flüge T, Hoymann HG, Hohlfeld J, Heinrich U, Fabel H, Wagner TOF, Forssmann WG. Type A natriuretic peptides exhibit different bronchoprotective effects in rats. Eur J Pharmacol 1994; 271:395–402.
91. Flüge T, Fabel H, Wagner TOF, Forssmann WG. Localization of the bronchodilator effect induced by type A natriuretic peptide in asthmatic subjects. Clin Invest 1994; 72:772–774.
92. Flüge T, Fabel H, Wagner TOF, Schneider B, Forssmann WG. Urodilatin (Ularitide, INN): a potent bronchodilator in asthmatic subjects. Eur J Clin Invest 1995; 25:728–736.
93. Flüge T, Fabel H, Wagner TOF, Schneider B, Forssmann WG. Bronchodilating effects of natriuretic and vasorelaxant peptides compared to salbutamol in asthmatics. Regulatory Peptides 1995; 59:357–370.
94. Moss RB, Golightly MG. In vitro enhancement of natural cytotoxicity by atrial natriuretic peptide fragment 4-28. Peptides 1991; 12:851–854.
95. Pella R. The protective effect of atrial natriuretic peptide (ANP) on cells damaged by oxygen radicals is mediated through elevated CGMP-levels, reduction of calcium-inflow and probably G-proteins. Biochem Biophys Res Commun 1991; 174:549–555.
96. Wiedermann CJ, Niedermuhlbichler M, Braunsteiner H. Priming of polymorphonuclear neutrophils by atrial natriuretic peptide in vitro. J Clin Invest 1992; 89:1580–1586.
97. Vollmar AM, Schulz R. Expression and differential regulation of natriuretic peptides in mouse macrophages. J Clin Invest 1995; 95:2442–2450.
98. Vollman AM, Schulz R. Gene expression and secretion of atrial natriuretic peptide by murine macrophages. J Clin Invest 1994; 94:539–545.
99. Vollmar AM, Schulz R. Effect of natriuretic peptides on nitric oxide synthesis in mouse macrophages (abstr). Naunyn-Schmiedeberg's Arch Pharmacol 1995; 351(suppl):R126.
100. Vollmar AM, Schulz R. Atrial natriuretic peptide inhibits nitric oxide synthesis in mouse macrophages. Life Sci 1995; 56:PL149–PL155.
101. Imamura T, Ohnuma N, Iwasa F, Furuya M, Hayashi Y, Inomata N, Ishihara T, Noguchi T. Protective effect of alpha-human atrial natriuretic polypeptide (alpha-hANP) on chemical-induced pulmonary edema. Life Sci 1988; 42:403.

102. Inomata N, Ohnuma N, Furuya M. Alpha-human atrial natriuretic peptide (a-hANF) prevents pulmonary edema induced by arachidonic acid treatment in isolated perfused lung from guinea pig. Jpn J Pharmacol 1987; 44:211–214.
103. Berthiaume Y, Suzuki S, Gutkowska J, Tremblay J. Modulation of Na/K/ATPase activity in adult alveolar type II cells by ANF (abstr). Am J Respir Crit Care Med 1995; 151 (4):A182.
104. Zimmermann RS, Trippodo NC, MacPhee AA, Mattinez AJ, Barbee RW. High-dose atrial natriuretic factor enhances albumin escape from systemic but not the pulmonary circulation. Circ Res 1990; 67:461–468.
105. Solomon R, Protter A, McEnroe G, Porter JG, Silva P. C-type natriuretic peptides stimulate chloride secretion in the rectal gland of Squalus acanthias. Am J Physiol 1992; 262:R707–R711.
106. Chrisman TD, Schulz S, Potter LR, Garbers DL. Seminal plasma factors that cause large elevations in cellular cyclic GMP are C-type natriuretic peptides. J Biol Chem 1993; 268:3698–3703.

AUTHOR INDEX

Italic numbers give the page on which the complete reference is listed.

SUBJECT INDEX

A

Activity-dependent neurotrophic factor (ADNF)
  release by VIP, 353, 355, 387, 388, 396
Acute inflammation
  attenuation of, 410
Acute renal failure
  URO and CDD/ANP as important mediators, 592
Acute respiratory distress syndrome (adult respiratory distress syndrome, ARDS)
  models of, 346–350
  neutrophils in, 10
Acquired Immune Deficiency Syndrome (AIDS)
  VIP protection against neuronal cell death, 396, 402
Adhesion molecules
  in inflammatory response, 10, 11

Airway peptidases
  degradation of tachykinins, 521
  degradation of VIP, 410
Airway smooth muscle
  contraction induced by endothelin-1, 211
  human, contraction by sensory neuropeptides, 95, 98, 117
  mitogenesis induced by endothelin-1, 214
  relaxation by VIP and analogs, 410
  relaxation induced by endothelin-1, 214
Airway epithelium
  inflammatory peptides, 251–252
Airways
  hyperresponsiveness and sensory neuropeptides, 89–146, 147–162, 103, 107, 109, 113, 125
  reactivity and sensitivity, 90
  sensory neuropeptides in, 45–68, 92
Alveolar wall, 218

Milton Keynes UK
Ingram Content Group UK Ltd.
UKHW021936071024
449327UK00022B/1829